ELEMENTS OF PETROLEUM GEOLOGY

THIRD EDITION

Petroleum Exploration: Past, Present, and Future

The Past: Petroleum geologist looking for oil in Persia (now Iran) in the early twentieth century. (Courtesy of the British Petroleum.)

The Present: Petroleum geoscientists looking for oil anywhere in the world in the late twentieth century. (Courtesy of Esso UK plc.)

The Future: Post-millennial cyber-cadet seeking abiogenic petroleum anywhere in the universe. (Courtesy of Paradigm Geophysical, created by Sachnowitz & Co.)

ELEMENTS OF PETROLEUM GEOLOGY

THIRD EDITION

RICHARD C. SELLEY
STEPHEN A. SONNENBERG

ELSEVIER

AMSTERDAM • BOSTON • HEIDELBERG • LONDON
NEW YORK • OXFORD • PARIS • SAN DIEGO
SAN FRANCISCO • SINGAPORE • SYDNEY • TOKYO
Academic Press is an imprint of Elsevier

ACADEMIC PRESS

Academic Press is an imprint of Elsevier
525 B Street, Suite 1800, San Diego, CA 92101-4495, USA
225 Wyman Street, Waltham, MA 02451, USA
The Boulevard, Langford Lane, Kidlington, Oxford OX5 1GB, UK
32 Jamestown Road, London NW1 7BY, UK

ISBN: 978-0-12-386031-6

British Library Cataloguing-in-Publication Data
A catalogue record for this book is available from the British Library

Library of Congress Cataloguing-in-Publication Data
A catalog record for this book is available from the Library of Congress

For information on all Academic Press publications
visit our web site at http://store.elsevier.com/

Contents

Preface to the Third Edition

The first edition of *Elements of Petroleum Geology* was published in 1985, some 30 years ago. The objective of the book was to describe the elements of petroleum geology. Beginning with the deposition and maturation of a source rock, followed by the migration of petroleum from the source into a porous permeable reservoir rock trapped beneath an impermeable seal. This book also described the science and technology of petroleum exploration and production, from the first geophysical surveys to the finale of enhanced recovery.

When the second edition was published in 1998 the fundamental elements of petroleum geology remain little changed, but the science and technology of petroleum exploration and production had evolved. For example, ever improving computer power enabled the development of 3D seismic surveys. The ability to interpret reflecting horizons progressed to interpreting the amplitude of individual seismic wave traces.

When Elsevier requested the production of a third edition RCS, by now in the spring time of his senility, felted daunted by the task. Elsevier suggested a co-author. SAS accepted the challenge and has made a major contribution to the revision. In the intervening years between the second and third editions the importance of unconventionally occurring petroleum has been realized. Technological advances, notably in horizontal drilling and hydraulic fracturing have enabled gas and oil to be produced from the very source rocks themselves, without migration and entrapment in a conventional reservoir. The production of gas and oil from shale, together with coal bed methane is accelerating. This shift has many economic and environmental benefits, not least because it diminishes reliance on burning coal leading to a cut in carbon dioxide emissions.

We hope that this third edition will, like its predecessors, be useful for student geoscientists and engineers preparing for a career in the energy industry, and also for mature practitioners in the upstream petroleum industry seeking a wide ranging account of a large area of science and engineering.

RCS
Royal School of Mines
Imperial College
London, UK

SAS
Colorado School of Mines
Golden
Colorado, USA

Acknowledgments

There are two main problems to overcome in writing a book on petroleum geology. The subject is a vast one, ranging from arcane aspects of molecular biochemistry to the mathematical mysteries of seismic data processing. The subject is also evolving very fast as new data become available and new concepts are developed. I am very grateful to the many people who read draft of the manuscript, pointed out errors of fact or emphasis, and suggested improvements. Much of this load was borne by staff at the Imperial College, London University. The geophysical sections were dealt with by Dr Thomas-Betts and the late Weildon and the late Williamson, geochemistry by the late Dr Kinghorn, petroleum engineering by the late Professor Wall, and most of the remaining topics by the late Professor Stoneley. Mr Maret of Schlumberger reviewed formation evaluation section.

For permission to use previously published illustrations I am grateful to the following: Academic Press, the American Association of Petroleum Geologists, Applied Science Publishers, Badley Earth Sciences Ltd., Blackwell Scientific Publications, BP Exploration, Gebruder Borntraeger, Geoexplorers International Inc., Cambridge University Press, the Canadian Association of Petroleum Geologists, Chapman and Hall, Esso UK plc., Coherence Technology Company, W.H. Freeman and Company, *Geology*, The Geologists Association of London, *Geological Magazine*, the Geological Society of London, the Geological Society of South Africa, Gulf Coast Association of Geological Societies, the Geophysical Development Corporation, GMG Europe Ltd., GVA International Consultants, the Institute of Petroleum, the *Journal of Geochemical Exploration*, the *Journal of Petroleum Geology*, *Marine and Petroleum Geology*, McGraw-Hill, the Norwegian Petroleum Society, NUMAR UK Ltd., the Offshore Technology Conference, Paradigm Geophysical Corporation, Princeton University Press, Sachnowitz & Co., Schlumberger Wireline Logging Services, Schlumberger Oil-field Review, the Society of Petroleum Engineers, the Society of Professional Well Log Analysts, Springer-Verlag, John wiley & Sons, World Geoscience UK Ltd., and World Oil.

CHAPTER

1

Introduction

And God said unto Noah...Make thee an ark of gopher wood; rooms shalt thou make in the ark, and shalt coat it within and without with pitch. *Genesis 6:13-14*

1.1 HISTORICAL REVIEW OF PETROLEUM EXPLORATION

1.1.1 Petroleum from Noah to Organization of Petroleum Exporting Countries

Petroleum exploration is a very old pursuit, as the preceding quotation illustrates. The Bible contains many references to the use of pitch or asphalt collected from the natural seepages with which the Middle East abounds. Herodotus, writing in about 450 BC, described oil seeps in Carthage (Tunisia) and the Greek island Zachynthus (Herodotus, c. 450 BC). He gave details of oil extraction from wells near Ardericca in modern Iran, although the wells could not have been very deep because fluid was extracted in a wineskin on the end of a long pole mounted on a fulcrum. Oil, salt, and bitumen were produced simultaneously from these wells. Throughout the first millenium AD, oil and asphalt were gathered from natural seepages in many parts of the world. The early uses of oil were for medication, waterproofing, and warfare. Oil was applied externally for wounds and rheumatism and administered internally as a laxative. From the time of Noah, pitch has been used to make boats watertight. Pitch, asphalt, and oil have long been employed in warfare. When Alexander the Great invaded India in 326 BC, he scattered the Indian elephant corps by charging them with horsemen waving pots of burning pitch. Nadir Shah employed a similar device, impregnating the humps of camels with oil and sending them ablaze against the Indian elephant corps in 1739 (Pratt and Good, 1950). Greek fire was invented by Callinicus of Heliopolis in AD 668. Its precise recipe is unknown, but it is believed to have included quicklime, sulfur, and naphtha and it ignited when wet. It was a potent weapon in Byzantine naval warfare.

Up until the mid-nineteenth century, asphalt, oil, and their by-products were produced only from seepages, shallow pits, and hand-dug shafts. In 1694, the British Crown issued a patent to Masters Eele, Hancock, and Portlock to "make great quantities of pitch, tarr, and oyle out of a kind of stone" (Eele, 1697). The stone in question was of Carboniferous age and occurred at the eponymous Pitchford in Shropshire (Torrens, 1994). The first well in the Western World that specifically sunk to search for oil (as opposed to water or brine) appears to have been at Pechelbronn, France, in 1745. Outcrops of oil sand were noted in this region, and Louis XV granted a license to M. de la Sorbonniere, who sank several borings and built a refinery in the same year (Redwood, 1913). The birth of the oil shale industry is

Elements of Petroleum Geology
http://dx.doi.org/10.1016/B978-0-12-386031-6.00001-1

credited to James Young, who began retorting oil from the Carboniferous shales at Torban, Scotland, in 1847. The resultant products of these early refineries included ammonia, solid paraffin wax, and liquid paraffin (kerosene or coal oil). The wax was used for candles and the kerosene for lamps. Kerosene became cheaper than whale oil, and therefore the market for liquid hydrocarbons expanded rapidly in the mid-nineteenth century. Initially, the demand was satisfied by oil shales and from oil in natural seeps, pits, hand-dug shafts, and galleries. Before exploration for oil began, cable-tool drilling was an established technique in many parts of the world in the quest for water and brine (Fig. 1.1). The first well to produce oil intentionally in the Western World was drilled at Oil Creek, Pennsylvania, by Colonel Drake in 1859 (Owen, 1975). Previously, water wells in the Appalachians and elsewhere produced oil as a contaminant. The technology for drilling Drake's well was derived from Chinese artisans who had traveled to the United States to work on the railroads.

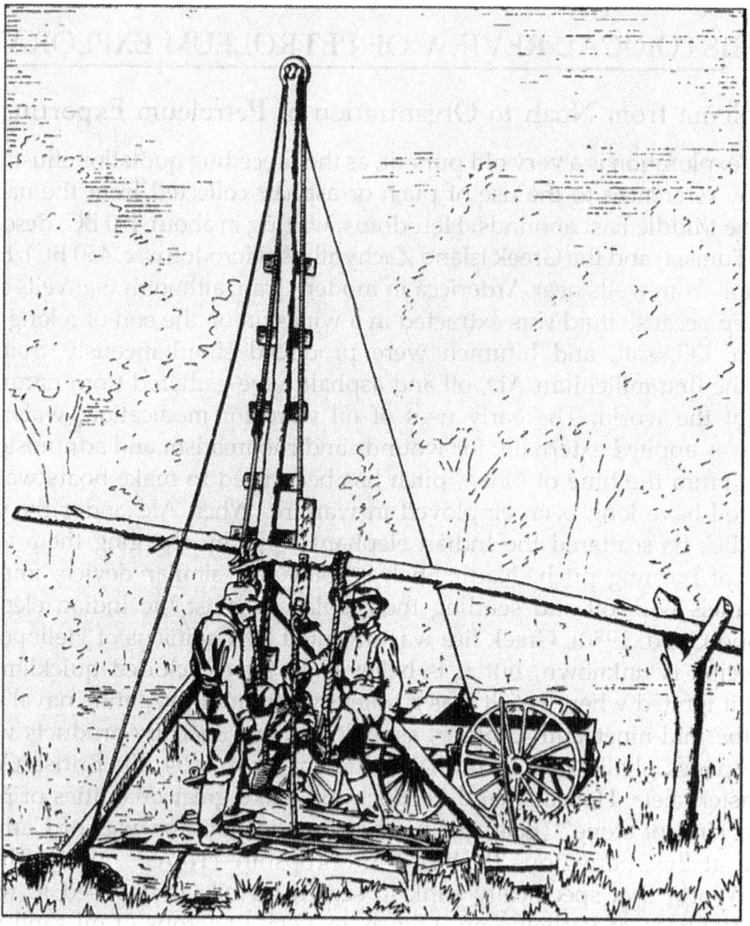

FIGURE 1.1 Early cable-tool rig used in America the motive power was provided by one man and a spring pole. *Courtesy of British Petroleum.*

Cable-tool drilling had been used in China since at least the first century BC, the drilling tools being suspended from bamboo towers up to 60 m high. In China, however, this drilling technology had developed to produce artesian brines, not petroleum (Messadie, 1995). The first "oil mine" was opened in Bobrka, Poland, in early 1854 by Ignacy Lukasiewicz (Frank, 2005; Wikipedia, 2014 "History of the Petroleum Industry"). Lukasiewicz was interested in using seep oil as an alternative to the more expensive whale oil and was the first in the world to distill kerosene from seep oil.

A rapid growth in oil production from subsurface wells soon followed, both in North America and around the world. A major stimulus to oil production was the development of the internal combustion engine in the 1870s and 1880s. Gradually, the demand for lighter petroleum fractions overtook that for kerosene. Uses were found, however, for all the refined products, from the light gases, via petrol, paraffin, diesel oil, tar, and sulfur, to the heavy residue. Demand for oil products increased greatly because of the First World War (1914–1918). By the 1920s, the oil industry was dominated by seven major companies, termed the "seven sisters" by Enrico Mattei (Sampson, 1975). These companies included:

- European:
  - British Petroleum
  - Shell
- American:
  - Exxon (formerly Esso)
  - Gulf
  - Texaco
  - Mobil
  - Socal (or Chevron)

British Petroleum and Shell found their oil reserves abroad from their parent countries, principally in the Middle and Far East, respectively. They were thus involved early in long-distance transport by sea, measuring their oil by the seagoing tonne. The American companies, by contrast, with shorter transportation distances, used the barrel as their unit of measurement. The American companies began overseas ventures, mainly in Central and South America, in the 1920s. In the 1930s, the Arabian–American Oil Company (Aramco, now Saudi Aramco) evolved from a consortium of Socal, Texaco, Mobil, and Exxon. Following the Second World War and the postwar economic boom, the idea of oil consortia became established over much of the free world. Oil companies risked the profits from one productive area to explore for oil in new areas. To take on all the risks in a new venture is unwise, so companies would invest in several joint ventures, or consortia. Table 1.1 shows some of the major consortia, demonstrating the stately dance of the seven sisters as they changed their partners around the world. In this process the major consortia shared a mutual love–hate relationship. The object of any business is to maximize profit. Thus, it was to their mutual benefit to export oil from the producing countries as cheaply as possible and to sell it in the world market for the highest price possible. The advantage of a cartel is offset by the desire of every company to enhance its sales at the expense of its competitors by selling its products cheaper.

In 1960, the Organization of Petroleum Exporting Countries (OPEC) was founded in Baghdad and consisted initially of Iraq, Iran, Kuwait, Saudi Arabia, and Venezuela (Martinez, 1969). It later expanded to include Algeria, Dubai, Ecuador, Gabon, Indonesia,

TABLE 1.1 Partners of Some of the Major Overseas Oil Consortia[a]

Companies	The Consortium, Iran	I.P.C., Iraq	Aramco, Saudi Arabia	Kuwait Oil Co., Kuwait	Admar, Abu Dhabi	A.D. P.C., Abu Dhabi	Oasis, Libya
B.P.	X	X		X	X	X	
Shell	X	X				X	X
Exxon	X	X	X			X	
Mobil	X	X	X			X	
Gulf	X			X			
Texaco	X		X				
Socal	X		X				
C.F.P.	X					X	
Conoco							X
Amerada							X
Marathon							X

[a]*Note that partners and their percentage interest varied over the lifetime of the various consortia.*

Libya, Nigeria, Qatar, and the United Arab Emirates. To qualify for membership, a country's economy must be predominantly based on oil exports; therefore, the United States and the United Kingdom do not qualify. By the mid-1970s, OPEC was producing two-thirds of the free world's oil. The object of OPEC is to control the power of the independent oil companies by a combination of price control and appropriation of company assets. For many OPEC countries, oil is their only natural resource. Once it is depleted, they will have no assets unless they can maximize their oil revenues and spend them in the development of other industries. The OPEC objective has been notably successful, although its large price increases in the early 1970s contributed to a global recession, which affected both the developed and the poorer Third World countries alike.

The idea of the producing state controlling the oil company's activities has now been exported from OPEC. Not only have state oil companies been formed in countries that formerly lacked indigenous oil expertise (e.g., Statoil in Norway and Petronas in Malaysia), but they have also been formed in those that had the expertise (Petrocan in Canada and the former Britoil in the United Kingdom). Formerly, the profit that the oil companies made in one country was the risk capital to be invested in the next country. With state oil companies the taxpayer shares the risk and the profit.

The most influential state oil and gas companies based in countries outside the Organization of Economic Co-operation and Development according to the Financial Times (2007) are:

- China National Petroleum Corporation (China)
- Gazprom (Russia)
- National Iranian Oil Company (Iran)
- Petrobras (Brazil)

- Petroleos de Venezuela S.A. (Venezuela)
- Petronas (Malaysia)
- Saudia Aramco (Saudi Arabia)

This group of state oil and gas companies has been labeled the "New Seven Sisters (Hoyos, 2007)." These largely state-owned companies are the new rule makers and control almost one-third of the world's oil and gas production and more than one-third of the world's total oil and gas reserves.

1.1.2 Evolution of Petroleum Exploration Concepts and Techniques

From the days of Noah to OPEC the role of the petroleum geologist has become more and more skilled and demanding. In the early days, oil was found by wandering about the countryside with a naked flame, optimism, and a sense of adventure. One major U.S. company, which will remain nameless, once employed a chief geologist whose exploration philosophy was to drill on old Indian graves. Another oil finder used to put on an old hat, gallop about the prairie until his hat dropped off, and start drilling where it landed. History records that he was very successful (Cunningham-Craig, 1912). One of the earliest exploration tools was "creekology." It gradually dawned on the early drillers that oil was more often found by wells located on river bottoms than by those on the hills (Fig. 1.2). The anticlinal theory of oil entrapment, which explained this phenomenon, was expounded by Hunt (1861). Up to the present day, the quest for anticlines has been one of the most successful exploration concepts.

Experience soon proved, however, that oil could also occur off structure. Carll (1880) noted that the oil-bearing marine Venango sands of Pennsylvania occurred in trends that reflected not structure, but paleoshorelines. Thus was borne the concept that oil could be trapped stratigraphically as well as structurally. Stratigraphic traps are caused by variations in deposition, erosion, or diagenesis within the reservoir.

Through the latter part of the nineteenth and the early part of the twentieth century, oil exploration was based on the surface mapping of anticlines. Stratigraphic traps were found accidentally by serendipity or by subsurface mapping and extrapolation of data gathered

FIGURE 1.2 Creekology—the ease of finding oil in the old days.

from wells drilled to test structural anomalies. Unconformities and disharmonic folding limited the depth to which surface mapping could be used to predict subsurface structure. The solution to this problem began to emerge in the mid-1920s, when seismic (refraction), gravity, and magnetic methods were all applied to petroleum exploration. Magnetic surveys seldom proved to be effective oil finders, whereas gravity and seismic methods proved to be effective in finding salt dome traps in the Gulf of Mexico coastal province of the United States. In the same period geophysical methods were also applied to borehole logging, with the first electric log run at Pechelbronn, France, in 1927. Further electric, sonic, and radioactive logging techniques followed. Aerial surveying began in the 1920s, but photogeology, which employs stereophotos, only became widely used after the Second World War. At this time aerial surveys were cheap enough to allow the rapid reconnaissance of large concessions, and photogeology was notably effective in the deserts of North Africa and the Middle East, where vegetation does not cover surface geology.

Pure geological exploration methods advanced slowly but steadily during the first half of the twentieth century. One of the main applications to oil exploration was the development of micropaleontology. The classic biostratigraphic zones, which are based on macrofossils such as ammonites, could not be identified in the subsurface because of the destructive effect of drilling. New zones had to be defined by microfossils, which were calibrated at the surface with macrofossil zones. The study of modern sedimentary environments in the late 1950s and early 1960s, notably on Galveston Island (Texas), the Mississippi delta, the Bahama Bank, the Dutch Wadden Sea, and the Arabian Gulf, gave new insights into ancient sedimentary facies and their interpretation. This insight provided improved prediction of the geometry and internal porosity and permeability variation of reservoirs.

The 1970s saw major advances on two fronts: geophysics and geochemistry. The advent of the computer resulted in a major quantum jump in seismic processing. Instead of seismologists poring painfully over a few bunched galvanometer traces, vast amounts of data could be displayed on continuous seismic sections. Reflecting horizons could be picked out in bright colors, first by geophysicists and later even by geologists. As techniques improved, seismic lines became more and more like geological cross sections, until stratigraphic and environmental concepts were directly applicable.

In the 1980s, increasing computing power led to the development of 3D seismic surveys that enabled seismic sections of the earth's crust to be displayed in any orientation, including horizontal. Thus, it is now possible to image directly the geometry of many petroleum reservoirs. Similarly enhanced processing methods made it possible to detect directly the presence of oil and gas. These improvements went hand in hand with enhanced borehole logging. It is now possible to produce logs of the mineralogy, porosity, and pore fluids of boreholes, together with images of the geological strata that they penetrate. These techniques are discussed and illustrated in detail in Chapter 3.

As the millenium approaches, one can only speculate on what new advances in petroleum exploration technology will be discovered. All techniques may be expected to improve. Remote sensing from satellites may be one major new tool, as might direct sensing from surface geochemical or geophysical methods. These latter methods generally involve the identification of gas microseeps and fluctuations in electrical conductivity of rocks above petroleum accumulations. Such methods have been around for half a century, but have yet to be widely accepted.

From the earliest days of scientific investigation the formation of petroleum had been attributed to two origins: inorganic and organic. Chemists, such as Mendeleyev in the nineteenth century, and astronomers, such as Gold and Hoyle in the twentieth, argued for an inorganic origin—sometimes igneous, sometimes extraterrestrial, or a mixture of both. Most petroleum geologists believe that petroleum forms from the diagenesis of buried organic matter and note that it is indigenous to sedimentary rocks rather than igneous ones. The advent of cheap and accurate chemical analytical techniques allowed petroleum source rocks to be studied. It is now possible to match petroleum with its parent shale and to identify potential source rocks, their tendency to generate oil or gas, and their level of thermal maturation. For a commercial oil accumulation to occur, five conditions must be fulfilled:

1. There must be an organic-rich source rock to generate the oil and/or gas.
2. The source rock must have been heated sufficiently to yield its petroleum.
3. There must be a reservoir to contain the expelled hydrocarbons. This reservoir must have *porosity*, to contain the oil and/or gas, and *permeability*, to permit fluid flow.
4. The reservoir must be sealed by an impermeable cap rock to prevent the upward escape of petroleum to the earth's surface.
5. Source, reservoir, and seal must be arranged in such a way as to trap the petroleum.
6. The timing of trap formation, petroleum generation, and accumulation must be in a favorable sequence.
7. The accumulation must be preserved or protected from breaching, flushing, aerobic bacteria, thermal degradation, etc. until exploitation.

Chapter 5 deals with the generation and migration of petroleum from source rocks. Chapter 6 discusses the nature of reservoirs, and Chapter 7 deals with the different types of traps.

1.2 THE CONTEXT OF PETROLEUM GEOLOGY

1.2.1 Relationship of Petroleum Geology to Science

Petroleum geology is the application of geology (the study of rocks) to the exploration for and production of oil and gas. Geology itself is firmly based on chemistry, physics, and biology, involving the application of essentially abstract concepts to observed data. In the past, these data were basically observational and subjective, but they are now increasingly physical and chemical, and therefore more objective. Geology, in general, and petroleum geology, in particular, still rely on value judgments based on experience and an assessment of validity among the data presented. The preceding section showed how petroleum exploration had advanced over the years with the development of various geological techniques. It is now appropriate to consider in more detail the roles of chemistry, physics, and biology in petroleum exploration (Fig. 1.3).

1.2.2 Chemistry and Petroleum Geology

The application of chemistry to the study of rocks (geochemistry) has many uses in petroleum geology. Detailed knowledge of the mineralogical composition of rocks is

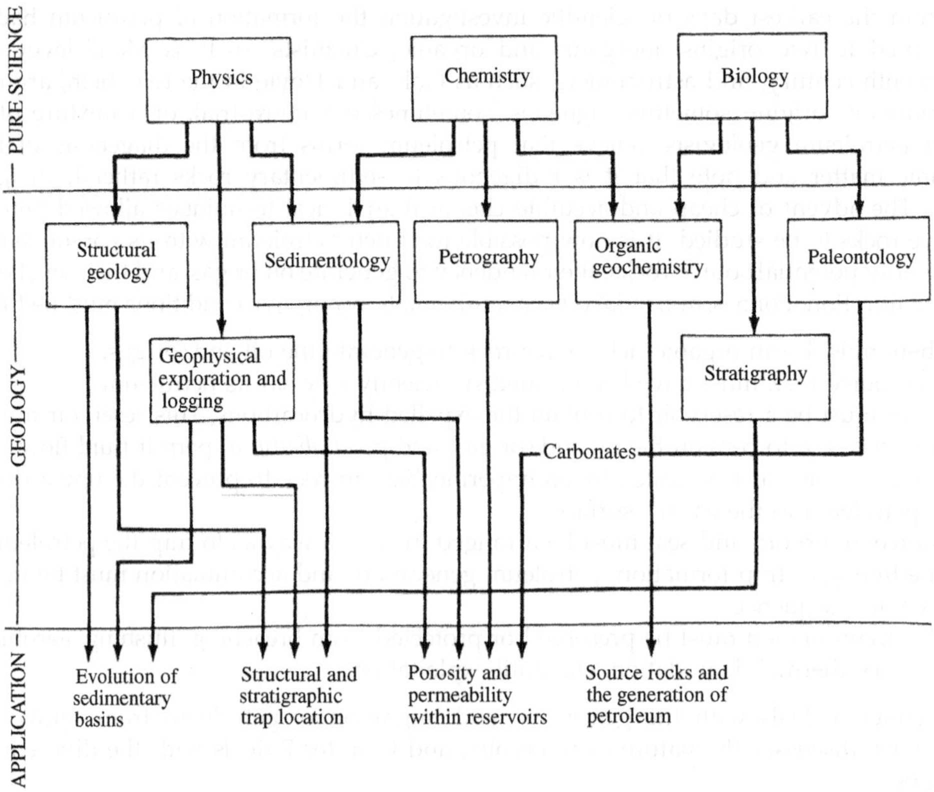

FIGURE 1.3 The relationship of petroleum geology to the pure sciences.

important at many levels. In the early stages of exploration, certain general conclusions as to the distribution and quality of potential reservoirs could be made from their gross lithology. For example, the porosity of sandstones tends to be facies related, whereas in carbonate rocks this is generally not so. Detailed knowledge of the mineralogy of reservoirs enables estimates to be made of the rate at which they may lose porosity during burial, and this detailed mineralogical information is essential for the accurate interpretation of geophysical well logs through reservoirs. Knowledge of the chemistry of pore fluids and their effect on the stability of minerals can be used to predict the places where porosity may be destroyed by cementation, preserved in its original form, or enhanced by the solution of minerals by formation waters. Organic chemistry is involved both in the analysis of oil and gas and in the study of the diagenesis of plant and animal tissues in sediments and the way in which the resultant organic compound, kerogen, generates petroleum.

1.2.3 Physics and Petroleum Geology

The application of physics to the study of rocks (geophysics) is very important in petroleum geology. In its broadest application geophysics makes a major contribution to understanding the earth's crust and, especially through the application of modern plate tectonic theory, to the genesis and petroleum potential of sedimentary basins. More specifically, physical concepts are required to understand folds, faults, and diapirs, and hence their roles in petroleum entrapment. Modern petroleum exploration is unthinkable without the aid of magnetic, gravity, and seismic surveys in finding potential petroleum traps. Nor could any finds be evaluated effectively without geophysical wireline well logs to measure the lithology, porosity, and petroleum content of a reservoir.

1.2.4 Biology and Petroleum Geology

Biology is applied to geology in several ways, notably through the study of fossils (paleontology), and is especially significant in establishing biostratigraphic zones for regional stratigraphical correlation. The way in which oil exploration shifted the emphasis from the use of macrofossils to microfossils for zonation has already been noted. Ecology, the study of the relationship between living organisms and their environment, is also important in petroleum geology. Carbonate sediments, in general, and reefs, in particular, can only be studied profitably with the aid of detailed knowledge of the ecology of modern marine fauna and flora. Biology, and especially biochemistry, is important in studying the transformation of plant and animal tissues into kerogen during burial and the generation of oil or gas that may be caused by this transformation.

1.2.5 Relationship of Petroleum Geology to Petroleum Exploration and Production

Geologists, in contrast to some nongeologists, believe that knowledge of the concepts of geology can help to find petroleum and, furthermore, often think that petroleum geology and petroleum exploration are synonymous, but they are not. Theories that petroleum is not formed by the transformation of organic matter in sediments have already been noted and are examined in more detail in Chapter 5. If the petroleum geologists' view of oil generation and migration is not accepted, then present exploration methods would need extensive modification.

Some petroleum explorationists still do not admit to a need for geologists to aid them in their search. In 1982, a successful oil finder from Midland, Texas, admitted to not using geologists because when his competitors hired them, all it did was increase their costs per barrel of oil found. The South African state oil company was under a statutory obligation imposed by its government to test every claim to an oil-finding method, be it dowsing or some sophisticated scientific technique. These examples are not isolated cases, and it has been argued that oil may better be found by random drilling than by the application of scientific principles.

Petroleum geology is only one aspect of petroleum exploration and production. Leaving aside atypical enterprises, petroleum exploration now involves integrated teams of people possessing a wide range of professional skills (Fig. 1.4). These skills include political and social expertise, which is involved in the acquisition of prospective acreage. Geophysical surveying is involved in preparing the initial data on which leasing and, later, drilling recommendations are based. Geological concepts are applied to the interpretation of the geophysical data once they have been acquired and processed. As soon as an oil well has been drilled,

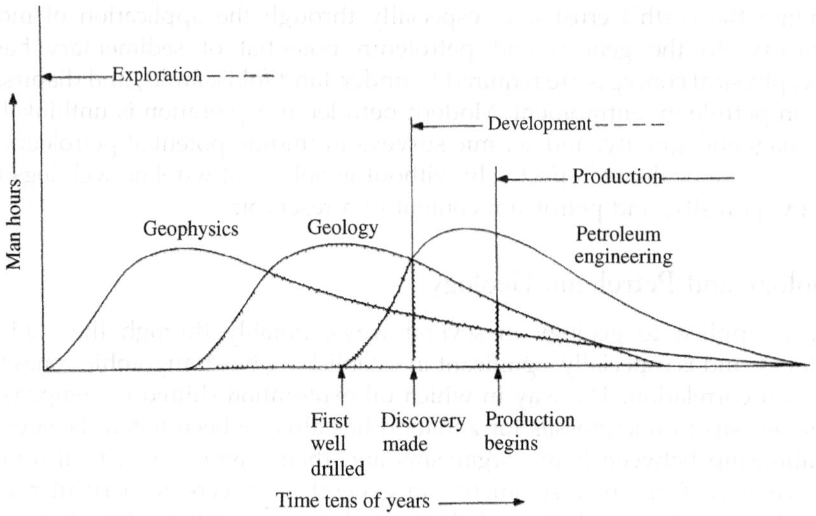

FIGURE 1.4 Graph showing how petroleum geology is part of a continuum of disciplines employed in the exploration and production of oil and gas. Note that geophysics now extends beyond the beginning of production. Repeated seismic surveys can monitor the migration of fluid interfaces within fields during their productive lifetime (4D seismic).

the engineering aspects of the discovery need appraisal. Petroleum engineering is concerned with establishing the reserves of a field, the distribution of petroleum within the reservoir, and the most effective way of producing it. Thus petroleum geology lies within a continuum of disciplines, beginning with geophysics and ending with petroleum engineering, but overlapping both in time and subject matter (Fig. 1.5).

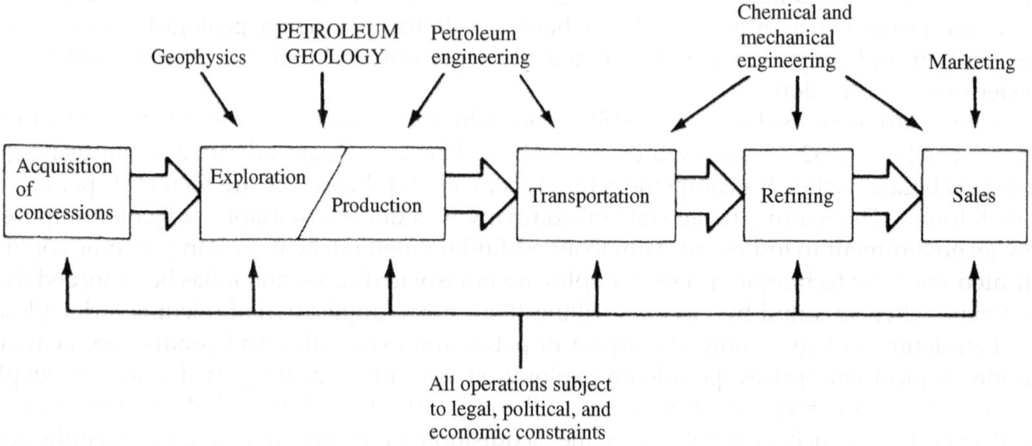

FIGURE 1.5 Flowchart showing how petroleum geology is only one aspect of petroleum exploration and production, and how these enterprises themselves are part of a continuum of events subjected to various constraints and expedited by many disciplines.

Underlying this sequence of events is the fundamental control of economics. Oil companies exist not only to find oil and gas but, like any business enterprise, to make money. Thus, every step of the journey, from leasing to drilling, to production, and finally to enhanced recovery, is monitored by accountants and economists. Activity in petroleum exploration and production accelerates when the world price of petroleum products increases, and it decreases or even terminates when the price drops. Petroleum geologists are in the unusual position of being subject to firing for either technical incompetence or excellence, but not for mediocrity. If they find no oil, they may be fired; similarly, if they find too much oil, then their presence on the company payroll is unnecessary. This has been demonstrated repeatedly since the oil industry began. Competent geologists are more important to small companies, for whom a string of dry holes spells catastrophe. Major companies can tolerate a fair degree of incompetence because they have the financial resources to withstand a string of disasters. Nowhere is this more true than in state oil companies. With an endless supply of taxpayers' money to sustain them, the political expediency of searching for indigenous petroleum reserves may outweigh any economic consideration.

References

Carll, J.F., 1880. The geology of the oil regions of Warren, Venango, Clarion and Butler Counties. Pa. Geol. Surv. 3, 482.
Cunningham-Craig, E.H., 1912. Oil-finding. Edward Arnold, London.
Eele, M., 1697. On making pitch, tar and oil out of a blackish stone in Shropshire. Philos. Trans. R. Soc. Lond. 19, 544.
Frank, A.F., 2005. Oil Empire: Visions of Prosperity in Austrian Galicia (Harvard Historical Studies). Harvard University Press, ISBN 0-674-01887-7.
Herodotus, H., c. 450 BC. The Histories. 9 books.
Hoyos, C., March 11, 2007. The New Seven Sisters: Oil and Gas Giants Dwarf Western Rivals. Financial Times. http://www.ft.com/cms/s/2/471ae1b8-d001-11db-94cb-000b5df10621.html#axzz2GpiGeMvd.
Hunt, T.S., March 1, 1861. Bitumens and Mineral Oils. Montreal Gazette.
Martinez, A.R., 1969. Chronology of Venezuelan Oil. Allen & Unwin, London.
Messadie, G., 1995. The Wordsworth Dictionary of Inventions. Chambers, Edinburgh.
Owen, E.W., 1975. Trek of the Oil Finders: A History of Exploration for Petroleum. Am. Assoc. Pet. Geol., Tulsa, OK.
Pratt, W.E., Good, D., 1950. World Geography of Petroleum. Princeton University Press, Princeton, NJ.
Redwood, Sir B., 1913. A Treatise on Petroleum, vol. 1. Griffin, London.
Sampson, A., 1975. The Seven Sisters. Hodden & Stoughton, London.
Torrens, H., 1994. 300 years of oil. In: The British Association Lectures 1993. The Geological Society, London, pp. 4–8.
Wikipedia, 2014, http://en.wikipedia.org/wiki/History_of_the_petroleum_industry

Selected Bibliography

For an account of the early historical evolution of the oil industry, see:
Redwood, S.B., 1913. A Treatise on Petroleum, vol. 1. Griffin, London.
For the evolution of geological concepts in petroleum exploration, see:
Dott, R.H., Reynolds, M.J., 1969. Mem. No. 5. In: Source Book for Petroleum Geology. Am. Assoc. Pet. Geol., Tulsa, OK.
For accounts of the evolution of the oil industry over the last century, see:
Owen, E.W., 1975. Trek of the Oil Finders: A History of Exploration for Petroleum. Am. Assoc. Pet. Geol., Tulsa, OK.
Pratt, W.E., Good, D., 1950. World Geography of Petroleum. Princeton University Press, Princeton, NJ.
Sampson, A., 1975. The Seven Sisters. Hodden & Stoughton, London.

CHAPTER

2

The Physical and Chemical Properties of Petroleum

Petroleum exploration is largely concerned with the search for oil and gas, two of the chemically and physically diverse group of compounds termed the *hydrocarbons*. Physically, hydrocarbons grade from gases, via liquids and plastic substances, to solids. The hydrocarbon gases include dry gas (methane) and the wet gases (ethane, propane, butane, etc.). Condensates are hydrocarbons that are gaseous in the subsurface, but condense to liquid when they are cooled at the surface. Liquid hydrocarbons are termed *oil, crude oil,* or just *crude,* to differentiate them from refined petroleum products. The plastic hydrocarbons include asphalt and related substances. Solid hydrocarbons include coal and kerogen. Gas hydrates are ice crystals with peculiarly structured atomic lattices, which contain molecules of methane and other gases. This chapter describes the physical and chemical properties of natural gas, oil, and the gas hydrates; it is a necessary prerequisite to Chapter 5, which deals with petroleum generation and migration. The plastic and solid hydrocarbons are discussed in Chapter 9, which covers the tar sands and oil shales.

The earth's atmosphere is composed of natural gas. In the oil industry, however, natural gas is defined as "a mixture of hydrocarbons and varying quantities of nonhydrocarbons that exist either in the gaseous phase or in solution with crude oil in natural underground reservoirs." The foregoing is the definition adopted by the American Petroleum Institute (API), the American Association of Petroleum Geologists (AAPG), and the Society of Petroleum Engineers (SPE). The same authorities subclassify natural gas into dissolved, associated, and nonassociated gas. Dissolved gas is in solution in crude oil in the reservoir. Associated gas, commonly known as gas cap gas, overlies and is in contact with crude oil in the reservoir. Nonassociated gas is in reservoirs that do not contain significant quantities of crude oil. Natural gas liquids, or NGLs, are the portions of the reservoir gas that are liquefied at the surface in lease operations, field facilities, or gas processing plants. NGLs include, but are not limited to, ethane, propane, butane, pentane, natural gasoline, and condensate. Basically, natural gases encountered in the subsurface can be classified into two groups: those of organic origin and those of inorganic origin (Table 2.1).

Gases are classified as *dry* or *wet* according to the amount of liquid vapor that they contain. A dry gas may be arbitrarily defined as one with less than 0.1 gal/1000 ft^3 of condensate; chemically, dry gas is largely methane. A wet gas is one with more than 0.3 gal/1000 ft^3 of

Elements of Petroleum Geology
http://dx.doi.org/10.1016/B978-0-12-386031-6.00002-3

TABLE 2.1 Natural Gases and Their Dominant Modes of Formation

Gas			Dominant source
Inert gases	Helium		Inorganic
Inert gases	Argon		Inorganic
Inert gases	Krypton		Inorganic
Inert gases	Radon		Inorganic
	Nitrogen		Inorganic
	Carbon dioxide		Mixed
	Hydrogen sulfide		Mixed
	Hydrogen		Mainly organic
Hydrocarbons	Methane—dry gas		Mainly organic
Hydrocarbons	Ethane	wet gases	Mainly organic
Hydrocarbons	Propane	wet gases	Mainly organic
Hydrocarbons	Butane	wet gases	Mainly organic

condensate; chemically, these gases contain ethane, butane, and propane. Gases are also described as *sweet* or *sour*, based on the absence or presence, respectively, of hydrogen sulfide.

2.1 NATURAL GASES

2.1.1 Hydrocarbon Gases

The major constituents of natural gas are the hydrocarbons of the paraffin series (Table 2.2). The heavier members of the series decline in abundance with increasing molecular weight. Methane is the most abundant; ethane, butane, and propane are quite common, and paraffins with a molecular weight greater than pentane are the least common. Methane (CH_4) is also known as *marsh gas* if found at the surface or *fire damp* if present down a coal mine. Traces of methane are commonly recorded as shale gas or background gas during the drilling of all but the driest of dry wells. Methane is a colorless, flammable gas, which is produced (along with other fluids) by the destructive distillation of coal. As such, it was commonly used for domestic purposes in Europe until replaced by natural gas, itself largely composed of methane. Methane is the first member of the paraffin series. It is chemically nonreactive, sparingly soluble in water, and lighter than air (0.554 relative density).

Methane forms in three ways. It may be derived from the mantle, it may form from the thermal maturation of buried organic matter, and it may form by the bacterial degradation of organic matter at shallow depths. Geochemical and isotope analysis can differentiate the source of methane in a reservoir. Mantle-derived methane is differentiated from biogenically sourced methane from the carbon 12:13 ratio. Methane occurs as a by-product of bacterial

TABLE 2.2 Significant Data of the Paraffin Series

Name	Formula	Molecular weight	Boiling point at atmospheric pressure (°C)	Solubility (g/10°g water)
Methane	CH_4	16.04	−162	24.4
Ethane	C_2H_6	30.07	−89	60.4
Propane	C_3H_8	44.09	−42	62.4
Isobutane	C_4H_{10}	58.12	−12	48.9
n-Butane	C_4H_{10}	58.12	−1	61.4
Isopentane	C_5H_{12}	72.15	30	47.8
n-Pentane	C_5H_{12}	72.15	36	38.5
n-Hexane	C_6H_{14}	86.17	69	9.5

decay of organic matter at normal temperatures and pressures. This biogenic methane has considerable potential as a source of energy. It has been calculated that some 20% of the natural gas produced today is of biogenic origin (Rice and Claypool, 1981). In the nineteenth century, eminent Victorians debated the possibility of lighting the streets of London with methane from the sewers. Today's avant-garde agriculturalists acquire much of the energy needed for their farms by collecting the gas generated by the maturation of manure. Methane generated by waste fills (garbage) is now pumped into the domestic gas grid in many countries. Biogenic methane is commonly formed in the shallow subsurface by the bacterial decay of organic-rich sediments. As the burial depth and temperature increase, however, this process diminishes and the bacterial action is extinguished. The methane encountered in deep reservoirs is produced by thermal maturation of organic matter. This process is discussed in detail later in Chapter 5.

The other major hydrocarbons that occur in natural gas are ethane, propane, butane, and occasionally pentane. Their chemical formulas and molecular structure are shown in Fig. 2.1. Their occurrence in various gas reservoirs is given in Table 2.3. Unlike methane these heavier members of the paraffin series do not form biogenically. They are only produced by the thermal maturation of organic matter. If their presence is recorded by a gas detector during the drilling of a well, it often indicates proximity to a significant petroleum accumulation or source rock.

2.1.2 Nonhydrocarbon Gases

2.1.2.1 Inert Gases

Helium is a common minor accessory in many natural gases, and traces of argon and radon have also been found in the subsurface. Helium occurs in the atmosphere at 5 ppm and has also been recorded in mines, hot springs, and fumaroles. It has been found in oil field gases in amounts of up to 8% (Dobbin, 1935). In North America helium-enriched natural gases occur in the Four Corners area and Texas panhandle of the United States and in Alberta

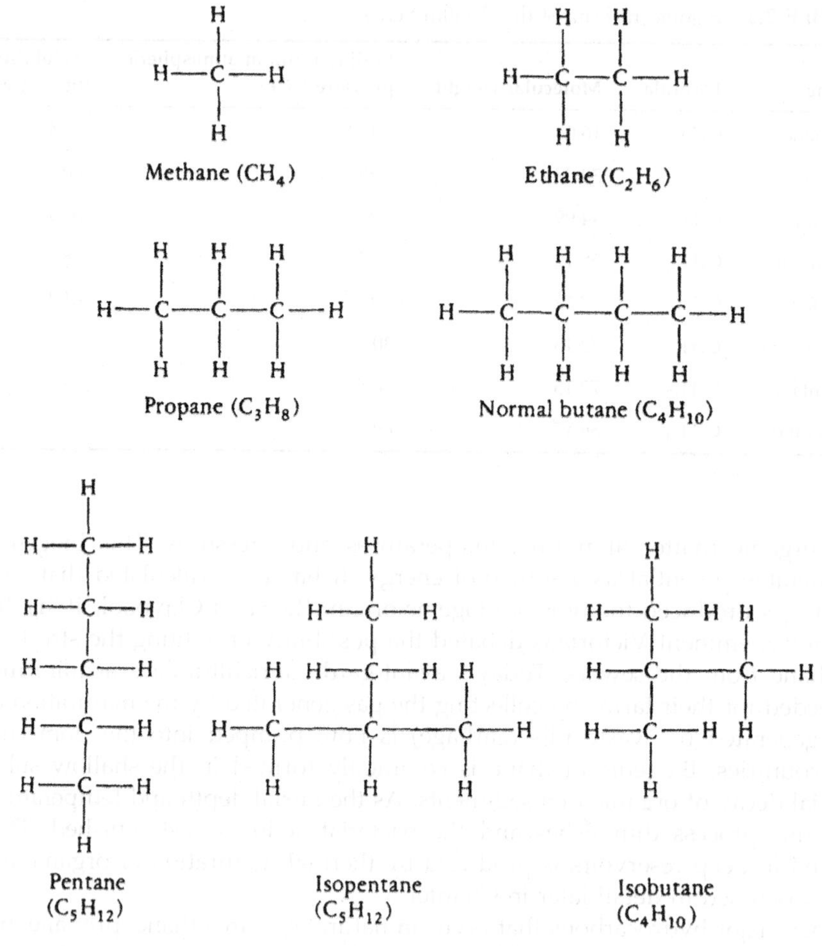

FIGURE 2.1 Molecular structures of the more common hydrocarbon gases.

and Saskatchewan, Canada (Lee, 1963; Hitchon, 1963). In Canada the major concentrations occur along areas of crustal tension, such as the Peace River and Sweetwater arches, and the foothills of the Rocky Mountains. Other regions containing helium-enriched natural gases include Poland, Alsace, and Queensland in Australia. Helium is known to be produced by the decay of various radioactive elements, principally uranium, thorium, and radium. The rates of helium production for these elements are shown in Table 2.4. Rogers (1921) calculated that between 282 and 1.06 billion ft^3 of helium are generated annually. Based on these data, the helium found in natural gas is widely believed to have emanated from deep-seated basement rocks, especially granite. Although the actual rate of production is slow and steady, the expulsion of gas into the overlying sediment cover may occur rapidly when the basement is subjected to thermal activity or fracturing by crustal arching.

Ideally it would be useful to demonstrate a correlation between helium-enriched natural gases and radioactive basement rocks. This correlation is generally difficult to establish

TABLE 2.3 Chemical Composition of Various Gas Fields

	Composition									
Field and area	**Methane**	**Ethane**	**Propane**	**Butane**	**Pentane^{+}**	**Co_2**	**N**	**H_2S**	**He**	**References**
SOUTHERN N. SEA BASIN										
Groningen	81.3	2.9	0.4	0.1	0.1	0.9	14.3	Tr[a]	Tr	Cooper (1977, 1975)
Hewett	83.2	5.3	2.1	0.4	0.5	0.1	8.4	Tr	Tr	Cumming and Wyndham (1975)
West Sole	94.4	3.1	0.5	0.2	0.2	0.5	1.1			Butler (1975)
ALGERIA										
Hassi-R'Mel	83.5	7.0	2.0	0.8	0.4	0.2	6.1			Maglione (1970)
FRANCE										
Lacq	69.3	3.1	1.1	0.6	0.7	9.6	0.4	15.2		Cooper (1977)
CANADA										
Turner Valley	92.6	4.1	2.5	0.7	0.13					Slipper (1935)
U.S.A.										
Panhandle	91.3	3.2	1.7	0.9	0.56	0.1			0.5	Cotner and Crum (1935)
Hugoton	74.3	5.8	3.5	1.5	0.6				Av.[b]	Keplinger et al. (1949)
TRINIDAD										
Barrackpore	95.65	2.25	1.55	0.80	0.25					Suter (1952)
VENEZUELA										
La Concepcion	70.9	8.2	8.2	6.2	3.7	2.8				Anon (1948)
Cumarebo	63.89	9.49	14.41	8.8	5.4					Payne (1951)
NEW ZEALAND										
Kapuni	46.2	5.2	2.0	0.6	44.9	1.0				Cooper (1975)
RUSSIA										
Baku	88.0	2.26	0.7	0.5	6.5					Shimkin (1950)
ABU DHABI										
Zakum	76.0	11.4	5.4	2.2	1.3	2.3	1.1	0.3.3		Cooper (1975)
IRAN										
Agha Jari	66.0	14.0	10.5	5.0	2.0	1.5	1.0			Cooper (1975)

[a] *Tr = trace.*
[b] *Av. = average.*

TABLE 2.4 Production Rates of Helium from Various Radioactive Elements

Radioactive substance	Helium production [g/(year)(mm^3)]
Uranium	2.75×10^{-5}
Uranium in equilibrium with its products	11.0×10^{-5}
Thorium in equilibrium with its products	3.1×10^{-5}
Radium in equilibrium with emanation, radium A, and radium C	158

From: Geology of Petroleum by Levorsen. © 1967 by W. H. Freeman and Company. Used with permission.

because helium tends to occur in deep, rather than shallow, wells, and there is seldom sufficient well control to map the geology of the basement. The major source of helium in the United States is the Panhandle Hugoton field in Texas. This field locally contains up to 1.86% helium, and an extraction plant has been working since 1929 (Pippin, 1970). It is significant that this field produces helium from sediments pinching out over a major granitic fault block. Helium also occurs from the breakdown of uranium ore bodies within sedimentary sequences, for example, at Castlegate in Central Utah. Apart from helium, traces of other inert gases have been found in the subsurface. Argon occurs in the Panhandle Hugoton field (as does radon) and in Japan. Argon and radon are by-products of the radioactive disintegration of potassium and radium, respectively, and are believed to have an origin similar to that of helium.

Helium is of considerable economic significance because it is lighter than air and, because it is inert, is safer than hydrogen for use in dirigibles. Argon and radon are of little economic significance. Radon is a considerable environmental hazard, however, because its inhalation is a cause of lung cancer. Thus, regional maps of radon gas are prepared, and steps are taken to ameliorate radon gas invasion into houses by the use of impermeable membrane foundations. Radon health hazards are highest in areas of basement in general, and granites in particular (Durrance, 1986). The highest radon reading in the United Kingdom was recorded in the lavatory of a health center on the Dartmoor Granite of Devon.

Nitrogen is another nonhydrocarbon gas that frequently occurs naturally in the earth's crust. It is commonly associated both with the inert gases just described and with hydrocarbons. Nitrogen is found in North America in a belt stretching from New Mexico to Alberta and Saskatchewan. The Rattlesnake gas field of New Mexico is a noted example (Hinson, 1947) and has the following composition:

Nitrogen	72.6%
Methane	14.2%
Helium	7.6%
Ethane	2.8%
Carbon dioxide	2.8%
	100.0%

Note the association of nitrogen with helium. Nitrogen is also a common constituent of the Rotliegendes gases of the southern North Sea basin. Published accounts show ranges from 1 to 14% in the West Sole and Groningen fields, respectively (Butler, 1975; Stauble and Milius, 1970). Individual wells in the German offshore sector are reported to have encountered far higher quantities. The origin of nitrogen is less straightforward than that of the inert gases. Nitrogen has been recorded from volcanic emanations, but some studies have suggested that it may form organically, for example, by the bacterial degradation of nitrates via ammonia. Although chemically feasible, such biogenic nitrogen is likely to be produced only in shallow conditions, whereas in nature it occurs in deep hydrocarbon reservoirs. It has also been suggested that the thermal metamorphism of bituminous carbonates could generate both nitrogen and carbon dioxide. This mechanism has been postulated for the Beaverhill Lake formation of Alberta and Saskatchewan (Hitchon, 1963).

Several pieces of information suggest that the major source of nitrogen is basement, or more specifically, igneous rock. First, the association of nitrogen with basement-derived inert gases has been already noted. Second, in the southern North Sea basin nitrogen content is unrelated to the variation of the coal rank of the Carboniferous beds that sourced the hydrocarbon gases (Eames, 1975). The amount of nitrogen in Rotliegendes reservoirs appears to increase eastward across the basin in the direction of increased igneous intrusives. Another significant case has been described from southwestern Saskatchewan by Burwash and Cumming (1974). Natural gas produced from basal Paleozoic clastics contains 97% nitrogen, 2% helium, and 1% carbon dioxide. No associated hydrocarbons are found. The accumulation overlies an upstanding volcanic plug in metamorphic basement (Fig. 2.2). Taken together, these various lines of evidence suggest that nitrogen in natural gases are predominantly of inorganic origin, although organic processes may be significant generators of atmospheric and shallow nitrogen. Furthermore, some atmospheric nitrogen may have been trapped in sediments during deposition, occurring now as connate gas.

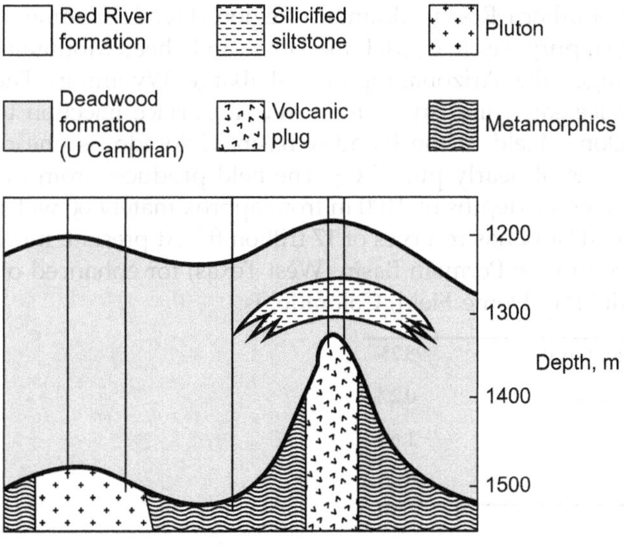

FIGURE 2.2 Cross section of the accumulation of nitrogen, helium, and carbon dioxide gases over a buried volcanic plug in southwestern Saskatchewan, Canada. *From Burwash and Cumming (1974), with permission from the Canadian Society of Petroleum Geologists.*

2.1.2.2 Hydrogen

Free hydrogen gas rarely occurs in the subsurface, partly because of its reactivity and partly because of its mobility. Nonetheless, one or two instances are known. More than 1.36 trillion ft^3 of hydrogen gas has been discovered in Mississippian rocks of the Forest City basin, Kansas. Analysis of the gas discovered revealed a composition of 40% hydrogen, 60% nitrogen, and traces of carbon dioxide, argon, and methane (Anonymous, 1948). Hydrogen is commonly dissolved in subsurface waters and in petroleum as traces, but it is seldom recorded in conventional analyses (Hunt, 1979, 1996). Subsurface hydrogen is probably produced by the thermal maturation of organic matter.

2.1.2.3 Carbon Dioxide

Carbon dioxide (CO_2) is often found as a minor accessory in hydrocarbon natural gases. It is also associated with nitrogen and helium, as shown by previously presented data. Natural gas accumulations in which carbon dioxide is the major constituent are common in areas of extensive volcanic activity, such as Sicily, Japan, New Zealand, and the Cordilleran chain of North America, from Alaska to Mexico. Levorsen (1967) cites a specific example from the Tensleep Sandstone (Pennsylvanian) of the Wertz Dome field, Wyoming:

Carbon dioxide	42.00%
Hydrocarbons	52.80%
Nitrogen	4.09%
Hydrogen sulfide	1.11%
	100.00%

Levorsen also quotes wells in New Mexico capable of producing up to 26 million ft^3/day of 99% pure carbon dioxide. Numerous natural accumulations of CO_2-dominant gas are located in the Colorado Plateau and Southern Rocky Mountain Region (Fig. 2.3). Some of these fields are produced for commercial purposes (e.g., McElmo Dome and Sheep Mountain, Colorado; Farnham Dome, Utah; Springerville, Arizona; Big Piney-LaBarge, Wyoming). The commercial uses of CO_2 include dry ice sales, industrial uses, and subsurface injection to enhance oil recovery. The McElmo Dome Field of the Paradox Basin, Colorado, is one of the world's largest known accumulations of nearly pure CO_2. The field produces from the Mississippian Leadville Limestone at average depths of 2100 m from approximately 60 wells. The field has produced over 3.3 trillion ft^3 and has reserves of 17 trillion ft^3. At present, more than 1.1 billion ft^3/day can be delivered to the Permian Basin (West Texas) for enhanced oil operations. The gas composition at McElmo Dome Field, Colorado, is:

Carbon dioxide	98.2%
Hydrocarbons	0.2%
Nitrogen	1.6%
	100.00%

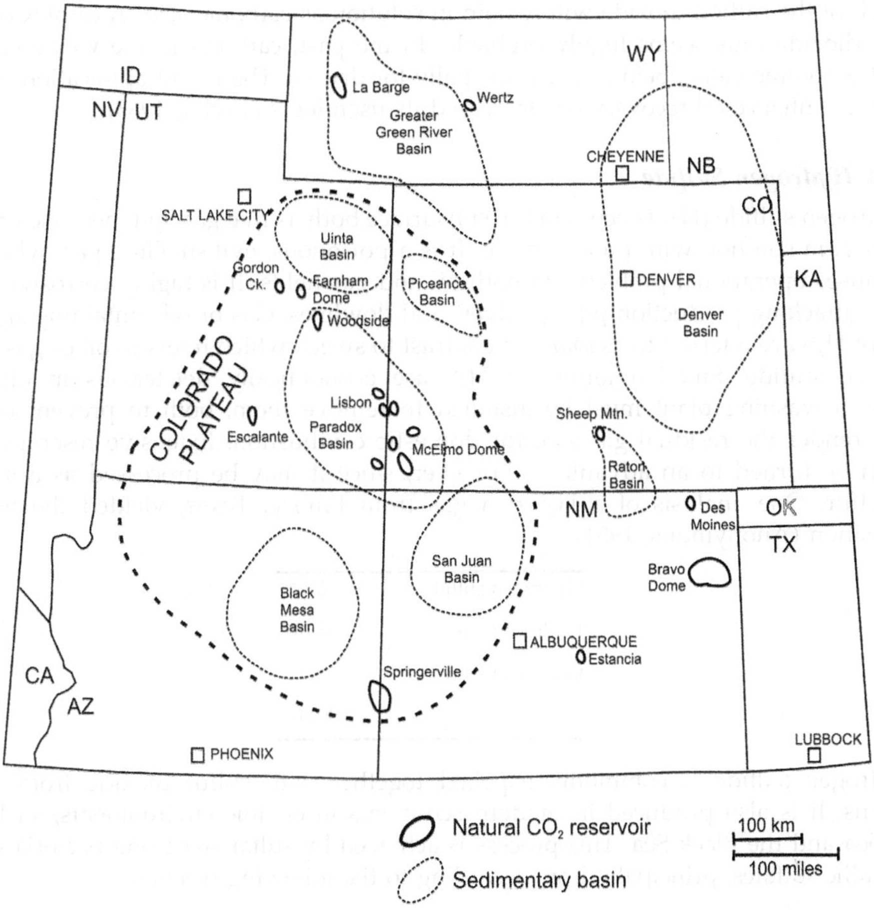

FIGURE 2.3 Natural CO_2 fields around the Colorado Plateau. *Modified from Allis et al. (2001).*

Both organic and inorganic processes can undoubtedly generate significant amounts of carbon dioxide in the earth's crust. Carbon dioxide is commonly recorded in natural gasess (Sugisaki et al., 1983). It may also be generated where igneous intrusives metamorphose carbonate sediments. Permeable limestones and dolomites can also yield carbon dioxide when they are invaded and leached by acid waters of either meteoric or connate origin. As described in Chapter 5, carbon dioxide is a normal product of the thermal maturation of kerogen, generally being expelled in advance of the petroleum. Carbon dioxide is also given off during the fermentation of organic matter as well as by the oxidation of mature organic matter due to either fluid invasion or bacterial degradation. Specifically, methane in the presence of oxygenated water may yield carbon dioxide and water:

$$3CH_4 + 6O_2 = 3CO_2 + 6H_2O.$$

Much of the carbon dioxide will remain in solution as carbonic acid. A duplex origin for carbon dioxide thus seems highly probable. In the past, carbon dioxide was generally of limited economic value, being used principally for dry ice. The recent application of carbon dioxide to enhance oil recovery has increased its usefulness (Taylor, 1983).

2.1.2.4 Hydrogen Sulfide

Hydrogen sulfide (H_2S) occurs in the subsurface both as free gas and, because of its high solubility, in solution with oil and brine. It is a poisonous, evil-smelling gas, whose presence causes operational problems in both oil and gas fields. It is highly corrosive to steel, quickly attacking production pipes, valves, and flowlines. Gas or oil containing significant traces of H_2S are referred to as *sour*—in contrast to sweet, which refers to oil or gas without hydrogen sulfide. Small amounts of H_2S are economically deleterious in oil or gas because a washing plant must be installed to remove them, both to prevent corrosion and to render the residual gas safe for domestic combustion. Extensive reserves of sour gas can be turned to an advantage, however, since it may be processed as a source of free sulfur. One analysis of a H_2S-rich gas from Emory, Texas, yielded the following composition (Anonymous, 1951):

Hydrogen sulfide	42.40%
Hydrocarbons	53.10%
Carbon dioxide	4.50%
	100.00%

Hydrogen sulfide is commonly expelled together with sulfur dioxide from volcanic eruptions. It is also produced in modern sediments in euxinic environments, such as the Dead Sea and the Black Sea. This process is achieved by sulfate-reducing bacteria working on metallic sulfates, principally iron, according to the following reaction:

$$2C + MeSO_4 + H_2O = MeCO_3 + CO_2 + H_2S,$$

where Me = metal. Note that sour gas generally occurs in hydrocarbon provinces where large amounts of evaporites are present. Notable examples include the Devonian basin of Alberta, the Paradox basin of Colorado, southwest Mexico, and extensive areas of the Middle East. Anhydrite may be converted to calcite in the presence of organic matter, the reaction leading to the generation of H_2S according to the following equation:

$$CaSO_4 + 2CH_2O = CaCO_3 + H_2O + CO_2 + H_2S,$$

where $2CH_2O$ = organic matter. There is a further interesting point about sour gas. Not only is it associated with evaporites but it is frequently associated with carbonates, generally reefal and lead–zinc sulfide ore bodies. A genetic link has long been postulated between these telethermal ores and evaporites (Davidson, 1965). Dunsmore (1973) has shown that the reduction of anhydrite to calcite and sour gas can occur inorganically, without the aid of bacteria. The reaction is an exothermic one, which may generate the high temperatures needed to mobilize the metallic sulfides.

2.2 GAS HYDRATES

2.2.1 Composition and Occurrence

Gas hydrates are compounds of frozen water that contain gas molecules (Kvenvolden and McMenamin, 1980). The ice molecules themselves are referred to as *clathrates* (Sloan, 1990). Physically, hydrates look similar to white, powdery snow (Fig. 2.4) and have two types of unit structure. The small structure with a lattice structure of 12 Å holds up to eight methane molecules within 46 water molecules. This clathrate may contain not only methane but also ethane, hydrogen sulfide, and carbon dioxide. The larger clathrate, with a lattice structure of 17.4 Å, consists of unit cells with 136 water molecules. This clathrate can hold the larger hydrocarbon molecules of the pentanes and *n*-butanes (Fig. 2.5).

Gas hydrates occur only in very specific pressure and temperature conditions. They are stable at high pressures and low temperatures, the pressure required for stability increasing logarithmically for a linear thermal gradient (Fig. 2.6). Gas hydrates occur in shallow arctic sediments and in deep oceanic deposits. Gas hydrates in arctic permafrost have been described from Alaska and Siberia. In Alaska they occur between about 750 and 3500 m (Holder et al., 1976). In Siberia, where the geothermal gradient is lower, they extend down to about 3100 m (Makogon et al., 1971; Makogan, 1981).

Hydrates have been found in the sediments of many of the oceans around the world. They have been recovered in the Deep Sea Drilling Project (DSDP) borehole cores, and their presence has been inferred from seismic data (e.g., Moore and Watkins, 1979; MacLeod, 1982). Specifically, they have been recognized from bright spots on seismic lines in water

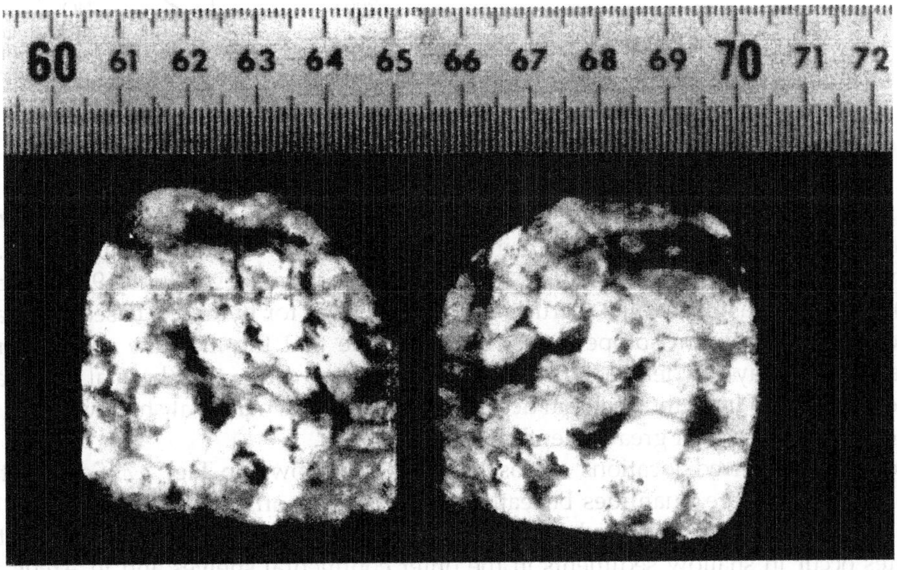

FIGURE 2.4 Samples of massive gas hydrate cored by the Deep Sea Drilling Project. *Courtesy of R. von Huene and J. Aubouin and the Deep Sea Drilling Project, Scripps Institute of Oceanography.*

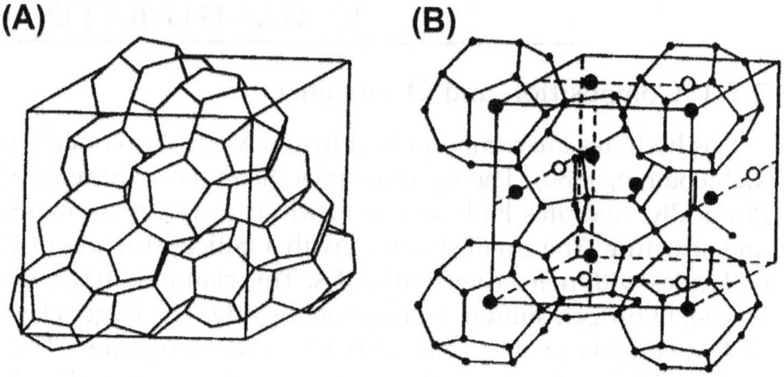

FIGURE 2.5 The two types of lattice structure for gas hydrates: (A) the larger structure with a 17.4-Å lattice and (B) the smaller one with a 12-Å lattice. *From Krason and Ciesnik (1985).*

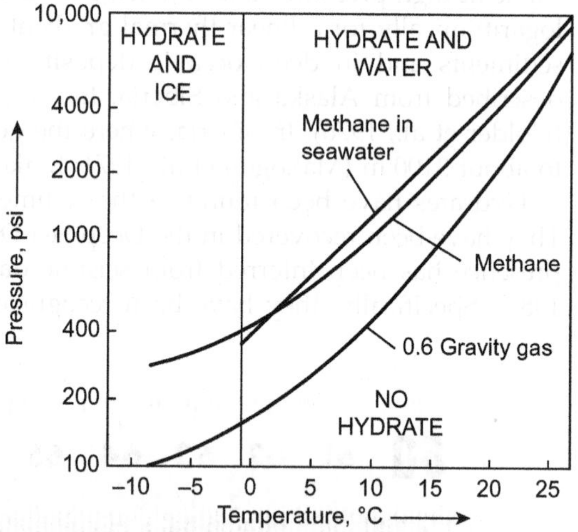

FIGURE 2.6 Pressure–temperature graph showing the stability field of gas hydrates. *After Hunt (1979, 1996), with permission from W. H. Freeman and Company.*

depths of 1000–2500 m off the eastern coast of North Island, New Zealand (Katz, 1981), and in water depths of 1000–4000 m in the western North Atlantic (Dillon et al., 1980). Gas hydrates have been attributed to a shallow biogenic origin (e.g., Hunt, 1979). MacDonald (1983), however, has postulated a crustal inorganic origin for hydrates, based on analysis of their carbon and helium isotope ratios. It is probable that the methane comes from three sources. Some may be derived from the mantle, some from the thermal maturation of kerogen, and some from the bacterial degradation of organic matter at shallow burial depths. This topic is discussed in greater depth in Chapter 5.

Known and inferred locations of gas hydrates are shown in Fig. 2.7. Sedimentary gas hydrates exist in large quantities beneath permafrost and offshore. Recent drilling activity in Japan (Nankai Trough), Canada, the United States, Korea, and India has shown that gas hydrates occur in shallow sediments in the outer continental shelves and in Arctic regions. Sand-dominated gas hydrate reservoirs are considered to be the most viable target for gas hydrate production (Collett et al., 2009; Boswell and Collett, 2011).

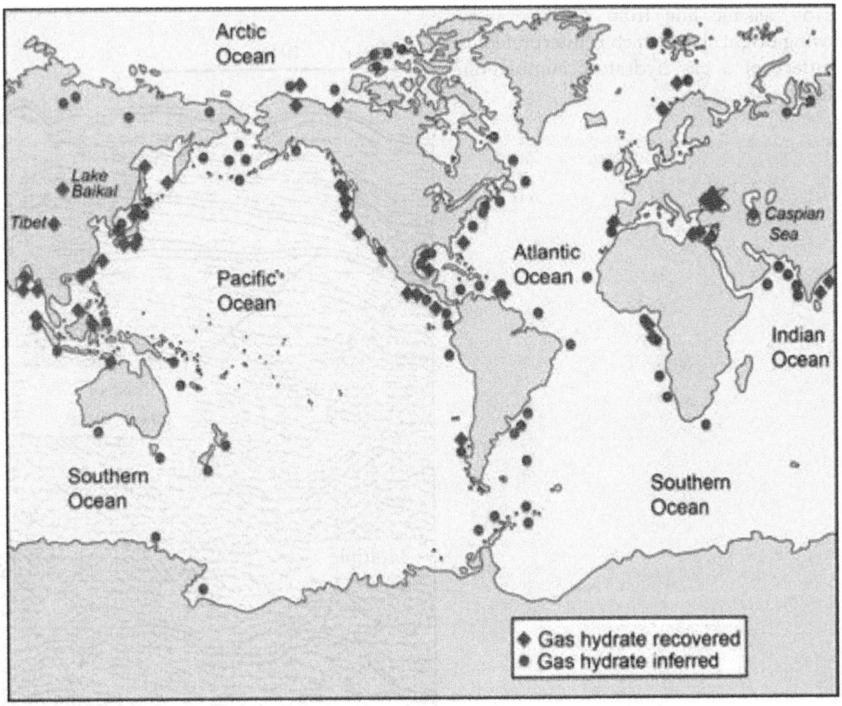

FIGURE 2.7 Known and inferred locations of gas hydrate occurrence. *From USGS gas hydrate primer (2014).*

2.2.2 Identification and Economic Significance

As just noted, the presence of gas hydrates can be suspected, but not proved, from seismic data. The lower limit of hydrate-cemented sediment is often concordant with bathymetry, because the velocity contrast between the gas-hydrate-cemented sediment and underlying noncemented sediment is large enough to generate a detectable reflecting horizon. The bottom simulating reflector may appear as a bright spot, which cross-cuts bedding-related reflectors. Care must be taken to distinguish gas hydrate bottom reflectors from ordinary seabed multiples (Fig. 2.8). Basal hydrate reflectors commonly increase in subbottom depth with increasing water depth because of the decreasing temperature of the water above the sea floor.

The presence of gas hydrates can only be proved, however, by engineering data. The penetration rate of the bit is low when drilling clathrate-cemented sediments. Pressure core barrels provide the ultimate proof, since gas hydrates show a characteristic pressure decline when pressure cells are brought to the surface and opened (Stoll et al., 1971). Gas hydrates also show certain characteristic log responses (Bily and Dick, 1974). They have high resistivity and acoustic velocity, coupled with low density (Fig. 2.9).

Large areas of the arctic permafrost and of the ocean floors contain vast reserves of hydrocarbon gas locked up in clathrate deposits. Clathrates can hold six times as much gas as can an open, free, gas-filled pore system and are a potential energy resource of great significance (McIver, 1981; Max and Lowrie, 1996). Detailed studies have been carried out on 12 selected areas of known gas hydrate occurrence (Finley and Krason, 1989; Krason, 1994). These

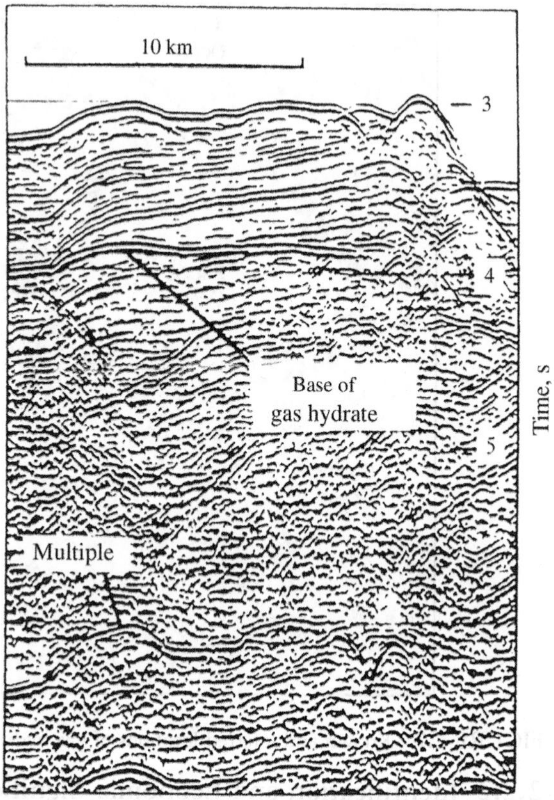

FIGURE 2.8 Seismic line from offshore New Zealand showing bright spot, which is interpreted as the lower surface of a gas hydrate accumulation. *From Katz (1981).*

studies have revealed that these 12 areas contain well in excess of 100,000 Trillion Cubic Feet (TCF) of gas within the hydrate-cemented sediment, and more than 4000 TCF of gas trapped beneath the hydrate seal (Table 2.5). Unfortunately, gas hydrates present considerable production problems that are yet to be overcome. These problems are due, in part, to the low permeability of the reservoir and, in part, to chemical problems concerning the release of gas from the crystals. Clathrate deposits may be of indirect economic significance, however, by acting as cap rocks. Because of their low permeability, they form seals that prevent the upward movement of free gas. Some gas is produced from gas hydrates in Western Siberia, where they pose some interesting engineering problems (Krason and Finley, 1992).

Production methods for producing gas hydrates include thermal stimulation, depressurization, and chemical inhibition (Fig. 2.10). Production tests at the permafrost Mallik and Mt. Elbert wells and laboratory simulations on sediment cores have produced important data on gas production (Ruppel, 2011). Thermal stimulation involves injection of heated fluids into the formation or directly heating the formation. Thermal stimulation introduces heat into the marine gas hydrate-bearing sediments (GHBS), changes the gas hydrate stability zone, and dissociates gas hydrate. Depressurization through drilling, perforating, and producing lowers the pressure in the gas hydrate-bearing sediments. This also leads to instability in the gas hydrate reservoir and causes gas hydrate to dissociate. Depressuring is the

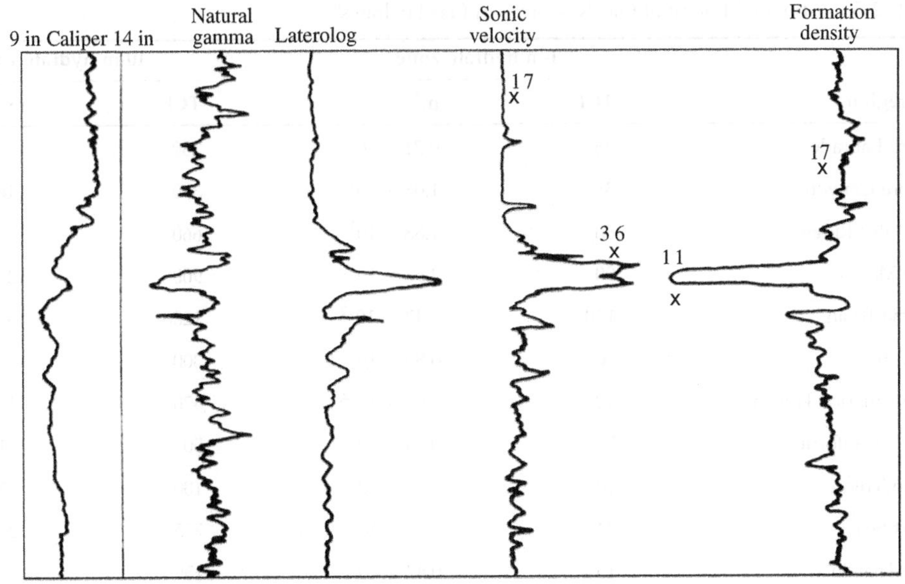

FIGURE 2.9 Wireline log of part of the Deep Sea Drilling Project Site 570, showing a gas hydrate zone. It is characterizd by high resistivity and acoustic velocity, and by low radioactivity and density. Crosses show velocities and densities in hydrate-bearing and normal intervals. *Courtesy of R. von Huene and J. Aubouin and the Deep Sea Drilling Project, Scripps Institute of Oceanography.*

most preferred and most economic method of producing gas from methane hydrates. Gas hydrate stability is inhibited in the presence of organic or ionic (seawater or brine) compounds. Inhibitors shift the gas hydrate stability boundary toward lower temperatures. Thus, injection of an inhibitor will dissociate gas hydrate near the wellbore.

It has also been suggested that gas hydrates may be of considerable importance in understanding climatic change, both the rapid rise in global temperature at the end of the Paleocene (Dickens et al., 1995) and more recent Quaternary climatic fluctuations (Nisbet, 1990). During glacial maxima vast quantities of methane, of whatever origin, become trapped within permafrost and submarine clathrate-cemented sediment. As the climate warms, a critical temperature will be reached at which point the clathrate will destabilize and huge amounts of methane gas will be released. This sudden release of gas may trigger mud volcanoes and pock marks on the sea bed, and pingos in permafrost. The huge increase of a greenhouse gas into the atmosphere may be responsible for the sudden increases in global temperatures observed at the end of glacial maxima.

2.3 CRUDE OIL

Crude oil is defined as "a mixture of hydrocarbons that existed in the liquid phase in natural underground reservoirs and remains liquid at atmospheric pressure after passing through surface separating facilities" (joint API, AAPG, and SPE definition). In appearance

TABLE 2.5 Estimated Potential Gas Resources in Gas Hydrates[a]

	I-m hydrate zone		10-m hydrate zone	
Study region	**TCF**	**m^3**	**TCF**	**m^3**
Offshore Labrador	25	0.71×10^{12}	250	7.1×10^{12}
Baltimore Canyon	38	1.08×10^{12}	380	10.8×10^{12}
Blake Outer Ridge	66	1.88×10^{12}	660	18.8×10^{12}
Gulf of Mexico	90	2.57×10^{12}	900	25.7×10^{12}
Colombia Basin	120	3.42×10^{12}	1200	34.2×10^{12}
Panama Basin	30	0.85×10^{12}	300	8.5×10^{12}
Middle America Trench	92	2.62×10^{12}	470	13.4×10^{12}
Northern California	5	0.14×10^{12}	50	1.4×10^{12}
Aleutian Trench	10	0.28×10^{12}	100	2.8×10^{12}
Beaufort Sea	240	6.85×10^{12}	725	20.7×10^{12}
Nankai Trough	15	0.42×10^{12}	150	4.2×10^{12}
Black Sea	3	0.08×10^{12}	30	0.8×10^{12}
Total	730	21.0×10^{12}	5200	149.0×10^{12}
		2.1×10^{13}		1.5×10^{14}

[a]*Estimated gas resources in gas hydrates in selected offshore areas. In most of the regions cited, the gas hydrate zone is much thicker than 10 m, and often extends down to several 100 m. Total estimated offshore gas hydrate resources potential is 100,000 TCF. Total estimated gas trapped beneath hydrates is 4000 TCF. TCF; Trillion Cubic Feet.*
From Krason (1994).

crude oils vary from straw yellow, green, and brown to dark brown or black in color. Oils are naturally oily in texture and have widely varying viscosities. Oils on the surface tend to be more viscous than oils in warm subsurface reservoirs. Surface viscosity values range from 1.4 to 19,400 cSt and vary not only with temperature but also with the age and depth of the oil.

Most oils are lighter than water. Although the density of oil may be measured as the difference between its specific gravity and that of water, it is often expressed in gravity units defined by the API according to the following formula:

$$^{\circ}\text{API} = \frac{141.5}{\text{specific gravity } 60/60^{\circ}\text{F}} - 131.5$$

where 60/60 °F is the specific gravity of the oil at 60 °F compared with that of water at 60 °F. Note that API degrees are inversely proportional to density. Thus light oils have API gravities of more than 40° (0.83 specific gravity), whereas heavy oils have API gravities of less than 10° (1.0 specific gravity). Heavy oils are thus defined as those oils that are denser than water. Variations of oil gravity with age and depth are discussed in Chapter 8. Oil viscosity and API gravity are generally inversely proportional to one another.

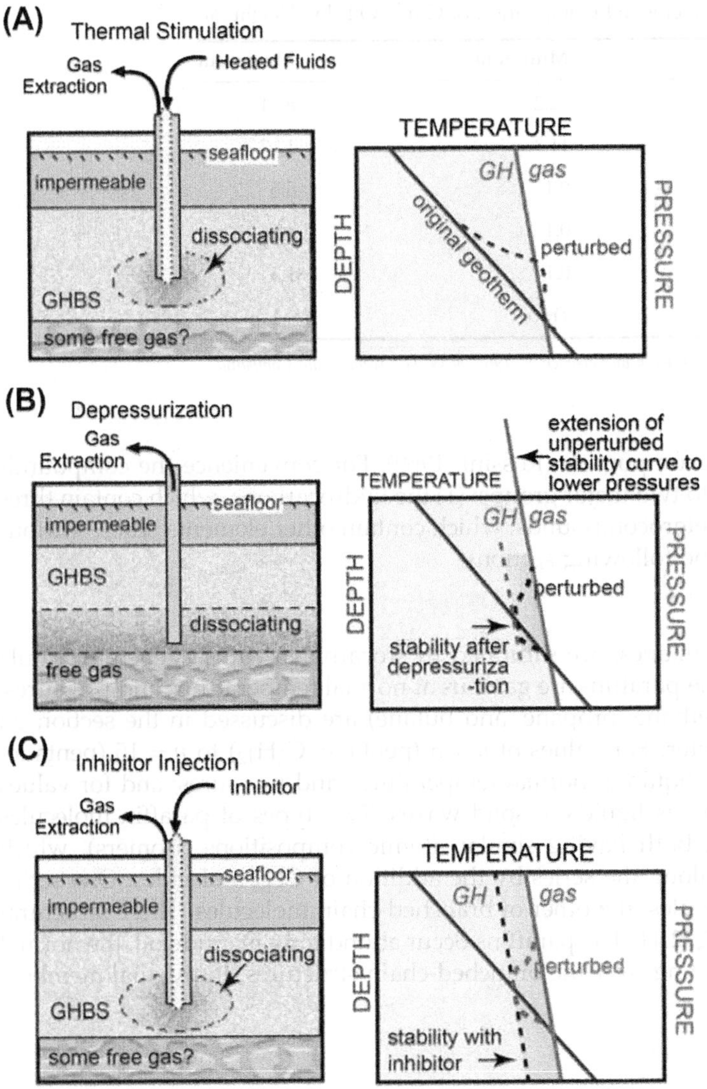

FIGURE 2.10 Possible methods for producing gas from marine gas hydrate-bearing sediments (GHBS). (A) Illustrates thermal stimulation. (B) Illustrates depressuring. (C) Illustrates inhibitor injection. In each case the stability field for the gas hydrate-bearing sediment is perturbed or shifted causing dissociation of gas. *From Ruppel (2011).*

2.3.1 Chemistry

In terms of elemental chemistry, oil consists largely of carbon and hydrogen with traces of vanadium, nickel, and other elements (see Table 2.6). Although the elemental composition of oils is relatively straightforward, there may be an immense number of molecular compounds. No two oils are identical either in the compounds contained or in the various proportions present. However, certain compositional trends are related to the age, depth, source, and geographical location of the oil. Conoco's Ponca City crude oil from Oklahoma, for

TABLE 2.6 Elemental Composition of Crude Oils by Weight %

Element	Minimum	Maximum
Carbon	82.2	87.1
Hydrogen	11.8	14.7
Sulfur	0.1	5.5
Oxygen	0.1	4.5
Nitrogen	0.1	1.5
Other	Trace	0.1

From: Geology of Petroleum by Levorsen. © 1967 by W. H. Freeman and Company. Used with permission.

example, contains at least 234 compounds (Rossini, 1960). For convenience the compounds found in oil may be divided into two major groups: (1) the hydrocarbons, which contain three major subgroups and (2) the heterocompounds, which contain other elements. These various compounds are described in the following sections.

2.3.1.1 *Paraffins*

The paraffins, often called alkanes, are saturated hydrocarbons, with a general formula C_nH_{2n+2}. For values of $n < 5$ the paraffins are gaseous at normal temperatures and pressures. These compounds (methane, ethane, propane, and butane) are discussed in the section on natural gas earlier in this chapter. For values of $n = 5$ (pentane, C_5H_{12}) to $n = 15$ (pentadecane, $C_{15}H_{32}$) the paraffins are liquid at normal temperatures and pressures; and for values of $n > 15$ they grade from viscous liquids to solid waxes. Two types of paraffin molecules are present within the series, both having similar atomic compositions (isomers), which increase in molecular weight along the series by the addition of CH_2 molecules. One series consists of straight-chain molecules, the other of branched-chain molecules. These structural differences are illustrated in Fig. 2.11. The paraffins occur abundantly in crude oil, the normal straight-chain varieties dominating over the branched-chain structures. Individual members

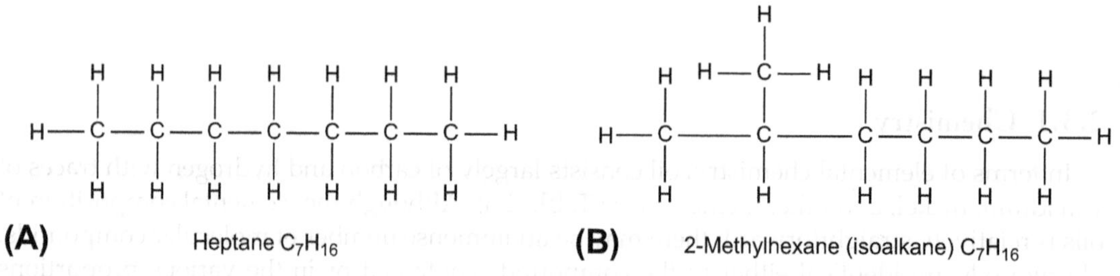

FIGURE 2.11 The isomeric structure of the paraffin series. Octane may have a normal chain structure (A) or a branched chain isomer (B).

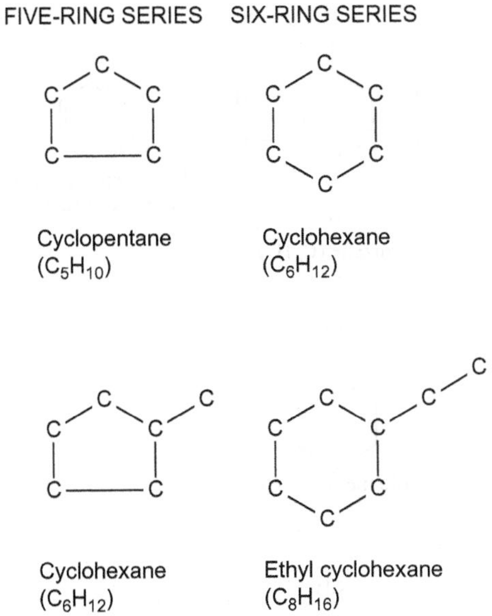

FIGURE 2.12 Examples of the molecular structures of five- and six-ring naphthenes (cycloalkanes). For convenience the accompanying hydrocarbon atoms are not shown.

of the series have been recorded up to $C_{78}H_{158}$. For a given molecular weight, the normal paraffins have higher boiling points than do equivalent weight isoparaffins.

2.3.1.2 *Naphthenes*

The second major group of hydrocarbons found in crude oils is the naphthenes or cycloalkanes. This group has a general formula C_nH_{2n}. Like the paraffins they occur in a homologous series consisting of five- and six-membered carbon rings termed the cyclopentanes and cyclohexanes, respectively (Fig. 2.12). Unlike the paraffins all the naphthenes are liquid at normal temperatures and pressures. They make up about 40% of both light and heavy crude oil.

2.3.1.3 *Aromatics*

Aromatic compounds are the third major group of hydrocarbons commonly found in crude oil. Their molecular structure is based on a ring of six carbon atoms. The simplest member of the family is benzene (C_6H_6), whose structure is shown in Fig. 2.13. One major series of the aromatic compounds is formed by substituting hydrogen atoms with alkane (C_nH_{2n+2}) molecules. This alkyl benzene series includes ethyl benzene (C_6H_5, C_2H_5) and toluene ($C_6H_5CH_3$). Another series is formed by straight- or branched-chain carbon rings. This series includes naphthalene ($C_{10}H_8$) and anthracene ($C_{14}H_{10}$). The aromatic hydrocarbons include asphaltic compounds. These compounds are divided into the resins, which are soluble in *n*-pentane, and the asphaltenes, which are not. The aromatic hydrocarbons are liquid at normal temperatures and pressures (the boiling point of benzene is 80.5 °C).

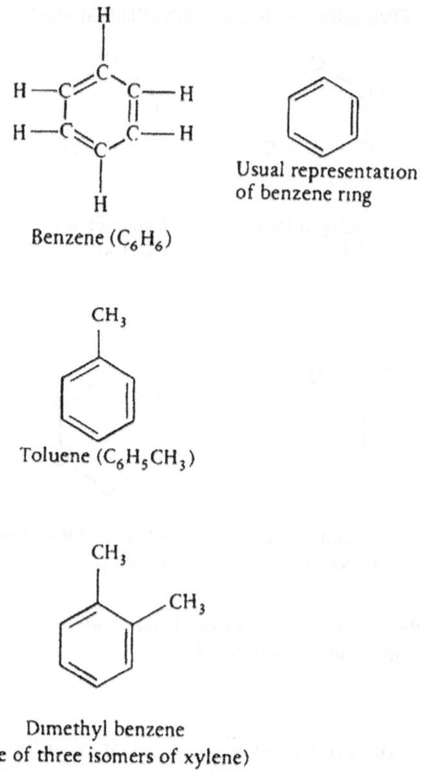

FIGURE 2.13 Molecular structures of aromatic hydrocarbons commonly found in crude oil.

They are present in relatively minor amounts (about 10%) in light oils, but increase in quantity with decreasing API gravity to more than 30% in heavy oils. Toluene ($C_6H_5CH_3$) is the most common aromatic component of crude oil, followed by the xylenes ($C_6H_4(CH_3)_2$) and benzene.

2.3.1.4 *Heterocompounds*

Crude oil contains many different heterocompounds that contain elements other than hydrogen and carbon. The principal ones are oxygen, nitrogen, and sulfur, together with rare metal atoms, commonly nickel and vanadium. The oxygen compounds range between 0.06% and 0.4% by weight of most crudes. They include acids, esters, ketones, phenols, and alcohols. The acids are especially common in young, immature oils and include fatty acids, isoprenoids, and naphthenic and carboxylic acids. The presence of steranes in some crudes is an important indication of their organic origin. Nitrogen compounds range between 0.01% and 0.9% by weight of most crudes. They include amides, pyridines, indoles, and pyroles.

Sulfur compounds range from 0.1% to 7.0% by weight in crude oils. This figure represents genuine sulfur-containing carbon molecules; it does not include H_2S gas or native sulfur.

Elemental sulfur occurs in young, shallow oils (above 100 °C sulfur combines with hydrogen to form hydrogen sulfide). Five main series of sulfur-bearing compounds in crude oil are: alkane thiols (mercaptans), the thio alkanes (sulfides), thio cycloalkanes, dithio alkanes, and cyclic sulfides.

Traces of numerous other elements have been found in crude oils, but determining whether they occur in genuine organic compounds or whether they are contaminants from the reservoir rock, formation waters, or production equipment is difficult. Hobson and Tiratsoo (1975) have tabulated many of these trace elements. Their lists include common rock-forming elements, such as silicon, calcium, and magnesium, together with numerous metals, such as iron, aluminum, copper, lead, tin, gold, and silver. Of particular interest is the almost ubiquitous presence of nickel and vanadium. Unlike the other metals it can be proved that nickel and vanadium occur, not as contaminants, but as actual organometallic compounds, generally in porphyrin molecules. The porphyrins contain carbon, nitrogen, and oxygen, as well as a metal radical. The presence of porphyrins in crude oil is of particular genetic interest because they may be derived either from the chlorophyll of plants or the hemoglobin of blood. Data compiled by Tissot and Welte (1978) show that the average vanadium and nickel content of 64 crude oils is 63 and 18 ppm, respectively. The known maximum values are 1200 ppm vanadium and 150 ppm nickel in the Boscan crude of Venezuela.

Metals tend to be associated with resin, sulfur, and asphaltene fractions of crude oils. Metals are rare in old, deep marine oils and are relatively abundant in shallow, young, or degraded crudes.

2.3.1.5 *Chemical Composition of Crude Oil*

Crude oil is separated into useful products by distillation. Refiners are interested in the various valuable petroleum products that can be formed in a refinery (e.g., naphtha, gasoline, diesel fuel, asphalt base, heating oil, kerosene, liquefied petroleum gas, etc.). Petroleum products have different boiling points that allow them to be separated by distillation. Modern refiners distill thousands of barrels of crude through distillation towers (Fig. 2.14). Crude oil is heated to very high temperatures until the petroleum is converted into a gas phase that contains a variety of hydrocarbons. The hydrocarbon gases rise in the distillation tower until they reach their unique condensing temperature. Vapor distilled from one of the chambers rises to the chamber above. The vapor passes through condensed liquid of the overlying chamber. Each overlying chamber condenses smaller and lighter molecules. Light gasoline is present at the top of the distillation tower whereas residuum is found at the base. Products are taken out at various levels in the distillation tower.

The composition of a typical crude oil is shown in Table 2.7. The typical oil has more aromatics and asphaltics in the residuum and more paraffins in the gasoline fraction.

Figure 2.15 illustrates the abundance of the various hydrocarbon compounds with boiling range occurring in naphthenic crude oils.

2.3.2 Classification

Many schemes have been proposed to classify the various types of crude oils. Broadly speaking, the classifications fall into two categories: (1) those proposed by chemical engineers

FIGURE 2.14 Distillation Tower. Vapor distilled from one of the chambers rises to the chamber above and passes through condensed liquid of the overlying chamber. Each overlying chamber in the tower condenses lighter and smaller molecules.

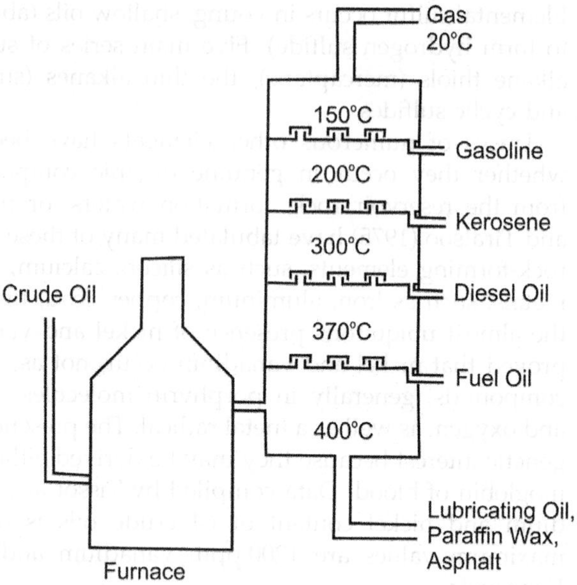

TABLE 2.7 Composition of a 35° API Gravity Crude Oil

Molecular size	Volume percent
Gasoline (C_5 to C_{10})	27
Kerosene (C_{11} to C_{13})	13
Diesel fuel (C_{14} to C_{18})	12
Heavy gas oil (C_{19} to C_{25})	10
Lubricating oil (C_{26} to C_{40})	20
Residuum (>C_{40})	18
Total	100
Molecular type	**Weight percent**
Paraffins	25
Naphthenes	50
Aromatics	17
Asphaltics	8
Total	100

API, American Petroleum Institute.

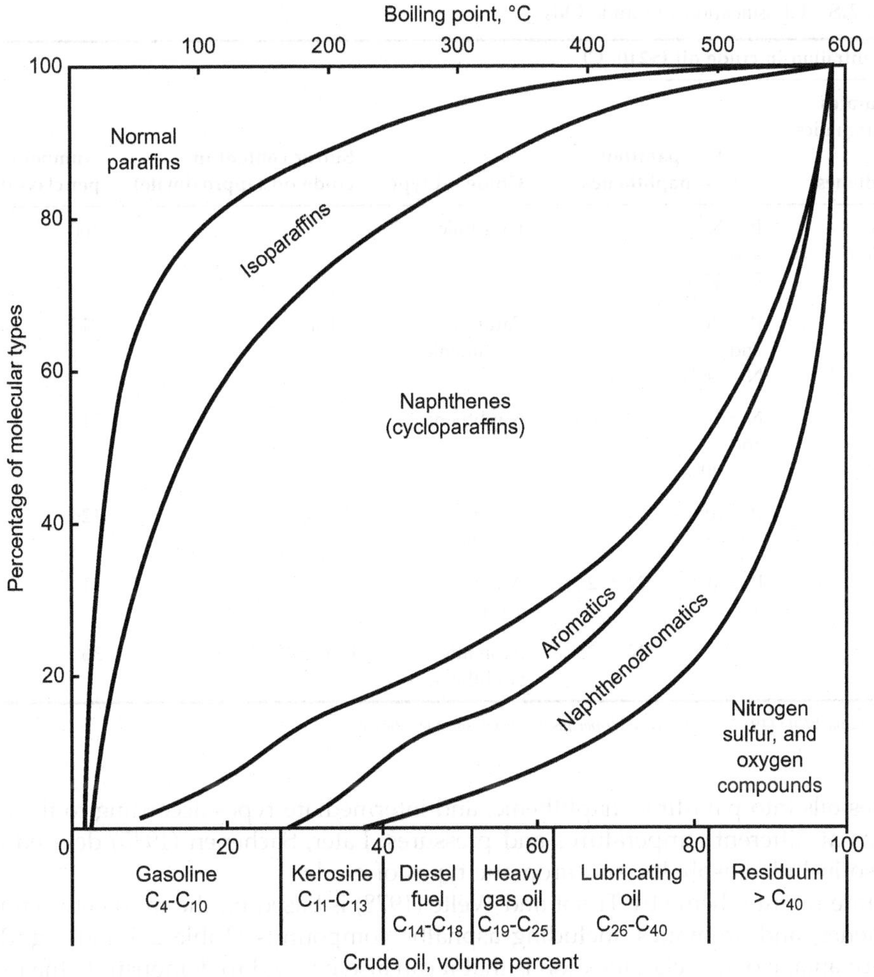

FIGURE 2.15 Chemical composition of naphthenic crude oil. *From: Petroleum Geochemistry and Geology 2/E by Hunt. © 1979 by W. H. Freeman and Company. Used with permission.*

interested in refining crude oil and (2) those devised by geologists and geochemists as an aid to understanding the source, maturation, history, or other geological parameters of crude oil occurrence.

The first type of classification is concerned with the quantities of the various hydrocarbons present in a crude oil and their physical properties, such as viscosity and boiling point. For example, the *n.d.M.* scheme is based on refractive index, density, and molecular weight (n = refractive index, d = density, and M = molecular weight).

Classificatory schemes of interest to geologists are concerned with the molecular structures of oils because these may be the keys to their source and geological history. One of the first schemes was developed in the US Bureau of Mines (Smith, 1927; Lane and Garton, 1935). It

TABLE 2.8 Classification of Crude Oils

Concentration in crude oil (>210 °C) S = saturates AA = aromatics + resins + asphaltenes	P = paraffins N = naphthenes		Crude oil type	Sulfur content in crude oil (approximate)	Number of samples per class (total = 541)
S > 50% AA < 50%	P > N and P > 40%		Paraffinic		100
	P ⩽ 40% and N ⩽ 40%		Paraffinic-naphthenic	<1%	217
	N > P and N > 40%		Naphthenic		21
S ⩽ 50% AA ⩾ 50%	P > 10%		Aromatic intermediate	>1%	126
	P ⩾ 10%	N ⩽ 25%	Aromatic asphaltic		41
		N ⩾ 25%	Aromatic naphthenic	Generally S < 1%	36

From Tissot and Welte (1978). Reprinted with permission from Springer-Verlag.

classifies oils into paraffinic, naphthenic, and intermediate types according to their distillate fractions at different temperatures and pressures. Later, Sachenen (1945) devised a scheme that also included asphaltic and aromatic types of crude.

A more recent scheme by Tissot and Welte (1978) is based on the ratio between paraffins, naphthenes, and aromatics, including asphaltic compounds (Table 2.8 and Fig. 2.16). The great advantage of this classification is that it can also be used to demonstrate the maturation and degradation paths of oil in the subsurface. A scheme based more on geological occurrence was proposed by Biederman (1965). This empirical classification is based on an arbitrarily defined depth and age:

1. Mesozoic and Cenozoic oil at less than 600 m
2. Mesozoic and Cenozoic oil at over 3000 m
3. Paleozoic oil at less than 600 m
4. Paleozoic oil at more than 3000 m

Essentially, this scheme recognized four classes of oil: young-shallow, young-deep, old-shallow, and old-deep. A detailed study of many oils by Martin et al. (1963) showed a number of statistically significant chemical and physical differences between these groups. Young-shallow oils tend to be heavy and viscous. They are generally sulfurous and relatively low in paraffins and rich in aromatics. Young-deep oils, on the other hand, are less viscous, of higher API gravity, more paraffinic, and low in sulfur content. Old-shallow oils are broadly

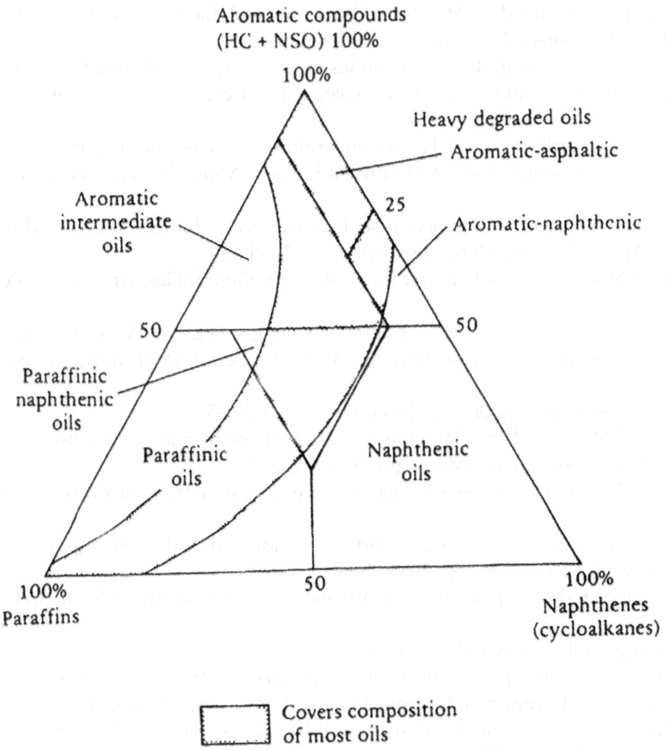

FIGURE 2.16 Ternary diagram showing the classification of oils proposed by Tissot and Welte (1978). *Reprinted with permission from Springer-Verlag.*

comparable to young-deep crudes in purity, viscosity, and paraffinic nature. Like young-shallow oil, however, they tend to be relatively sulfurous. Old-deep oils tend to have the lowest viscosity, density, and sulfur content of the four groups.

This system is very much a rule-of-thumb classification, and the variations of crude oils based on geological parameters are discussed further in Chapter 8. In particular, note that oils vary not only with age and depth, but also with variations in their source rock and the degree of degradation to which they have been subjected. This degradation causes very wide variations, especially in shallow oils.

References

Allis, R., Chidsey, T., Gwynn, W., White, S., Moore, J., 2001. Natural CO_2 reservoirs on the Colorado Plateau – candidates for CO_2 sequestration. http://www.geology.utah.gov/emp/co2sequest/pdf/reservoirs.pdf.

Anonymous, 1948. Oil fields of Royal Dutch-Shell group in western Venezuela. Am. Assoc. Pet. Geol. Bull. 32, 595.

Anonymous, May 24, 1951. Sour gas discovery at Emory field, Texas. Oil Gas J. 82.

Biederman, E.W., 1965. Crude oil composition—a clue to migration. World Oil 161, 78–82.

Bily, C., Dick, J.W.L., 1974. Naturally occurring gas-hydrates in the Mackenzie Delta, N.W.T. Bull. Can. Pet. Geol. 22, 340–352.

Boswell, R., Collett, T.S., 2011. Current perspectives on gas hydrate resources. Energy Environ. Sci. 4, 1206–1215.

Burwash, R.A., Cumming, G.L., 1974. Helium source-rock in southwestern Saskatchewan. Bull. Can. Pet. Geol. 22, 405–412.

Butler, J.B., 1975. The west sole gas field. In: Woodland, A.W. (Ed.), Petroleum and the Continental Shelf of Northwest Europe, vol. 1. Applied Science Publishers, London, pp. 213–222.
Collett, T.S., et al., 2009. Natural Gas hydrates: a review. In: Collett, T., Johnson, A., Knapp, C., Boswell, R. (Eds.), Natural Gas Hydrates—Energy Resource Potential and Associated Geologic Hazards, Am. Assoc. Pet. Geol. Memoir, vol. 89, pp. 146–219.
Cooper, B.S., Coleman, S.H., Barnard, P.C., Butterworth, J.S., 1975. Palaeotemperatures in the northern North Sea. In: Woodland, A.W. (Ed.), Petroleum and the Continental Shelf of Northwest Europe. Applied Science Publishers, London, pp. 487–492.
Cooper, B.S., 1977. Estimation of the maximum temperature attained in sedimentary rocks. In: Hobson, G.D. (Ed.), Developments in Petroleum Geology 1. Applied Science Publishers, London, 127–146.
Cotner, V., Crum, H.E., 1935. Natural gas in Amarillo District, Texas. In: Geology of Natural Gas. Am. Assoc. Pet. Geol., Tulsa, OK, pp. 409–427.
Cumming, A.D., Wyndham, C.L., 1975. The geology and development of the Hewett gas field. In: Woodland, A.W. (Ed.), Petroleum and the Continental Shelf of Northwest Europe. Applied Science Publishers, London, pp. 313–326.
Davidson, C.F., 1965. A possible mode of strata-bound copper ores. Econ. Geol. 60, 942–954.
Dickens, G.R., O'Neil, J.R., Rea, D.K., Owen, R.M., 1995. Dissociation of oceanic methane hydrate as a cause of the carbon isotope excursion at the end of the Paleocene. Paleooceanography 10, 965–971.
Dillon, W.P., Grow, J.A., Paull, C.K., January 7, 1980. Unconventional gas hydrate seals may trap gas off Southeast U.S.A. Oil Gas J. 124–130.
Dobbin, C.E., 1935. Geology of natural gases rich in helium, nitrogen, carbon dioxide and hydrogen sulphide. In: Geology of Natural Gas. Am. Assoc. Pet. Geol., Tulsa, OK, pp. 1053–1064.
Dunsmore, H.E., 1973. Diagenetic processes of lead-zinc emplacement in carbonates. Trans. Inst. Mm. Metall. Sect. B 82, 168–173.
Durrance, E.M., 1986. "Radioactivity in Geology. Ellis Horwood, Chichester.
Eames, T.D., 1975. Coal rank and gas source relationships—Rotliegendes reservoirs. In: Woodland, A.W. (Ed.), Petroleum and the Continental Shelf of Northwest Europe, vol. 1. Applied Science Publishers, London, pp. 191–204.
Finley, P.D., Krason, J., 1989. Summary Report. U.S. Geological Evolution and Analyis of Confirmed or Suspected Gas Hydrate Localities, vol. 15. Department of Energy, Morgantown, WV.
Hinson, H.H., 1947. Reservoir characteristics of Rattlesnake oil and gas field, San Juan County, N. Mexico. Am. Assoc. Pet. Geol. Bull. 31, 731–771.
Hitchon, B., 1963. Geochemical studies of natural gas. Part III. Inert gases in western Canada natural gases. J. Can. Pet. Technol. 2, 165–174.
Hobson, D.G., Tiratsoo, E.N., 1975. Introduction to Petroleum Geology. Scientific Press, Beaconsfield.
Holder, G.D., Katz, D.L., Hand, J.H., 1976. Hydrate formation in subsurface environments. AAPG Bull. 60, 981–988.
Hunt, J.M., 1979. Petroleum Geochemistry and Geology. Freeman, San Francisco.
Hunt, J.M., 1996. Petroleum Geochemistry and Geology, second ed. Freeman, San Francisco.
Katz, H.R., 1981. Probable gas hydrate in continental slope east of the North Island, New Zealand. J. Pet. Geol. 3, 315–324.
Keplinger, C.H., Wanemacher, C.H., Burns, K.R., January 6, 1949. Hugoton, worlds largest dry gas field in amazing development. Oil Gas J. 84–88.
Krason, J., August 1994. Study of 21 marine basins indicates wide prevalence of hydrates. Offshore 34–35.
Krason, J., Ciesnik, J., 1985. Geological Evolution and Analysis of Confirmed or Suspected Gas Hydrate Localities. Vol. 5. Gas Hydrates in the Russian Literature. U.S. Department of Energy, Morgantown, WV.
Krason, J., Finley, P.D., 1992. Messoyakh Gas Field-Russia. In: Beaumont, E.A., Foster, N.H. (Eds.), Treatise on Oil and Gas Fields, vol. 7. Am. Assoc. Pet. Geol., Tulsa, OK, pp. 197–220.
Kvenvolden, K.A., McMenamin, M.A., 1980. Hydrates of natural gas: a review of their geological occurrence. U.S. Geol. Circ. 825, 1–11.
Lane, E.C., Garton, E.L., 1935. 'Base' of a Crude Oil. Rep. Invest.—U.S. Bur. Mines. RI-3279.
Lee, H., 1963. The technical and economic aspects of helium production in Saskatchewan. J. Can. Pet. Technol. 2, 16–27.
Levorsen, A.I., 1967. Geology of Petroleum, second ed. Freeman, San Francisco.
MacDonald, G.J., 1983. The many origins of natural gas. J. Pet. Geol. 5, 341–362.
MacLeod, M.K., 1982. Gas hydrates in ocean bottom sediments. AAPG Bull. 66, 2649–2662.
Maglione, P.R., 1970. Triassic gas field of Hassi er R'Mel, Algeria. Mem.—Am. Assoc. Pet. Geol. 14, 489–501.

Makogon, Y.F., 1981. Hydrates of Natural Gas. Penn Well Publ. Co., Tulsa, OK.
Makogon, Y.F., Trebin, F.A., Trofimuk, A.A., Tsarev, V.P., Cherskiy, N.V., 1971. Detection of a pool of natural gas in a solid (hydrated gas) state. Dokl. Akad. Nauk SSSR 196, 197–200.
Martin, R.L., Winters, J.C., Williams, J.A., 1963. Composition of crude oil by gas chromatography: geological significance of hydrocarbon distribution. In: Proc.—World Pet. Congr. 6; Sect. 5, Pap. 13.
Max, M.D., Lowrie, A., 1996. Oceanic methane hydrates: a "frontier" gas resource. J. Pet. Geol. 19, 95–112.
McIver, R.D., 1981. Gas hydrates. In: Mayer, R.G., Olson, J.C. (Eds.), Long Term Energy Resources. Pitman, Boston, pp. 713–726.
Moore, J.C., Watkins, J.S., 1979. Middle america trench. Geotimes 24, 20–22.
Nisbet, E.G., 1990. The end of the ice-age. Can. J. Earth Sci. 27, 148–157.
Payne, A.L., 1951. Cumarebo oil field, Falcon, Venezuela. Am. Assoc. Pet. Geol. Bull. 35, 1870.
Pippin, L., 1970. Panhandle-Hugoton field, Texas-Oklahoma-Kansas—the first fifty years. Mem.—Am. Assoc. Pet. Geol. 14, 204–222.
Rice, D.D., Claypool, G.E., 1981. Generation, accumulation and resource potential of biogenic gas. AAPG Bull. 65, 1–25.
Rogers, G.S., 1921. Helium-bearing natural gas. Geol. Surv. Bull. (U.S.) 1–113.
Rossini, F.D., 1960. Hydrocarbons in petroleum. J. Chem. Educ. 37, 554–561.
Ruppel, C., 2011. Methane Hydrates and the Future of Natural Gas. Supplementary Paper #4, The Future of Natural Gas, MIT Energy Initiative study, 25 pp.
Sachenen, A.N., 1945. Chemical Constituents of Petroleum. Rheinhold, New York.
Shimkin, D.B., December 21, 1950. Is petroleum a Soviet weakness? Oil Gas J. 214–226.
Slipper, S.E., 1935. Natural gas in Alberta. Am. Assoc. Pet. Geol. Bull. 51, 25–37.
Sloan, E.D., 1990. Clathrate Hydrates of Natural Gases. Marcel Decker, New York.
Smith, N.A.C., 1927. The Interpretation of Crude Oil Analyses. Rep. Invest.—U.S. Bur. Mines. RI-2806.
Stauble, A.J., Milius, B., 1970. Geology of Groningen gas field, Netherlands. Mem.—Am. Assoc. Pet. Geol. 14, 359–369.
Stoll, R.D., Ewing, J., Bryan, G.M., 1971. Anomalous wave velocities in sediments containing gas hydrates. J. Geophys. Res. 76, 2090–2094.
Sugisaki, R., Id, M.T., Takeda, H., Isobe, Y., 1983. Origin of hydrogen and carbon dioxide in fault gases and its relation to fault activity. J. Geol. 91, 239–258.
Suter, H.H., 1952. The general and economic geology of Trinidad. Colon. Geol. Miner. Resour. 4, 4–28.
Taylor, G., June 1, 1983. CO_2 projects to test recovery theories. AAPG Explorer 20–21.
Tissot, B.P., Welte, D.H., 1978. Petroleum Formation and Occurrence. Springer-Verlag, Berlin.
USGS Gas Hydrates Primer, 2014. http://woodshole.er.usgs.gov/project-pages/hydrates/primer.html.

Selected Bibliography

Hunt, J.M., 1979. Petroleum Geochemistry and Geology. Freeman, San Francisco.
Hunt, J.M., 1996. Petroleum Geochemistry and Geology, second ed. Freeman, San Francisco.
Neumann, 1, Pazynska-Lahme, B., Severin, D., 1981. Composition and Properties of Petroleum. Wiley, London.
Tissot, B.P., Welte, D.H., 1978. Petroleum Formation and Occurrence. Springer-Verlag, Berlin. Part IV, Chapter I (pp. 333–368) gives a detailed account of the chemistry of oil and gas.

CHAPTER

3

Methods of Exploration

3.1 WELL DRILLING AND COMPLETION

In the earliest days of oil exploration, oil was collected from surface seepages. Herodotus, writing in about 450 BC, described oil seeps in Carthage (Tunisia) and the Greek island Zachynthus. He gave details of oil extraction from wells near Ardericca in modern Iran, although, as mentioned in Chapter 1, the wells could not have been very deep, because fluid was extracted in a wineskin on the end of a long pole mounted on a fulcrum. Oil, salt, and bitumen were produced simultaneously from these wells. In China, Burma, and Romania, mine shafts were dug to produce shallow oil. Access was gained by ladders or a hoist, and air was pumped down to the mines through pipes. Oil seeped into the shaft and was lifted to the surface in buckets. This type of technology was not conducive to a healthy life and long retirement for the miners.

Oil has also been mined successfully in several parts of the world by driving horizontal adits into reservoirs. Oil dribbles down the walls onto the floor and flows down to the mine entrance. This technique has been used in the North Tisdale field of Wyoming (Dobson and Seelye, 1981). Conventionally, however, oil and gas are located and produced by drilling boreholes. Before exploration for oil began, cable-tool drilling was an established technique in many parts of the world in the quest for water and brine (Fig. 3.1). The first well to produce oil intentionally in the Western World was drilled at Oil Creek, Pennsylvania, by Colonel Drake in 1859 (Owen, 1975). Previously, water wells in the Appalachians and elsewhere produced oil as a contaminant. The technology for drilling Drake's well was derived from Chinese artisans who had traveled to the United States to work on the railroads. Cable-tool drilling had been used in China since at least the first century b.c., the drilling tools being suspended from bamboo towers up to 60 m high. In China, however, this drilling technology had developed to produce artesian brines, not petroleum (Messadie, 1995). The two methods, cable-tool and rotary-tool drilling, are briefly described in the following sections.

3.1.1 Cable-Tool Drilling

Cable-tool drilling seems to have developed spontaneously in several parts of the world. In the early nineteenth century, the Chinese were sinking shafts to depths of some 700 m using an 1800-kg bit suspended from a rattan cord. Hole diameters were of the order of 10–15 cm, and the rate of penetration was about 60–70 cm/day. The wells were cased with bamboo or

Elements of Petroleum Geology
http://dx.doi.org/10.1016/B978-0-12-386031-6.00003-5

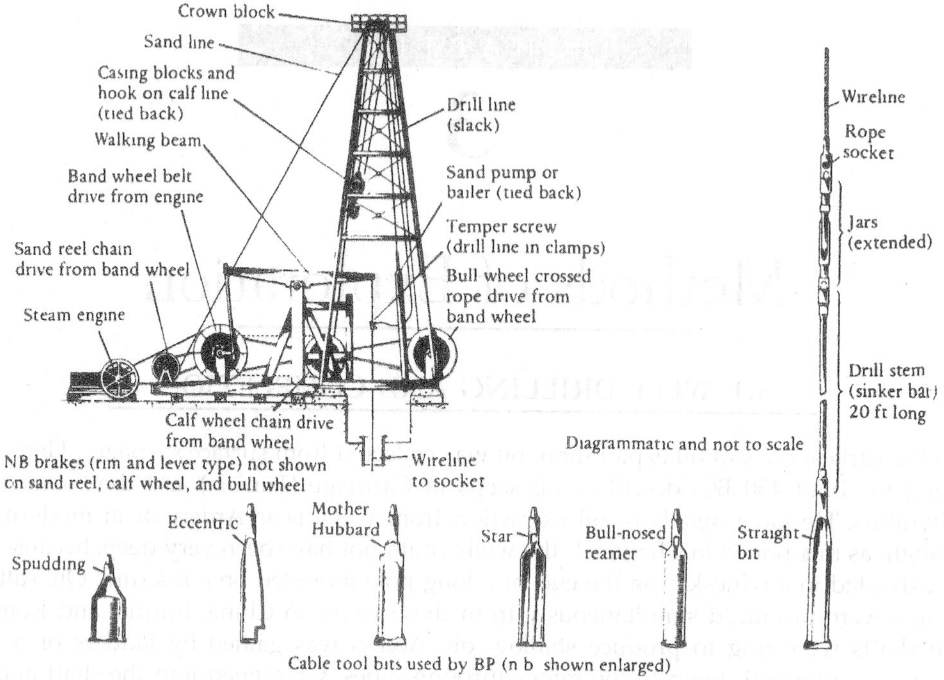

FIGURE 3.1 A steam-driven cable-tool rig and equipment. *Courtesy of British Petroleum.*

hollow cypress trunks (Imbert, 1828; Coldre, 1891). These wells were sunk in the search for freshwater and for brines from which salt could be extracted. In modern cable-tool drilling, a heavy piece of metal, termed the *bit,* is banged up and down at the end of a cable on the bottom of the hole. The bit is generally chisel shaped. This repeated percussion gradually chips away the rock on the bottom of the hole. Every now and then, the bit is withdrawn to the surface, and a bailer is fitted to the end of the cable. This bailer is a steel cylinder with a one-way flap at the bottom. As the bailer is dropped on the floor of the hole, chips of rock are forced into it through the trap flap. When the bailer is lifted prior to the next drop, the rock cuttings are retained in the trap (Fig. 3.1). When the cuttings have been removed from the borehole, the bailer is drawn out and emptied. The bit is then put back on the end of the cable, and percussion recommences. As the hole is gradually deepened, the sides have a natural tendency to cave in. This tendency is counteracted by lining the hole with steel casing.

In the early days of oil exploration, the percussive power of the cable tool was provided by a man or men pulling on the rope or, later, aided by a spring-pole. In more recent times, however, motive power was provided by a steam or internal combustion engine.

The mechanical cable-tool drilling method evolved toward the end of the eighteenth century. It was then used primarily for the sinking of water wells. Occasionally, such wells would find water contaminated with oil, to the displeasure of the driller. When the economic uses of oil were discovered in the mid-nineteenth century, however, the cable-tool rig became the prime method of sinking oil wells, and it remained so for some 80 years.

Cable-tool drilling has several major mechanical constraints. First, the depth to which one may drill is severely limited. The deeper the hole, the heavier the cable. There comes a point, therefore, when the cable at the well head is not strong enough to take the combined weight of the bit and the downhole cable. Although cable-tool rigs have drilled to >3000 m, the average capability is about 1000 m. This capability is adequate for most water wells, but too shallow for the increased depths required for oil exploration. A further limitation of the cable-tool method is that it can only work in an open hole. The cable must be free to move, so it is not possible to keep the hole full of fluid. The bailer removes water that oozes into the hole. Thus, when the bit breaks through into a high-pressure formation, the oil or gas shoots up to the surface as a gusher. Because of these limitations of penetration depth and safety, cable-tool drilling is of limited use in petroleum exploration.

3.1.2 Rotary Drilling

Because of the greater safety and depth penetration of rotary drilling, it has largely superseded the cable-tool method for deep drilling in the oil industry (Fig. 3.2). In this technique, the bit is rotated at the end of a hollow steel tube called the *drill string*. Many types of bit are used, but the most common consists of three rotating cones set with teeth (Fig. 3.3). The bit is rotated, and the teeth gouge or chip away the rock at the bottom of the borehole. Simultaneously, mud or water is pumped down the drill string, squirting out through nozzles in the bit and flowing up to the surface between the drill string and the wall of the hole. This circulation of the drilling mud has many functions: It removes the rock cuttings from the bit; it removes cavings from the borehole wall; it keeps the bit cool; and, most importantly, it keeps the hole safe. The hydrostatic pressure of the mud generally prevents fluid from moving into the hole, and if the bit penetrates a formation with a high-pore pressure, the weight of the mud may prevent a gusher. A gusher can also be prevented by sealing the well head with a series of valves termed the *blowout preventers*.

As the bit deepens the hole, new joints of drill pipe are screwed on to the drill string at the surface. The last length of drill pipe is screwed to a square-section steel member called the kelly, which is suspended vertically in the kelly bushing, a square hole in the center of the rotary table. Thus, rotation of the table by the rig motors imparts a rotary movement down the drill string to the bit at the bottom of the hole. As the hole deepens, the kelly slides down through the rotary table until it is time to attach another length of drill pipe (Fig. 3.4). When the bit is worn out, which depends on the type of bit and the hardness of the rock, the drill pipe is drawn out of the hole and stacked in the derrick. When the bit is brought to the surface, it is removed and a new one is fitted.

After drilling for some depth, the borehole is lined with steel casing, and cement is set between the casing and the borehole wall. Drilling may then recommence with a narrower gauge of bit. The diameters of bits and casing are internationally standardized. Depending on the final depth of the hole, several diameters of bit will be used with the appropriate casing (Fig. 3.5). The average depth of an oil well is between 1 and 3 km, but depths of up to 11.5 km can be penetrated.

FIGURE 3.2 Simplified sketch of an onshore derrick for rotary drilling. *Courtesy of British Petroleum.*

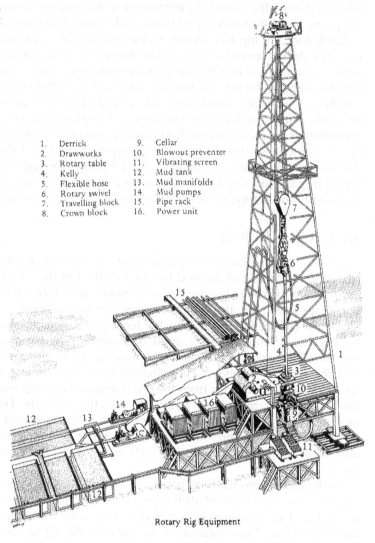

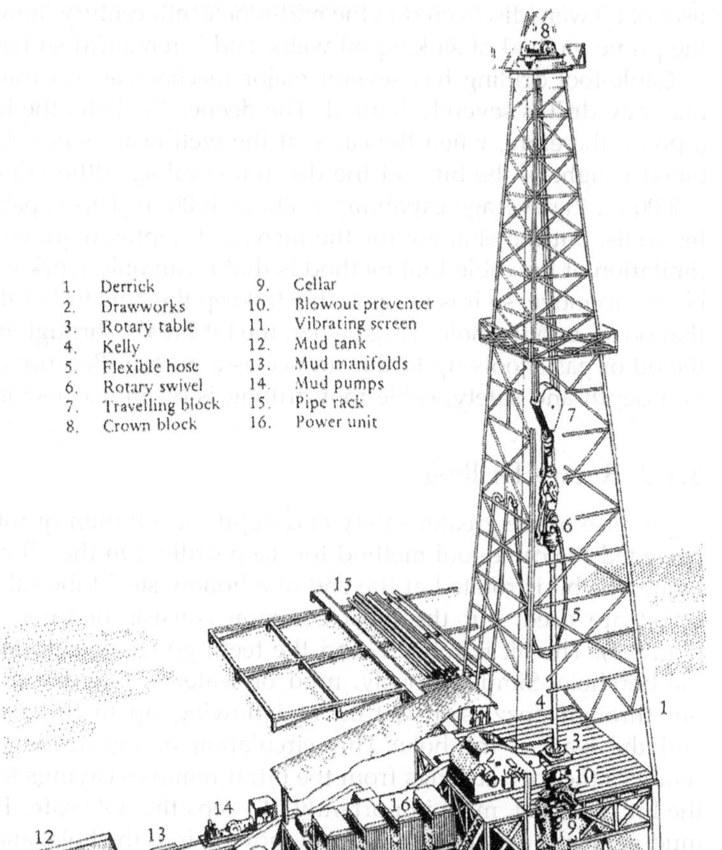

When drilling into a reservoir, an ordinary bit may be removed and replaced by a core barrel. A core barrel is a hollow steel tube with teeth, commonly diamonds, at the downhole end. As the core barrel rotates, it cuts a cylinder of rock and descends over it as the hole deepens. When the core barrel is withdrawn to the surface, the core of rock is retained in the core barrel by upward-pointing steel springs. Coring is slower than drilling with an ordinary bit and is thus more expensive. It is only used sparingly in hydrocarbon exploration to collect large, intact rock samples for geological and engineering information.

Geologists are involved to varying degrees in drilling wells. In routine oil field development wells, where the depth and characteristics of the reservoir are already well known,

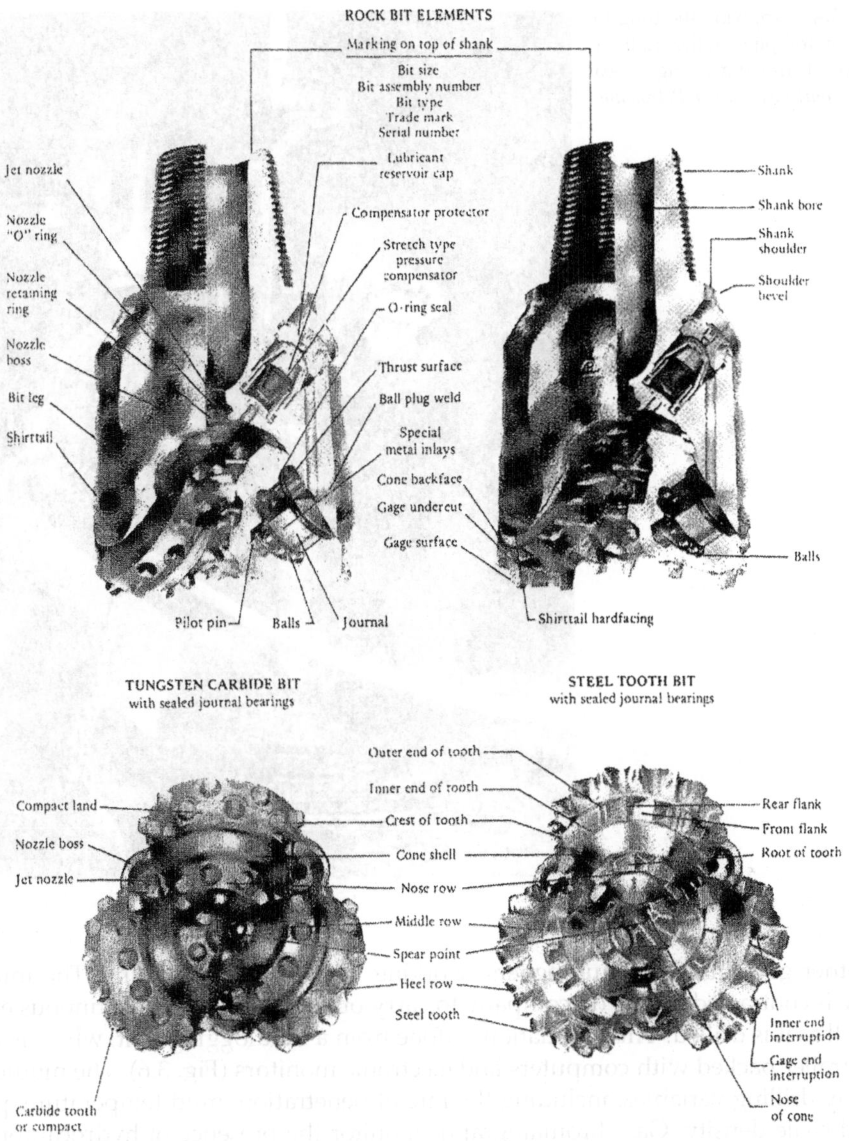

FIGURE 3.3 Two types of tricone bit used for rotary drilling. *Courtesy of Hughes Tool Co.*

geologists may not be present at the well site, although they will monitor the progress of the well from the office. On an offshore wildcat well, however, there may be five or more geologists. The oil company for whom the well is being drilled will have one of their own well site geologists on board. His or her duties include advising both the driller and the operational headquarters on the formations, fluids, and pressures to be anticipated; picking casing and coring points; deciding when to run wireline logs; supervising logging and interpreting the end result; and, most importantly, identifying and evaluating hydrocarbon shows in the well.

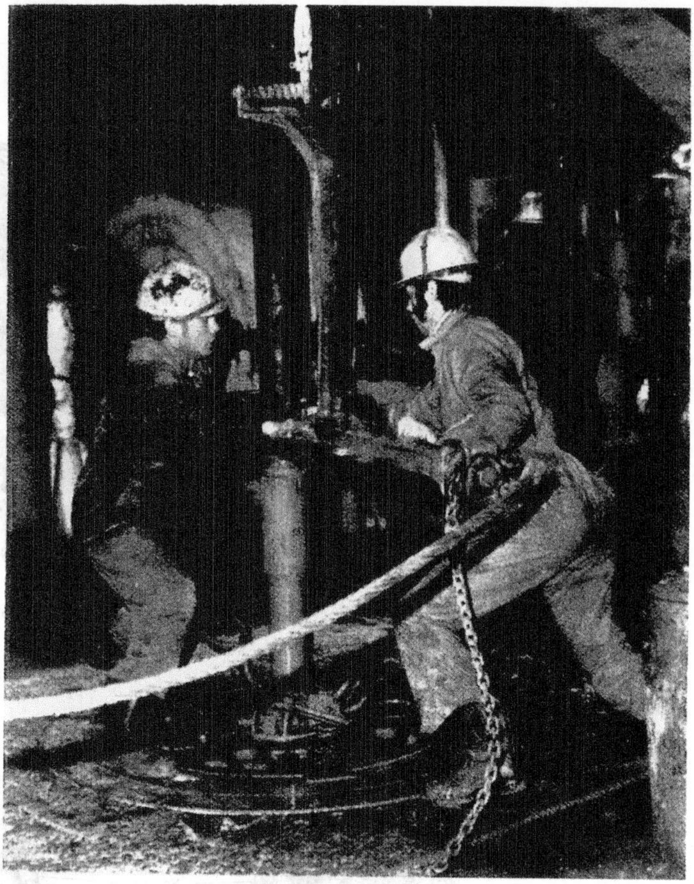

FIGURE 3.4 Applying the tong to make up the drill pipe on the rig floor of British Petroleum's rig, Sea Conquest. *Courtesy of British Petroleum.*

The other geologists are mudloggers, working in pairs on 12-h shifts. The mudlogging company is contracted by the oil company to carry out a complex and continuous evaluation of the well as it is drilled. This evaluation is done from a mudlogging unit, which is a cabin or caravan trailer packed with computers and electronic monitors (Fig. 3.6). The mudloggers record many drilling variables, including the rate of penetration, mud temperature, pore pressure, and shale density. Gas chromatographs monitor the presence of hydrocarbon gases in the mud. Samples of the well cuttings are collected at specified depth intervals, checked with ultraviolet light and other tests for the presence of oil, identified, and described. These data are continuously plotted on a mud log (Fig. 3.7). A useful account of the geological aspects of mudlogging and associated rig activities is given by Dickey (1979). Nowadays, however, it is also possible to run a series of geophysical logs while the well is actually being drilled, the sondes being lowered inside the drill pipe. Geophysical logs are described in more detail later in the chapter; for now, note that MWD (measurement while drilling) is an important aspect of formation evaluation, providing early warnings of rock type, hydrocarbon saturation, and potentially perilous zones of high pressure.

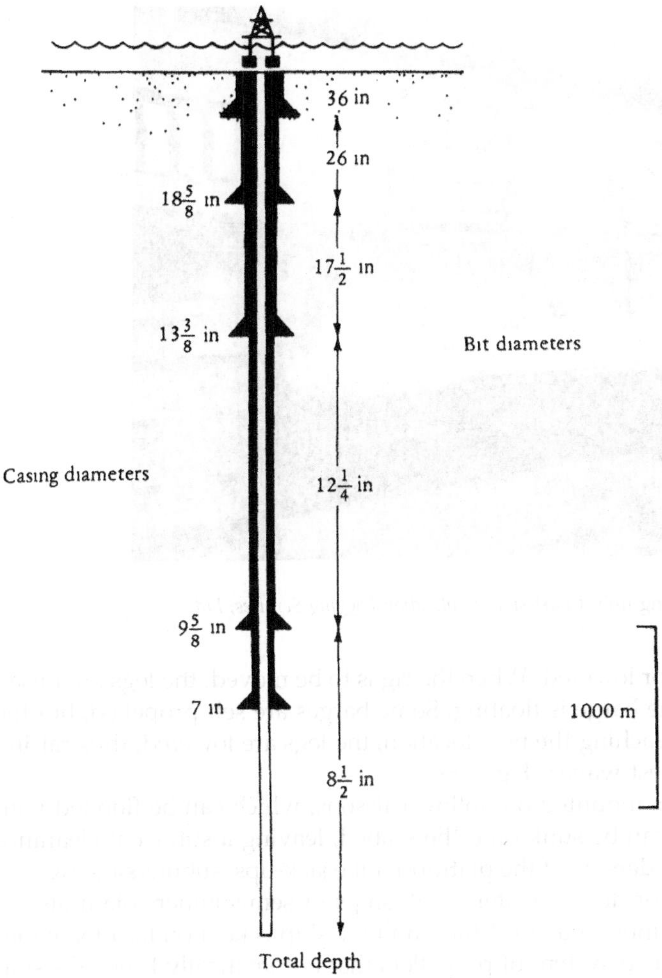

FIGURE 3.5 The casing and bit diameters of a typical well.

3.1.3 Various Types of Drilling Units

The derrick for rotary drilling may be used on land or at sea. On land, prefabricated rigs are used; they can be transported by skidding, by vehicle, or, in the case of light rigs, by helicopter. Once a well has been drilled, the derrick is dismantled and moved to the next location, whether the well is productive or barren of hydrocarbons (colloquially termed a *dry hole* or *duster*). Thus, modern oil fields are not marked by a forest of derricks as shown in old photographs and films.

In offshore drilling, the derrick can be mounted in various ways. For sheltered inland waterways, it may be rigged on a flat-bottomed barge. This technique has been employed, for example, in the bayous of the Mississippi delta for >60 years. In water depths of up to about 100 m, jack-up rigs are used. In jack-up rigs, the derrick is mounted on a flat-bottomed barge

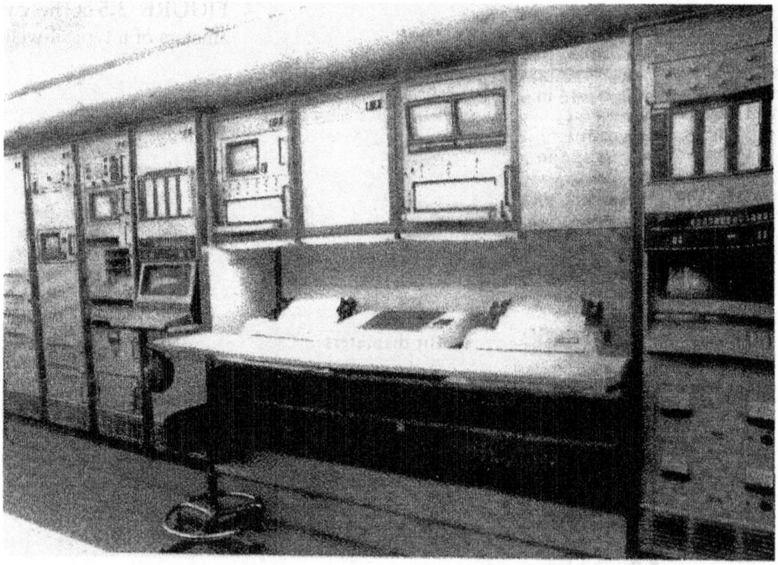

FIGURE 3.6 The inside of a mudlogging unit. *Courtesy of Exploration Logging Services, Ltd.*

fitted with legs that can be raised or lowered. When the rig is to be moved, the legs are raised (i.e., the barge is lowered) until the barge is floating. Some barges are self-propelled, but the majority are moved by tugs. On reaching the new location, the legs are lowered, thus raising the barge clear of all but the highest waves (Fig. 3.8).

Submersible units are platforms mounted on hollow caissons, which can be flooded with seawater. The platform supports can be sunk onto the seabed, leaving a sufficient clearance between the sea surface and the underside of the platform. Like jack-ups, submersible rigs are only suitable for shallow water. For deeper waters, drill ships or semisubmersible units are used. On drill ships, the derrick is mounted amidships, and the ship is kept on location either by anchors or by a specially designed system of propellers that automatically keeps the ship in the same position regardless of wind, waves, and currents. Semisubmersibles are floating platforms having three or more floodable caisson legs (Fig. 3.9). With the aid of anchors and judicious flooding of the legs, the unit can be stabilized, although still floating, with the rotary table 30 m or so above the sea. Some semisubmersibles are self-propelled, but the majority are towed by tugs from one location to another (Fig. 3.10). In shallow water Arctic conditions, where there is extensive pack ice, drill ships, jack-ups, and semisubmersibles are unsuitable. In very shallow water, artificial islands are made from gravel and ice. Monopod rigs are also used, which, as their name implies, balance on one leg around which the pack ice can move with minimum obstruction.

3.1.4 Vertical, Directional, and Horizontal Drilling

Most wells drilled in the United States for oil and natural gas historically have been vertical wells (Fig. 3.11). A dramatic change started in 2004 when a switch occurred to horizontal

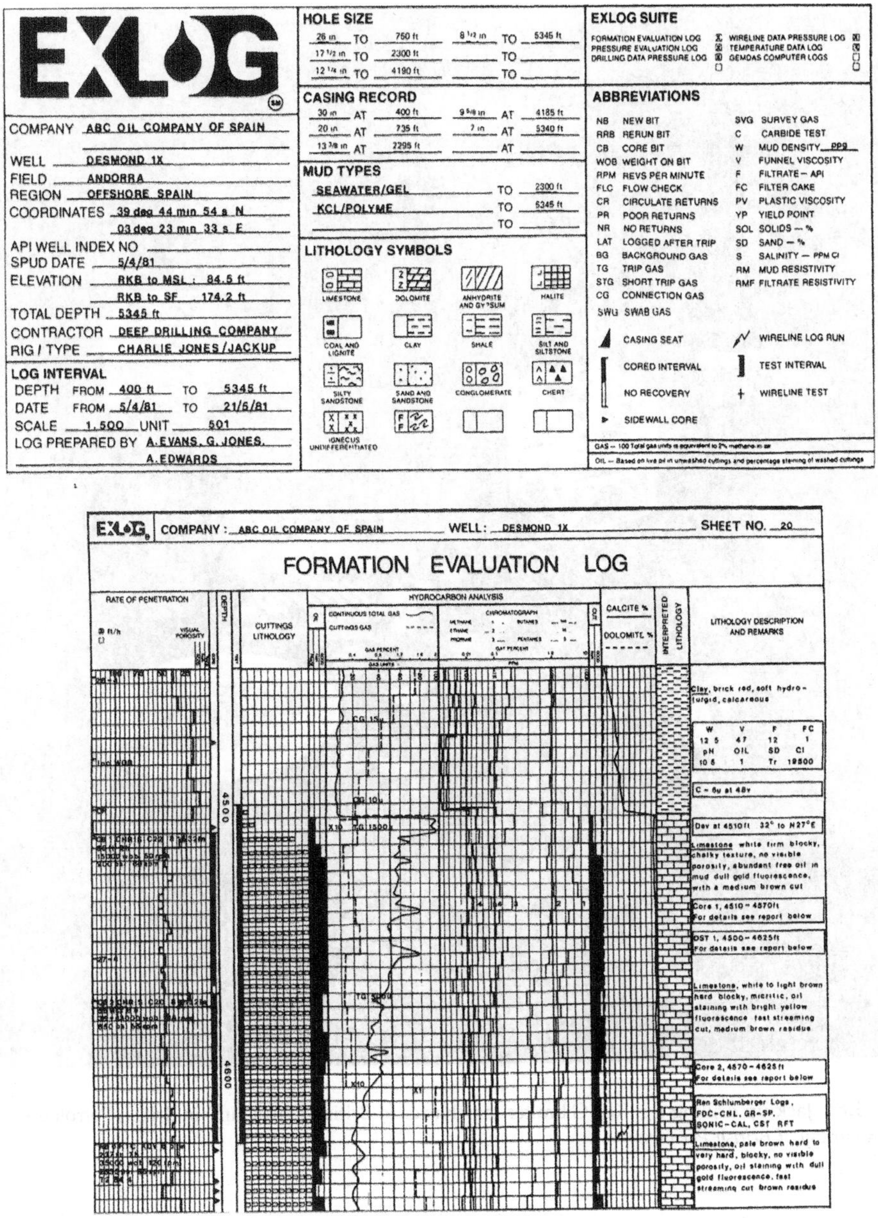

FIGURE 3.7 Example of part of a typical mud log, showing the range of drilling and geological data recorded. *Courtesy of Exploration Logging Services, Ltd.*

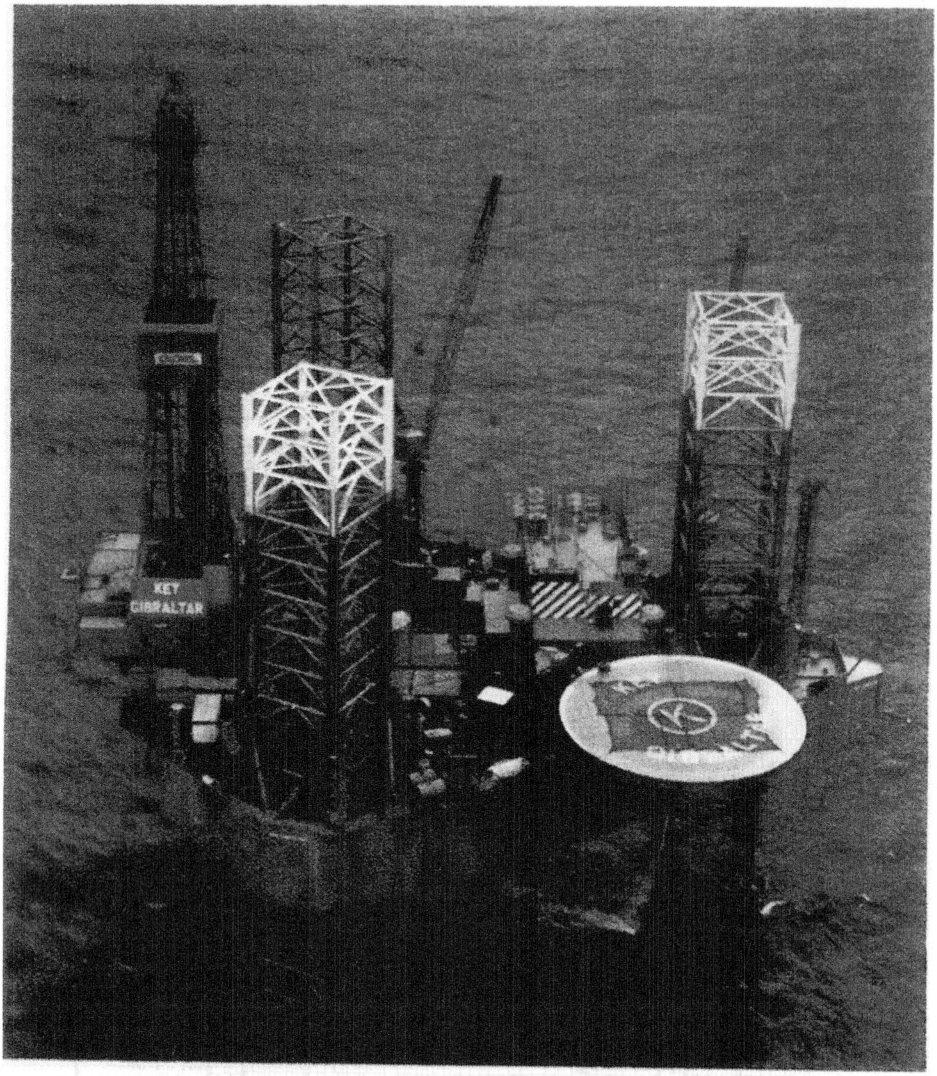

FIGURE 3.8 Jack-up drilling rig, Key Gibralter (on charter to British Petroleum) in British Petroleum's West Sole gas field. *Courtesy of British Petroleum.*

wells. The 2014 percentage of wells drilled in the United States is as follows: vertical 20%, horizontal 67%, and directional 12%.

Drilling wells directionally or horizontally has several advantages over vertical wells:

1. target areas under cities or lakes where surface occupancy is impossible;
2. drain a large area with pad drilling (several wells from one pad to reduce surface footprint);

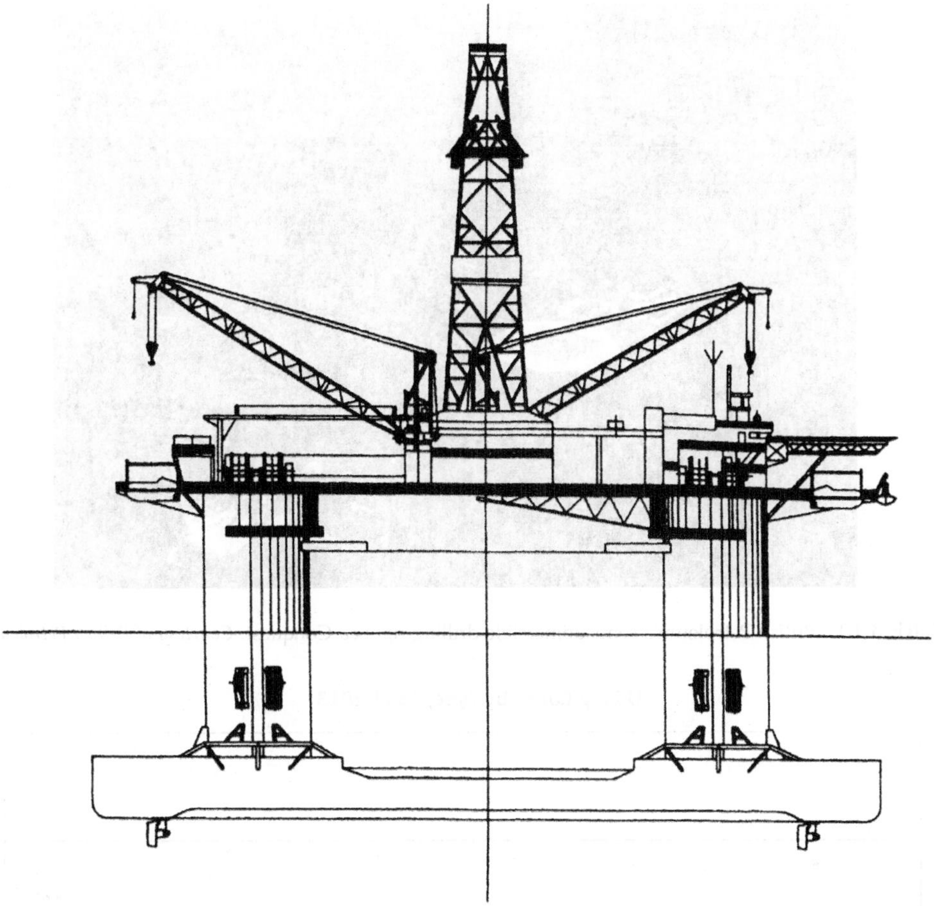

FIGURE 3.9 Diagram of a modem semisubmersible drilling rig, the GVA 8400. *Courtesy of GVA Consultants.*

3. increase the contact area of the pay zone with the well bore;
4. improve productivity in fractured reservoirs by contacting more fractures;
5. relieve pressure or seal "out-of-control wells."

Figure 3.12 illustrates vertical, slant-hole, and horizontal wells. Slant holes in this example target a combination of lenticular (e.g., channel sandstones) and blanket deposits (e.g., shoreline sandstones). Horizontal wells target a single horizon and are generally drilled in a direction to maximize natural fracture intersections or selected reservoir compartments.

3.1.5 Horizontal Drilling and Multistage Hydraulic Fracture Stimulation

Most tight reservoirs (e.g., shale gas, tight gas, tight oil, and coal seam gas) require hydraulic fracture stimulation in order for them to produce. The fracture stimulation process involves injecting water, chemicals, and proppant (sand or ceramics) at high pressures into

FIGURE 3.10 British Petroleum's semisubmersible drilling rig, Sea Conquest. *Courtesy of British Petroleum.*

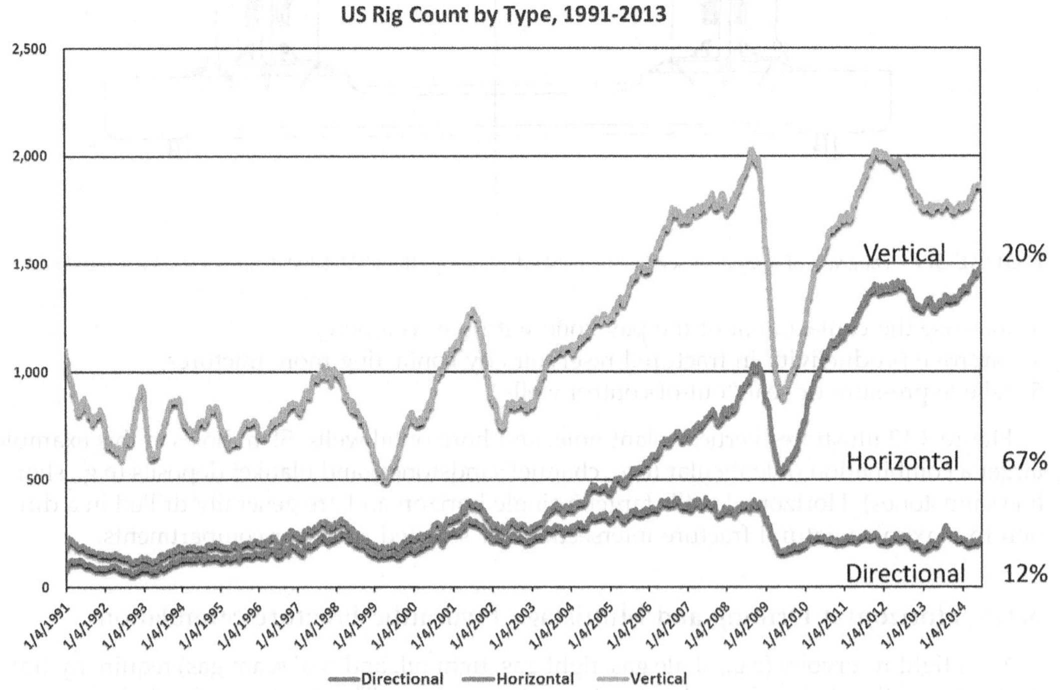

FIGURE 3.11 US rig count by type, 1991–2013. *Data from Baker-Hughes.*

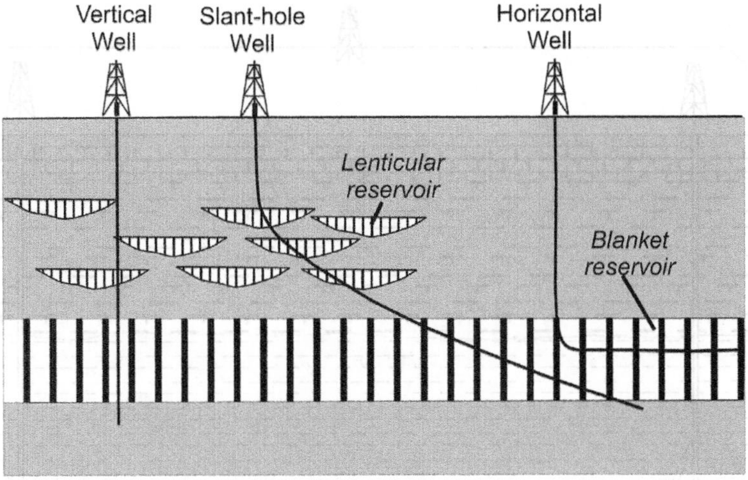

FIGURE 3.12 Diagrammatic sketch illustrating vertical, slant-hole, and horizontal wells. Fractures are the vertical lines in the sketch.

a formation, which creates small fractures in the surrounding rock and allows natural gas or oil to flow from the rock fractures to the production well. Chemicals used in this process generally include gelling agents, crosslinkers, friction reducers, corrosion inhibitors, scale inhibitors, biocide, and even diesel fuel. Proppants are used to keep the fractures open. Hydraulic fracture stimulation of wells has occurred since the 1940s but increased dramatically in 2003, when energy companies began drilling shale formations horizontally and completing the wells with multistage fracturing. Hydraulic fracturing is applied to tight formations to enhance well performance, minimize drilling, and recover otherwise inaccessible resources.

Figure 3.13 illustrates horizontal and vertical wells completed with hydraulic fracture stimulation. The advantage of horizontal drilling with multistage hydraulic fracturing over vertical wells with hydraulic fracturing is dramatically increasing the contact area of the formation with the well bore. Hydraulic fractures lengths are described by Davies et al. (2012) and Fisher and Warpinski (2011).

3.1.6 Various Types of Production Units

As previously noted, when a well has been drilled, the derrick is moved to the next location. Oil and gas are produced from wells in several ways. Details of the various drive mechanisms of reservoirs are outlined in Chapter 6. Actual production is achieved in one of two ways, depending on whether or not the oil flows to the surface without the use of a pump.

In most cases, the reservoir has sufficient pressure for oil or gas to flow to the surface. In this situation, casing is generally run below the producing zone, and perforations are shot through it by explosive charges opposite the hydrocarbon pay interval. Steel tubing is hung from the well head to the producing zone, and the annulus between tubing and casing

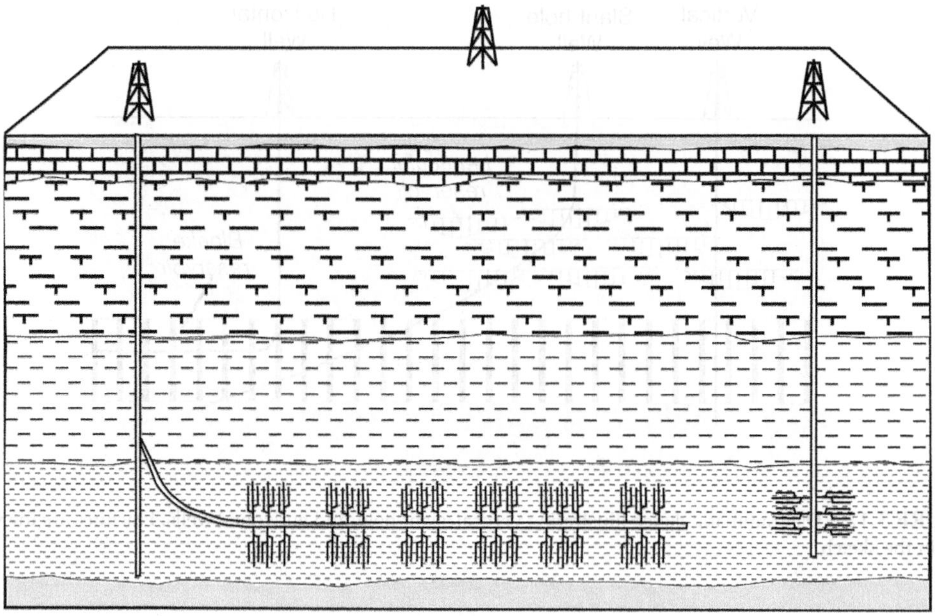

FIGURE 3.13 Sketch illustrating horizontal and vertical wells completed with hydraulic fracture stimulation.

sealed off with a packer (Fig. 3.14). The productivity of many wells may be stimulated, either initially or worked over later during their life. Typical stimulation techniques involve fracturing the formation, generally by high-pressure pumping of metallic or plastic shrapnel into fractures to wedge them open. For a finale, hydrochloric or some other acid may then be injected to enlarge the fractures and hence increase permeability and productivity. These techniques are largely, but not exclusively, applied to carbonate reservoirs.

At the well head, a system of valves, termed the Christmas tree, is installed from which a flowline leads to a tank in which produced oil and gas are separated at atmospheric pressure.

If the reservoir pressure is too low for oil to flow to the surface, a pumping device is used: either a nodding donkey at the well head, or a downhole pump installation. Figure 3.15 illustrates a well-head pump. A diesel or electrically powered beam engine raises and lowers a connected string of sucker rods, which are connected to a piston at the base of the borehole. Alternatively, an electrically driven centrifugal pump may be installed at the bottom of the hole. This device is more effective than the nodding donkey and less ecologically shocking. Maintenance costs may be higher because of difficulties of access, and reliability may be lower.

Onshore development wells are drilled on various types of geometric arrangements at spacings determined by the permeability of the reservoir and radius of drainage anticipated for each well. Offshore, however, this procedure is not feasible. Various types of production facilities are now used. These include floating buoys, ships, floating production platforms (Fig. 3.16), and fixed production platforms (Figs 3.17 and 3.18). Fixed platforms have been used for water depths of up to some 450 m; below that floating production facilities are

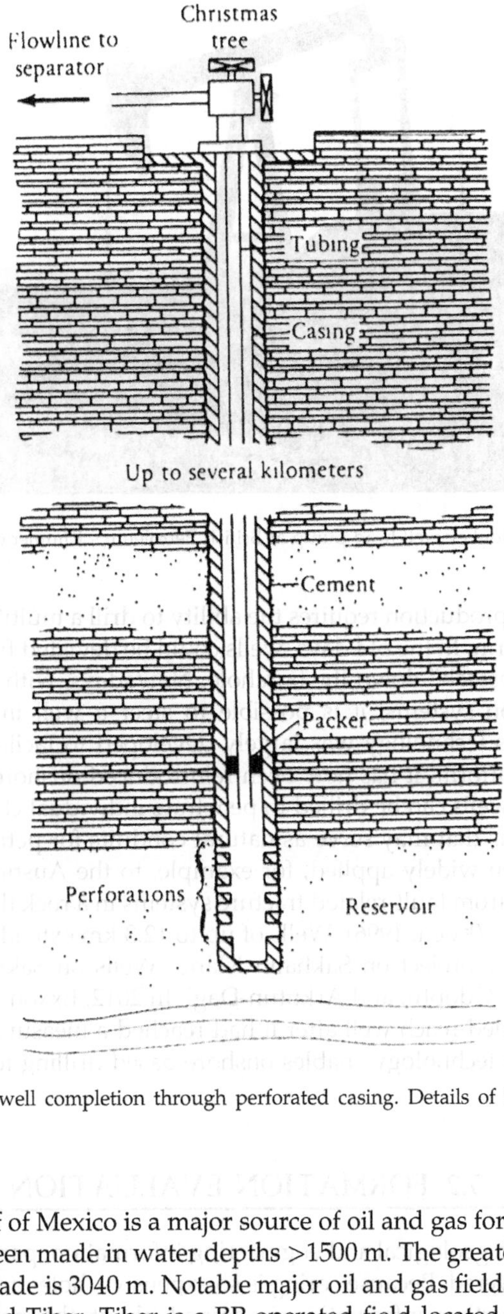

FIGURE 3.14 A typical well completion through perforated casing. Details of the various casing strings are omitted (Fig. 3.5).

used. The offshore Gulf of Mexico is a major source of oil and gas for the United States. Sixty-five discoveries have been made in water depths >1500 m. The greatest water depth in which a discovery has been made is 3040 m. Notable major oil and gas fields include Mensa, Eugene Island 330, Atlantis, and Tiber. Tiber is a BP operated field located in the Keathley Canyon Block 102. The field is reported to have four to six billion barrels in place and an estimated ultimate recovery of 600–900 million barrels of oil. The deepest well in the field was drilled to 10,685 m under 1260 m of water.

FIGURE 3.15 Nodding donkey well-head pump at British Petroleum's Kimmeridge Bay field, Dorset.

Offshore petroleum production requires the ability to drill a multitude of wells that radiate out from a single platform. In the old days, wells could be deviated from vertical by dropping a steel wedge or "whipstock" down the borehole. Nowadays, with motorized drill bits and sophisticated navigation systems, it is possible to steer a well in any orientation that is required. Although this technology was largely developed to facilitate drilling many wells off a single offshore platform, it has now been reintroduced onshore and applied to drilling horizontal wells. Such wells can be drilled to penetrate individual channel reservoir sands or to crosscut fault systems that may serve as natural conduits for petroleum production. Horizontal drilling has been widely applied, for example, to the Austin Chalk of Texas, where petroleum is produced from fault-related fracture systems in a rock that has porosity, but normally lacks permeability (Koen, 1996). Wells of up to 12.3 km extended-reach have now been drilled in the Sakhalin-1 project on Sakhalin Island. Wells on Sakhalin Island target three offshore fields: Chayvo, Odoptu, and Arkutun-Dagi. In 2012, Exxon Neftegas Ltd. completed the Z-44 Chayvo extended-reach well after it had reached a measure depth of 12,376 m. The extended-reach drilling technology enables onshore-based drilling to reach petroleum reservoirs far offshore.

3.2 FORMATION EVALUATION

Boreholes yield much geological and engineering information, particularly when extensive lengths of core are recovered. Because coring is so expensive, holes are usually drilled with an ordinary rotary bit. Further information is gained about the penetrated formations by measuring their geophysical properties with the aid of wireline logs. Formerly, geophysical logs were run after drilling a section of the well and before setting casing, so several log runs are necessary during the drilling of a single well. Nowadays, however, it is possible

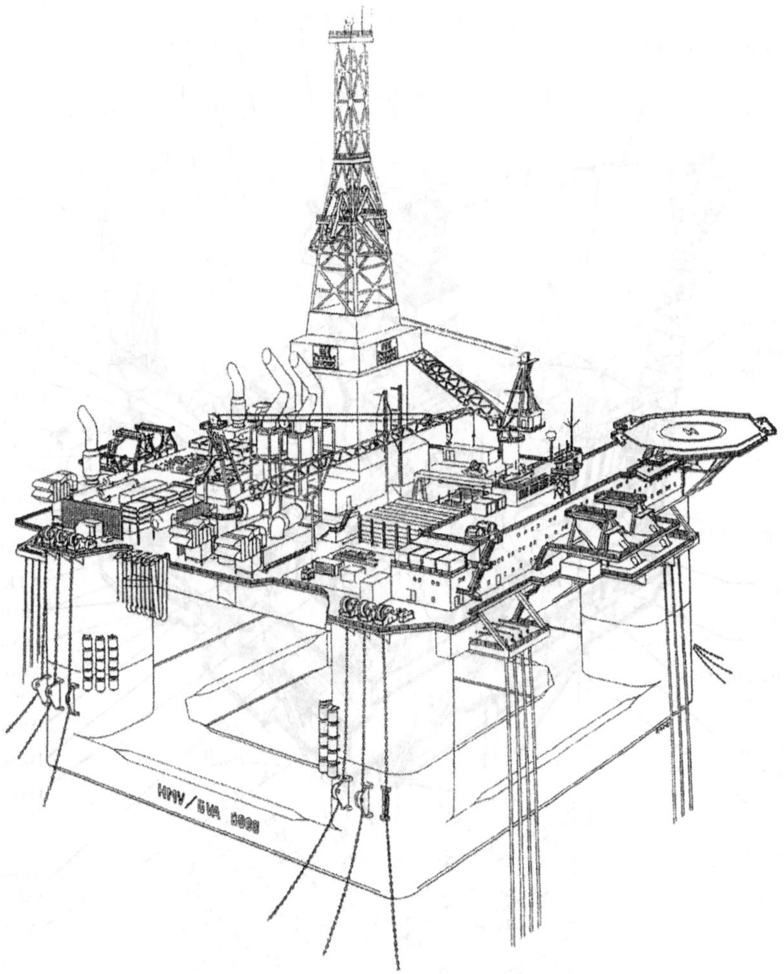

FIGURE 3.16 A typical modern floating production platform, the HMV/GVA 8000. *Courtesy of GVA Consultants.*

to run some geophysical logs within the drill pipe while the well is actually being drilled. MWD is an important advance in formation evaluation, providing early warnings of rock type, hydrocarbon saturation, and potentially perilous zones of high pressure.

The equipment that measures the geophysical properties of the penetrated rocks is housed in a cylindrical sonde, which is lowered down the borehole on a multiconductor electric cable (Fig. 3.19). On reaching the bottom of the hole, the recording circuits are switched on and the various properties measured as the sonde is drawn to the surface. The measurements are transmitted up the cable and recorded on film or magnetic tape in the logging unit. In onshore operations, the recorders are generally mounted on a truck (Fig. 3.20).

Many different parameters of the rock can be recorded, such as formation resistivity, sonic velocity, density, and radioactivity. The recorded data can then be interpreted to determine

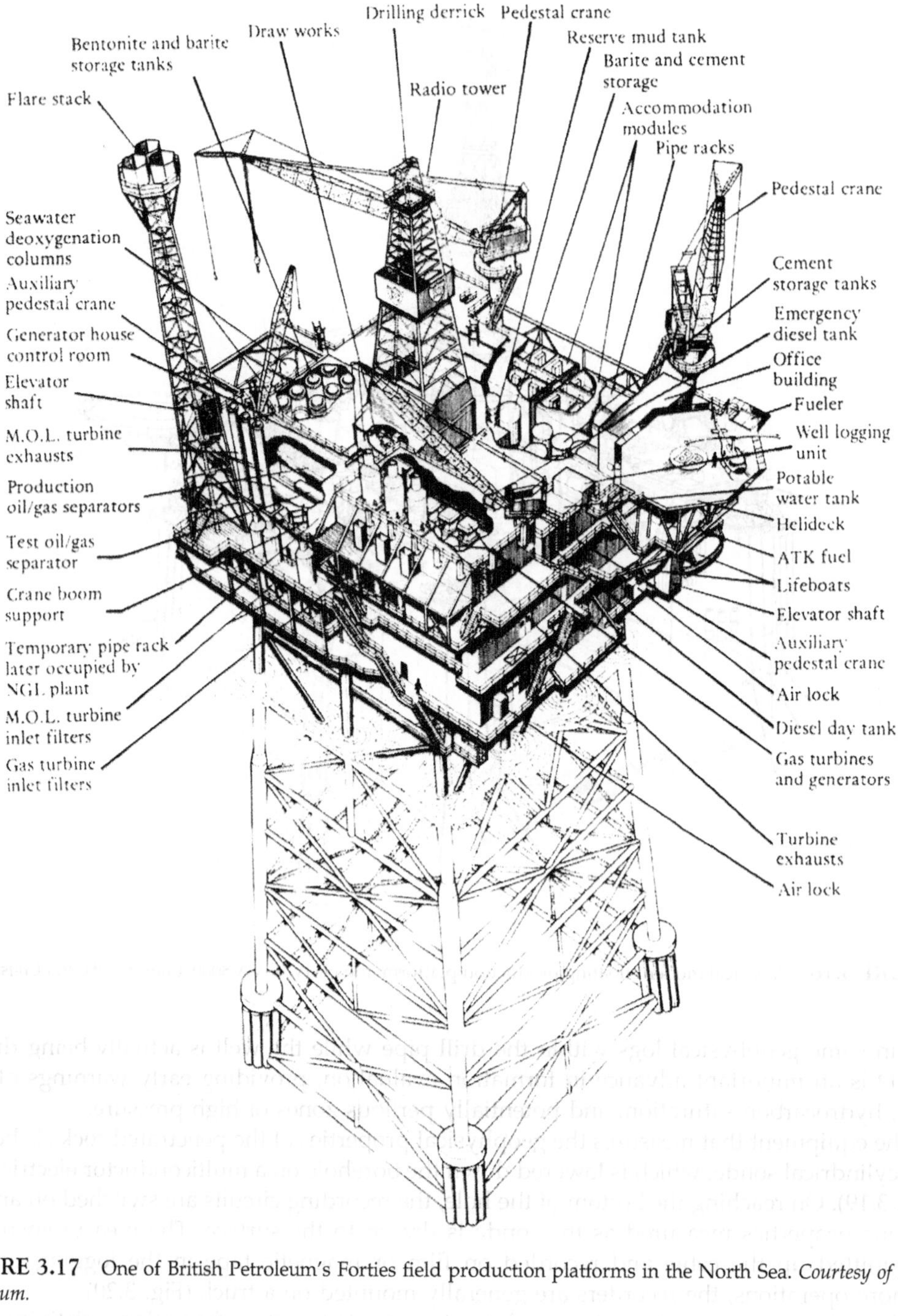

FIGURE 3.17 One of British Petroleum's Forties field production platforms in the North Sea. *Courtesy of British Petroleum.*

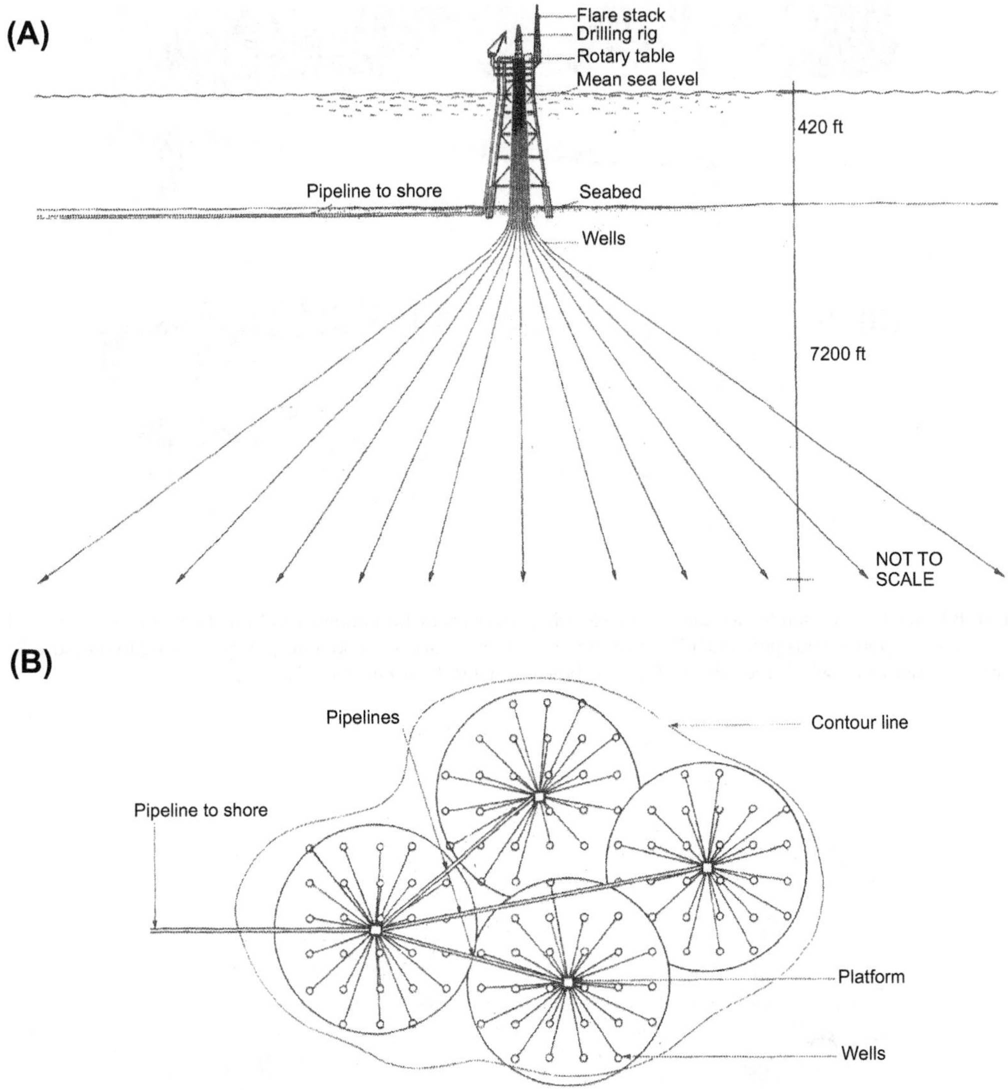

FIGURE 3.18 (A) Cross-section showing how deviated wells are drilled from a platform to penetrate the reservoir at the required depth (B). *Courtesy of British Petroleum.*

the lithology and porosity of the penetrated formations and also the type and quantity of fluids (oil, gas, or water) within the pores.

Formation evaluation is a large and complex topic. Many oil companies employ full-time log analysts. The following account is only intended as an introduction to the topic. The various types of logs are now described, and the principles of log analysis are reviewed. Petroleum geoscientists generally attend courses on log analysis early in their professional career. These courses provide up-to-date and detailed information on new tools and

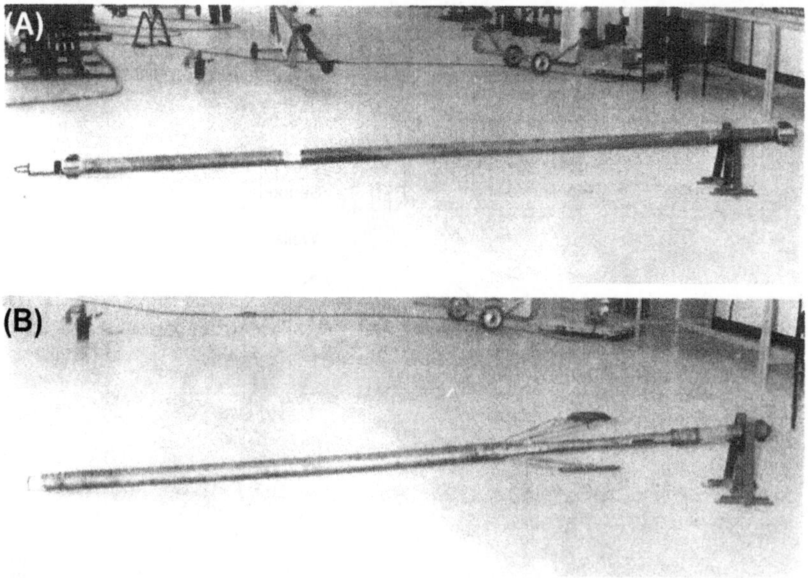

FIGURE 3.19 Two sondes for measuring electric properties of formations: (A) Induction resistivity tool, which combines the spontaneous potential (SP), a 16-in normal to measure R_{xo}, and a deep induction log to measure R_t. (B) A microfocused log, which records SP, R_{xo}, and hole diameter. *Courtesy of Tesel Services, Ltd.*

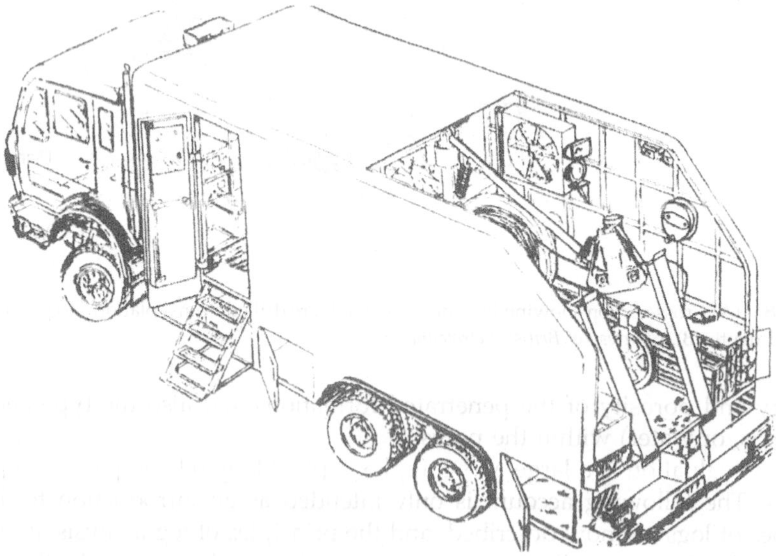

FIGURE 3.20 A typical truck for onshore well-logging operations. *Courtesy of Tesel Services Ltd.*

analytical methods and software programs. When the author attended his first logging course some 30 years ago, calculations were done by slide rules, aided by curious graphs and nomograms. Nowadays, of course, it is all done by computer. Even so, it is important to understand the fundamental principles that underlie the complex calculations, statistical gymnastics, and brightly colored displays that are an integral part of modern formation evaluation.

3.2.1 Electric Logs

3.2.1.1 The Spontaneous Potential, Self-Potential, or SP Log

The spontaneous potential, or self-potential, log is the oldest type of geophysical log in use. The first one was run in 1927. The spontaneous log records the electric potential set up between an electrode in a sonde drawn up the borehole and a fixed electrode at the earth's surface (Fig. 3.21). It can only be used in open (i.e., uncased) holes filled with conductive mud. Provided that there is a minimum amount of permeability, the SP response is dependent primarily on the difference in salinity between drilling mud and the formation water.

The electric charge of the SP is caused by the flow of ions (largely Na^+ and Cl^-) from concentrated to more dilute solutions. Generally, this flow is from salty formation water to fresher drilling mud (Fig. 3.22). This naturally occurring electric potential (measured in millivolts) is basically related to the permeability of the formation. Deflection of the log from an arbitrarily determined shale baseline indicates permeable and therefore porous sandstones or carbonates. In most cases, this deflection, termed a *normal* or *negative* SP deflection, is to the left of the baseline. Deflection to the right of the baseline, termed *reversed* or *positive* SP, occurs when formation waters are fresher than the mud filtrate. A poorly defined or absent SP deflection occurs in uniformly impermeable formations or where the salinities of mud and formation water are comparable (Fig. 3.23). In most cases, with a normal SP, the curve can

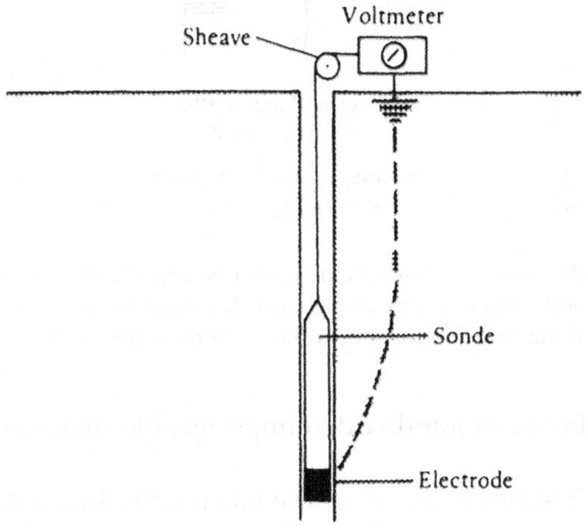

FIGURE 3.21 Basic arrangement for the SP log.

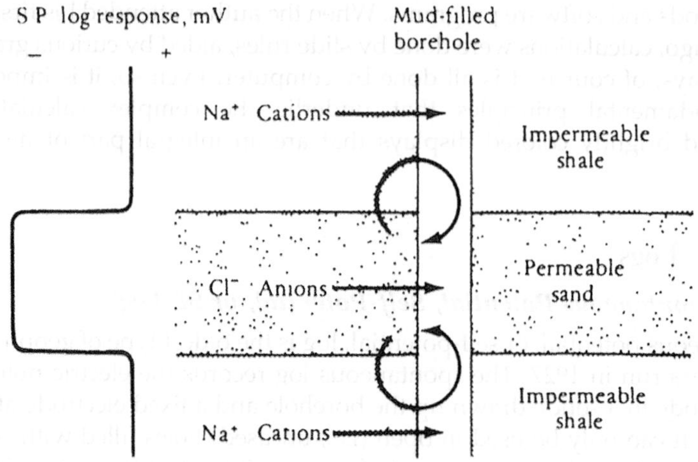

FIGURE 3.22 Diagram showing how ionic diffusion causes the spontaneous potential effect. Looped arrows show the direction of a positive current flow. Log response is for the situation in which the salinity of the formation water is greater than that of the drilling mud.

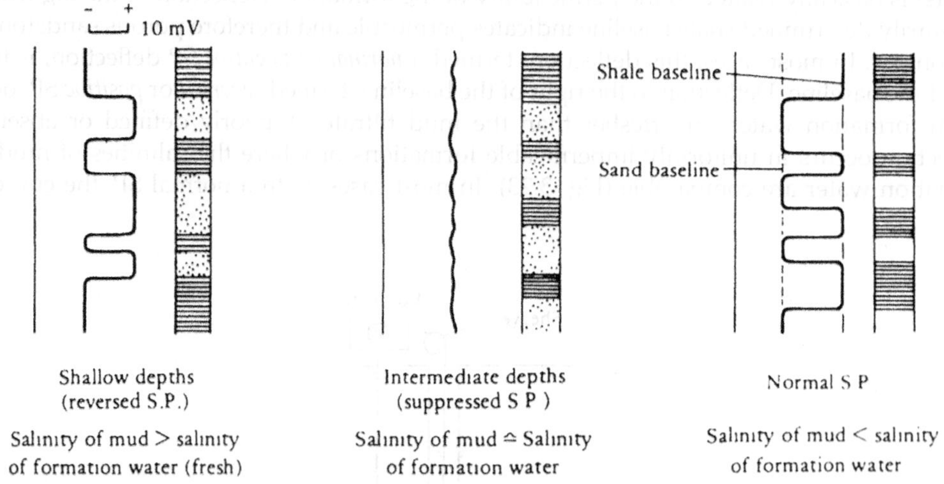

FIGURE 3.23 Schematic SP logs for different salinity contrasts of mud and formation water. Reversed SP logs are very rare. Suppressed SP logs occur where salt muds are used. The usual response, in which the salinity of the drilling mud is less than the salinity of the formation water, is shown in the right-hand log.

be used to differentiate between interbedded impermeable shales and permeable sandstones or carbonates.

Note that the millivolt scale on the SP curve has no absolute value. The logging engineer shifts the baseline as the curve gradually drifts across the scale during the log run. The SP deflections of adjacent wells are not comparable. Similarly, although local deflections on the curve are caused by vertical variations in permeability, no actual millidarcy values can be measured.

The amount of the current and, hence the amplitude of deflection on the SP curve, is related not only to permeability but also to the contrast between the salinity of the drilling mud and the formation water. Specifically, it is the contrast between the resistivity of the two fluids. Empirically, it has been found that

$$E = K \log \frac{R_{mf}}{R_w}$$

where E is the SP charge (millivolts); K is a constant, which is generally taken as $65 + 0.24T$ (degrees centigrade) or $61 + 0.133T$ (degrees Farenheit); R_{mt} is the resistivity of mud filtrate (ohms per meter); and R_w is the resistivity of formation water (ohms per meter).

Details of this relationship are found in Wyllie (1963) and the various wireline-logging service company manuals. Note that since the resistivity of salty water varies with temperature, it is necessary to allow for that factor when solving the equation. The resistivity of the filtrate of the drilling mud may be measured at the surface and, if the bottom hole temperature is known, it can be recalculated for the depth of the zone at which the SP charge is measured. The equation may then be solved, and R_w, the resistivity, and hence the salinity of the formation water can be determined.

In summary, the SP log may be used to delineate permeable zones, and hence, it aids lithological identification and well-to-well correlation. The SP log can also be used to calculate R_w, the resistivity of the formation water. The SP is limited by the fact that it cannot be run in cased holes and is ineffective when R_{mf} is approximately equal to R_w. This situation occurs with many offshore wells drilled using saltwater-based drilling muds.

3.2.1.2 *Resistivity Logs*

The three main ways of measuring the electrical resistivity of formations penetrated by boreholes are the *normal log*, *laterolog*, and *induction log* techniques. With the normal, or conventional resistivity, log an electric potential and flow of current is set up between an electrode on the sonde and an electrode at the surface. A pair of electrodes on the sonde is used to measure the variation in formation resistivity as the sonde is raised to the surface. The spacing between the current electrode and the recording electrode can be varied, as shown in Fig. 3.24(A). The three electrode spacings usually employed are 16 in (short normal), 64 in (long normal), and 8 ft 8 in (long lateral). They can generally be run simultaneously with an SP log.

Normal resistivity devices, although largely superseded by more sophisticated types, may be encountered on old well logs. For low-resistivity salty muds, laterologs, or guard logs, are now generally used (Fig. 3.24(B)). In these systems, single electrodes cause focused current to flow horizontally into the formation. This horizontal flow is achieved by placing two guard electrodes above and below the current electrode. By balancing the guard electrode current with that of the central generating electrode, a sheet of current penetrates the formation. The potential of the guard and central electrodes is measured as the sonde is raised. As with conventional resistivity logs, various types of laterologs can be used to measure

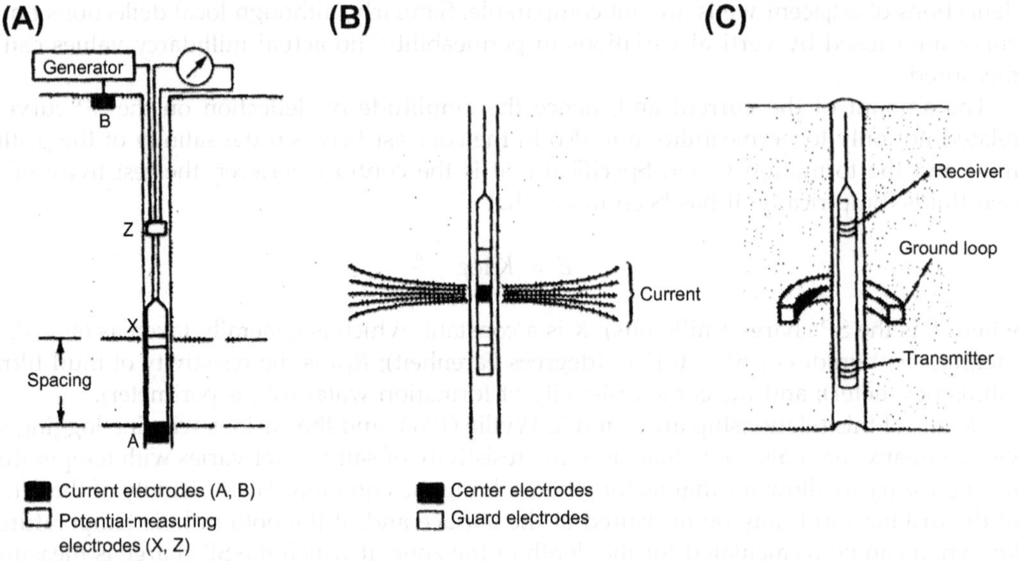

FIGURE 3.24 Illustrations of the three types of resistivity logging devices. (A) The normal resistivity logging device. Variation of the spacing between A and X determines the distance away from the borehole at which the resistivity is measured. (B) Diagram of the laterolog method. (C) Diagram of the induction principle. For explanations see the text.

resistivity at different distances from the borehole. This measurement is achieved by changing the geometry of the focusing electrodes.

For freshwater or oil-based muds, which have a low resistivity, a third type of device is used. This device is the induction log, in which transmitter and receiver coils are placed at two ends of a sonde and are used to transmit a high-frequency alternating current. This current creates a magnetic field, which, in turn, generates currents in the formation. These currents fluctuate according to the formation resistivity and are measured at the receiver coil, as shown in Fig. 3.24(C).

The electrical resistivity of formations varies greatly. Solid rock is highly resistive, as is porous rock saturated in freshwater, oil, or gas. Shale, on the other hand, and porous formations saturated with salty water or brine have very low resistivities. When run simultaneously, SP and resistivity logs enable qualitative interpretations of lithology and the nature of pore fluids to be made (Fig. 3.25). One of the functions of drilling mud is to prevent fluids from flowing into the borehole from permeable formations. At such intervals, a cake of mud builds up around the borehole wall, and mud filtrate is squeezed into the formation. Thus, the original pore fluid is displaced, be it connate water, oil, or gas. So a circular invaded, or flushed, zone is created around the borehole with a resistivity (referred to as R_{xo}) that may be very different from the resistivity of the uninvaded zone (R_t). A transition zone separates the two. This arrangement is shown in Fig. 3.26. As already noted, various types of resistivity logs are not only adapted for different types of mud but also for measuring resistivity of both the uninvaded zone (R_t) and the flushed zone (R_{xo}). The latter

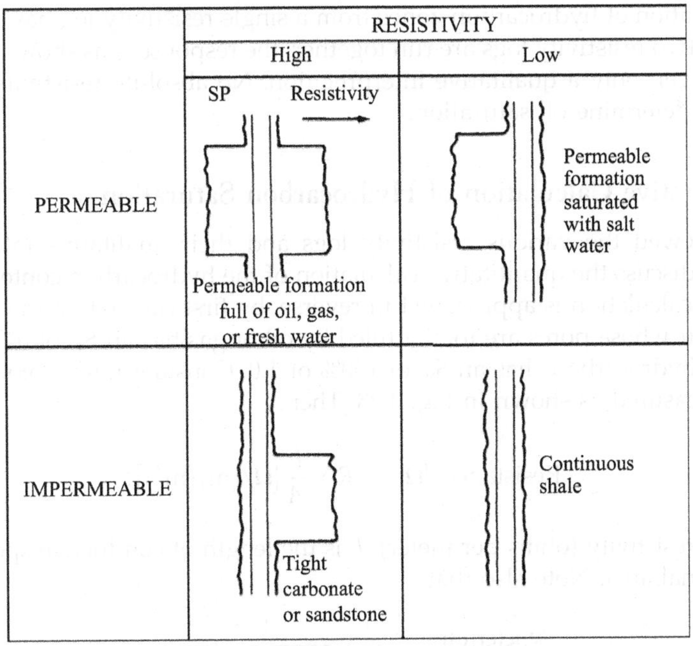

RESISTIVITY
High
Low
SP
Resistivity
PERMEABLE
Permeable formation full of oil, gas, or fresh water
Permeable formation saturated with salt water
IMPERMEABLE
Tight carbonate or sandstone
Continuous shale

FIGURE 3.25 The four basic responses for SP and resistivity logs for a bed between impermeable formations.

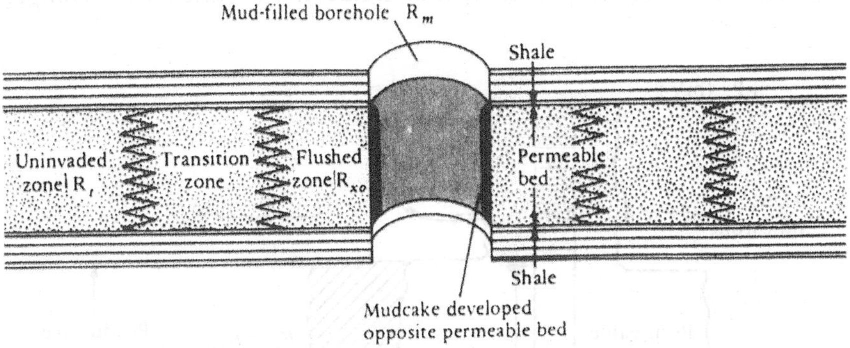

FIGURE 3.26 The situation around a borehole adjacent to a permeable bed. Resistivity of flushed zone, R_{xo}; resistivity of uninvaded zone, R_t; resistivity of mud filtrate, R_{mf}; and resistivity of formation water, R_w.

are generally referred to as *microresistivity logs*, of which there are many different types with different trade names. Figure 3.24 shows the responses of R_t and R_{xo} (deep and shallow) resistivity logs in various water-bearing reservoirs. Note the convention that the deep-penetrating log is shown by a dashed curve and the shallow log by a continuous line. Where formations are impermeable, there is no separation between the two logs, since there is no flushed zone. A single sonde can simultaneously record the SP and more than one resistivity curve (Fig. 3.19).

The identification of hydrocarbon zones from a single resistivity log has already been discussed. Where two resistivity logs are run together, the response is as shown in Fig. 3.27. This figure is, however, only a qualitative interpretation. No absolute resistivity cut-off reading can be used to determine oil saturation.

3.2.2 Quantitative Calculation of Hydrocarbon Saturation

Having reviewed the various resistivity logs and their qualitative interpretation, it is appropriate to discuss the quantitative calculation of the hydrocarbon content of a reservoir. Generally, this calculation is approached in reverse, by first calculating the water saturation (S_W). A reservoir whose pores are totally filled by oil or gas has an S_W of 0% or 0.00; a reservoir devoid of hydrocarbons has an S_W of 100% or 1.0. Consider now a block of rock whose resistivity is measured as shown in Fig. 3.28. Then,

$$\text{Resistance } [\Omega] = R \times \frac{L}{A}\left[\Omega/(\text{m})(\text{m}^2)\right],$$

where R is the resistivity (ohms per meter), L is the length of conductive specimen, and A is the cross-sectional area. Note also that

$$\text{Resistivity} = \frac{1}{\text{conductivity (mhos)}}.$$

Most reservoirs consist of mineral grains that are themselves highly resistive, but between them are pores saturated with fluids whose conductivity varies with composition and

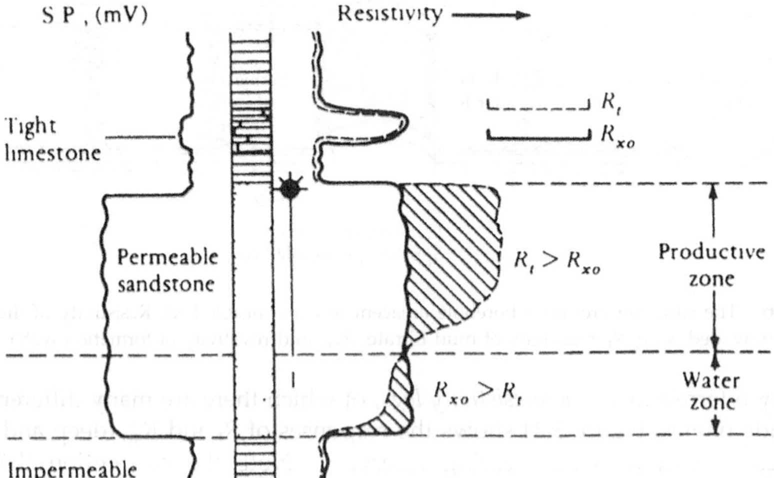

FIGURE 3.27 SP and resistivity logs through a hydrocarbon reservoir showing typical responses for the situation where $R_{mf} > R_w$.

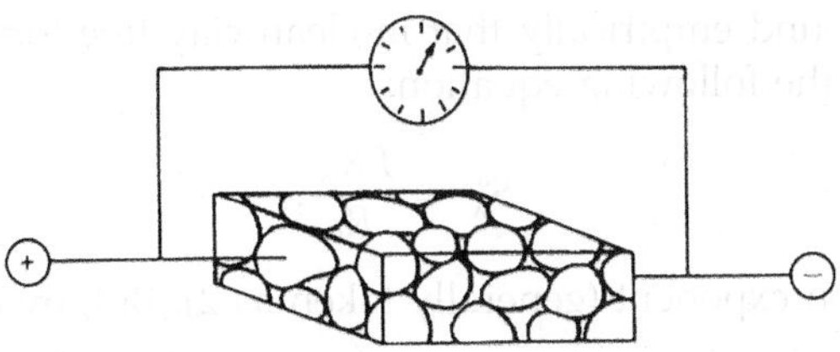

FIGURE 3.28 Sketch showing how the resistivity of a rock is measured.

temperature. An important number in log interpretation therefore is the particular formation resistivity factor (F), defined as

$$\text{Formation resistivity factor (F)} = \frac{R_o}{R_w}$$

where R_o is the resistivity of rock 100% saturated with water of resistivity R_W. Note that R_W decreases with increasing salinity and temperature, as, therefore, does R_o. The value of F increases with decreasing porosity. Archie (1942) empirically derived what is now termed the Archie formula:

$$F = \frac{a}{\phi^m},$$

where a is a constant, ϕ is the porosity, and m is the cementation factor. The values of a and m depend on formation parameters that are difficult to measure, being influenced by the tortuosities among interconnected pores.

Several values for a and m have been found that empirically match various reservoirs. For sands,

$$F = \frac{0.81}{\phi^2}.$$

For compacted formations, notably carbonates,

$$F = \frac{1}{\phi^2}.$$

The Shell formula for low-porosity nonfractured carbonates is

$$F = \frac{1}{\phi^m},$$

where

$$m = 1.87 + \frac{0.019}{\phi}.$$

The Humble formula (suitable for soft formations) is

$$F = \frac{0.62}{\phi^{2.15}}.$$

Again, it has been found empirically that in clean clay-free formations, water saturation can be calculated from the following equation:

$$S_w^n = \frac{FR_w}{R_t},$$

where n is the saturation exponent (generally taken as 2). But, by definition,

$$F = \frac{R_o}{R_w}$$

or

$$R_o = FR_w.$$

Thus,

$$S_w^2 = \frac{R_o}{R_t}$$

or

$$S_W = \sqrt{\frac{R_o}{R_t}}.$$

Using a log reading deep resistivity, R_t can be measured in the suspected oil zone and R_o in one that can be reasonably assumed to be 100% water saturated. Thus, S_w may be calculated throughout a suspected reservoir. This is the simplest method and is only valid for clean (i.e., clay-free) reservoirs. If a clay matrix is present, the resistivity of the reservoir is reduced, and oil-saturated shaley sands may be missed. This method also assumes that the value for F is the same in the oil and water zones, which is seldom true because oil inhibits cementation, which may continue in the underlying water zone. To overcome some of these problems, the ratio method may be used. This method is based on the assumption that

$$S_{xo} = S_w^{0\,2}.$$

From this equation, the following relationship may be derived:

$$S_w = \frac{(R_{xo}/R_t)^{5/8}}{R_{mf}/R_w},$$

where R_{xo} can be measured from a microlog; R_t from a deep resistivity log; R_{mt} from the drilling mud at the surface (corrected for temperature at the zone of interest); and R_w from the SP log, as shown in Fig. 3.29.

3.2.3 Radioactivity Logs

3.2.3.1 *The Gamma-Ray Log (and Caliper)*

Three types of logs that measure radioactivity are commonly used: the gamma-ray, neutron, and density logs. The gamma-ray log, or gamma log, uses a scintillation counter

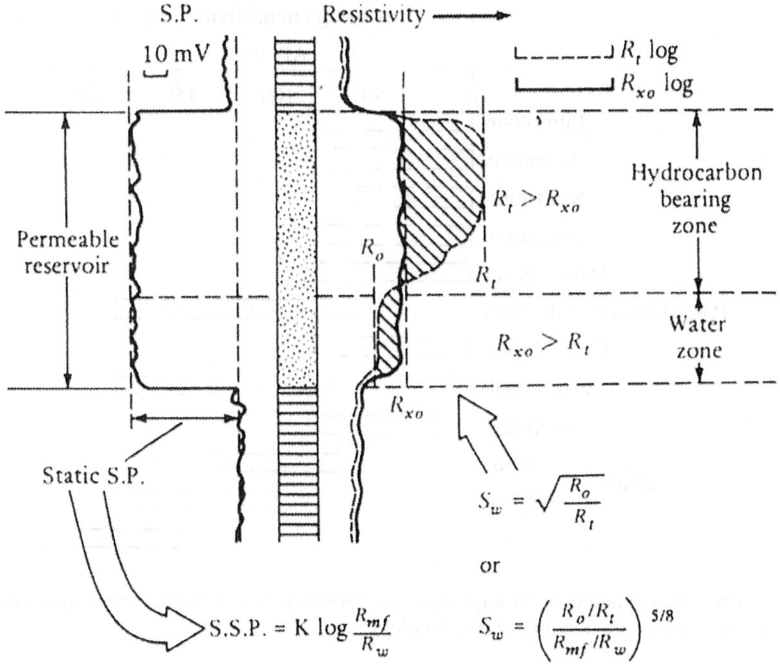

FIGURE 3.29 SP and resistivity logs for a hydrocarbon zone showing where readings are taken to solve the R_w and S_w calculations. For an explanation, see the text.

to measure the natural radioactivity of formations as the sonde is drawn up the borehole. The main radioactive element in rocks is potassium, which is commonly found in illitic clays and to a lesser extent in feldspars, mica, and glauconite. Zirconium (in detrital zircon), monazite, and various phosphate minerals are also radioactive. Organic matter commonly scavenges uranium and thorium, and thus oil source rocks, oil shales, sapropelites, and algal coals are radioactive. Humic coals, on the other hand, are not radioactive. The gamma log is thus an important aid to lithological identification (Fig. 3.30). The radioactivity is measured in API (American Petroleum Institute) units and generally plotted on a scale of 0–100 or 0–120 API.

Conventionally, the natural gamma reading is presented on the left-hand column of the log in a similar manner to, and often simultaneously with, the SP log. The gamma log can be used in much the same way as an SP, with a shale baseline being drawn. Deflection to the left of this line does not indicate permeability, but rather a change from shale to clean lithology, generally sandstone or carbonate. The gamma reading is affected by hole diameter, so it is generally run together with a caliper log, a mechanical device that records the diameter of the borehole. The caliper log shows where the hole may be locally enlarged by washing out or caving and hence deviating the expected gamma-ray and other log responses. The hole may also be narrower than the gauge of the bit where *bridging* occurs. Bridging is caused by either sloughing of the side of the hole and incipient collapse or a build-up of mud cake opposite permeable zones. Although the gamma log is affected by hole diameter,

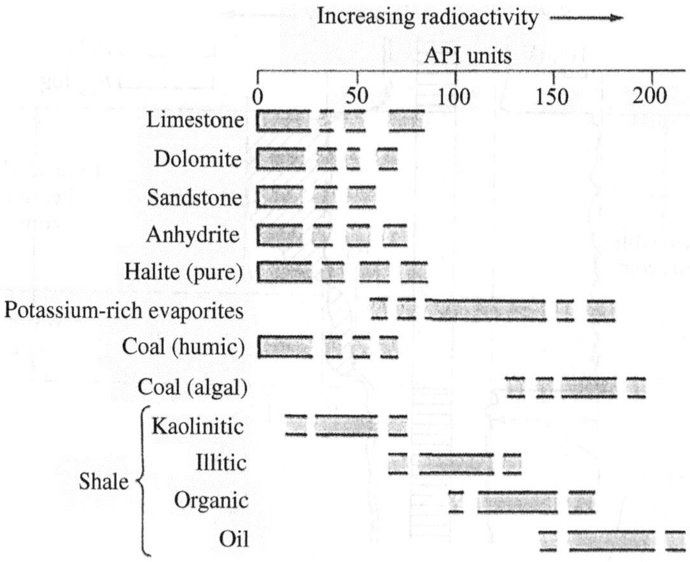

FIGURE 3.30 The approximate gamma log ranges for various rocks. Note that small quantities of radioactive clay, for example, can increase the reading of any lithology.

it has the important advantage that it can be run through casing. The gamma log is important for identifying lithology, calculating the shaliness of reservoirs, and correlating between adjacent wells.

3.2.3.2 *The Natural Gamma-Ray Spectrometry Tool*

One of the limitations of the standard gamma-ray log is that it is unable to differentiate among various radioactive minerals causing the gamma response. This lack of differentiation causes a severe problem when the log is used to measure the clay content of a reservoir in which either the clay is kaolin (potassium-free and nonradioactive) or other radioactive minerals are present, such as mica, glauconite, zircon, monazite, or uranium adsorbed on organic matter.

By analyzing the energy wavelength spectrum of detected gamma radiation, the refined gamma-ray spectrometry tool measures the presence of three commonly occurring radioactive decay series, whose parent elements are thorium, uranium, and potassium (Hassan et al., 1976). This information can be used for detailed mineral identification and, in particular, it allows a study of clay types to be made. An example of such an application is shown in Fig. 3.31, where the potassium–thorium ratio indicates the trend from potassic feldspar to kaolinite clays. Gamma-ray spectrometry logging is also important in source rock evaluation because it can differentiate detrital radioactive minerals containing potassium and thorium from organic matter with adsorbed uranium.

3.2.3.3 *The Neutron Log*

The neutron log, as its name suggests, is produced by a device that bombards the formation with neutrons from an Americium beryllium or other radioactive source. Neutron bombardment causes rocks to emit gamma rays in proportion to their hydrogen content.

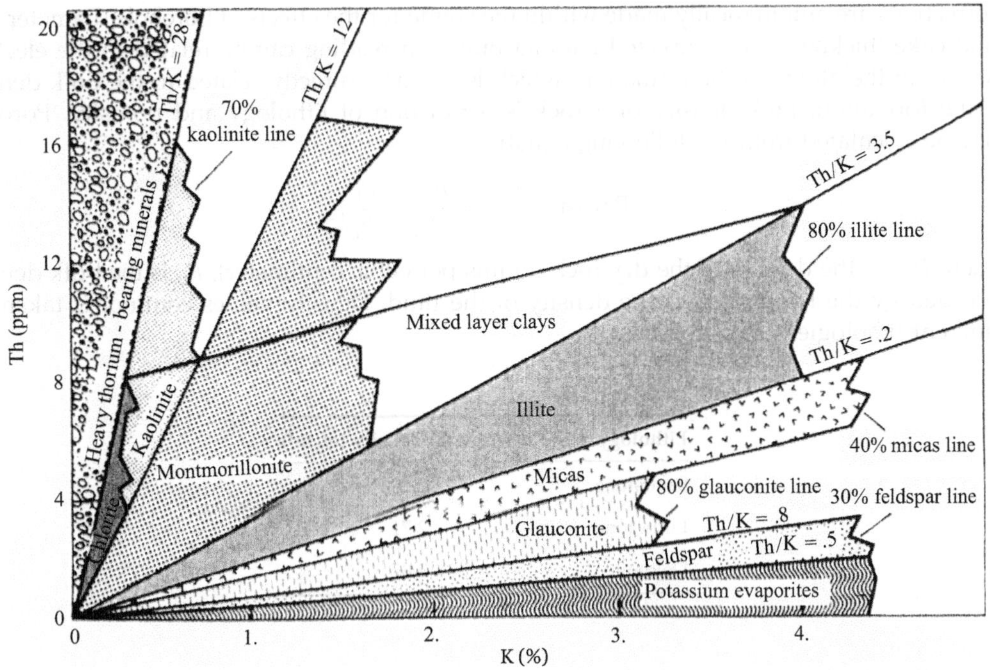

FIGURE 3.31 Crossplot of thorium and potassium showing how the different types of clays (and other minerals) can be identified by the gamma-ray spectrometry log.

This gamma radiation is recorded by the sonde. Hydrogen occurs in all formation fluids (oil, gas, or water) in reservoirs, but not in the minerals. Thus, the response of the neutron log is essentially correlative with porosity. Hole size, lithology, and hydrocarbons, however, all affect the neutron log response. The effect of variation in hole size is overcome by simultaneously running a caliper with an automatic correction for bit gauge. In the early days, the neutron log was recorded in API units. Because it is so accurate for clean reservoirs, the neutron log is now directly recorded in either limestone or sandstone porosity units (LPUs and SPUs, respectively). As with all porosity logs, the log curve is presented to the right of the depth scale, with porosity increasing to the left. Because shales always contain some bonded water, the neutron log will always give a higher apparent porosity reading in dirty reservoirs than actually exists. The hydrogen content of oil and water is about equal, but is lower than that of hydrocarbon gas. Thus, the neutron log may give too low a porosity reading in gas reservoirs. As shown in the following section, this fact can be turned into an advantage. The neutron log can be run in cased holes because its neutron bombardment penetrates steel.

3.2.3.4 The Density Log

The third type of radioactivity tool measures formation density by emitting gamma radiation from the tool and recording the amount of gamma radiation returning from the formation (Fig. 3.32). For this reason, the device is often called the gamma–gamma tool.

Corrections are automatically made within the sonde for the effects of borehole diameter and mud cake thickness. The corrected gamma radiation reading can be related to the electron density of the atoms in the formation, which is, in turn, directly related to the bulk density of the formation. Bulk density of a rock is a function of lithology and porosity. Porosity may be calculated from the following equation:

$$\text{Porosity}(\phi) = \frac{P_{ma} - P_b}{P_{ma} - P_f},$$

where P_{ma} is the density of the dry rock (grams per cubic centimeter), P_b is the bulk density recorded by the log, and P_f is the density of the fluid. Density values commonly taken for different lithologies are as follows:

Lithology	Density (g/cm^3)
Sandstone	2.65
Limestone	2.71
Dolomite	2.87

The fluid in the pores near the borehole wall is generally the mud filtrate. Because the tool has only a shallow depth of investigation and effectively "sees" only that part of the formation invaded by filtrate from the drilling mud, it reads this value for the porosity. Thus, the density of the fluid may vary from 1.0 g/cm^3 for freshwater mud to 1.1 g/cm^3 for salty mud.

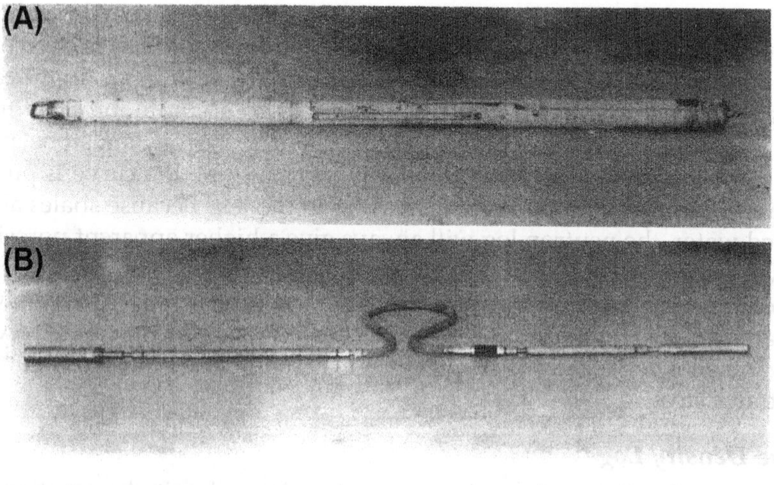

FIGURE 3.32 (A) A density tool, which records the porosity of the formation, making automatic corrections for hole diameter and mud cake thickness. (B) A sonic sonde for measuring the acoustic velocity of formations. *Courtesy of Tesel Services, Ltd.*

Shale also affects the accuracy of the density-derived porosity of the reservoir. Also, several minerals have anomalous densities, the log traces of which may affect porosity values. Notable among these minerals are mica, siderite, and pyrite. The presence of oil has little effect on porosity values, but gas lowers the density of a rock and thus causes the log to give too high a porosity. This effect can be turned to an advantage, however, when combined with the information derived from the neutron log.

3.2.3.5 *The Lithodensity Log*

Improvements in density logging techniques include the addition of a new parameter: the photoelectric cross-section, commonly denoted P_e, which is less dependent on porosity than is the formation density and is particularly useful in analyzing the effects of heavy minerals on log interpretation. The P_e records the absorption of low-energy gamma rays by a formation in units of barns per electron. The logged value is a function of mineralogy and the aggregate atomic number of the elements in the formation. Common reservoir mineral reference values are quartz 1.81; dolomite 3.14; calcite 5.08 b/electron. Coals typically are <1, and typical shales are approximately 3 b/electron (can be distinguished from dolomite by high gamma ray log readings). Typical log scale for a P_e curve is 0–10 b/electron. The P_e curve has a finer resolution (about half a foot) than the neutron/density curves (~2 ft). Thus, the curve can help resolve lithology in thin bedded units.

An application that is particularly useful involves combining P_e with the thorium–potassium ratio from the gamma-ray spectrometry device, as indicated in Fig. 3.33.

3.2.4 The Sonic, or Acoustic, Log

A third way of establishing the porosity of a rock is by measuring its acoustic velocity by the sonic, or acoustic, log (Fig. 3.33(B)). In this technique, interval transit times are recorded of clicks emitted from one end of the sonde traveling to one or more receivers at the other end.

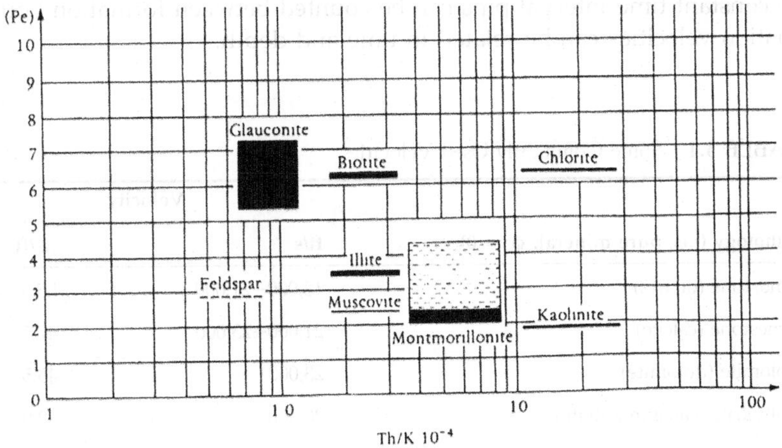

FIGURE 3.33 Crossplot showing mineral identification using the spectrometry and lithodensity logs. *Courtesy of Schlumberger.*

Sound waves generally travel faster through the formation than through the borehole mud. The interval transit time (Δt), which is measured in microseconds per foot, can then be used to calculate porosity according to the following equation (Wyllie et al., 1956, 1958):

$$\phi = \frac{\Delta t_{\text{log}} - \Delta t_{\text{ma}}}{\Delta t_{\text{f}} - \Delta t_{\text{ma}}},$$

where Δt_{log} is the interval transit time recorded on the log, Δt_{ma} is the velocity of the rock at $\phi = 0$, and Δt_{f} is the velocity of the pore fluid. Some commonly used velocities are shown in Table 3.1.

The sonic log can be used only in open, uncased holes. The circuitry associated with the receiver has to be carefully adjusted for sensitivity so that it does not trigger on spurious noise, yet picks up the first arrival from the signal sound wave. A rapid to-and-fro log trace, termed *cycle skipping*, occurs if the sensitivity is insufficient or if the returned signal is very weak. It results from triggering on a later part of the sound wave pulse, which causes an erroneously long computed transit time. Cycle skip occurs in undercompacted formations (especially if gas filled), fractured intervals, and areas where the hole is enlarged and out of gauge so that a wider than normal width of borehole mud is traversed before the signal pulse enters the formation. Advances with computer-controlled logging devices greatly reduce the problem of balancing trigger sensitivity, which previously required continuous monitoring by the logging engineer.

The sonic method is the least accurate of the three porosity logs because it is the one most affected by lithology. On the other hand, for this very reason, it is widely employed as a means of lithology identification and hence for correlation from well to well. The sonic log is also extremely useful to geophysicists because it can be used to determine the interval velocities of formations and thus relate timing of seismic reflectors to actual rocks around a borehole by means of computed time–depth conversions. For this reason, the sonic log also records the total travel time in milliseconds along its length by a process of integration, the result being recorded as a series of pips along the length of the log. The cumulative number of these constant time interval pips can be counted between formation boundaries, and hence, formation velocities can be related in time and depth.

TABLE 3.1 Some Commonly Used Velocities

	Velocity	
Lithology (i.e., pure mineral, ϕ = 0)	**ft/s**	**μs/ft**
Sandstone (quartz)	18,000–21,000	55.5–51.3
Limestone (calcite)	21,000–23,000	47.5
Dolomite (dolomite)	23,000	43.5
Anhydrite (calcium sulfate)	20,000	50.0
Halite (sodium chloride)	15,000	67.0
Fluid (freshwater or oil)	5300	189.0

3.2.5 Porosity Logs in Combination

The three porosity logs are influenced by formation characteristics other than porosity, notably by their lithology, clay content, and the presence of gas. When used in combination rather than individually, the logs give a more accurate indication of porosity and extract much other useful information. It has already been noted that in gas zones the neutron log indicates too low a porosity and the density log, too high a porosity. These different porosity values can be turned to advantage if the two curves are calibrated so that they track each other on the log. In limestones, the curves are calibrated so that 0 LPU = 2.71 g/cm^3; in sandstones, 0 SPU = 2.65 g/cm^3. If this calibration is done, separation between the log traces indicates the presence of gas (Fig. 3.34). Note that this phenomenon only applies in reservoir zones. Separation will commonly be seen in the shale sections, but in the reverse direction.

Accurate identification of porosity and lithology can be determined by crossplotting the readings of two porosity logs, using the chart books provided by the particular service company that ran the logs. Figure 3.35 shows one such graph for Schlumberger's formation density and compensated neutron logs run in freshwater holes. These crossplots are satisfactory only in the presence of limestone, sandstone, or dolomite, but are inaccurate in the presence of clay or other anomalous minerals.

A more sophisticated method of lithology identification is the M–N crossplot (Burke et al., 1969). The two formulas used in this method take the readings from the three porosity logs and remove the effect of porosity, thereby leaving only the lithological effect. For example, for water-saturated formations,

$$M = \frac{\Delta t_f - \Delta t_{log}}{\rho_b - \rho_f} \times 0.01,$$

$$N = \frac{(\phi_N)f - \phi_N}{\rho_b - \rho_f}.$$

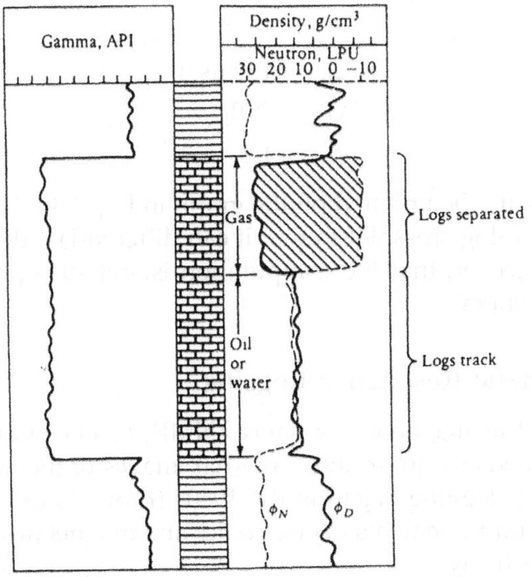

FIGURE 3.34 Log showing how separation of the density and neutron logs may indicate gas-bearing intervals.

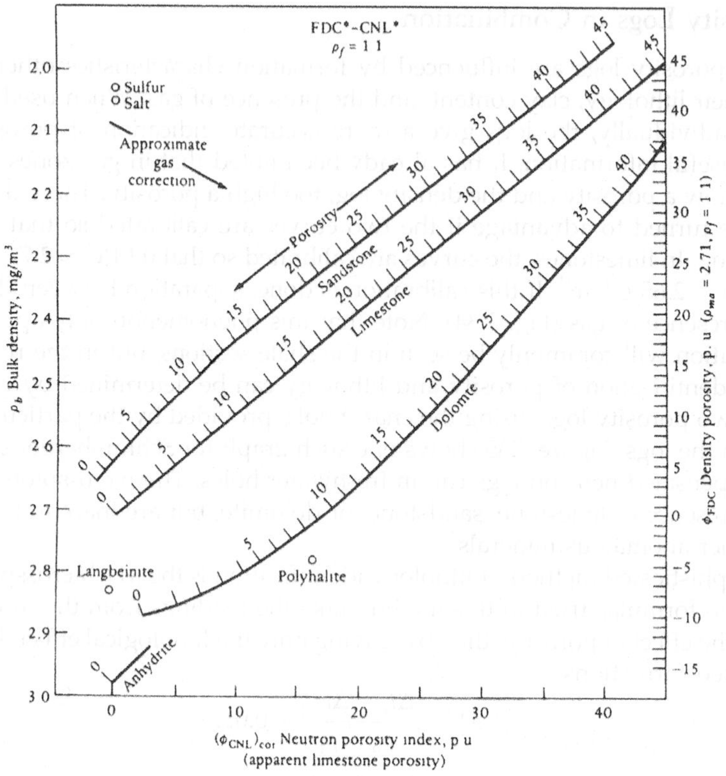

FIGURE 3.35 Density-neutron crossplot for determining porosity and lithology. For an explanation, see the text. *Courtesy of Schlumberger.*

For freshwater muds,

$$\Delta_t = 189\ \mu s/ft,$$
$$\phi_N = 1.0,$$
$$\rho_b = 1.0.$$

The *M* and *N* values are then plotted on the graph in Fig. 3.36. This method has an advantage over the simple two-log crossplot in that it can differentiate the proportions of multimineralic rocks. Note, however, that the computation is still susceptible to the effects of gas, clay, and anomalous minerals.

3.2.6 Nuclear Magnetic Resonance Logging

The principle of nuclear magnetic resonance (NMR) has been understood for >25 years, and has been used in medicine to produce internal images of the human body. The application of NMR to borehole logging began in the 1950s (Brown and Gammon, 1960). It is only recently, however, that improvements in electronics and magnet design have made it possible for it to be applied effectively.

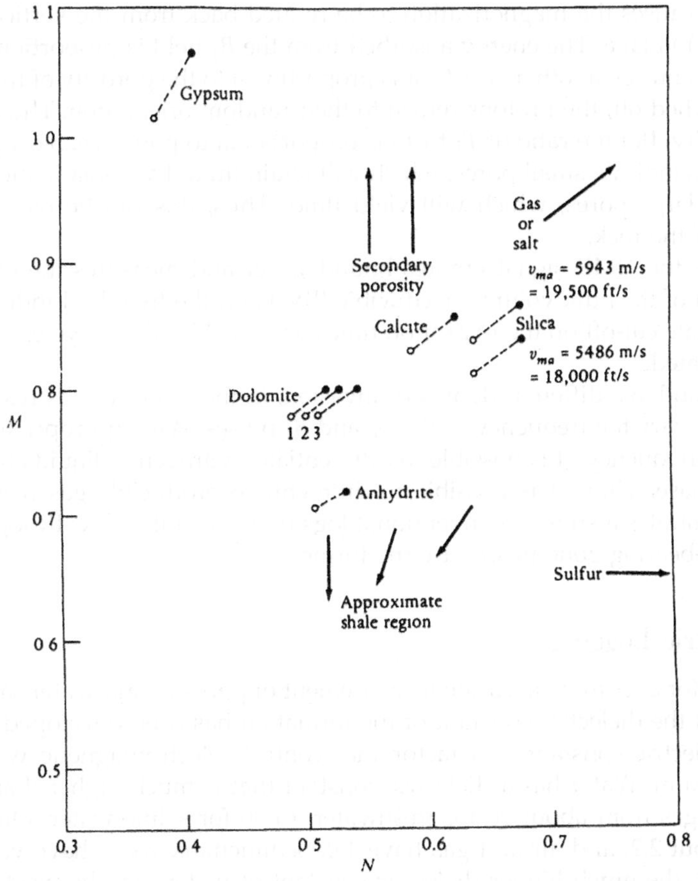

FIGURE 3.36 *M–N* crossplot for complex lithology identification. For an explanation, see the text. *Courtesy of Schlumberger.*

NMR logging is an extremely effective way of measuring porosity and permeability. It can differentiate movable from immovable water. When correctly calibrated, it can be used to identify gas reservoirs that would have been missed by conventional logging arrays (Coope, 1994). To really enjoy NMR logging, a PhD in nuclear physics is essential. For simple geologists without such a qualification, an elementary account follows based on Camden (1994).

In rocks, hydrogen occurs principally in the fluids, water, oil, and/or gas that fills pores. Hydrogen nuclei act as randomly aligned positive magnets. When they are exposed to a magnetic field, termed the B_0 field, they will become aligned to it and proceed around it at a velocity proportional to the field strength, known as the *Larmor Frequency*. The orientation of the field may be termed the Z axis, and the perpendicular plane normal to it the X–Y axis. The time taken for the hydrogen nuclei to orient themselves to the magnetic field is termed the polarization time (T_1). When the field is removed, the nuclei will return to their original random orientation. The time taken for this to happen is known as the relaxation time (T_2). The NMR is measured by exposing the hydrogen atoms to pulses of a second magnetic field (B_1). This causes the magnetization to be rotated back from the vertical (Z) axis to the horizontal ($X-Y$) plane. The energy absorbed from the B_1 field is proportional to the density of the hydrogen nuclei, in other words, it is proportional to the porosity of the rock. When the B_1 field is switched off, the protons return to their random orientation. Thus, it is possible to measure porosity. But the ratio of T_1 to T_2 is proportional to pore size. Slow polarization and relaxation times indicate small pores, which will retain fluid. Fast polarization and relaxation times indicate large pores, which will yield fluid. Thus, this can be used to measure the permeability of the rock.

Fortuitously, the NMR signal ignores bound water and measures the effective porosity. This is the sum of the bulk volume irreducible (BVI) and the free-fluid index (FFI). Selection of the appropriate cut-off on the relaxation time curve enables these two essential parameters to be differentiated.

NMR logs can have different depths of invasion, in the same way as resistivity logs. It is also possible to vary the frequency of the B_o and B_1 pulses. With appropriate depths of invasion and pulse frequency, it is possible to differentiate hydrogen in liquids from hydrogen in the gaseous phase. Thus, it is possible to differentiate producible gas from bound water. Figure 3.37 illustrates a suite of conventional logs beside a suite of NMR logs that identified a hydrocarbon-bearing zone missed by the former.

3.2.7 Dielectric Logging

In the quest for ever more accurate measurement of porosity and water saturation logging the variation of the dielectric constant of the formation has been developed (Wharton et al., 1980). The dielectric constant is a factor that controls electromagnetic wave propagation through a medium. Water has a dielectric constant that is much higher than of other fluids or rocks. It ranges from about 50 for freshwater, to 80 for saline water. Oil has a dielectric constant of about 2.2, and air and gas have 1.0. Sedimentary rocks have values of between 4 and 10. Thus, the much higher dielectric constant of water may be used, in combination with other logs, to measure porosity and S_w.

Several tools are available for measuring the dielectric constant. These include the electromagnetic propagation tool, the dielectric constant log (DCL), and the deep propagation tool. The difference between total porosity, measured by the other porosity logs, and the porosity measured by the DCL will indicate the hydrocarbon saturation of the reservoir. Dielectric logs are particularly useful in low R_w situations, and also where there are rapid vertical variations in R_w. As with all tools, there are problems with the dielectric log method. Dielectric logs respond to water, whether it is connate water, mud filtrate, or water bound to mineral grains. If the depth of investigation is shallow, it may record high readings where mud filtrate has invaded permeable hydrocarbon-bearing zones. This problem may be overcome, as with resistivity logging, by running shallow and deep investigative dielectric logs together.

3.2.8 Logging or Measurement while Drilling

The foregoing account of logging related to logs run in open or cased holes after drilling sections of a well. Since the early 1980s, however, it has been possible to run some logs within drill pipes while the well is actually being deepened. The main problem that had

(A)

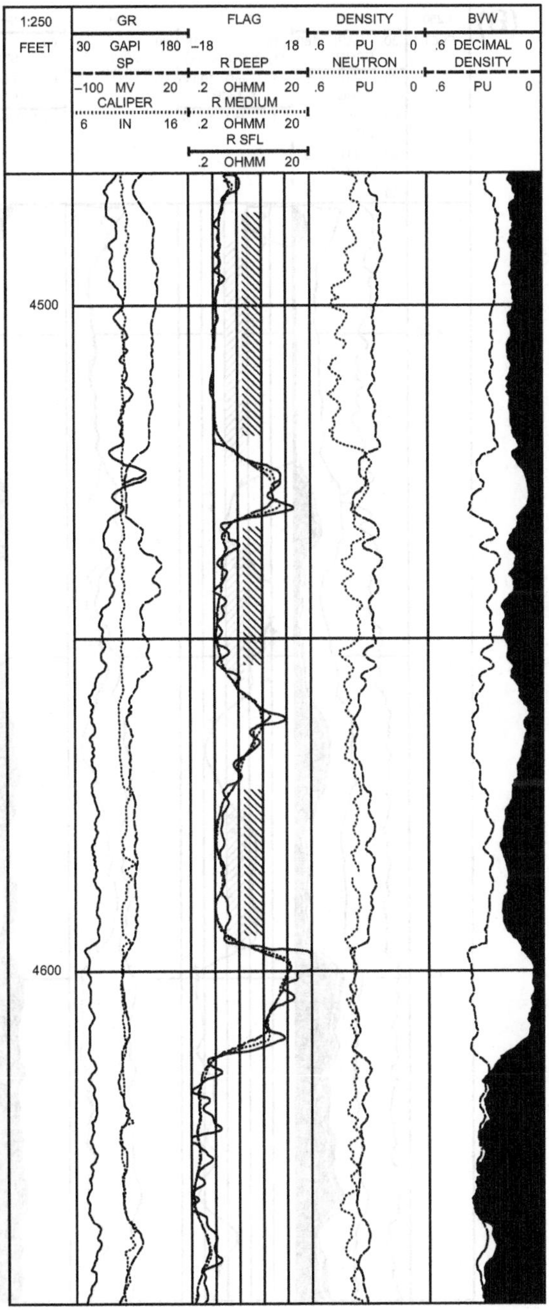

FIGURE 3.37 (A) Tracks 1, 2, and 3 illustrate a conventional suite of logs. Track 4 shows water saturation and porosity calculated from them. Three high-resistivity hydrocarbon-bearing zones are clearly visible. The low-resistivity zones (crosshatched in track 2) appear to be water bearing. In reality, the whole interval tested oil, with no water production. How can this be so? (B) MRIL (NMR) logs over the same interval. Effective (MPHI), irreducible (BVI), and free-fluid (FFI) porosity are shown in track 3. Assuming that the rock is water wet, then the BVI (light gray area) indicates irreducible water. This shows that most of the water in the low-resistivity zone is irreducible, and not producible. Hence, the whole interval produced oil, even though conventional logs showed high water saturation in some intervals QED. *Courtesy D. Coope and Numar (UK) Ltd.*

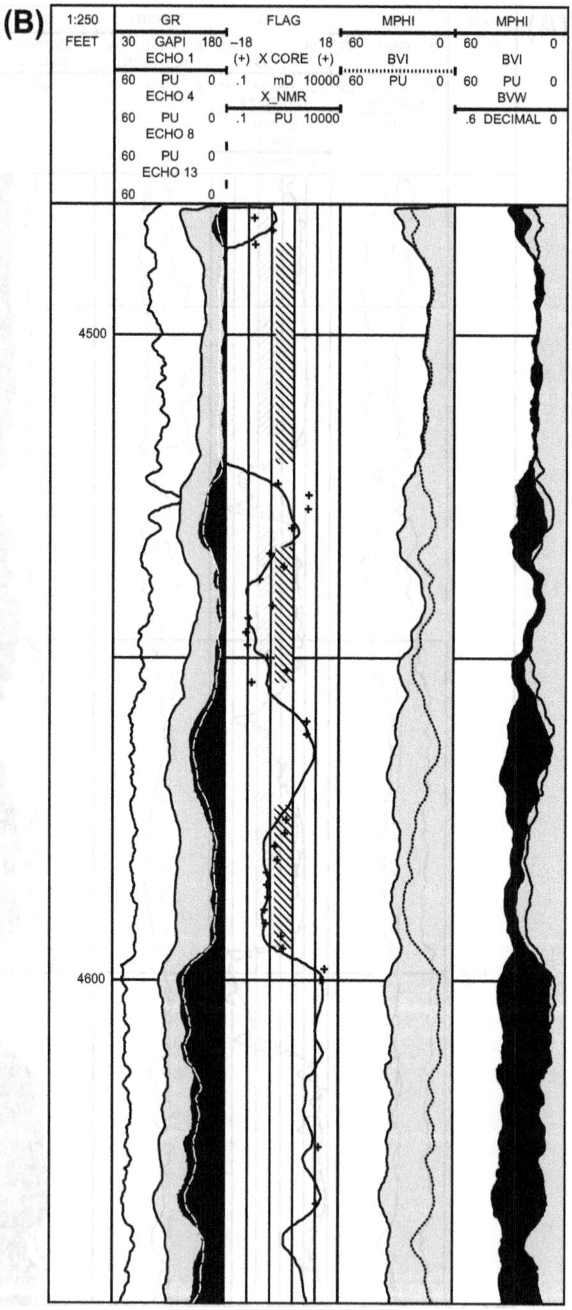

FIGURE 3.37 cont'd

to be overcome was the telemetric link between the surface and the sonde adjacent to the drill bit. Various techniques were tried, including electromagnetic, acoustic, mud pulse, and wire telemetry. Only the last two of these have proved to be effective. Running logs while drilling is variously termed *measurement while drilling* or *logging while drilling* (LWD).

In the early days, only simple resistivity and gamma logs could be run (Fontenot, 1986). Their usefulness was largely restricted to correlating the drilling well with adjacent previously drilled wells. Today, density and neutron logs can also be used, together with a range of resistivity logs (Bonner, 1992, 1996). Both laterologs and induction logs are available for dealing with a range of formation and drilling mud resistivities. MWD/LWD now has many more uses than heretofore. As mentioned earlier, offshore petroleum exploration has led to the increased need to drill lengthy directional holes. It is thus extremely important not only to know where the drill bit is underground but also to have real-time information on rock type, hydrocarbon occurrences, and signs of overpressure.

3.2.9 Dip Meter Log and Borehole Imaging

The dip meter is a device for measuring the direction of dip of beds adjacent to the borehole. It is essentially a multiarm microresistivity log (Fig. 3.38). Three or four spring-loaded arms record separate microresistivity tracks, while, within the sonde, a magnetic compass records the orientation of the tool as it is drawn up the hole. A computer correlates deviations, or kicks, on the logs and calculates the amount and direction of bedding dip and assesses dip reliability (Fig. 3.39). Many different computer programs can be applied. Some programs essentially smooth the dips over large intervals; others use statistical techniques to discriminate against minor bedding features so as to display only large-scale dip trends. Both these types are suitable for determining structural dip. Some programs calculate dips over intervals of only a few centimeters. This type of program (after removal of structural dip) can be used to discover small-scale sedimentary dips, such as crossbedding.

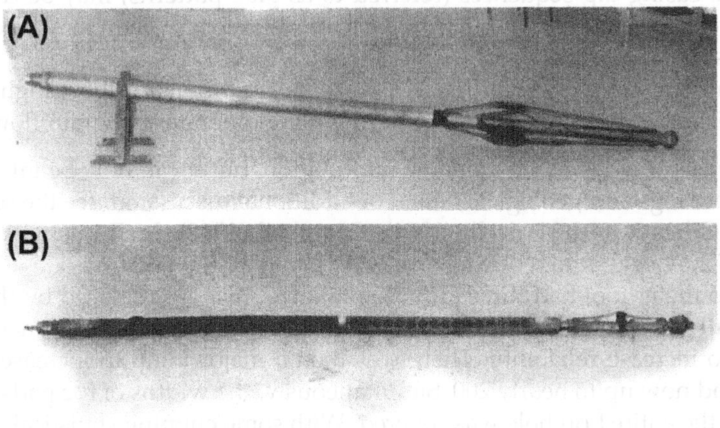

FIGURE 3.38 (A) Three-arm dip meter sonde. (B) Sidewall core gun. This device fires cylindrical steel bullets, which are attached to the gun by short cables, into the side of a borehole. Small samples of rock may thus be collected from known depths. *Courtesy of Tesel Services, Ltd.*

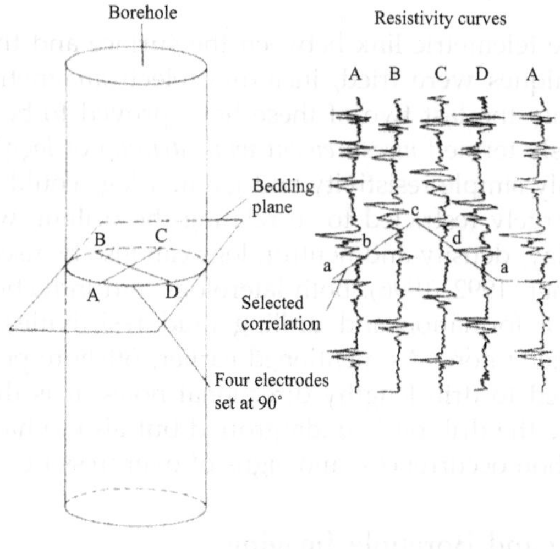

FIGURE 3.39 Sketch of an antique four-pad four-track dip meter showing how the direction of dip around a borehole may be calculated. Today, the simple dip meter shown here has been superseded by multitrack imaging tools, but the basic principle of dip calculation is still as shown here.

Calculated dips are usually presented on a tadpole plot (Fig. 3.40). Four basic types of motif are commonly identifiable:

1. Uniformly low dips (referred to as green patterns) are generally seen in shales and indicate the structural dip of the formation.
2. Upward declining dip sequences (referred to as red patterns) may be caused by the drape of shales over reefs or sandbars; by the infilling of sandstones within channels; or by the occurrence of folds, faults, or unconformities.
3. Upward increasing dip sequences (referred to as blue patterns) may be caused by sedimentary progrades in reefs, submarine fans, or delta lobes. They may also be caused by folds, faults, or unconformities.
4. Random bag o' nails motifs can reflect poor hole conditions or they might be geologically significant, indicating fractures, slumps, conglomerates, or grain flows.

The dip meter provides much valuable information, but it can only be interpreted fully in the light of other logs and geological data. For additional discussion, see the service company manuals and Gilreath and Maricelli (1964), Campbell (1968), McDaniel (1968), Jageler and Matuszak (1972), Goetz et al. (1978), Serra (1985), and Selley (1996).

The first dip meter tool had three arms 120° apart. This was replaced by the four-arm dip meter. Originally having only four micrologs, the number was eventually increased to 8, four on each pad to increase reliability. There was then a major jump to increase the number of tracks to 25, and now up to nearly 200. Simultaneously, the widths of the pads were increased so that almost the entire borehole was covered. With some cunning statistical gymnastics, it is now possible to process this multiplicity of microresistivity log data to produce incredible pseudovisual images of the borehole (Fig. 3.41). These tools, made by several companies with several trade names, can be used, both to measure the direction and amount of dip of

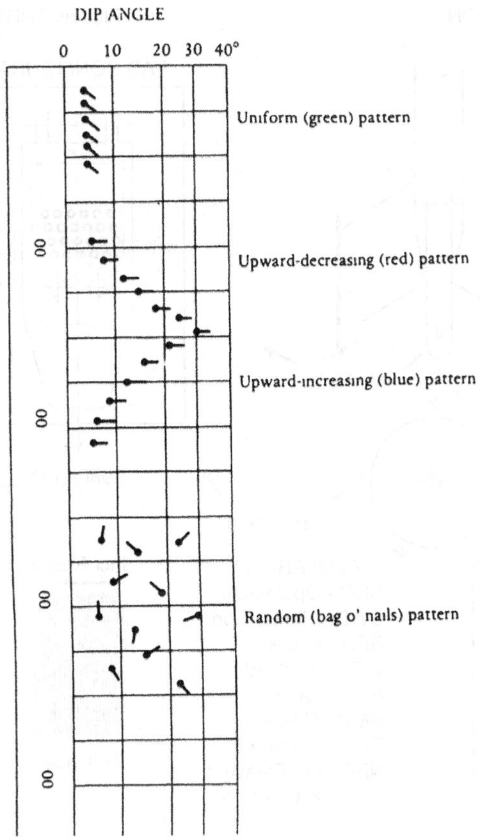

FIGURE 3.40 Conventional dip meter tadpole plot showing the four common dip motifs. Each motif can be produced by several quite different geological phenomena. The head of the tadpole shows the amount of dip. The tail of the tadpole points in the direction of dip.

strata around the borehole, and also to identify faults, fractures, unconformities, and sedimentary structures (Fig. 3.42).

Acoustic borehole imaging tools have also been developed. They are generally not as effective as resistivity-based tools. The best results are normally seen in very hard items such as basement, quartzites, and carbonates, where there is a large contrast in the acoustic velocity between the rock and open fractures and vugs.

3.2.10 Uses of Logs in Petrophysical Analysis: Summary

The previous sections provide a short, and therefore simplified, review of a major group of borehole logging techniques that are crucial to the exploration for and production of hydrocarbons. New tools and techniques are continuously being introduced, but the basic principles remain constant. Figure 3.43 provides a summary of the more common geophysical logs and their principal uses. Figures 3.44 and 3.45 provide typical examples of suites of

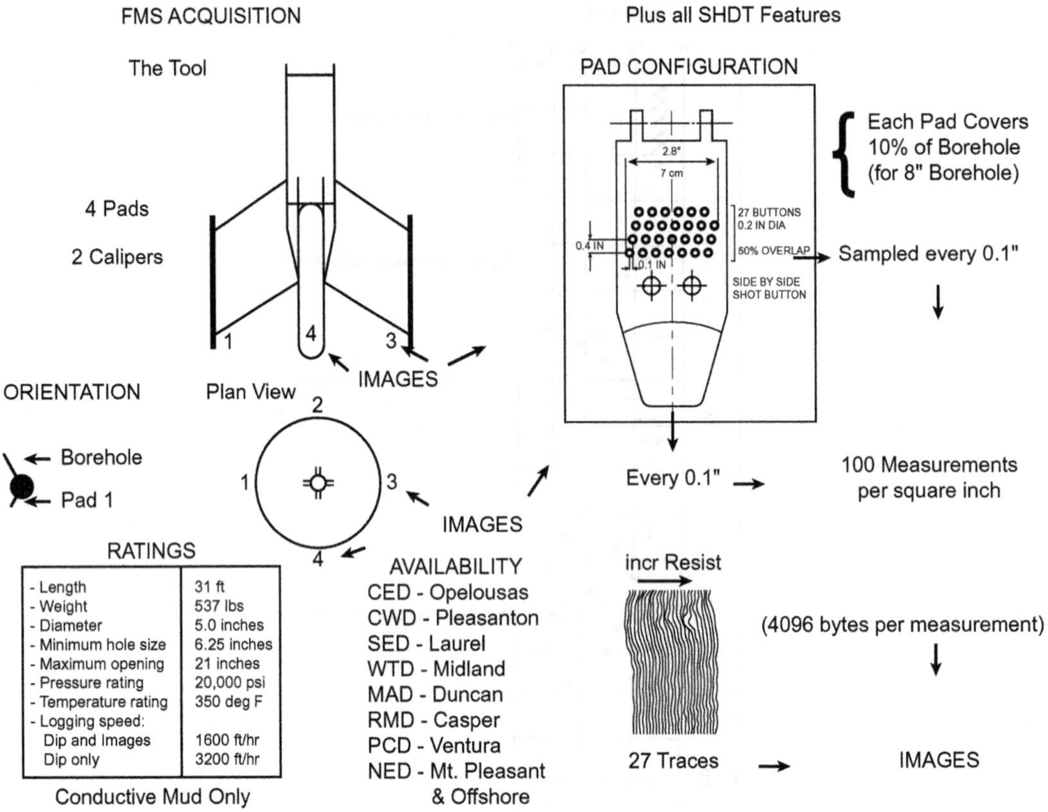

FIGURE 3.41 Diagrams to show the workings of the Schlumberger Formation MicroScanner. This was the first borehole imaging tool. It is now superseded by the Fullbore MicroImager. *Courtesy of Schlumberger.*

logs with their responses for parts of two boreholes. In the old days, the calculations of R_w, S_w, ϕ, and so forth were made using service company chart books and a calculator. Today, however, computer-processed interpretations, which give continuous readings of lithology, porosity, and the percentages of the various fluids, are preferable (Fig. 3.46). Although computer outputs look very convincing, especially when brightly colored, the old adage "rubbish in, rubbish out" must be remembered (Plate 3.1). Failure to check the quality of the original logs or failure to note the presence of an unusual or aberrant mineral will invalidate the whole procedure. For example, abundant mica flakes in a sandstone will read as "clay" on logs, giving a false value of porosity. This is because of the way in which small amounts of anomalous minerals can alter log readings that geochemical logging programs have been developed. These involve very complex algorithms to identify and measure the vertical variations of such minerals. Once this has been done, it is possible to calculate the essential reservoir parameters of porosity and hydrocarbon saturation (Wendtland and Bhuyan, 1990; Harvey and Lovell, 1992). These logs, together with conventional logs, can also be used to identify and understand diagenetic phenomena within reservoirs. Thus, it has been

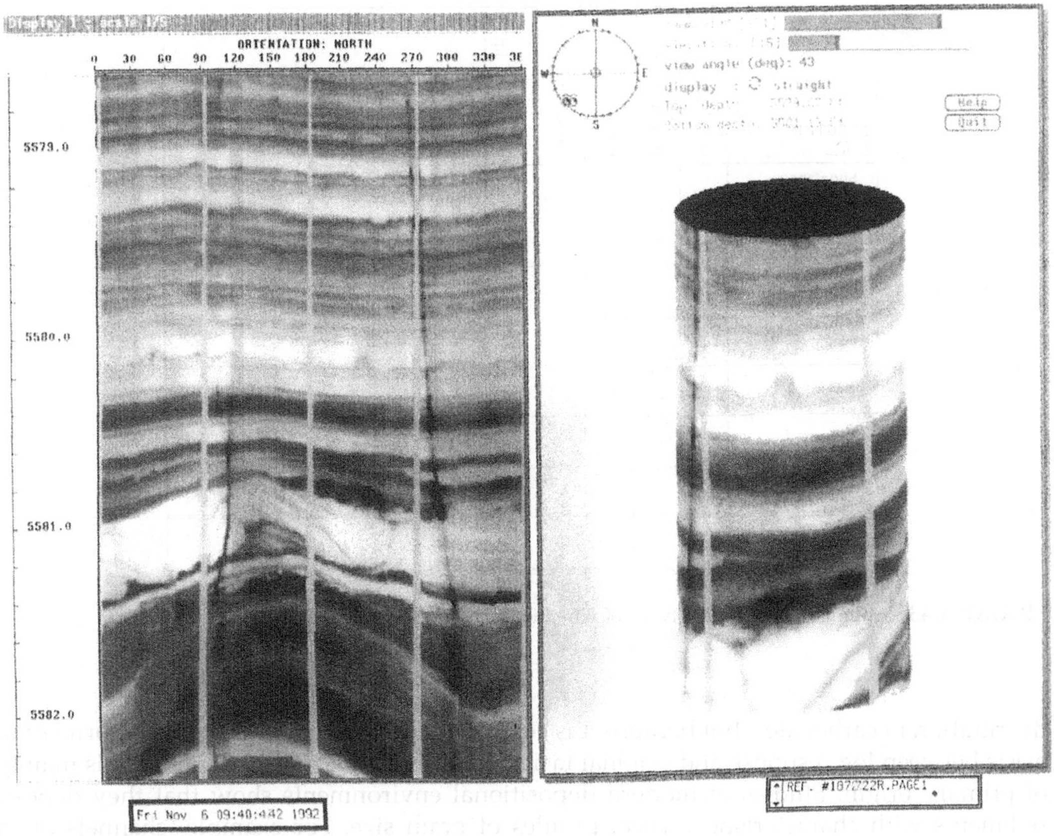

FIGURE 3.42 An example of a borehole image produced by the Formation MicroImager. Note that the borehole can be displayed in both cylindrical and unrolled formats. *Courtesy of Schlumberger.*

argued that formation evaluation is now entering its third age (Selley, 1992). The first age was the application of logs to petrophysical analysis, the second age to facies analysis (discussed later), and the third to reservoir diagenesis (discussed in Chapter 6).

3.2.11 Applications of Logs in Facies Analysis

The previous sections reviewed the various types of geophysical logs and their application in identifying lithology, porosity, and hydrocarbon saturation. Logs are also used in other ways. When the stratigraphy of a well has been worked out, whether it be the lithostratigraphy or the biostratigraphy (in which paleontological zones are introduced), adjacent wells can be stratigraphically correlated. Logs can also be used to determine the facies and depositional environment of a reservoir. From such studies, the geometry and orientation of reservoirs may be predicted. Carbonate facies are largely defined, and their depositional environments deduced, from petrography. Geophysical logs show porosity

ELECTRIC LOGS	Lithology	Hydrocarbons	Porosity	Pressure prediction	Structural and sedimentary dip	OTHER
Spontaneous potential	✓					Calculation of R_w, and bed shaliness, Qualitative identification of permeability
Resistivity	✓	✓		✓		Calculation of R_{xo}, R_t, and hence S_w
RADIOACTIVE LOGS						
Gamma	✓					Calculation of bed shaliness and organic content
Neutron		✓	✓ Gas effect			Lithology identification by cross-plots
Density		✓	✓			
SONIC	✓	✓	✓	✓		Calibration of formations with seismic data
DIPMETER					✓	

FIGURE 3.43 Summary of the main types of wireline logs and their major applications.

distribution in carbonates, but because it is commonly of secondary origin, little correlation exists between log response and original facies. In sandstones, however, porosity is mainly of primary origin. Studies of modern depositional environments show that they deposit sediments with characteristic vertical profiles of grain size. For example, channels often fine up, from a basal conglomerate, via sand, to silt and clay. Conversely, prograding deltas and barrier islands deposit upward-coarsening grain-size profiles. Grain-size profiles may thus be of use in facies analysis. Both the SP and gamma logs may indicate grain-size profiles in sand–shale sequences. As discussed earlier, the deflection on the SP is locally controlled by permeability, with the maximum leftward deflection occurring in the most permeable interval. Permeability, however, increases with grain size. Therefore, the SP log is generally a vertical grain-size log in all but the most cemented sand–shale sections.

The gamma log may be used in a similar way, since the clay content (and hence radioactivity) of sands increases with declining grain size. Exceptions to this general statement may be caused by intraformational clay clast conglomerates and the presence of anomalous radioactive minerals, such as glauconite, mica, and zircon (Rider, 1990). Gamma and SP logs often show three basic motifs:

1. Sands that fine upward gradually from a sharp base (bell motifs)
2. Sands that coarsen up gradually toward a sharp top (funnel motifs)
3. Clean sands with sharp upper and lower boundaries (boxcar, or blocky motifs)

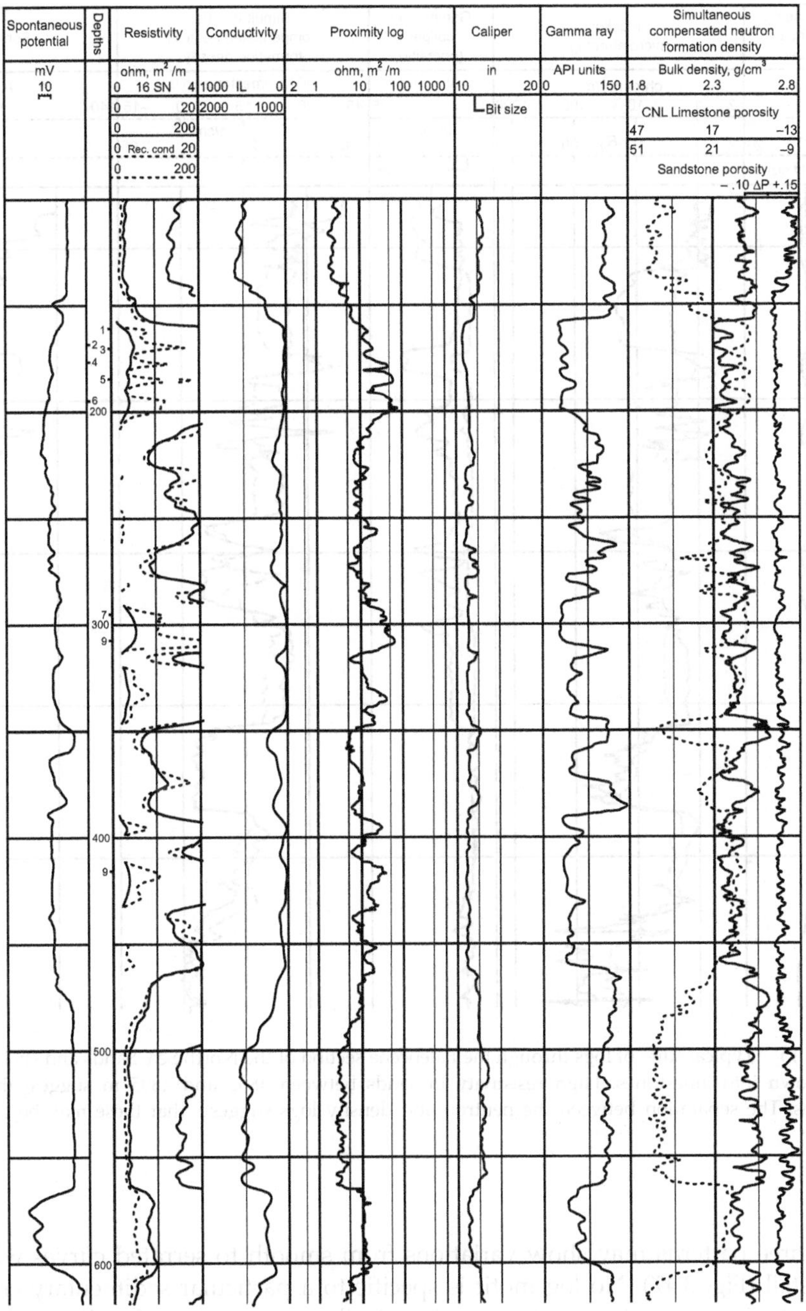

FIGURE 3.44 Typical suite of logs through a sand–shale sequence in the Niger delta. Note high-resistivity hydrocarbon-bearing zones at 160–210 ft and below. *Courtesy of Schlumberger.*

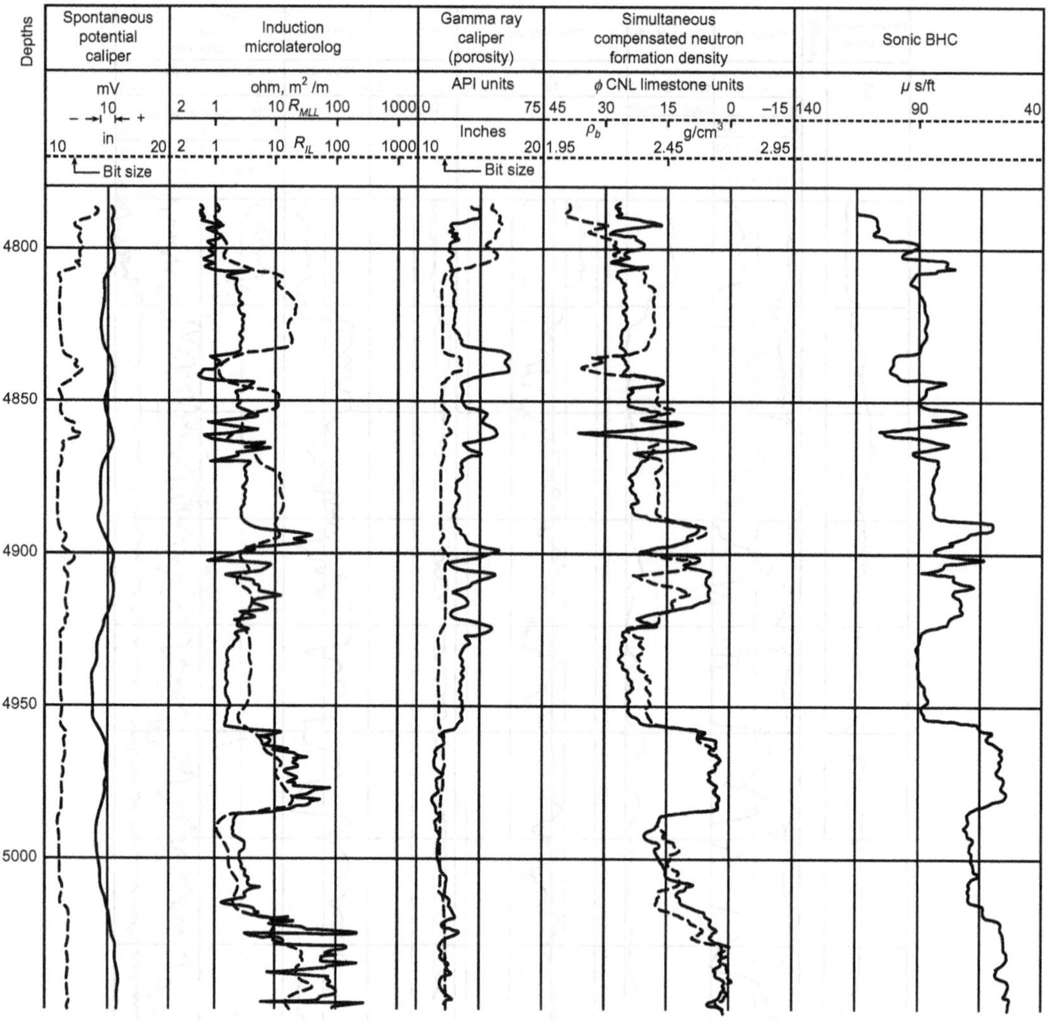

FIGURE 3.45 Typical suite of logs through the Paleocene section of the North Sea. Sands and shales in the upper part pass down into limestones. High resistivity in sands between 4805 and 4895 m suggest the presence of hydrocarbons. The separation between the neutron and density logs suggests that these may be gas. *Courtesy of Schlumberger.*

These three patterns may show variations from smooth to serrated curves when considered in detail (Fig. 3.47). No log motif is specific to a particular sedimentary environment, but by combining an analysis of log profile with the composition of well samples, an interpretation of environment can be attempted. Constituents to look for in the cuttings are glauconite, shell debris, carbonaceous detritus, and mica. Glauconite forms during the early diagenesis of shallow marine sediments. Once formed, it is stable in the marine realm, but

(A)

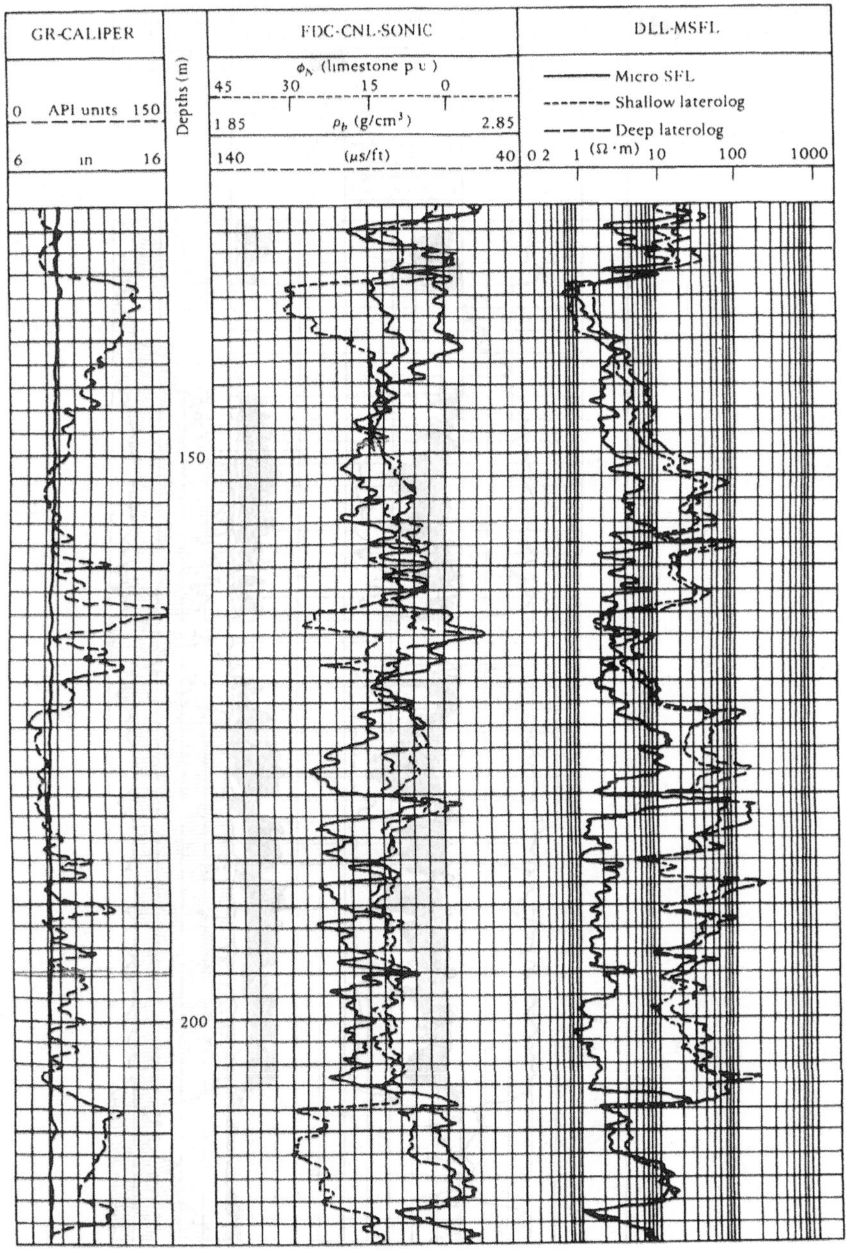

FIGURE 3.46 (A) A "raw" unprocessed suite of logs, compare with (B), a computer-calculated interpretation. This shows the percentage of porosity and hydrocarbon saturation, together with the variation in clay and matrix. Note that engineers use the term *matrix* to describe grains, not in the geological sense of the finer sediment (silt and clay) that is syndepositionally deposited along with the framework grains. *Courtesy of Schlumberger.*

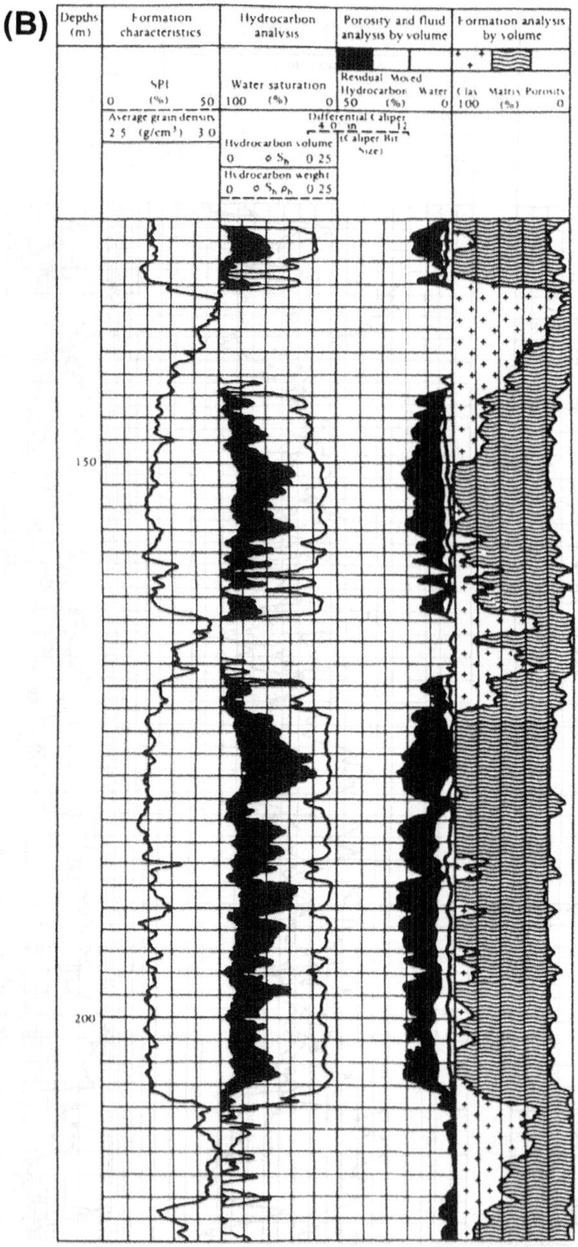

FIGURE 3.46 cont'd

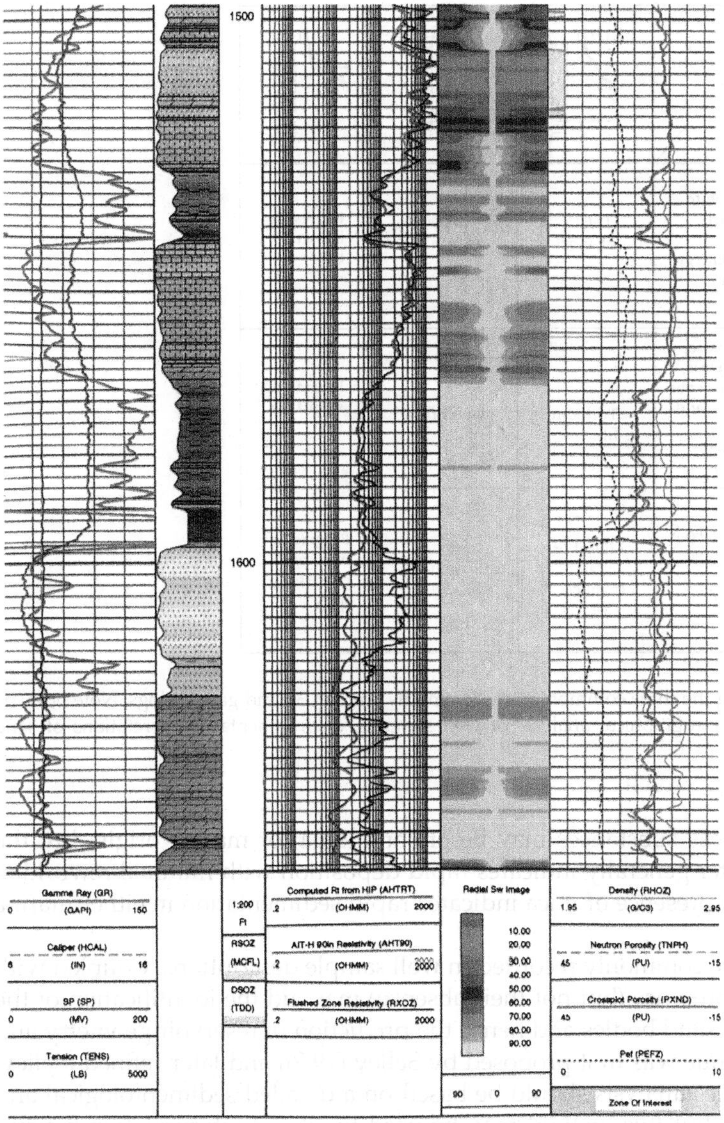

LITHOLOGY with AITH Sw Processing

PLATE 3.1 An example of a suite of well logs. Gamma, self-potential, and caliper logs in left-hand column help to indicate lithology. Resistivity logs in track to immediate right of depth column calculate hydrocarbon saturation in right-hand colored column. Porosity logs in the far right column. Reservoir zone at the top of the section is highlighted by pink bar at right of radial S_w log (See the color plate). *Courtesy of Schlumberger Wireline and Testing.*

can be transported landward on to beaches or basinward on to deep sea fans. Glauconite is readily oxidized at outcrop, however, so reworked second-cycle glauconite is virtually unknown. Thus, the presence of glauconite grains in a sandstone indicates a marine environment, although its absence indicates nothing.

Shell-secreting invertebrates live in freshwater and seawater environments, but shelly sands tend to be marine rather than nonmarine. Diagnosis is obviously enhanced if fragments of specific marine fossils can be identified. Carbonaceous detritus includes coal, plant

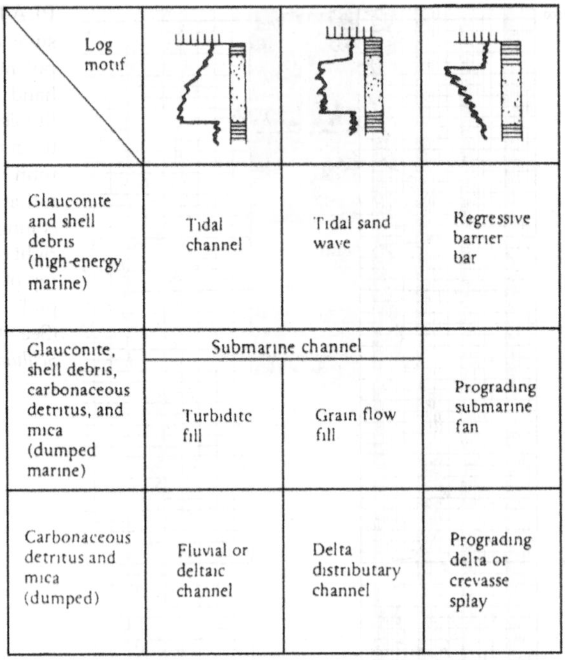
Log motif
Glauconite and shell debris (high-energy marine)
Tidal channel
Tidal sand wave
Regressive barrier bar
Glauconite, shell debris, carbonaceous detritus, and mica (dumped marine)
Submarine channel
Turbidite fill
Grain flow fill
Prograding submarine fan
Carbonaceous detritus and mica (dumped)
Fluvial or deltaic channel
Delta distributary channel
Prograding delta or crevasse splay

FIGURE 3.47 Diagram showing the three common grain-size motifs seen on SP and gamma logs. Note how, by combining these with the visual petrographical analysis of well samples, environmental interpretations of sand bodies can be made.

fragments, and kerogen. These substances may be of continental or marine origin, but the preservation of organic matter generally indicates rapid deposition with minimal reworking and oxidation. Similarly, the presence of mica indicates rapid sedimentation in either marine or continental environments.

These four constituents are commonly recorded in well sample descriptions. Coupled with a study of log motifs, their presence (but not their absence) may aid the identification of the depositional environment of sand bodies and hence the prediction of reservoir geometry and trend (Fig. 3.47). This technique was first proposed by Selley (1976) and later refined (Selley, 1996). In an ideal world, facies analysis should be based on a detailed sedimentological and petrographic study of cores, but this method may be used to extend interpretation beyond cored wells or in regions where cores are not available. Figures 3.48 and 3.49 show the sort of facies synthesis that can be produced by integrating log and rock data. Real examples of some of these motifs are illustrated in this book, especially in Chapters 6 and 7.

Facies analysis becomes easier still when grain-size profiles from logs are combined with facsimiles of sedimentary structures from borehole imaging tools. Then, the orientation of sedimentary structures, notably crossbedding, may be used to predict paleocurrent direction, and hence the orientation of reservoir units. As seismic becomes ever more effective at imaging the shapes of reservoirs, however, facies analysis using these techniques becomes increasingly unnecessary.

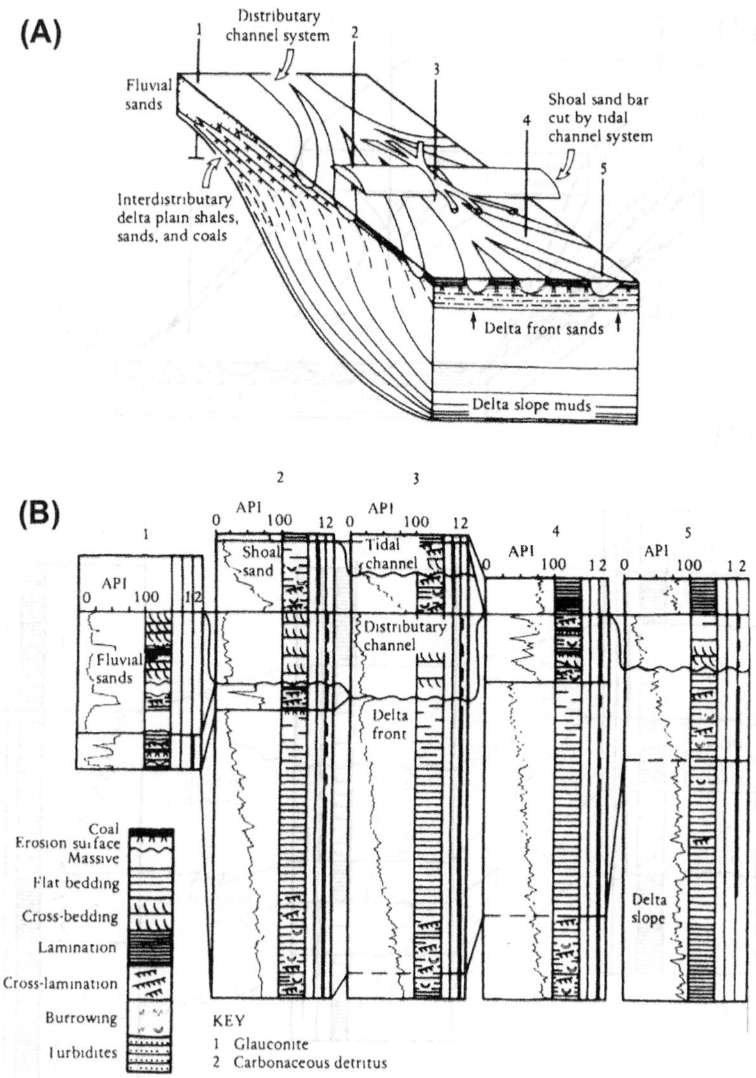

FIGURE 3.48 (A) Sand body geometry and (B) log responses for a delta system. Note how carbonaceous detritus and mica occur in the progradational phase of the delta, and are replaced by glauconite and shell debris in the transgressive sands that were deposited following delta abandonment. *From Selley (1996). Reprinted with permission from Chapman & Hall.*

3.3 GEOPHYSICAL METHODS OF EXPLORATION

Petroleum exploration and production are largely concerned with the geological interpretation of geophysical data, especially in offshore areas. Petroleum geologists need to be well acquainted with the methods of geophysics. For many years, a large communication barrier

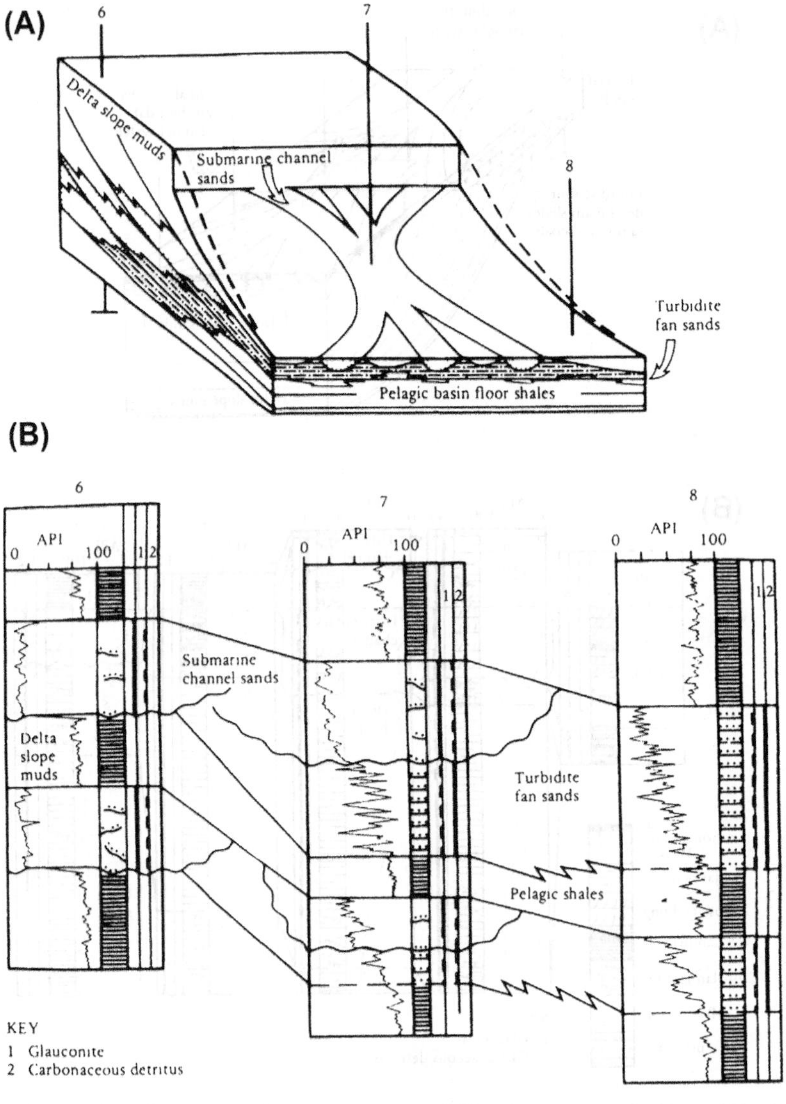

FIGURE 3.49 (A) Sand body geometry and (B) log responses for submarine channel and fan deposits. Note how deep marine sediments may be characterized by an admixture of land-derived carbonaceous detritus and mica together with shelf-derived glauconite and shell fragments. The log profiles are not dissimilar from those produced by a prograding delta, but the petrography may be used to differentiate them. *From Selley (1996). Reprinted with permission from Chapman & Hall.*

existed in many oil companies between the two groups, which were usually organized in different departments. This separation has now largely disappeared, as a new breed of petroleum geoscientist appears—half geologist and half geophysicist. Unfortunately, a new division has developed within the geophysical world between the mathematicians, physicists, and computer programmers (who acquire and process geophysical data) and those who use geological concepts to interpret this information.

The following account of geophysical methods of petroleum exploration has two objectives. It seeks to explain the basic principles and to illustrate the wonders of modern geophysical display. What it does not pretend to do, however, is explain the arcane mathematics and statistical gymnastics that are used to get from first principles to the finished brightly colored geophysical image. This account is probably sufficient for petroleum reservoir engineers seeking to know the origins of the maps by which fields are found and reserves assessed. For petroleum geologists, however, it is only an introduction and will be followed by courses in geophysics in general and in seismic interpretation in particular.

Three main geophysical methods are used in petroleum exploration: magnetic, gravity, and seismic. The first two of these methods are used only in the predrilling exploration phase. Seismic surveying is used in both exploration and development phases and is by far the most important of the three methods (Fig. 3.50).

3.3.1 Magnetic Surveying

3.3.1.1 Methodology

The earth generates a magnetic field as if it were a dipole magnet. Lines of force radiate from one magnetic pole and converge at the other. The inclination of the magnetic field varies from vertical at the magnetic poles to horizontal at the magnetic equator. The magnetic axis of the earth moves about within a circle of some 10–15 degrees of arc from the rotational axis. The force between two magnetic poles may be expressed as follows:

$$F = \frac{AM_1M_2}{r^2},$$

where F is the force; A is the constant, generally unity; M_1, M_2 are strengths of the respective poles; and r is the distance between the poles.

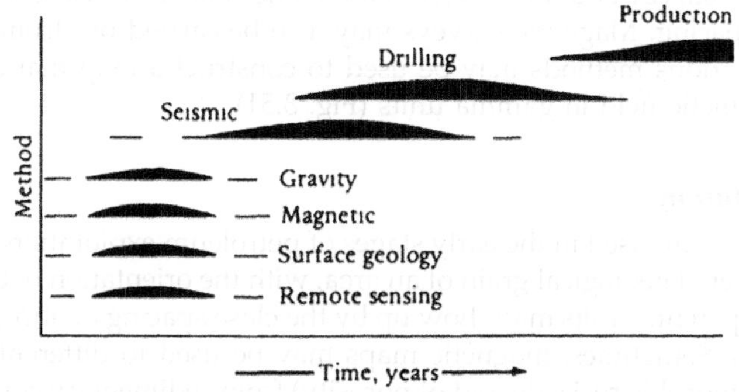

FIGURE 3.50 Sketch showing the times at which different geophysical methods are used in petroleum exploration.

Magnetic field strength is measured either in gammas or in oersteds. One oersted is the field that exerts a force of 1 dyne on a unit magnetic pole. One gamma is 10^{-5} Oe. The magnetic field of the earth varies from about 60,000 gamma (0.6 Oe) at the poles to about 35,000 gamma (0.35 Oe) at the Equator. By way of contrast, a schoolboy's horseshoe magnet has a field of about 350 Oe.

The magnetic field of the earth shows considerable time-varying fluctuations, with periods ranging from hundreds of years to about a second. Diurnal variations are the most significant for magnetic surveys and have to be corrected before interpreting the results. Less easy to allow for, however, are large amplitude fluctuations in the earth's magnetic field during magnetic storms, which last for several hours and are caused by sun spot activity.

The intensity of magnetization of a magnetic mineral will normally be related to the regional field strength according to the following relation:

$$J = kH,$$

where J is the intensity of magnetization, defined to be the magnetic moment per unit volume of the rock; k is the magnetic susceptibility of the rock; and H is the intensity of the magnetic field inducing the magnetism. The intensity of magnetization may also have increments that have been acquired and retained at earlier stages in the history of the rock and may have to be taken into account in the interpretation. The magnetic susceptibility of rocks is very variable, ranging from <10 to 4 emu/cm^3 for sedimentary rocks to between 10^{-3} and 10^{-2} emu/cm^3 for iron-rich basic igneous rocks.

At any point above the earth, the measured geomagnetic field will be the sum of the regional field and the local field produced by the magnetic rocks in the vicinity. The purpose of magnetic surveys, therefore, is to measure the field strength over the area of interest. The recorded variations of the field strength will be due to changes in the earth's magnetic field and to the volume and magnetic susceptibility of the underlying rocks. These variations can be eliminated by removing the fluctuations recorded by a fixed station in the survey area. The residual values are then directly related to the rocks beneath. Magnetic surveys can be carried out on the ground or on board ship, but taking the readings from the air is far more efficient. In aeromagnetic surveys, a plane trails a magnetometer in a sonde attached to a cable and flies in a grid pattern over the survey area. The spacing of the survey lines will vary according to the nature of the survey and the budget, but the lines may be as close as 1 km with crosstie lines on a 5-km spacing. Magnetic surveys may also be carried out from satellites. Data acquired by these various methods may be used to construct a map that contours anomalies in the earth's magnetic field in gamma units (Fig. 3.51).

3.3.1.2 *Interpretation*

Magnetic surveys are used in the early stages of petroleum exploration. Magnetic anomaly maps show the overall geological grain of an area, with the orientation of basement highs and lows generally apparent. Faults may show up by the close spacing or abrupt changes in orientation of contours. Sometimes, magnetic maps may be used to differentiate basement (i.e., igneous and metamorphic rocks devoid of porosity) from sedimentary cover that may be prospective. However, this differentiation is only possible where there is a sharp contrast between the magnetic susceptibility of basement and cover, and where this contrast is

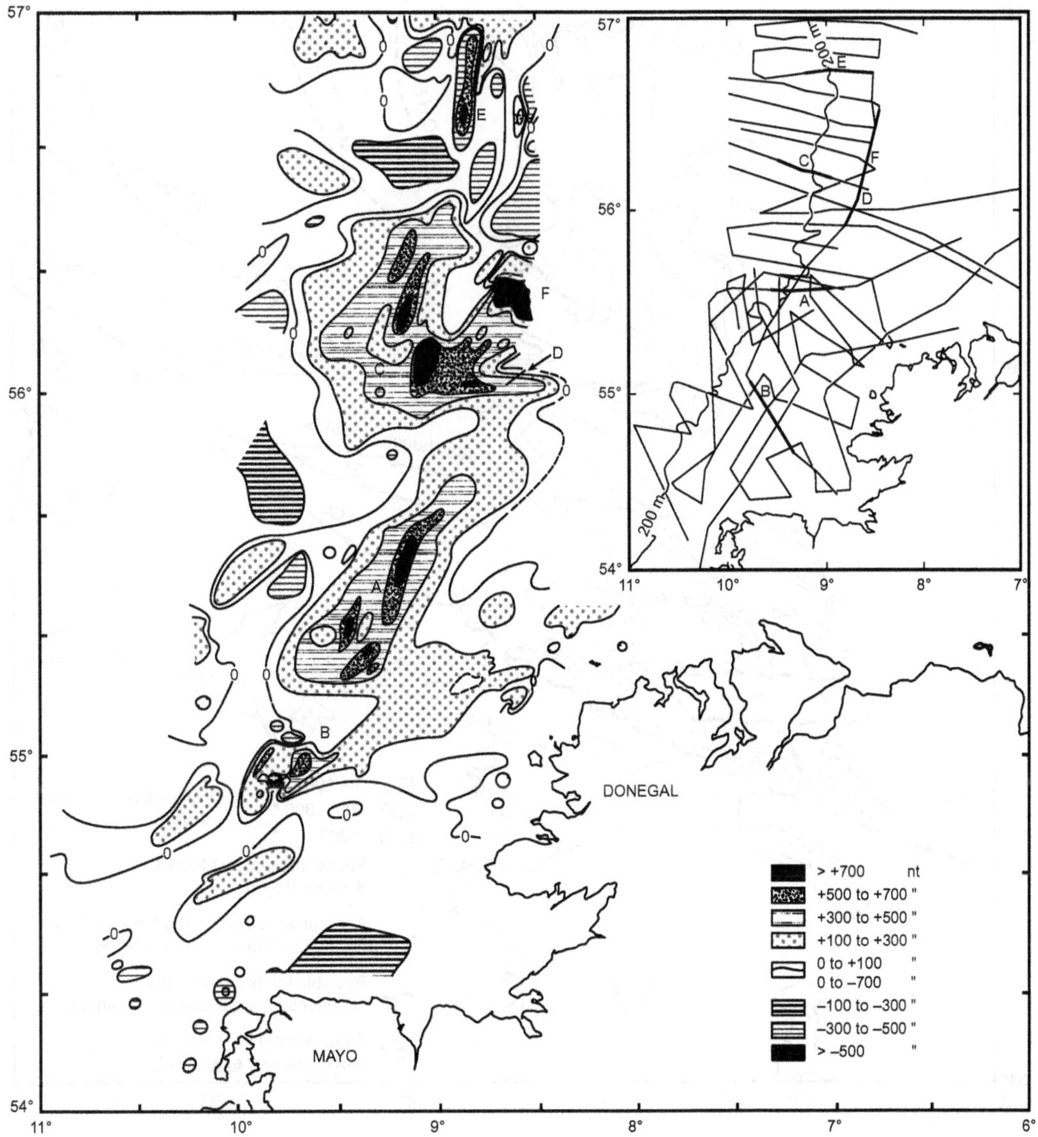

FIGURE 3.51 An example of a total magnetic intensity map of offshore northern Ireland. Location of survey lines shown in the inset map. Compare with the refined, processed version in Fig. 3.49. *From Bailey et al. (1975).*

laterally persistent. An original magnetic survey can be interpreted and presented as a depth-to-magnetic basement map (Fig. 3.52). This procedure can be done when there is some well control on the basement surface, backed by measured rock parameters, and some knowledge of the regional geology. Magnetic anomaly maps are also useful to petroleum exploration in indicating the presence of igneous plugs, intrusives, or lava flows, areas normally to be

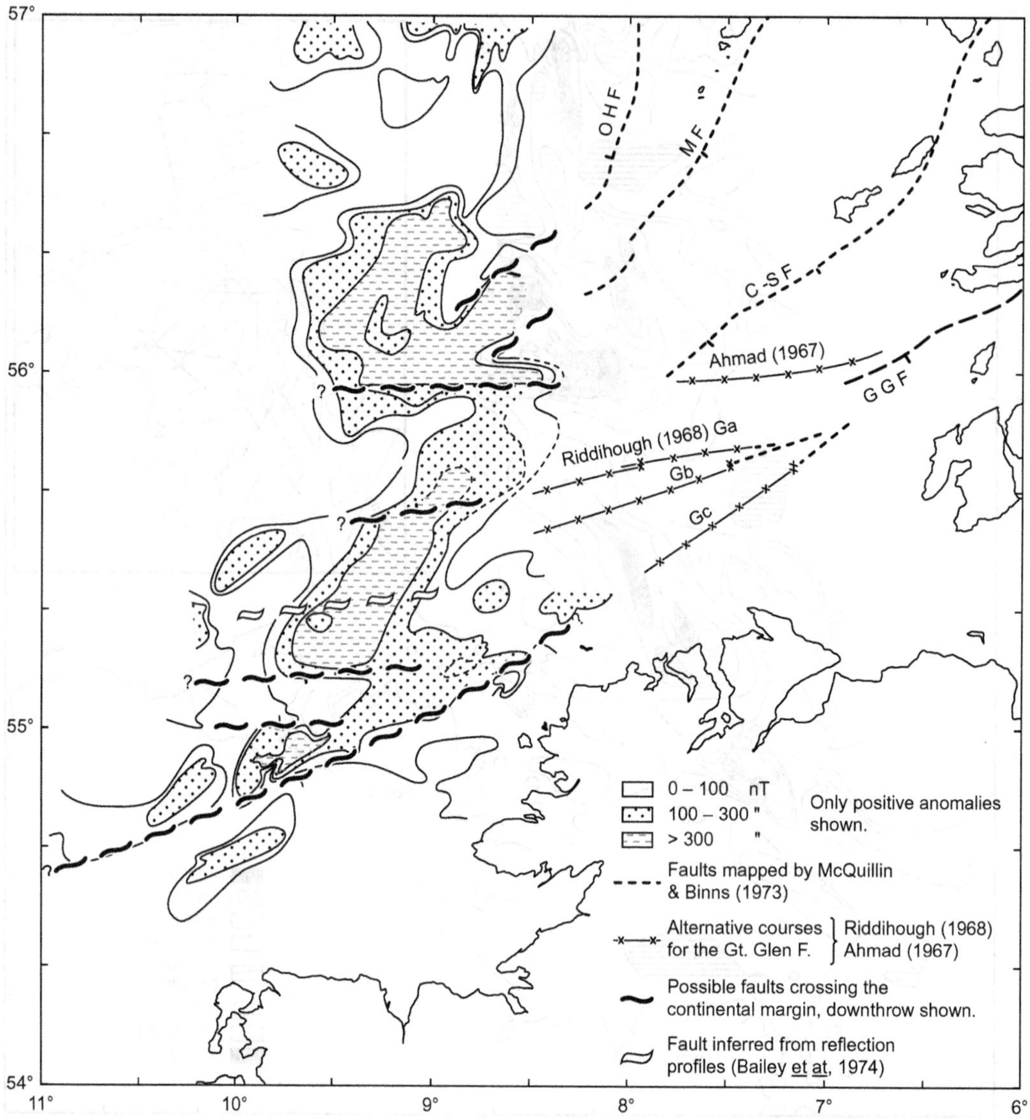

FIGURE 3.52 An interpretation of the magnetic intensity map of offshore northern Ireland shown in Fig. 3.48. *From Bailey et al. (1975).*

avoided in the search for hydrocarbons. Seismically delineated "reefs" have often been drilled and discovered to be igneous intrusions or volcanic plugs. A quick check of the aeromagnetic map would have shown the magnetic anomaly produced by the igneous material instantly.

In recent years, closer grid spacing and improved computer power have lead to a renaissance in magnetic surveying, producing what is now described as high-resolution aeromagnetics. Computer software programs have been devised that process the data quickly and display the results in attractive polychromatic displays (Plate 3.2). High-resolution aeromagnetics becomes still more effective when integrated with other geophysical surveying methods, such as gravity and seismic.

In conclusion, magnetic surveys are a quick and cost-effective way of defining broad-basin architecture. Although they can seldom be used to locate drillable petroleum prospects, they can sometimes differentiate genuine from phoney prospects.

3.3.2 Gravity Surveying

3.3.2.1 *Methodology*

According to Newton's law of gravitation,

$$F = \frac{GM_1M_2}{d^2},$$

where F is the gravitational force between two point masses, M_1 and M_2; d is the distance between M_1 and M_2; G is the universal gravitational constant, usually taken as 6.670×10^{11} $m^3/(kg)(s^2)$ or 6.670×10^{-8} in centimeter–gram–second units.

According to Newton's second law of motion, the acceleration a when a point, mass M_1, is attracted to another point, mass M_2, may be expressed as

$$a = \frac{F}{M_1} = \frac{GM_2}{r^2}.$$

If M_2 is considered to be the mass of the earth and r its radius, then a is the gravitational acceleration on the earth's surface. Were the earth a true sphere of uniform density, then a would be a constant anywhere on the earth's surface. The value of a, however, varies from place to place across the earth. This variation is due to the effect of latitude, altitude, topography, and geology. These first three variations must be removed before the last residual can be detected.

The earth is not a true sphere, but is compressed at the poles such that the polar radius is 21 km less than the equatorial radius. Thus, the force of gravity increases with increasing latitude. Further, gravity is affected by the speed of the earth's rotation, which increases the effect of latitudinal variation.

The correction for latitude is generally calculated using the international gravity formula of 1967, which supersedes an earlier version of 1930:

$$G = 978031.8\left[1/(0.0053024\ \sin^2 y - 0.0000058\ \sin^2 2y)\right],$$

where y is the latitude. The acceleration due to gravity is measured in gals (from Galileo): 1 gal is an acceleration of 1 cm/s^2. For practical purposes, the milligal (mgal), which is 1 thousandth of 1 gal, or the gravity unit (gu), which equals 0.00001 gal, is generally used.

When allowance has been made for variations in G due to latitude, the gravity value still varies from place to place. This variation occurs because the force of gravity is also affected by local variations in the altitude of the measuring station and topography. Gravity decreases with elevation, that is, distance from the center of the earth, at a rate of 0.3086 mgal/m. Thus, a free-air correction must be made to compensate for this effect. A second correction

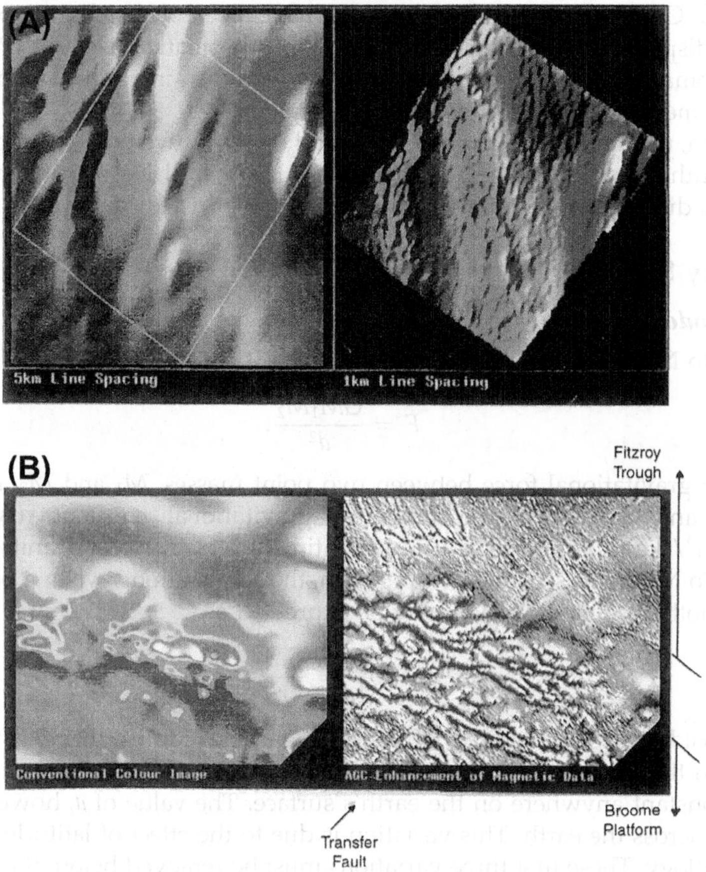

PLATE 3.2 Examples of aeromagnetic maps. Note how the increase in line spacing enhances resolution (A), and how data can be manipulated for further resolution enhancement (B) (See the color plate). *Courtesy of World Geoscience.*

must be made to compensate for the force due to the mass of rock between the survey station and a reference datum, generally taken as sea level. This difference, the Bouguer anomaly, is 0.04191 d mgal/m, where d is the density of the rock.

The free-air and Bouguer effects (E) can be removed simultaneously:

$$E = (0.3086 - 0.4191d)h,$$

where d is the density of rock and h is the height in meters above reference datum.

In mountainous terrain, a further correction is necessary to compensate for the gravitational pull exerted by adjacent cliffs or mountains. When the free-air, Bouguer, and terrain corrections have been made for the readings at each station in a gravity survey, they can be plotted on a map and contoured in milligals (Fig. 3.53).

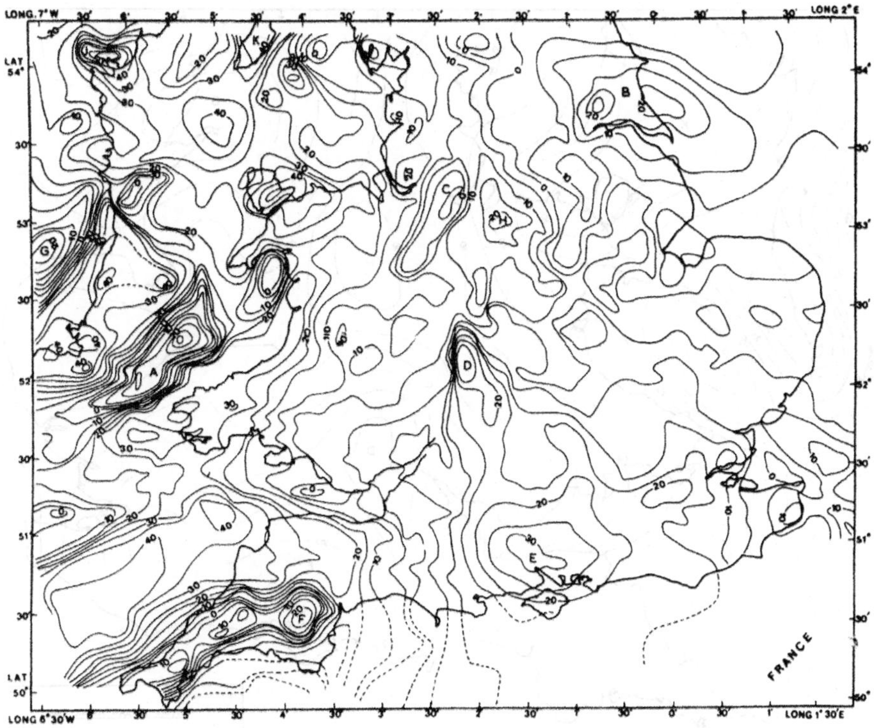

FIGURE 3.53 Bouguer anomaly map of southern Great Britain. Contours in milligals. *From Maroof (1974).*

Gravity surveys can be carried out on land and at sea, both on the sea floor and aboard a ship. In the latter case, corrections must be made for the motion of the ship and for the density and depth of the seawater. Airborne gravity surveys are also now feasible.

3.3.2.2 Interpretation

The interpretation of gravity maps presents many problems, the simplest of which are caused by different subsurface bodies producing the same anomaly on the surface. For example, distinguishing between a small sphere of large density and a large sphere of low density at similar depths is impossible.

The precise cause of an individual anomaly can be tested by a series of models, each model being posited on various depths, densities, or geometries for the body. Gravity maps are seldom used for such detailed interpretation, however, because seismic surveys are generally more useful for small areas. Like magnetic maps, gravity maps are more useful for showing the broad architecture of a sedimentary basin. In general, terms depocenters of low-density sediment appear as negative anomalies, and ridges of dense basement rock show up as positive anomalies (Fig. 3.54).

In some circumstances, gravity maps may indicate drillable prospects by locating salt domes and reefs. Salt is markedly less dense than most sediments and, because of this lower

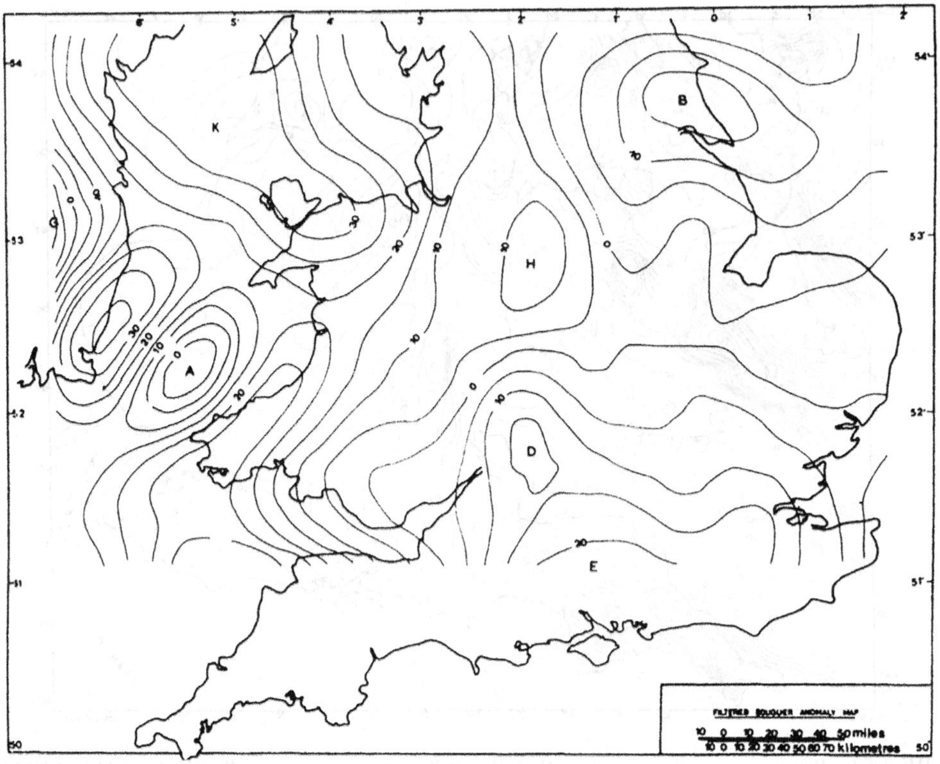

FIGURE 3.54 Filtered Bouguer anomaly map of southern Great Britain computed from the raw data shown in Fig. 3.53. *From Maroof (1974).*

density, often flows up in domes. These salt domes can generate hydrocarbon traps in petroliferous provinces. Because of their low-density, salt domes may often be located from gravity maps. Likewise, reefs may trap hydrocarbons, and they may also show up as gravity anomalies because of the density contrast between the reef limestone and its adjacent sediments (Ferris, 1972).

3.3.3 Magnetic and Gravity Surveys: Summary

Magnetic and gravity surveys are seldom sufficiently responsive to small-scale geological variations so that they can be used to locate individual petroleum prospects. Although gravity surveys can sometimes locate reef and salt dome traps, aeromagnetic surveys can identify igneous bodies (Fig. 3.55).

Both magnetic and gravity surveys are, however, cost-effective methods of reconnoitering large areas of the earth's surface onshore and offshore before lease acquisition. Their main use is in defining the limits and scale of sedimentary basins and the internal distribution of structural highs and lows. When used in combination, a far more accurate basement map can be prepared than when they are interpreted separately. Aeromagnetic and gravity surveys can

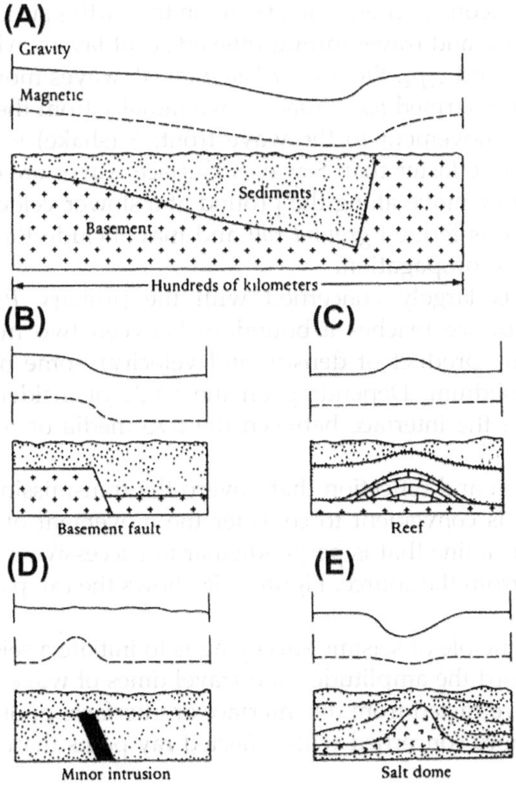

FIGURE 3.55 Gravity and magnetic responses for various geological features. (A) Regional responses across a sedimentary basin. (B)–(E) Local anomalies that may be present on a scale of a few kilometers.

be integrated with one another, or with seismic surveys. Once well control becomes available, the true depth to basin is known. Then, the surveys can be accurately calibrated and more reliable depth-to-basement maps redrawn.

Both magnetic and gravity surveys are undergoing a renaissance in 2014 for several reasons. Magnetic and gravity surveys are now being run concurrently with offshore seismic surveys. Ever-improving computing power, coupled with extremely accurate satellite navigation systems, enables integrated geophysical studies to be carried out that lead to much improved imaging of the geological structure. The sum is greater than the individual parts (George, 1993).

3.3.4 Seismic Surveying

3.3.4.1 *Basic Principles*

Seismic surveying is the most important of the three main types of geophysical prospecting. To understand the method, a review of the physical principles that govern the movement of acoustic or shock waves through layered media is necessary.

Consider a source of acoustic energy at a point on the earth's surface. Three types of waves emanate from the surface and travel through the adjacent layers, which have acoustic velocities and densities v_1p_1, and v_2p_2. *Surface*, or *longitudinal*, waves move along the surface. The two other types of waves termed *body waves* move radially from the energy source: *P* (push) waves impart a radial movement to the wave front; *S* (shake) waves impart a tangential movement. *P* waves move faster than *S* waves. Surface waves are of limited significance in seismic prospecting. They move along the ground at a slower velocity than do body waves. The surface disturbance is termed *ground roll* and may include Rayleigh and other vertical and horizontal modes of propagation.

Seismic surveying is largely concerned with the primary *P* waves. When a wave emanating from the surface reaches a boundary between two media that have different acoustic impedance (the product of density and velocity), some of the energy is reflected back into the upper medium. Depending on the angle of incidence, some of the energy may be refracted along the interface between the two media or may be refracted into the lower medium.

The laws of refraction and reflection that govern the transmission of light also pertain to sound waves. Thus, it is convenient to consider the movement of waves in terms of their ray paths. A ray path is a line that is perpendicular to successive wave fronts as an acoustic pulse moves outward from the source. Figure 3.56 shows the ray paths for the three types of wave just described.

The fundamental principle of seismic surveying is to initiate a seismic pulse at or near the earth's surface and record the amplitudes and travel times of waves returning to the surface after being reflected or refracted from the interface or interfaces of one or more layers of rock. Seismic surveying is more concerned with reflected ray paths than refracted ones.

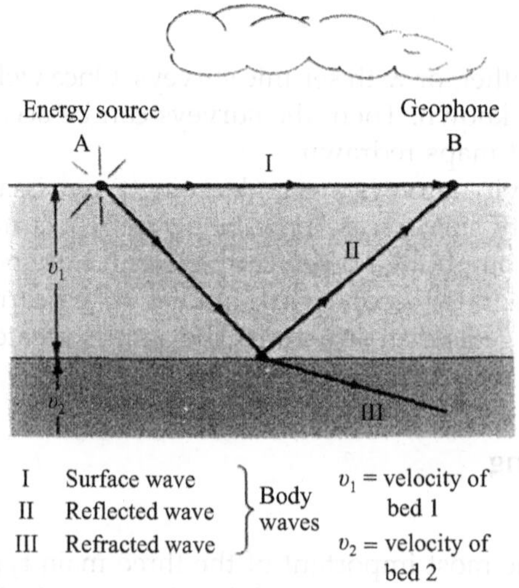

FIGURE 3.56 Cross-section illustrating various seismic wave paths.

If the average acoustic velocity of the rock is known, then it is possible to calculate the depth D to the interface:

$$D = \frac{vt}{2},$$

where v is the acoustic velocity and t is the two-way (i.e., there and back) travel time.

The acoustic velocity of a rock varies according to its elastic constants and density. As sediments compact during burial, their density (and hence their acoustic velocities) generally increases with depth. Theoretically, the velocity of a P wave can be calculated as follows:

$$v = \sqrt{\left(k + \frac{4}{3}h\right)/p},$$

where k is the bulk modulus, h is the shear modulus, and p is the density. The relationships between the velocity and density of common sedimentary rocks are shown in Fig. 3.57. The product of the velocity and density of a rock is termed the *acoustic impedance*. The ratio of the amplitudes of the reflected and incident waves is termed the *reflectivity* or *reflection coefficient*:

$$\text{Reflection coefficient} = \frac{\text{amplitude of reflected wave}}{\text{amplitude of incident wave}}$$

$$= \frac{\text{change in acoustic impedance}}{2 \times \text{mean acoustic impedance}}$$

$$= \frac{P_2 v_2 - p_1 v_1}{p_2 v_2 + p_1 v_1},$$

where p_1 is the density of the upper rock; P_2 is the density of the lower rock; v_1 is the acoustic velocity of the upper rock; and v_2 is the acoustic velocity of the lower rock.

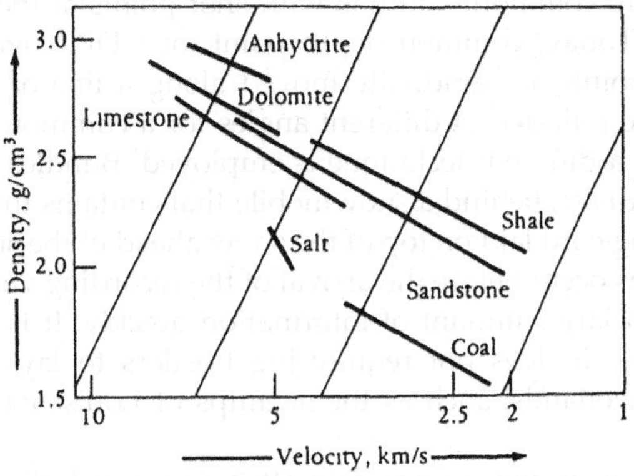

FIGURE 3.57 The velocity–density relationships for sedimentary rocks. Note how velocity increases with density for most rocks except salt. Diagonals are lines of equal acoustic impedance. *After Gardner et al. (1974).*

Because velocity and density generally increase with depth, the reflection coefficient is usually positive. Negative coefficients occur where there is a downward decrease in velocity and density, for example, at the top of an overpressured shale diapir. A reflection coefficient that is positive for a downgoing wave is negative for an upgoing wave and, of course, vice versa.

Velocity and density generally increase with depth for a homogenous lithology, but at the boundary between two units we can, and usually will, get a lithological contrast. Therefore, the positive and negative reflection coefficients tend to appear with approximately equal frequency (taking the case of a downgoing wave only). Seismic processing is often based on the assumption that the earth's reflection coefficient series is random, which is empirically true. This assumption would not be the case if most reflection coefficients were positive. Negative coefficients occur where there is a decrease in acoustic impedance across the reflecting interface. Reflection coefficients have a maximum range limited by +1 and −1. Values of +0.2 or −0.2 would occur for very strong reflectors. In practice, most reflection coefficients lie between −0.1 and +0.1. The boundary between a porous sand and a dense, tight limestone will have a high reflection coefficient and show up as a prominent reflecting surface. By contrast, the interface between two shale formations of similar impedance will have a negligible reflection coefficient and will reflect little energy.

It is now appropriate to see how these general principles are used in seismic surveying. Traditionally, seismic exploration involves three steps: data acquisition, data processing, and interpretation.

3.3.4.2 Data Acquisition

Seismic surveys are carried out on land and at sea in different ways. On land, the energy source may be provided by detonating explosives buried in shot holes, by dropping a heavy weight off the back of a lorry (the thumper technique is actually a rather sophisticated procedure), or by vibrating a metal plate on the ground (Vibroseis). The returning acoustic waves are recorded on geophones arranged in groups. The signals are transmitted from the geophones along cables to the recording truck. Equipment in this truck controls the firing of the energy source and records the incoming signals from the geophones on magnetic tapes.

The shot points and the receiving geophones may be arranged in many ways. Many groups of geophones are commonly on line with shot points at the end or in the middle of the geophone spread. Today, common depth point, or CDP, coverage is widely used. In this method the shot points are gradually moved along a line of geophones. In this way, up to 48 signals may be reflected at different angles for a common depth point.

In Arctic conditions, a different technique is employed. Bundles of geophones are rowed on a cable, termed a *streamer*, behind a snowmobile that contains the recording gear. The energy source is detonating cord laid on top of the snow ahead of the snowmobile. Care is taken to ensure that explosions occur before the arrival of the recording vehicle. Used carefully, this technique can acquire a large amount of information quickly. It is faster than conventional onshore seismic because it does not require jug hustlers to lay and pick up geophones (Rygg et al., 1992). In wetlands, such as the swamps of Louisiana, hovercraft are used for seismic surveys.

The basic method of acquiring seismic data offshore is much the same as that of onshore, but it is simpler, faster, and, therefore, cheaper. A seismic boat replaces a truck as the controller and recorder of the survey. This boat trails an energy source and a cable of hydrophones, again termed a *streamer* (Fig. 3.58). It is possible for one boat to operate several energy sources, but experience has shown that, in this instance, more bangs is not necessarily the best. Streamer lengths can extend for up to 6000 m to the annoyance of fisher folk. Currently, survey vessels, such as the *Ramform Explorer*, can operate up to three energy sources whose signals are received by hydrophones on 8–12 streamers, up to 3000 m in length, with a total survey width of 800 m.

In marine surveys, dynamite is seldom used as an energy source. For shallow, high-resolution surveys, including sparker and transducer surveys, high-frequency waves are used. In sparker surveys, an electric spark is generated between electrodes in a sonde towed behind the boat. Every time a spark is generated, it implodes after a few milliseconds. This implosion creates shock waves, which pass through the sea down into the strata. Transducers alternately transmit and receive sound waves. These high-resolution techniques generally only penetrate up to 1.0-s two-way time. They are useful for shallow geological surveys (e.g., to aid production platform and pipeline construction) and can sometimes indicate shallow gas accumulations and gas seepages (Fig. 3.59).

For deep exploration, the air gun is a widely used energy source. In this method, a bubble of compressed air is discharged into the sea; usually a number of energy pulses are triggered simultaneously from several air guns. The air guns can emit energy sufficient to generate signals at between 5- and 6-s two-way travel time. Depending on interval velocities, these signals may penetrate to >5 km.

The reflected signals are recorded by hydrophones on a cable towed behind the ship. The cable runs several meters below sea level and may be up to 6 km in length. As with land surveys, the CDP method is employed, and many recorders may be used. The reflected signals are transmitted electronically from groups of hydrophones along the cable to the recording unit on the survey ship. Other vital equipment on the ship includes a fathometer and position fixing devices. The accurate location of shot points at sea is obviously far more difficult than it is on land. Formerly this was done either by radio positioning, or by getting fixes on two or more navigation beacon transmitters from the shore. Nowadays, satellite navigation systems enable pinpoint accuracy to be achieved.

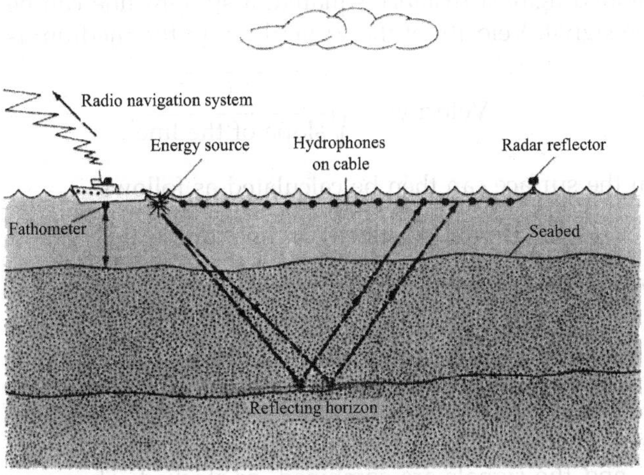

FIGURE 3.58 Sketch showing how seismic data are acquired at sea.

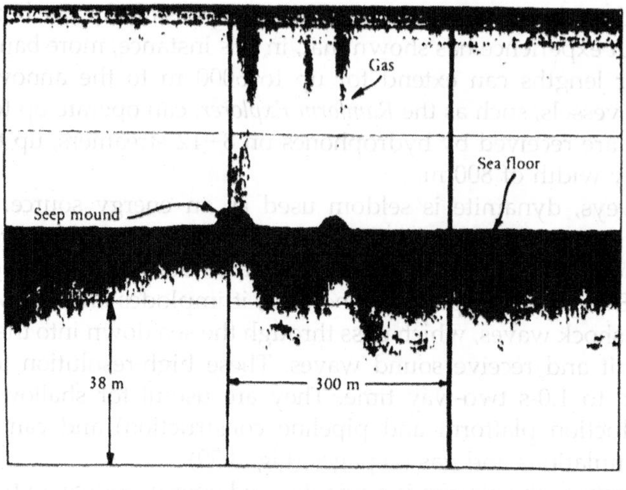

FIGURE 3.59 Transducer record showing gas seeps and mounds in offshore Texas. *From H. C. Sieck and G. W. Self, namely Fig. 4, p. 358 in AAPG Memoir Series No. 26, © 1977;* Analysis of High-Resolution Seismic Data.

3.3.4.3 *Data Processing*

Once seismic data have been acquired, they must be processed into a format suitable for geological interpretation. This process involves the statistical manipulation of vast numbers of data using mathematical techniques far beyond the comprehension of geologists. Seismic data processors include mathematicians, physicists, electronic engineers, and computer programmers.

Consider the signals of three receivers arranged in a straight line from a shot point. The times taken for a wave to be reflected from a particular layer will appear as a wiggle on the receiver trace. The arrival times will increase with the distance of receivers from the shot point. Thus, a time–distance graph may be constructed as shown in Fig. 3.60. When time squared is plotted against distance squared, a straight line can be drawn through the arrival times of the signal. Velocity of the wave through the medium is

$$\text{Velocity} = \sqrt{\frac{1}{\text{slope of the line}}}.$$

The depth from the surface can then be calculated as follows:

$$\text{Depth} = \text{velocity} \times \text{time (one way)}.$$

But since the travel time recorded is two way (there and back),

$$\text{Depth} = \text{velocity} \times \frac{\text{time (one way)}}{2}.$$

With common depth point shooting (discussed previously), reflections from the same subsurface points are recorded with a number of different combinations of surface source and receiver positions, and the signals are combined, or stacked. Thus, wave–time graphs can

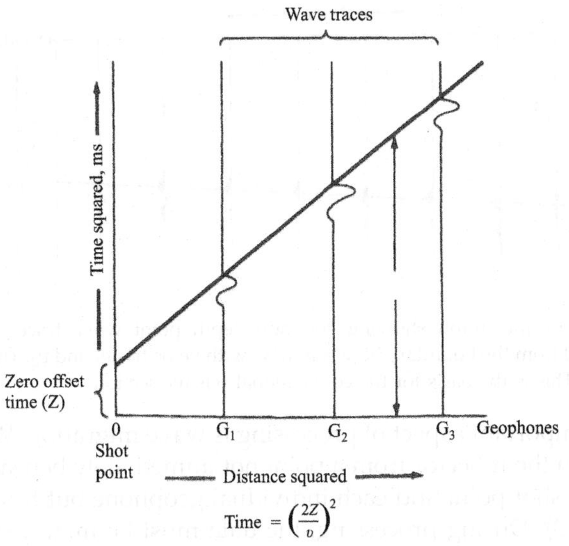

FIGURE 3.60 Travel–time graph showing how velocity may be calculated by measuring the length of time taken for a wave to travel from a shot point to geophones placed known distances away.

be displayed in a continuous seismic section, and a single seismic reflecting horizon can be traced across it. Wave traces are displayed in several ways. They may be shown as a straightforward graph or wiggle trace. Reflectors show up better on a variable area display in which deviations in one direction (positive or negative polarity) are shown in black (Fig. 3.61). This type of display is the basis for the conventional seismic section used today.

Four main steps are involved in the processing of raw seismic data before the production of the final seismic section:

1. Conversion of field magnetic tape data into a state suitable for processing
2. Analysis of data to select optimum processing parameters (e.g., weathering correction and seawater velocity)
3. Processing to remove multiple reflectors and enhance primary reflectors
4. Conversion of data from digital to analog form and printout on a graphic display (i.e., final seismic line)

During processing, many unwanted effects must be filtered out. All seismic records contain both the genuine signal from the rocks and the background "noise" due to many extraneous events, which may range from a microseism to a passing bus. The signal-to-noise ratio is a measure of the quality of a particular survey. Noise should obviously be filtered out. The frequency and amplitude of the signal are the result of many variables, including the type of energy source used and its resultant wave amplitude, frequency, and shape, and the parameters of the various rocks passed through. Deconvolution is the reverse filtering that counteracts the effects of earlier undesirable filters (Fig. 3.62).

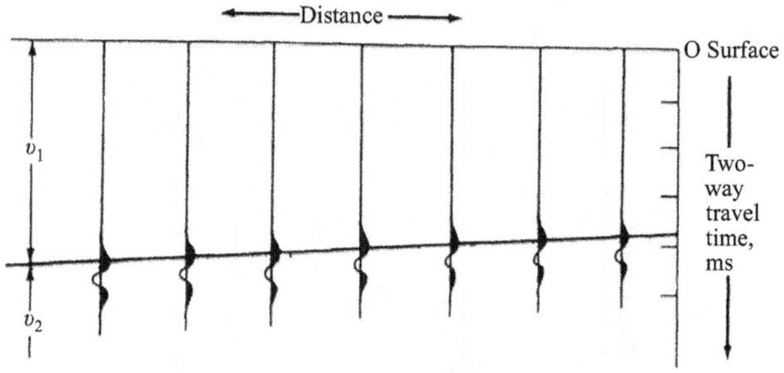

FIGURE 3.61 Distance–time graph showing common depth point wave traces for a series of geophones recording a signal reflected from the boundary between rocks with velocities v_1 and v_2. The wave traces are shaded in the variable area format. This is the basis for the conventional seismic section.

One particularly important aspect of processing is wave migration. When beds dip steeply, the wave returns from the reflector from a point not immediately beneath the surface location midway between the shot point and each individual geophone but from a point up-dip from this position (Fig. 3.63). During processing, the data must be migrated to correct this effect. This migration causes several important modifications in the resultant seismic section. Anticlines become sharper, synclines gentler, and faults more conspicuous.

The preceding discussion briefly covered some aspects of seismic processing. The main point to note is that digitally recorded seismic data can be processed time and time again, either by using new, improved computer programs or by tuning the signal to bring out aspects of particular geological significance.

3.3.4.4 3D Seismic Surveying

In the old days, seismic surveys were shot on loose grids several kilometers wide. Increased computer power, however, has made it possible to handle ever larger amounts of data faster and cheaper. If seismic surveys are shot on a grid of only 50 m or so, then a three-dimensional matrix of data is acquired that enables seismic displays to be produced, not only along the survey lines shot but also in any other orientation. This is called 3D seismic, naturally. The 3D seismic is very appealing because horizontal displays, or time slices, can be produced (Plate 3.3). Not only can these reveal structural features, such as faults and salt domes, but also stratigraphic ones, such as reefs and channels (Brown, 1991, 2004).

3.3.5 Interpretation of Seismic Data

3.3.5.1 First Principles

Geologists looking at seismic lines and maps inevitably tend to see them as representations of rock and forget that they do, in fact, represent travel time.

According to Tucker and Yorston (1973), three main groups of pitfalls in seismic interpretation are to be avoided:

1. Pitfalls due to processing
2. Pitfalls due to local velocity anomalies
3. Pitfalls due to rapid changes in geometry

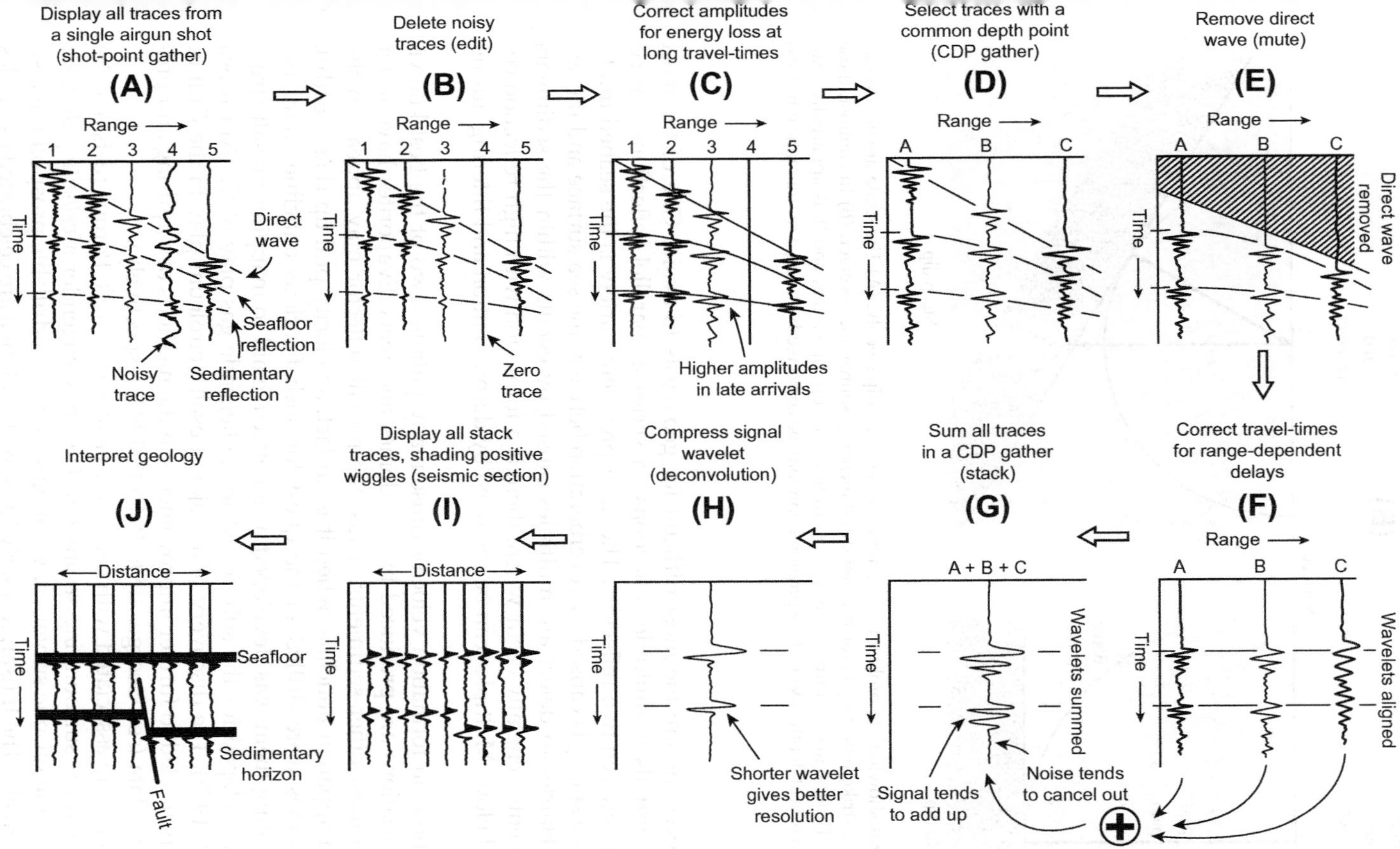

FIGURE 3.62 Simplified seismic processing sequence. The traces illustrated in part (A) result from a single shot, and were recorded by a single receiver. The direct wave (refracted wave) appears at a time proportional to the receiver distance from the shot, while the reflections lie on a hyperbolic travel–time curve. Any bad or noisy traces are deleted (B). The seismic amplitudes recorded decrease with increasing travel–time because the reflectors are further away, so the weaker amplitudes are boosted (C). Then, the traces are resorted so that all traces with an identical source–receiver midpoint are gathered together in (D), one trace each from (A), (B), and (C). The direct wave is removed (the shaded area in (E)), and then each trace is separately corrected for the time delay appropriate to its source–receiver offset (F). This is known as the *normal move-out*, or NMO, correction. The wavelets from each reflection are now lined up, as if each trace had been recorded with coincident source and receiver (i.e., without source–receiver offset delays), and all the traces in the gather can be summed up (stacked) to give the single trace in (G) with a higher signal-to-noise ratio. If the wavelet is rather reverberatory, or lengthy, then its resolution is poor, but can be shortened by a digital filtering technique called *deconvolution* (H). Finally, all the traces are displayed as a seismic section and may then be interpreted. *From Klemperer and Peddy (1992).* © *Cambridge University Press, reprinted with permission.*

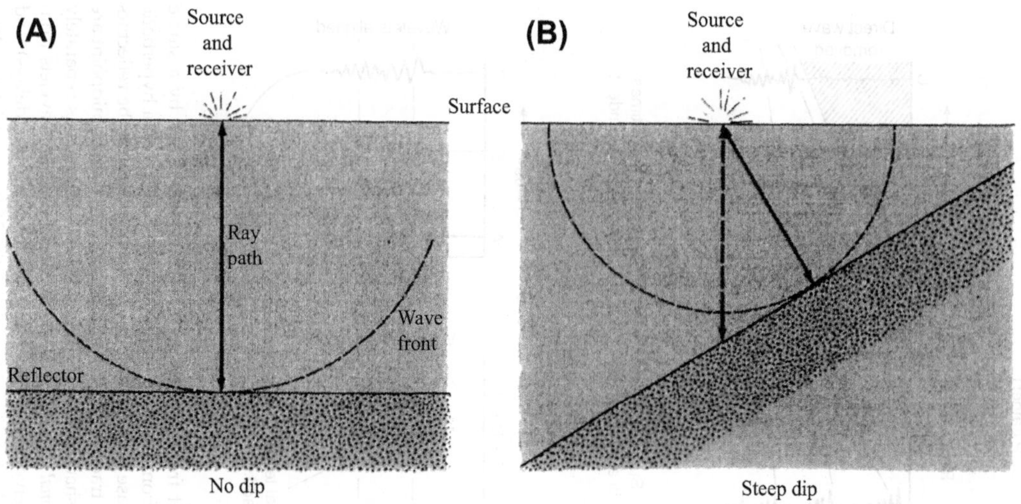

FIGURE 3.63 Diagrams showing the problem encountered when beds dip steeply. (A) The strata are horizontal. Thus, the wave fronts first strike reflectors vertically beneath the energy source and receiver. (B) In contrast, these beds are steeply dipping. Thus, signals return to the receiver from points up-dip earlier, and thus apparently shallower, than from the true vertical depth (TVD). This phenomenon may be corrected by using a mathematical process termed *migration*.

Pitfalls due to processing are the most difficult for geologists to avoid. One of the most common processing pitfalls is multiple reflections—a series of parallel reflections caused by the reverberation between two reflectors. The most prevalent variety is the seabed multiple, which, as its name states, is caused by reverberation between the sea surface and the sea floor (Fig. 3.64(A)). Harder to detect are multiples caused by events within the sediments themselves. These events generally occur where there is a formation with high reflection coefficients above and below. Multiples can be removed by deconvolution and filtering during processing.

Rapid local variations in formation velocity cause many pitfalls. Two of the best-known examples of these variations are produced by salt domes and reefs. Evaporites have faster travel times than do most other sediments. Thus, the presalt reflector may appear on the seismic section as an apparent anticline, when it is, in fact, a velocity pull-up (Fig. 3.64(B)). A number of salt domes were drilled in the quest for such fictitious anticlines until the method of undershooting them was developed as an effective aid to mapping presalt structure. This was first developed for the sub-Zechstein Rotliegende gas play of the North Sea (Krey and Marschall, 1975). The discovery of a major subsalt petroleum play in the Gulf of Mexico in recent years has led to further improvements in the methods of acquiring and processing subsalt seismic data (see George, 1996; Ratcliff, 1993; respectively).

Other velocity pitfalls are associated with carbonate *build-ups* or *reefs*. Porous reefs may be acoustically slower than the surrounding sediments. This may engender a *velocity pull-down* beneath the reef. Thus, the seismic survey may not only identify a drillable prospect but also indicate that it has porosity. The Intisar, formerly Idris, reefs of Libya are celebrated examples

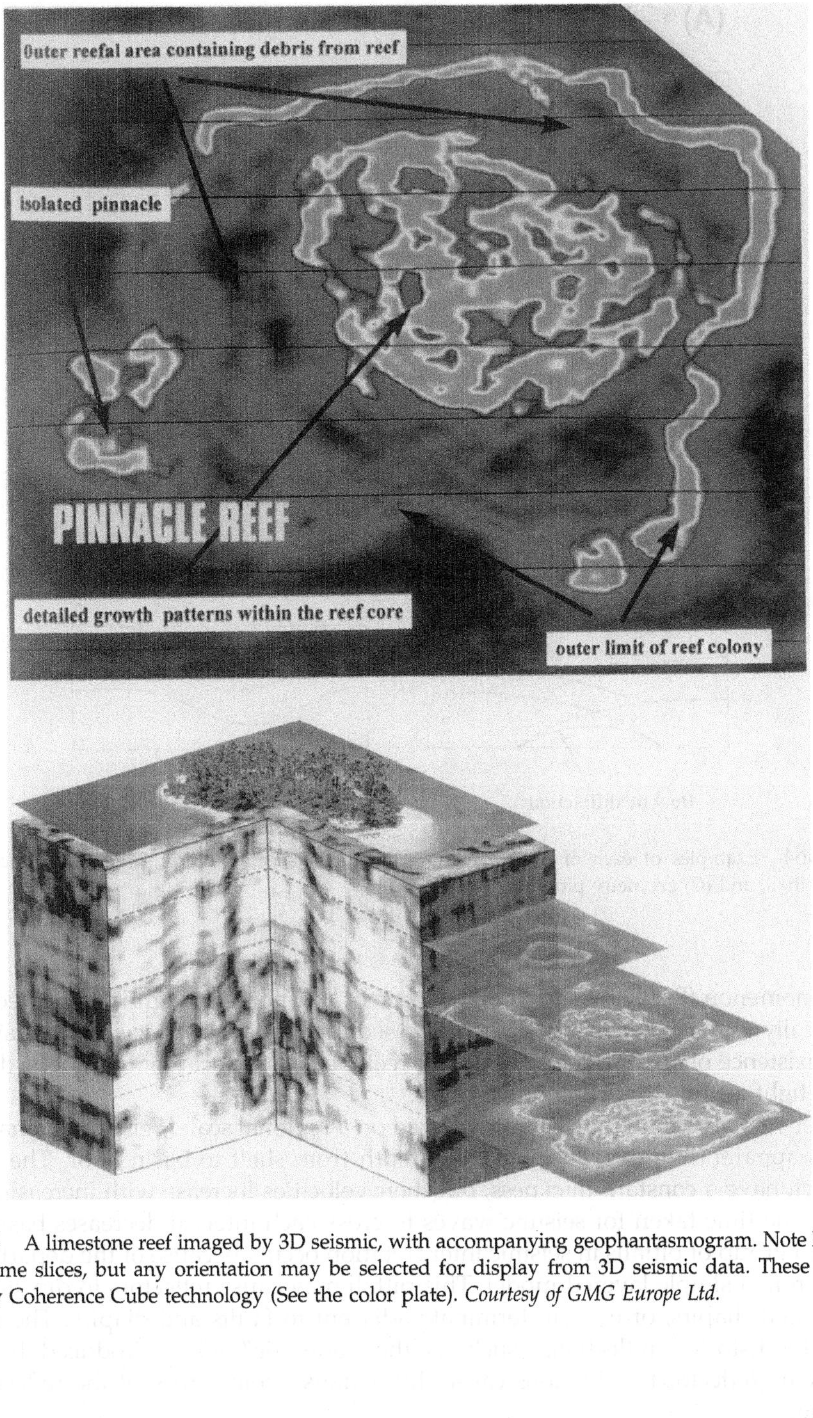

PLATE 3.3 A limestone reef imaged by 3D seismic, with accompanying geophantasmogram. Note how not only horizontal time slices, but any orientation may be selected for display from 3D seismic data. These images were produced by Coherence Cube technology (See the color plate). *Courtesy of GMG Europe Ltd.*

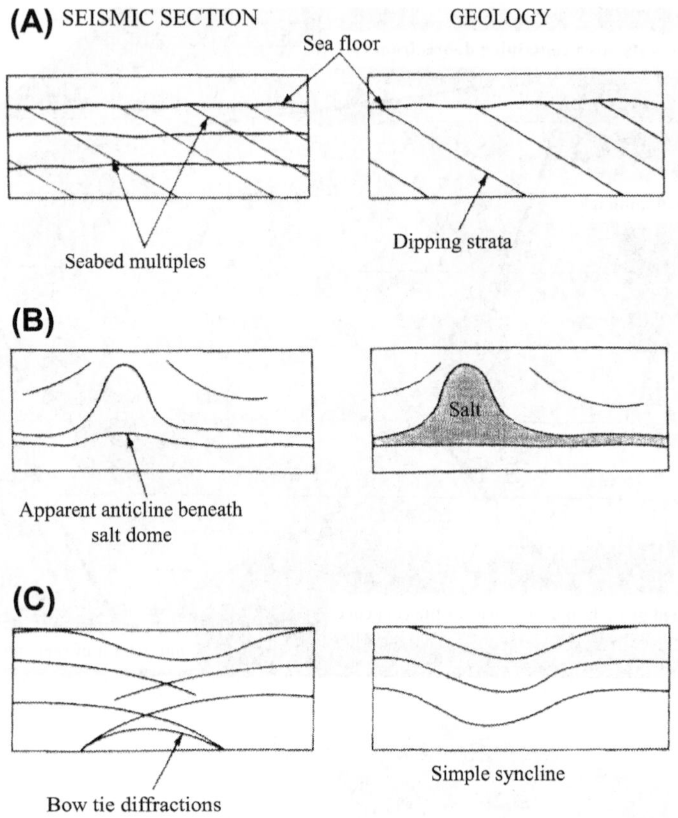

FIGURE 3.64 Examples of each of the three main pitfalls in seismic interpretation: (A) processing pitfall, (B) velocity pitfall, and (C) geometry pitfall.

of this phenomenon (Brady et al., 1980). Conversely low-porosity, tightly cemented reefs may be acoustically much faster than the enclosing sediments. The seismic section may show the apparent existence of an anticline beneath the reef. Such prospects should not be drilled since the reef is tight and the anticline nonexistent.

A further velocity pitfall may be encountered on a regional scale. Seismically mapped units sometimes apparently thin with increasing depth from shelf to basin floor. The formations may, in fact, have a constant thickness, but their velocities increase with increasing compaction. Thus, the time taken for seismic waves to cross each interval decreases basinward.

The third group of pitfalls in seismic interpretation occurs because of the departure of rock geometry from a simple layered model. This pitfall may cause reflectors to dip steeply, as in tight folds and diapirs, or even to terminate adjacent to faults and diapirs. The former can give rise to distorted reflections, such as the "bow tie" effect produced by synclines (Fig. 3.61(C)). Reflector terminations cause diffractions, which crosscut the reflectors on the seismic line.

When the seismic data have been correctly interpreted, avoiding the pitfalls just discussed, the various reflecting horizons can be mapped. In the old days, reflectors were mapped from hard copy prints using brightly colored pencils. Today, computers loaded with appropriate software packages enable reflectors to be mapped on monitors with a mouse. Since seismic lines are generally shot on an intersecting grid, reflectors can be tied from line to line. For large surveys, reflectors can be traced from line to line and tied in a loop to check that the reflector has been correctly picked and that the interpreter has not jumped a reflecting horizon, colloquially referred to as a *leg*, in the process.

The times for the reflectors for each shot point may then be tabulated. Contour maps may then be drawn for each horizon (red, blue, green, etc.). Note that these are time maps, not depth maps. They show isochrons (lines of equal two-way travel time), not structure contours. Figure 3.65 shows, in a very simple way, how reflectors may be picked on two intersecting seismic lines and used to construct isochron maps.

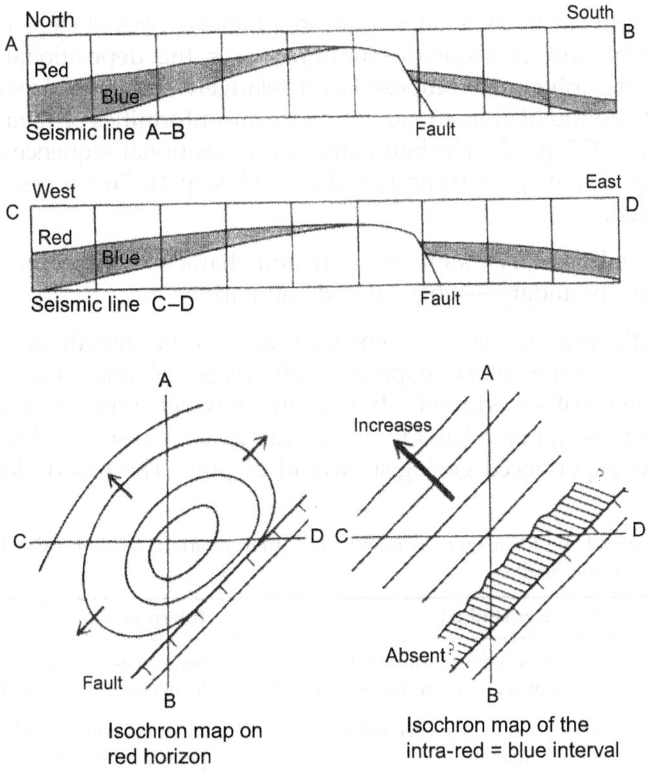

FIGURE 3.65 Simplified sketch showing how reflectors may be picked on two intersecting seismic lines (A–B) and (C–D), and used to construct isochron maps of an individual reflector (bottom left), or the interval between two reflectors (bottom right).

3.3.5.2 *Geological Application*

Seismic surveying is an essential part of the whole cycle of petroleum exploration and production. It is applied in six main ways: regional mapping, prospect mapping, reservoir delineation, seismic modeling, direct hydrocarbon detection, and the monitoring of petroleum production. Pacht et al. (1993) have proposed a threefold hierarchy of seismic interpretation, in terms of the parameter that is analyzed, and the interpretations that may be drawn from the analysis; these are seismic sequence analysis, seismic facies analysis, and seismic attribute analysis (Table 3.2).

3.3.5.2.1 SEISMIC SEQUENCE ANALYSIS

In the early stages of reconnaissance mapping of a new sedimentary basin, a broad seismic grid will be shot and integrated with magnetic and gravity surveys, as discussed earlier. The advent of high-quality offshore seismic data in the 1960s and 1970s led to a whole new way of looking at sedimentary sequences, termed *sequence stratigraphy* (Vail et al., 1977). This is now a very important and integral part of petroleum exploration. It is briefly expounded here. For further details, see Neidell and Poggiagliolmi (1977), Payton (1977), Vail et al. (1991), Van Wagoner et al. (1988), Weimer (1992), Neal et al. (1993), and Steel et al. (1995). Seismic sequence stratigraphic analysis is carried out in a logical series of steps (Fig. 3.66).

The fundamental unit of sequence stratigraphy is the depositional sequence. This is defined as "a stratigraphic unit composed of a relatively conformable succession of genetically related strata bounded at its top and base by unconformities or their correlative conformities" (Vail et al., 1977, p. 53). The boundaries of depositional sequences may be associated with onlap, toplap, downlap, or truncation (Fig. 3.63, step 1). Two types of sequence boundaries are recognized:

Type 1: Sequence boundary—subaerial only and characterized by channeling
Type 2: Sequence boundary—subaerial and submarine

The age and lithology of a seismic sequence may not be directly apparent without well control, but they are, nonetheless, mappable units (Figs 3.67 and 3.68). Vail and colleagues argue for the existence of a worldwide stratigraphy in which a regular sequence of characteristic seismic sequences can be related to global fluctuations in sea level (Fig. 3.69). These are old ideas that few experienced geologists would dispute. The broad global changes of sea

TABLE 3.2 Summary of the Parameters Examined and Interpretations Made for the Threefold Hierarchy of Seismic Data Analysis

Seismic analysis	Parameter studied	Interpretation
Sequence analysis	Sequences defined and mapped; nature of sequence contacts recorded	Sequences correlated with global sea level changes (if possible) and time calibrated
Facies analysis	Description of sequence-defined seismic character	Identification of vertical mounded and pelagic sediments
Attribute analysis	Analysis of wave shape, amplitude, polarity, and continuity	Identification of vertical changes in rock properties, including grain-size profiles and direct hydrocarbon indicators (DHIs)

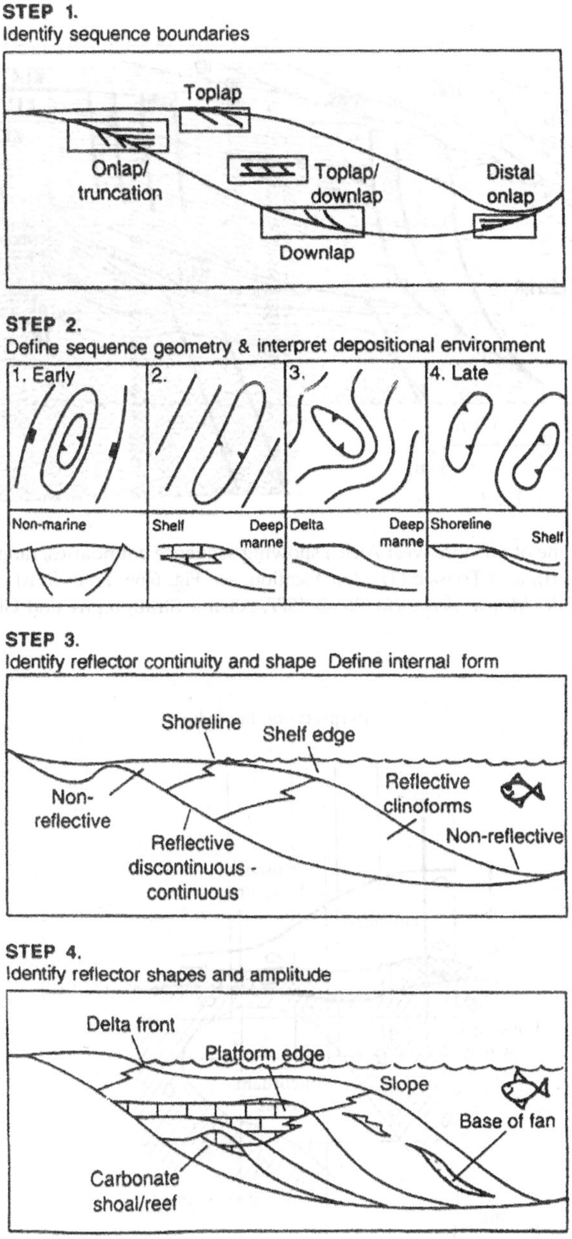

FIGURE 3.66 Illustration to show the terminology of reflection conjunctions (step I) and the subsequent steps to be taken in seismic sequence analysis. *Based on Hubbard et al. (1985).*

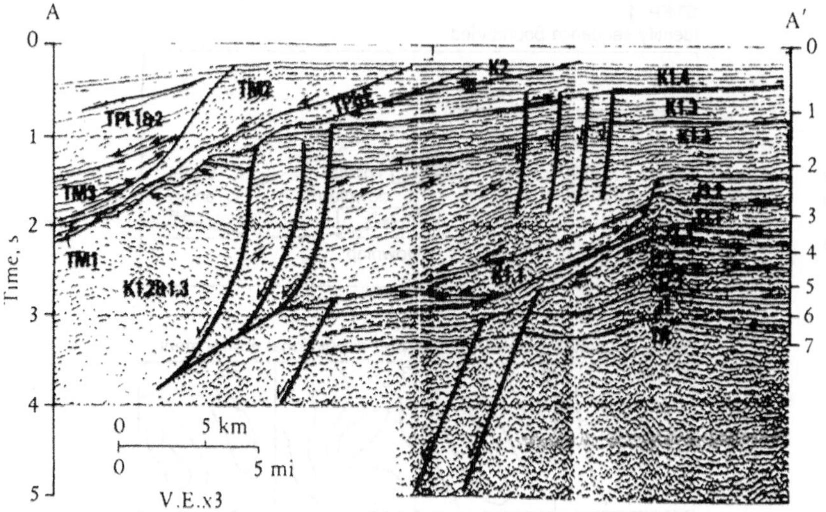

FIGURE 3.67 Seismic line of offshore west Africa showing sequence boundaries, ranging in age from Tertiary (T) to Cretaceous (K), Jurassic (J), and Triassic (Tr). For location, see Fig. 6.66. *From R. M. Mitchum, Jr., and P. R. Vail, namely Fig. 1, p. 137 in AAPG Memoir Series No. 26, © 1977;* Seismic Stratigraphy and Global Changes of Sea Level, Part 7.

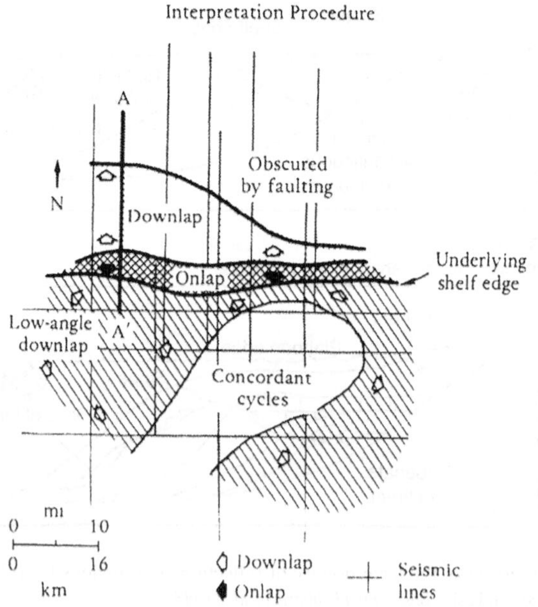

FIGURE 3.68 Map showing reflection patterns at lower surface of Lower Cretaceous sequence, offshore west Africa. A–A′ is the seismic line shown in Fig. 3.64. *From R. M. Mitchum, Jr., and P R. Vail, namely, Fig. 1, p. 137 in AAPG Memoir Series No. 26, © 1977,* Seismic Stratigraphy and Global Changes of Sea Level, Part 7.

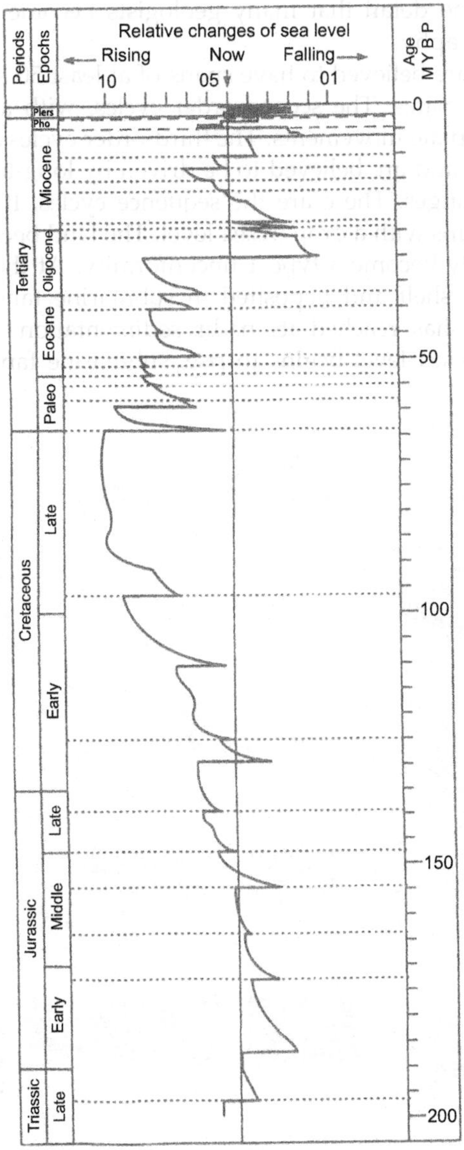

FIGURE 3.69 A global sea level curve for the last 200 million years, or so. *For sources, see Vail et al. (1977) and Haq et al. (1988).*

level were remarked by Stille (1924), and the remarkable uniformity of sedimentary facies by Ager (1973, 1993). Vail and colleagues further argue, however, that seismic surveys can now be used to erect a sequence stratigraphy with a hierarchy of cycles on four different scales. These cycles can be correlated universally because seismic reflectors are chronostratigraphic horizons. It is in this fine detail that many geologists become nervous about seismically defined sequence stratigraphy.

The first-order cycles are believed to have spans of at least 50 my, and to be driven by the break-up of continental plates. The second-order cycles, with spans of 3–50 my are also believed to be driven by plate movements. The third-order cycles are deemed to be on a scale of 500,000 years to 3 my and are believed to be driven by long-term tectonic processes, and shorter-term climatic changes. These are the sequence cycles. It is argued that each third-order sequence cycle begins with a drop in sea level. The land becomes emergent and incised by channels, to ultimately become a type 1 unconformity. At the same time, sand is transported off the emergent shelf and deposited in submarine fans on the basin floor (Plate 3.4(A)). When sea level has reached its nadir sedimentation, slows down to deposit a condensed sequence. As sea level begins to rise, submarine fan deposition encroaches up

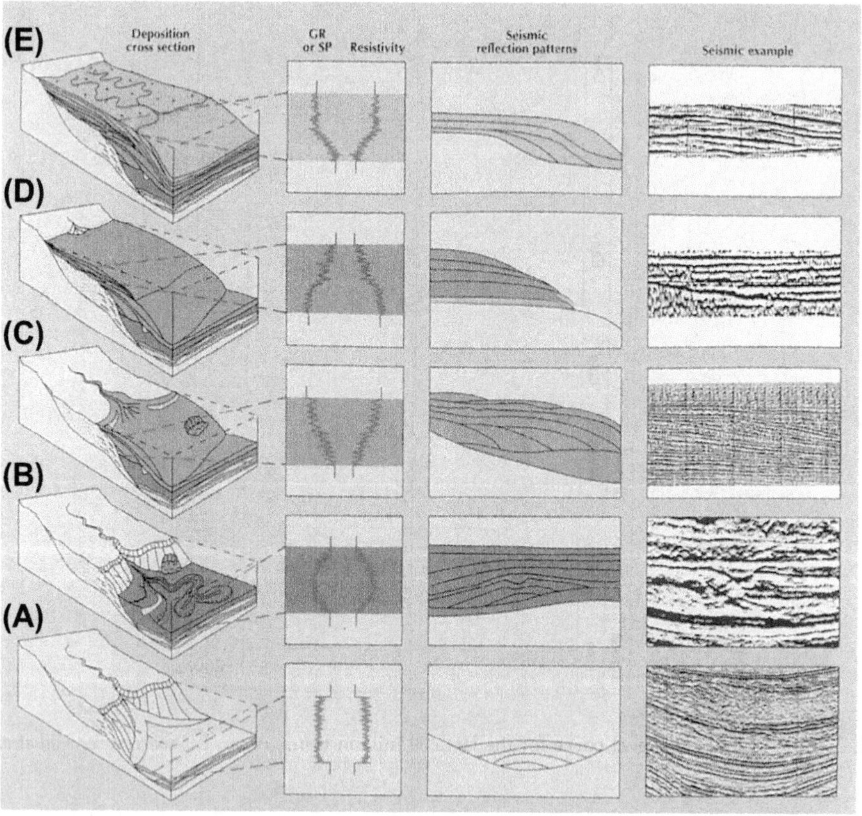

PLATE 3.4 Geophantasmograms, log motifs, seismic reflection patterns, and examples to illustrate a third-order stratigraphic sequence. A = Low stand sea level. B = Rising sea level floods the continental slope. C = Sea floods shelf providing accommodation space for prograding deltas. D = Rapid flooding of the shelf permits deposition of a transgressive systems tract. E = Slow rising sea level permits the slow basinward progradation of thin high stand system tracts (See the color plate). *From Neal et al. (1993); courtesy of Schlumberger oil field review.*

onto the slope (Plate 3.4(B)). A continually rising sea level favors the progradation of clastic wedges out into the basin (Plate 3.4(C)). As sea level continues to rise, it will brim over the continental shelf edge rapidly flooding the continental shelf itself. The horizon at which the shelf is drowned is termed the *main* or *maximum flooding surface.* This may then be succeeded by a thin sequence of reworked shallow marine sands (Plate 3.4(D)). As the rate of rise of sea level begins to slow down toward the end of the cycle a second episode of progradation may occur, but this is thinner than the earlier basinfill phase, and restricted to the shelf itself (Plate 3.4(E)). Note how each stage in the cycle has a characteristic wireline log motif, and a characteristic seismic reflection pattern. The sediments deposited during the cycle are termed *systems tracts,* qualified by transgressive, high stand, shelf edge, and low stand, as appropriate (Fig. 3.70).

The fourth-order cycles, termed *parasequences,* are believed to span 10,000–50,000 years and are supposed to be driven by climatic cyclic events. Parasequences are what used to be termed *genetic increments,* and *parasequence* sets are what used to be called genetic sequences (Busch, 1971).

While many geologists have happily pedaled the cycles of sequence stratigraphy, some have viewed the underlying tenets of sequence stratigraphy critically. The two main areas of concern are the recognition of the hierarchy of cycles, and the dogma that seismic reflectors are of chronostratigraphic significance. These are now considered in turn.

Geologists are trained in pattern recognition. We see what we want to see. This was demonstrated in a rotten trick played on geologists by Zeller (1964). He generated an artificial sequence of coal measure sediments using random numbers taken from the last digit of the telephone directory of Lawrence, Kansas (i.e., 1 = coal, 2 = sandstone, 3 = siltstone, etc.). He gave these artificially generated stratigraphic sequences to geologists, who had no difficulty

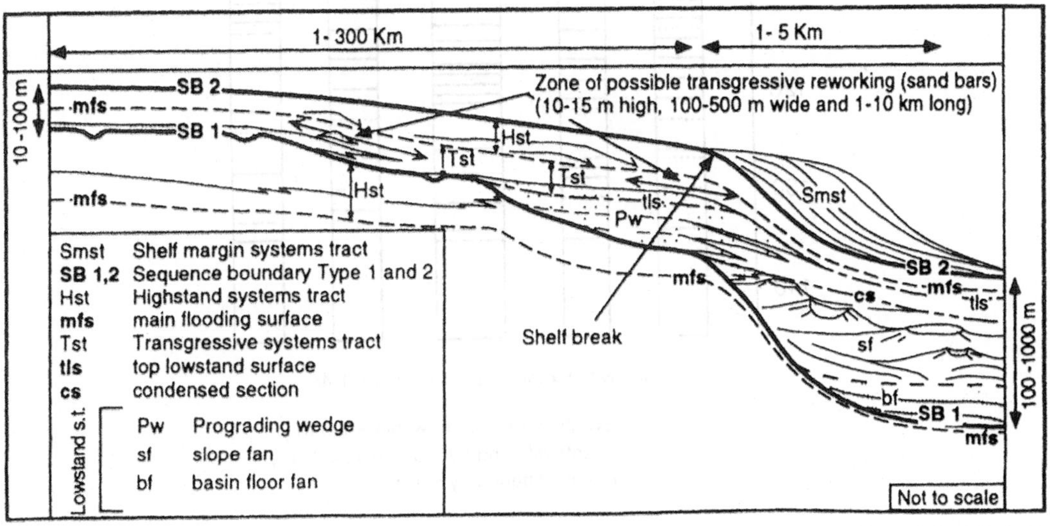

FIGURE 3.70 Sketch to show the terminology of sequence boundaries and systems tracts for a third-order cycle on a clastic continental shelf. *Reprinted from* Marine and Petroleum Geology, *Vol. 13, Reymond and Stempli, pp. 41–60. Copyright 1996, with kind permission from Elsevier Science Ltd.*

in correlating coal measure cyclothems from section to section. More recently, and more relevant to the present discussion, both Embry (1995) and Miall (1992) have queried the validity of the events on the Exxon sea level curve, arguing that they are too subjective, and based on circular arguments. Miall (1992) generated several artificial sequences, with random events. He then tested the statistical significance of these with the event boundaries on the Exxon sea level chart. A minimum of 77% successful correlations of random events with the Exxon curve was achieved (Fig. 3.71).

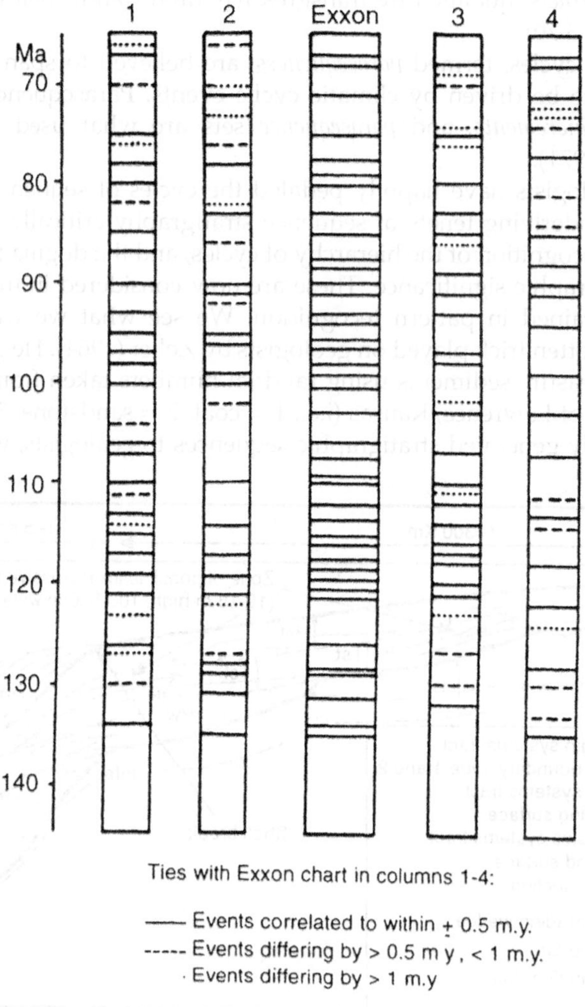

FIGURE 3.71 (Middle column) Event boundaries in the Exxon global cycle chart (Haq et al., 1988). Adjacent columns show events generated randomly. In the worst example in column 3, a 77% successful correlation of random events was achieved for intervals ±1 my from an Exxon event. *From Geology. Miall. Reproduced with permission of the publisher, The Geological Society of America, Boulder, Colorado USA. Copyright © 1992.*

Some geologists find it hard to accept that sea level changes are global and therefore can be correlated. Geologists know that the advances or retreat of a coastline may be caused by three processes. It may indeed be related to eustatic changes caused by fluctuations of the earth's temperature. It may also, however, be due to local tectonic processes, or even to changes in the rate and locus of deposition. Neotectonics shows how rapid and localized tectonic uplift and subsidence may be on a geological timescale. Today, for instance, the eastern coast of England is subject to severe coastal erosion and encroachment by the North Sea, while the western coasts show evidence of uplift. Synchronous advance and retreat of coastlines within a few kilometers of one another have been described from Spitzbergen (Martinsen, 1995), from New Zealand (Leckie, 1994), and are particularly well known in Louisiana. The modern Mississippi delta advances into the Gulf of Mexico, depositing a low stand systems tract. At the same time, however, only a few kilometers to the north, the sea advances across the subsiding older St Bernard delta, depositing a transgressive systems tract (Coleman and Gagliano, 1964). In some published sequence stratigraphic studies, which shall remain nameless, authors admit that the third-order cycles are not ubiquitous, and that they are identified in different wells by different criteria and are below the limits of seismic resolution.

The second tenet of sequence stratigraphy that makes geologists uneasy is that seismic reflectors are time horizons. Hubbard et al. (1985) have demonstrated that this is false when horizons are traced on a basinwide scale. On a smaller scale, Thorne (1992) and Tipper (1993) have demonstrated that this is erroneous on a smaller scale. Essentially, seismic waves are reflected from tangible changes in lithology, not from intangible time lines. The diachronous nature of formation boundaries is too well known to geologists for it is to be believed that lithological boundaries are time horizons. Tipper modeled the seismic responses for sand–shale bodies of different thickness, and for different seismic wavelength. He showed how seismic reflectors may clearly crosscut time lines (Fig. 3.72).

Thus, some of the fundamental assumptions of sequence stratigraphy are open to question, notably the tetracyclic nature of global transgressions and regressions, and the synchroneity of seismic reflectors. Nonetheless, sequence stratigraphy has made geologists question some of the fundamental assumptions about sedimentary processes. The main contribution of sequence stratigraphy to petroleum exploration has been that it has drawn attention to the fact that high stands of the sea favor the deposition of source rocks, for reasons discussed in Chapter 5. Conversely, a drop in sea level leads to the transport of sand off the continental shelf, to be deposited as submarine fans on adjacent basin floors (Shanmugam and Moiola, 1988).

3.3.5.2.2 SEISMIC FACIES ANALYSIS

Seismic facies analysis is the description and geological interpretation of seismic reflectors between sequence boundaries (Sieck and Self, 1977). It includes the analysis of parameters such as the configuration, continuity, amplitude, phase, frequency, and interval velocity. These variables give an indication of the lithology and sedimentary environment of the facies. Large-scale sedimentary features that may be recognized by the configuration of seismic reflectors include prograding deltas, carbonate shelf margins, and submarine fans (Sheriff, 1976). There are also characteristic features of reflectors that correlate with geophysical log motifs in the different systems tracts. Plate 3.4 illustrates examples of some of these.

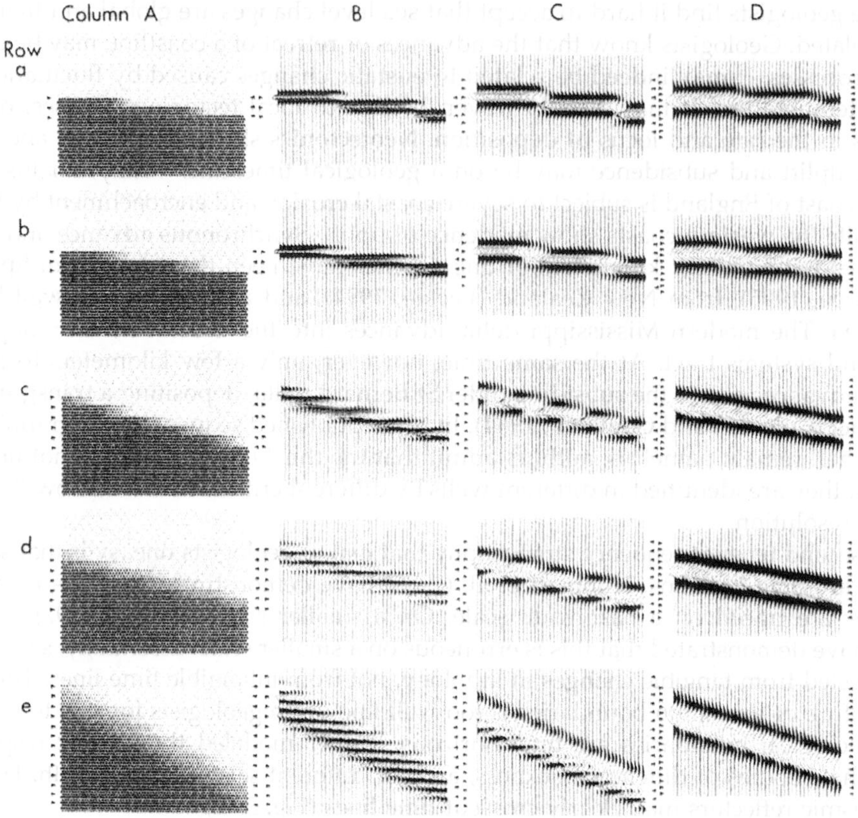

FIGURE 3.72 Synthetic seismic models for different sand:shale bodies using different wavelengths. Column A shows different shale (light tone) on sand (dark tone) transitions for uniform sedimentation, so that horizontal surfaces are synchronous. Rows a–e are for lithological transition belt width to lateral shift ratios of 1:4, 1:2, 1:1, 2:1. and 4:1., respectively. Columns B–D model these situations for different wavelength/bed thickness ratios of 1/2, 1/4, and 1/8, respectively. They have been depth converted, so that vertical intervals equal time. This study clearly shows instances where seismic reflections of lithological boundaries cross horizontal time surfaces. Some instances image the episodic nature of the lithological shift, but this is by no means ubiquitous. *From Tipper (1993). © Cambridge University Press, reprinted with permission.*

3.3.5.2.3 SEISMIC ATTRIBUTE ANALYSIS

The third and final stage of the hierarchy of seismic interpretation is seismic attribute analysis (Taner and Sheriff, 1977). This is concerned with study of wave shape, polarity, continuity, dip, frequency, phase, and amplitude. Seismic attribute analysis provides information related to structure, stratigraphy, and reservoir properties. Such analysis may give an indication of the thickness and of the nature of the upper and lower contacts of a sand bed. Comparison of observed seismic waves with synthetic traces computed from a geological model may give some insight into the depositional environment of the sand, and hence help to predict its geometry and internal reservoir characteristics (Fig. 3.73). Because of the advances

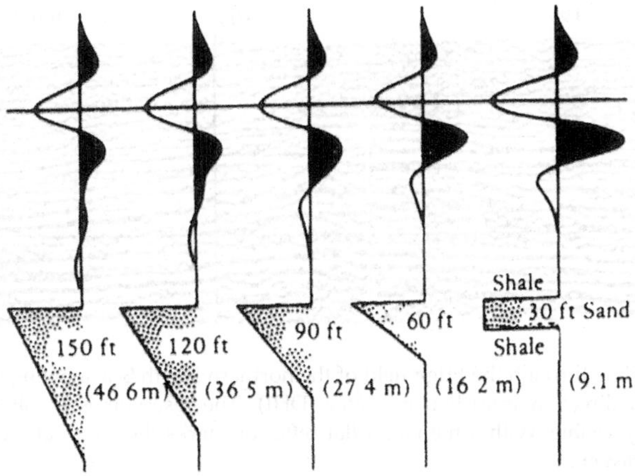

FIGURE 3.73 Seismic responses for different types of sand-shale sequences. These signals may give some indication of the depositional environment of a sand body. Those signals to the left suggest progradational deposition. The right-hand example is more typical of some channels. *From N. S. Neidell and E. Poggiagliomi, namely Fig. 17, p. 404 in AAPG Memoir Series No. 26, © 1977;* Stratigraphic Modeling and Interpretation–Geophysical Principles and Techniques.

made in 3D seismic methods, illustrated earlier, this type of application is of less use than heretofore.

The extraction of seismic attributes, such as amplitude, dip, curvature, and azimuth can produce remarkable 3D images of rock formations (e.g., Rijks and Jauffred, 1991). Refer again to Plate 3.3.

The simplest attribute is seismic amplitude generally reported as maximum (positive or negative) amplitude along a horizon picked. Amplitude in many cases corresponds directly to the porosity or saturation of the formation picked (oil/water vs gas).

Time/horizon attributes include coherence, dip, azimuth, and curvature. Coherence measures trace-to-trace similarity of the seismic waveform within a small analysis window. It is used to image faults, channels, and other discontinuous events. Dip and azimuth measure the direction of trace offset for maximum similarity. Curvature is a measure of how bent a surface is at a particular point and is closely related to the second derivative of a curve defining a surface.

Frequency attributes commonly separate seismic events based on their frequency content. The application of this attribute is called spectral decomposition. This attribute can be used to help determine bed thickness, presence of hydrocarbons, and fracture zones.

One application of attribute analysis that is particularly important, however, is in the direct recognition of hydrocarbons. Sometimes a single reflector cross cuts many parallel dipping reflectors. This is termed a *flat spot* (Backus and Chen, 1975). Flat spots are commonly discovered to be reflectors generated from petroleum/water contacts. Normally, they are produced by gas/water contacts, rather than by oil/water contacts, because of the greater impedance of the former. Figure 3.74 illustrates a celebrated flat spot produced by the gas/liquid contact of the Frigg field in the North Sea.

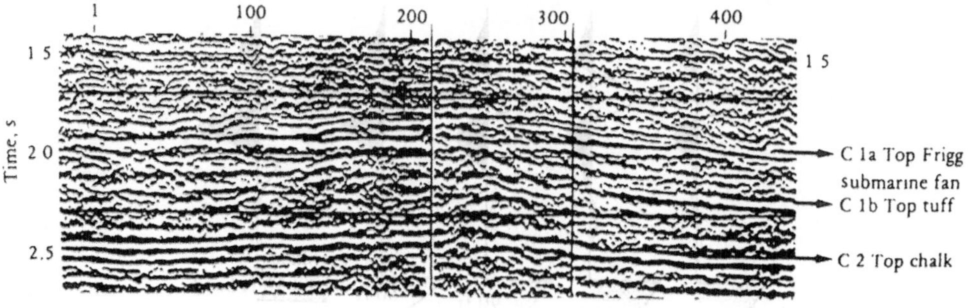

FIGURE 3.74 Seismic line through the Frigg field of the northern North Sea showing a famous example of a flat spot, a particular type of direct hydrocarbon indicator (DHI). The reservoir is a submarine fan, whose arched paleotopography is clearly visible. Within the fan, a flat reflector marks the gas: water contact. *From Hëritier et al. (1979). Reprinted with permission.*

Occasionally, however, structures exhibiting flat spots fail to reveal the presence of hydrocarbons. A possible explanation for these disappointments is that these are "ghosts" of old fluid contacts in traps that have leaked. As discussed in Chapter 6, petroleum inhibits cementation, but cementation of a reservoir commonly continues in the water zone beneath a hydrocarbon accumulation. Thus, if the petroleum leaks out of a trap, there may be a sufficient velocity contrast at the old fluid contact between cemented and uncemented sand to generate a reflecting horizon.

A flat spot is normally the result of a rapid increase in velocity caused by a seismic wave crossing from acoustically slow gas-saturated sand to faster water-saturated sand. Given the conventional display, whereby a downward increase in velocity generates a strong deflection of the seismic trace to the right, a flat spot appears as a black high amplitude reflector termed a *bright spot*. By contrast, the reflecting horizon over the crest of a gas sand may show a reversal of polarity, as the wave passes from faster to slower rock, the reverse of the normal situation (Fig. 3.75).

Thus, they calculate out as

$$A = \frac{(3,200 \times 2.15) - (2,300 \times 2.2)}{(3,200 \times 2.15) + (2,300 \times 2.2)} = +0.152,$$

$$B = \frac{(1,900 \times 2.1) - (2,300 \times 2.2)}{(1,900 \times 2.1) + (2,300 \times 2.2)} = -0.118,$$

$$C = \frac{(3,200 \times 2.\ 15) - (1,900 \times 2.1)}{(3,200 \times 2.15) + (1,900 \times 21)}.$$

The reverse of a bright spot is termed a *dim spot*. This is where a positive reflector locally reverses polarity. This might be seen, for example, over the crest of a limestone reef, and could be interpreted as indicating that the limestone becomes acoustically slower due to increased porosity, over the crest of the reef. This would obviously be an important indicator of improved reservoir characteristics due to solution or fracture-generated porosity.

One particular technique of verifying whether a bright spot is really a gas anomaly has been the analysis of the amplitude variation with offset (colloquially referred to as AVO

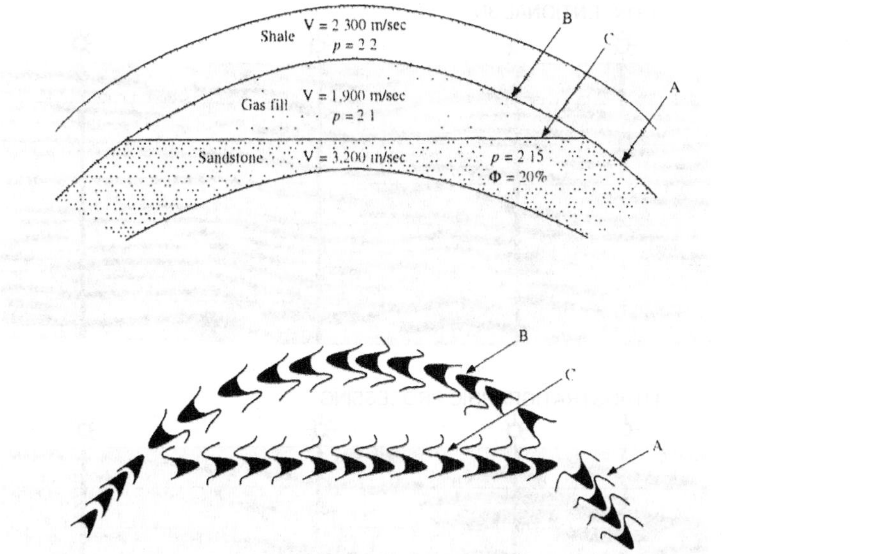

FIGURE 3.75 Illustration demonstrating the theory behind the formation of flat spots, and the reversal of polarity over a gas-saturated sand *(After Stone, 1977)*. A = reflection coefficient of shale on water-saturated sand, B = reflection coefficient of shale on gas-saturated sand, and C = reflection coefficient of gas-saturated sand on water-saturated sand. The reflection coefficient, described earlier in the text is $p_2v_2 - p_1v/p_1v_1 + p_2v_2$.

analysis). AVO analysis is a particular type of attribute analysis that measures variations in the Poisson ratio of different rock formations (D'Angelo, 1994). It consists of examining the amplitude of a reflection at increasing source–receiver distances. Gas sands commonly have a low Poisson ratio (0.1–0.2) compared with other lithologies such as shale (0.3–0.35). A large difference between the Poisson ratio of gas sands and shales produces a dramatic anomalous pattern of amplitudes versus offset (Fig. 3.76). These AVO sections are also referred to as *lithostratigraphic sections* (Hilterman and Sherwood, 1996).

Modern 3D seismic and inversion algorithms can yield information concerning rock strength parameters such as Young's modulus and Poisson's ratio. Principal stresses, vertical, maximum horizontal, and minimum horizontal, can also be analyzed.

Shale reservoir characterization is also the focus of 3D seismic interpretation. Thickness, fracture permeability, brittleness, organic richness, pore pressure, and more can be determined from seismic data (Chopra et al., 2014). Thickness can be determined by seismic stratigraphic interpretation and conversion from time data to depth. Fracture permeability can be determined from seismic discontinuity attributes (coherence and curvature). Fractures are correlated to areas of high strain as measured by coherence and curvature on poststack seismic data. Brittleness is exhibited by areas of high Young's modulus and low Poisson's ratio. Organic richness can be determined with the use of simultaneous inversion. Variations in Young's modulus and Poisson's ratio are expected due to variations in lithology, porosity, fluid content, and cementation. The density attribute helps with total organic carbon (TOC) characterization. Low densities imply high TOC.

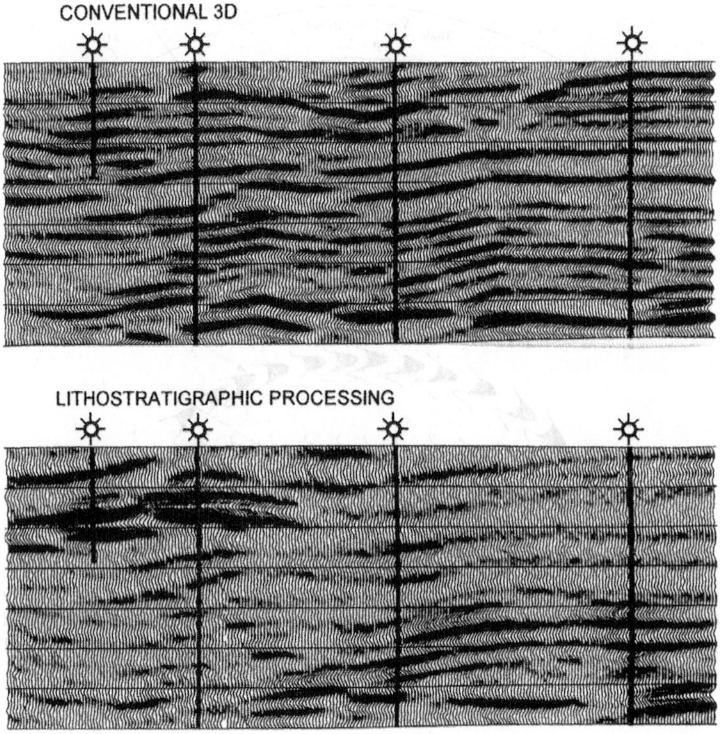

FIGURE 3.76 Comparison of conventional 3D and amplitude versus offset attribute lithostratigraphic processing. The latter highlights gas sands in the shallower parts of the two left-hand wells, and in the lower part of the two right-hand wells. *Courtesy of F. Hilterman and Geophysical Development Corporation.*

3.4 BOREHOLE GEOPHYSICS AND 4D SEISMIC

The early part of this chapter discussed the one-dimensional imaging of the subsurface using boreholes and geophysical well logs. The last section described the two- and three-dimensional imaging of the subsurface using seismic surveying. This section integrates geophysical logs with seismic surveys in what is now generally referred to as *borehole geophysics*. The integration of these two technologies opens up the feasibility of imaging petroleum in four dimensions, the fourth dimension being time.

3.4.1 The Vertical Seismic Profile

Earlier, it was shown how the sonic log could be used to calibrate seismic reflectors with the rock boundaries penetrated by a borehole. This is the starting point for an evolutionary sequence of techniques termed *vertical seismic profiles* (colloquially VSPs).

For more than half a century, it has been standard practice to place receivers at strategic depths in a borehole, and to record the time taken for sound to travel from an energy source at the surface. These velocity check shots are more reliable than counting the pips on a sonic log. In VSPs, the downhole geophones are more closely spaced than for the earlier check shots. Indeed the recovered data are sufficiently abundant that a simulated seismic section of the strata adjacent to the borehole can be generated (Fitch, 1987).

The simplest instance of this technique is the zero-offset VSP in which there is a single energy source at the surface located immediately adjacent to the borehole (Fig. 3.77). Over the years, however, this has been superseded by a spectrum of improvements (Christie et al., 1995). These include the offset VSP, in which the energy source is removed some distance from the well. For theoretical reasons, this will provide a maximum distance of coverage of half the offset distance. Next comes the walkaway VSP in which there is a series of surface energy sources. These can be arranged in a straight line away from the borehole, or may be arranged in radial patterns like the spokes of an umbrella. Walkaway VSPs can thus produce a 3D image of the strata adjacent to the borehole that can be calibrated with conventional 2D and 3D surveys. Further, not only can they image geology away from the side of the borehole but they can also image geology below the bottom of the borehole. Walkaway VSPs make it possible to image geological discontinuities, such as faults, the margins of salt domes and reefs. When run over deviated wells these surveys are referred to as walk-above VSPs.

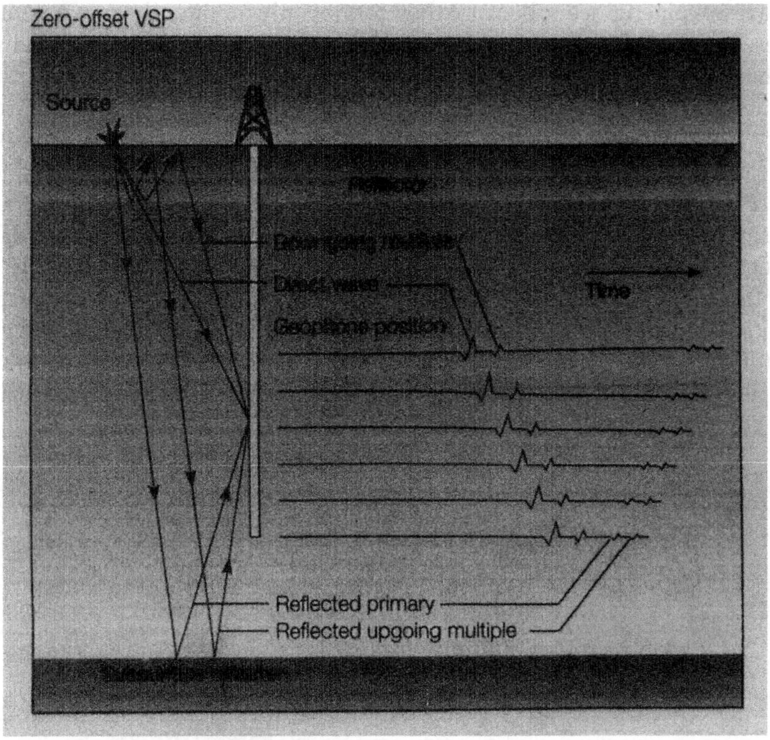

FIGURE 3.77 Illustration of the arrangement for a zero-offset vertical seismic profile (VSP). *Courtesy of Schlumberger oil field review.*

Specialized variants of the walkaway VSP have been developed for imaging salt domes, for measuring shear waves, and for listening to drilling noise (Fig. 3.78).

Traditionally, VSP surveys were run after a well was drilled. Now, however, they are run during the actual drilling of a borehole. Because VSP surveys record data adjacent to and below a borehole, it is possible to recalibrate continually estimates of the depth to geological horizons ahead of the drill bit (Borland et al., 1996). Data can be acquired by using a conventional energy source to generate a signal and also by using the noise generated by the drill bit, a technique termed the reverse VSP. The latter produces a continuous signal, and is of course cheaper than a conventional VSP. This technique has important implications, not only for predicting the depth to petroleum reservoirs but also more importantly for the depth of potentially perilous intervals of formation overpressure (Fig. 3.79). Thus, the acronym LWD is joined by seismic while drilling.

3.4.2 4D Seismic

Now, to add the fourth dimension, time, it is possible to shoot both 3D seismic and VSP surveys at regular intervals during the lifetime of a producing field. In fields where petroleum/water contacts may be imaged it thus becomes possible to produce what is, in effect, a time lapse film recording the movement of fluids during the life of a field (Archer et al., 1993). This is extremely useful for the reservoir engineers, because they cannot only monitor the movement of petroleum within the reservoir but also the movement of fluids that may have been injected into the field as part of an enhanced recovery scheme (as described in Chapter 6). Figure 3.80 illustrates the theory, showing computer-generated seismic sections before and after simulated fluid production, and the difference in signal extracted from the "before" and "after" surveys. Figure 3.81 illustrates a VSP shot for a well in the Frigg field some years after it was discovered and production commenced. Plate 3.5 shows VSPs over a field shot some 18 months apart. These last two examples both clearly demonstrate the upward migration of the gas/liquid contact. 3D examples of time lapse 4D surveys have been published by Anderson et al. (1996).

Once an offshore field has been discovered, it is now becoming common practice to permanently position geophones on the seabed. Then, seismic surveys are shot before production begins and throughout the life of the field. Thus, a preproduction image of the fluid boundaries of the field may be mapped, and deviations from the primary model can be used to

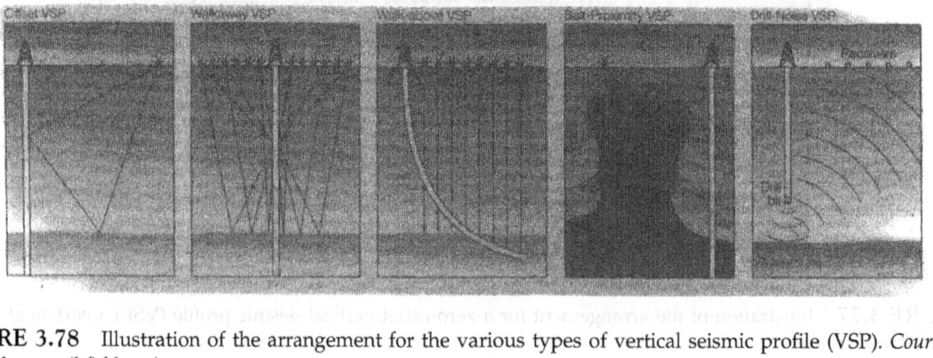

FIGURE 3.78 Illustration of the arrangement for the various types of vertical seismic profile (VSP). *Courtesy of Schlumberger oil field review.*

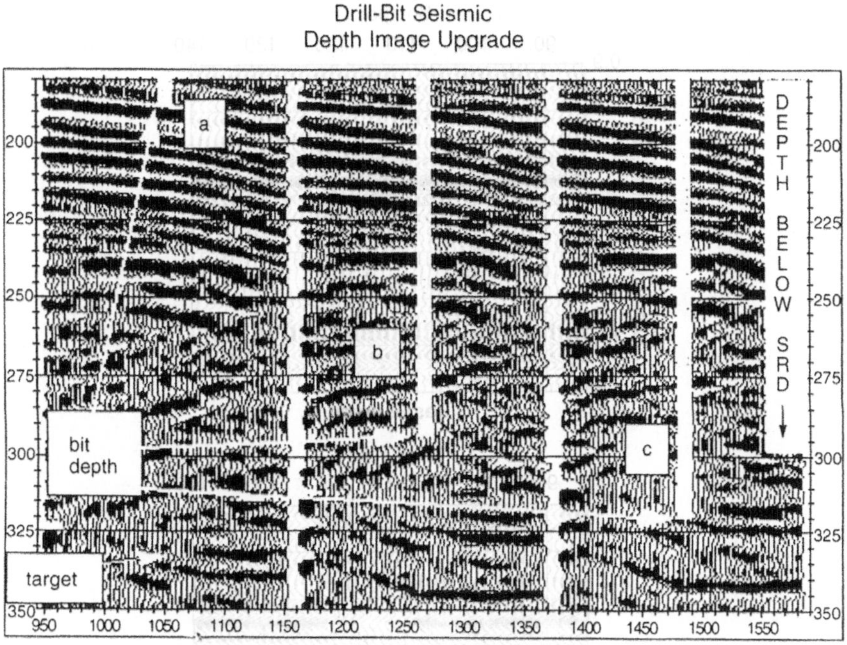

FIGURE 3.79 An example of how vertical seismic profiles (VSPs) shot at three times during the drilling of a well (A, B, and C) produce increasingly accurate predictions of a reflector marking the top of a zone of overpressure. *From Borland et al. (1996); courtesy of Schlumberger.*

interpret changes in the rock properties and fluid geometries during the life of the field. Examples of this practice from fields in the North Sea and West Shetlands are cited in Berteussen (1994) and George (1995) respectively.

3.5 SUBSURFACE GEOLOGY

Successful petroleum exploration involves the integration of the wireline logs and geophysical surveys, described earlier in this chapter, with geological data and concepts. This section is concerned with the manipulation and presentation of subsurface geological information. It is written on the assumption that the reader is familiar with the basic principles of stratigraphy, correlation, and contour mapping. Many of the methods to be discussed are not, of course, peculiar to petroleum exploration, but have wider geological applications. Essentially, the two methods of representing geological data are cross-sections and maps. These methods are discussed in the following sections.

3.5.1 Geological Cross-Sections

Vertical cross-sections are extremely important in presenting geological data. This account begins with small-scale sections and well correlations and proceeds to regional sections.

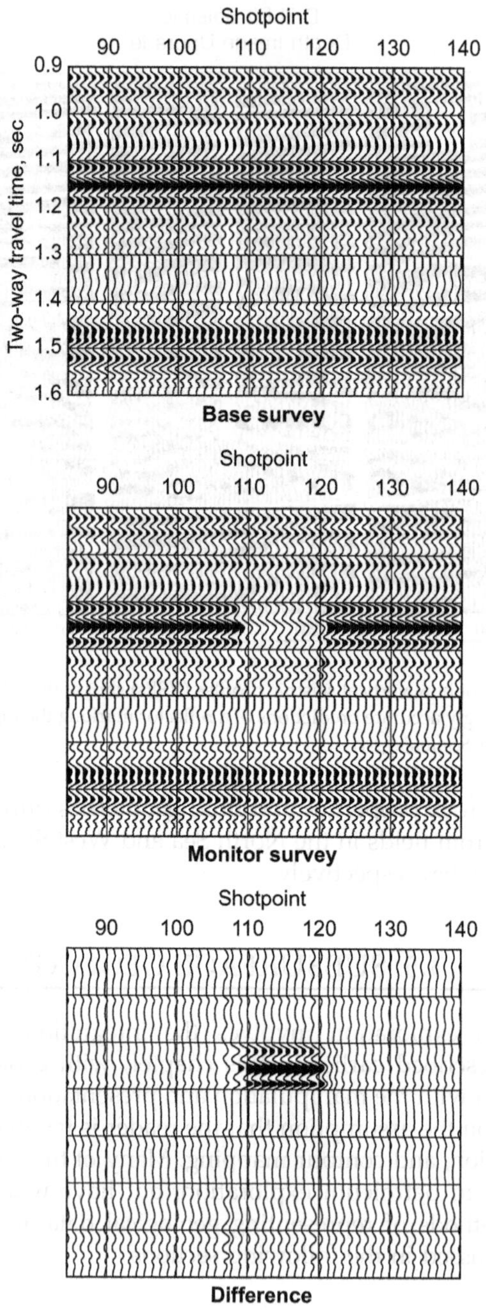

FIGURE 3.80 Computer-modeled seismic lines to show the response produced by the simulated production of petroleum from a reservoir horizon. *From Albright et al. (1994); courtesy of Schlumberger oil field review.*

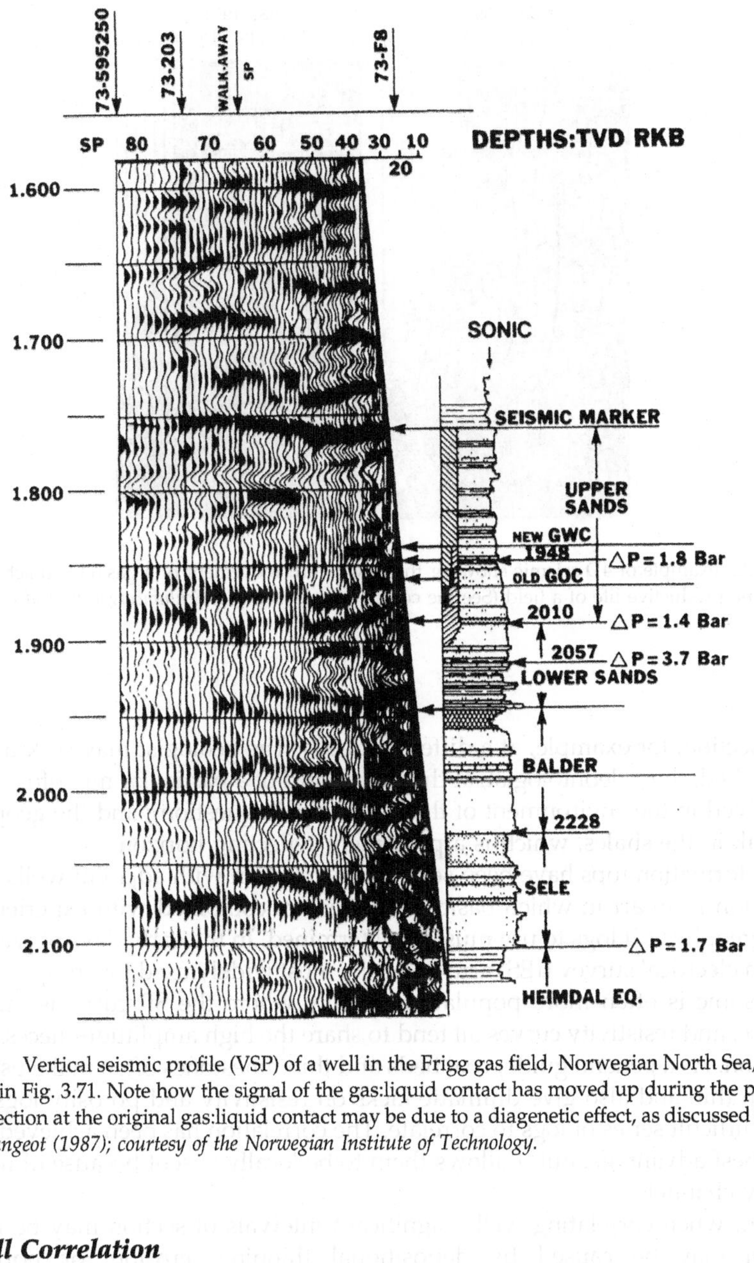

FIGURE 3.81 Vertical seismic profile (VSP) of a well in the Frigg gas field, Norwegian North Sea, whose flat spot was illustrated in Fig. 3.71. Note how the signal of the gas:liquid contact has moved up during the productive life of the field. A reflection at the original gas:liquid contact may be due to a diagenetic effect, as discussed in the text. *From Brewster and Jeangeot (1987); courtesy of the Norwegian Institute of Technology.*

3.5.1.1 Well Correlation

The starting point for detailed cross-section construction, as for example within a field, is consideration of well correlation. When a well has been drilled and logged, a composite log is prepared. This log correlates the geological data gathered from the well cuttings with that of the wireline logs. The formation tops then have to be picked, which is not always an easy task. The geologist, the paleontologist, and the geophysicist may all pick the top of the

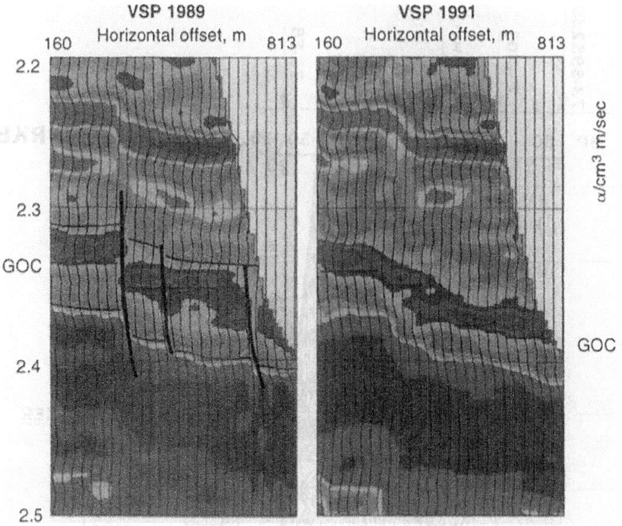

PLATE 3.5 An example of 4D seismic showing the downward movement of a gas:oil contact over a 22-month period during the productive life of a field (See the color plate). *Courtesy of Schlumberger oil field review.*

Cretaceous section, for example, at a different depth. The geologist may pick the top at, say, the first sand bed; the paleontologist, at the first record of a particular microfossil (which may not have thrived in the environment of the sediments in question); and the geophysicist, at a velocity break in the shales, which is a prominent reflecting horizon.

When the formation tops have been selected, correlation with adjacent wells can be made. This correlation is an art in which basic principles are combined with experience. First, the most useful geophysical logs to use must be determined. In the early days of well correlation, the induction electrical survey (IES) array was the most widely used for composite logs. Now, the gamma-sonic is often more popular, especially where the SP curve is ineffective. The gamma, sonic, and resistivity curves all tend to share the high amplitudes necessary for effective correlation. As a general guide, coal beds and thin limestones often make useful markers: they are both thin and also give dramatic kicks on resistivity and porosity logs. Figure 3.82 illustrates a difficult series of logs to correlate. The correlation has been achieved by using the coals to the best advantage, but it allows them to be locally absent because of nondeposition or erosion by channels.

Sometimes, when correlating wells, significant intervals of section may be missing. This phenomenon may be caused by depositional thinning, erosion, or normal faulting (Fig. 3.83). Examination of the appropriate seismic lines will generally reveal which of these three possibilities is the most likely. Repetition of sections may sometimes be noted. Repetition may be caused by reverse faulting, but this possibility should only be considered in regions known to have been subjected to compressional tectonics. An alternative explanation is that sedimentation is cyclic, causing the repetition of log motifs.

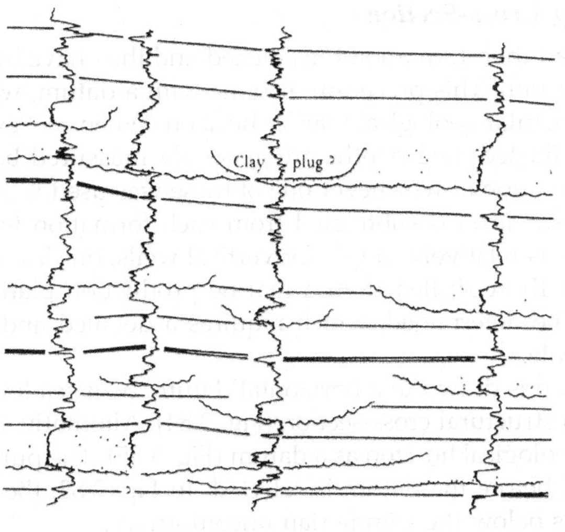

FIGURE 3.82 Correlation of wells in a complex deltaic reservoir. Note that the coals provide the best markers. They have been identified not only from the gamma logs shown here but also from porosity logs that are not illustrated in this figure. *After Rider and Laurier (1979).*

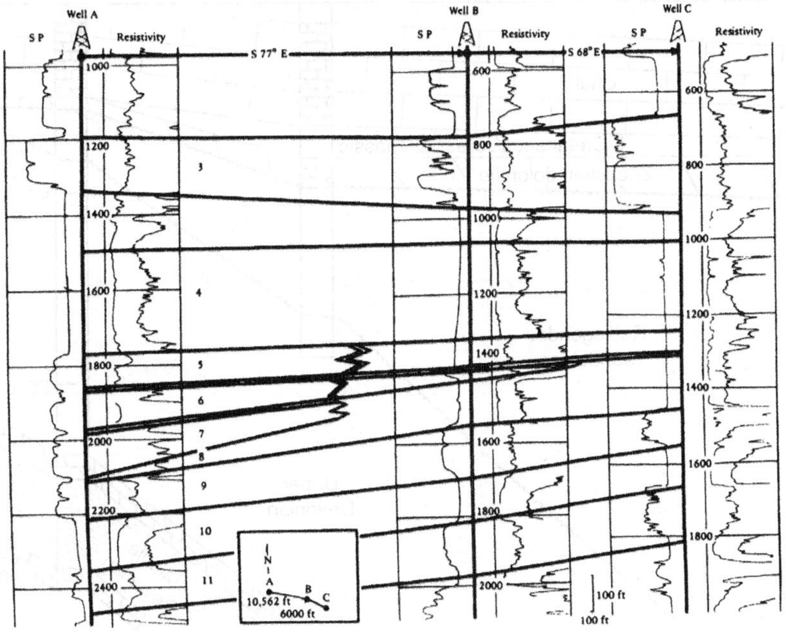

FIGURE 3.83 Correlation of three wells in the Niger delta showing lateral thinning and facies change. *Courtesy of Schlumberger.*

3.5.1.2 Constructing Cross-Sections

Once wells have had their formation tops picked and they have been correlated, they may be hung on a cross-section. This procedure is done with a datum, which may be sea level, a fluid contact, or a particular geological marker horizon. When sea level is used, the elevation of the log must be adjusted. Log depths are generally measured below the rotary table or kelly bushing. The elevation of whichever one of these was used is given on the log heading. The elevation above sea level is subtracted from each formation top to find its altitude or depth. This procedure is relatively simple for vertical wells, bur less so for deviated and horizontal wells, such as those drilled from a marine production platform. For these types of wells the TVD must be determined, which requires a detailed and accurate knowledge of the path of the borehole.

When a cross-section is drawn to a horizontal datum, be it sea level or a petroleum:water contact, the result is a structural cross-section (Fig. 3.84). Alternatively, a cross-section can be constructed using a geological horizon as a datum (Fig. 3.85). The purpose of the cross-section should be considered before the datum is selected. In Fig. 3.82, the purpose is to show the truncation of horizons below the Cimmerian unconformity.

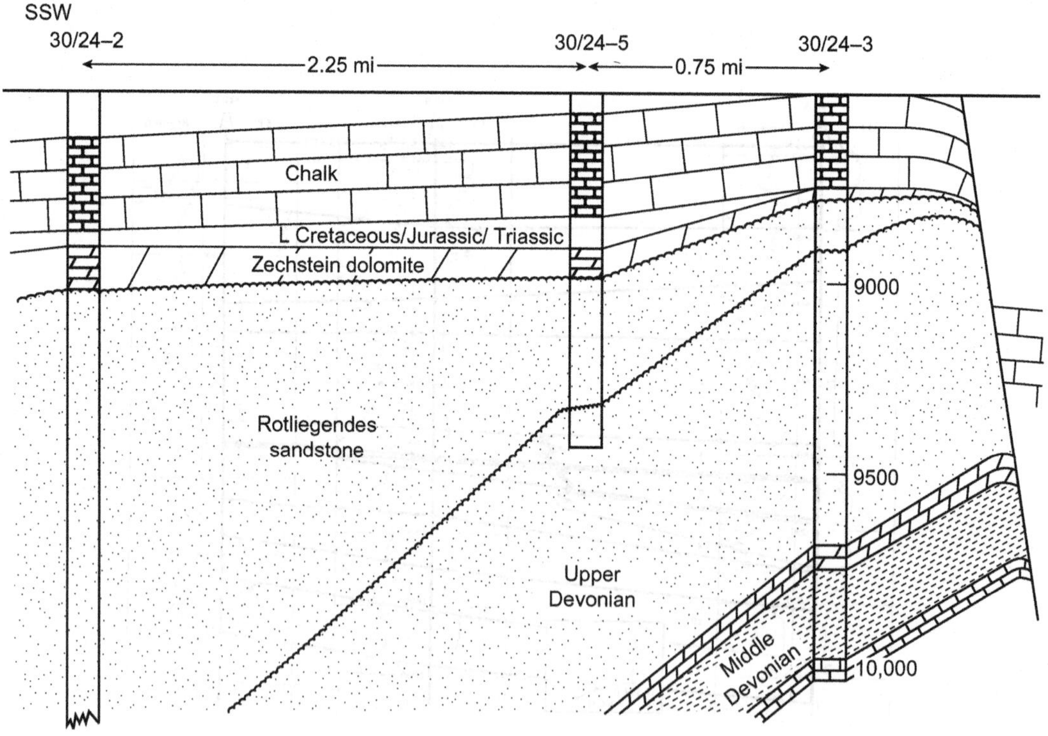

FIGURE 3.84 Structural cross-section through the Argyll field, North Sea. *After Pennington (1975).*

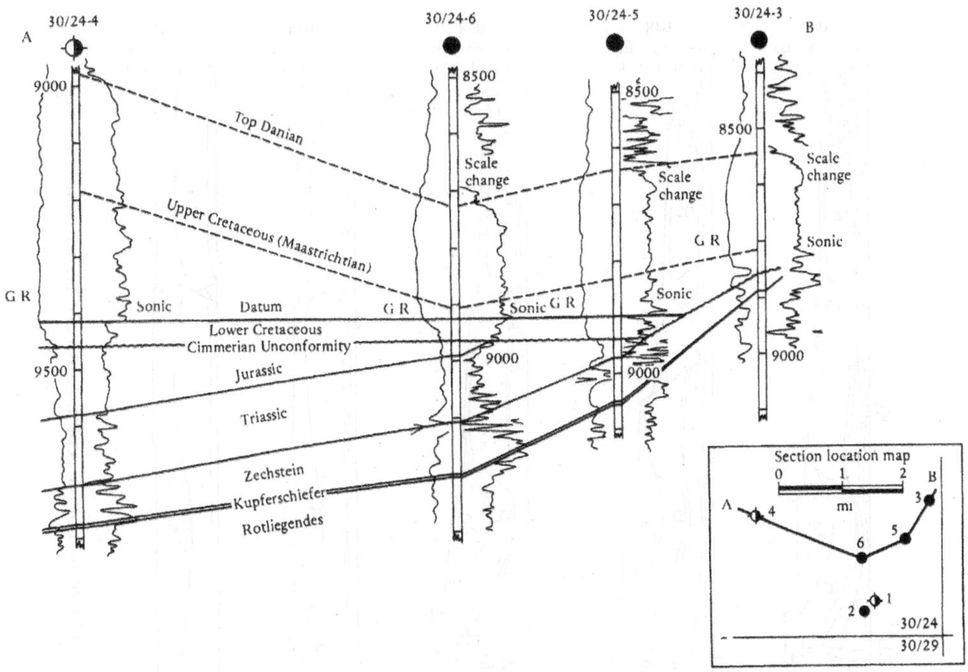

FIGURE 3.85 Cross-section of wells through the Argyll field of the North Sea, using the Cimmerian unconformity as a datum. This figure usefully demonstrates the pre-Lower Cretaceous truncation of strata. *After Pennington (1975).*

Cross-sections of fields are generally based on well control. For regional studies, a combination of seismic and well data is used. A development of the single cross-section is a series drawn using a sequence of different data horizons. Where these data horizons are selected to span a number of markers up to the present day, they can be used to document the evolution of a basin or an individual structural feature (Fig. 3.86). More accurate, but complex, reconstructions can be made by decompacting the shale formations (selecting an appropriate decompaction formula), and adding in water depth across the line of section. This technique of "backstripping" cross-sections is used for basin modeling (Chapter 8).

Although in olden days cross-sections were constructed by geologists leaning over drawing tables with well logs, tracing paper, rulers, pencils, and erasers, nowadays, it is all done by computer (Fang et al., 1992).

3.5.2 Subsurface Geological Maps

Many different types of subsurface geological maps are used in oil exploration. Only a few of the more important kinds are reviewed in this section. The simplest, and probably most

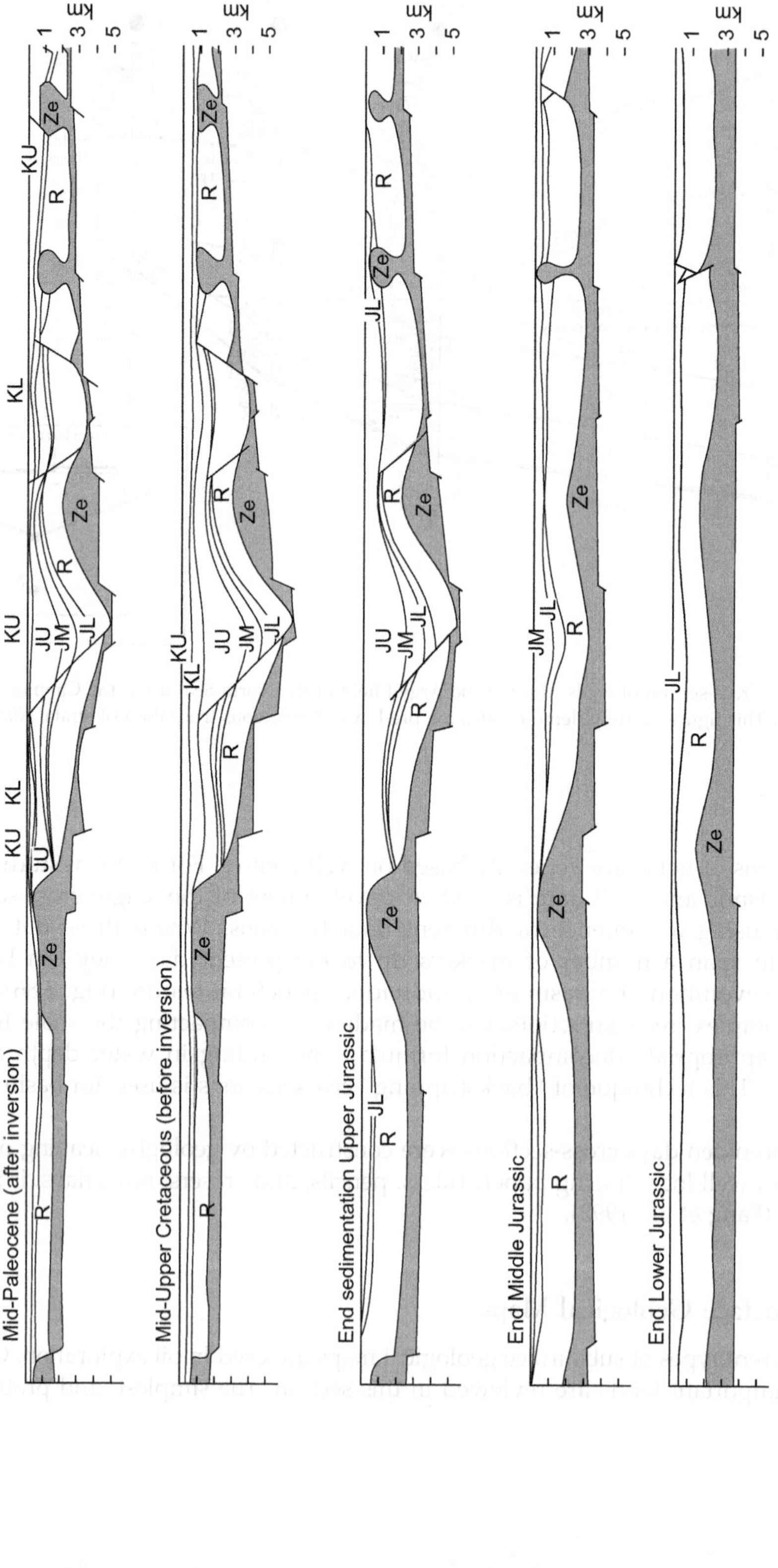

FIGURE 3.86 Regional cross-sections showing the structural evolution of the southern part of the central North Sea graben. *After Heybroek (1975).*

important, subsurface map is the structure contour map. This map shows the configuration of a particular horizon with respect to a particular datum, generally sea level. Structure contour maps may be regional or local, indicating the morphology of basins and traps, respectively. This information may be based on seismic data, well control, or, most effectively, a combination of both. Structure contour maps delineate traps and are essential for reserve calculations.

Next in importance are isopach maps, which record the thickness of formations. As with structural maps, they may be regional or local and tend to be most reliable when seismic and well data are integrated. Care must be taken when interpreting isopach maps. There is a natural tendency to think that thickness increases with basin subsidence, which is not necessarily so. Both terrigenous and carbonate sedimentation rates are often at a maximum a short way into a basin. Beds thin toward the shore, because of erosion, and basinward because of nondeposition or slow sedimentation rates. Likewise, local or regional truncation may result in formation thickness being unrelated to synde-positional basin morphology.

An isochore map is a particular type of isopach map that indicates the thickness of the interval between the oil:water contact and the cap rock of a trap. A related type of subsurface geological map is the net pay map, which contours the ratio of gross pay to net pay within a reservoir. Sand:shale ratio maps can be drawn, which are very useful both on a local and a regional scale, because they may indicate the source of sand and, therefore, areas where good reservoirs may be found. Net sand and net pay maps are essential for detailed evaluation of oil and gas fields. Field development also requires maps that contour porosity and permeability across a field in percent and millidarcies, respectively. These maps are generally based on well information, but, as seismic data improve, net sand and porosity maps can be constructed without well control. Obviously, their accuracy is enhanced when they can be calibrated with log information. Figure 3.87 illustrates the various types of maps just discussed for the Lower Permian (Rotliegendes) of the southern North Sea basin.

Net sand maps may often be combined with paleogeographic maps. Paleogeographic maps should be based on seismic and well data from which the depositional environment has been interpreted. These data may then be used to delineate the paleogeography of the area during the deposition of the sediments studied. Such maps can be used to predict the extent and quality of source rocks and reservoirs across a basin (Fig. 3.88).

Another useful type of map is the subcrop, or preunconformity map, which can be constructed from seismic and well data (Fig. 3.89). Regional subcrop maps are useful for showing where reservoirs are overlain by source rocks and vice versa. On a local scale, subcrop maps may be combined with isopach maps to delineate truncated reservoirs. Subcrop maps also give some indication of the structural deformation imposed on the underlying strata.

Finally, maps for all of the types previously discussed may be used in combination to delineate plays and prospects. A play map shows the probable geographic extent of oil or gas fields of a particular genetic type (i.e., a reef play, a rollover-anticline play, etc.). Figure 3.90 shows the extent of the Permian gas play of the southern North Sea. It selectively overlays several of the maps previously illustrated. Subject to a structure being present, the parameters that define this play include subcropping Carboniferous coal beds (the source rock), Lower Permian Rotliegendes of eolian type (the other facies lack adequate porosity and permeability), and capping Upper Permian Zechstein evaporites to prevent the gas from escaping. Prospect maps are also based on a judiciously selected combination of the

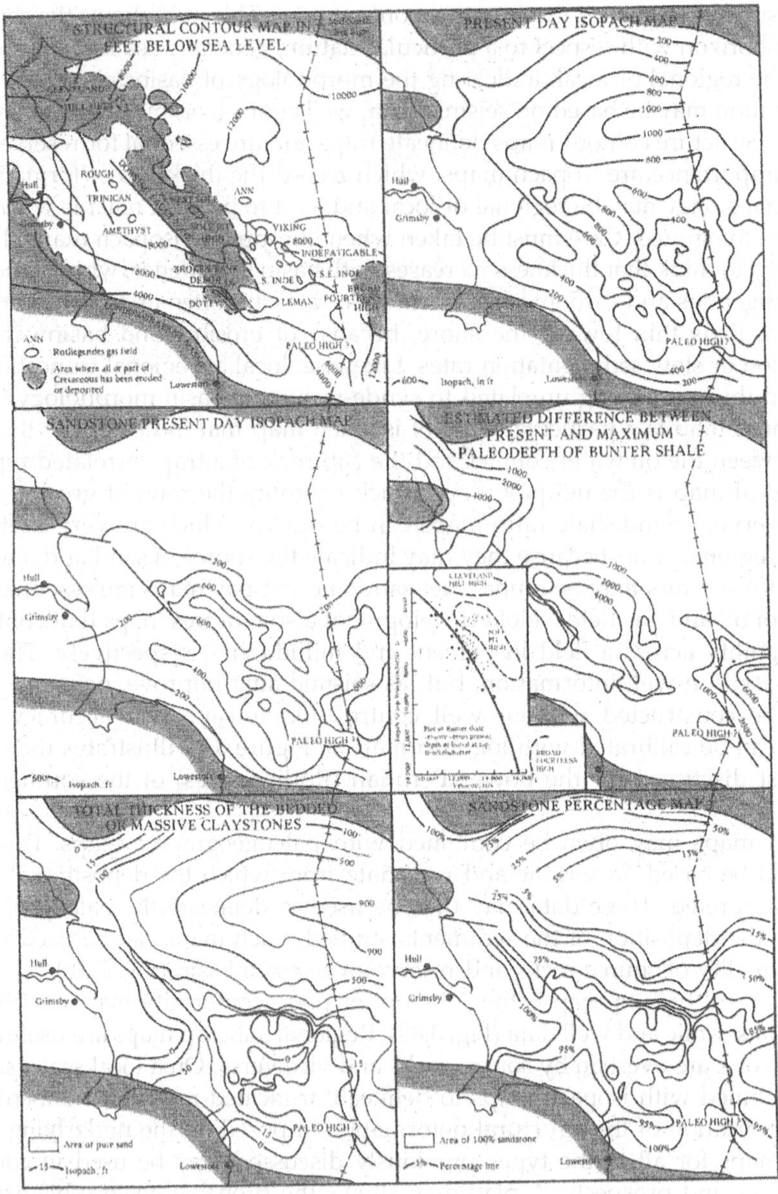

FIGURE 3.87 Maps of the Lower Permian (Rotliegendes) all constructed from the same basic data set. *After Marie (1975).*

previously discussed types of maps, but will, of necessity, contain a structure contour map to define the prospect in question.

The preceding review shows some of the subsurface geological maps that may be constructed and some of the problems of making and interpreting them. Examples of these maps are used throughout the following chapters.

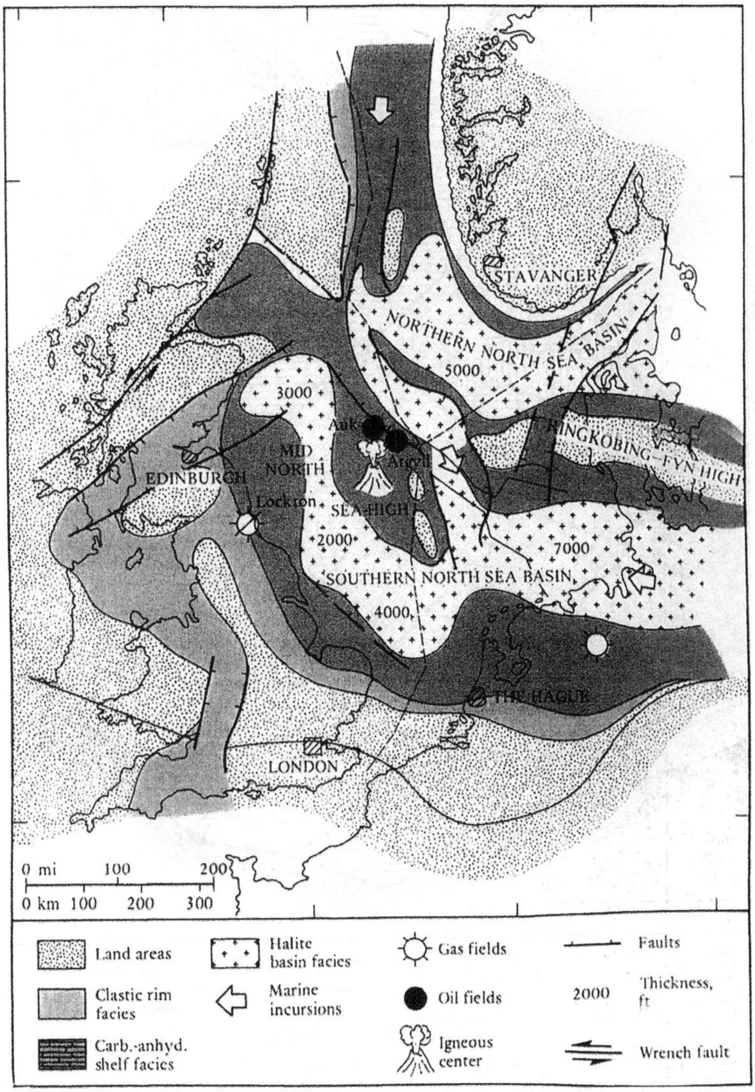

FIGURE 3.88 Paleogeographic map for the Zechstein (Upper Permian) of the North Sea. *After Ziegler (1975).*

3.6 REMOTE SENSING

Remote sensing is the collection of data without actual contact with the object being studied. Thus, in terms of geology, remote sensing includes aeromagnetic and gravity geophysical surveys, which have already been described. This section is concerned only with remote sensing using electromagnetic waves.

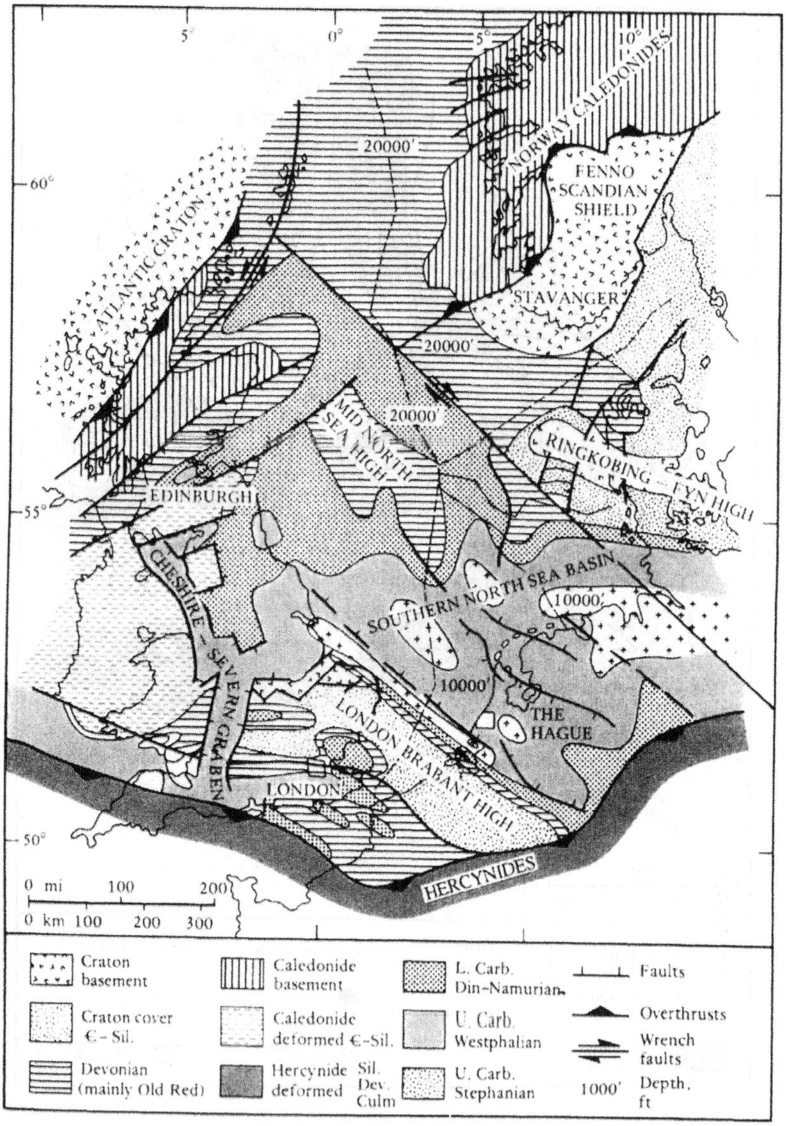

FIGURE 3.89 Pre-Permian subcrop map of the North Sea showing the nature and extent of pre-Permian structural deformation and erosion. It also indicates the extent of the Upper Carboniferous coal measures, which are the source for gas trapped in overlying Permian sands. *After Ziegler (1975).*

The electromagnetic spectrum includes visible light, but ranges from cosmic radiation to radiowaves (Fig. 3.91). Remote sensing can be carried out from various elevations, ranging from treetop height to satellites in the upper atmosphere. Thus, the two crucial parameters to remote sensing using electromagnetic waves are elevation and the wavelength analyzed. However, these techniques contain many common factors. An energy source is essential; it

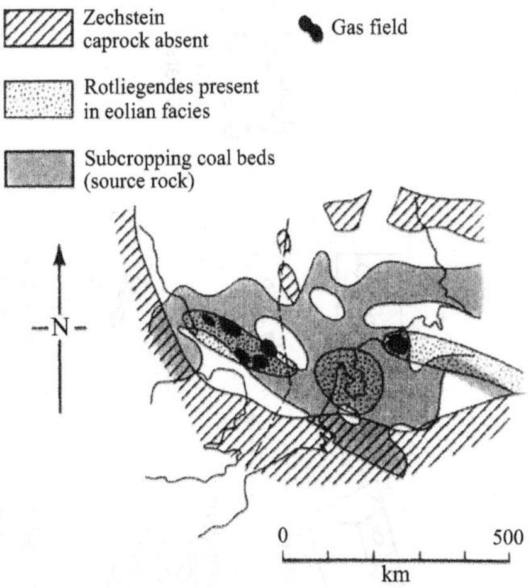

FIGURE 3.90 Play map showing the factors that control the habitat of the Permian Rotliegende gas fields of the southern North Sea.

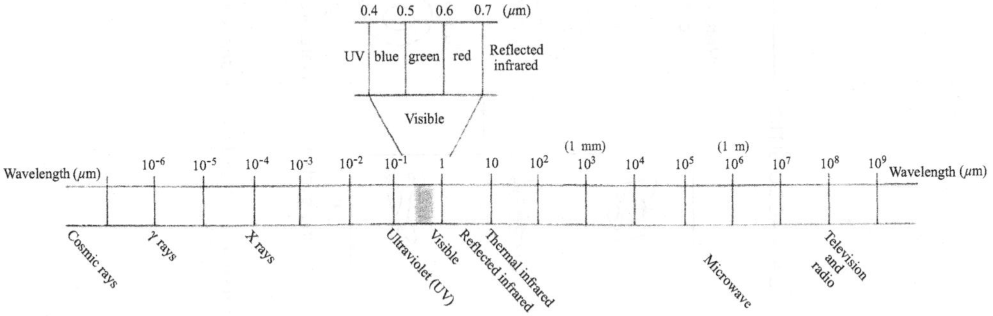

FIGURE 3.91 The electromagnetic spectrum. *After Lillesand and Kiefer (1994).* Remote Sensing and Image Interpretation, *© John Wiley & Sons, Inc.*

may be induced (as with microwave radiation used in radar) or natural (as with solar radiation). The sun's electromagnetic waves are filtered through the atmosphere and either absorbed by or reflected from the earth's surface. Reflected waves are modified by surface features of the earth. Some types of wave vary according to thermal variance of the surface, vegetation cover, geology, and so on. These wave variations can be measured photographically or numerically. They are then analyzed, either visually and subjectively or by more sophisticated computer processing. The results may then be interpreted (Fig. 3.92).

The field of remote sensing is large and rapidly expanding. Many new techniques are evolving and are being applied not only to petroleum exploration but to other geological surveys and the study of many other aspects of the earth's surface. The main remote sensing

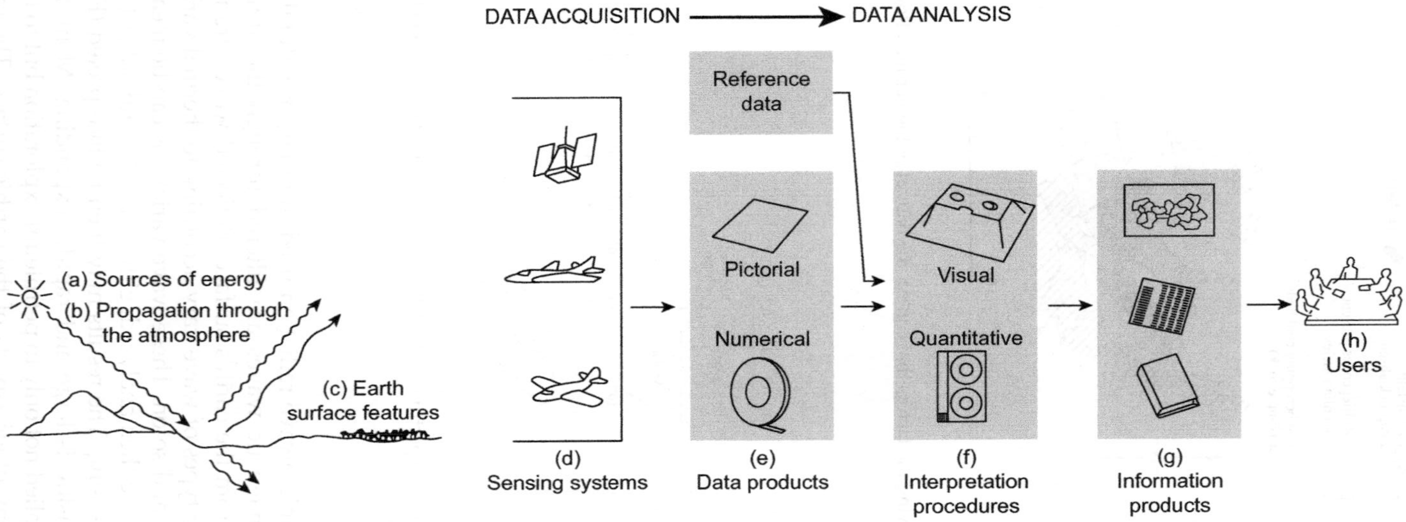

FIGURE 3.92 Electromagnetic sensing of earth resources. *After Lillesand and Keifer (1994).* Remote Sensing and Image Interpretation, © *John Wiley & Sons, Inc.*

techniques applied to petroleum exploration are visual, radar, and multispectral. These techniques are described in the following sections. For further details, see Lillesand and Kiefer (1994) and Sabins (1996).

3.6.1 Visual Remote Sensing

Conventional aerial photography is one of the oldest applications of remote sensing to petroleum exploration. The first aerial photographs were taken from balloons in the late nineteenth century. In the early part of the twentieth century, photographs were taken from planes, mainly as isolated snapshots used for map making. The discipline of aerial photography advanced rapidly for military purposes during the Second World War. By this time, a basic technique was established by which specially designed cameras took a sequence of photos at regular intervals along a plane's carefully controlled flight path, which must be accurately mapped. Photographs are taken at a sufficiently close interval to produce a 50–60% overlap (endlap within a line and sidelap between adjacent lines).

The photographs may be used in two ways. In one method, they can be fitted together in a mosaic, with the unwanted overlapping photograph carefully removed. The resultant mosaic may then be rephotographed for ease of reproduction. Air photograph mosaics may be used as a base for both topographic and geological surveys. Before a final map is prepared, however, the mosaic requires a number of corrections. The actual scale of the mosaic must be calibrated by reference to locations whose separation has been measured on the ground. The scale may not be constant over the mosaic if altitude has altered during the flight or if the elevation of the land surface varies. The terrain elevation is generally made with reference to sea level. The exact orientation and location of the mosaic on the earth's surface must also be determined. This task may be done by making astrofixes of ground locations, which can be clearly identified on the photograph.

A second technique widely employed in aerial photography is to use the overlap of adjacent photographs. Two overlaps can be viewed side by side stereoscopically (Fig. 3.93). Thus, the topography can actually be seen in three dimensions. In the early days of aerial photography, the photographs were viewed with a simple binocular stereoscope. In 2014, far more sophisticated viewers are available. Accurately measuring ground elevation with respect to a datum is now possible. Thus, spot heights may be digitized and located on a base map, or contours can be drawn straight from the photographs. Therefore, extremely accurate topographic and geological maps can be prepared.

Since topography is closely related to geology, aerial photographs can also be used in geological mapping. The boundaries of various rock units may be delineated, but, more importantly, the actual dip and strike of strata can be measured from air photographs, so the regional structure can be mapped. Further, linear features show up from the air far better than on the ground. Major structural lineaments, invisible on the ground, can be seen and traced for hundreds of kilometers using air photographs. At the other extreme of the scale, fracture systems may be detected from air photographs and their orientation and frequency numerically analyzed.

The preceding account pertains to the interpretation of air photographs shot from planes at elevations of only a few kilometers. Satellite photographs are useful in a rather different way.

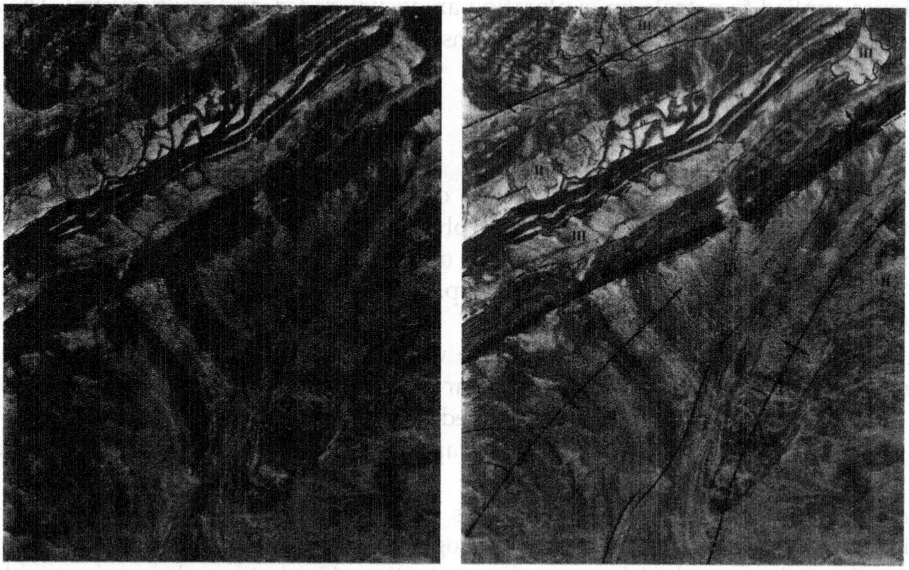

FIGURE 3.93 Stereo pair off air photographs with interpretation (right-hand photograph). The scale is approximately 1:30,000. I represents the oldest unit exposed, and consists of well-bedded limestones. II consists of sands and shales, and III represents superficial deposits. Several folds and faults can be mapped. *Courtesy of Huntings Geology and Geophysics Ltd.*

Because of the great height, stereoscopic mapping is not feasible. On the other hand, satellite photographs may show the presence of major lineations invisible from the ground or from the lower elevations of aircraft. Similarly, because of the large area covered, whole sedimentary basins may be seen in a single photograph, complete with concentric centripetally dipping strata.

3.6.2 Radar

Another method of remote sensing is provided by radar (an acronym for RAdio Detection And Ranging). Although aerial photography records light reflected from the sun, radar uses an energy source on the plane or satellite that emits microwave radiation. The returning radiation is recorded and displayed in a format similar to an aerial photograph.

Radar has several advantages over light photographs. Microwaves penetrate cloud cover and haze and can be used at night as well as during the day. Thus, radar surveys can be carried out with fewer operational constraints than can photographic surveys. A second advantage is that radar records data continuously, rather than in a series of separate photographs, so many of the problems of overlap and matching are eliminated. The spatial resolution of radar imagery is largely related to the diameter of the antenna. The larger the antenna, the better the resolution for a given wavelength. Because of the difficulty in mounting a rotating antenna on a plane, it is generally mounted underneath and pointed out to the side. Thus, most radar systems are side-looking airborne radar, referred to as SLAR or, more simply, SLR.

Radar images can, of course, not only be recorded from planes but also from satellites (Radarsat). The recent development of synthetic aperture radar imaging is now being deployed in satellites to produce maps for use in petroleum exploration surveys (Tack, 1996).

3.6.3 Multispectral Scanning

A conventional aerial photograph measures reflected light. Measuring the thermal radiation of the earth's surface using a thermal sensor is also possible. Multispectral scanning extends these concepts, measuring several different energy wavelengths simultaneously. Instead of recording light on film via an optic lens, the data are recorded electronically. This type of recording has the great advantage of eliminating the need to have the information transported to the ground in a film that has to be processed. Multispectral scanning can be carried out from planes, and is also extensively employed in satellites, notably the LANDSAT and SEASAT systems. The data can be transmitted back to a receiver on earth by radiowaves. The information can then be displayed visually in false-color photographs, which show variations not of light, but of whatever wavelength was selected for recording and analysis. Although light photographs show wavelengths of 0.3–0.9 μm, multispectral scanning can extend this value to some 14 μm.

Because the information is recorded electronically rather than optically, it can be processed by computers in many ways. Three principal steps are used in processing before the data can be interpreted. First, the image must be restored as closely as possible to the original. Image defects and the effects of haze are corrected. The location of the image with respect to the earth's surface must be exactly fixed, and adjacent images fitted into a mosaic. In the second stage of processing, the image is improved by enhancing the edges of different types of surface by central stretching, density slicing, and selecting colors for various spectral bands. In the third and final stage of processing, the information is extracted and displayed in hard copy form using false colors.

Multispectral scanning provides far more data than a conventional photograph does. Because such a wide range of spectra can be recorded and because the data can be processed in so many ways, the information is open to a great diversity of interpretation. Thus, it is possible to identify petroleum seeps from space by remote sensing (Plate 3.6).

3.6.4 Conclusion

Remote sensing is an invaluable technique in petroleum exploration. Indirectly, it is useful for topographic mapping; more directly, aerial photography has been widely used in geological mapping, especially in desert areas, where the effect of vegetation is minimal. The application of multispectral sensing from satellites requires great care to be taken in its interpretation. Thus, Halbouty (1980) showed how 15 giant oil and gas fields around the world are not revealed by LANDSAT imagery. Nonetheless, as Plate 3.8 shows, when appropriately processed, it may identify surface petroleum

Seeps on LANDSAT TM - Sulphur Springs (blue), NW Iraq

PLATE 3.6 A remotely sensed image of northwest Iraq from Landsat thematic mapper. False-color display. Fold structures can be seen crosscutting the central area from the northwest to the southeast. Petroleum-related sulfur springs are visible in blue (See the color plate). *Courtesy of World Geoscience Ltd.*

seepages. As with all exploration tools, multispectral sensing is open to abuse and misinterpretation when used on its own. When used in conjunction with other techniques, such as gravity and magnetics, it may delineate anomalies that deserve further attention on the ground.

References

Ager, D.V., 1973. The Nature of the *Stratigraphical* Record, first ed. Wiley, Chichester.

Ager, D.V., 1993. The Nature of the *Stratigraphical* Record, third ed. Wiley, Chichester.

Albright, J., Cassell, B., Dangerfield, J., Deflandre, J.P., Johnstad, S., Withers, W., January 1994. Seismic surveillance for monitoring reservoir changes. Schlumberger Oilfield Rev. 6, 4–14.

Anderson, R.N., Boulanger, A., He, Wei, Sun, Y.F., Xu, Liqing, Sibley, D., Austin, J., Woodhams, R.R., Andre, R., Rinehart, N.K., May 20 1996. Gulf of Mexico reservoir management. Oil & Gas J. 41–46.

Archer, S.H., King, G.A., Seymour, R.H., Uden, R.C., 1993. Seismic reservoir modelling— the potential. First Break 11, 391–397.

Archie, G.F., 1942. The electrical resistivity log as an aid in determining some reservoir characteristics. J. Pet. Technol. 5. Tech. Pap. No. 1422.

Backus, M.M., Chen, R.L., 1975. Flat spot exploration. Geophys. Prospect 23, 533–577.

Bailey, R.J., Buckley, J.S., Kielmas, M.M., 1975. Geomagnetic reconnaissance on the continental margin of the British Isles between 54° and 57° N. J. Geol. Soc. London 131, 275–282.

Berteussen, K.A., 1994. The use of permanent or semipermanent sea bottom acoustic sensors for reservoir seismic and reservoir surveillance. In: Aason, J. (Ed.), North Sea Oil and Gas Reservoirs III. Norwegian Petroleum Institute, Trondheim, pp. 125–129.
Bonner, S., 1992. Logging while drilling—images from the string. Schlumberger Oilfield Rev. 4, 4–21.
Bonner, S., 1996. Resistivity while drilling: a three year perspective. Schlumberger Oilfield Rev. 8, 4–19.
Borland, W.H., Hayashida, N., Kusaka, H., Leancy, W.S., Nakanashi, S., 1996. Integrating high resolution look-ahead VSP & drill bit seismic data to aid drilling decisions in the Tho Tinh structure—Nam Con Son Basin—Vietnam. In: Proc. 40th Anniv. Geophy., Vietnam Conf., Hanoi. Petroleum Exploration Society of Vietnam, pp. 1–16.
Brady, T.J., Campbell, N.D.H., Maher, C.E., 1980. Intisar 'D' oil field, Libya. AAPG Mem. 30, 543–564.
Brewster, J., Jeangeot, G., 1987. The production geology of the Frigg field. In: Kleppe, J. (Ed.), North Sea Oil and Gas Reservoirs. Graham & Trotman, London, pp. 75–88.
Brown, A.R., 1991. Interpretation of Three-Dimensional Seismic Data. AAPG Mem. 42, 1–341.
Brown, A.R., 2004. Interpretation of Three-Dimensional Seismic Data, sixth ed. AAPG Mem. 42. 1–541.
Brown, R.J.S., Gammon, B.W., 1960. Nuclear magnetism logging. J. Pet. Technol. 12, 199–201.
Burke, J.A., Campbell, R.L., Schmidt, A.T., 1969. The lithoporosity cross plot. Log Anal. 1–29. November–December.
Busch, D.A., 1971. Genetic units in delta prospecting. Am. Assoc. Pet. Geol. Bull. 55, 1137–1154.
Camden, D., 1994. NMR theory made simple. Dialog. London Petrophys. Soc. 2 (4), 4–5.
Campbell, R.L., 1968. Stratigraphic applications of dipmeter data in mid-continent. Am. Assoc. Pet. Geol. Bull. 52, 1700–1719.
Chopra, S., Sharma, R., Marfurt, K., 2014. Shale gas reservoir characterization workflows. AAPG Search Discov. 1–40. Article # 41266. http://www.searchanddiscovery.com/pdfz/documents/2014/41266chopra/ndx_chopra.pdf.html.
Christie, P., Ireson, D., Rutherford, J., Smith, N., Dodds, K., Johnson, L., Schaffner, J., 1995. Borehole seismic data sharpen the reservoir image. Oilfield Rev. 18–31. Winter.
Coldre, M.L., 1891. Les salines et les puits de feu de la province du Se-Tehouan. Ann. Mines 8 (19), 441–528.
Coleman, J.M., Gagliano, S.M., 1964. Cyclic sedimentation in the Mississippi river deltaic plain. Trans.—Gulf Coast Assoc. Geol. Soc. 14, 67–80.
Coope, D.F., 1994. Petrophysical applications of NMR. Dialog. London Petrophys. Soc. 2 (4), 8–10.
D'Angelo, R.M., 1994. An AVO study of North Sea data. In: Aason, J. (Ed.), North Sea Oil and Gas Reservoirs III. Norwegian Petroleum Institute, Trondheim, pp. 109–114.
Davies, R., Mathias, S.A., Moss, J., Hustoft, S., Newport, L., 2012. Hydraulic fractures: how far can they go? Mar. Pet. Geol. 37, 1–6.
Dickey, P.A., 1979. Petroleum Development Geology. PPC Books, Tulsa, OK.
Dobson, W.F., Seelye, D.R., 1981. Mining technology assists oil recovery from Wyoming field. J. Pet. Technol. 34, 259–265.
Embry, A.F., 1995. Sequence boundaries and sequence hierarchies: problems and perspectives. In: Steel, R.J., Felt, V.L., Johannsen, E.P., Mathieu, C. (Eds.), Sequence Stratigraphy on the Northwest European Margin. Elsevier, Amsterdam, pp. 1–11. NPF Spec. Pub. No. 5.
Fang, J.H., Chen, H.C., Shulz, A.W., Mahmoud, W., 1992. Computer-aided well log correlation. AAPG Bull. 76, 307–317.
Ferris, C., 1972. Use of gravity meters in search for traps. Mem.—Am. Assoc. Pet. Geol. 16, 252–270.
Fisher, K., Warpinski, N., 2011. Hydraulic Fractures—Height Growth. Real Data. Society of Petroleum Engineers, 145949.
Fitch, A.A., 1987. Use of the VSP (vertical seismic profile) in reservoir description. In: Kleppe, J. (Ed.), North Sea Oil and Gas Reservoirs. Graham & Trotman, London, pp. 109–120.
Fontenot, J.E., February 1986. Measurement while drilling—A new tool. J. Pet. Technol. 38, 128–130.
Gardner, G.H.F.L., Gardner, L.W., Gregory, A.R., 1974. Formation velocity and density—the diagnostic basics of stratigraphic traps. Geophysics 39, 770–780.
George, D., August 1993. Gravity and magnetic surveys compliment 3D seismic acquisitions. Offshore 58–64.
George, D., July 1995. First deepwater 4D seismic to be acquired over Foinaven. Offshore 31–32.
George, D., June 1996. Longer cable multi-streamer acquisitions key to greater accuracy in 3D seismic data. Offshore 64–69.
Gilreath, J.A., Maricelli, J.J., 1964. Detailed stratigraphic control through dip computation. Am. Assoc. Pet. Geol. Bull. 48, 1902–1910.
Goetz, J.F., Prins, W.J., Logar, J.F., 1978. Reservoir delineation by wireline techniques. In: Proc. Annu. Conv.—Indones. Pet. Assoc. 6, pp. 1–40.

Halbouty, M.T., 1980. Geologic significance of Landsat data for 15 giant oil and gas field. AAPG Bull. 64, 8–36.
Haq, B.U., Hardenbohl, J., Vail, P., 1988. Mesozoic and cenozoic chronostratigraphy and eustatic cycles. In: Wilgus, C.K., Posamentier, H., Ross, C.A., Kendall, C. G. St C. (Eds.), Sea-Level Changes: An Integrated Approach. Soc. Econ. Pal. Min. Spec. Publ. 42, pp. 71–108.
Harvey, P.K., Lovell, M.A., 1992. Downhole mineralogy logs: Mineral inversion methods and the problem of compositional colinearity. In: Hurst, A., Griffiths, C.M., Worthington, P.F. (Eds.), Geological Aspects of Wireline Logging 2. Geol. Soc. London, London, pp. 361–368. Spec. Publ.
Hassan, M., Hossin, A., Combaz, A., 1976. Fundamentals of the differential gamma ray log. In: Trans. Eur. Symp. Soc. Prof. Well Log Anal., 4th, Paper F, Society of Professional Well Log Analysts, pp. 1–18.
Heritier, F.E., Lossel, P., Wathne, E., 1979. Frigg field—large submarine fan trap in lower Eocene rocks of North sea Viking graben. AAPG Bull. 63, 1999–2020.
Heybroek, P., 1975. On the structure of the Dutch part of the Central North Sea Graben. In: Woodland, A.W. (Ed.), Petroleum and the Continental Shelf of North West Europe. Applied Science Publishers, London, pp. 339–352.
Hilterman, F.J., Sherwood, J.W.C., 1996. Identification of lithology in Gulf of Mexico Miocene rocks. OTC 7961, 83–90.
Hubbard, R.J., Pape, J., Roberts, D.G., 1985. Depositional sequence mapping as a technique to establish tectonic and stratigraphic framework and evaluate hydrocarbon potential on passive continental margin. AAPG Mem. 39, 79–92.
Imbert, P., 1828. Letters describing brine wells and natural gas of Szechuan province, China. Ann. Assoc. Propag. Foi. 3, 369–381.
Jageler, A.H., Matuszak, D.R., 1972. Use of well logs and dipmeters in stratigraphic trap exploration. Mem.—Am. Assoc. Pet. Geol. 10, 107–135.
Klemperer, S.L., Peddy, C., 1992. Chapter 13. In: Brown, G.C., Hawkesworth, C.J., Wilson, R.C.L. (Eds.), Understanding the Earth. Cambridge University Press, Cambridge, UK, pp. 35–74.
Koen, A.D., April 29 1996. Horizontal technology helps spark Louisiana's Austin Chalk trend. Oil Gas J. 15–19.
Krey, T., Marschall, R., 1975. Undershooting salt-domes in the North Sea. In: Woodland, A.W. (Ed.), Petroleum and the Continental Shelf of North West Europe. Applied Science Publishers, London, pp. 265–274.
Leckie, D.A., 1994. Canterbury Plains, New Zealand—Implications for sequence stratigraphic models. AAPG Bull. 78, 1240–1256.
Lillesand, T.M., Kiefer, R.W., 1994. Remote Sensing and Image Interpretation, third ed. Wiley, New York.
Marie, J.P.P., 1975. Rotliegendes stratigraphy and diagenesis, 1975. In: Woodland, A.W. (Ed.), Petroleum Geology and the Continental Shelf of North–West Europe. Applied Science Publishers, London, pp. 205–210.
Maroof, S.I., 1974. A Bouguer anomaly map of southern Great Britain and the Irish Sea. J. Geol. Soc. London 130, 471–474.
Martinsen, O.J., 1995. Sequence stratigraphy, three dimensions and philosophy. In: Steel, R.J., Felt, V.L., Johannsen, E.P., Mathieu, C. (Eds.), Sequence Stratigraphy on the Northwest European Margin. Elsevier, Amsterdam, pp. 23–29. NPF Spec. Pub. No. 5.
McDaniel, G.A., 1968. Application of sedimentary directional features and scalar properties to hydrocarbon exploration. Am. Assoc. Pet. Geol. Bull. 52, 1689–1699.
Messadie, G., 1995. Dictionary of Inventions. Wordsworth Reference. Chambers, Edinburgh.
Miall, A.D., 1992. Exxon global cycle chart: an event for every occasion? Geology 20, 787–790.
Neal, J., Risch, D., Vail, P., 1993. Sequence stratigraphy—A global theory for local success. Schlumberger Oilfield Rev. 5 (1), 51–62.
Neidell, N.S., Poggiagliolmi, E., 1977. Stratigraphic models from seismic data. Mem.—Am. Assoc. Pet. Geol. 26, 389–416.
Owen, E.W., 1975. Trek of the Oil Finders: A History of Exploration for Petroleum. Am. Assoc. Pet. Geol. Tulsa, OK.
Pacht, J.A., Bowen, B., Schaffer, B.L., Pottorf, W.R., 1993. Systems tracts, seismic facies and attribute analysis within a sequence stratigraphic framework—example from the offshore Gulf coast. In: Rhodes, E.G., Moslow, T.F. (Eds.), Marine Clastic Reservoirs. Springer-Verlag, Berlin, pp. 3–21.
Payton, C.E., 1977. Seismic stratigraphy—Applications to hydrocarbon exploration. Mem.—Am. Assoc. Pet. Geol. 26, 1–516.
Pennington, J.J., 1975. The geology of the Argyll field, 1975. In: Woodland, A.W. (Ed.), Petroleum and the Continental Shelf of North West Europe. Applied Science Publishers, London, pp. 285–294.
Ratcliff, D.W., September 27 1993. New technologies improve seismic images of salt bodies. Oil Gas J. 41–49.
Rider, M.H., 1990. Gamma-ray log shape used as a facies indicator: a critical analysis of an oversimplified methodology. Spec. Publ.—Geol. Soc. London 48, 27–37.

Rider, M.H., Laurier, D., 1979. Sedimentology using a computer treatment of well logs. In: Trans. Eur. Symp. Soc. Prof. Well Log Anal., 6th, J, pp. 1–12.

Rijks, E.J.H., Jauffred, J.C.E.M., September 1991. Attribute extraction: an important application in any detailed 3-D interpretation study. Geophysics 11–18.

Rygg, E., Riste, P., Nottvedt, A., Rod, K., Kristoffersen, Y., 1992. The Snowstreamer—a new device for acquisition of seismic data on land. In: Vorren, T.O., Bergsager, E., Dahl-Stamnes, O.A., Holter, E., Johansen, B., Lie, E., Lund, T.B. (Eds.), Arctic Geology and Petroleum Potential. Elsevier, Amsterdam, pp. 703–709. NPF Spec. Publ. No. 2.

Sabins, F.F., 1996. Remote Sensing: Principles and Interpretation, third ed. Freeman, San Francisco.

Selley, R.C., 1976. Subsurface environmental analysis of North sea sediments. AAPG Bull. 60, 184–195.

Selley, R.C., 1992. The third age of log analysis: application to reservoir diagenesis. In: Hurst, A., Griffiths, C.M., Worthington, P.F. (Eds.), Geological Aspects of Wireline Logging 2. Geol. Soc. London, London, pp. 377–387. Spec. Publ.

Selley, R.C., 1996. Ancient Sedimentary Environments, fourth ed. Chapman & Hall, London.

Serra, O., 1985. Sedimentary Environments from Wireline Logs. Schlumberger.

Shanmugam, G., Moiola, R.J., 1988. Submarine fans: characteristics, models, classification and reservoir potential. Earth Sci. Rev. 24, 383–428.

Sheriff, R.E., 1976. Inferring stratigraphy from seismic data. AAPG Bull. 60, 528–542.

Sieck, H.C., Self, G.W., 1977. Analysis of high resolution seismic data. In: Payton, C.E. (Ed.), Seismic Stratigraphy, Mem.—Am. Assoc. Pet. Geol., pp. 26,353–386.

Steel, R.J., Felt, V.L., Johannsen, E.P., Mathieu, C. (Eds.), 1995. Sequence Stratigraphy on the Northwest European Margin. Elsevier, Amsterdam. NPF Spec. Publ. No. 5.

Stille, H., 1924. Grundfragen der vergleichenden Tekronik. Borntraeger, Berlin.

Stone, C.S., 1977. 'Bright spot' techniques. In: Hobson, G.D. (Ed.), Developments in Petroleum Geology, vol. 1. Applied Science Publishers, London, pp. 275–292.

Tack, R.E., July 15 1996. Canada's commercially oriented radarsat returns SAR data for oil, gas exploration. Oil Gas J. 70–73.

Taner, M.T., Sheriff, R.E., 1977. Application of amplitude, frequency, and other attributes to stratigraphic and hydrocarbon determination. Mem.—Am. Assoc. Pet. Geol. 26, 301–327.

Thorne, J.A., 1992. An analysis of the implicit assumptions of the methodology of seismic sequence stratigraphy. AAPG Mem. 53, 375–394.

Tipper, J.C., 1993. Do seismic reflectors necessarily have chronostratigraphic significance? Geol. Mag. 130, 47–55.

Tucker, P.M., Yorston, H.J., 1973. "Pitfalls in Seismic Interpretation," Monogr. No. 2. Soc. Explor. Geophys.

Vail, P.R., Mitchum, R.M., Todd, R.G., Widmeir, J.M., Thompson, S., Sangree, J.B., Bubb, J.N., Hatledid, W.G., 1977. Seismic stratigraphy and global changes of sea level. Mem.—Am. Assoc. Pet. Geol. 26, 49–212.

Vail, P., Audemard, F., Bowman, S.A., Eisner, P.N., Perez-Cruz, C., 1991. The stratigraphic signatures of tectonics, eustasy and sedimentology—an overview. In: Einsele, G., et al. (Eds.), Cycles and Event Stratigraphy. Springer-Verlag, Berlin, pp. 617–659.

Van Wagoner, J.C., Posamentier, H.W., Mitchum, R.M., Vail, P.R., Sarg, J.F., Loutit, T.S., Hardenbol, J., 1988. An overview of the fundamentals of sequence stratigraphy and key definitions. Spec. Publ.—Soc. Econ. Paleontol. Mineral 42, 39–46.

Weimer, R.J., 1992. Developments in sequence stratigraphy: Foreland and Cratonic Basins. AAPG Bull. 76, 965–982.

Wendtland, R.F., Bhuyan, K., 1990. Estimation of mineralogy and lithology from geochemical log measurements. AAPG Bull. 74, 837–856.

Wharton, R.P., Hazen, G.A., Rau, R.N., Best, D.L., 1980. Electromagnetic propagation logging: advances in technique and interpretation. In: Soc. Pet. Eng. Pap. No. 9267, pp. 1–12.

Wyllie, M.R.J., 1963. The Fundamentals of Electric Log Interpretation, third ed. Academic-Press, New York.

Wyllie, M.R.J., Gregory, A.R., Gardner, G.H.F., 1956. Elastic wave velocities in heterogenous porous media. Geophysics 21, 41–70.

Wyllie, M.R.J., Gregory, A.R., Gardner, G.H.F., 1958. An experimental investigation of factors affecting elastic wave velocities in porous media. Geophysics 23, 459–493.

Zeller, E.J., 1964. Cycles and psychology. Geol. Surv. Kans. Bull. 169, 631–636.

Ziegler, W.H., 1975. Outline of the geological history of the North Sea. In: Woodland, A.W. (Ed.), Petroleum and the Continental Shelf of North West Europe, vol. 1. Applied Science Publishers, London, pp. 165–190.

Selected Bibliography

Oil Well Drilling and Production

Hyne, N.H., 1995. Nontechnical Guide to Petroleum Geology, Exploration, Drilling and Production. Penn Well Publ. Co., Tulsa, OK.

Formation Evaluation

Books and chart books of the various wireline service companies are essential (and free), but some assume an understanding of the subject. Other useful books include

Bateman, R.M., 1995. Open-Hole Log Analysis and Formation Evaluation. IHRDC, Boston.

Hurst, A., Lovell, M.A., Moreton, A.C. (Eds.), 1990. Geological Application of Wireline Logs. Geol. Soc. London, London. Spec. Publ. No. 48.

Hurst, A., Griffiths, C.M., Worthington, P.F. (Eds.), 1992. Geological Applications of Wireline Logs 2. Geol. Soc. London, London. Spec. Publ. No. 65.

Geophysics

Books that cover gravity, magnetic, and seismic surveying:

Backus, G., Parker, R., Constable, C., 1996. Foundations of Geomagnetism. Cambridge University Press, Cambridge, UK.

Milsom, J., 1991. Field Geophysics. Wiley, Chichester.

Telford, W.M., Geldart, L.P., Sheriff, R.E., 1991. Applied Geophysics, second ed. Cambridge University Press, Cambridge, UK.

For seismic prospecting only, see:

Doyle, H., 1995. Seismology. Wiley, Chichester.

Sheriff, R.E., Geldert, L.P., 1995. Exploration Seismology. Cambridge University Press, Cambridge, UK.

Remote Sensing

Lillesand, T.M., Kiefer, R.W., 1994. Remote Sensing and Image Interpretation, third ed. Wiley, New York.

Sabins, F.F., 1996. Remote Sensing: Principles and Interpretation, third ed. Freeman, San Francisco.

CHAPTER

4

The Subsurface Environment

Petroleum geology is largely concerned with the study of fluids, not just the oil and gas discussed in Chapter 2, but the waters with which they are associated and through which they move. Before examining the generation and migration of oil and gas in Chapter 5, the subsurface environment in which these processes operate should be considered.

This chapter begins with an account of subsurface waters, and then considers pressure and temperature and their effects on the condensation and evaporation of gas and oil. The chapter concludes by putting these ingredients together and discussing the dynamics of fluids in basins.

4.1 SUBSURFACE WATERS

Two types of water can be recognized in the subsurface by their mode of occurrence:

1. Free water
2. Interstitial, or irreducible, water

Free water is free to move in and out of pores in response to a pressure differential. Interstitial water, on the other hand, is bonded to mineral grains, both by attachment to atomic lattices as hydroxyl radicals, and as a discrete film of water. Interstitial water is often referred to as *irreducible water* because it cannot be removed during the production of oil or gas from a reservoir.

4.1.1 Analysis

Subsurface waters are analyzed for several specific reasons, apart from general curiosity. As discussed in Chapter 3, the measurement of the resistivity of formation water (R_w) is essential for the accurate assessment of S_w, and hence hydrocarbon saturation. Of course, R_w is closely related to salinity. Salinity varies both vertically and laterally across a basin. Salinity often increases with proximity to hydrocarbon reservoirs. Therefore, regional salinity maps may be an important exploration tool. Similarly, subsurface waters contain traces of dissolved hydrocarbon gases, whose content increases with proximity to hydrocarbon accumulations.

Subsurface waters can be analyzed in two ways. The total concentration of solids, or salinity, can be calculated from R_w by using the S.P. log, as discussed in Chapter 3.

Elements of Petroleum Geology
http://dx.doi.org/10.1016/B978-0-12-386031-6.00004-7

Alternatively, samples can be obtained from drill stem tests or during production. Care has to be taken when interpreting samples from drill stem tests because of the likelihood of contamination by mud filtrate.

4.1.2 Genesis

Traditionally, four types of subsurface water can be defined according to their genesis:

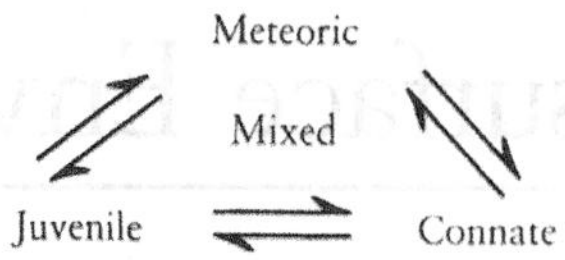

Meteoric waters occur near the earth's surface and are caused by the infiltration of rainwater. Their salinity, naturally, is negligible, and they tend to be oxidizing. Meteoric waters are often acidic because of dissolved humic, carbonic, and nitrous acid (from the atmosphere), although they may quickly become neutralized in the subsurface, especially when they flow through carbonate rocks. Connate waters are harder to define. They were originally thought of as residual seawater that was trapped during sedimentation. Current definitions proposed for connate waters include "interstitial water existing in the reservoir rock prior to disturbance by drilling" (Case, 1956) and "waters which have been buried in a closed hydraulic system and have not formed part of the hydraulic cycle for a long time" (White, 1957). Connate waters differ markedly from seawater both in concentration and chemistry.

Juvenile waters are of primary magmatic origin. It may be difficult to prove that such hydrothermal waters are indeed primary and have received no contamination whatsoever from connate waters. These three definitions lead naturally to the fourth class of subsurface waters—those of mixed origin. The mixed waters may be caused by the confluence of juvenile, connate, or meteoric waters. In most basins a transition zone exists between the surface aquifer and the deeper connate zone. The effect of this transition zone on the S.P. curve was discussed in Chapter 3.

4.1.3 Chemistry of Subsurface Waters

The four characteristics of subsurface water to consider are Eh, pH, concentration, and composition.

4.1.3.1 Eh and pH

The current data on the Eh and pH of subsurface waters are summarized in Fig. 4.1 (see Krumbein and Garrels (1952), Pirson (1983), Friedman and Sanders (1978)). The data show that rainwater is oxidizing and acidic. It generally contains oxygen, nitrogen, and carbon dioxide in solution, together with ammonium nitrate after thunderstorms.

As rainwater percolates into the soil it undergoes several changes as it becomes meteoric water. It tends to become reducing as it oxidizes organic matter. The pH of meteoric water may remain low in swampy environments because of the humic acids, but it increases in arid climates. If meteoric waters flow deep into the subsurface, they gradually dissolve salts

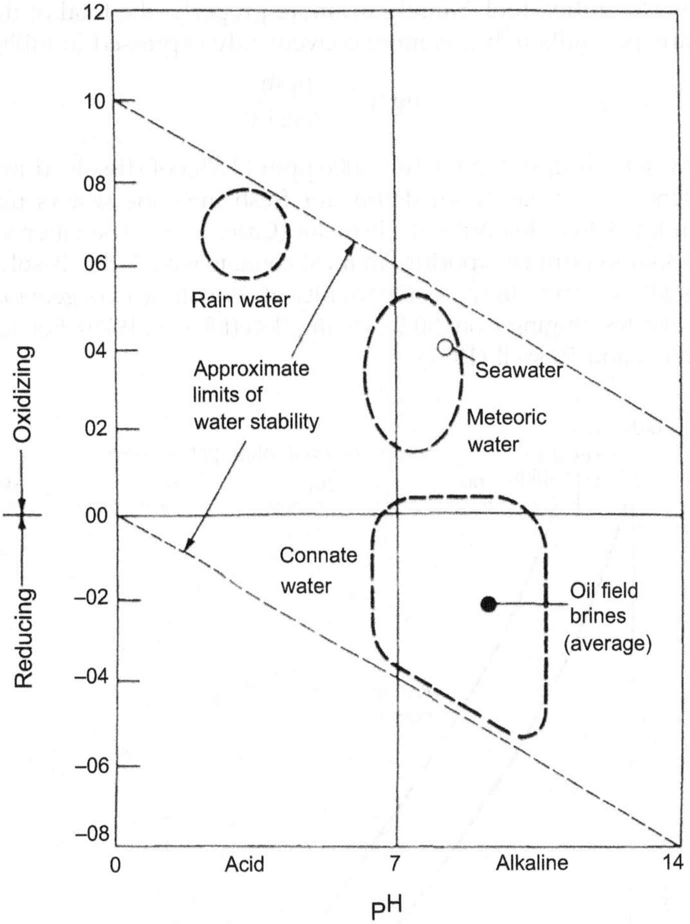

FIGURE 4.1 Oxidation:reduction potential—pH graph showing the approximate distribution of the various types of subsurface fluid.

and increase in pH. Deep connate waters show a wide range of Eh and pH values depending on their history, and particularly on the extent to which they have mixed with meteoric waters or contain paleoaquifers trapped beneath unconformities.

Oil field brines tend to be alkaline and strongly reducing. For further details on the Eh of subsurface fluids and its significance in petroleum exploration, see Pirson (1983). The Eh and pH of pore fluids control the precipitation and solution of clays and other diagenetic mineral cements. Obviously, the study of their relationship with diagenesis and porosity evolution is important, and is discussed in Chapter 6 in further detail.

4.1.3.2 Concentration

The significance of measuring the concentration of salts in subsurface waters has already been mentioned. Not only is it important for well evaluation, but the data may be plotted regionally as an exploration tool. Salinity or, more properly, the total of dissolved solids, is measured in parts per million, but is more conveniently expressed in milligrams per liter:

$$\mathrm{mg/l} = \frac{\mathrm{ppm}}{\mathrm{density}}.$$

Average seawater has approximately 35,000 ppm (3.5%) of dissolved minerals. Values in subsurface waters range from about 0 ppm for fresh meteoric waters up to 642,798 ppm for a brine from the Salina dolomite of Michigan (Case, 1956). The latter value is extremely high because of solution from evaporites. In most connate waters the dissolved solids content seldom exceeds 350,000 ppm. In sands the salinity of connate waters generally increases with depth (Fig. 4.2) at rates ranging from 80 to 300 mg/l m (Dickey, 1979). For sources of data, see Dickey (1966, 1969) and Russell (1951).

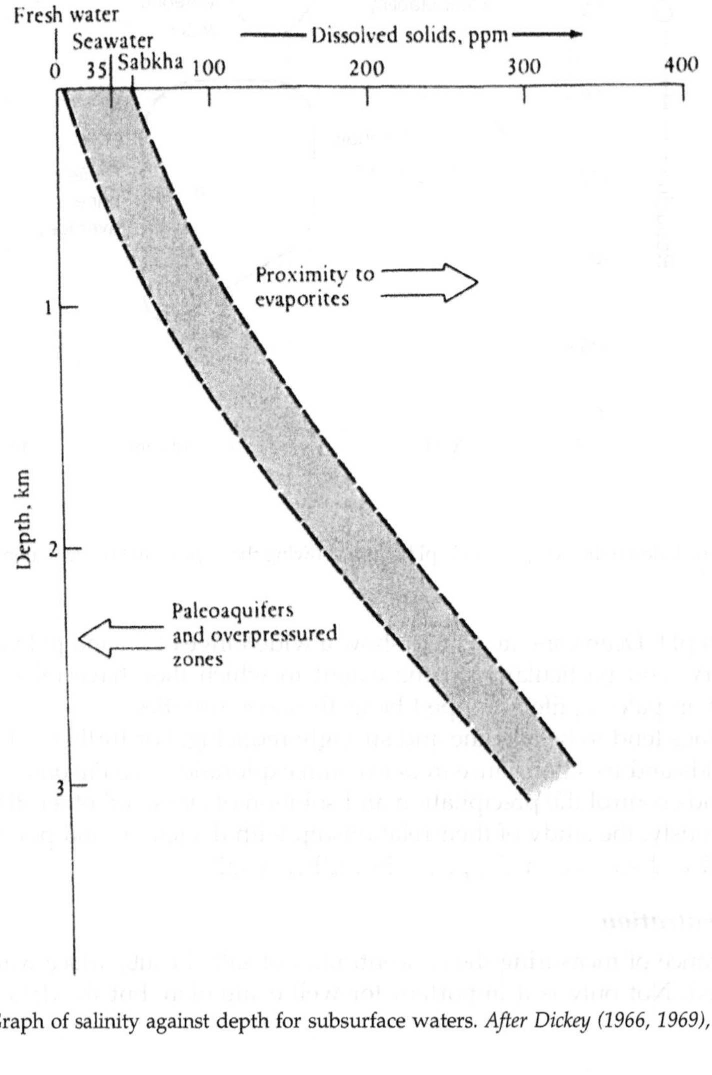

FIGURE 4.2 Graph of salinity against depth for subsurface waters. *After Dickey (1966, 1969), Russell (1951).*

Local reversals of this general trend have been seen and are attributable to two causes. Meteoric water may sometimes be trapped beneath an unconformity and is thus preserved as a paleoaquifer. A noted example of this occurrence has been documented from the sub-Cretaceous unconformity of Israel (Bentor, 1969). Here the beds above the unconformity have salinities of 60,000 ppm. The salinity drops to approximately 20,000 ppm beneath the unconformity before gradually increasing again with depth to more than 40,000 ppm.

Similarly in the North Sea there is good evidence that Cretaceous rainwater still lies, with little modification, as a paleoaquifer in Jurassic sands truncated by the Cimmerian unconformity (Macaulay et al., 1992). Reversals of increasing salinity with depth are also noted in zones of overpressure. This phenomenon is discussed in more detail later, but basically it reflects the fact that overpressured waters are trapped and cannot circulate. In shales, however, the increase in salinity with depth is less marked. The salinity of sand is often about three times that of the shales with which it may be interbedded (Dickey, 1979). This difference, together with the overall increase in salinity with depth, has been attributed to salt sieving (De Sitter, 1947). Shales may behave like semipermeable membranes. As water moves upward through compacting sediments, the shale prevents the salt ions from escaping from the sands, with the net result that salinity increases progressively with depth (see Magara (1977, 1978), for further details). Based on studies in the Gulf of Mexico and the Mackenzie Delta, Overton (1973) and Van Elsberg (1978) have defined four major subsurface environments:

Zone 1: Surface, depth of about 1 km; zone of circulating meteoric water. Salinity is fairly uniform.
Zone 2: Depth of about 1–3 km; salinity gradually increases with depth. Saline formation water is ionized (Fig. 4.2).
Zone 3: Depth greater than 3 km; chemically reducing environment in which hydrocarbons form. Salinity is uniform with increasing depth; may even decline if overpressured.
Zone 4: Incipient metamorphism with recrystallization of clays to micas.

Having finished our discussion of vertical salinity variations, it is now appropriate to consider lateral salinity changes. Salinity has long been known to tend to increase from the margins of a basin toward its center. Figure 4.3 illustrates a regional salinity map for the Mississippian Madison limestone aquifer in the Williston Basin of North Dakota and surrounding areas. Many instances of this salinity increase have been documented (Case, 1945; Youngs, 1975). This phenomenon is easy to explain: The basin margins are more susceptible to circulating meteoric water than is the basin center, where flow is negligible or coming from below. Regional isohaline maps can be a useful exploration tool, indicating areas of anomalously high salinity. These areas are presumably stagnant regions unaffected by meteoric flow, where oil and gas accumulations may have been preserved.

4.1.3.3 Composition

Subsurface waters contain varying concentrations of inorganic salts together with traces of organic compounds, including hydrocarbons. Table 4.1 presents analyses of many subsurface waters. For additional data, see Krejci-Graf (1962, 1978) and Collins (1975).

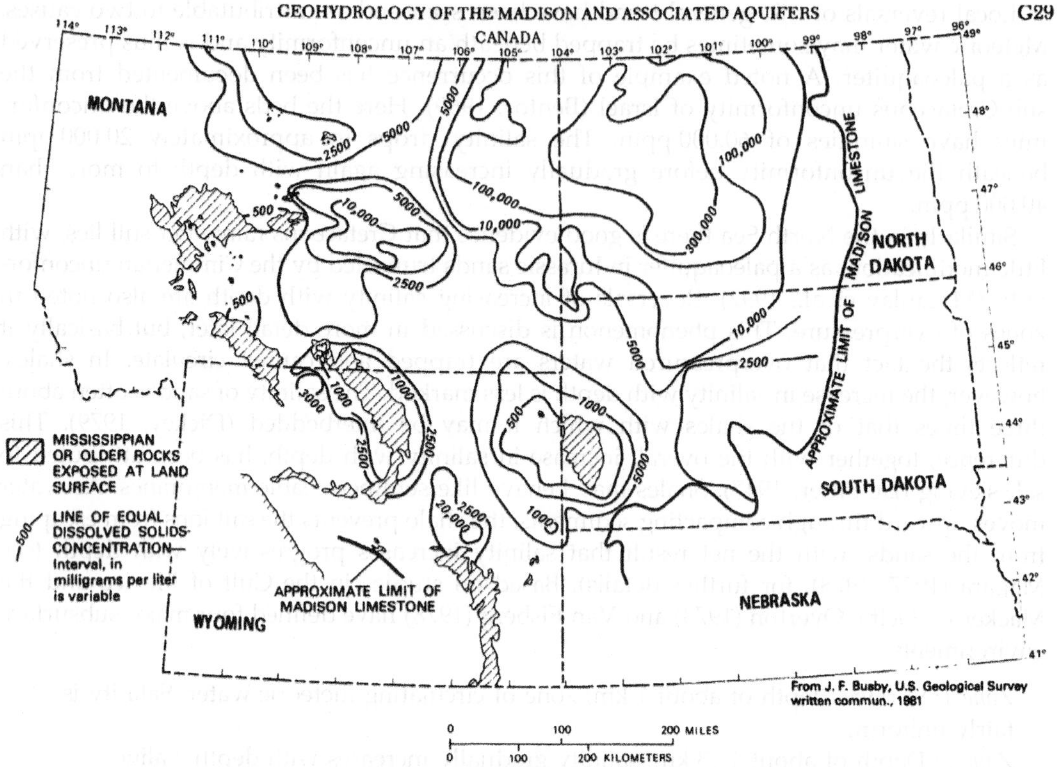

FIGURE 4.3 Regional salinity map for the Mississippian Madison limestone aquifer in the Williston Basin. Contour values are in parts per million of total dissolved solids; contour interval is variable. *From Downey (1984).*

Meteoric waters differ from connate waters not only in salinity but also in chemistry. Meteoric waters have higher concentrations of bicarbonate and sulfate ions and lower amounts of calcium and magnesium. These differences are the basis for the classification of subsurface waters proposed by Sulin (1946), which has been widely adopted by geologists (Table 4.2). For reviews of this classification and others, see Ostroff (1967).

Connate waters differ from seawater not only because they contain more dissolved solids but also in their chemistry. Connate waters contain a lower percentage of sulfates, magnesium, and often calcium (possibly caused by the precipitation of anhydrite, dolomite, and calcite) and a higher percentage of sodium, potassium, and chlorides than does seawater.

Because of the complex composition of subsurface waters, they can best be displayed and compared graphically. Schemes have been proposed by Tickell (1921) and Sulin (1946). The composition of seawater and typical connate waters are plotted according to these schemes in Fig. 4.4.

Connate waters also contain traces of dissolved hydrocarbons. The seminal work on this topic was published by Buckley et al. (1958). Basing their work on a study of hundreds of drill stem tests from the Gulf Coast, they found methane dissolved in subsurface waters at

TABLE 4.1 Representative Oil Field Water Analyses (ppm)

Pool	Reservoir rock, age	Cl^-	SO_4^-	CO_3^-	HCO_3^-	$Na^+ + K^+$	Ca^{2+}	Mg^{2+}	Total (ppm)
Seawater, ppm		19,350	2690	150		11,000	420	1300	35,000
Seawater, %		55.3	7.7	0.2		31.7	1.2	3.8	
Lagunillas, Western Venezuela	2000–3000 ft, Miocene	89	—	120	5263	2003	10	63	7548
Conroe, Texas	Conroe sands, Eocene	47,100	42	288		27,620	1865	553	77,468
East Texas	Woodbine sand, Upper Cretaceous	40,598	259	387		24,653	1432	335	68,964
Burgan, Kuwait	Sandstone, Cretaceous	95,275	198	–	360	46,191	10,158	2206	154,388
Rodessa, Texas, LA	Oolitic limestone, Lower Cretaceous	140,063	284	–	73	61,538	20,917	2874	225,749
Davenport, OK	Prue sand, Pennsylvanian	119,855	132	–	122	62,724	9977	1926	194,736
Bradford, PA	Bradford sand, Devonian	77,340	730	–	–	32,600	13,260	1940	125,870
Oklahoma City, OK	Simpson sand, Ordovician	184,387	286	–	18	91,603	18,753	3468	298,497
Garber, OK	Arbuckle limestone, Ordovician	139,496	352	–	43	60,7333	21,453	2791	224,868

From Geology of Petroleum by Leworsen. © 1967 by W. H. Freeman and Company. Used with permission.

concentrations of up to 14 standard cubic feet per barrel. They also recorded ethane and propane and very minor concentrations of heavier hydrocarbons. The amount of dissolved gases correlated with salinity, increasing with depth and from basin margin to center. Halos of gas-enriched connate waters were recorded around oil fields.

TABLE 4.2 Major Classes of Water by Sulin Classification

Types of water (V. A. Sulin)	Ratios of concentrations, expressed as milliequivalent percent $\frac{Na}{Cl}$	$\frac{Na-Cl}{So_4}$	$\frac{Cl-Na}{Mg}$
Meteoric			
Sulfate-sodium	>1	<1	<0
Bicarbonate-sodium	>1	>1	<0
Connate			
Chloride-magnesium	<1	<0	<1
Chloride-calcium	<1	<0	>1

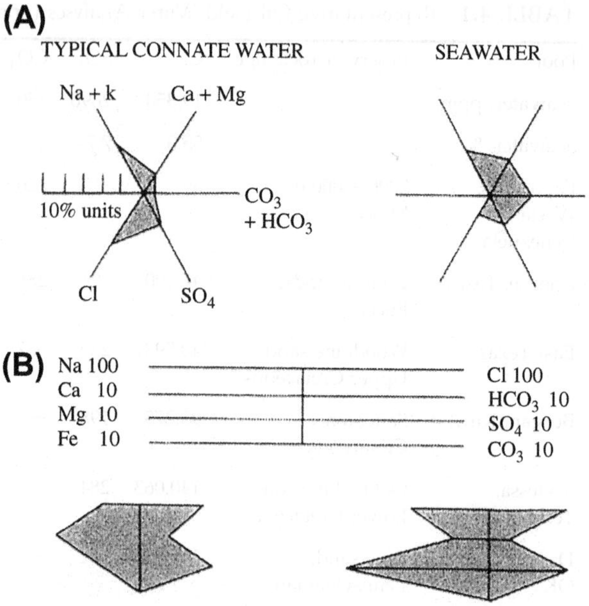

FIGURE 4.4 Methods of plotting water chemistry. (A) Tickell plots and (B) Sulin plots for a typical connate water sample (left) and seawater (right).

These data are of great significance for two reasons. They raise the possibility of regionally mapping dissolved gas content as a key to locating new oil and gas fields. These data also have some bearing on the problems of the migration of oil and gas (for an example, see Price (1980)). This topic is discussed in more detail in Chapter 5.

4.2 SUBSURFACE TEMPERATURES

4.2.1 Basic Principles

Temperature increases from the earth's surface toward its center. Bottom hole temperatures (BHTs) can be recorded from wells and are generally taken several times at each casing point. As each log is run, the BHT can be measured. It is important to take several readings at each depth because the mud at the bottom of the hole takes hours to warm up to the ambient temperature of the adjacent strata. Thus BHTs are recorded together with the number of hours since circulation. Figure 4.5 shows a BHT buildup curve. The true stabilized temperature can be determined from the Horner plot (Fertl and Wichmann, 1977). In this method the recorded temperature is plotted against the following ratio:

$$\frac{\Delta t}{(t + \Delta t)},$$

where Δt = number of hours since circulation and logging and t = hours of circulation at that depth. An example of a Horner plot is shown in Fig. 4.6. For a more detailed analysis of this topic, see Carstens and Finstad (1981).

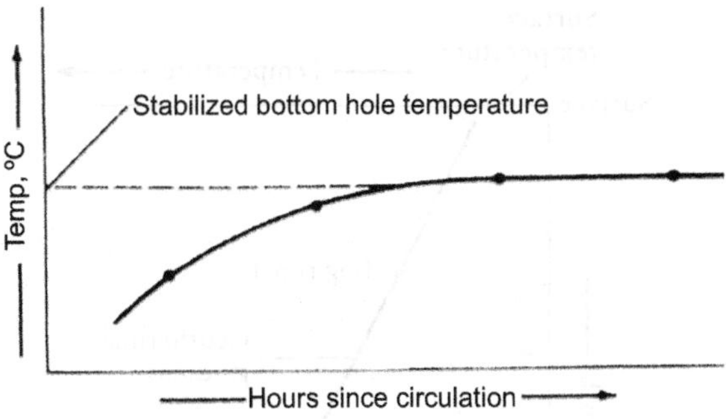

FIGURE 4.5 Graph showing how true BHT can only be determined from several readings taken many hours apart.

Once several corrected BHTs have been obtained as a well is drilled, they can be plotted against depth to calculate the geothermal gradient (Fig. 4.7). These values range from approximately 1.8 to 5.5 °C/100 m. The global average is about 2.6 °C/100 m. When several BHTs are plotted against depth for a well they may show that the geothermal gradient is not constant with depth. This discrepancy is generally caused by variations in the thermal conductivity of the penetrated strata. This relationship may be expressed as follows:

Table 4.3 shows the thermal conductivity of various sediments. Where sediments of different thermal conductivity are interbedded, the geothermal gradient will be different for each formation.

$$\text{Heat flow} = \text{geothermal gradient} \times \text{thermal conductivity of rock.}$$

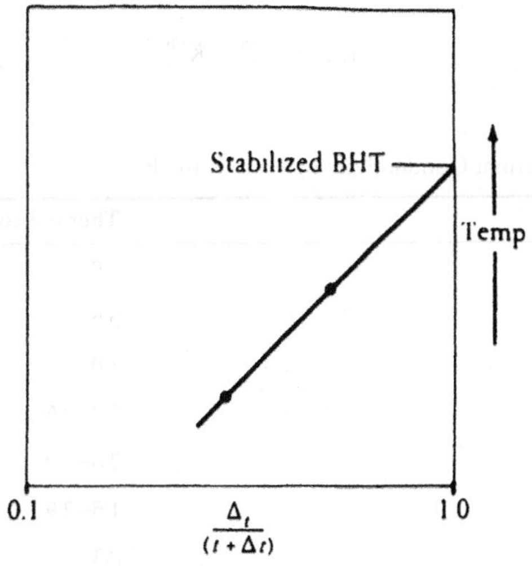

FIGURE 4.6 Horner plot showing how true BHT can be extrapolated from two readings.

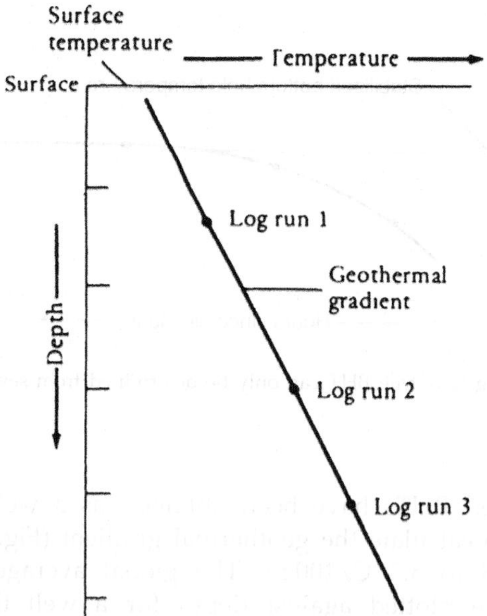

FIGURE 4.7 Sketch showing how geothermal gradient may be determined from two or more BHTs taken at different log runs.

A wide range of values can be expected for sands, shales, and limestones because of porosity variations. Because thermal conductivity is lower for water than it is for minerals, it increases with decreasing porosity and increasing depth of burial according to the following formula:

$$K_{pr} = K_{w}^{\Phi},\ K_{r}^{1\Phi},$$

TABLE 4.3 Thermal Conductivity of Various Rocks

Lithology	Thermal conductivity (W/m °C)
Halite	5.5
Anhydrite	5.5
Dolomite	5.0
Limestone	2.8–3.5
Sandstone	2.6–4.0
Shale	1.5–2.9
Coal	0.3

From Evans (1977), Oxburgh and Andrews-Speed (1981).

where

K_{pr} = bulk saturated conductivity
K_w = conductivity of pore fluid
K_r = conductivity of the rock at zero porosity
Φ = porosity

Figure 4.8 illustrates the vertical variations in conductivity, porosity, and geothermal gradient for a well in the North Sea. Note how conductivity increases and gradient decreases with depth and declining porosity. Local vertical variations are due to lithology, especially the high thermal conductivity of the Zechstein evaporites.

Once the geothermal gradient is established, isotherms can be drawn (Fig. 4.9). Note that the vertical spacing of isotherms decreases with conductivity and increasing geothermal gradient.

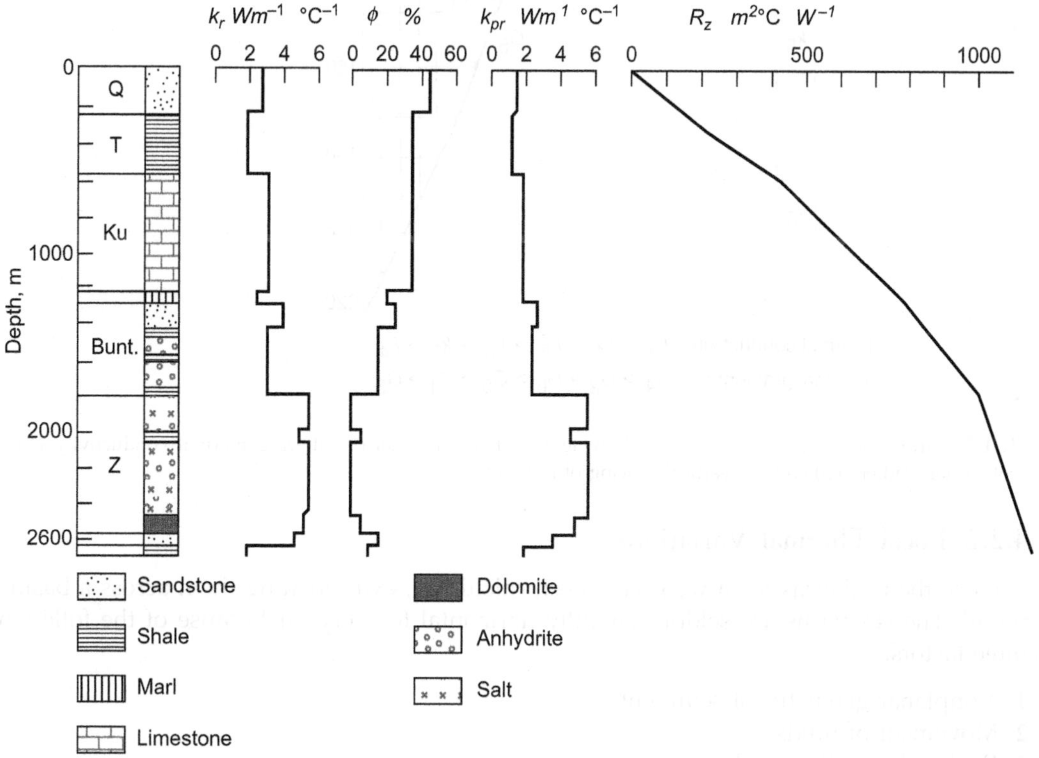

FIGURE 4.8 Variations in thermal conductivity, porosity, and geothermal gradient for a well in the North Sea. Note how thermal conductivity and gradient gradually increase with depth and declining porosity. Note also the local fluctuations due to lithology, especially the high conductivity of the Zechstein evaporites. *After Oxburgh and Andrews-Speed (1981).*

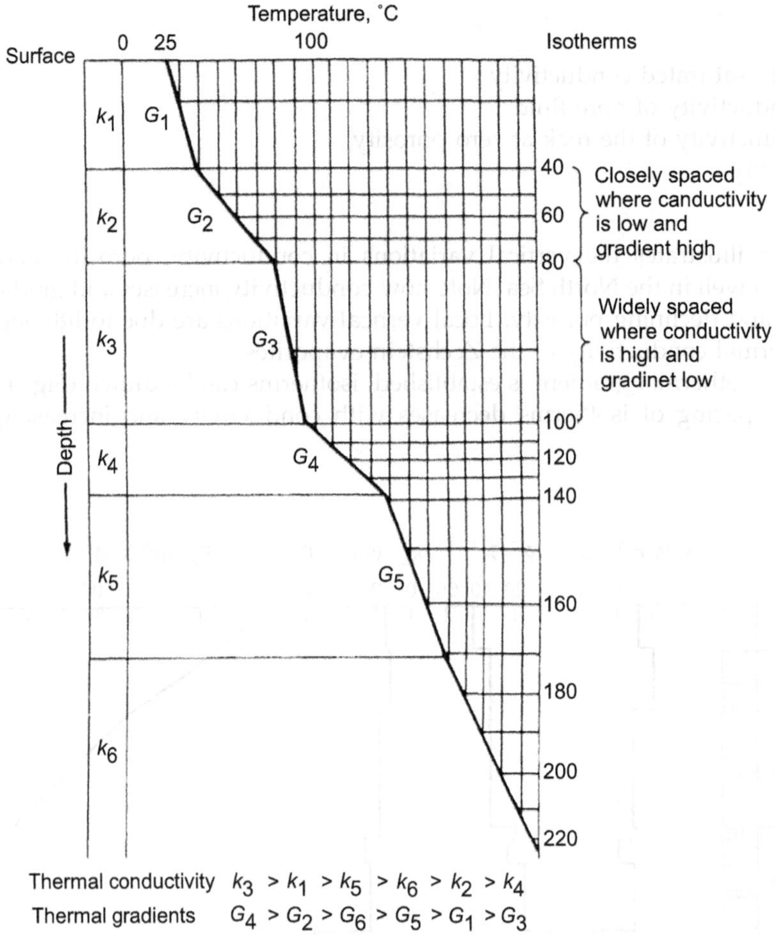

FIGURE 4.9 Depth–temperature plot showing the effect of rocks of differing thermal conductivity (K) on geothermal gradient (G) and the vertical spacing of isotherms.

4.2.2 Local Thermal Variations

Once the isotherms for a well have been calculated, extrapolating them across a basin is useful. The isotherms are seldom laterally horizontal for very far because of the following three factors:

1. Nonplanar geometry of sediments
2. Movement of fluids
3. Regional variations in heat flow

When strata are folded or where formations are markedly lenticular, anomalies are likely to occur in the geometry of isotherms. Two local variations are of considerable interest. Table 4.3 shows that salt has a far higher conductivity than most sediments (in the order of 5.5 W/m °C).

Evans (1977), in a study of North Sea geothermal gradients, noted how isotherms dome up over salt diapirs, and are depressed beneath them, because of the high thermal conductivity of evaporites (Fig. 4.9). This work has been confirmed in studies in other parts of the world. Rashid and McAlary (1977) and Rashid (1978) studied the thermal maturation of kerogen in two wells on the Grand Banks of Newfoundland. The Adolphus 2K-41 on the crest of a salt structure contained kerogen with a higher degree of maturation than the Adolphus D-50 some 3 km down flank. Conversely, sandstones lose their porosity in response to many variables, but heat is one of most important. Thus sandstones may be more cemented above a salt dome than are adjacent sands at comparable depths. Figure 4.10 shows that isotherms are depressed beneath a salt dome. Thus sandstones may have higher porosities than their lateral equivalents, and source rocks may be less mature than their lateral equivalents. This phenomenon has been noted in the presalt plays of the Gulf of Mexico, Brazil, and West Africa (Mello et al., 1995).

In contrast to salt, mud diapirs of highly porous overpressured clay have an anomalously low thermal conductivity. The isotherms within the clay will be closely spaced and depressed over the dome. Extensive overpressured clay formations act as an insulating blanket, trapping thermal energy and aiding source rock maturation.

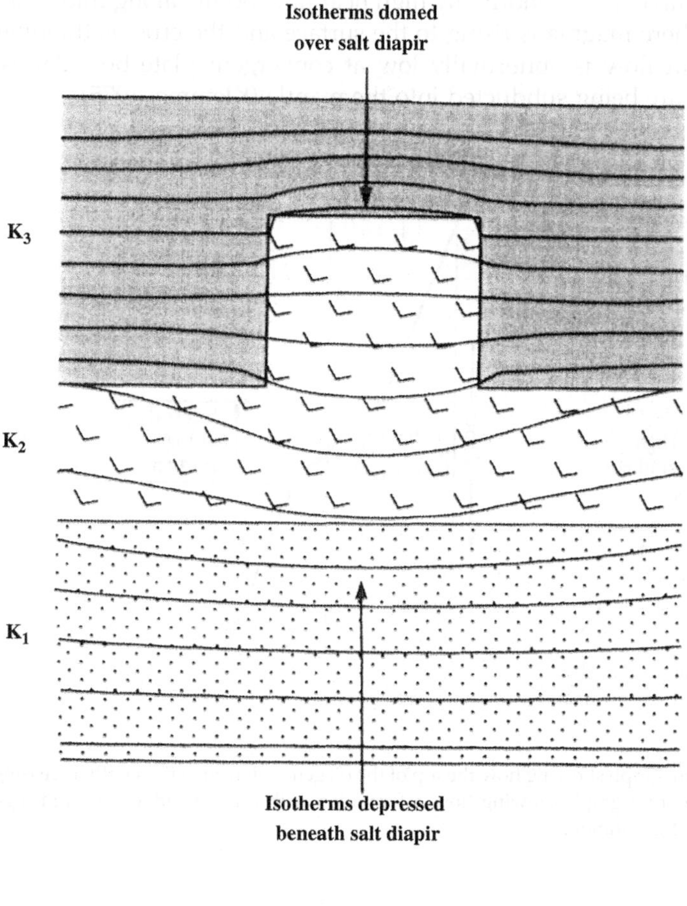

FIGURE 4.10 Isotherms modeled for a salt dome. Contours at 20 °C. Conductivities are as follows: postsalt sediments, $K_3 = 1.74$ W/m °C; Zechstein (Permian) salt, $K_2 = 4.22$ W/m °C; pre-Permian Carboniferous clastics, $K_1 = 2.82$ W/m °C. Note how source rocks will be abnormally mature above a salt dome and abnormally immature beneath one. Conversely, reservoir sands may be abnormally cemented above and abnormally porous beneath a diapir. *Developed from Evans (1977).*

Igneous intrusions are a further cause of local perturbations in heat flow, since they may cause positive heat flow anomalies long after the magma was emplaced (Sams and Thomas-Berts, 1988). The resultant thermal chimney may result in the development of a convection cell that draws connate water into the flanks of the intrusion and expels it from the crest. This mechanism may be responsible for the emplacement of the petroleum that is found in fractured reservoirs in many granites. It has been invoked, for example, to explain the petroleum reservoirs in fractured basement of offshore Vietnam (Schnip, 1992; Dmitriyevsky, 1993) and in the granites of southwest England (Selley, 1992).

A further cause of local perturbations in heat flow results when waters are rapidly discharged along open faults. This phenomenon has been described for some of the Viking Graben boundary faults of the North Sea (Cooper et al., 1975) and from growth faults of the Gulf Coast (Jones and Wallace, 1974).

4.2.3 Regional Thermal Variations

The heat flow of the earth's crust has previously been defined as the product of the geothermal gradient and the thermal conductivity. Data on global heat flow and discussions of its regional variation have been given by Lee (1965) and Sass (1971). The global average heat flow rate is of the order of 1.5 μcal/cm^2 s. Abnormally high heat flow occurs along midocean ridges and intracratonic rifts, where magma is rising to the surface and the crust is thinning and separating. Conversely, heat flow is abnormally low at convergent plate boundaries, where relatively cool sediments are being subducted into the mantle (Klemme, 1975).

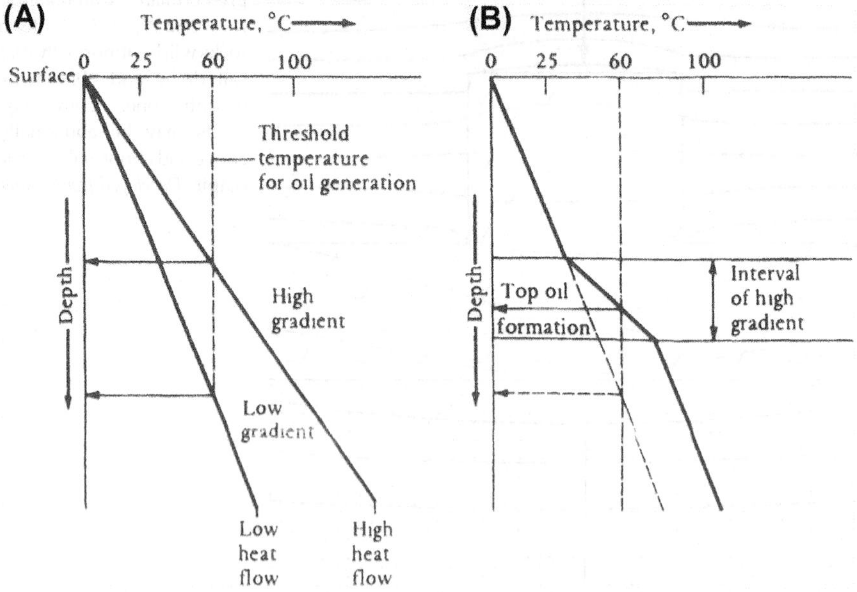

FIGURE 4.11 (A) Depth–temperature graph showing how the top of the oil generation zone rises with increasing geothermal gradient. (B) Depth–temperature graph showing how a formation with a low conductivity and high gradient raises the threshold depth of oil generation.

Regional variations in heat flow affect petroleum generation, as discussed in the next chapter. Data showing that oil generation occurs between temperatures of 60 and 120 °C will be presented. In areas of high heat flow, and hence high geothermal gradient, the optimum temperature will be reached at shallower depths than in areas of low heat flow and geothermal gradient (Fig. 4.11(A)). Note also the effect of low conductive formations. With their high geothermal gradients they raise the depth at which the oil window is entered (Fig. 4.11(B)). Thus oil generation begins at greater depths in subductive troughs than in rift basins. Layers of low-conductivity rock may raise the depth at which oil generation begins.

4.3 SUBSURFACE PRESSURES

4.3.1 Measurement

Subsurface pressures can be measured in many ways. Some methods indicate pressure before a well is drilled, some operate during drilling, and others operate after drilling.

In wildcat areas with minimal well control the interval velocities calculated from seismic data may give a clue to pressure, or at least overpressure. Velocity generally increases with depth as sediments compact. A reversal of this trend may indicate undercompacted and hence overpressured shales (Fig. 4.12).

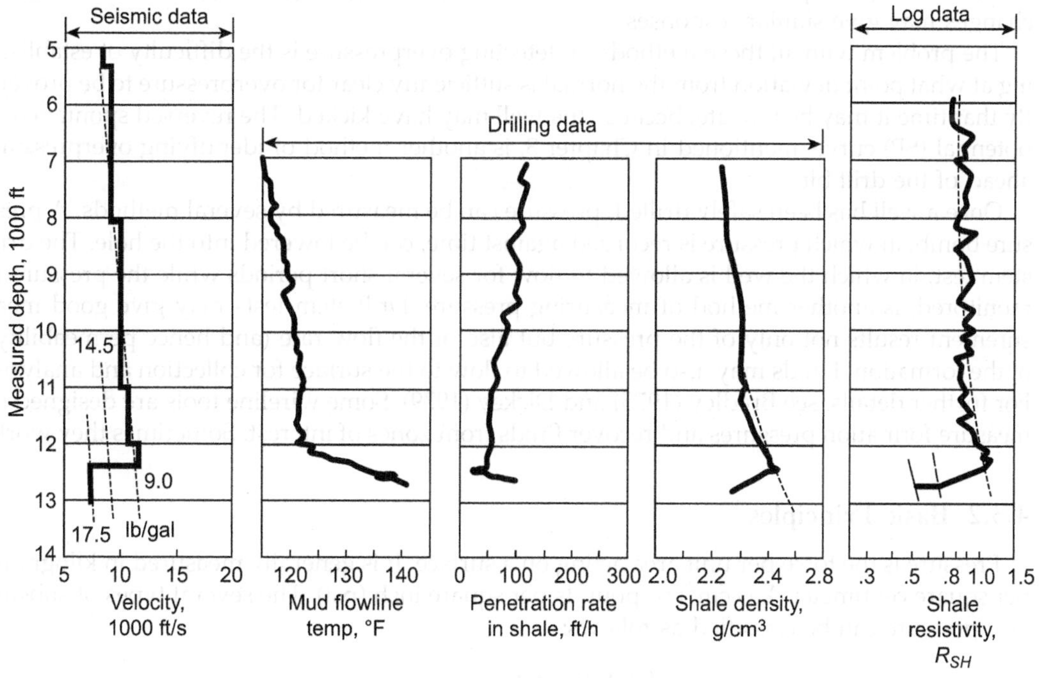

FIGURE 4.12 Various indicators of overpressure detection. In this well a zone of overpressure is present below 12,400 ft. *After Fertl and Chilingarian (1976) © SPE-AIME.*

While a well is being drilled, a number of parameters may indicate abnormal pressure. These parameters include a rapid increase in the rate of penetration, a rapid increase in the temperature of the drilling mud, and a decrease in the density of shale cuttings. A particularly useful method is the drilling (*d*) exponent (Jordan and Shirley, 1966). This method takes into account that the rate of penetration of the bit reflects not only the degree of compaction of the sediments but also the weight on the bit and the rotary speed:

$$d \text{ exponent} = \frac{\log(R/60N)}{\log(12W/106D)},$$

where

R = rate of penetration (ft/h)
N = rotary speed (rpm)
W = weight on bit (lb)
D = diameter of borehole (in.)

The *d* exponent is plotted against depth as the well is drilled. It will decrease linearly until reaching the top of abnormal pressure, at which depth it will increase.

When these methods suggest that a zone of overpressure has been penetrated, it may be wise to stop drilling and run logs. Sonic, density, and neutron logs may all show a sudden increase in porosity, whereas resistivity may sharply decrease (Hottman and Johnson, 1965). Since logs respond to more than one variable, no single log is diagnostic. Lithological changes may give similar responses.

The problem with all these methods of detecting overpressure is the difficulty of establishing at what point deviation from the normal is sufficiently clear for overpressure to be proven. By that time it may be too late, because the well may have kicked. The reversed spontaneous potential (SP) curve, mentioned in Chapter 3, is another method of identifying overpressure ahead of the drill bit.

Once a well has been safely drilled, pressure can be measured by several methods. A pressure bomb, in which pressure is recorded against time, can be lowered into the hole. The drill stem test, in which the well is allowed to flow for several short periods while the pressure is monitored, is another method of measuring pressure. Drill stem tests may give good measurement results not only of the pressure but also of the flow rate (and hence permeability) of the formation. Fluids may also be allowed to flow to the surface for collection and analysis. For further details, see Bradley (1975) and Dickey (1979). Some wireline tools are designed to measure formation pressures and recover fluids from zones of interest. Sometimes they work.

4.3.2 Basic Principles

Pressure is the force per unit area acting on a surface. It is generally measured in kilograms per square centimeter (kg/cm^2) or pounds per square inch (psi). The several types of subsurface pressure can be classified as follows:

Overburden pressure
- 1. Lithostatic
- 2. Fluid pressure
  - (a) Hydrostatic
  - (b) Hydrodynamic

The lithostatic pressure is caused by the pressure of rock, which is transmitted through the subsurface by grain-to-grain contacts. The lithostatic pressure gradient varies according to depth, the density of the overburden, and the extent to which grain-to-grain contacts may be supported by water pressure. It often averages about 1 psi/ft.

The fluid pressure is caused by the fluids within the pore spaces. According to Terzaghi's law (Terzaghi, 1936; Hubbert and Rubey, 1959).

$$s = p + o,$$

where $s =$ overburden pressure, $p =$ lithostatic pressure, and $o =$ fluid pressure. As fluid pressure increases in a given situation, the forces acting at sediment–grain contacts diminish and lithostatic pressure decreases. In extreme cases this effect may transform the sediment into an unstable plastic state.

In the oil industry fluid pressure is generally calculated as follows:

$$p = 0.052 \times wt \times D,$$

where $p =$ hydrostatic pressure (psi), $wt =$ mud weight (lb/gal), and $D =$ depth (ft).

The two types of fluid pressure are hydrostatic and hydrodynamic. The hydrostatic pressure is imposed by a column of fluid at rest. For a column of fresh water (density 1.0) the hydrostatic gradient is 0.433 psi/ft, or 0.173 kg/cm^2 m. For water with 55,000 ppm of dissolved salts the gradient is 0.45 psi/ft; for 88,000 ppm of dissolved salts the gradient is about 0.465 psi/ft. These values are, of course, temperature dependent. Figure 4.13 shows the relationship between lithostatic and hydrostatic pressure.

The second type of fluid gradient is the hydrodynamic pressure gradient, or fluid potential gradient, which is caused by fluid flow. When a well is drilled, pore fluid has a natural tendency to flow into the well bore. Normally, this flow is inhibited by the density of the drilling mud. Nonetheless, the ability to measure the level to which the fluid would rise if the well were open is important. This level is termed the *potentiometric* or *piezometric* level and is calculated as follows:

$$\text{Elevation to potentiometric level} = \frac{P}{W} - (D - E),$$

where

$P =$ bottom hole pressure (psi)
$W =$ weight of fluid (psi/ft)
$D =$ depth (ft)
$E =$ elevation of kelly bushing above sea level (ft)

The potentiometric level of adjacent wells may be contoured to define the potentiometric surface. If this surface is horizontal, then no fluid flows across the region and the fluid pressure is purely hydrostatic. If the potentiometric surface is tilted, then fluid is moving across the basin in the direction of dip of the surface and the fluid pressure is caused by both hydrostatic and hydrodynamic forces (Fig. 4.14). Formation water salinity commonly increases in the direction of dip of the potentiometric surface (Fig. 4.15). Where the pressure gradient is hydrostatic (approximately 0.43 psi/ft), the pressure is termed normal. Abnormal gradients

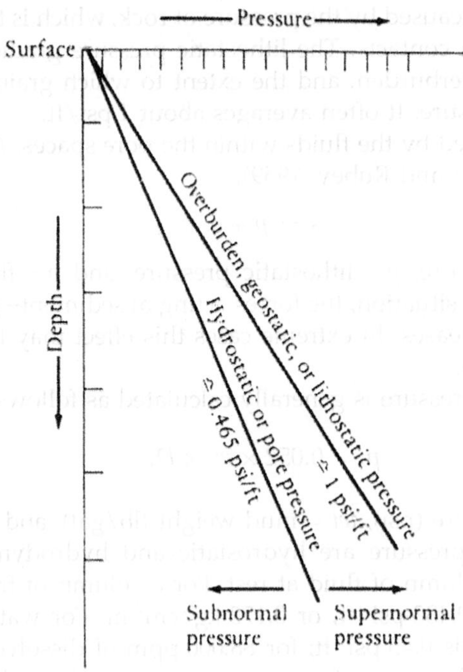

FIGURE 4.13 Depth–pressure graph illustrating hydrostatic and lithostatic (geostatic) pressures and concepts.

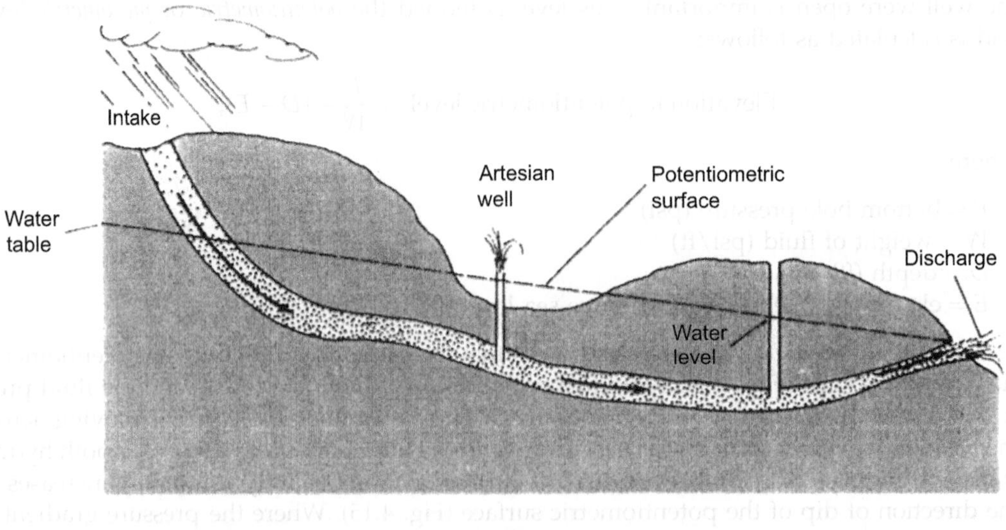

FIGURE 4.14 Sketch illustrating the concept of the potentiometric, or piezometric, surface.

may be subnormal (less than hydrostatic) or supernormal (above hydrostatic). The level to which formation fluid would rise or fall to attain the potentiometric surface is expressed as the fluid potential. Hubbert (1953) showed that this potential could be calculated as follows:

$$\text{Fluid potential} = Gz + \frac{P}{\rho},$$

where

$G =$ the acceleration due to gravity
$z =$ datum elevation at the site of pressure measurement
$P =$ static fluid pressure
$\rho =$ density of the fluid

The fluid potential is also directly related to the head of water, h:

$$\text{Fluid potential} = hG.$$

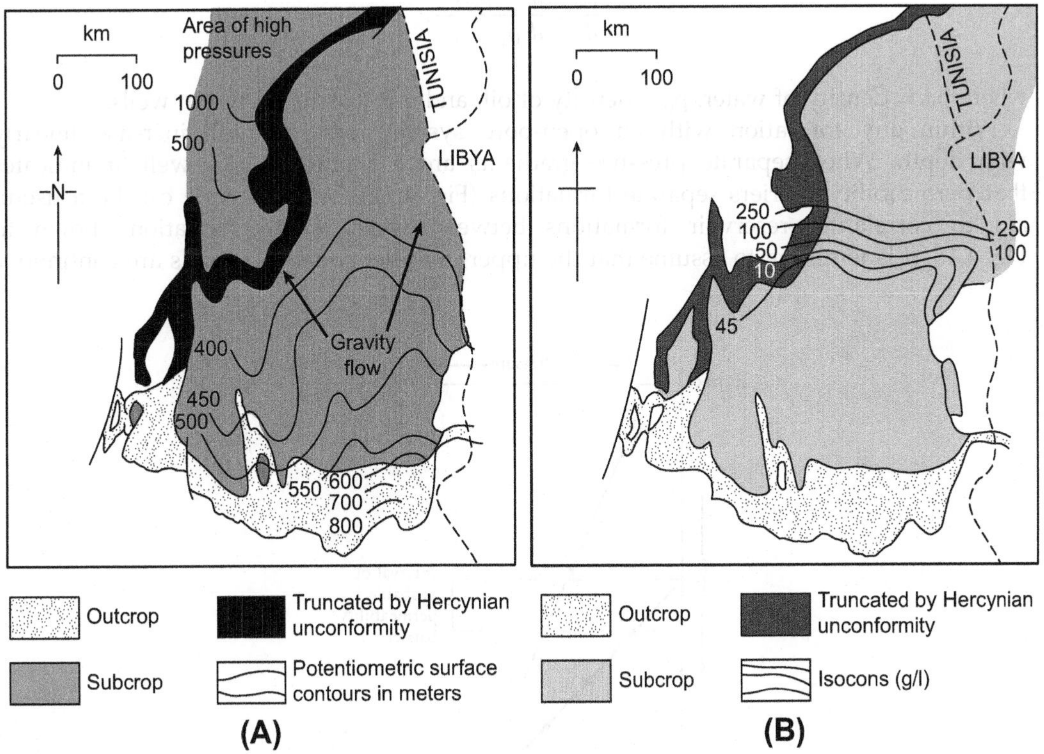

FIGURE 4.15 (A) Potentiometric and (B) isocon maps of the Silurian–Lower Devonian (Acacus and Tadrart) reservoir sandstones of eastern Algeria. Note how the potentiometric surface drops northward with increasing formation water salinity. *Modified from Chiarelli (1978); reprinted by permission of the American Association of Petroleum Geologists.*

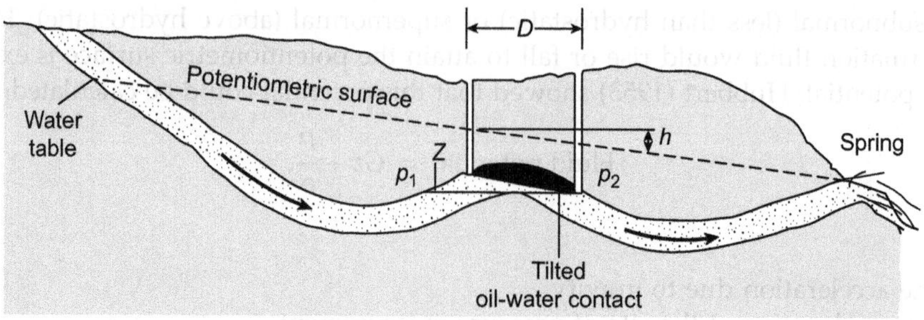

FIGURE 4.16 Sketch showing how the slope of an oil–water contact may be due to hydrodynamic flow.

Sometimes oil/water contacts are not horizontal, but tilted. Hydrodynamic flow is one of several causes of such contacts. Hubbert (1953) showed how in such cases the slope of the contact is related to the fluid potential (Fig. 4.16):

$$\frac{z}{d} = \frac{h}{d}\frac{(\rho_w)}{\rho_w - \rho_o},$$

where ρ_w = density of water, ρ_o = density of oil, and d = distance between wells.

Within any formation with an open-pore system, pressure will increase linearly with depth. When separate pressure gradients are encountered in a well, it indicates that permeability barriers separate formations (Fig. 4.17). This principle can be a useful aid to correlating reservoir formations between wells. In the situation shown in Fig. 4.18, it is tempting to assume that the upper, middle, and lower sands are continuous

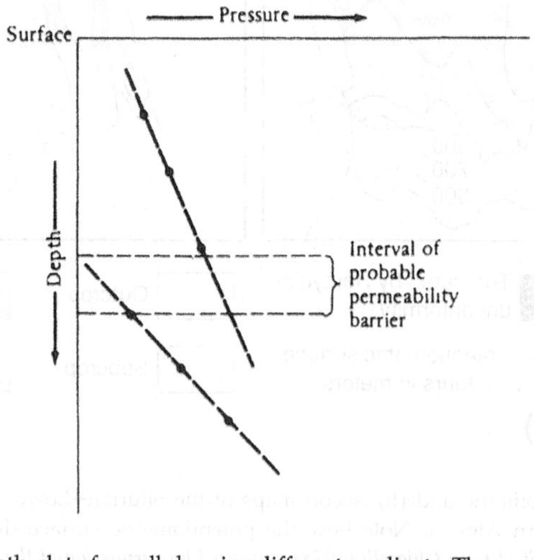

FIGURE 4.17 Pressure–depth plot of a well showing different gradients. These gradients suggest the existence of a permeability barrier between the two intervals with uniform, but different, gradients.

from well to well. Pressure gradient plots may actually reveal a quite different correlation. This difference is particularly common in stacked regressive barrier bar sands. Borehole pressure data are also used to construct regional potentiometric maps, which may be used to locate hydrodynamic traps.

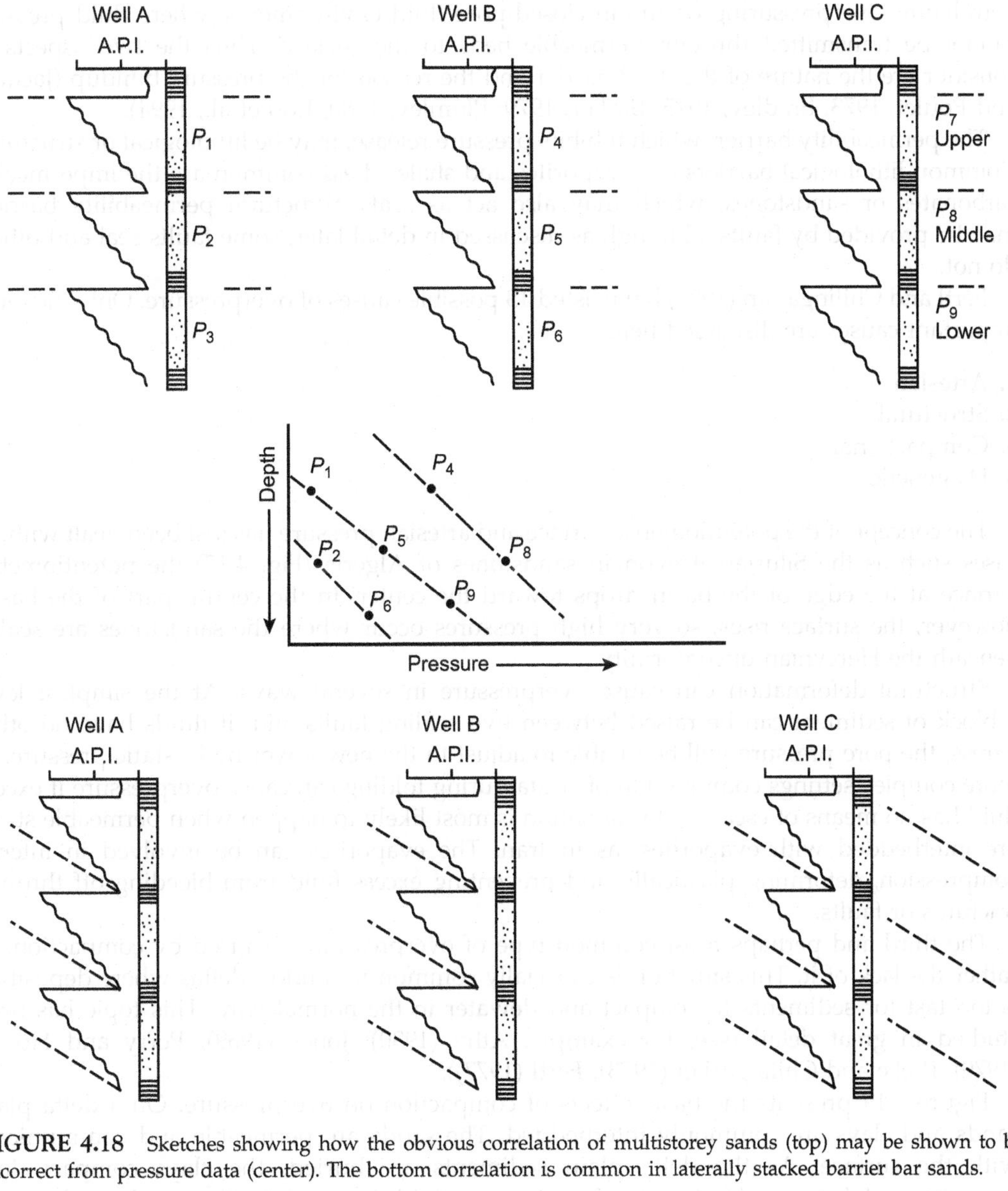

FIGURE 4.18 Sketches showing how the obvious correlation of multistorey sands (top) may be shown to be incorrect from pressure data (center). The bottom correlation is common in laterally stacked barrier bar sands.

4.3.3 Supernormal Pressure

Supernormal pressures are those pressures that are greater than hydrostatic. They are found in sediments ranging in age from Cambrian to recent, but are especially common in Tertiary deltaic deposits, such as in the North Sea, Niger Delta, and Gulf Coast of Texas. Study of the causes and distribution of overpressuring is very important because overpressuring presents a hazard to drilling and is closely related to the genesis and distribution of petroleum. Overpressuring occurs in closed-pore fluid environments, where fluid pressure cannot be transmitted through permeable beds to the surface. Thus the two aspects to consider are the nature of the fluid barrier and the reason for the pressure buildup (Jacquin and Poulet, 1973; Bradley, 1975; Barker, 1979; Plumley, 1980; Luo et al., 1994).

The permeability barrier, which inhibits pressure release, may be lithological or structural. Common lithological barriers are evaporites and shales. Less common are the impermeable carbonates or sandstones, which may also act as seals. Structural permeability barriers may be provided by faults, although as discussed in detail later, some faults seal and others do not.

Fertl and Chilingarian (1976) have listed 13 possible causes of overpressure. Only the more important causes are discussed here:

1. Artesian
2. Structural
3. Compactional
4. Diagenetic

The concept of the potentiometric surface and artesian pressure has just been dealt with. In cases such as the Silurian–Devonian sandstones of Algeria (Fig. 4.15) the potentiometric surface at the edge of the basin drops toward the center. In the central part of the basin, however, the surface rises, so very high pressures occur where the sandstones are sealed beneath the Hercynian unconformity.

Structural deformation can cause overpressure in several ways. At the simplest level a block of sediment can be raised between two sealing faults and, if fluids have no other egress, the pore pressure will be unable to adjust to the new lower hydrostatic pressure. In more complex settings compression of strata during folding can cause overpressure if excess fluid has no means of escape. This situation is most likely to happen when permeable strata are interbedded with evaporites, as in Iran. The evaporites can be involved in intense compression, deforming plastically and preventing excess fluid from bleeding off through fractures or faults.

The third and perhaps most common type of overpressure is caused by compaction, or rather the lack of it. This situation is especially common in muddy deltas where deposition is too fast for sediments to compact and dewater in the normal way. This topic has been studied in great detail (see, for example, Athy (1930), Jones (1969), Perry and Hower (1970), Rieke and Chilingarian (1973), Fertl (1977)).

Figure 4.19 presents the basic effects of compaction on overpressure. On a delta plain sands and clays are commonly interbedded. The sands are permeable and communicate with the surface. As the delta plain sediments are buried, the clays compact, lose porosity, and increase in density. Excess pore fluids move into the sands and escape

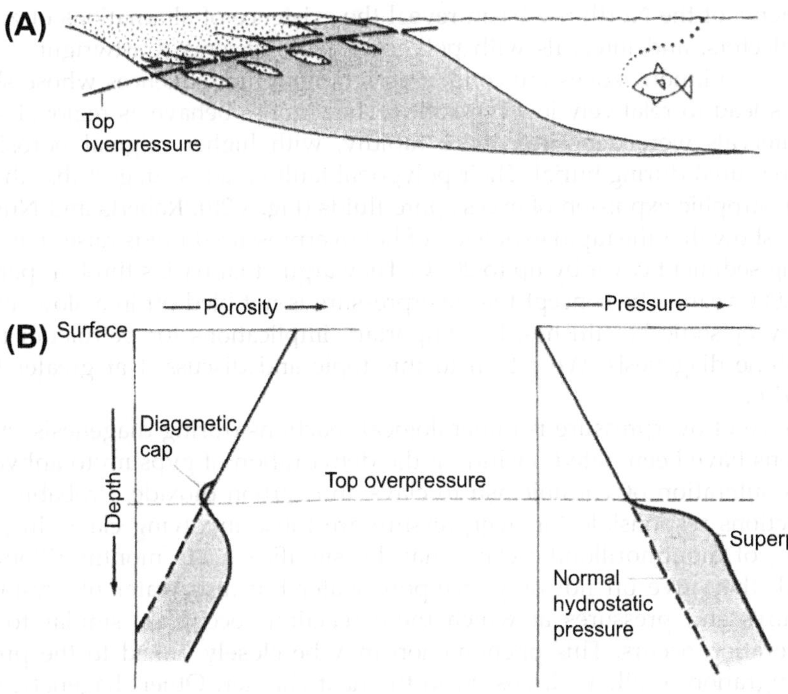

FIGURE 4.19 (A) Cross-section through a delta to show that overpressuring occurs beneath the break-up zone where there are no longer any sand beds to act as conduits to permit muds to dewater. (B) depth–porosity and depth–pressure curves, indicating how overpressuring may occur due to undercompaction.

to the surface. Thus although pore pressure increases with burial, it remains at the hydrostatic level. On the seaward side of the delta, however, sands become discontinuous and finally die out. As the delta progrades basinward, the delta plain sediments overlie the prodelta muds. The overburden pressure increases on the prodelta muds, but, lacking permeable conduits, they cannot dewater. Porosity remains high, and pore pressure exceeds hydrostatic. This type of overpressure has several important implications for petroleum geology. First, overpressured shales are a drilling hazard. Methods of predicting overpressure ahead of the drill bit were discussed in Chapter 3. Secondly, overpressured shales have lower densities and lower thermal conductivities than normally pressured shales.

The lower density of overpressured shales compared with normally compacted sediments means that they tend to flow upward and be replaced by denser, normally compacted sediments from above. This flow causes growth faults and clay diapirs, which can form important hydrocarbon traps, as discussed and illustrated later. The significance of the low thermal conductivity of overpressured shales was noted earlier. By increasing the geothermal gradient, oil generation can occur at shallower depths.

There is now a growing body of evidence that the dewatering of overpressured shale is a catastrophic and episodic process (Capuano, 1993). For example, 3D seismic studies of the

Tertiary sediments of the North Sea basin reveal the existence of alternations of zones with continuous reflectors, and intervals with polygonal fault systems (Cartwright, 1994). It is argued that the continuous zones are condensed stratigraphic sequences whose slow sedimentation rates lead to relatively low porosities. These zones behave as regional seals. The intervening intervals were deposited more rapidly, with higher original porosities, and became overpressured during burial. Their polygonal fault systems suggest that these zones underwent catastrophic expulsion of excess pore fluids (Fig. 4.20). Roberts and Nunn (1995) present data to show that the rapid expulsion of hot overpressured fluids raises temperatures in the overlying sediment cover by up to 20 °C. They argue that basins throb at periodicities of 10,000–500,000 years. The concept that overpressure is not bled off in a slow and steady manner, but by episodic hot flushes, has important implications for petroleum migration and for sandstone diagenesis. We return to this topic and discuss it at greater length in Chapters 5 and 6.

A fourth cause of overpressure is mineralogical reactions during diagenesis. A number of such reactions have been noted, including the dehydration of gypsum to anhydrite and water, and the alteration of volcanic ash to clays and carbon dioxide. Probably the most important reactions responsible for overpressure are those involving clays. In particular the dewatering of montmorillonitic clays may be significant. As montmorillonitic clays are compacted, they give off not only free-pore water but also water of crystallization. The temperatures and pressures at which these reactions occur are similar to those at which oil generation occurs. This phenomenon may be closely linked to the problem of the primary migration of oil, as discussed in the next chapter. Other diagenetic reactions occur in shales, which may not only increase pore pressure but also decrease permeability. In particular, a cap of carbonate-cemented shale often occurs immediately above an overpressured interval.

Finally, overpressuring can be caused during production either by fluid injection schemes or by faulty cementing jobs in which fluids move from an overpressured formation to a normally pressured one. In some presently normally pressured basins the presence of fibrous calcite along veins and fractures may be a relict indicator of past overpressures (Stoneley, 1983).

4.3.4 Subnormal Pressure

Subnormal pressures are those pressures that are less than the hydrostatic pressure. Examples and causes of subnormal pressures have been given by Fertl and Chilingarian (1976), Dickey and Cox (1977), and Bachu and Underschultz (1995).

Subnormal pressures will only occur in a reservoir that is separated from circulating groundwater by a permeability barrier; were this not so, the reservoir would fill with water and rise to hydrostatic pressure (Fig. 4.21). Subnormal pressures can be brought about by producing fluids from reservoirs, especially where they lack a water drive.

Naturally occurring subnormal pressures are found around the world in both structural and stratigraphic traps. Fertl and Chilingarian (1976) cite pre-Pennsylvanian gas fields of the Appalachians and the Morrow sands of trap in northwestern Oklahoma. Dickey and Cox (1977) cite subnormal pressures in the Cretaceous barrier bars from the Viking sands of Canada to the Gallup Sandstone of New Mexico.

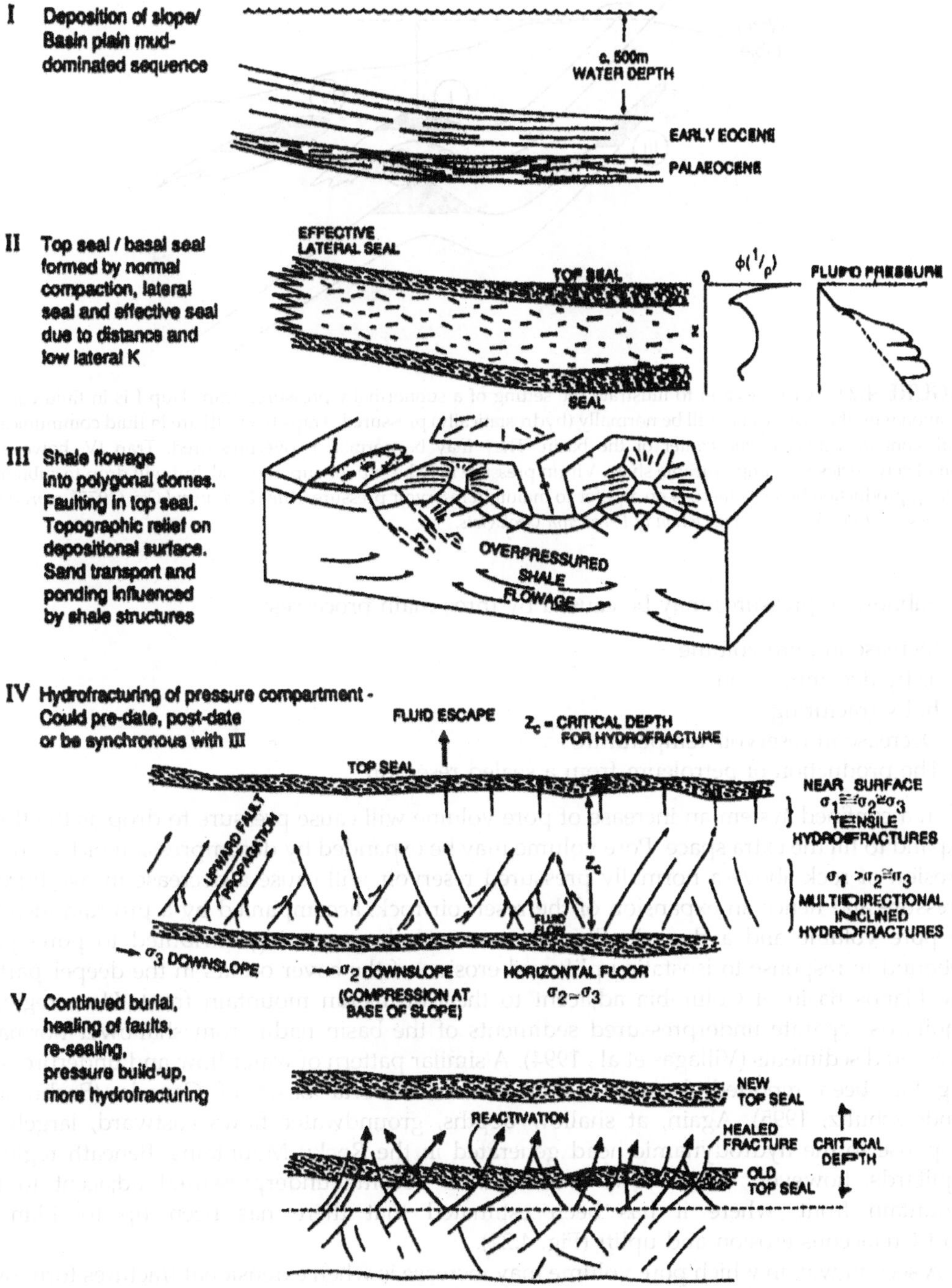

FIGURE 4.20 Geophantasmograms to explain how polygonal fault systems in the Tertiary section of the North Sea Basin may indicate the episodic expulsion of overpressured fluids. This mechanism has important implications for petroleum migration and for sandstone diagenesis. *Reprinted from Marine and Petroleum Geology, vol. II, Cartwright, pp. 587–607. Copyright 1994, with kind permission from Elsevier Science Ltd.*

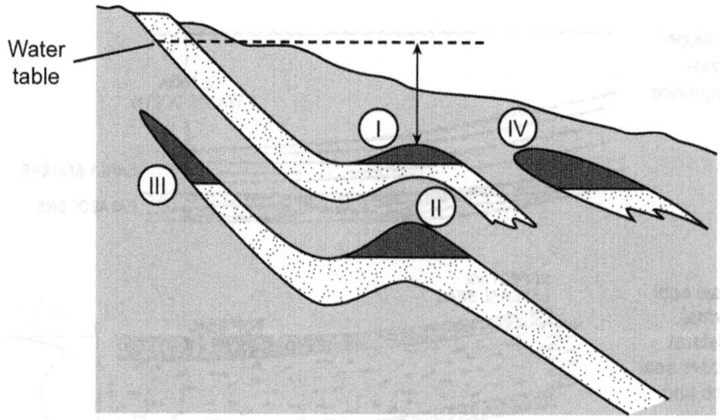

FIGURE 4.21 Cross-section to illustrate the setting of a subnormally pressured trap. Trap I is in fluid communication with the surface and will be normally (hydrostatically) pressured. Traps II and III are in fluid communication with connate waters in the center of the basin. They may be normal or overpressured. Trap IV, however, is completely enclosed in impermeable shale. Virgin pressure may be sub- or supernormal, but will drop to subnormal during production because there is no aquifer to maintain reservoir pressure. *After Dickey and Cox (1977), reprinted by permission of the American Association of Petroleum Geologists.*

Subnormal pressures may be caused by three main processes:

1. Increase in pore volume
   a. by decompression
   b. by fracturing
2. Decrease in reservoir temperature
3. The production of petroleum from a sealed reservoir

In a confined system an increase of pore volume will cause pressure to drop as the fluids expand to fill the extra space. Pore volume may be expanded by decompression or tectonism. Erosion of rock above a normally pressured reservoir will cause a decrease in overburden pressure and hence an expansion of the reservoir rock, accompanied by a pro rata increase in pore volume and a drop in fluid pressure. Underpressuring attributed to pore-space rebound in response to isostatic uplift and erosion of the cover occurs in the deeper part of the Llanos Basin of Columbia adjacent to the Cordilleran mountain front. Here regional aquitards separate underpressured sediments of the basin nadir from shallower normally pressured sediments (Villagas et al., 1994). A similar pattern of water flow and underpressuring has been reported from the southwestern Alberta Basin of Canada (Bachu and Underschultz, 1995). Again, at shallow depths, groundwater flows eastward, largely in response to the hydrodynamic head generated in the Rocky Mountains. Beneath regional aquitards, however, water flows westward to become underpressured adjacent to the mountain front, where it has been estimated that there has been up to 4 km of post-Cretaceous erosion and uplift (Fig. 4.22).

A second way in which pore volume may increase is when extensional fractures form over the crest of an anticline. This mechanism has been invoked to explain the subnormal pressure of the Kimmeridge Bay field of Dorset (Brunstrom, 1963).

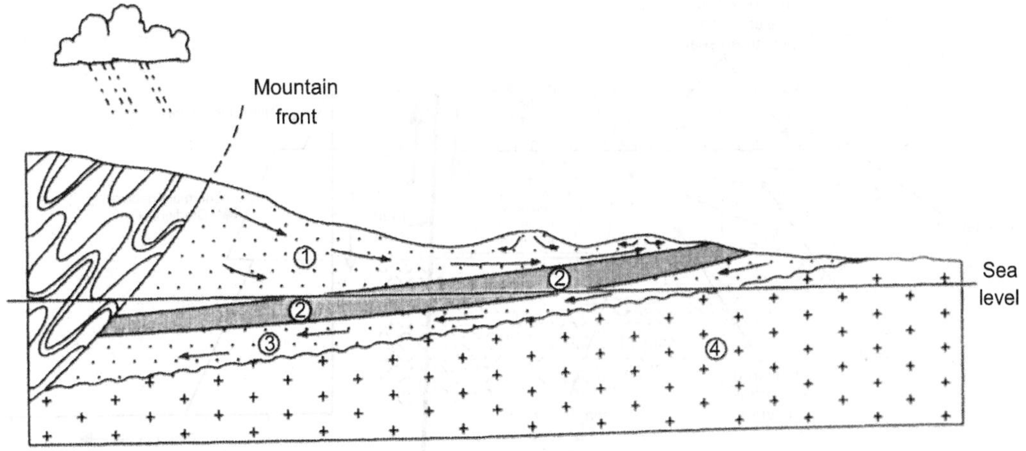

FIGURE 4.22 Cross-section through a back arc basin to show hydrologic regimes and water flow directions (per arrows). Based on the Alberta Basin of Canada and the Llanos Basin of Columbia (see Bachu and Underschultz (1995), and Villagas et al. (1994); respectively.) (1) Shallow zone where flow is driven across the basin away from the hydrodynamic head generated by the mountain range. Petroleum may be trapped in hydrostatic or hydrodynamic traps. (2) Zone of potential aquitards that may separate shallow from deep hydrological regimes. (3) Deep zone where water flows from the basin edge toward the mountain front. Underpressured reservoirs may occur adjacent to the mountain front due to isostatic uplift, erosion, and pore-space rebound. (4) Impermeable basement.

The second main cause of subnormal pressure is a decrease in geothermal gradient, which will cause reservoir fluids to cool and shrink, and thus decrease in pressure.

Finally, underpressure may develop where a petroleum reservoir occurs completely enclosed in impermeable formations, such as a reef enclosed by evaporites, or a channel or shoestring sand enclosed by shales. In these situations virgin pressure may be at or above hydrostatic. As the reservoir is produced, however, there will be no aquifer to allow water to flow in to replace the produced petroleum. Thus the time may come when the pressure in the reservoir drops below hydrostatic.

4.3.5 Pressure Compartments

Many sedimentary basins contain a layered arrangement of two superimposed hydrogeological systems. The shallow system extends down to a depth of approximately 3000 m or subsurface temperatures of 90 to 100 °C. Pore waters freely migrate in the shallow system.

The deeper seal-bounded fluid pressure system or compartments is where much of the world's oil and gas has been generated (Hunt, 1990). The seal for the compartments appears to be caused by carbonate mineralization along a thermocline (90 to 100 °C). The deep compartments are not basinwide and are generally overpressured due to generation of oil and gas and thermal expansion of pore fluids. This seal-bounded compartment is a restricted flow system. Some compartments that are underpressured may be caused by thermal contraction of confined fluids as buried rocks cool during continued uplift or erosion at

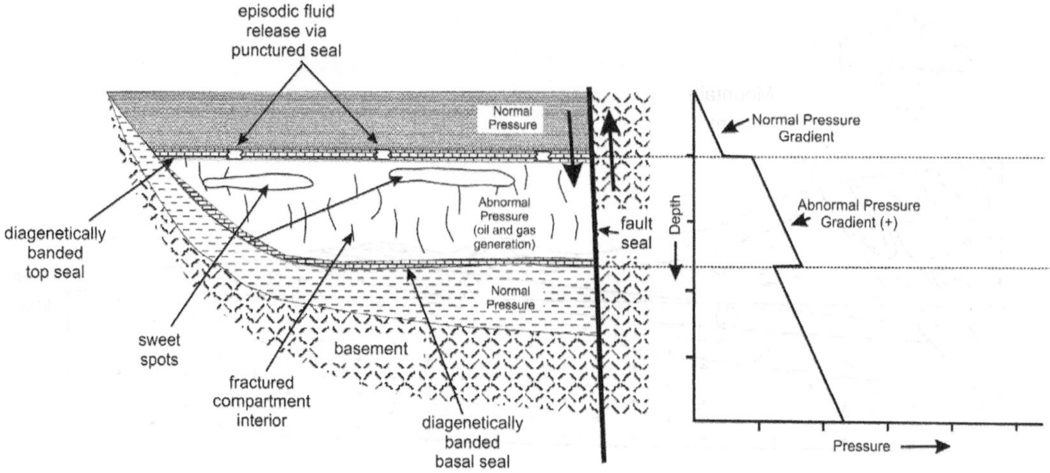

FIGURE 4.23 Diagrammatic basin-scale pressure compartment defined by diagenetically banded transbasinal top seal, stratum-localized bottom seal, fault side seal, and a fractured interior *(modified from Ortoleva, 1994)*. Pressure–depth plot illustrates abnormal pressure compartment.

the surface. Compartments have great longevity and they can undergo changes from overpressure to normal pressure to underpressure as basins go through stages of sinking and deposition, to quiescence, to basin uplift and erosion. The fluid pressure compartment has the unique characteristic of exhibiting internal dilation fractures.

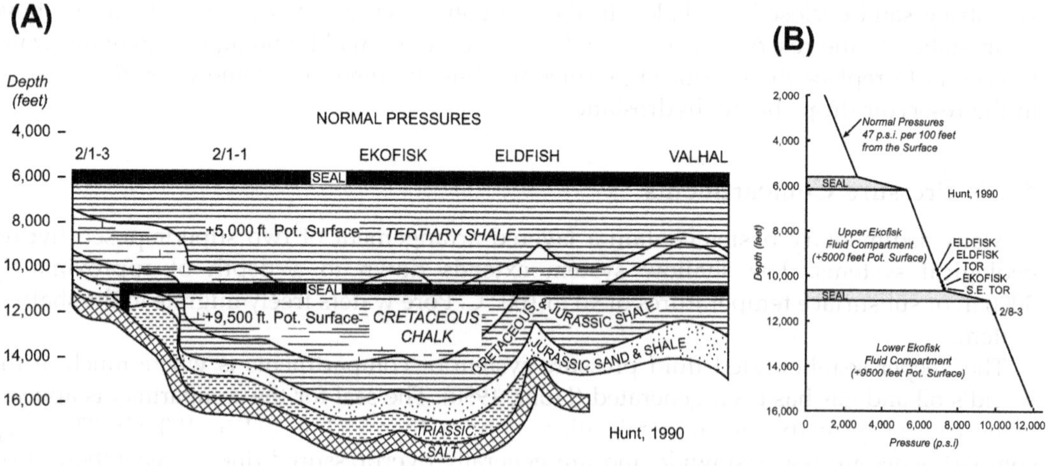

FIGURE 4.24 (A) Approximate position of seals of upper and lower Ekofisk pressure compartments in southern Central Graben, North Sea Basin (Hunt, 1990). (B) Pressure–depth plot for Ekofisk and nearby fields in Central Graben, North Sea Basin (Hunt, 1990).

The generation of oil and gas within compartments plus thermal expansion of pore fluids eventually causes fracturing at the top of the compartment seal. Oil and gas can migrate into structural and stratigraphic traps in the shallow normally pressured regime. Compartments are thought to reseal and pressure can build up again and the process will repeat itself.

Some basins have a third system or compartment that is normally pressured (Fig. 4.23). Figure 4.24 illustrates pressure compartments in the southern Central Graben, North Sea basin. Beneath each seal the pressure gradient increases.

4.4 SUBSURFACE FLUID DYNAMICS

4.4.1 Pressure–Temperature Relationships

Temperature and pressure in the subsurface have been reviewed separately; they are now examined together. More detailed accounts will be found in petroleum reservoir engineering textbooks, such as Archer and Wall (1986). From the laws of Boyle and Charles, the following relationship exists:

$$\frac{\text{Pressure} \times \text{volume}}{\text{Temperature}} = \text{a constant.}$$

This basic relationship governs the behavior of fluids in the subsurface, as elsewhere, and is particularly important in establishing the formation volume factor in a reservoir. For a given fluid at a constant pressure there is a particular temperature at which gas bubbles come out of liquid and at which they condense as temperature decreases. Similarly, at a uniform temperature there is a particular pressure at which liquid evaporates as pressure drops and gas condenses as pressure increases.

A pure fluid may exist in either the liquid or gaseous state, depending on the pressure and temperature (Fig. 4.25). Above the critical point (*c*), however, only one phase can exist. Subsurface fluids are mixtures of many compounds: connate water contains traces of hydrocarbons in solution; petroleum is a mixture of many different hydrocarbons in liquid and/or gaseous states.

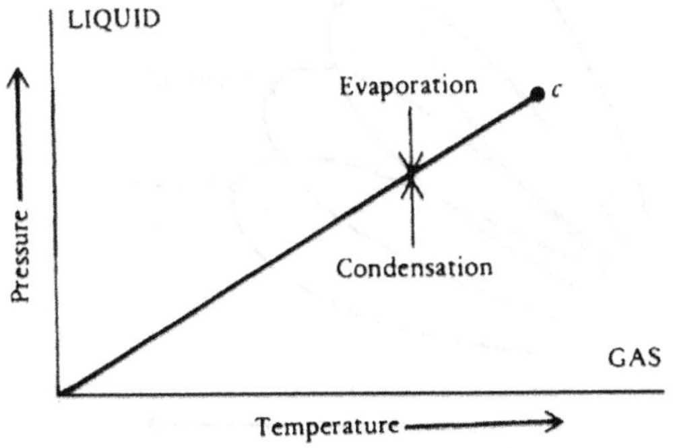

FIGURE 4.25 Pressure–temperature graph for a pure fluid. Above the critical point (*c*) only one phase can exist.

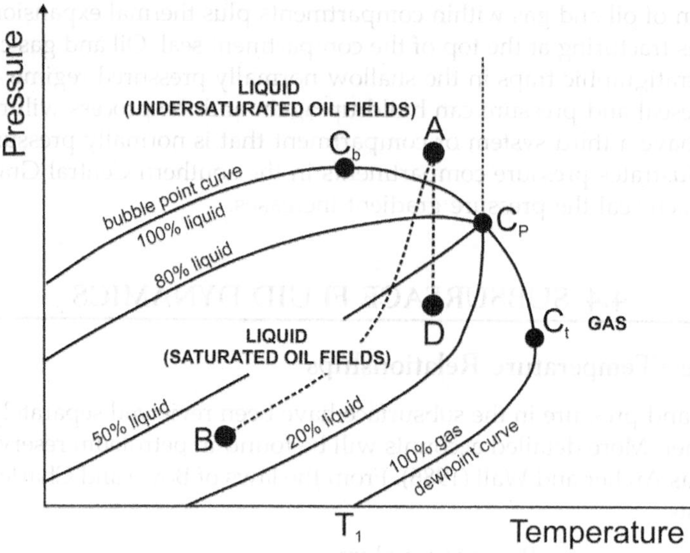

FIGURE 4.26 Pressure–temperature phase diagram of a petroleum mixture. Cb is maximum bubble point called the cricondenbar; Cp is the critical point where bubble point curve and dew point curve coincide; Ct is the cricondentherm, dew point curve maximum temperature value. *(Modified from North (1985).)* A to B symbolizes change from reservoir conditions to near-surface conditions during production; A to D symbolizes pressure drop in a reservoir from production.

A pressure–temperature phase diagram for a petroleum mixture is shown in Fig. 4.26. This figure shows three different fluid phases: liquid, liquid and vapor, and gas. The bubble point line marks the boundary at which gas begins to bubble out of liquid. The dew point line marks the boundary at which gas condenses. These two lines join the critical point (P). Point

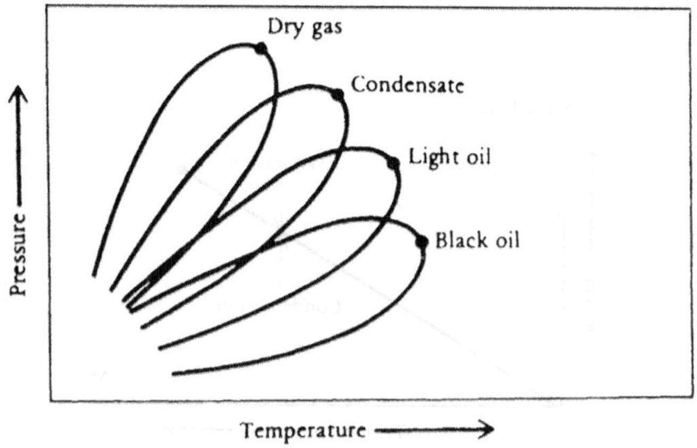

FIGURE 4.27 Pressure–temperature graph showing phase behavior for hydrocarbons of different gravity.

A represents an oil pool with high pressure and moderate temperature. All the gas is in solution. The oil is termed undersaturated. Point B represents the same pool but at lower pressure and the same temperature (production lowers pressure). The pressure is not high enough to keep all the gas in solution in the oil so free gas bubbles come out of the oil and a gas cap may form. This is called the bubble point. Above this point only liquid is present in the reservoir. As pressure continues to drop, the amount of gas released increases and the percentage of liquid oil falls progressively. The oil is termed saturated.

Figure 4.27 shows pressure–temperature curves for various mixtures of hydrocarbons. Several general conclusions can be drawn from these graphs. The amount of gas that can be dissolved in oil increases with pressure and therefore depth. The lower the density (higher the API gravity) of an oil, the more dissolved gas it contains. Within a reservoir with separate oil and gas columns, the two zones are in thermodynamic equilibrium. As production begins and pressure drops, gas will come out of solution and the volume of crude oil will decrease.

4.4.2 Secondary Migration of Petroleum

The effects of pressure and temperature on petroleum are obviously extremely important. They control both the way in which petroleum behaves in a producing reservoir and its migration from source rock to trap. Petroleum production from reservoirs is discussed in Chapter 6. Two types of petroleum migration are recognized.

Primary migration is the movement of hydrocarbons from the source rock into permeable carrier beds (Hunt, 1996). Secondary migration is the movement of hydrocarbons through the carrier beds to the reservoir. Primary migration is still something of a mystery and is reviewed in Chapter 5.

Between leaving the source rocks and filling the pores of a trap, oil must exist as droplets. Therefore, as long as the diameter of the droplets is less than that of the pore throats, buoyancy will move the droplets until they reach a throat whose radius is less than that of the droplet. Further movement can only occur when the displacement pressure of the oil exceeds the capillary pressure of the pore. Capillary pressure is discussed in more detail in Chapter 6, but basically the following relationship exists:

$$\text{Capillary pressure} = \frac{2\,t\cos\theta}{r},$$

where i = interfacial tension between the two fluids, θ = angle of contact, and r = radius of pore. Neither buoyancy nor hydrodynamic pressure alone can exceed the displacement pressure. As the oil droplets build up beneath the narrow throat, however, they increase the pressure at the throat until a droplet is squeezed through the opening. Thus the process will continue until the oil reaches the sediment whose pores are so small that the pressure of the column of oil beneath is insufficient to force further movement. The oil has thus become trapped beneath a cap rock.

The height of the oil column necessary to overcome displacement pressure has been shown by Berg (1975) to be:

$$Z_o = 2y\left(\frac{1}{r_t} - \frac{1}{r_p}\right) \Big/ G(\rho_w - \rho_o)$$

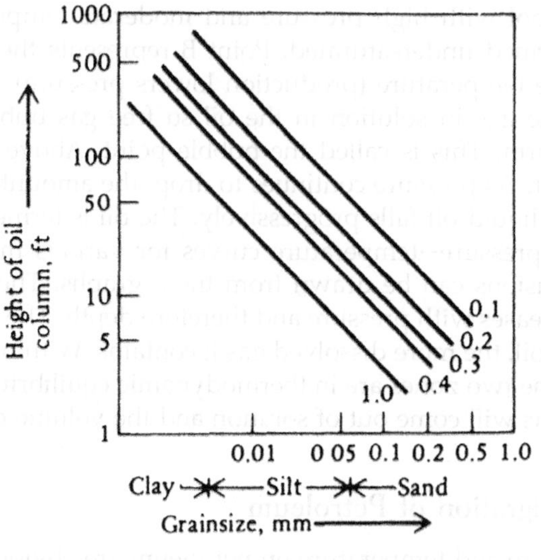

FIGURE 4.28 Graph showing the column of oil trapped by various grades of sediment for oil/water systems of various density differences. Basically, the finer the grain size (i.e., the smaller the pore diameter), the thicker the oil column that it may trap. *Modified from Berg (1975); reprinted by permission of the American Association of Petroleum Geologists.*

where

Z_o = height of oil column
y = interfacial tension between oil and water
r_t = radius of pore throat in the cap rock
r_p = radius of pore throat in the reservoir rock
G = gravitational constant
ρ_w = density of water
ρ_o = density of oil

Figure 4.28 shows the relationship between the height of oil column and grain size for oil–water systems of various density contrasts (a difference of about 0.3 is usual). Oil is

TABLE 4.4 Documented Examples of Long-Distance Lateral Petroleum Migration

Example	Distance km	Distance (mi)	Reference
Phosphoria shale, Wyoming and Idaho	400	(250)	Claypool et al. (1978)
Gulf Coast, USA Pleistocene	160	(100)	Hunt (1996)
Pennsylvanian, northern Oklahoma	120	(75)	Levorsen (1967)
Magellan Basin, Argentina	100	(60)	Zielinski and Bruchhausen (1983)
Athabasca tar sands, Canada	100	(60)	Tissot and Welte (1978)

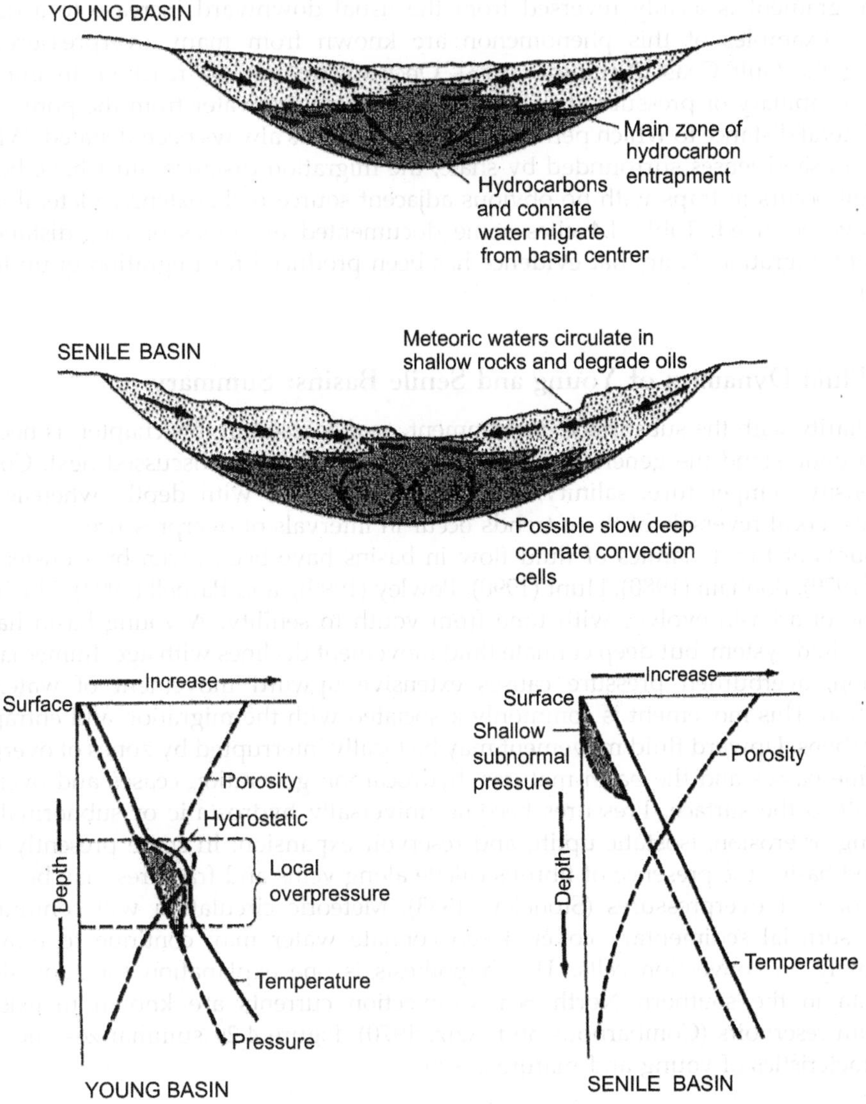

FIGURE 4.29 Cross-sections and depth curves contrasting the fluid dynamics of young and senile sedimentary basins.

thus retained beneath a fine-grained cap rock by capillary pressure in what may be referred to as a *capillary seal*. A capillary seal is likely to be more effective for oil than for the more mobile gases. Gas entrapment may be due not so much to capillary seals but to pressure seals (Magara, 1977). A pressure seal occurs when the pressure due to the buoyancy of the hydrocarbon column is less than the excess pressure of the shales above hydrostatic. Pressure seals are common where overpressured clays overlie normally pressured sands; that is, the pressure gradient is locally reversed from the usual downward increase to a downward decrease. Examples of this phenomenon are known from many overpressured basins, including the Gulf Coast (Schmidt, 1973). Once oil or gas has reached an impermeable seal, be it capillary or pressure, it will horizontally displace water from the pores.

The lateral distance to which petroleum can migrate has always been debated. Where oil is trapped in sand lenses surrounded by shale, the migration distance must have been short. Where oil occurs in traps with no obvious adjacent source rock, extensive lateral migration must have occurred. Table 4.4 cites some documented examples of long-distance lateral petroleum migration. Note that evidence has been produced for migration of up to 400 km (250 mi).

4.4.3 Fluid Dynamics of Young and Senile Basins: Summary

Familiarity with the subsurface environment, as discussed in this chapter, is necessary in order to understand the generation and migration of oil, to be discussed next. Commonly, rock density, temperature, salinity, and pressure increase with depth, whereas porosity decreases. Local reversals of these trends occur in intervals of overpressure.

Accounts of the dynamics of fluid flow in basins have been given by Cousteau (1975), Neglia (1979), Bonham (1980), Hunt (1990), Powley (1990),, and Parnell (1994). The hydrologic regime of a basin evolves with time from youth to senility. A young basin has a very dynamic fluid system, but deep connate fluid movement declines with age. Immediately after deposition, overburden pressure causes extensive upward movement of water due to compaction. This movement is commonly associated with the migration and entrapment of hydrocarbons. Upward fluid movement may be locally interrupted by zones of overpressure.

As time passes and the basin matures, hydrocarbon generation ceases and overpressure bleeds off to the surface. Pressures become universally hydrostatic or subnormal because of cooling or erosion, isostatic uplift, and reservoir expansion. In some presently normally pressured basins the presence of fibrous calcite along veins and fractures may be a relict indicator of past overpressures (Stoneley, 1983). Meteoric circulation will continue in the shallow surficial sedimentary cover. Deep connate water may continue to move, albeit very slowly, in convection cells. This hypothesis is one explanation for anomalous heat flow data in the southern North Sea. Convection currents are known to exist within petroleum reservoirs (Combarnous and Aziz, 1970). Figure 4.29 summarizes and contrasts the characteristics of young and mature basins.

References

Archer, J.S., Wall, C.G., 1986. Petroleum Engineering Principles and Practice. Graham & Trotman, London.

Athy, L.F., 1930. Density, porosity and compaction of sedimentary rocks. Am. Assoc. Pet. Geol. Bull. 14, 1–24.

Bachu, S., Underschultz, J.R., 1995. Large-scale underpressuring in the Mississippian—Cretaceous succession, southwestern Alberta Basin. AAPG Bull. 79, 989–1004.

Barker, C., 1979. Role of temperature and burial depth in development of sub-normal and abnormal pressures in gas reservoirs. AAPG Bull. 63, 4–14.

Bentor, Y.K., 1969. On the evolution of subsurface brines in Israel. Chem. Geol. 4, 83–110.

Berg, R.R., 1975. Capillary pressures in stratigraphic traps. AAPG Bull. 59, 939–956.

Bonham, L.C., 1980. Migration of hydrocarbons in compacting basins. AAPG Bull. 64, 549–567.

Bradley, J.S., 1975. Abnormal formation pressure. AAPG Bull. 59, 957–973.
Brunstrom, R.G.W., 1963. Recently discovered oil fields in Britain. In: Proc—World Pet. Congr. 6, Sect. 1, Pap. 49, pp. 11–20.
Buckley, S.E., Hocutt, C.R., Taggart, M.S., 1958. Distribution of dissolved hydrocarbons in subsurface waters. In: Weeks, L.G. (Ed.), The Habitat of Oil. Am. Assoc. Pet. Geol., Tulsa, OK, pp. 850–882.
Capuano, R.M., 1993. Evidence of fluid flow in microfractures in geopressured shales. AAPG Bull. 77, 1303–1314.
Carstens, H., Finstad, K., 1981. Geothermal gradients of the northern North Sea basin, 59–62°N. In: Illing, L.V., Hobson, G.D. (Eds.), Petroleum Geology of the Continental Shelf of North West Europe. Heyden Press, London, pp. 152–161.
Cartwright, J.A., 1994. Episodic basin-wide hydrofracturing of overpressured Early Cenozoic mudrock sequences in the North Sea Basin. Mar. Pet. Geol. 11, 587–607.
Case, L.C., 1945. Exceptional Silurian brine near Bay City, Michigan. AAPG Bull. 29, 567–570.
Case, L.C., 1956. The contrast in initial and present application of the term "connate water." J. Pet. Technol. 8, 12.
Chiarelli, A., 1978. Hydrodynamic framework of eastern Algerian Sahara—influence on hydrocarbon occurrence. AAPG Bull. 62, 667–685.
Claypool, G.E., Love, A.H., Maugham, E.K., 1978. Organic geochemistry, incipient meta-morphism, and oil generation in black shale members of Phosphoria formation, western interior United States. AAPG Bull. 62, 98–120.
Collins, A.G., 1975. Geochemistry of Oilfield Waters. Elsevier, Amsterdam.
Combarnous, M., Aziz, K., 1970. Influence de la convection naturelle dan les reservoirs d'huille ou de gas. Rev. Inst. Pr. Pet. 25, 1335–1354.
Cooper, B.S., Coleman, S.H., Barnard, P.C., Butterworth, J.S., 1975. Palaeotemperatures in the northern North Sea basin. In: Woodland, A.W. (Ed.), Petroleum and the Continental Shelf of North West Europe, vol. I. Applied Science Publishers, London, pp. 487–492.
Cousteau, H., 1975. Classification hydrodynamiques des bassins sedimentaires. Proc. World Pet. Congr. 9 (2), 105–119.
De Sitter, L.V., 1947. Diagenesis of oilfield brines. AAPG Bull. 31, 2030–2040.
Dickey, P.A., 1966. Patterns of chemical composition in deep subsurface waters. AAPG Bull. 50, 2472–2478.
Dickey, P.A., 1969. Increasing concentration of subsurface brines with depth. Chem. Geol. 4, 361–370.
Dickey, P.A., 1979. Petroleum Development Geology. Petroleum Publishing Co., Tulsa, OK.
Dickey, P.A., Cox, W.C., 1977. Oil and gas in reservoirs with subnormal pressures. AAPG Bull. 61, 2134–2142.
Downey, J.S., 1984. Geohydrology of the Madison and Associated Aquifers in Parts of Montana, North Dakota, South Dakota, and Wyoming. U.S. Geological Survey Professional Paper 1273-G, 47 p.
Dmitriyevsky, A.N., 1993. Hydrothermal origin of oil and gas reservoirs in basement rocks of the south Vietnamese continental shelf. Int. Geol. Rev. 35, 621–630.
Evans, T.R., 1977. Thermal properties of North Sea rocks. Log Analyst 18 (2), 3–12.
Fertl, W.H., 1977. Shale density studies and their application. In: Hobson, G.D. (Ed.), Developments in Petroleum Geology, vol. 1. Applied Science Publishers, London, pp. 293–328.
Fertl, W.H., Chilingarian, G.V., 1976. Importance of abnormal pressures to the oil industry. Soc. Pet. Eng. Pap. 5946, 1–11.
Fertl, W.H., Wichmann, P.A., 1977. How to determine static BHT from well log data. World Oil 184 (1), 105–106.
Friedman, G.M., Sanders, J.E., 1978. Principles of Sedimentology. Wiley, New York.
Hottman, C.E., Johnson, R.K., June 17, 1965. Estimation of formation pressures from log-derived shale properties. J. Pet. Technol. 717–722.
Hubbert, M.K., 1953. Entrapment of petroleum under hydrodynamic conditions. AAPG Bull. 37, 1954–2026.
Hubbert, M.K., Rubey, W.W., 1959. Role of fluid pressure in mechanics of overthrust faulting. AAPG Bull. 70, 115–166.
Hunt, J.M., 1996. Petroleum Geochemistry and Geology, second ed. Freeman, San Francisco.
Hunt, J.M., 1990. Generation and migration of petroleum from abnormally pressured fluid compartments. AAPG Bull. 74, 1–12.
Jacquin, C., Poulet, M., 1973. Essai de restitution des conditions hydrodynamiques regnant dans un basin sedimentaire au cours de son évolution. Rev. Inst. Fr. Pet. 28, 269–298.
Jones, P.H., 1969. Hydrodynamics of geopressure in the northern Gulf of Mexico basin. J. Pet. Technol. 21, 803–810.

Jones, P.H., Wallace, R.H., 1974. Hydrogeologic aspects of structural deformation in the northern Gulf of Mexico basin. J. Res. U.S. Geol. Surv. 2, 511–517.

Jordan, J.R., Shirley, O.J., 1966. Application of drilling performance data to overpressure detection. J. Pet. Technol. 18, 1387–1394.

Klemme, H.D., 1975. Geothermal gradients, heat flow and hydrocarbon recovery. In: Fisher, A.G., Judson, S. (Eds.), Petroleum and Global Tectonics. Princeton University Press, Princeton, NJ, pp. 251–304.

Krejci-Graf, K., 1962. Oilfield waters. Erdoel Kohle, Erdgas, Petrochem. 15, 102.

Krejci-Graf, K., 1978. Data on the Geochemistry of Oil Field Waters. K. Krejci, Frankfurt.

Krumbein, W.C., Garrels, R.M., 1952. Origin and classification of chemical sediments in terms of pH and oxidation–reduction potentials. J. Geol. 60, 1–33.

Lee, W.H.K. (Ed.), 1965. Terrestrial Heat Flow. Am. Geophys. Union, Washington, DC. Monogr. No. 8.

Levorsen, A.I., 1967. Geology of Petroleum. Freeman, London.

Luo, M., Baker, M.R., Le Mone, D.V., 1994. Distribution and generation of the overpressure system, eastern Delaware basin, western Texas and southern New Mexico. AAPG Bull. 78, 1386–1405.

Macaulay, C.Y., Haszeldine, R.S., Fallick, A.E., 1992. Diagenetic pore waters stratified for at least 35 million years: Magnus oil field, North Sea. AAPG Bull. 76, 1625–1634.

Magara, K., 1977. Petroleum migration and accumulation. In: Hobson, G.D. (Ed.), Developments in Petroleum Geology, vol. 1. Applied Science Publishers, London, pp. 83–126.

Magara, K., 1978. Compaction and Fluid Migration. Elsevier, Amsterdam.

Mello, U.T., Karner, G.T., Anderson, R.N., 1995. Role of salt in restraining the maturation of subsalt source rocks. Mar. Pet. Geol. 12, 697–716.

Neglia, S., 1979. Migration of fluids in sedimentary basins. AAPG Bull. 63, 573–597.

North, F.K., 1985. Petroleum Geology. Unwin Hyman Inc., Winchester, Mass, 631 p.

Ostroff, A.G., 1967. Comparison of some formation water classification systems. AAPG Bull. 51, 404.

Ortoleva, P.J., 1994. Basin Compartments and seals. AAPG Memoir 61, 39–51.

Overton, H.L., 1973. Water chemistry analysis in sedimentary basins. In: 14th Annu. Logging Symp. Soc. Prof. Well Log Analysts Assoc., Paper L.

Oxburgh, E.H., Andrews-Speed, C.P., 1981. Temperature, thermal gradients and heat flow in the southwestern North Sea. In: Iling, L.V., Hobson, G.D. (Eds.), Petroleum Geology of the Continental Shelf of Northwest Europe. Heyden Press, London, pp. 141–151.

Parnell, J. (Ed.), 1994. Geofluids: Origin, Migration and Evolution of Fluids in Sedimentary Basins. Spec. Publ. No. 78. Geol. Soc. London, London.

Perry, E.D., Hower, J., 1970. Burial diagenesis in Gulf Coast pelitic sediments. Clays Clay Miner. 18, 165–177.

Pirson, S.J., 1983. Geological Well Log Analysis, third ed. Gulf Publishing Co., Houston, TX.

Plumley, W.J., 1980. Abnormally high fluid pressure: survey of some basic principles. AAPG Bull. 64, 414–430.

Powley, D.E., 1990. Pressures and Hydrogeology in petroleum basins. Earth Sci. Rev. 29, 215–226.

Price, L.C., 1980. Aqueous solubility of crude oil at 400°C and 2,000 bars pressure in the presence of gas. J. Pet. Geol. 4, 195–223.

Rashid, M.A., 1978. The influence of a salt dome on the diagenesis of organic matter in the Jeanne d'Arc Subbasin of the northeast Grand Banks of Newfoundland. Org. Geochem. 1, 67–77.

Rashid, M.A., McAlary, J.D., 1977. Fairly maturation of organic matter and genesis of hydrocarbons as a result of heat from a shallow piercement salt dome. J. Geochem. Explor. 8, 549–569.

Rieke, H.H., Chilingarian, G.V., 1973. Compaction of Argillaceous Sediments. Elsevier, Amsterdam.

Roberts, S.J., Nunn, J.A., 1995. Episodic fluid expulsion from geopressured sediments. Mar. Pet. Geol. 12, 195–204.

Russell, W.L., 1951. Principles of Petroleum Geology. McGraw-Hill, New York.

Sams, M.S., Thomas-Betts, A., 1988. Models of convective heat flow and mineralization in south west England. J. Geol. Soc. London 145, 809–818.

Sass, J.H., 1971. The earth's heat and internal temperature. In: Gass, I.G., Smith, P.J., Wilson, R.C.L. (Eds.), Understanding the Earth. Artemis Press, Sussex, UK, pp. 80–87.

Schmidt, G.W., 1973. Interstitial water composition and geochemistry of deep Gulf Coast shales and sandstones. AAPG Bull. 57, 321–337.

Schnip, O.A., 1992. Reservoirs in fractured basement on the continental shelf of Vietnam. J. Pet. Geol. 15, 451–464.

Selley, R.C., 1992. Petroleum seepages and impregnations in Great Britain. Mar. Pet. Geol. 9, 226–244.
Stoneley, R., 1983. Fibrous calcite veins, overpressures and primary oil migration. AAPG Bull. 67, 1427–1428.
Sulin, V.A., 1946. Vody neftyanykh mestorozhdenn v sisteme priroduikhvod (Waters of Oil Reservoirs in the System of Natural Waters). Gostoptedkhizdat, Moscow.
Terzaghi, K., 1936. The shearing resistance of saturated soils. In: Proc. Int. Conf. Soil Mech., 1st, Harvard, vol. 1, pp. 54–56.
Tickell, F.G., 1921. Summary of operations. Bull. Calif. Div. Mines Geol. 6, 7–16.
Tissot, B.P., Welte, D.H., 1978. Petroleum Formation and Occurrence: A New Approach to Oil and Gas Exploration. Springer-Verlag, Berlin.
Van Elsberg, J.N., 1978. A new approach to sediment diagenesis. Bull. Can. Pet. Geol. 26, 57–86.
Villagas, M.E., Bachu, S., Ramon, J.C., Underschultz, J.R., 1994. Flow of formation waters in the cretaceous-miocene succession of the Llanos Basin, Columbia. AAPG Bull. 78, 1843–1862.
White, D.E., 1957. Magmatic, connate and metamorphic waters. Geol. Soc. Am. Bull. 68, 1659–1682.
Youngs, B.G., 1975. The hydrology of the Gidgealpa formation of the western and central Cooper Basin. Rep. Invest. South Aust. Geol. Surv. 43, 1–35.
Zielinski, G.W., Bruchhausen, P.M., 1983. Shallow temperature and thermal regime in the hydrocarbon province of Tierra del Fuego. AAPG Bull. 67, 166–177.

Selected Bibliography

Chemistry of Subsurface Fluids

Collins, A.G., 1975. Geochemistry of Oilfield Waters. Elsevier, Amsterdam.
Drever, J.I., 1982. The Geochemistry of Natural Waters. Prentice-Hall, Englewood Cliffs, NJ.
Hanor, J.S., 1994. Physical and chemical controls on the composition of waters in sedimentary basins. Mar. Pet. Geol. 11, 31–45.

Subsurface Temperatures

Klemme, H.D., 1975. Geothermal gradients, heat flow and hydrocarbon recovery. In: Fisher, A.G., Judson, S. (Eds.), Petroleum and Global Tectonics. Princeton University Press, Princeton, NJ, pp. 251–304.

Subsurface Pressures

Vockroth, G.B. (Ed.), 1974. Abnormal Subsurface Pressure. Reprint Ser. No. 11. Am. Assoc. Pet. Geol., Tulsa, OK.

Subsurface Pressure Compartments

Ortoleva, P.J. (Ed.), 1994. Basin Compartments and Seals. AAPG Memoir 61, 486 pp.

Subsurface Fluid Dynamics

Dahlberg, E.C., 1995. Applied Hydrodynamics in Petroleum Exploration, second ed. Springer-Verlag, Heidelberg.
Neglia, S., 1979. Migration of fluids in sedimentary basins. AAPG Bull. 63, 573–597.
Parnell, J. (Ed.), 1994. Geofluids: Origin, Migration and Evolution of Fluids in Sedimentary Basins. Spec. Publ. No. 78. Geol. Soc. London, London.
Powley, D.E., 1990. Pressures and hydrogeology in petroleum basins. Earth Sci. Rev. 29, 215–226.
Young, A., Galley, I.E., 1965. Fluids in subsurface environments. Mem.—Am. Assoc. Pet. Geol. 4, 1–414. Dated, but with a mass of data and several seminal papers.

CHAPTER

5

Generation and Migration of Petroleum

The preceding chapters discussed the physical and chemical properties of crude oil and natural gas and the behavior of fluids in the subsurface. The object of this chapter is to discuss the genesis, migration, and maturation of petroleum. We now have a good understanding of what oil and gas are made up of and plenty of empirically derived data that show how they occur. Thus, we know how to find and use oil and gas. However, the precise details of petroleum generation and migration are still debatable. Recent advances in geochemistry, especially in analytical techniques, have resulted in rapid progress on this front, but many problems remain to be solved. Any theory of petroleum generation must explain two sets of observations: geological and chemical.

A theory of petroleum genesis must explain the following *geological facts*:

1. Major accumulations of hydrocarbons characteristically occur in sedimentary rocks (Table 5.1). In the words of Pratt (1942): "I believe that oil in the earth is far more abundant and far more widely distributed than is generally realized. Oil is characteristic of unmetamorphosed marine rocks of shallow water origin. In its native habitat in the veneer of marine sediments that came with time to be incorporated into the earth's crust, oil is a normal constituent of that crust; a creature of the direct reaction of common earth forces on common earth materials." Half a century later, few geologists would disagree with this statement, except perhaps to note in passing that continental and deep marine sediments are also petroliferous.
2. Numerous examples of hydrocarbon accumulations in sandstone and limestone reservoirs are totally enclosed above, below, and laterally by impermeable rocks. The Devonian reefs of Alberta and the Cretaceous shoestring sands of the Rocky Mountain foothill basins are examples of this phenomenon.
3. Other geological occurrences to consider are
   a. Commercial accumulations of hydrocarbons have been found in basement rocks, but a young geologist's career is not enhanced by recommending that management drill such prospects.
   b. Traces of indigenous hydrocarbons have been found in igneous and metamorphic rocks. Commercial accumulations in basement, however, are always in lateral fluid continuity with sedimentary rocks.
   c. Traces of hydrocarbons occur in stony chondritic meteorites.

Elements of Petroleum Geology
http://dx.doi.org/10.1016/B978-0-12-386031-6.00005-9

TABLE 5.1 Amount of Organic Matter in Sedimentary Rocks

	Mass in (10^8 g)	
Organic sediments	1	
Petroleum	0.5	
Asphalts		
Coal	15	
Lignite		
Peat		
		16.5
Organic matter disseminated in sedimentary rock		
Hydrocarbons	200	
Asphalts	275	
Kerogen (nonextractable organic matter)	12,000	
		12,475
	Total	12,491.5

From: Hunt (1979). Used with permission.

A theory of petroleum genesis must also explain the following *chemical facts*:

1. Crude oils differ from shallow hydrocarbons in 2014.
   a. They show a preference for even-numbered carbon chains (modern hydrocarbons tend to occur in odd-numbered chains).
   b. They contain >50% light hydrocarbons, which are rare or absent in modern sediments.
2. Crude oils show the following affinities with modern organic hydrocarbons:
   a. Young oils show levorotation. This property of optical activity is characteristic of hydrocarbons produced biosynthetically.
   b. They contain certain complex molecules that occur either in modern organic matter or as a product of their degradation (e.g., the porphyrins and steroids).
   c. It is possible to use biomarkers to "fingerprint" the depositional environment within which a particular petroleum originated (Volkman, 1988).
3. Correlation between source rock and reservoir oil can be carried out by fingerprinting using gas chromatography (Bruce and Schmidt, 1994).

In discussing theories that seek to explain the preceding geological and chemical facts, it is very important to make several distinctions. Whether petroleum is of organic (biological) or of inorganic (cosmic or magmatic) origin has been debated often. How petroleum is formed from its parent material has also been debated. Origin and mode of formation are two related but quite distinct problems. Most geologists are clear in their own minds that oil is organic in origin, although most have only the vaguest ideas as to how it forms. Ignorance of the mode of formation of oil need not invalidate the fact that its nature is organic. In an appropriate analogy, ignorance of the principles of brewing need not prevent one from knowing that beer can be made from yeast, hops, and malted barley.

The next section discusses whether petroleum is organic or inorganic in origin. This section is followed by a discussion of the mode of formation of petroleum.

5.1 ORIGIN OF PETROLEUM: ORGANIC OR INORGANIC

Early theories of petroleum generation postulated an inorganic origin (e.g., Berthelot, 1860; Mendele'ev, 1877, 1902). Jupiter, Saturn, and some of the satellites of the outer planets are known to contain methane. A particular class of meteorites, the carbonaceous chondrites, contains traces of various hydrocarbons, including complex amino acids and the isoprenoids phytane and pristane (Mueller, 1963; Epstein et al., 1987; Pillinger, 1987; Chyba, 1990). Extraterrestrial hydrocarbons are believed by most authorities to have formed abiogenically (Studier et al., 1965). The opposite view has been argued by Link (1957) and Zahnle and Grinspoon (1990) who suggest that the presence of cosmic hydrocarbons provides evidence of extraterrestrial life.

Some astronomers have argued that the earth's oil was of cosmic origin. Thus, in 1955, Professor Sir Frederic Hoyle wrote: "The presence of hydrocarbons in the bodies out of which the Earth is formed would certainly make the Earth's interior contain vastly more oil than could ever be produced from decayed fish—a strange theory that has been in vogue for many years.... If our prognostication that the oil deposits have also been squeezed out from the interior of the Earth is correct, then we must, I think, accept the view that the amount of oil still present at great depths vastly exceeds the comparatively tiny quantities that man has been able to recover" (Hoyle, 1955).

Let us return to earth to examine this theory. Until the turn of the century, many distinguished scientists believed in the magmatic origin of hydrocarbons. Among the earliest and greatest of these scientists were the geographer Alexander von Humboldt and the great chemist Gay-Lussac (Becker, 1909). This theory was adopted by Mendele'ev, who suggested that the mantle contained iron carbide. This iron carbide could react with percolating water to form methane and other oil hydrocarbons (Mendele'ev, 1902), which is analogous to the reaction in which acetylene is produced by carbide and water:

$$CaC_2 + 2H_2O = C_2H_2 + Ca(OH)^2$$

There is little evidence for the existence of iron carbide in the mantle. Yet the belief in a deep, inorganic origin for hydrocarbons has been widely held by many scientists, chemists, and astronomers, but not, however, by geologists (see Dott and Reynolds, 1969; for a historical review). Vestigial traces of this belief still survive today in Russia (Porfir'ev, 1974) and even in the United States. If this theory is true, one would expect hydrocarbons to be commonly associated with igneous rocks and areas of deep crustal disturbance and faulting. We now examine evidence for the exhalation of hydrocarbons from volcanoes, their occurrence in congealed magma, and their association with faults.

Gaseous hydrocarbons have been recorded to emanate from volcanoes in many parts of the world. White and Waring (1963) have reviewed some of these occurrences. Methane is the most commonly found gaseous hydrocarbon, generally present at $<1\%$, although readings of $\geq 15\%$ are noted. In all these instances, however, the volcanoes erupt through a cover of sedimentary rocks, from which methane may have been derived by the heating of organic matter. Data have accumulated over the years describing the occurrence of hydrocarbons within igneous rocks. Three genetic types of igneous hydrocarbons may be classified:

1. Hydrocarbon gases, bitumens, and liquids occurring within vesicles andicroscopic inclusions and as irregular disseminated masses within igneous and metamorphic rocks
2. Hydrocarbons trapped where igneous rocks intrude sediments
3. Hydrocarns in weathered igneous basement trapped beneath unconformits

Solid and liquid hydrocarbons of the first type have been widely reported in Precambrian basement rocks in Karelia and the Kola peninsula of northern Russia (Filippov, 1995; Gottikh, 1995; Morakhovsky, 1995; Pushkarev, 1995). One of the most celebrated cases occurs in the Khibiny massif of the Kola peninsula in northern Russia (Petersil'ye, 1962; Ikorskiy, 1967). Here, alkaline intrusives penetrate a series of volcanic and volcaniclastic sediments. Ninety percent of the intrusives are composed of nepheline syenite. Gaseous and liquid hydrocarbons occur within inclusions in the nepheline crystals. Some 85–90% of the hydrocarbons consist of methane, but there are traces of ethane, propane, and *n*-butane and isobutane, together with hydrogen and carbon dioxide. Traces of bitumens have also been recorded, with long-chain alkanes and aromatics. Even though the quantity of hydrocarbons present in the Khibiny intrusives is infinitesimal, their diversity is impressive.

Further Russian occurrences of igneous oil are reviewed by Porfir'ev (1974). There is an ongoing debate as to whether these occurrences formed from abiogenic mantle sources, or whether they are metamorphosed late Precambrian organic-rich sediments. The debate is not an idle one, as deep (>5530-m) wells in the Republic of Tarastan have located extensive fracture systems below 5000 m (Nazipov and Iskanderov, 1995). Deep wells have been drilled in Russia in an endeavor to locate commercial accumulations of hydrocarbons in Precambrian basement.

A second celebrated case of igneous oil has been described by Evans et al. (1964) from Dyvika in the Arendal area of Norway. Here, oil occurs in vesicles in a dolerite dyke. The vesicles are rimmed with crystals of calcite, analcite, chalcedony, and quartz. Paraffinic hydrocarbons occur in the center of the vesicles. The dolerite dyke intruded Precambrian gneisses and schists.

Somewhat different to the examples just discussed are the widespread reports of subcommercial quantities of hydrocarbons in Phanerozoic intrusives (Robinson et al., 1986). The Cornubian granites of southwest England are one of many instances where subcommercial quantities of bitumen and free oil have been reported (Selley, 1992). These granites are extensively mineralized. Field evidence shows that the petroleum was emplaced at a late stage in the cooling of the magma together with lead–zinc sulfide ores at temperatures of some 150 °C. It is suggested that the petroleum and the mineralizing fluids were transported into the granites by convective flow from sedimentary cover rocks that have since eroded (Fig. 5.1).

The second type of occurrence of hydrocarbons in basement is far more common and contains many commercial accumulations of oil and gas. Oil sometimes occurs where sediments, especially organic shales, have been intruded by igneous material. Instances occur, for example, in the Carboniferous oil shales of the Midland Valley of Scotland. Commercial accumulations of this type occur in the Jatibonico field of Jatibonico, Cuba, and in the interior coastal plain of Texas (Collingwood and Rettger, 1926). In these instances, oil occurs in various hydrothermally altered basic intrusions. Such fields lack water and,

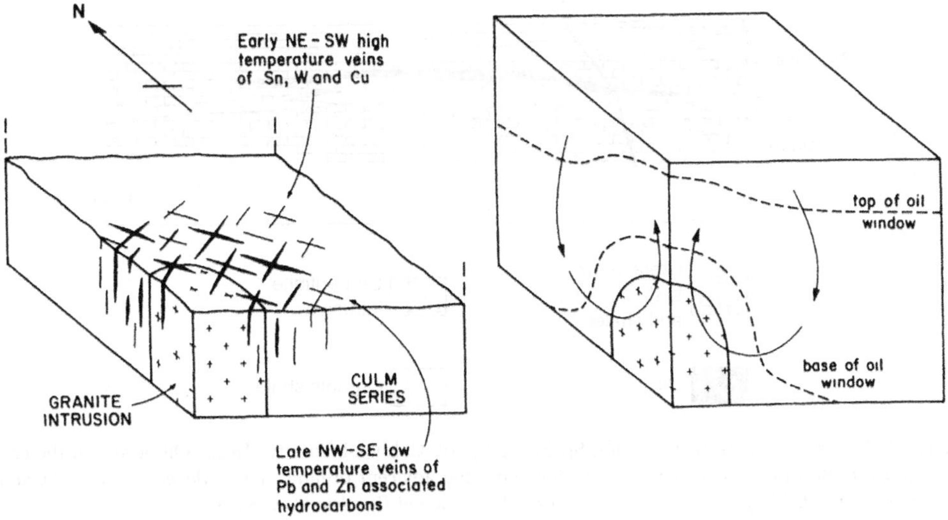

FIGURE 5.1 Illustrations showing how hydrocarbons may have been emplaced into the Cornubian granites of southwest England by convective flow during late mineralization. Subsequent erosion has removed the sedimentary cover that provided the source for the hydrocarbons and, most probably, the minerals. This mechanism has been advanced to explain many, though not all, occurrences of hydrocarbons in igneous rocks. *Reprinted from Selley (1992). Copyright 1992, with kind permission from Elsevier Science Ltd.*

although they may show very high initial production rates because of oil or gas expansion drive, often deplete rapidly. The Texas intrusives, of which >17 fields have been found, intrude Cretaceous limestones and shales (Fig. 5.2).

The third type of oil accumulation in igneous rock is frequently commercially significant. This type is where oil or gas occurs in fractures and solution pores within igneous and metamorphic rocks beneath unconformities. The porosity is obviously due to weathering. The basement rocks are overlain and onlapped by sediments that include organic-rich clays. This occurs, for example, in the oil and gas fields of offshore Vietnam (Schnip, 1992; Dmitriyevsky, 1993), in the Augila-Nafoora fields of Libya, and in northern China (Guangming and Quanheng, 1982).

Let us now consider the origins of these three genetic types of oil in igneous rock. For the first type, where hydrocarbons occur in vesicles or inclusions, there are two possible explanations. These occurrences could have originated from abiogenic hydrocarbons derived from the mantle. Alternatively, they could have resulted from the metamorphism of late Precambrian organic-rich sediments. Contamination of magma as it intruded Phanerozoic organic sediments could be postulated in some cases, but not in those where sedimentary rocks are absent.

The second type of occurrence, where hydrocarbons occur at the margin of igneous intrusives, is genetically equivocal. It could be argued that either the oil was genetically related to the magma or it was formed by thermal metamorphism of kerogen in the intruded sediments. Geochemical fingerprinting, however, demonstrates that the petroleum could be correlated with petroleum extracted from kerogen in adjacent sediments.

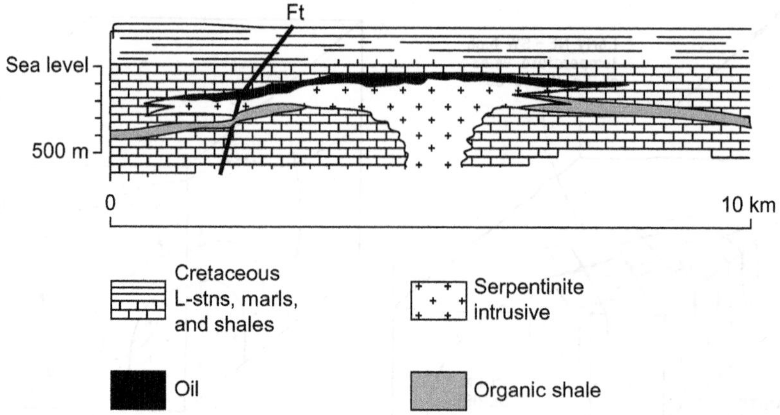

FIGURE 5.2 Cross-section of the Lytton Springs oil pool, Caldwell County, Texas. Oil occurs in the crest of a circular serpentinite plug. There is no free water. Note the adjacent organic shale. *Modified from Collingwood and Rettger (1926), reprinted by permission of the American Association of Petroleum Geologists.*

The third type, where oil occurs in weathered basement beneath unconformities, can best be explained as being caused by emigration from the enveloping sediments. Genesis synchronous with intrusion is hard to accept, because oil emplacement must have postdated both the weathering of the basement and the deposition of the sediments that provide the seal. Again, geochemical fingerprinting demonstrates that the petroleum correlates with petroleum extracted from kerogen in adjacent sediments.

In both of the preceding instances, it is probable that the petroleum was transported into the granites by convective flow from sedimentary cover rocks, as discussed in Chapter 4.

5.1.1 Hydrocarbons from the Mantle

As discussed earlier, faults play an important part in the migration and entrapment of petroleum. Although some faults are impermeable and act as seals to petroleum migration, thus causing fault traps, others allow the upward movement of oil and gas. Some faults exhale flammable gases at the surface of the earth during earthquakes. Exhalation is preceded by bulging of the ground and accompanied by loud bangs, flashing lights, and visible ground waves. An ensuing evil smell is sometimes noted, due to the presence of hydrogen sulfide gas. These phenomena are all well documented and certainly demonstrate that some faults are conduits for flammable gas (Gold, 1979). Detailed analyses of these gases reveal the presence of methane, carbon dioxide, hydrogen sulfide, and traces of noble gases (Sugisaki et al., 1983). Peyve (1956), Subbotin (1966), and Gold and Held (1987) have argued that subcrustal abiogenic hydrocarbon and other gases migrate up major faults to be trapped in sedimentary basins or dissipated at the earth's surface. Porfir'ev (1974) cited the flanking faults of the Suez, Rhine, Baikal, and Barguzin grabens as examples of such petroleum feeders.

This school of thought has been supported in the West (Gold, 1979, 1984; Gold and Soter, 1982). Gold has argued that earthquake outgassing along faults allows methane to escape from the mantle. This process gives rise to deep gas reservoirs and, by polymerization, to oil at shallower horizons. To most geologists, these ideas seem improbable and are largely rejected out of hand. Gold's earthquake outgassing theory implies that to find limitless quantities of oil and gas, all that needs to be done is to drill adjacent to faults deep enough and often *enough*.

The classic geologist's response has been cogently argued by North (1982). Because porosity and permeability decrease with depth, there can be no deep reservoirs to contain and disburse any gas that might exist. On a global scale, gas reservoirs are closely related to coal-bearing formations of Late Carboniferous (Pennsylvanian) and Cretaceous to Paleocene ages (the gas-generating potential of humic kerogen is discussed later).

Note that many of the most dramatic reports of earthquake outgassing occur in faulted basins with thick sequences of young sediments. These loosely packed sands are bound to be charged with biogenic methane and hydrogen sulfide. A seismic shock will cause the packing of the sand to tighten, resulting in a decrease in porosity and a violent expulsion of excess pore fluid. It is possible, however, to differentiate biogenic from abiogenic methane. Carbon has two isotopes: C^{12} and C^{13}. The C^{12}:C^{13} ratio varies for different compounds, and this ratio is especially large for the different types of methane. Biogenic methane is enriched in C^{12}; abiogenic methane contains a higher proportion of C^{13} (Hoefs, 1980). The C^{12}:C^{13} ratios of methane from the "hot spots" of the Red Sea, Lake Kivu (East Africa), and the East Pacific Rise suggest an abiogenic origin (MacDonald, 1983).

Gold's theory has been tested by wells drilled on the Siljan ring, a meteorite impact structure in Sweden dated at some 360 mybp. This structure occurs in Precambrian basement. Within the impact crater, there are Ordovician and Silurian limestones and shales that contain petroleum (Fig. 5.3). Gold argued that this was just the place to drill for oil, because the impact of the meteorite would have shattered the crust, thus allowing the petroleum to escape to the surface, as proved by the seeps. The first well was drilled to 6957 m. Drilling problems were encountered due to lost circulation and due to drilling mud being lost into the formation. This indicates the existence of deep porosity and permeability. Gaseous and liquid hydrocarbons were encountered, and up to 85 barrels of oil were produced (Vlierbloom et al., 1986). Unfortunately, however, when the well lost circulation, diesel oil was added to the drilling mud system. It is argued, however, that the recovered oil is different from diesel. Enthusiasts for the abiogenic theory were disappointed to learn, however, that the black gelatinous hydrocarbons recovered from the bottom of the well were actually an artifact of the drilling process, due to the mixing of diesel with caustic soda (Jeffrey and Kaplan, 1988, 1989).

5.1.2 Genesis of Petroleum by Fischer–Tropsch Synthesis

There is a third theory for the origin of petroleum that is intermediate between the abiogenic mantle theory, just discussed, and the shallow biogenic theory. The Fischer–Tropsch reaction is a well-known industrial process that is used to produce synthetic hydrocarbons. It was used by Germany in World War II, and more recently by South Africa to produce petroleum and related products. In the Fischer–Tropsch reaction, carbon dioxide and

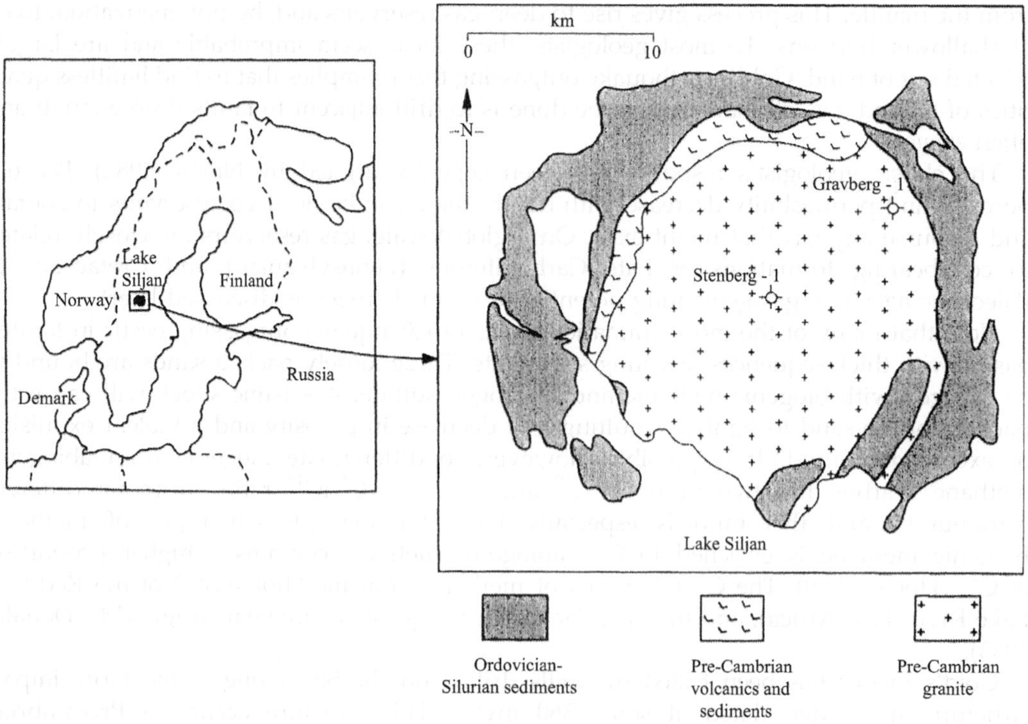

FIGURE 5.3 Map of the Siljan ring meteorite impact crater, Sweden. This was drilled in 1986 to test the abiogenic mantle-derived theory of petroleum genesis. It proved the existence of deep porosity and permeability and recovered oil and gas. The oil proved to have been produced by the alteration of diesel in the mud system caused by the action of the drill bit on hot rock.

hydrogen are passed over a catalyst of hematite and magnetite at temperatures of >500 °C. According to Gould (1991), the reactions involved are probably as follows:

$$C + Fe + H + OH \rightarrow CH_4 + C_2H_6 + C_3H_8 + C_4H_8 + FeO \tag{5.1}$$

$$CO_2 + HOH + Fe \rightarrow CH_4 + FeO \tag{5.2}$$

$$C + HOH + Fe \rightarrow CH_4 + CO_2 + H_2 \tag{5.3}$$

$$CO_2 + H_2 \rightarrow CH_4 + H_2O \tag{5.4}$$

Robinson (1963) suggested that this process could produce methane and liquid petroleum in nature. More recently, Szatmari (1986, 1989) discussed this process in the light of plate tectonics. He argued that hydrocarbons could be generated by the Fischer–Tropsch process at convergent plate boundaries where sedimentary rocks and oceanic crust undergo subduction. The carbon dioxide could be formed from the metamorphism of carbonates, and the hydrogen by the serpentinization of ophiolites. The latter would provide the iron oxides necessary to catalyze the reaction. A critique of this theory can be found in Le Blanc (1991).

5.1.3 Abiogenic versus Biogenic Genesis of Petroleum: Conclusion

In concluding this review of theories for the abiogenic origin of petroleum hydrocarbons, it must be stressed that the apparently unquestionable instances of indigenous oil in basement are rare and not commercially important. Not only are the volumes of hydrocarbons trapped this way insignificant but the "reservoirs" are impermeable unless fractured.

Commercial accumulations of hydrocarbons in igneous rocks only occur where the igneous rocks intrude or are unconformably overlain by sediments.

There is now clear evidence for the origin of abiogenic hydrocarbons in the deep crust or mantle, and for its emergence along faults and fractures, notably in midoceanic ridges and intracontinental rifts. Geologists will not fail to note, however, that commercial accumulations of oil are restricted to sedimentary basins. Petroleum seeps and accumulations are absent from the igneous and metamorphic rocks of continental shields, both far from and adjacent to faults. It is now routine to use gas chromatography (fingerprinting) to match the organic matter in shales with the petroleum in adjacent reservoirs (Flory et al., 1983; Hunt, 1996). Geologists thus conclude that commercial quantities of petroleum are formed by the thermal maturation of organic matter. This thesis is now examined, beginning with an account of the production and preservation of organic matter on the earth's surface and then charting its evolution when buried with sediments.

5.2 MODERN ORGANIC PROCESSES ON THE EARTH'S SURFACE

The total amount of carbon in the earth's crust has been estimated to weigh 2.65×10^{20} g (Hunt, 1977). Some 82% of this carbon is locked up as CO^3 in limestones and dolomites. About 18% occurs as organic carbon in coal, oil, and gas (Schidlowski et al., 1974). The key reaction is the conversion of inorganic carbon into hydrocarbons by photosynthesis. In this reaction, water and atmospheric carbon dioxide are converted by algae and plants into water and glucose:

$$6CO_2 + 12H_2O = C_6H_{12}O_6 + 6H_2O + 6O_2$$

Glucose is the starting point for the organic manufacture of polysaccharides and other more complex carbon compounds. This polysaccharide production may happen within plants or within animals that eat the plants. In the natural course of events, the plants and animals die and their organic matter is oxidized to carbon dioxide and water. Thus, the cycle is completed (Fig. 5.4). In certain exceptional circumstances, however, the organic matter may be buried in sediments and preserved, albeit in a modified state, in coal, oil, or gas. Gas chromatography can be used to "fingerprint" petroleum and to correlate it with the source rock from which it was derived (Bruce and Schmidt, 1994) and to determine the type of organism and depositional environment in which it lived (Volkman, 1988).

To trace the evolution of living organic matter into crude oil and gas, it is necessary to begin with an examination of its chemistry. The major groups of chemicals that occur in organic matter are proteins, carbohydrates, lipids, and lignin. The proteins are found largely in animals and to a lesser extent in plants. They contain the elements hydrogen, carbon, oxygen, and nitrogen, with some sulfur and phosphorus. This combination of

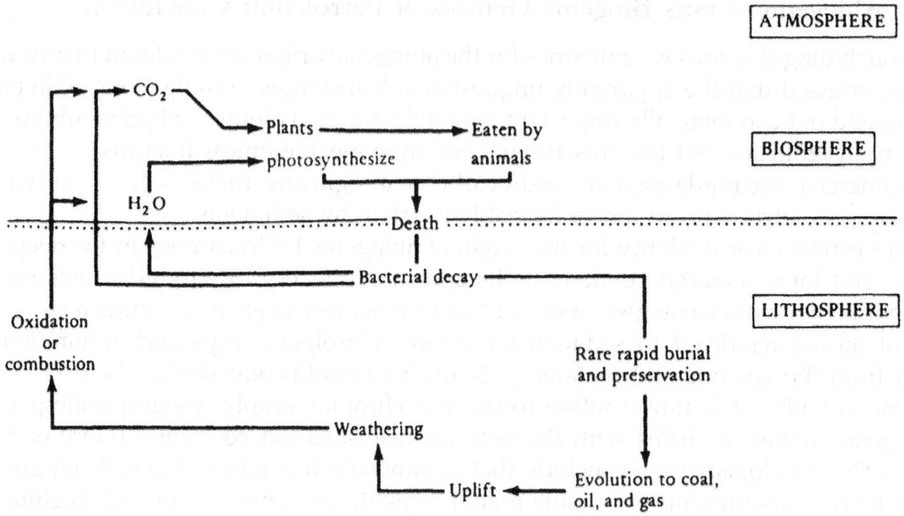

FIGURE 5.4 The cycle of organic carbon within the earth's crust. The primary source for the carbon is from the weathering of crustal rocks, together with mantle-derived carbon dioxide and methane.

elements occurs in the form of amino acids. The carbohydrates are present in both animals and plants. They have the basic formula $C_n(H_2O)_n$ and include the sugars, such as glucose, and their polymers—cellulose, starch, and chitin. The lipids are also found in both animals and plants. They are basically recognized by their insolubility in water and include the fats, oils, and waxes. Chemically, the lipids contain carbon, hydrogen, and oxygen atoms. The basic molecule of the lipids is made up of five carbon atoms (C_5H_8). From this base, the steroids are built. The last of the four major groups of organic compounds is lignin, which is found only in the higher plants. Lignin is a polyphenol of high molecular weight, consisting of various types of aromatic carbon rings. Figure 5.5 shows the distribution of these four groups of compounds in living organisms and in recent shallow sediments. Table 5.2 shows their abundance in different groups of animals and plants.

5.2.1 Productivity and Preservation of Organic Matter

As previously discussed, the amount of organic matter buried in sediments is related to the ratio of organic productivity and destruction. Generally, organic matter is destroyed on the earth's surface, and only minor amounts are preserved. The deposition of an organic-rich sediment is favored by a high rate of production of organic matter and a high preservation potential. These two factors are now discussed. Determining production and preservation for the present day is relatively easy; extrapolating back in time is harder. This problem is especially true for continental environments, whose production–destruction ratio of organic matter is largely related to the growth of land plants. Therefore, marine and continental environments should be considered separately.

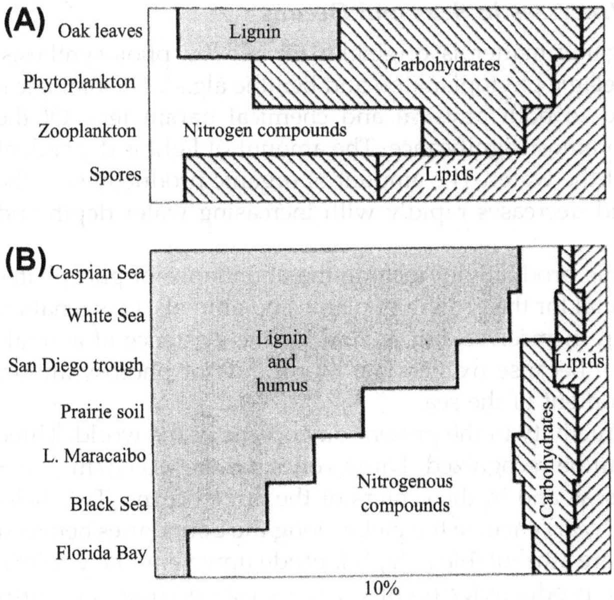

FIGURE 5.5 Composition of organic matter in (A) organisms and (B) shallow Recent sediments. *From: Hunt (1979). Used with permission.*

TABLE 5.2 Chemical Composition of Various Groups of Animals and Plants

	Weight percent of major constituents (ash-free)			
Substance	**Proteins**	**Carbohydrates**	**Lignin**	**Lipids**
Plants				
Spruce	1	66	29	4
Oak leaves	6	52	37	5
Pine needles	8	47	17	28
Phytoplankton	23	66	0	11
Diatoms	29	63	0	8
Lycopodium spores	8	42	0	50
Animals				
Zooplankton	60	22	0	18
Copepods	65	25	0	10
Higher invertebrates	70	20	0	10

From: Hunt (1979). Used with permission.

5.2.1.1 *Organic Productivity and Processes in Seas and Oceans*

In the sea, as on the land, all organic matter is originally formed by photosynthesis. The photosynthesizers in the sea are pelagic phytoplankton and benthic algae. The biological productivity of these plants is related to both physical and chemical parameters. Of the former, temperature and light are of foremost significance. The amount of light is dependent on the depth, latitude, and turbidity of the water. The amount of organic productivity is the highest in the shallow photic zone and decreases rapidly with increasing water depth and decreasing light and temperature.

Chemical conditions favoring organic productivity include the abundance of phosphates and nitrates. These chemicals are essential for the growth of plants and animals. Oxygenation is not important for the phytoplankton themselves, but is vital for the existence of animals that later form links in the food chain. Because oxygen is a by-product of photosynthesis, phytoplankton increases the oxygen content of the sea.

Figure 5.6 shows the biological productivity in the present-day oceans of the world. Three major areas of organic productivity can be recognized. Large zones, where $<50\ g/m^2/year$ are produced, are found in the polar seas and in the centers of the large oceans. Two belts of higher productivity ($200-400\ g/m^2/year$) encircle the globe along the boundaries between the polar and equatorial oceans. These are attributable to high-latitude upwelling (Hay, 1995). Three zones of above average organic productivity occur off the western coasts of North America, South America, and Africa. These zones are where deep, cold ocean currents well up from great depths (the California, Humboldt, and Benguela currents, respectively). The nutrients that these currents contain, principally nitrates and phosphates, are responsible for the high organic productivity of these zones.

It is important to stress that the areas of high oceanic productivity are not necessarily the areas where organic matter is best preserved. The preservation of organic matter is favored by anaerobic bottom conditions and a rapid sedimentation rate. Johnson Ibach (1982) described a detailed study of the relationship between rate of sedimentation and total organic carbon (TOC) in cores recovered from the ocean floors by the Deep Sea Drilling Project. The samples ranged in age from the Jurassic to Recent. Figure 5.7 illustrates the results. Black shale, siliceous, and carbonate sediments are shown separately. For all three sediments, however, the TOC content increases with sedimentation rate for slow rates of deposition, but

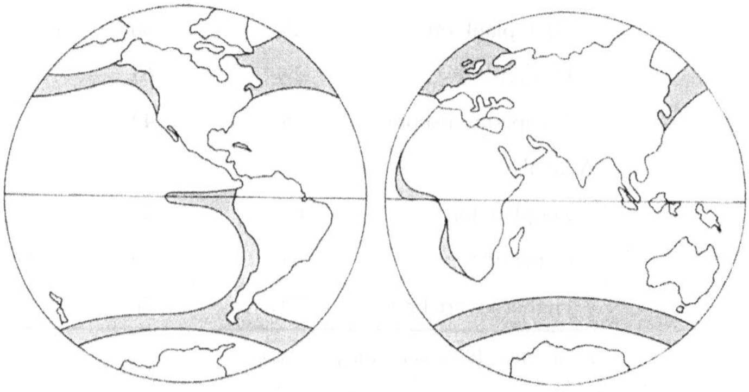

FIGURE 5.6 Production of organic matter in the present-day world oceans. Note the low productivity in polar regions and the high productivity along eastward sides of the oceans. *For sources, see Debyser and Deroo (1969), Degens and Mopper (1976), Calvert (1987), and Hay (1995).*

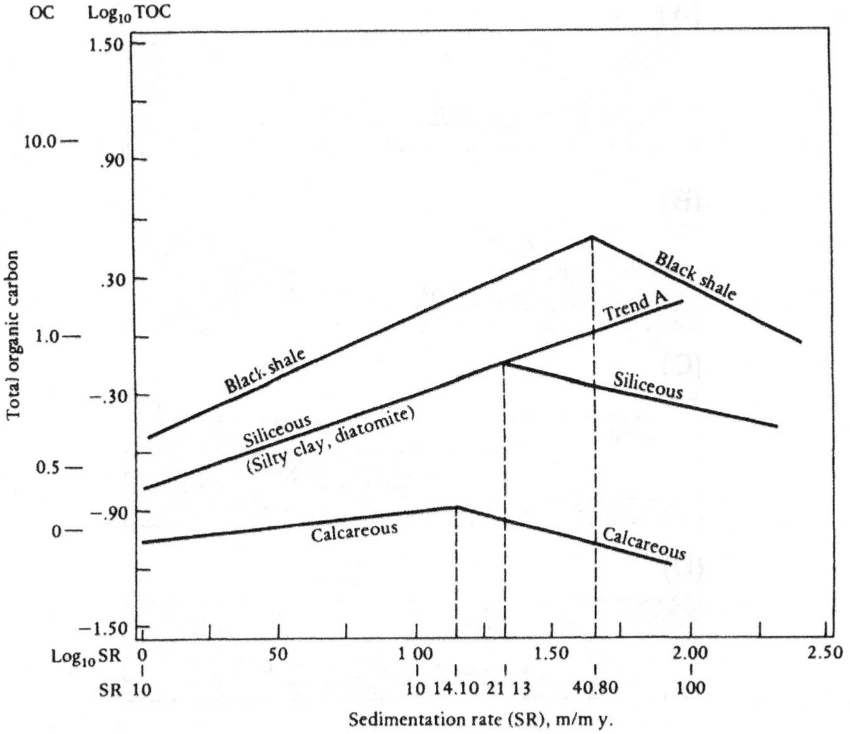

FIGURE 5.7 Graph showing the relationship between total organic carbon and sedimentation rate for Jurassic to Recent cores recovered by the Deep Sea Drilling Project. The graph shows that each different type of sediment has an optimum sedimentation rate for the maximum preservation of organic matter. At too slow a rate, oxidation occurs; at too high a rate, organic carbon is diluted by sediment. *Modified from Johnson Ibach (1982), reprinted by permission of the American Association of Petroleum Geologists.*

decreases with sedimentation at high rates of deposition. The switchover occurs at slightly different sedimentation rates for the three lithologies. Johnson Ibach interprets these results as showing that at slow sedimentation rates organic matter is oxidized on or near the seabed. Rapid deposition effectively dilutes the organic content of the sediment. For each lithology, there appears to be an optimum rate for the preservation of organic matter; this rate is a balance between dilution and destruction of the organic fraction. Nonetheless, more recent studies of both modern and ancient organic-rich sediments have led to the conclusion that organic productivity is the dominant controlling factor in its preservation (Pederson and Calvert, 1990; Parrish, 1995).

Another important condition favoring the preservation of organic matter is the presence of stratification within the waters from which sedimentation occurs. Demaison and Moore (1979, 1980) recognize four major settings for the formation of organic-rich sediments in anoxic environments (Fig. 5.8).

The first of these settings is the freshwater lake, in which thermally induced stratification of water develops, with warm water near the surface and cooler, denser water deeper down

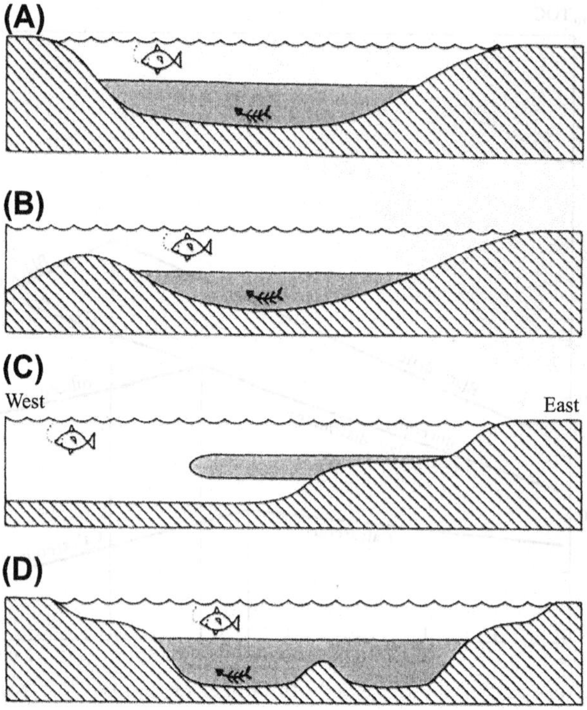

FIGURE 5.8 Illustrations of the four main settings for anoxic environments favorable for the preservation of organic matter in sediments as classified by Demaison and Moore (1980): (A) a lake (thermally induced stratification), (B) a barred basin (stratification due to salinity contrast), (C) a continental shelf with upwelling, and (D) an ocean basin anoxic event.

(Fig. 5.8(A)). Life will thrive in the upper zone, with phytoplankton photosynthesizing and oxygenating the photic zone. In the deeper layer, oxygen, once used up, will not be replenished because photosynthesis does not occur in the dark, and the density layering inhibits circulation. Sporadic storms may stimulate water circulation and cause mass mortality by asphyxiation of life throughout the lake, enhancing the amount of organic detritus deposited in the lake bed (Fleet et al., 1988). This situation is well known in modern lakes, such as those of East Africa (Degens et al., 1971, 1973). Similar lakes in ancient times include the series of Tertiary lakes of Colorado, Utah, and Wyoming, in which the Green River formation oil shales were deposited (Bradley, 1948; Robinson, 1976; Katz, 1995).

A second setting for the deposition of organic matter and its preservation in an anoxic environment occurs in barred basins (Fig. 5.8(B)). In this setting, seawater enters semirestricted lagoons, gulfs, and seas. In arid climates, evaporation causes the water level to drop and salinity to increase. Less dense water from the open sea enters across the bar and floats on the denser indigenous water. The same bar prevents the escape of the denser water to the sea. Thus, a water stratification analogous to that of the freshwater lake develops. But in this case, the density layering is due to differences not of temperature, but of salinity. Similar anoxic bottom conditions develop, however, and organic matter may be preserved. The Black Sea is a recent example of an anoxic barred basin (Deuser, 1971; Degens and Ross, 1974). The organic-rich Liassic shales of northwest Europe may have been deposited in this type of barred basin (Hallam, 1981; Bessereau et al., 1995).

A third situation favorable to the deposition and preservation of organic matter occurs today on the western sides of continents in low latitudes. The high productivity of these regions has already been described, and attributed to the upwelling of waters rich in nutrients in general, and phosphates in particular (Fig. 5.8(C)). In these settings, the seawater is commonly oxygen-deficient between 200 and 1500 m (650 and 5000 ft) (Fig. 5.9). The Phosphoria formation (Permian) and the Chattanooga–Woodford shales (Devonian–Mississippian) were probably deposited in analogous settings (Claypool et al., 1978; Duncan and Swanson, 1965; Parrish, 1982, 1995). Both are relatively thin, but laterally extensive, formations that cover ancient continental shelves. The deposition of such organic-rich sediments is favored by a rise in sea level, possibly caused by a global temperature rise, so that the anoxic zone rises to extend across the continental shelves. Thus, this type of source rock commonly blankets an unconformity and occurs during the high stand phase of the sequence stratigraphic cycle. Specifically, it overlies the maximum flooding surface and occurs within the lower part of the high stand system tract (Creaney and Passey, 1993; Bessereau et al., 1995).

The fourth setting for the deposition of organic sediment is in an anoxic ocean basin. There is no known Recent instance of this situation. Most modern ocean floors are reasonably well oxygenated. Deep oceanic circulation is caused by density currents, where cold polar waters flow beneath warmer low-latitude waters. It has been argued, however, that deep oceanic circulation may not have existed in past periods when the earth had a uniform, equable climate without the polar ice caps of today; this seems to have been particularly true of

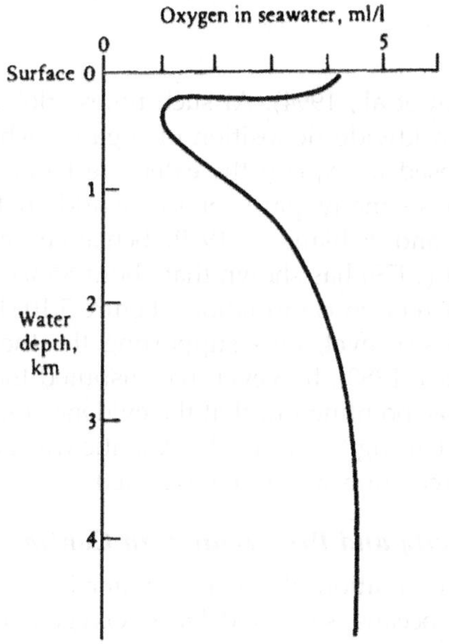

FIGURE 5.9 Variation of oxygen content against depth in modern oceans. *After Emery (1963), and Dietrich (1963).*

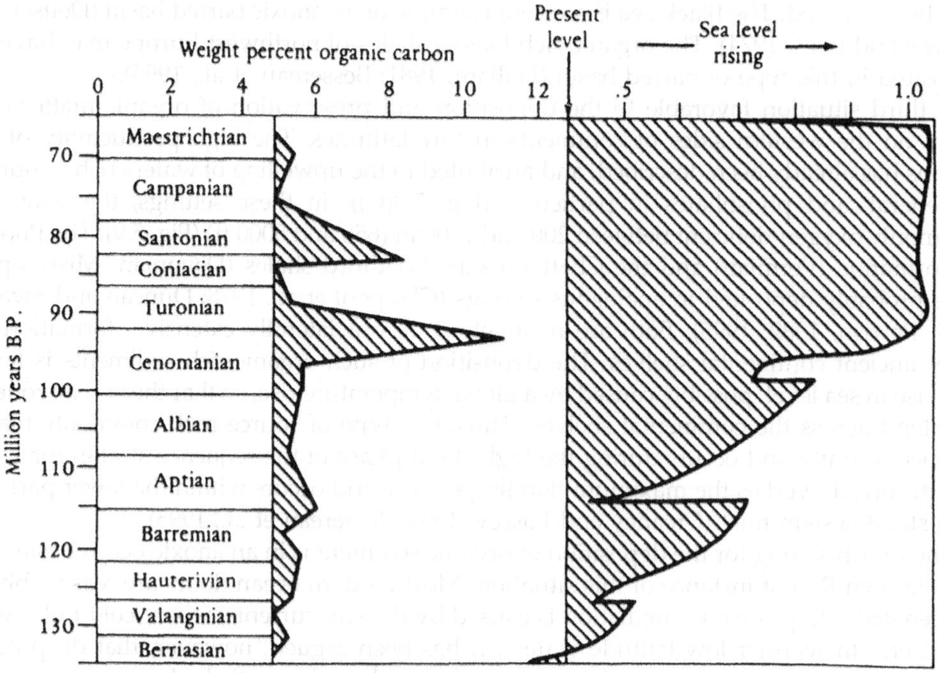

FIGURE 5.10 Relationship between source rocks and estimated sea level. Left: Graph of average values of organic carbon in Deep Sea Drilling Project cores in the North Atlantic. *(After Arthur and Schlanger (1979).)* Right: Global sea level curve. Modified from Vail et al. (1977), reprinted by permission of the American Association of Petroleum Geologists.

the late Mesozoic Era (Allen et al., 1994). At such times, global "anoxic" events may have been responsible for the worldwide deposition of organic-rich sediment (Fig. 5.8(D)). This mechanism has been proposed to explain the extensive Upper Jurassic–Lower Cretaceous black shales that occur in so many parts of the world, notably in the Atlantic Ocean and its environs (Arthur and Schlanger, 1979; Schlanger and Cita, 1982; Stein, 1986; Summerhayes, 1987). Tissot (1979) has shown that about 85% of the world's oil was sourced from Late Jurassic to Early Cretaceous formations. Figure 5.10 shows the correlation between organic sedimentation and sea level, thus supporting the thesis that transgressions favor anoxic environments. Waples (1983), however, has disputed the existence of a global anoxic event in the early Cretaceous, pointing out that the evidence mainly comes from the Atlantic Ocean and its borderlands. During this time, the Atlantic was but a narrow seaway in which local restriction of water circulation was to be expected.

5.2.1.2 *Organic Productivity and Preservation in Continental Environments*

The productivity and preservation of organic matter in the continental realm are rather different from those in the oceans, seas, and lakes. Oxygen content is constant. Important variables are water and the number of growing days per year, which is a function of temperature and daylight hours. Thus, the organic productivity of the polar regions is low. Given adequate humidity, it increases with increasing temperature and daylight toward the equator.

Significant destruction of organic matter does not occur in upland areas below a mean temperature of approximately 10 °C and, in swamps, of approximately 15 °C (Gordon et al., 1958). Since any sediment with or without organic matter is unlikely to be preserved in upland environments, only swamp deposits need concern us. Gordon et al. (1958) also showed that the rate of production versus decay of organic matter increases from polar temperatures up to about 25 °C. At this point, organic productivity begins to decline, although the rate of decay continues to increase. Thus, the preservation potential of organic detritus declines above 25 °C. These data are of relatively little importance as far as the generation of fluid hydrocarbons are concerned, but are of considerable interest regarding the formation of coal. The same conclusion is reached for continental realms as for oceans, namely, that the rate of decay of organic matter is more significant than its rate of formation.

5.2.2 Preservation of Organic Matter in Ancient Sediments

Extrapolating these data from the present day back in geological time is very difficult. Whereas the oceans appear to have contained diverse plant and animal life since way back in the Precambrian, plants only began to colonize the land toward the end of the Silurian Period, and extensive coal formation did not begin until the Carboniferous Period. In addition to the possibility of fluctuation of the rate of production of organic matter in the past, the chemical composition of the various plants may have changed as well. In particular, the amount of lignin may have increased in proportion to lipids, carbohydrates, and proteins.

Similarly, many paleontologists have traced the abundance of different groups of fossils back through geological time. Most studies use the number of species of a particular group as a criterion for their abundance. Although extrapolating from this information to actual numbers and biomass is hard, studies such as that by Tappan and Loeblich (1970) on phytoplankton may be particularly significant. They postulate a gradual increase in phytoplankton abundance from the Precambrian to 2014, with major blooms in the late Ordovician to early Silurian, and in the late Cretaceous.

This review shows that the preservation of organic matter in sediments is favored by high organic productivity; by an optimum, though not necessarily slow, sedimentation rate; and by anoxic conditions. These conditions may occur in diverse settings, from localized lakes to whole oceans. Gas chromatography can be used to fingerprint petroleum and to correlate it with the source rock from which it was derived (Bruce and Schmidt, 1994) and to determine the type of organism and depositional environment in which it lived (Volkman, 1988).

Many geologists have documented the distribution of organic-rich sediments in space and time (North, 1979, 1980; Grunau and Gruner, 1978; Grunau, 1983; Klemme and Ulmishek, 1991). Tissot (1979) and Klemme (1993) have pointed out the concentration of source rocks in Late Jurassic to Early Cretaceous sediments (Fig. 5.11).

Three old-established themes of geological research can now be brought together. Since the days of Du Toit and Wegener, geologists have been reconstructing the movement of the continents through time. It is now possible to use computers to model not only the wanderings of the continents but also the patterns of water circulation in the ancient oceans through

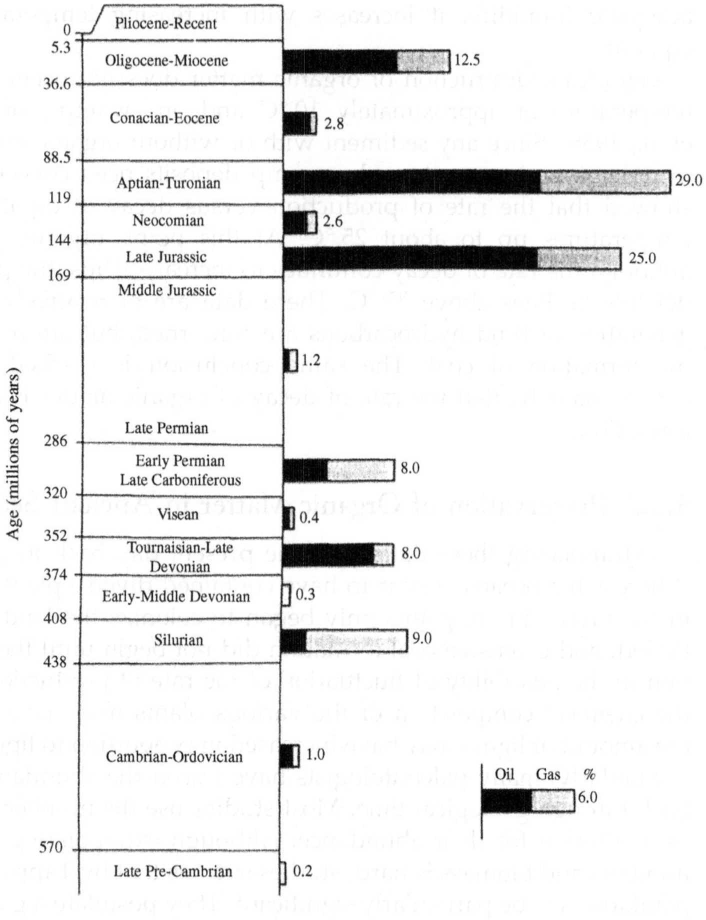

FIGURE 5.11 Histograms showing the distribution of the world's currently discovered oil and gas reserves through geological time. Note the concentration in rocks of Late Jurassic to Early Cretaceous ages, times of known organic-rich sediment deposition. *After Klemme and Ulmishek (1991).*

which they drifted, together with climatic variations. These can be integrated to map the global distribution of petroleum source rocks through the history of the earth (Huc, 1995).

5.3 FORMATION OF KEROGEN

Having now discussed the generation and preservation of organic matter at the earth's surface, it is appropriate to consider what happens to this organic matter when buried in a steadily subsiding sedimentary basin. As time passes, burial depth increases, exposing the sediment to increased temperature and pressure.

Tissot (1977) defined three major phases in the evolution of organic matter in response to burial:

1. *Diagenesis*: This phase occurs in the shallow subsurface at near normal temperatures and pressures. It includes both biogenic decay, aided by bacteria, and abiogenic reactions. Methane, carbon dioxide, and water are given off by the organic matter, leaving a

complex hydrocarbon termed *kerogen* (to be discussed in much greater detail shortly). The net result of the diagenesis of organic matter is the reduction of its oxygen content, ling the hydrogen:carbon ratio largely unaltered.

2. *Catagenesis*: This phase occurs in the deeper subsurface as burial continues and temperature and pressure increase. Petroleum is released from kerogen during catagenesis—first oil and later gas. The hydrogen:carbon ratio declines, with no significant change in the oxygen:carbon ratio.
3. *Metagenesis*: This third phase occurs at high temperatures and pressures verging on metamorphism. The last hydrocarbons, generally only methane, arxpelled. The hydrogen:carbon ratio declines until only carbon is left in the form of graphite. Porosity and permeability are now negligible.

The evolution of surface organic matter into kerogen and the ensuing generation of petroleum are discussed in detail in the following section.

5.3.1 Shallow Diagenesis of Organic Matter

The pH and Eh adjacent to the sediment:water interface are controlled by chemical reactions in which bacteria play an important role. In the ordinary course of events, circulation continually mixes the oxygenated and oxygen-depleted waters. As previously discussed, stratified water bodies inhibit this circulation. An oxygenated zone (+Eh) overlies a reducing zone (−Eh). Depending on circumstances, the interface between oxidizing and reducing conditions may occur within the water, at the sediment:water interface, or, if the sediments are permeable, below the sea bottom. In the lower reducing zone, anaerobic sulfate-reducing bacteria of the genus *Desulfovibrio* remove oxygen from sulfate ions, releasing free sulfur. In the upper oxygenated zone, the bacteria *Thiobacillus* oxidize the sulfur again:

$$SO_4 \xrightleftharpoons[\textit{Thiobacillus}]{\textit{Desulfovibrio}} S + 2O_2$$

In the reducing zone, the sulfur may combine with iron in ferrous hydroxide to form pyrite:

$$Fe(OH)_2 + 2S = FeS_2 + H_2O$$

Sulfate ions may also react with organic matter to form hydrogen sulfide:

$$SO_4 + \underset{\substack{\textit{Organic}\\\textit{matter}}}{2CH_2O} = 2HCO_3 - +H_2S$$

This formula for organic matter is very simple, and it may be more appropriate to assume that the average content of marine organic matter has a carbon: nitrogen:phosphorus ratio of 106:16:1 (Goldhaber, 1978).

The first stage of biological decay is oxidation, which generates water, carbon dioxide, nitrates, and phosphates:

$$(CH_2O)_{106}(NH_3)_{16}H_3PO_4 + 138O_2 = 106CO_2 + 16NHO_3 + H_3PO_4 + 122H_2O$$

In the next stage, reduction of nitrates and nitrites occurs:

$$(CH_2O)_{106}(NH_3)_{16}H_3PO_4 + 94{\cdot}4HNO_3 = 106CO_2 + 55{\cdot}2N_2 + 177{\cdot}2H_2O + H_3PO_4$$

This process is followed in turn by sulfate reduction, which results in the generation of hydrogen sulfide and ammonia:

$$(CH_2O)_{106}(NH_3), 6H_3PO_4 + 52SO_4^{2-} = 106HCO_3^- + 53H_2S + 16NH_3 + H_3PO_4$$

These equations are, of course, very simplified. As has been shown earlier, the organic compounds that are the starting point of this sequence of reactions are diverse and complex, consisting essentially of proteins, carbohydrates, lipids, and lignin; the relative rate of decay being in this order, with proteins being the least stable and woody lignin the most resistant. These compounds are all acted on by the enzymes of microbes to produce various biomonomers. For example, carbohydrates, such as starch and cellulose, are broken down to sugars. Cellulose is also converted into methane and carbon dioxide:

$$(C_6H_{10}O_5)_n \rightarrow CO_2 + CH_4$$

Methane is a major by-product of the bacterial decay of not only cellulose but also many other organic compounds. As this biogenic methane moves upward from the decaying organic matter, it may cross the −Eh:+Eh surface and be oxidized:

$$3CH_4 + 6O_2 = 3CO_2 + 6H_2O$$

In environments where the deposition of organic matter and its rate of decay are rapid, free methane may seep to the surface as bubbles of marsh gas. This gas occasionally ignites spontaneously, giving rise to the will o' the wisp phenomenon, which terrifies the uninformed in bogs and graveyards.

The proteins in decaying organic matter are degraded into amino acids and peptides. The lipid waxes and fats degrade to glycerol and fatty acids. The lignin of woody tissues breaks down to phenols, aromatic acids, and quinones. Plant material is degraded into fulvic acids, humic acids, and humin. These changes all take place within the top few meters of sediment. With gradually increasing burial depth, the physical environment of the sediment changes, and chemical reactions alter in response to these changes. The increase in overburden pressure results in compaction of the sediment. Numerous studies of compaction have been made and are discussed elsewhere. At this point, it is sufficient to state that within the first 300 m of burial, the porosity of a clay typically diminishes from some 80% down to only about 30 or 40%. During compaction, water escapes. Part of this water is primary pore water, but some of it is of biogenic origin. It contains in solution carbon dioxide, methane, hydrogen sulfide, and other organic compounds of decay, loosely termed *humic acids*.

At this time, several important inorganic reactions may be taking place, causing the formation of early authigenic minerals. The generation of pyrite has already been discussed, but in carbonate-rich clays siderite ($FeCO_3$) is also common. Early calcium carbonate cementation may occur by direct precipitation or, in evaporitic environments, by the interreaction of calcium sulfate and organic matter (Friedman, 1972). Sulfate ions react with organic matter to produce hydrogen sulfide (as previously discussed) and HCO_3. The latter reacts with the calcium to form calcite, water, and carbon dioxide. Essentially,

$$\underset{}{CaSO_4} + \underset{\textit{Organic matter}}{2CH_2O} \rightarrow CaCO_3 + H_2O + CO_2 + H_2S$$

With increasing depth of burial, the ambient temperature increases and the role of bacteria in biogenic reactions gradually declines as they die out (although some thiophilic bacteria can occur deep in the subsurface). At the same time, however, increasing temperature allows inorganic reactions to accelerate. The generation of hydrocarbons now declines as the production of biogenic methane ceases. There are occasional reports of the shallow occurrence of biogenic hydrocarbons other than methane. One of the best known is the occurrence of up to 200 ppm of aromatics and paraffinic naphthenes in sand less than 10,000 years old and only 30 m deep off the Orinoco delta (Kidwell and Hunt, 1958). More recently, lipid-rich liquid hydrocarbons have been recorded from ocean floor samples. This protopetroleum is formed by the minor breakdown of organic oils. It is chemically distinct from crude oil, has not undergone thermal maturation, and does not appear to be present in commercial accumulations (Welte, 1972). Important changes continue to take place, however, within the preserved organic matter. Water and carbon dioxide continue to be expelled as the formation of kerogen begins.

Figure 5.12 illustrates the transformation of living material into kerogen and various types of hydrocarbons.

5.3.2 Chemistry of Kerogen

The etymology and original definition of kerogen as recognized in oil shale is discussed in Chapter 9. In 2014, kerogen is the term applied to disseminated organic matter in sediments that is insoluble in normal petroleum solvents, such as carbon bisulfide. This insolubility distinguishes it from bitumen. Chemically, kerogen consists of carbon, hydrogen, and oxygen, with minor amounts of nitrogen and sulfur (Table 5.3).

Three basic types of kerogen are generally recognizable. The differences are chemical and are related to the nature of the original organic matter. Because these three kerogen types generate different hydrocarbons their distinction and recognition are important (Tissot, 1977; Dow, 1977). Figure 5.13 illustrates the three kerogen types and depositional environments.

Type I kerogen is essentially algal in origin (Plate 5.6). It has a higher proportion of hydrogen relative to oxygen than the other types of kerogen have (H:O ratio is about 1.2–1.7). The H:C ratio is about 1.65 (Table 5.3). Lipids are the dominant compounds in this kerogen, with derivates of oils, fats, and waxes. This kerogen is particularly abundant in algae such as *Bottryococcus*, which occurs in modern Coorongite and ancient oil shales. Similar algal kerogen is characteristic of many oil shales, source rocks, and the cannel or boghead coals. The basic chemical structure of type I algal kerogen is shown in Fig. 5.14(A).

Type II, or liptinitic, kerogen is of intermediate composition (Plate 5.6). Like algal kerogen, it is rich in aliphatic compounds, and it has an H:C ratio of >1. The chemical structure is shown in

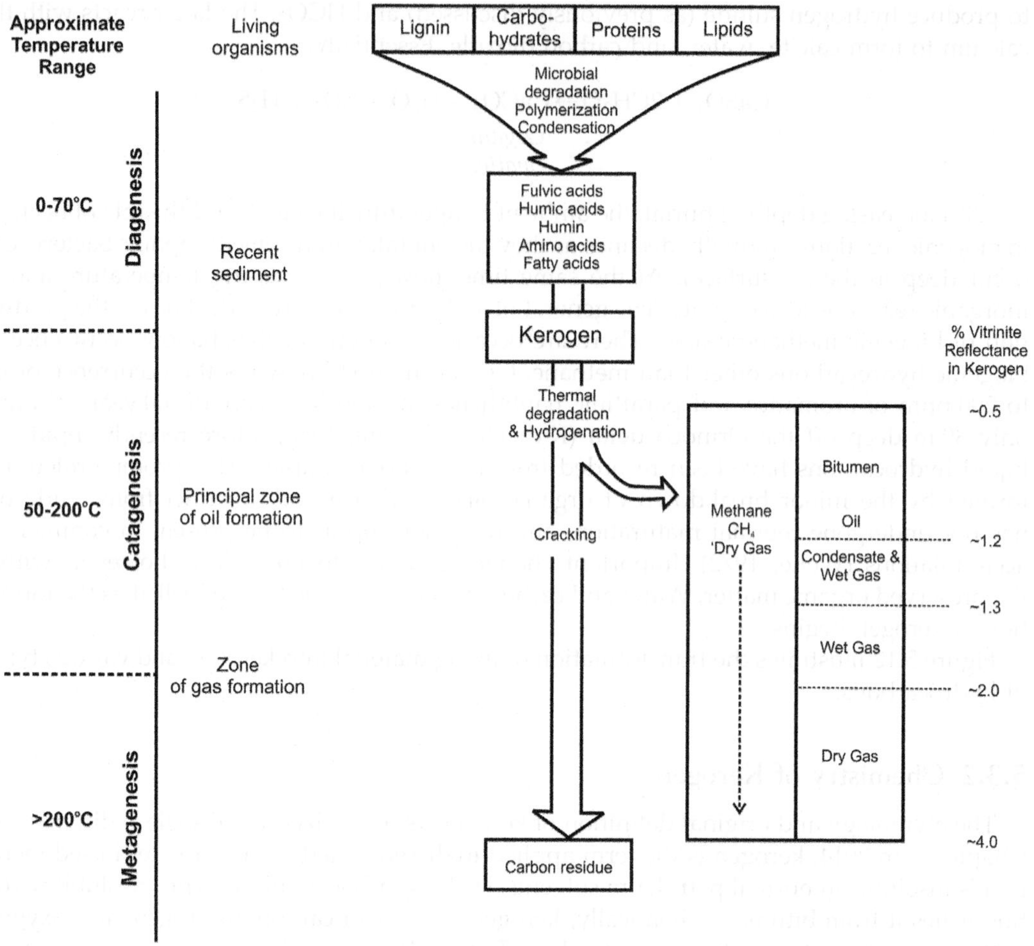

FIGURE 5.12 Alteration of organic matter in the subsurface. *Modified from Tissot and Welte (1984) and Meissner (1991).*

Fig. 5.14(B). The original organic matter of type II kerogen consisted of algal detritus, and also contained material derived from zooplankton and phytoplankton. The Kimmeridge clay of the North Sea and the Tannezuft shale (Silurian) of Algeria are of this type.

Type III, or humic, kerogen has a much lower H:C ratio (<0.84). Chemically, it is low in aliphatic compounds, but rich in aromatic ones (Plate 5.6). The basic molecular structure is shown in Fig. 5.14(C). Humic kerogen is produced from the lignin of the higher woody plants, which grow on land. It is this humic material that, if buried as peat, undergoes diagenesis to coal. Type III kerogen tends to generate largely gas and little, if any, oil. Nonmarine basins were once thought to be gas prone because of an abundance of humic kerogen, whereas marine basins were thought to be oil provinces because of a higher proportion of algal kerogen. This type of generalization is not valid. Many continental basins contain

TABLE 5.3 Chemistry of Kerogens

	Weight percent					Ratios		
	C	H	O	N	S	H–C	O–C	Petroleum type
Type I	75.9	8.84	7.66	1.97	2.7	1.65	0.06	Oil
Algal	(7 samples)							
Type II	77.8	6.8	10.5	2.15	2.7	1.28	0.1	Oil and gas
Liptinic	(6 samples)							
Type III	82.6	4.6		2.1	0.1	0.84	0.13	Gas
Humic	(3 samples)							

Weight percent data from Tissot and Welte (1978), ratios from Dow (1977). Reprinted with permission from Springer-Verlag.

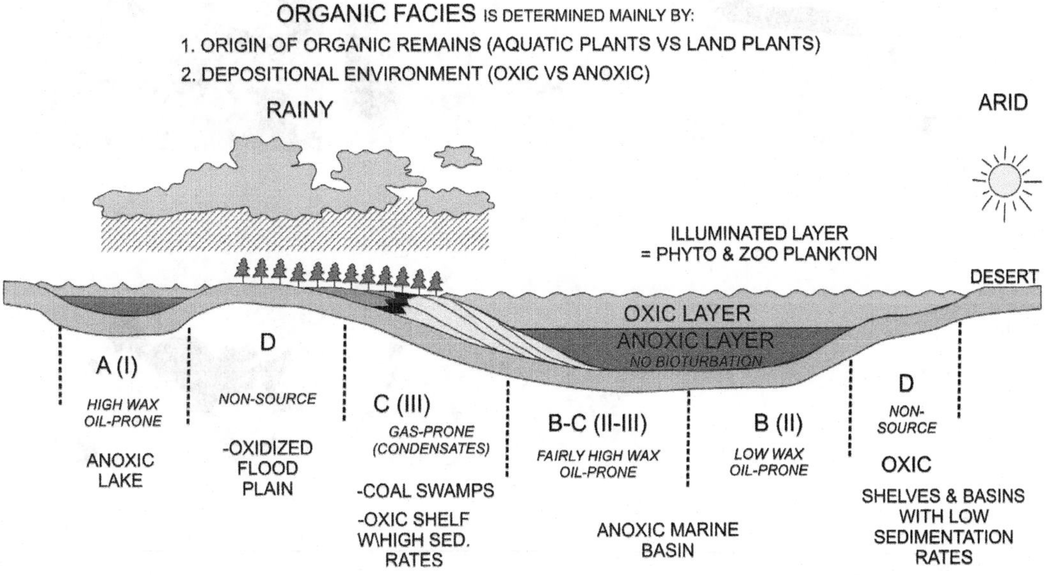

FIGURE 5.13 Organic facies and kerogen type. *Modified from Jones and Demaison (1980).*

lacustrine shales rich in algal kerogen. The Green River formation of Colorado, Utah, and Wyoming is a noted example; other examples occur in China (Li et al., 1982). Further details on kerogen can be found in Tissot and Welte (1978), Hunt (1996), Katz (1995), Brooks and Fleet (1987), and Durand (1980).

This review of the three basic types of kerogen shows the importance of identifying the nature of the organic matter in a source rock so as to assess accurately its potential for generating hydrocarbons. A second important factor to consider is not only the quality of kerogen but also the quantity necessary to generate significant amounts of oil and gas suitable for

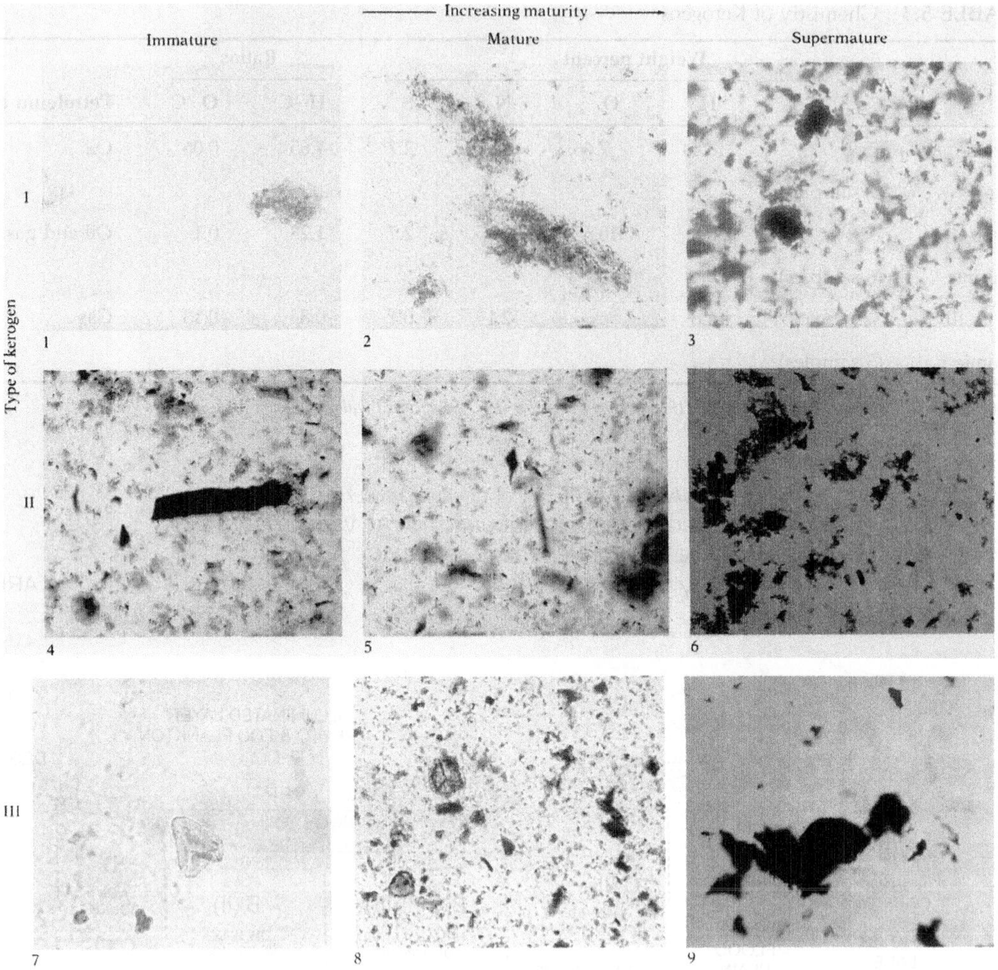

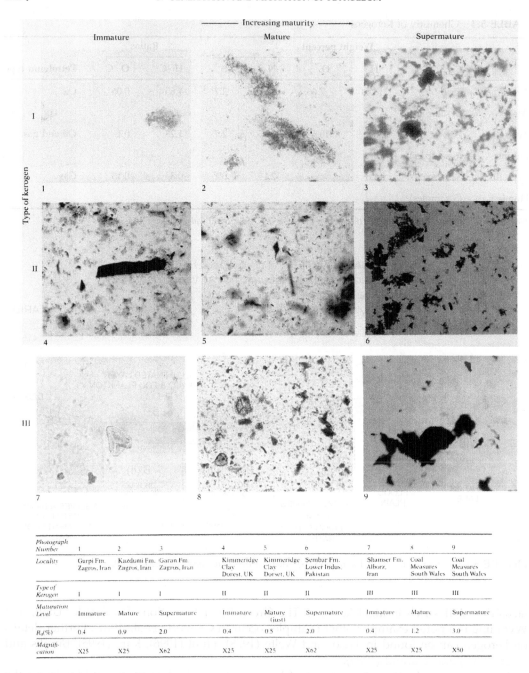

Photograph Number	1	2	3	4	5	6	7	8	9
Locality	Gurpi Fm. Zagros, Iran	Kazdumi Fm. Zagros, Iran	Garan Fm. Zagros, Iran	Kimmeridge Clay Dorest, UK	Kimmeridge Clay Dorset, UK	Sembar Fm. Lower Indus, Pakistan	Shamser Fm. Alborz, Iran	Coal Measures South Wales	Coal Measures South Wales
Type of Kerogen	I	I	I	II	II	II	III	III	III
Maturation Level	Immature	Mature	Supermature	Immature	Mature (just)	Supermature	Immature	Mature	Supermature
R_0(%)	0.4	0.9	2.0	0.4	0.5	2.0	0.4	1.2	3.0
Magnification	X25	X25	X62	X25	X25	X62	X25	X25	X50

PLATE 5.6 Photomicrographs of different types of kerogen at various maturation stages (See the color plate). *Courtesy of Kinghorn and Rahman.*

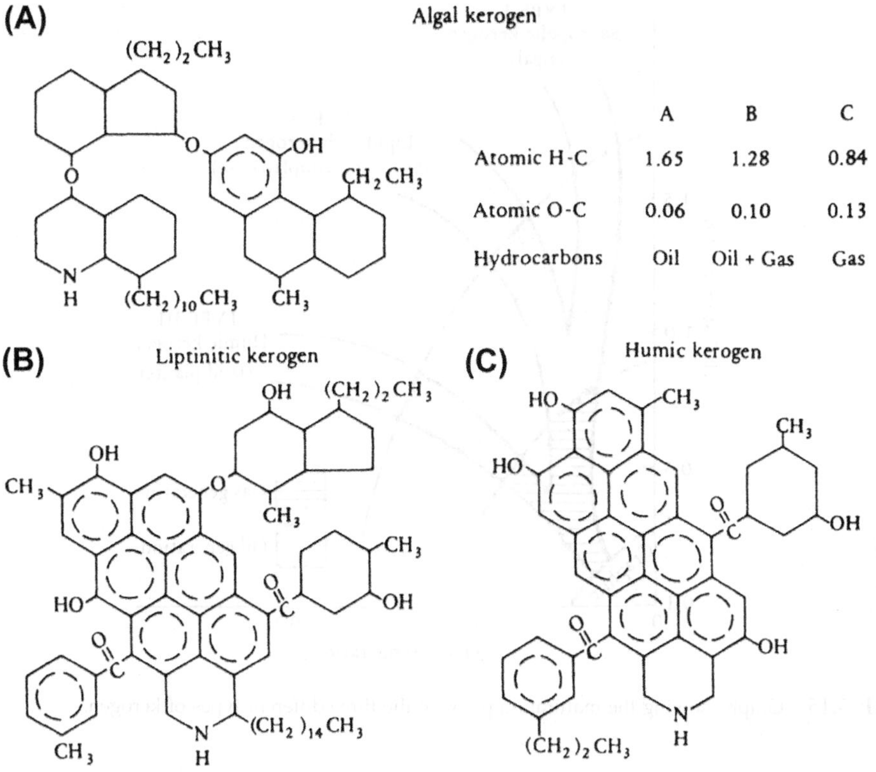

	A	B	C
Atomic H-C	1.65	1.28	0.84
Atomic O-C	0.06	0.10	0.13
Hydrocarbons	Oil	Oil + Gas	Gas

FIGURE 5.14 The molecular structure of (A) type I, or algal, kerogen; (B) type II, or liptinitic, kerogen; and (C) type III, or humic, kerogen. *Reprinted from Dow (1977), with kind permission of Elsevier Science, The Netherlands.*

commercial production. Several separate items are to be considered here, including the average amount of kerogen in the source bed, the bulk volume of the source bed, and the ratio of emigrated to residual hydrocarbons. The total organic matter in sediments varies from 0% in many Precambrian and continental shales to nearly 100% in certain coals. A figure of 1500 ppm TOC is sometimes taken as the minimum requirement for further exploration of a source rock (Pusey, 1973a).

5.3.3 Maturation of Kerogen

During the phase of catagenesis, kerogen matures and gives off oil and gas. Establishing the level of maturation of kerogen in the source rocks of an area subject to petroleum exploration is vital. When kerogen is immature, no petroleum has been generated; with increasing maturity, first oil and then gas are expelled; when the kerogen is overmature, neither oil nor gas remains. Figure 5.15 shows the maturation paths for the different types of kerogen. The maturation of kerogen can be measured by several techniques to be described shortly. The rate of maturation may be dependent on temperature, time, and, possibly, pressure.

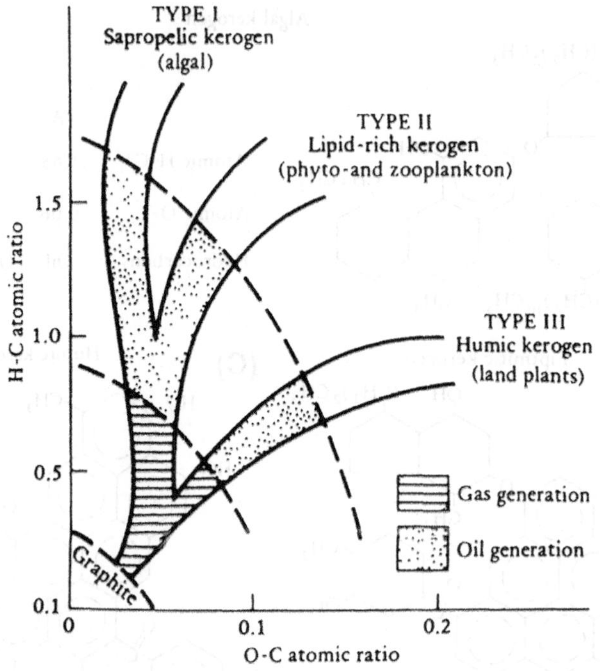

FIGURE 5.15 Graph showing the maturation paths of the three different types of kerogen.

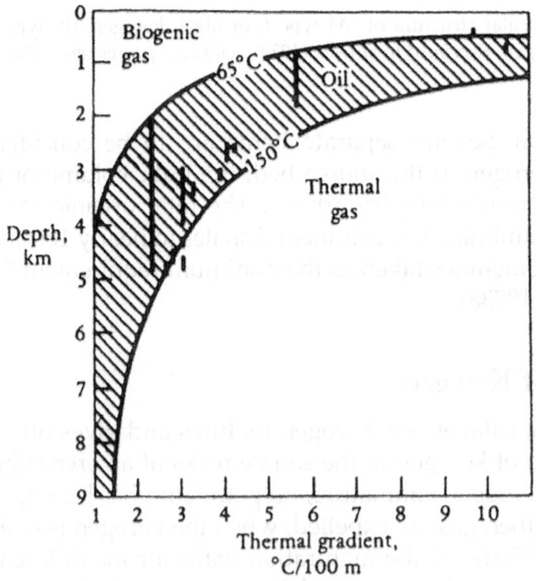

FIGURE 5.16 Graph of depth versus thermal gradient showing how oil occurs in a liquid window of between 65 and 150 °C, which extends from shallow depths with high thermal gradients to deep basins with lower gradients. *After Pusey (1973a).*

A number of workers have documented an empirical correlation between temperature and petroleum generation (Pusey, 1973a,b; Philippi, 1975; Hunt, 1996). Significant oil generation occurs between 60 and 120 °C, and significant gas generation between 120 and 225 °C. Above 225 °C, the kerogen is inert, having expelled all hydrocarbons; only carbon remains as graphite (Fig. 5.16). The temperatures just cited are only approximate boundaries of the oil and gas windows.

5.3.3.1 *The Chemical Laws of Kinetics*

The chemical laws of kinetics, as expressed in the Arrhenius equation, state that the rate of a chemical reaction is related to temperature and time. Reaction rate generally doubles for each 10 °C increase. Thus, many geologists have considered that kerogen maturity is a function of temperature and time (Fig. 5.17). Petroleum may therefore have been generated from old, cool source rocks as well as from young, hot ones (e.g., Connan, 1974; Erdman, 1975; Cornelius, 1975). Several techniques have been developed that try to quantify the relationship of temperature and time to kerogen maturity. These techniques are generally based on a burial history curve (Van Hinte, 1978), which is a graph on which burial depth is plotted against geological time for a particular region (Fig. 5.18).

Two commonly used maturation indices are the time–temperature index (TTI) and the level of organic maturation (LOM). The TTI was first proposed by Lopatin (1971) and developed by Waples (1980, 1981). The TTI is calculated from a formula that integrates temperature with the time spent in each temperature (in increments of 10 °C) as a source rock is buried. The length of time spent in each temperature increment is established from the burial curve. The LOM was first proposed by Hood et al. (1975) and developed by Royden et al. (1980) and Cohen (1981). The LOM is based on the assumption that reaction rate doubles for each 10 °C increment of temperature. Oil generation occurs between LOM values of 7 and 13, and gas generation occurs between values of 13 and 18.

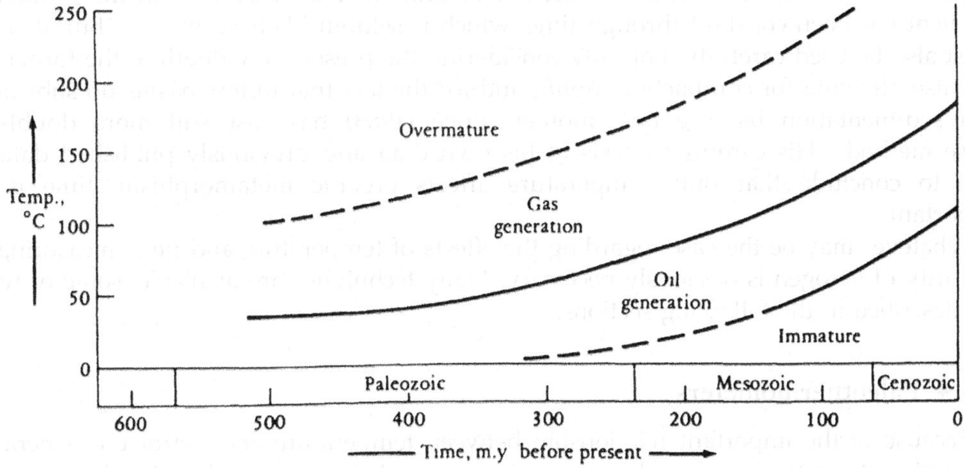

FIGURE 5.17 Graph of temperature against time of oil and gas formation. For the problems of interpreting this figure, see the text. *After Connan (1974), Erdman (1975) and Cornelius (1975).*

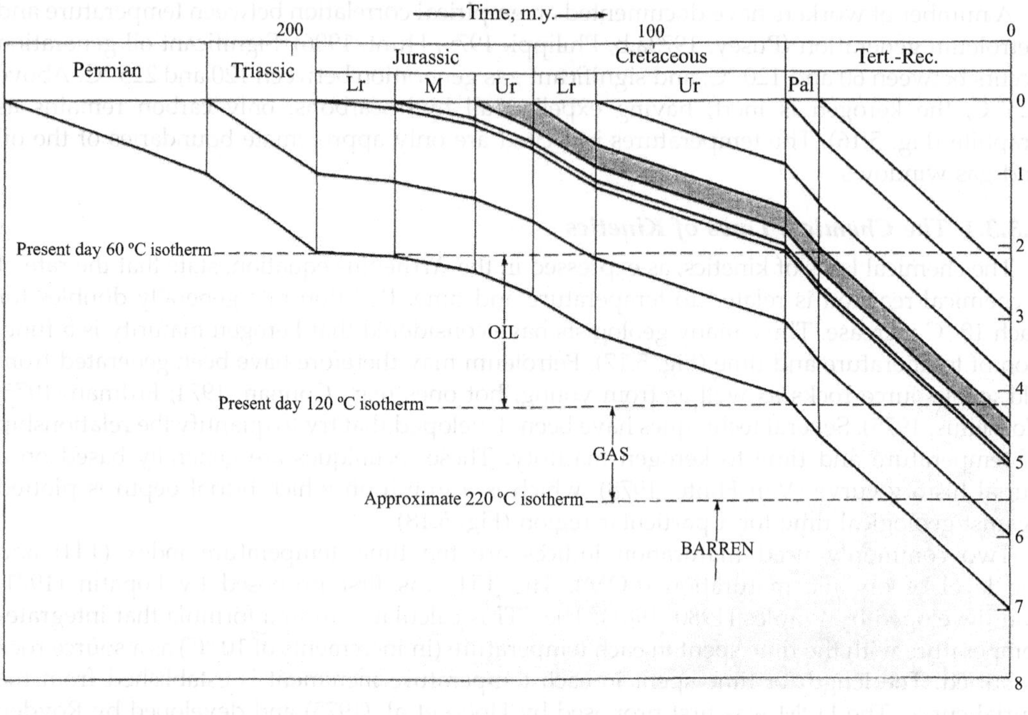

FIGURE 5.18 Burial curve for the Central Graben of the North Sea. Major Kimmeridge clay source rock shown in black. Since present-day isotherms are used, the graph only gives a rough idea of when the source rock entered the oil and gas windows. *After Selley (1980).*

Both the TTI and LOM must be used with caution. They assume that the geothermal gradient has been constant through time, which is seldom likely to be true. Burial curves must also be used carefully, not only considering the present-day depth of the formations but also allowing for compaction, uplift, and for the fact that in few basins do subsidence and sedimentation balance one another. Price (1983) has cast still more doubts on these methods. His careful analysis of his own data and previously published data led him to conclude that only temperature affects organic metamorphism; time is not important.

Whatever may be the case regarding the effects of temperature and time, measuring the maturity of kerogen is obviously necessary. Many techniques are available, some of which are described in the following sections.

5.3.4 Paleothermometers

Because of the important relationship between temperature and petroleum generation, measuring the maturity of the kerogen is important. It is not enough to be able to answer these questions: Are there organic-rich source rocks in the basin? Are they present in large-enough volumes? Are they oil prone or gas prone? It is also necessary to know whether or not the source rocks have matured sufficiently to generate petroleum or whether they are supermature and barren.

To measure only the bottom hole temperature from boreholes does not answer the question of kerogen maturity. This measurement only indicates the present-day temperature, which may be considerably lower than that of the past. It is necessary to have a paleothermometer, which can measure the maximum temperature to which the source rock was ever subjected. Several such tools are available, each varying in efficiency and each requiring expensive laboratory equipment and a considerable degree of technical expertise. Broadly speaking, two major groups of techniques are used for measuring the maximum paleotemperature to which a rock has been heated (Cooper, 1977):

1. Chemical paleothermometers
   a. Organic
      - Carbon ratio
      - Electron spin resonance (ESR)
      - Pyrolysis
      - Gas chromatography

   b. Inorganic
      - Clay mineral diagenesis
      - Fluid inclusions
2. Biological paleothermometers
   a. Pollen coloration
   b. Vitrinite reflectance

5.3.4.1 *Chemical Paleothermometers*

One of the oldest and most fundamental of maturation indices is the carbon ratio method. Obviously, as organic matter matures, organic carbon is lost; first as carbon dioxide during decarboxylation and later during petroleum generation. The net result of kerogen maturation is thus a gradual decrease in the proportion of movable organic carbon and a relative increase in the proportion of fixed carbon. The carbon ratio method compares the TOC (C_T) in the sample with the residual carbon (C_R) after pyrolysis to 900 °C. The ratio C_rC_I increases with maturation (Gransch and Eisma, 1970). This paleothermometer is not easy to calibrate, however, because of the variations in the original carbon content in the different types of kerogen and the risky assumption that their carbon ratios change at the same rate.

A second chemical thermometer is the ESR technique developed by Conoco (Pusey, 1973b). The ESR method is based on the fact that the atomic fraction of kerogen contains free electrons, whose number and distribution are related to the number and distribution of the benzene rings. These electron characteristics change progressively as the kerogen matures (Marchand et al., 1969). Kerogen can be extracted from the source rock and placed in a magnetic field whose strength can be controlled. If the sample is then exposed to microwaves, the free electrons resonate (hence the term ESR) and, in so doing, alter the frequency of the microwave. With an electrometer, certain properties of the electrons can be measured, including their number, signal width, and the location of the resonance point. As already stated, these parameters have been shown to vary with kerogen maturation.

The ESR technique was originally developed as a paleothermometer in the Gulf Coast of the United States (Pusey, 1973b) and later in the North Sea (Cooper et al., 1975). The method has the advantage of being quick (once a laboratory is established), and it can be used on minute and impure samples of kerogen. As with most paleothermometers, great care must be taken that the shale samples gathered during drilling are not caved; ideally, core or sidewall core material should be used. Again, as with the carbon ratio method, the ESR has certain problems caused by original variations in kerogen composition and the presence of recycled organic detritus (Durand et al., 1970).

Pyrolysis, the heating of kerogen or source rock, was the process developed to produce petroleum from oil shales. In 2014, the pyrolysis of kerogen samples is an important laboratory technique in source rock analysis (Espitalie et al., 1977; Tissot and Welte, 1978; Peters, 1986; Hunt, 1996). Pyrolysis is carried out with a flame ionization detector. Initially, it was necessary to separate the kerogen from the source rock. Nowadays, however, analysis can be carried out on whole rock samples using the Rock–Eval pyroanalyzer designed by the Institut Français du Petrole. The sample is heated in a stream of helium. The temperature is gradually raised at a carefully measured rate, and the expelled hydrocarbon gases recorded with a hydrogen flame ionization detector. At relatively low temperatures (200–300 °C), any free hydrocarbons in the sample are volatilized. These hydrocarbons are referred to as S_1. With increasing temperature, hydrocarbons are expelled from the kerogen itself. These hydrocarbons are termed S_2. With further heating to some 390 °C, carbon dioxide is expelled to generate a third peak, S_3. Pyrolysis generally continues up to 800–900 °C. The temperature at which the S_2 peak occurs is termed T_{max}. The three readings can be used to determine the maturation level of the source rock. Where migration has not occurred, the ratio $S_1/(S_1 + S_2)$ shows the amount of petroleum generated compared with the total amount capable of being generated. This ratio is referred to as the transformation ratio or production index (PI). When plotted against depth, PI will generally show a gradual increase as the source rock matures.

Another important parameter established by pyrolysis is the hydrogen index (HI); this indicates the amount of pyrolyzable organic compounds (hydrocarbons) from the S_2 peak, relative to the TOC (or C_{org}). Similarly, the oxygen index indicates the amount of carbon dioxide generated compared to the TOC. Figure 5.19 illustrates a typical pyrolysis well log. A critique of the pyrolysis method can be found in Snowdon (1995).

Gas chromatography is another method of assessing the maturity of a source rock. This is used to study the distribution of n-alkanes, since Allan and Douglas (1977) have shown that there is a correlation between the shift in the pattern of these when compared with other paleothermometers such as vitrinite reflectance (to be discussed shortly). Gas chromatography shows that n-alkanes evolve with increasing temperature and burial depth. The gas chromatographs of immature source rocks of n-alkanes show a broad "whaleback," with a bimodal distribution of unsolved components. As the temperature increases and the source rock matures, the whaleback submerges and the unsolved components form a single peaky mode (Fig. 5.20). Degraded oils are characterized by a broad low whaleback that is devoid of spikes.

Clay mineral analysis is one of the oldest of the inorganic paleothermometers, having been studied quite independently of kerogen maturation. Newly deposited clay minerals include smectites (montmorillonite), illite, and kaolin. The varying abundance of different mineral

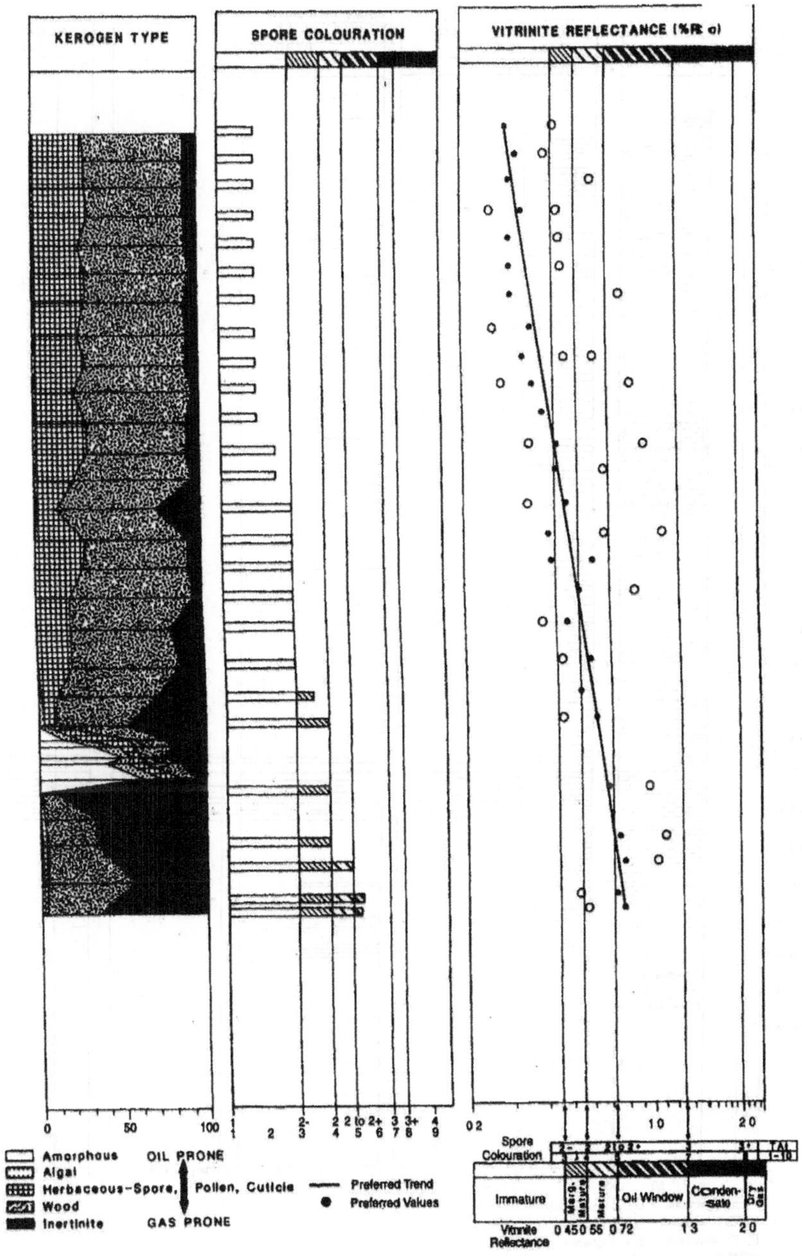

FIGURE 5.19 An example of a geochemical well log compiled to document the quantity, quality, and maturation level of a source rock. *Left*: From left to right, logs illustrate the lithology, TOC content, and Rock–Eval parameters of S2 and HI as explained in the text. *Right* From left to right, logs of kerogen type give an indication of the oil- or gas-prone nature of the source rock and two paleothermometers, spore coloration, and vitrinite reflectance, as discussed in the text. *Courtesy of Geochem, Ltd.*

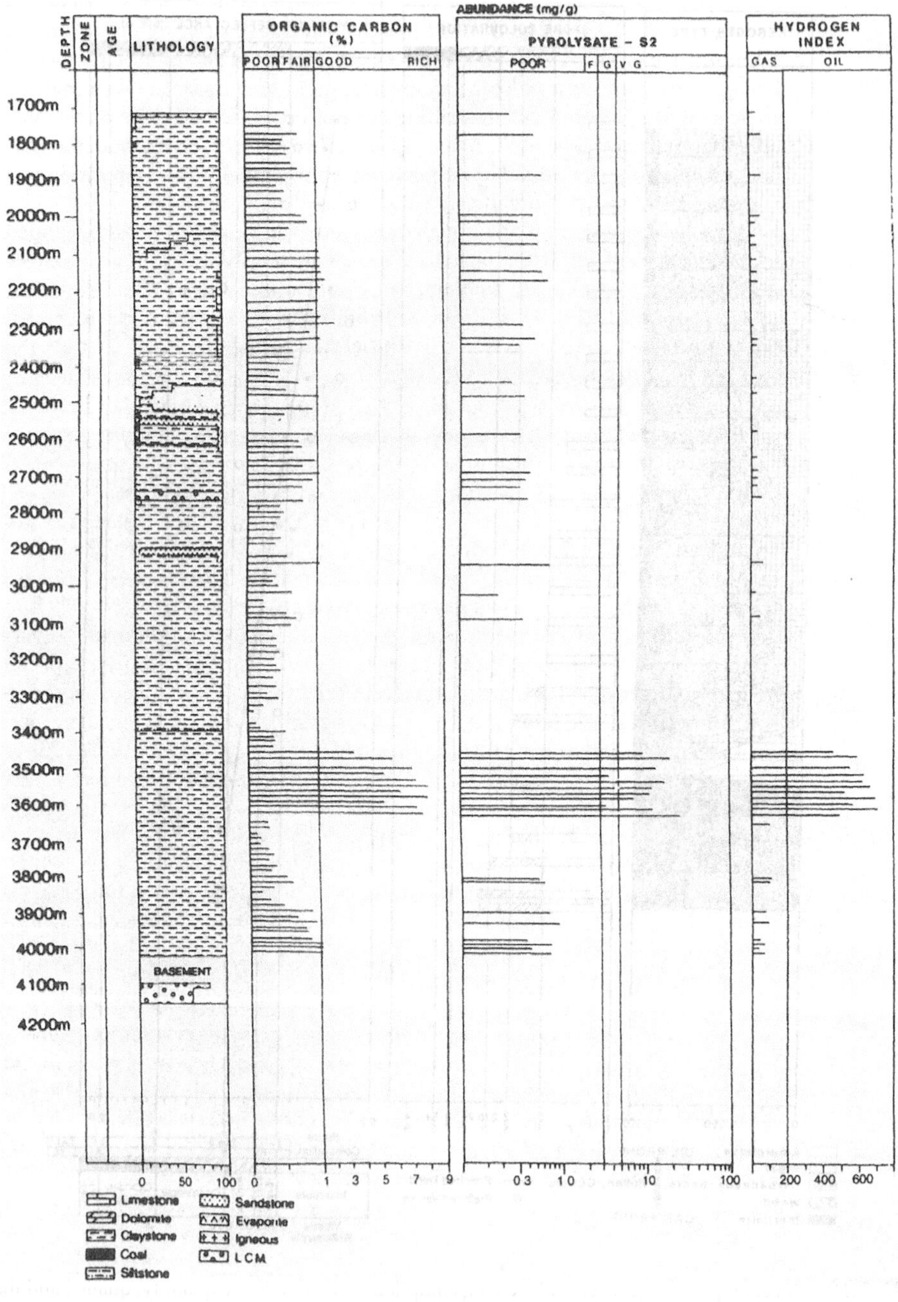

FIGURE 5.19 Cont'd

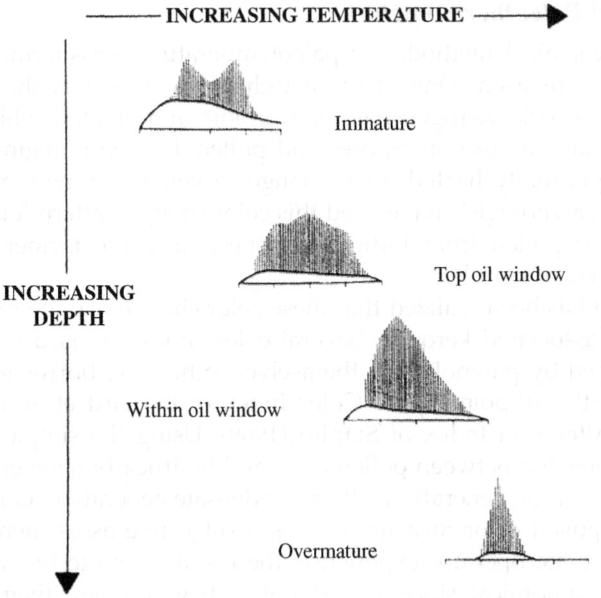

FIGURE 5.20 Diagram of gas chromatographs that show the evolution of *n*-alkanes with increasing temperature and burial depth. Each specimen is plotted on a carbon number scale of 0–40. Developed from Kinghorn (1982–1983). Immature source rocks show a broad whaleback, with a bimodal distribution of unsolved components. As the temperature increases and the source rock matures, the whaleback submerges and the unsolved components form a single peaky mode. Degraded oils are characterized by a broad low whaleback that is devoid of spikes.

types depends partly on the source area and partly on the depositional environment. As clays are buried, they undergo various diagenetic changes. Illite and kaolin authigenesis is largely related to the pH of the pore fluid. With increasing depth of burial, higher temperatures recrystallize the clays. Of particular interest is the behavior of the smectites, which dewater and change to illite within the optimum temperature for oil generation (80–120 °C). This phenomenon is considerably important to oil migration (Powers, 1967) and is discussed later in this chapter.

As temperatures increase to the upper limits at which oil and gas form, another major change in clay takes place. Kaolin and illite recrystallize into micas, whereas in ferromagnesian-rich environments they form chlorites. A number of studies have documented these changes (e.g., Fuchtbauer, 1967). These calibration points are rather imprecise to pick on the kerogen maturation scale. A greater degree of sophistication was achieved, however, by Gill et al. (1979), who showed how the crystallinity of kaolin and illite could be mapped across the South Wales coal basin and how those variations correlated with coal rank. This technique can undoubtedly be applied to other regions.

Another inorganic paleothermometer is provided by the study of fluid inclusions. This method has been used for many years by hard rock geologists interested in finding the temperature of formation of crystalline minerals (Roedder, 1984). It is now widely used in the oil industry, but principally as an aid to understanding diagenetic processes in petroleum reservoirs (Rankin, 1990). Thus, it is discussed in Chapter 6.

5.3.4.2 *Biological Paleothermometers*

In addition to chemical methods for paleotemperature measurement, several biological paleothermometers are used. One of these techniques measures the color of the organic matter in the source rock. Kerogen has many colors and shades, which are dependent on both maturation and composition. Spores and pollen, however, begin life essentially colorless. As they are gradually heated, they change to yellow, orange, brown (light to dark), and then to black. Palynologists have used this color change pattern for years to differentiate recycled spores and pollen from indigenous ones, since the former will be darker than average for a particular sample.

More recently, it has been realized that these color changes can be related to the degree of maturation of the associated kerogen. Several color codes are used by palynologists; some have been developed by palynologists themselves, others are borrowed from brewers. One popular system is the 10-point Spore Color Index of Barnard et al. (1978); another is the 5-point Thermal Alteration Index of Staplin (1969). Using the simple brewer's color code, the following relationship between pollen color and hydrocarbon generation has been noted: water: immature; lager: oil generation; bitter: condensate generation; Guinness: dry gas. It has been argued that pollen color measurement is a subjective assessment, but with reference standards and adequate operator experience, the results seem to be valid. The technique is relatively fast and economical, since the palynologists will review their samples as a routine check for contamination and recycling anyway.

The last paleothermometer to consider is vitrinite reflectance (Dow, 1977; Dow and O'Connor, 1979; Cooper, 1977), a very well-established technique used by coal petrographers to assess the rank of coal samples. Basically, the shininess of coal increases with rank from peat to anthracite. This shininess, or reflectance, can be measured optically. Vitrain, the coal maceral used for measurement, occurs widely, if sparsely, throughout sedimentary rocks. Kerogen, which includes vitrain, is separated from the sample by solution in hydrofluoric and hydrochloric acids. The residue is mounted on a slide, or in a resin block with a low-temperature epoxy resin and then polished. A reflecting-light microscope is then used to measure the degree of reflectivity, termed R_o. This step requires great care and can only be done by a well-trained operator. The prepared sample may contain not only indigenous kerogen but also caved and recycled material, as well as organic material from drilling mud (lignosulfonate). Reflectivity may also vary from the presence of pyrite or asphalt. A number of reflectivity readings are taken for each sample and plotted on a histogram, reference being made to standard samples of known R_o values. R_o may be measured for a series of samples down a borehole, and a graph of reflectivity versus depth may be plotted.

An empirical relationship has been noted between vitrinite reflectance and hydrocarbon generation. Crude oil generation occurs for R_o values between 0.6 and 1.5. Gas generation takes place between 1.5 and 3.0; at values above 3.0 the rocks are essentially graphitic and devoid of hydrocarbons. These values are only approximate because, as already discussed, the oil- or gas-generating potential of kerogen varies according to its original composition.

Reflectivity is generally assumed to increase with temperature and time. McTavish (1978) and Fang et al. (1995) have shown, however, the retarding effect of overpressure and, as

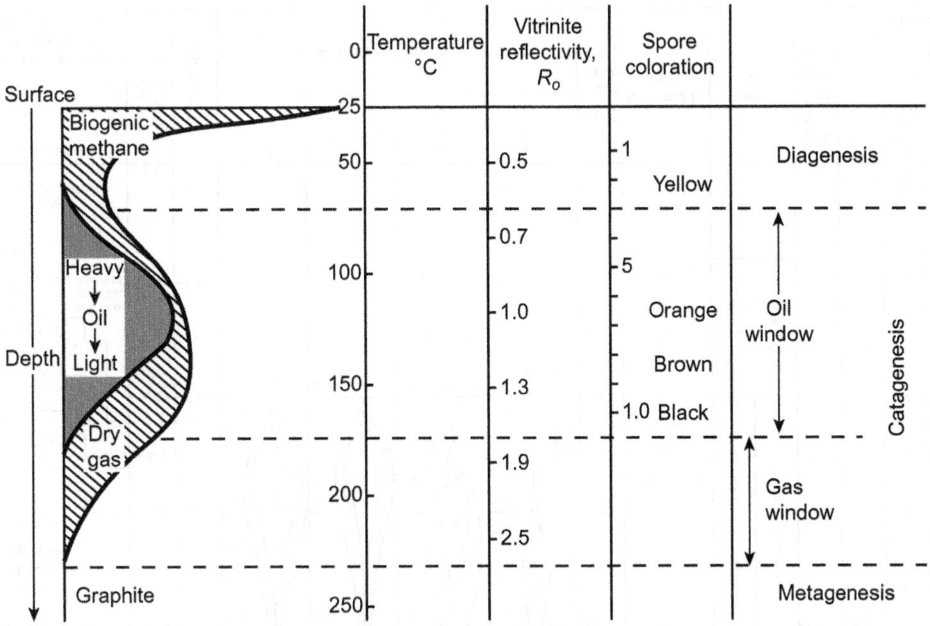

FIGURE 5.21 Correlation between hydrocarbon generation, temperature, and some paleothermometers.

already noted, Price (1983) considers temperature to be the only important variable. The measurement of vitrinite reflectivity is one of the most widely used techniques for establishing the maturity of a potential source rock in an area. Among other uses, borehole logs that plot R_o against depth indicate the intervals in which oil or gas may have been generated. Abrupt shifts in the value of R_o with depth may indicate faults or unconformities. An abrupt increase in R_o with depth followed by a return to the previous gradient may be caused by igneous intrusives.

This review of paleothermometers shows that a wide range of techniques is available. All the methods require elaborate laboratory equipment, carefully trained analytical operators, and skilled interpretation of the resultant data. There are many pitfalls for the inexperienced interpreter. Probably, pollen coloration and vitrinite reflectivity are the most reliable and widely used maturation indices at the present time. Figure 5.21 shows the relation between R_o, spore coloration, and hydrocarbon generation. Figure 5.22 shows a maturation range chart with various types of kerogen.

5.4 PETROLEUM MIGRATION

The preceding sections of this chapter showed that petroleum is of organic origin; discussed the production of organic matter and its preservation in sediments; and traced the diagenesis of organic matter into kerogen and the maturation of kerogen as petroleum is expelled. The object of this section is to discuss how fluid hydrocarbons may emigrate from the source rock to the reservoir.

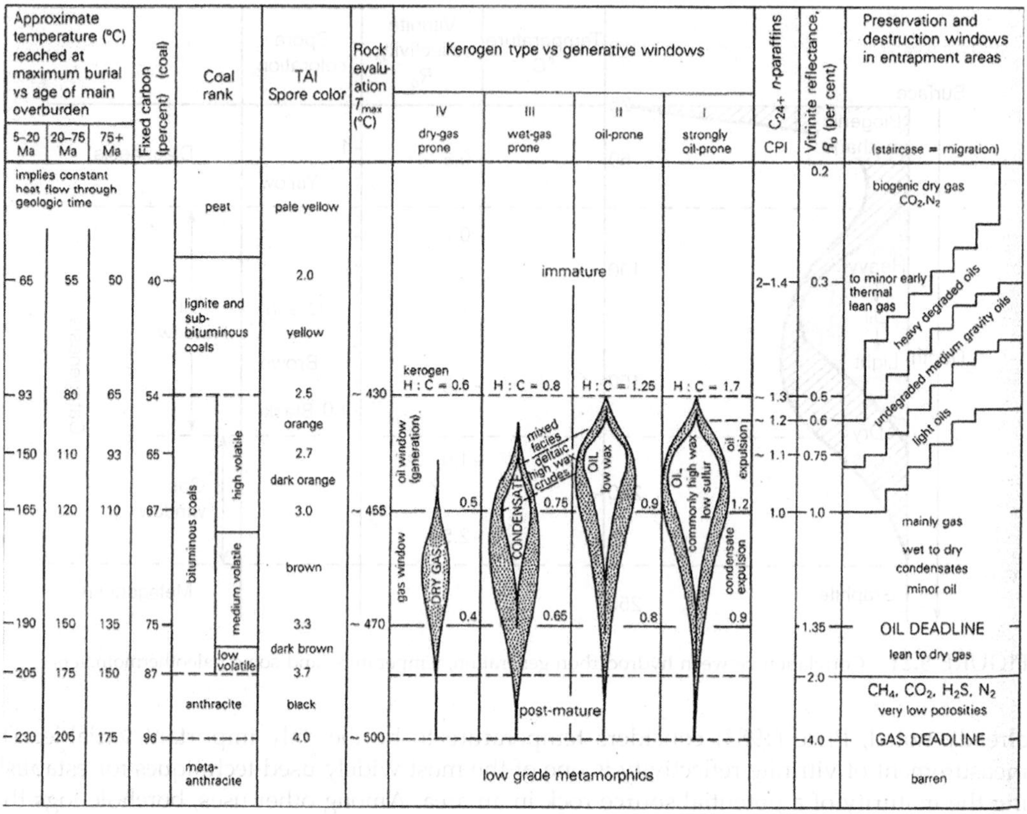

FIGURE 5.22 Maturation range chart. *Modified from Demaison (1980), North (1985).*

A number of lines of observational evidence show that oil and gas do not generally originate in the rock in which they are found, but that they must have migrated into it from elsewhere. This theory is proved by the following observations:

1. As previously discussed, organic matter is easily destroyed by oxidization in porous, permeable sediments at the earth's surface. It must therefore have invaded the reservoir rock after considerable burial and raised temperature.
2. Oil and gas often occur in solution pores and fractures that must have formed after the burial and lithification of the host rock.
3. Oil and gas are trapped in the highest point (structural culmination, or stratigraphic pinchout) of a permeable rock unit, which implies upward and lateral migration.
4. Oil, gas, and water occur in porous, permeable reservoir rock stratified according to their relative densities. This stratification implies that they were, and are, free to migrate vertically and laterally within the reservoir.

These observations all point to the conclusion that hydrocarbons migrate into reservoir rocks at a considerable depth below the surface and some time after burial.

An important distinction is made between primary and secondary migration. Primary migration is understood as the emigration of hydrocarbons from the source rock (clay or shale) into permeable carrier beds (generally sands or limestones). Secondary migration refers to subsequent movement of oil and gas within permeable carrier beds and reservoirs (Showalter, 1979; England, 1994). There is consensus that secondary migration occurs when petroleum is clearly identifiable as crude oil and gas, and, although gas may be dissolved in oil, their solubility in connate water is negligible. Secondary migration occurs by buoyancy due to the different densities of the respective fluids and in response to differential pressures, as discussed in Chapter 4.

Primary migration, the emigration of hydrocarbons from source bed to carrier, is still a matter for debate. The object of this section is to examine the problem of primary hydrocarbon migration, which is to many people the last great mystery of petroleum geology. Seminal works on this topic are by Magara (1977), Momper (1978), Tissot and Welte (1978), McAuliffe (1979), Hunt (1979, 1996), Roberts and Cordell (1980), Lewan (1987), England and Fleet (1991), Verweij (1993), Mann (1994), and Pepper and Corvi (1995a,b,c). The following account draws freely on these works.

A number of basic questions need to be answered before the problem of primary migration can be fully understood. These questions include "When did migration occur?" "What was the physical environment (temperature, pressure, permeability, porosity) of the source bed at that moment?" "What was the chemical composition of the source bed at that moment (clay mineralogy may be one critical factor; water content another)?" "What was the nature of the hydrocarbons?" "Were they in the form of protopetroleum, some nebulous substance transitional between kerogen and crude oil or gas?" "Did oil and gas emigrate in discrete phases, in solution (one in the other), in pore water, or in complex organic compounds in the pore fluid?" These questions are numerous and complex. They require much geological data that are difficult to acquire, and they are hard to attack experimentally because of the time, temperatures, and pressures involved. In addition, a knowledge of physics and chemistry beyond that of the many older geologists is required.

The study of primary migration contains a major paradox. Oil and gas are trapped in porous, permeable reservoirs. The source rocks from which they emigrated can be identified (Welte et al., 1975; Hunt, 1979; Magara, 1979; Flory et al., 1983). Yet these same source rocks are impermeable shales. How then did the fluids emigrate? It would be nice to believe that oil and gas were squeezed from the source clay during early burial before compaction destroyed permeability. This process cannot be so, however, because the temperatures necessary for hydrocarbon generation are not reached until compaction has greatly diminished permeability and water saturation.

At this point, a review of the relationship between clay porosity, permeability, compaction, water loss, and hydrocarbon migration is appropriate. Numerous compaction curves for argillaceous sediments have been published (Fig. 5.23). These curves show that most water expulsion by compaction occurs in the upper 2 km of burial. Pore water expelled by compaction is minimal below this depth. Note that for an average geothermal gradient (25 °C/km), oil generation begins below the depth at which most of the compactional pore water has been expelled. The migration of oil by the straightforward flushing of pore water is not therefore a viable proposition.

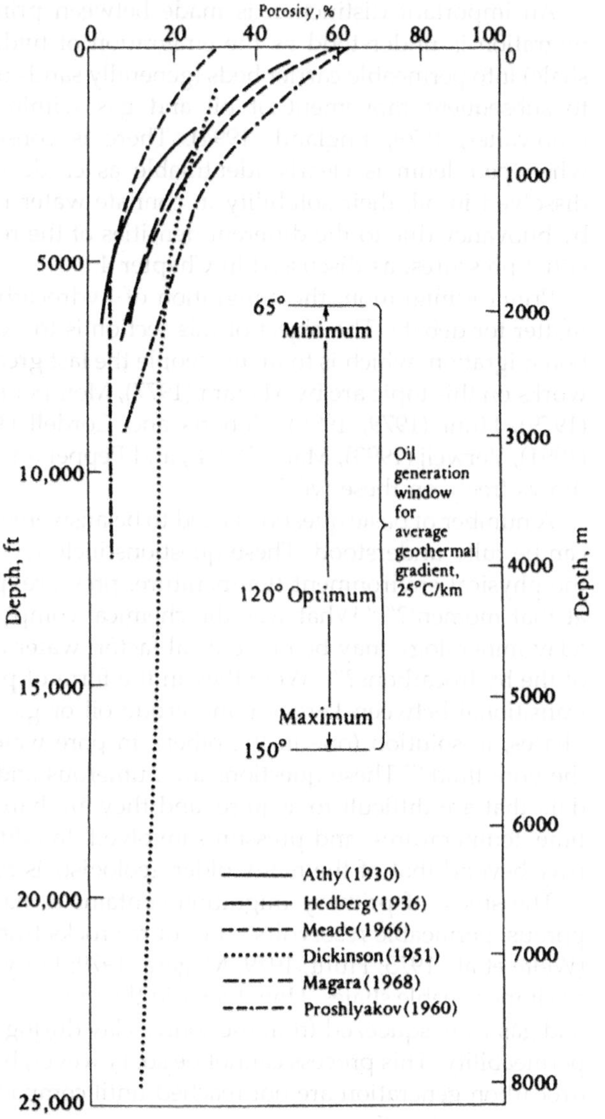

FIGURE 5.23 Shale compaction curves from various sources. Note that there is minimal water loss through compaction over the depth range of the oil window.

Other factors, however, are relevant to this problem, including consideration of the role of supernormal temperatures and pressures and the diverse kinds of water present in clays. Many major hydrocarbon provinces have been widely noted to be areas of supernormal temperatures and pressures. The Tertiary clastic prism of the Gulf of Mexico in the United States, which has undergone intense detailed studies, is a case in point.

Powers (1967) pointed out that there are two types of water in clays: normal pore water and structured water that is bonded to the layers of montmorillonitic clays (smectites), as

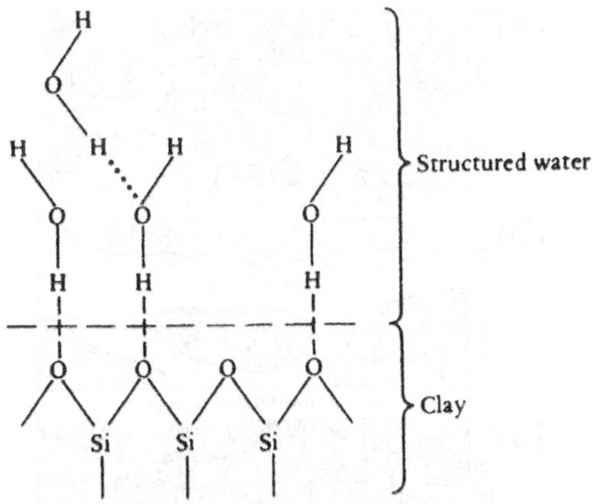

FIGURE 5.24 The relationship of structured water molecules to clay mineral lattice. *Modified from Barker (1978), reprinted by permission of the American Association of Petroleum Geologists.*

shown in Fig. 5.24. When illitic or kaolinitic clays are buried, a single phase of water emission occurs because of compaction in the first 2 km of burial. When montmorillonitic-rich muds are buried, however, two periods of water emission occur: an early phase and a second, quite distinct phase when the structured water is expelled during the collapse of the montmorillonite lattice as it changes to illite (Fig. 5.25). Further work by Burst (1969) detailed the transformation of montmorillonite to illite and showed that this change occurred at an average temperature of some 100–110 °C, right in the middle of the oil generation window (Fig. 5.26). The actual depth at which this point is reached varies with the geothermal gradient, but Burst (1969) was able to show a normal distribution of productive depth at some 600 m above the clay dehydration level (Fig. 5.27). By integrating geothermal gradient, depth, and the clay change point, it was possible to produce a fluid redistribution model for the Gulf Coast area. Similar studies have been reported from other regions (Foscolos and Powell, 1978).

Barker (1980) has pursued this idea, showing that not only water but also hydrocarbons may be attached to the clay lattice (Fig. 5.28). Obviously, the hydrocarbons will be detached from the clay surface when dehydration occurs. The exact physical and chemical process whereby oil is expelled from the source rock is still not clear, but Fig. 5.24 demonstrates an empirical relationship between clay dehydration and hydrocarbon accumulation. Regional mapping of the surface at which this change occurs is thus a valid exploration tool, although the processes responsible for the relationship may not be fully understood.

Several qualifications must be placed on this technique. These data pertain to the Tertiary Gulf Coast of the United States. In many other hydrocarbon provinces in the world, smectitic clays are largely absent. The dewatering of clay cannot therefore be advocated as the dominant process of emigrating hydrocarbons from source rocks. Overpressure is obviously a factor that may aid petroleum generation by maintaining porosity and permeability and inhibiting the formation of a rigid framework to the rock. Several geologists have suggested that fluid emigration from clays is aided by the development of microfractures (e.g., Palciauskus and Domenico, 1980). These

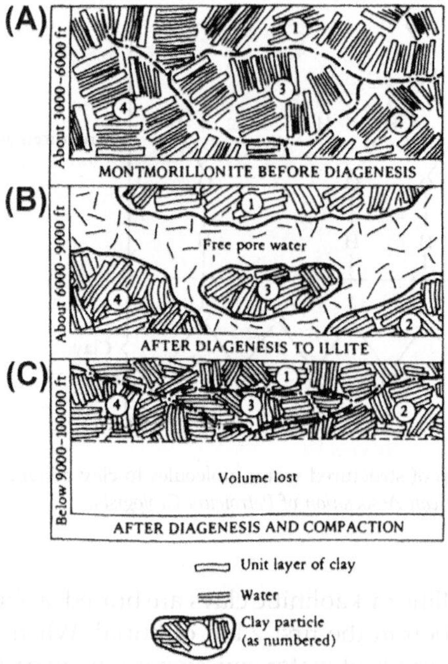

FIGURE 5.25 The two-stage dewatering of montmorillonitic clay. *From Powers (1967), namely Fig. I, p. 1242.*

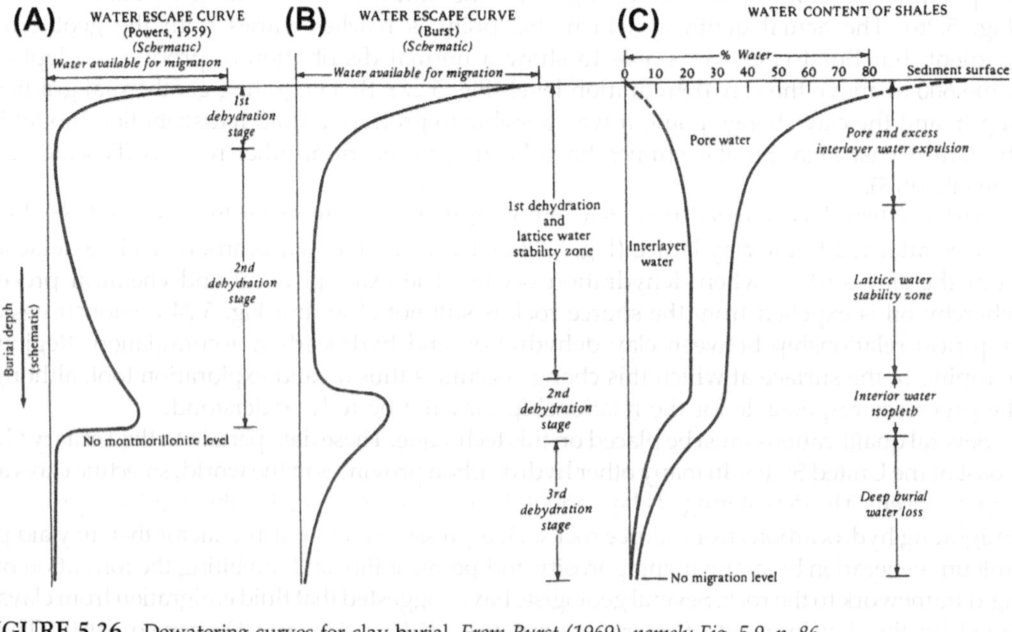

FIGURE 5.26 Dewatering curves for clay burial. *From Burst (1969), namely Fig. 5.9, p. 86.*

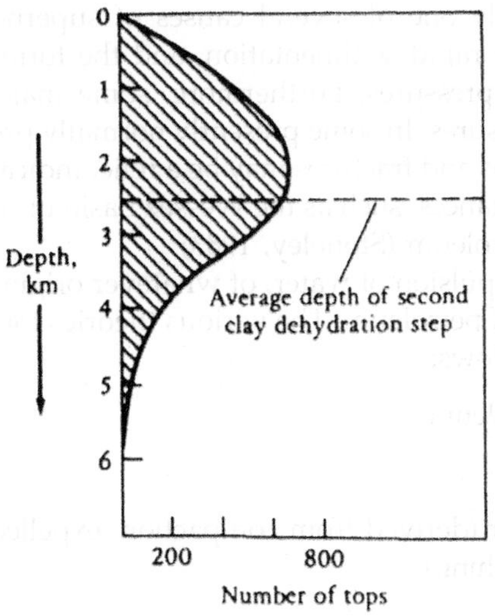

FIGURE 5.27 Depth–frequency curve of the top of 5368 oil reservoirs in the Gulf Coast of the United States. This curve shows that the peak production depth is at about 600 m above the average depth at which the second phase of clay dehydration occurs. *Modified from Burst (1969), reprinted by permission of the American Association of Petroleum Geologists.*

microfractures would cause a marked increase in permeability and thus allow fluid to escape. The microfractures would then close as pore pressure dropped. It has been suggested that petroleum globules could migrate by shouldering aside the unfixed clay grains. This process was introduced in Chapter 4 where evidence was presented that overpressured basins bleed of excess fluid, not in a slow steady discharge, but in episodic hot flushes (Capuano, 1993; Cartwright, 1994; Miller, 1995; Roberts and Nunn, 1995).

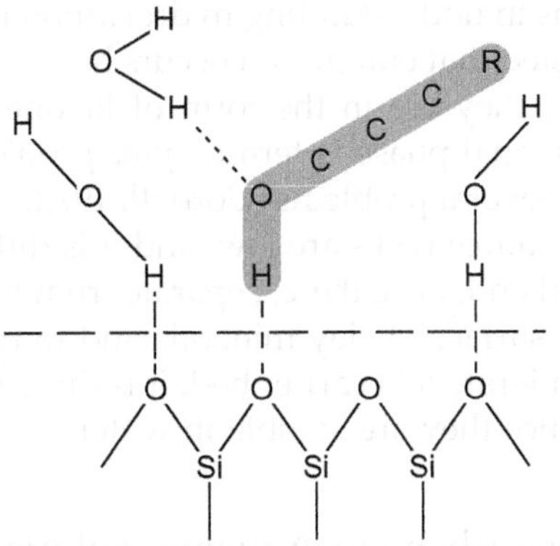

FIGURE 5.28 Sketch showing how hydrocarbon molecules (in this case an R–C–C–COH alcohol) may be attached to a clay mineral lattice together with water molecules. *Modified from Barker (1978), reprinted by permission of the American Association of Petroleum Geologists.*

Clay dehydration is only one of several causes of supernormal pressure. Inhibition of normal compaction due to rapid sedimentation, and the formation of pore-filling cements, can also cause high pore pressures. Furthermore, some major hydrocarbon provinces do not have supernormal pressures. In some presently normally pressurized basins, the presence of fibrous calcite along veins and fractures may be a relict indicator of past overpressures. It is significant that in some instances, such as the Wessex basin of southern England, these calcite veins contain traces of petroleum (Stoneley, 1983).

Nevertheless, the late expulsion of water, of whatever origin, must play an important role in the primary migration of petroleum. The various theories for primary hydrocarbon migration can be grouped as follows:

1. Expulsion as protopetroleum
2. Expulsion as petroleum
   a. In solution
      - Dissolved in water (derived from compaction, expelled from clays, or dissolved from meteoric flushing)
      - Within micelles
      - Solution of oil in gas
   b. Globules of oil in water
   c. Continuous phase

The merits and limitations of these various mechanisms are described and discussed in the following sections.

5.4.1 Expulsion of Hydrocarbons as Protopetroleum

To study hydrocarbons actually in the act of primary migration is, of course, extremely difficult. The evolutionary sequence from kerogen to crude oil and/or gas is very complex. Assessing whether this transformation is completed before, during, or after migration from source rock to carrier bed is very difficult.

One of the major problems in understanding hydrocarbon migration is their low solubility in water. Hunt (1968) suggested that emigration occurs before the hydrocarbons are recognizable crude oil, that is, while they are in the form of ketones, acids, and esters, which are soluble in water. This transitional phase is termed *protopetroleum*.

This mechanism contains several problems (Cordell, 1972). The observed concentrations of ketones, acids, and esters in source rocks are low, and it is difficult to see how they can actually migrate to the carrier bed and, once there, separate from the water. These compounds are likely to be adsorbed on the surface of clay minerals and to resist expulsion from the source rock. If, however, they do emigrate to a carrier bed, it is difficult to envisage how they evolve into immiscible crude oil, since they are soluble in water.

5.4.2 Expulsion of Hydrocarbons in Aqueous Solution

One obvious possibility to consider is that the hydrocarbons emigrate from the source bed fully formed, yet dissolved in water. The solubility of hydrocarbons is negligible at the earth's

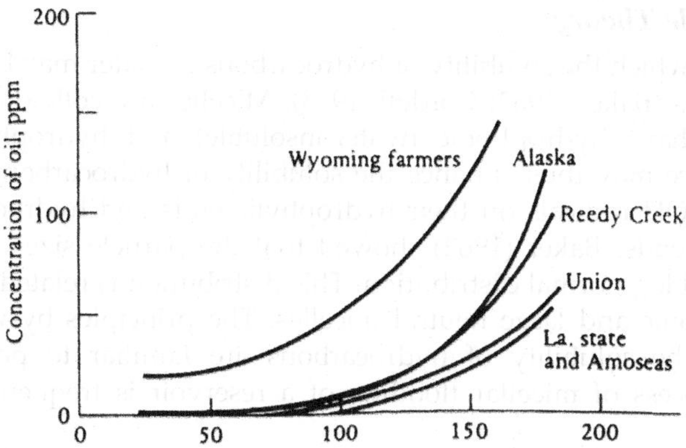

FIGURE 5.29 Graph showing the solubilities of various crude oils plotted against temperature. *From Price (1976), namely, Fig. 5.4, p. 218.*

surface, but may be enhanced by temperature or the presence of micelles. These two mechanisms are examined in the following sections.

5.4.2.1 The Hot Oil Theory

Figure 5.29 shows the solubility of various crude oils plotted against temperature. This graph shows that solubilities are negligible below about 150 °C and do not become significant until about 150 °C. It is worth remembering that paleotemperature analysis shows that optimum oil generation occurs at about 120 °C; at this temperature experimental data suggest solubilities of the order of 10–20 ppm.

This information does not seem very helpful. Hydrocarbon solubility is worth looking at in more detail. Figure 5.30 shows the solubility of hydrocarbons plotted according to hydrocarbon numbers. This graph shows that hydrocarbon solubility increases with decreasing carbon number for both the normal alkanes (paraffins) and the aromatic series. Note particularly that solubilities are very high, several hundred to several thousand milligrams per liter, for the paraffin gases (C_1–C_5). Perhaps this information has solved part of the problem of understanding primary migration. It seems quite easy for the gaseous hydrocarbons to emigrate from the source bed dissolved in pore water. Exsolution may then occur when they reach the lower temperatures and pressures of the carrier bed.

Only about 25% of a typical crude oil is composed of the moderately soluble alkanes, and another 25% is composed of the heavy naphthenoaromatics and resins, which are virtually insoluble in water. Nonetheless, there are advocates for primary oil emigration by solution in water (e.g., Price, 1976, 1978, 1980, 1981a,b). Experimental data show that solubilities of as much as 10 wt.% occur for hydrocarbons in pressured salty water. Solubility is enhanced by the presence of gas. Even so, Price advocates temperatures of between 300 and 350 °C for oil generation. As already seen, these temperatures are well above those that paleothermometers indicate as optimal for oil generation.

Further, at these temperatures, diagenesis will probably have destroyed the porosity and permeability of source and carrier beds alike. The stability of hydrocarbons at these temperatures has also been queried (Jones, 1978).

5.4.2.2 The Micelle Theory

Another way in which the solubility of hydrocarbons in water may be enhanced is by the presence of micelles (Baker, 1962; Cordell, 1973). Micelles are colloidal organic acid soaps whose molecules have hydrophobic (water-insoluble) and hydrophylic (water-soluble) ends. Their presence may thus enhance the solubility of hydrocarbons in water by acting as a link between OH radicals on their hydrophylic ends and hydrocarbon molecules on their hydrophobic ends. Baker (1962) showed that the particle sizes of micelles in crude oils have a bimodal log-normal distribution. This distribution is related to two basic micelle types—the small ionic and large neutral micelles. The principles by which soaps may be used to enhance the solubility of hydrocarbons are familiar to petroleum production engineers. The process of micellar flooding of a reservoir is frequently used to enhance recovery.

If this micelle theory is correct, then the proportions of different hydrocarbons in a crude oil should be related to their micellar solubility. Experimental data show this to be the case. The naphthenic and aromatic fractions are present in monomodal log-normal frequency distributions. By contrast the paraffins, both normal and branched, show a frequency distribution suggesting that they are solubilized by both ionic and neutral micelles. Hydrocarbons of low atomic weight (C_5 or less) show a normal frequency distribution. As already shown, this distribution is probably related to the fact that they are naturally soluble in water and do not require the assistance of micelles.

Numerous writers have reviewed the micelle theory. Major objections listed by McAuliffe (1979) include the requirement of a very high ratio of micelles to hydrocarbons to provide an effective process. Micelles are commonly only present in traces. In addition, the size of micelle

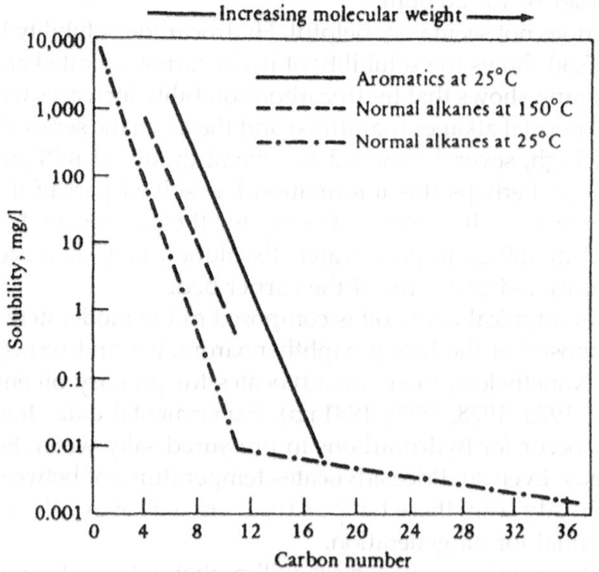

FIGURE 5.30 Solubility of hydrocarbons with different carbon numbers. Solubility decreases with increasing molecular weight. *From McAuliffe (1979), namely Fig. I, p. 762.*

molecules is greater than the diameter of pore throats in clay source beds. The processes whereby the micelles lose their hydrocarbon molecules and presumably break down are not clear.

5.4.3 Expulsion of Oil in Gaseous Solution

Other theories of hydrocarbon migration have considered the role of gases in acting as catalysts or transporting media. Momper (1978), in particular, has discussed the role of carbon dioxide, which is known to be driven off during kerogen maturation. Directly and indirectly, carbon dioxide may have considerable influence on hydrocarbon migration. Indirectly, by combining with calcium ions it precipitates calcite cement, which diminishes pore volume and thus increases pore pressure. The presence of carbon dioxide gas in solution lowers the viscosity of oil, thus increasing its mobility. It causes the precipitation of the N–S–O heavy ends of oil, thus making the residual oil lighter and increasing the gas:oil ratio.

On the debit side, however, the precipitation of calcite cement may diminish permeability. Further, the main phase of decarboxylation of kerogen is known to occur before hydrocarbon generation. Concentrations of carbon dioxide at the time of petroleum generation may thus be too low to assist migration in the ways previously outlined.

Gases may play a direct role in oil migration in another way. Hydrocarbon gases in deep wells are well known to carry oil in gaseous solution. The oil condenses when pressure and temperature drop as the hydrocarbon gases are brought to the surface. Therefore, oil may possibly migrate in this manner from the source bed. This mechanism quite satisfactorily explains the secondary migration of condensates through carrier and reservoir beds. As McAuliffe (1979) points out, however, it does not quite satisfactorily explain the secondary migration of normal crudes, or the primary migration of any oil. The solubility of the heavier naphtheno-aromatics and resins appears to be negligible. The expulsion of a gas bubble through a shale pore throat is as difficult as it is for an oil droplet. As is shown shortly, this presents considerable problems for theories of primary oil phase migration.

5.4.4 Primary Migration of Free Oil

A whole spectrum of theories postulates that oil emigrates from the source bed not in any kind of solution but as a discrete oil phase. There are two major types of such migration: the expulsion of discrete droplets associated with pore water and the expulsion of a three-dimensional continuous phase of oil (Magara, 1981). Hobson and Tiratsoo (1981) have shown mathematically how in water-wet pores it is impossible for a globule of oil to be squeezed through a pore throat in a clay source rock, either by virtue of its own properties (e.g., buoyancy or immiscibility) or by the assistance of flowing water. This mathematical hypothesis is based on the assumption that the diameter of the oil globule is greater than that of the pore throat. Hobson and Tiratsoo (1981) cite pore diameters of the order of 50–100 Å for shales at 2000 m. Individual molecules of asphaltenes and complex ring-structured hydrocarbons have diameters of 50–100 Å and 10–30 Å, respectively (Tissot and Welte, 1978). Figure 5.31 illustrates the interrelationship of various physical parameters (porosity, temperature, fluid pressure, shale pore diameters) with increasing depths of burial for shale-type sediments. Figure 5.32 illustrates the sizes of hydrocarbon molecules and pore throats in shales. Therefore, it is hard to see how discrete oil globules

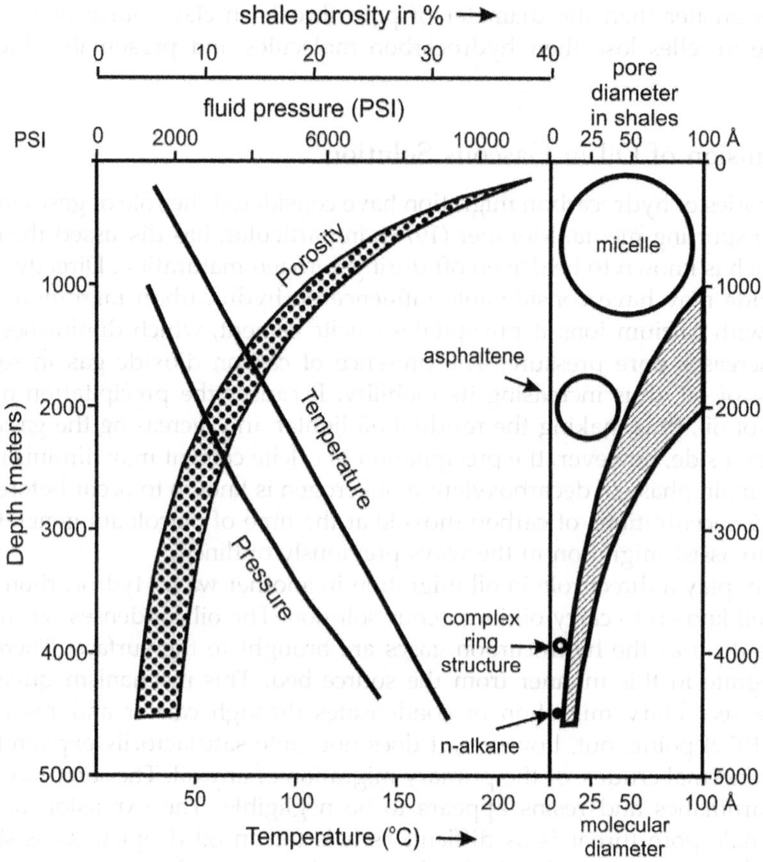

FIGURE 5.31 Interrelationship of various physical parameters (Pressure, Temperature, Porosity, and Shale pore diameters) with increasing depths of burial for shale-type sediments. *Modified from Tissot and Welte (1984).*

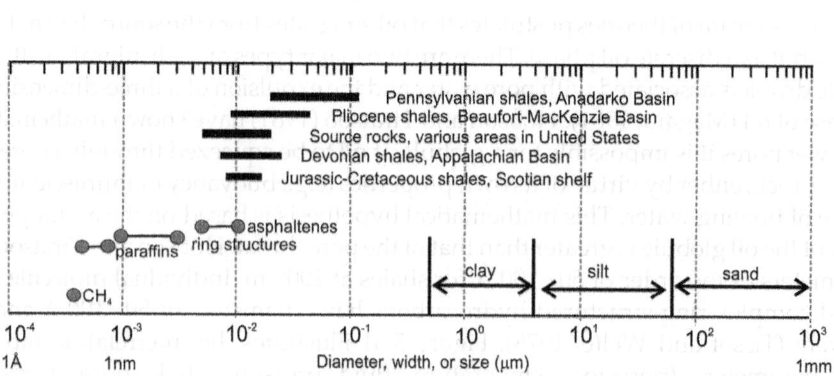

FIGURE 5.32 Sizes of hydrocarbon molecules and pore throats in shales on a logarithmic scale covering several orders of magnitude. *Modified from Nelson (2009).*

could have diameters less than those of pore throats. It is possible, however, that organic-rich source rocks are not water wet, but oil wet. In this situation, petroleum would not migrate as discrete globules of oil in water, but as a continuous three-dimensional phase. This mechanism is the so-called "greasy wick" theory, pointing to the analogy of molten wax moving through the fiber of a candle wick. Such a mechanism may work for rich source rock, but is less likely in leaner source rocks, which are probably water wet.

Hydrous pyrolysis and petrographic thin section work by Lewan (1985) illustrates in a laboratory setting how oil generation and expulsion occurs. These analyses are thought to be similar to actual oil generation. The following was observed by Lewan on samples from the Woodford Shale of Oklahoma. The samples analyzed had amorphous masses of kerogen dispersed in the shale. As thermal stress increased, viscous bitumen generated from the kerogen impregnated the planar bedding fabric of the shale. The bitumen was water soluble and polar-rich. This was followed by a continuous bitumen network forming when the amount and distribution of organic matter were favorable (>2.5 wt.% TOC). Increasing thermal stress resulted in generation of oil from the bitumen. Dissolved water in the bitumen acts as a source of hydrogen for the oil. The oil generated impregnated micropores and microfractures and was expelled. Any oil or bitumen that remained in the shales was carbonized to pyrobitumen.

For most petroleum accumulations, micropressuring is probably a major mechanism for primary migration. The micropressure is created by a net volume increase of organic matter within a confined matrix space (Fig. 5.32). The increase in volume and pressure is caused by thermal expansion of the generated liquid organic phase and overall decrease in the density of the reaction products (Lewan, 1985).

The pressure build-up, microfracturing, subsequent pressure release, expansion of a fluid or gas is an episodic process (Figs 5.33, 5.34, 5.35). Meaningful oil or gas migration requires this process be repeated many times. Fluid or gas escapes after a fracture is opened followed by the fracture closing because of overburden stress. The process then repeats itself. Figure 5.35 illustrates the interrelationship of porosity, pore pressure, and hydrocarbon generation with depth of burial for shale-type sediments (Momper, 1981). Pore pressure created by hydrocarbon generation and aquathermal expansion of any pore water may create pore pressures that exceed the fracture gradient of the shale.

Tissot and Welte (1984) state that the pressure build-up inside a pore network of a shale source rock does not always create microfractures. If a continuous organic phase and/or three-dimensional network of kerogen or if capillary pressures are not too high because pore throats are not too narrow exists, then oil or gas can be driven out of the kerogen without microfracturing.

Thus, pressure-driven hydrocarbon movement with or without microfracturing is an important part of primary migration.

5.4.5 The Importance of Hydrocarbon Adsorption in Source Rocks

Shale oil and gas reservoirs are becoming an important target for the petroleum industry as the switch to unconventional reservoirs. Shale oil and gas account for a large majority of the remaining oil and gas resources in the world.

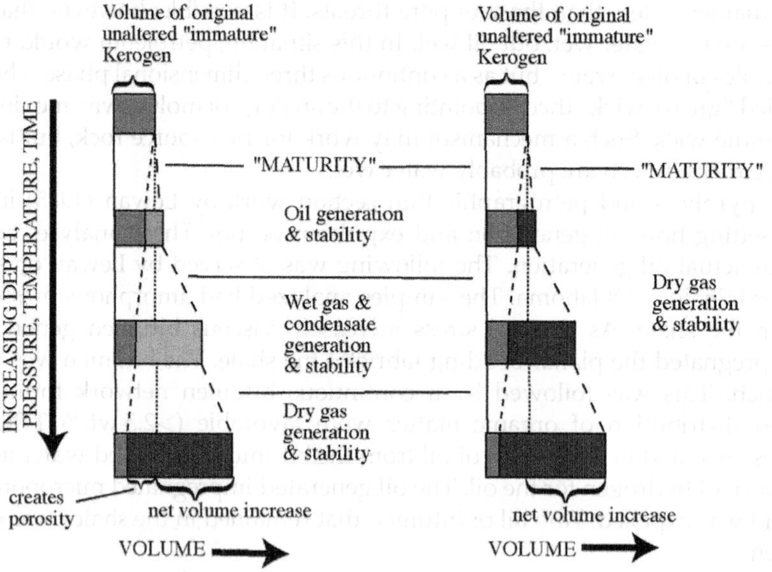

FIGURE 5.33 Illustration of petroleum volume increases associated with both oil-prone (left) and gas-prone (right) kerogens. *From Meissner (1980).*

The geochemistry of the shales is an important aspect to understand. Parameters like thermal maturity and TOC content need to be assessed. TOC is responsible for the volumes of hydrocarbons generated. Thermal maturity of kerogen is a measure of the amount of kerogen that has to be converted to hydrocarbons. Thus, a play is dependent on a combination of thermal maturity and TOC content.

Shales can act as both source rock and reservoir rock. Hydrocarbons are stored in shales in the matrix pore volume, and they are adsorbed onto the surface areas of the pores.

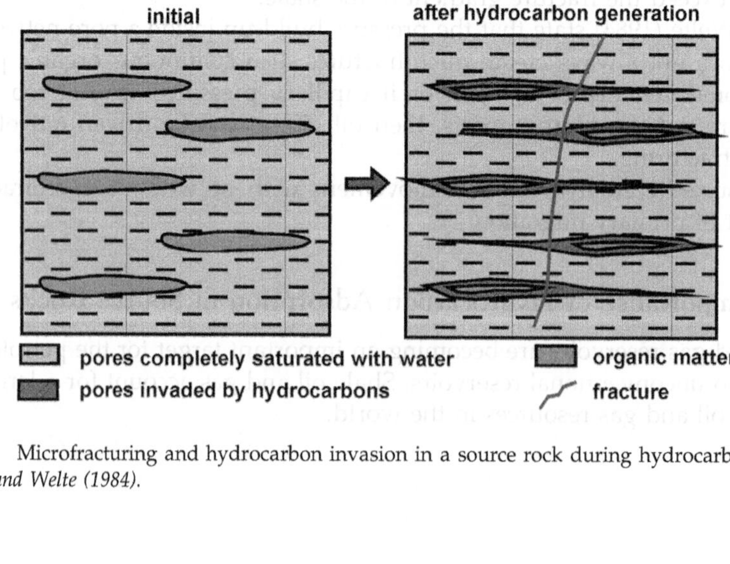

FIGURE 5.34 Microfracturing and hydrocarbon invasion in a source rock during hydrocarbon generation in a kerogen. *Tissot and Welte (1984).*

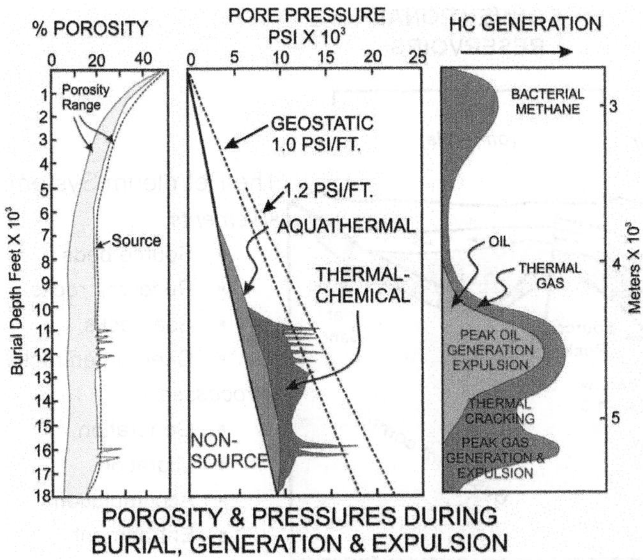

FIGURE 5.35 Interrelationship of porosity, pore pressure, and hydrocarbon generation with depth of burial for shale-type sediments. *Modified from Momper (1981).*

Adsorbed hydrocarbons plus interstitial hydrocarbons (free hydrocarbons) equals total hydrocarbons in place.

Adsorption is the adhesion of atoms, ions, or molecules from a gas, liquid, or dissolved solid (adsorbate) to a solid surface (adsorbent). Molecules attach themselves on a surface in two ways: (1) weak van der Walls attraction of the adsorbate to the solid surface; (2) chemisorption where adsorbate sticks to solid surface by formation of chemical bond with surface. This type of bond is much stronger than the van der Walls attraction.

To properly assess a hydrocarbon shale play, the adsorbed and free hydrocarbons must be assessed.

5.5 THE PETROLEUM SYSTEM

Having established that commercial quantities of petroleum are of organic origin, and having discussed the primary migration of petroleum from the source rock into the carrier bed, it is appropriate to conclude this chapter with a discussion of the petroleum system, that is to say, the integration of petroleum migration with the thermal and tectonic evolution of a sedimentary basin. This involves consideration of the distance of secondary migration of petroleum, and the mathematical modeling of the time and amounts of petroleum that may have been generated within a given sedimentary basin.

The petroleum system is defined as a hydrocarbon fluid system that encompasses a pod of active source rock and all related oil and gas accumulations (Magoon and Dow, 2000). It includes the essential elements (source bed, reservoir rock, seal rock, and overburden rock) and processes (generation, migration, accumulation, preservation, and entrapment) needed for oil and gas accumulations to exist. Understanding petroleum systems reduces exploration

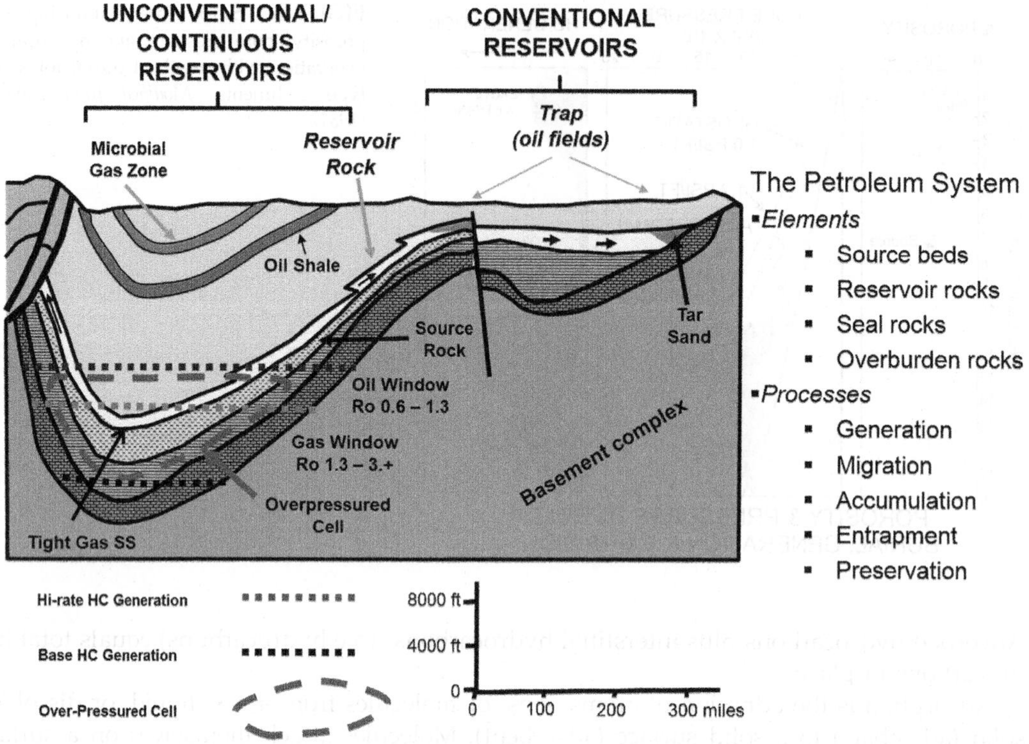

FIGURE 5.36 Cross-section of hypothetical basin illustrating aspects of a petroleum system. *Modified from (Magoon, 1988). See Chapter 9 for a discussion of unconventional reservoirs.*

risk in the search for undiscovered petroleum resources. Figures 5.36 and 5.37 illustrate the petroleum system. Unconventional reservoirs are discussed in Chapter 10.

5.5.1 Measurement of the Distance of Petroleum Migration

The lateral distance to which petroleum can migrate has always been debated. It is a difficult parameter to measure. Traditionally, it is done by physically measuring the distance between the petroleum accumulation and the nearest mature source rock; however, maturity and source rock might be defined. Both two- and three-dimensional studies have been attempted. Where oil is trapped in sand lenses surrounded by shale, the migration distance must have been short. Where oil occurs in traps with no obvious adjacent source rock, extensive lateral migration must have occurred. Correlation between source rock and reservoir oil can be carried out by fingerprinting using gas chromatography (Bruce and Schmidt, 1994).

Table 5.4 cites some documented examples of long-distance lateral petroleum migration. The record for the longest distance of oil migration is held by the West Canadian basin, where

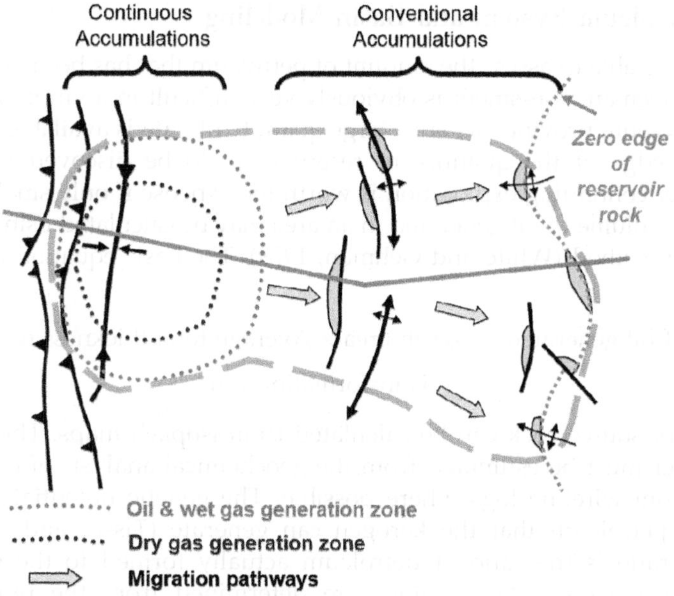

FIGURE 5.37 Map view of a petroleum system. *Magoon (1988).*

a migration distance of more than 1000 km has been calculated (Garven, 1989). A new geochemical method for calculating migration distances has been developed by Larter et al. (1996). This is based on the regional variation of traces of nonalkylated benzocarbazoles. The method is apparently effective, irrespective of the maturity of the oils.

Accurate estimates of the distance from the "devil's kitchen" to the petroleum trap is an essential part of basin modeling discussed in the next section. For a further discussion of this important parameter, see Hunt (1996, pp. 281–286).

TABLE 5.4 Some Published Accounts of Long-Distance Petroleum Migration[a]

Basin	Migration distance (km)	Reference
Athabasca Tar Sands, Canada	100	Tissot and Welte (1978)
Magellan basin, Argentina	100	Zielinski and Bruchhausen (1983)
Pennsylvanian Oklahoma, United States	120	Levorsen (1967)
Gulf Coast, Pleistocene	160	Hunt (1979)
Illinois basin, United States	200	Bethke et al. (1991)
Paris basin	200	Gaullier et al. (1991)
Phosphoria formation, Wyoming and Idaho, United States	400	Claypool et al. (1978)
Alberta basin, Canada	1000	Garven (1989)

[a]*Previously based in geometric analysis, it is now possible to use geochemistry to measure migration distance.*

5.5.2 The Petroleum System and Basin Modeling

It is useful to be able to assess the amount of petroleum that has been generated in a sedimentary basin. Such an assessment is obviously very difficult in a virgin area with no data. In a mature petroleum province where a large quantity of data is available, it is considerably easier. A knowledge of the quantity of reserves yet to be discovered is important for deciding whether continuing exploration is worth the expense if only small reserves remain to be found. The volume of oil generated in an area may be calculated using the geochemical material balance method (White and Gehman, 1979). The basic equation may be expressed as follows:

$$\text{Volume of oil generated} = \text{Basin area} \times \text{Average total thickness of source rock} \times \text{Transformation ratio.}$$

The volume of source rock can be calculated from isopach maps. The average amount of organic matter must be estimated from the geochemical analysis of cores and cuttings, extrapolating from wireline logs where possible. The genetic potential of a formation is the amount of petroleum that the kerogen can generate (Tissot and Welte, 1978). The transformation ratio is the ratio of petroleum actually formed to the genetic potential, and, as described earlier, these values are determined from the pyrolysis of source rock samples.

Estimates of the volumes of petroleum generated in various basins have been published by Hunt and Jamieson (1956), Hunt (1961), Conybeare (1965), Pusey (1973a), Fuller (1975), and Goff (1983). These studies show that the transformation ratio needs to be >0.1 for significant oil generation and is usually in the range of 0.3–0.7 in major petroleum provinces.

The concept of the petroleum system is old, though it has become very fashionable now because it facilitates the modeling of sedimentary basins as a means of finding out how much petroleum they may have generated and where it may be located. The petroleum system integrates the sedimentary and structural history of a basin with its petroleum characteristics, in terms of the richness, volume, and maturity of source rocks. The petroleum system has been variously defined as "a dynamic petroleum generating and concentrating physico-chemical system functioning in a geologic space and time" (Demaison and Huizinga, 1991, 1994), or as the relationship between "a pod of active source rock and the resulting oil and gas accumulations" (Magoon and Dow, 1994). The latter authors see the petroleum system as part of a hierarchy; thus,

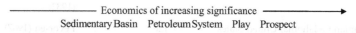

——— Economics of increasing significance ———→

Sedimentary Basin Petroleum System Play Prospect

Demaison and Huizinga (1991, 1994) classify petroleum systems according to three parameters: the charge factor, the style of migration (vertical or lateral), and the entrapment style. The charge factor is calculated on the basis of the richness and volume of the source rocks in a basin. This is measured according to the source potential index; thus,

$$\text{SPI} = \frac{h(S_1 + S_2)p}{1000},$$

where

SPI is the maximum quantity of hydrocarbons that can be generated within a column of source rock under 1 m^2 of surface area
h is the thickness of source rock (meters)
S_1+S_2 is the average genetic potential in kilograms hydrocarbons/metric ton of rock
p is the source rock density in metric tons/cubic meter.

The style of migration drainage is subdivided according to whether migration is principally lateral or vertical, though this can, of course, vary in time and space within the history of one basin. The style of entrapment is dependent on the length and continuity of carrier beds, the distribution and effectiveness of seals, and tectonic style. All these together control the degree of resistance of the basin to the dispersion of the petroleum charge. Thus entrapment style may be categorized as high or low impedance. Table 5.5 documents examples of the different types of petroleum system classified according to these three parameters.

The concept of the petroleum system can be usefully applied to the computer modeling of sedimentary basins. The object of this discipline is to try to discover how much petroleum a sedimentary basin may have generated and where it may be located. Basin modeling may take place in one, two, or three dimensions. Basin modeling is done by a computer. Several software packages are available. New versions and new packages are appearing constantly (Welte and Yukler, 1984; Waples et al., 1992a,b; Waples, 1994).

One-dimensional modeling involves no more than the construction of a burial history curve for a particular point in a basin, such as a well location (refer to Fig. 5.16). This may be used to establish the maturity of a source rock interval, either using a modern geothermal gradient data (if the well has been drilled) or developing a geothermal history based on the tectonic regime of the location.

A two-dimensional model consists of a cross-section. This may be constructed by "back-stripping" the geological history, based on a seismic section, calibrated with well data, if available. To do this, we need to estimate the depositional depth and compaction history of the sedimentary sequence. Then, a geohistory scenario can be plugged in. This can then be used to establish the pressure system of the section and the migration and entrapment history of petroleum (Plate 5.5). Examples of 2D modeling using TEMISPACK have been

TABLE 5.5 Examples of the Different Types of Depositional Systems Classified According to the System of Demaison and Huitzinga (1991)

Type of drainage	Charge	Impedance	Example
Lateral	Super	High	North slope of Alaska, United States
	Normal	Low	Williston basin, United States
Lateral	Super	Low	Venezuela foreland basin
Vertical	Super	High	Central Graben, United Kingdom North sea
Vertical	Normal	High	Tertiary Niger Delta
Vertical	Super	High	Los Angeles basin, United States

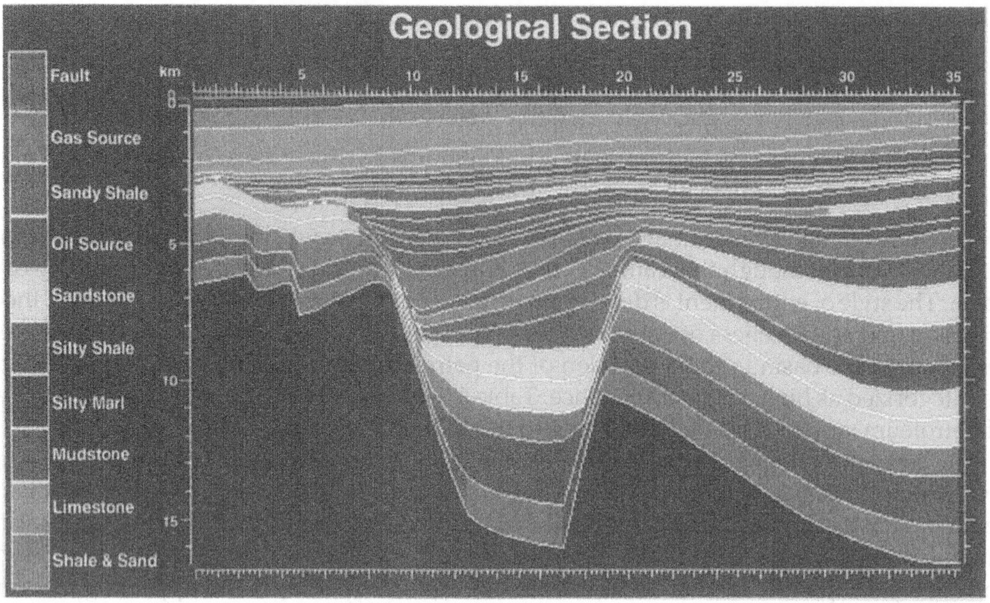

PLATE 5.5 Three computer-generated cross-sections to illustrate how a sedimentary basin may be modeled to predict the distribution of pressure and petroleum (See the color plate). *Courtesy of A. Vear and British Petroleum.*

published from the eastern offshore Canada and Williston basins by Forbes et al. (1992) and Burrus et al. (1996) respectively.

A three-dimensional model involves the same operations as just described, however, not for a cross-section, but for a volume of rock. An example of 3D modeling in the Gulf of Mexico basin has been published by Anderson et al. (1991).

5.5.3 Hydrocarbon Generation and Migration: Summary

Understanding the primary migration of hydrocarbons is one of the last problems of petroleum geology. Research in this field is currently very active, so any review is bound to date quickly; nonetheless, a summary of this complex topic is appropriate:

1. Commercial quantities of oil and gas form from the metamorphism of organic matter.
2. Kerogen, a solid hydrocarbon disseminated in many shales, is formed from buried organic detritus and is capable of generating oil and gas.
3. Three types of kerogen are identifiable: type I (algal), type II (liptinitic), and type III (humic). Type I tends to generate oil; type III, generates gas.
4. The maturation of kerogen is a function of temperature and, to a lesser extent, time. Oil generation occurs between 60 and 120 °C, and gas generation between 120 and 225 °C.
5. Source rocks generally contain >1500 ppm organic carbon, but yield only a small percentage of their contained hydrocarbon.
6. Several techniques may be used to measure the maturity of a source rock.

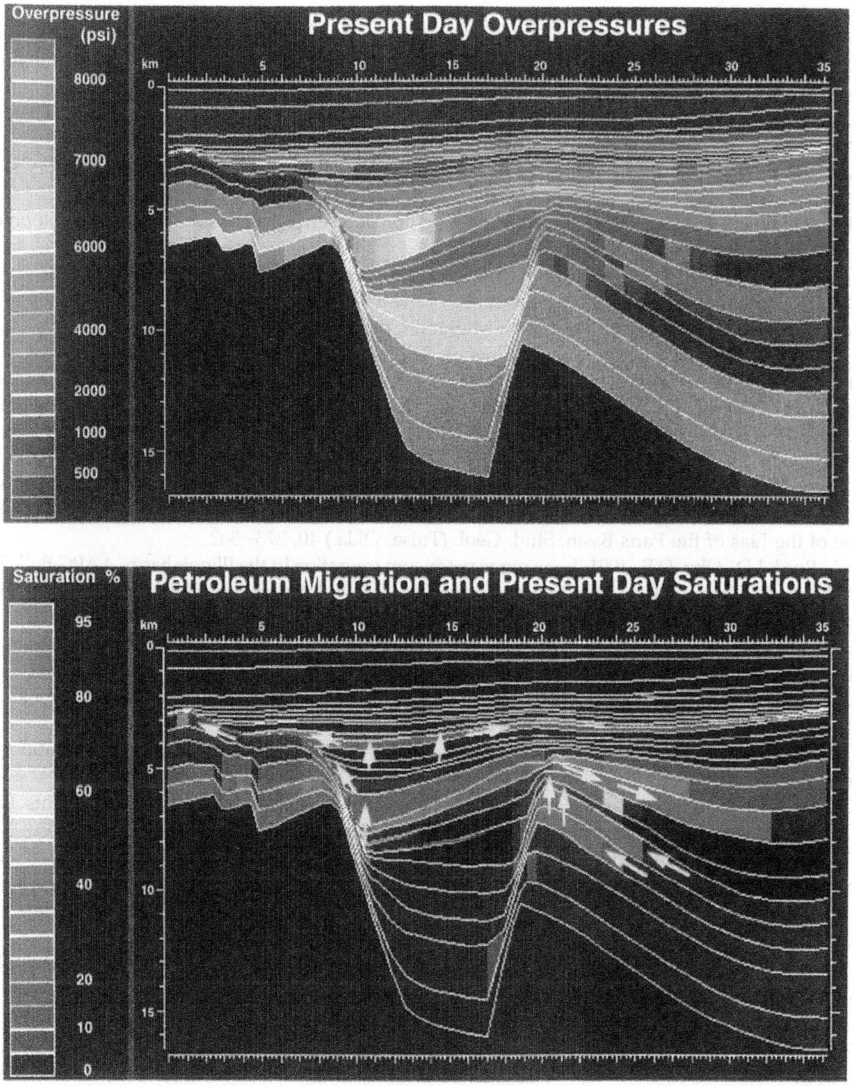

PLATE 5.5 cont'd

7. The exact process of primary migration, whereby oil and gas migrate from source beds, is unclear. Solubility of hydrocarbons in water is low, but, for the lighter hydrocarbons, is enhanced by high temperature and the presence of soapy micelles. An empirical relationship between oil occurrence and clay dehydration suggests that the flushing of water from compacting clays plays an important role in primary migration. There is much evidence to suggest that the episodic fshing of water from overpressured shales by hydraulic fracturing is an important pross in allowing petroleum to emigrate from source rocks.

References

Allan, J., Douglas, A.G., 1977. Variations in content and distribution of *n*-alkanes in a series of carboniferous vitrinites and sporinites of bituminous rank. Geochim. Cosmochim. Acta 40, 1223–1234.

Allen, J.R.L., Hoskins, B.J., Valdes, P.J., Sellwood, B.W., Spicer, R.A., 1994. Palaeoclimates and Their Modelling. Chapman & Hall, Andover, MA.

Anderson, R.N., Cathies, L.M., Nelson, H.R., November 1991. 'Data Cube' depicting fluid flow history in Gulf Coast sediments. Oil Gas J. 4, 60–65.

Arthur, M.A., Schlanger, S.O., 1979. Cretaceous 'oceanic anoxic events' as causal factors in development of reef-reservoired giant oil fields. AAPG Bull. 63, 870–885.

Baker, E.G., 1962. Distribution of hydrocarbons in petroleum. Am. Assoc. Pet. Geol. Bull. 46, 76–84.

Barker, C., 1980. Primary Migration: The Importance of Water-Mineral-Organic matter Interactions in the Source Rock. AAPG Sp. Pub, Problems of Petroleum Migration, pp. 19–31.

Barker, C., 1978. Physical and Chemical Constraints on Petroleum Migration. AAPG, Tulsa.

Barnard, P.C., Cooper, B.S., Fisher, M.J., 1978. Organic maturation and hydrocarbon generation in the Mesozoic sediments of the Sverdrup Basin, Arctic Canada. In: Proc. Int. Palynol. Congr., 4th, Lucknow, pp. 163–175.

Becker, G.F., 1909. Relations between local magnetic disturbances and the genesis of petroleum. Geol. Surv. Bull. (U.S.) 401, 1–23.

Berthelot, M., 1860. Sur 1'origine de carbures et des combustibles minéraux. C. R. Hebd. Seances Acad. Sci. 63, 949–951.

Bessereau, G., Gillocheau, F., Huc, A.-Y., 1995. Source rock occurrence in a sequence stratigraphic framework: the example of the Lias of the Paris Basin. Stud. Geol. (Tulsa, Okla.) 40, 273–302.

Bethke, C.M., Reed, J.D., Oltz, D.F., 1991. Long range petroleum migration in the Illinois basin. AAPG Bull. 75, 925–945.

Bradley, W.H., 1948. Limnology and the Eocene lakes of the Rocky Mtn. region. Bull. Geol. Soc. Am. 59, 635–648.

Brooks, J., Fleet, A.J. (Eds.), 1987. Marine Petroleum Source Rocks. Geol. Soc. London, London. Spec. Publ. No. 26.

Bruce, L.G., Schmidt, G.W., 1994. Hydrocarbon fingerprinting for application in forensic geology: review with case studies. AAPG Bull. 78, 1692–1710.

Burrus, J., Osadetz, K., Wolfe, S., Dologez, B., Visser, K., Dearborn, D., 1996. A two-dimensional regional basin model of Williston Basin hydrocarbon systems. AAPG Bull. 80, 265–291.

Burst, J.F., 1969. Diagenesis of Gulf Coast clay sediments and its possible relation to petroleum migration. Am. Assoc. Pet. Geol. Bull. 53, 73–93.

Calvert, S.E., 1987. Oceanographic controls on the accumulation of organic matter in marine sediments. In: Brooks, J., Fleet, A.J. (Eds.), Marine Petroleum Source Rocks. Blackwell Scientific, London, pp. 137–151.

Capuano, R.M., 1993. Evidence of fluid flow in microfractures in geopressured shales. AAPG Bull. 77, 1303–1314.

Cartwright, J.A., 1994. Episodic basin-wide hydrofracturing of overpressured Early Cenozoic mudrock sequences in the North Sea Basin. Mar. Pet. Geol. 11, 587–607.

Chyba, C.F., 1990. Extraterrestrial aminoacids and terrestrial life. Nature (Lond.) 358, 113–114.

Claypool, G.E., Love, A.H., Maughan, E.K., 1978. Organic geochemistry, incipient metamorphism, and oil generation in black shale members of phosphoria formation, western interior. United States. AAPG Bull. 62, 98–120.

Cohen, C.R., 1981. Time and temperature in petroleum formation: application of Lopatin's method to petroleum exploration: discussion. AAPG Bull. 65, 1647–1648.

Collingwood, D.M., Rettger, R.E., 1926. The Lytton springs oil field, Caldwell county, Texas. Am. Assoc. Pet. Geol. Bull. 10, 1199–1211.

Connan, J., 1974. Time–temperature relation in oil genesis. AAPG Bull. 58, 2516–2521.

Conybeare, C.E.B., 1965. Hydrocarbon-generation potential and hydrocarbon yield capacity of sedimentary basins. Bull. Can. Pet. Geol. 13, 509–528.

Cooper, B.S., 1977. Estimation of the maximum temperatures attained in sedimentary rocks. In: Hobson, G.D. (Ed.), Developments in Petroleum Geology, vol. 1. Applied Science Publishers, London, pp. 127–146.

Cooper, B.S., Coleman, S.H., Barnard, P.C., Butterworth, J.S., 1975. Palaeotemperatures in the northern North Sea basin. In: Woodland, A.W. (Ed.), Petroleum and the Continental Shelf of Northwest Europe. Applied Science Publishers, London, pp. 487–492.

Cordell, R.J., 1972. Depths of oil origin and primary migration: a review and critique. Am. Assoc. Pet. Geol. Bull. 56, 2029–2067.

Cordell, R.J., 1973. Colloidal soap as a proposed primary migration medium for hydrocarbons. Am. Assoc. Pet. Geol. Bull. 57, 1618–1643.

Cornelius, C.D.G., 1975. Geothermal aspects of hydrocarbon exploration in the North Sea area. In: Proc. Bergen Conf. Hydrocarbon Geol. Nor. Geol. Undersok, pp. 26–67.

Creaney, S., Passey, Q.R., 1993. Recurring patterns of total organic carbon and source rock quality within a sequence stratigraphic framework. AAPG Bull. 77, 386–401.

Debyser, J., Deroo, G., 1969. Faits t'observation sur la génèse du petrole. I. Facteurs controlant la repartition de la matière organiques dans les sediments. Rev. Inst. Fr. Pet. 24, 21–48.

Degens, E.T., Mopper, K., 1976. Factors controlling the distribution and early diagenesis of organic material in marine sediments. Chem. Oceanogr. 6, 59–113.

Degens, E.,T., Ross, D.A. (Eds.), 1974. The Black Sea—Geology, Chemistry and Biology. Am. Assoc. Pet. Geol., Tulsa, OK. Mem. No. 20.

Degens, E.T., von Hertzen, R.P., Wong, H.K., 1971. Lake Tanganika: water chemistry, sediments, geological structure. Naturwissenschaften 58, 224–291.

Degens, E.T., von Hertzen, R.P., Wong, H.K., 1973. Lake Kivu: structure, chemistry and biology of an East African rift lake. Geol. Rundsch. 62, 245–277.

Demaison, G., Huizinga, H.J., 1991. Genetic classification of petroleum systems. AAPG Bull. 75, 1626–1643.

Demaison, G., Huizinga, H.J., 1994. Genetic classification of petroleum systems using three factors: charge, migration and entrapment. In: Magoon, L.B., Dow, W.G. (Eds.), The Petroleum System—from Source to Trap. AAPG, pp. 73–92. Mem. No. 60.

Demaison, G.J., Moore, G.T., 1979. Anoxic environments and oil source bed genesis. Org. Geochem. 2, 9–13.

Demaison, G.J., Moore, G.T., 1980. Anoxic environments and oil source bed genesis. AAPG Bull. 64, 1179–1209.

Deuser, W.G., 1971. Organic-carbon budget of the Black Sea. Deep-Sea Res. 18, 995–1004.

Dietrich, G., 1963. General Oceanography, An Introduction. Wiley, New York.

Dmitriyevsky, A.N., 1993. Hydrothermal origin of oil and gas reservoirs in basement rocks of the South Vietnamese continental shelf. Int. Geol. Rev. 35, 621–630.

Dott, R.H., Reynolds, M.J., 1969. Sourcebook for Petroleum Geology. Mem. No. 5. Am. Assoc. Pet. Geol., Tulsa, OK.

Dow, W.G., 1977. Kerogen studies and geological interpretation. J. Geochem. Explor. 7, 79–100.

Dow, W.G., O'Connor, D.L., 1979. Vitrinite reflectance—what, how and why. AAPG Bull. 63, 441.

Duncan, D.C., Swanson, V.E., 1965. Organic-rich shale of the United and World Land Areas. Geol. Surv. Circ. (U.S.) 523, 1–30.

Durand, B., 1980. Kerogen—Insoluble Organic Matter from Sedimentary Rocks. Graham & Trotman, London.

Durand, B., Marchand, A., Amiell, J., Combaz, A., 1970. Etude de Kérogènes par RPE. In: Campos, R., Goni, J. (Eds.), Advances in Organic Geochemistry. Pergamon, Oxford, pp. 154–195.

Emery, K.O., 1963. Oceanic factors in accumulation of petroleum. In: Proc.—World Pet. Congr. 6, Sect. I, Pap. 42, PD2, pp. 483–491.

England, W.A., 1994. Secondary migration and accumulation of hydrocarbons. AAPG Mem. 60, 211–217.

England, W., Fleet, A.J. (Eds.), 1991. Petroleum Migration. Geol. Soc. London, London. Spec. Publ. No. 59.

Epstein, S., Krishnamurthy, R.V., Cronin, J.R., Pizzarello, S., Guen, G.U., 1987. Unusual stable isotope ratios in amino acid extracts from the Murchison meteorite. Nature (Lond.) 326, 477–479.

Erdman, J.G., 1975. Geochemical formation of oil. In: Fischer, A.G., Judson, S. (Eds.), Petroleum and Global Tectonics. Princeton University Press, Princeton, NJ, pp. 225–250.

Espitalié, J., Laporte, J.L., Madec, M., Marquis, F., Leplat, P., Paulet, J., Boutefou, A., 1977. Méthode rapide de caractérisation des roches meres de leur potentiel petrolier et de leur degré d'évolution. Rev. Inst. Fr. Pet. 3, 23–42.

Evans, W.D., Morton, R.D., Cooper, B.S., 1964. Primary investigations of the oleiferous dolerite of Dyvika. In: Colombo, U., Hobson, G.D. (Eds.), Advances in Geochemistry. Pergamon, Oxford, pp. 264–277.

Fang, H., Yongchuan, S., Sitian, L., Qiming, Z., 1995. Overpressure retardation of organic matter maturation and petroleum generation: a case study from the Yinggehai and Qiongdongan basins, South China Sea. APPG Bull. 79, 551–562.

Filipov, M.M., 1995. Proterozoic shungite-bearing rocks of Karelia. In: PreCambrian of Europe: Stratigraphy, Structure, Evolution and Mineralization. Russian Academy of Sciences, St. Petersburg, pp. 27–28.

Fleet, A.J., Kelts, K., Talbot, M.R. (Eds.), 1988. Lacustrine Petroleum Source Rocks. Geol. Soc. London, London. Spec. Publ. No. 40.

Flory, D.A., Lichtenstein, H.A., Biemann, K., Biller, J.E., Barker, C., 1983. Computer process uses entire GC-MS data. Oil Gas J. 17, 91–98.

Forbes, P.L., Ungerer, P., Mudford, B.S., 1992. A two-dimensional model of overpressure development and gas accumulation in venture field, Eastern Canada. AAPG Bull. 76, 318–338.

Foscolos, A.E., Powell, T.G., 1978. Mineralogical and geochemical transformation of clays during burial-diagenesis: relation to oil generation. In: Mortland, M.M., Farner, V.C. (Eds.), International Clay Conference. Elsevier, Amsterdam, pp. 261–270.

Friedman, G.M., 1972. Significance of Red Sea in problem of evaporites and basinal limestones. Am. Assoc. Pet. Geol. Bull. 56, 1072–1086.

Fuchtbauer, H., 1967. Influence of different types of diagenesis on sandstone porosity. In: Proc.—World Pet. Congr., vol. 7, pp. 353–369.

Fuller, J.G.C., 1975. Jurassic source-rock potential; and hydrocarbon correlation, North Sea. In: Finstad, K., Selley, R.C. (Eds.), Jurassic Northern North Sea Symposium. Norwegian Petroleum Society, Oslo, pp. 1–18.

Garven, G., 1989. A hydrogeologic model for the formation of the giant oil sand deposits of the western Canada sedimentary basin. Am. J. Sci. 289, 105–166.

Gaullier, J.M., Burrus, J., Barlier, J., Poulet, M., 1991. Paris basin petroleum systems revisited by two-dimensional modelling. AAPG Bull. 75, 578–579.

Gill, W.D., Khalaf, F.L., Massoud, M.S., 1979. Organic matter as indicator of the degree of metamorphism of the carboniferous rocks in the South Wales coalfields. J. Pet. Geol. 1, 39–62.

Goff, J.C., 1983. Hydrocarbon generation and migration from Jurassic source rocks in the E. Shetland Basin and Viking Graben of the northern North Sea. J. Geol. Soc. Lond. 140, 445–474.

Gold, T., 1979. Terrestrial sources of carbon and earthquake outgassing. J. Pet. Geol. 1, 3–19.

Gold, T., 1984. Contributions to the theory of an abiogenic origin of methane and other terrestrial hydrocarbons. Oil & Gas Fields. In: Proc. Int. Geol. Congr., 27th, Moscow, vol. 13, pp. 413–442.

Gold, T., Held, M., 1987. Helium–nitrogen-methane systematics in natural gases of Texas and Kansas. J. Pet. Geol. 10, 415–424.

Gold, T., Soter, S., 1982. Abiogenic methane and the origin of petroleum. Energy Explor. Exploit. 1, 89–104.

Goldhaber, M., 1978. Euxinic facies. In: Fairbridge, R.W., Bourgeois, J. (Eds.), Encyclopedia of Sedimentology. Dowden, Hutchinson & Ross, Stroudsburg, PA, pp. 296–300.

Gordon, M., Tracey, R.L., Ellis, M.W., 1958. Geology of the Arkansas Bauxite region. Geol. Surv. Prof. Pap. (U.S.) 299, 1–268.

Gottikh, R.P., 1995. Bituminogenesis and reduction degassing within the limits of the east of the Russian plate. In: PreCambrian of Europe: Stratigraphy, Structure, Evolution and Mineralization. Russian Academy of Sciences, St. Petersburg, pp. 40–41.

Gould, M.L., April 1991. Theory of methane origin and movement within a hydrocarbon basin. Offshore 26.

Gransch, J.A., Eisma, E., 1970. Characterization of the insoluble organic matter of sediments by pyrolysis. In: Hobson, G.D., Speers, G. (Eds.), Advances in Organic Geochemistry. Pergamon, Oxford, pp. 407–426.

Grunau, H.R., 1983. Abundance of source rocks for oil and gas worldwide. J. Pet. Geol. 6, 39–54.

Grunau, H.R., Gruner, U., 1978. Source rock and origin of natural gas in the Far East. J. Pet. Geol. 1, 3–56.

Guangming, Z., Quanheng, Z., 1982. Buried-hill oil and gas pools in the North China basin. AAPG Mem. 32, 317–336.

Hallam, A., 1981. Facies Interpretation and the Stratigraphic Record. Freeman, San Francisco.

Hay, W.H., 1995. Paleoceanography of marine organic-carbon-rich sediments. Stud. Geol. (Tulsa, Okla.) 40, 21–60.

Hobson, G.D., Tiratsoo, E.N., 1981. Introduction to Petroleum Geology. Scientific Press, Beaconsfield, IA.

Hoefs, J., 1980. Stable Isotope Geochemistry. Springer-Verlag, New York.

Hood, A., Gutjahr, C.C.M., Heacock, R.L., 1975. Organic metamorphism and the generation of petroleum. AAPG Bull. 59, 986–996.

Hoyle, F., 1955. Frontiers of Astronomy. Heinemann, London.

Huc, A.-Y. (Ed.), 1995. Paleogeography, Paleoclimates, and Source Rocks. Am. Assoc. Pet. Geol., Tulsa, OK. Stud. Geol. No. 40.

Hunt, J.M., 1961. Distribution of hydrocarbons in sedimentary rocks. Geochim. Cosmochim. Acta 22, 37–49.

Hunt, J.M., 1968. How gas and oil form and migrate. World Oil 167 (N4), 140–150.
Hunt, J.M., 1977. Distribution of carbon as hydrocarbons and asphaltic compounds in sedimentary rocks. AAPG Bull. 61, 100–116.
Hunt, J.M., 1979. Petroleum Geochemistry and Geology. Freeman, San Francisco.
Hunt, J.M., 1996. Petroleum Geochemistry and Geology, second ed. Freeman, San Francisco.
Hunt, J.M., Jamieson, G.W., 1956. Oil and organic matter in source rocks of petroleum. Am. Assoc. Pet. Geol. Bull. 40, 477–488.
Ikorskiy, S.V., 1967. Organic Substances in Minerals of Igneous Rocks in the Khibina Massif (from the Russian by L. Shapiro and I. A. Breger, Trans.). Clark Co., McLean, VA.
Jeffrey, A., Kaplan, I., 1988. Hydrocarbons and inorganic gases in the Gravberg-1 well, Siljan ring, Sweden. Chem. Geol. 71, 237–255.
Jeffrey, A., Kaplan, I., 1989. Asphaltine-like material in Siljan ring well suggests mineralized altered drilling fluid. J. Pet. Technol. 41, 1262–1313.
Johnson Ibach, L.E., 1982. Relationship between sedimentation rate and total organic carbon content in ancient marine sediments. AAPG Bull. 66, 170–188.
Jones, R.W., 1978. Some mass balance and geologic constraints on migration mechanisms. In: Roberts, W.H., Cordell, R.J. (Eds.), Physical and Chemical Constraints on Petroleum Migration. Am. Assoc. Pet. Geo, OK, pp. A1–A43. Course Notes, No. 8.
Jones, R.W., Demaison, G.J., 1980. Organic facies—stratigraphic concept and exploration tool. AAPG Bull. 64, 729.
Katz, B.J., 1995. Factors controlling the development of lacustrine petroleum source beds—an update. Stud. Geol. (Tulsa, Okla.) 40, 61–80.
Kidwell, A.L., Hunt, J.M., 1958. Migration of oil in recent sediments of Pedernales, Venezuela. In: Weeks, L.G. (Ed.), The Habitat of Oil. Am. Assoc. Pet. Geol., Tulsa, OK, pp. 790–817.
Kinghorn, R.R.F., 1982–1983. An Introduction to the Physics and Chemistry of Petroleum. Wiley, Chichester.
Klemme, H.D., November 1993. World petroleum systems with Jurassic source rocks. Oil Gas J. 8, 96–99.
Klemme, H.D., Ulmishek, G.F., 1991. Effective petroleum source rocks of the world: stratigraphic distribution and controlling depositional factors. AAPG Bull. 75, 1809–1851.
Larter, S.R., Bowler, B.F.J., Li, M., Chen, M., Brincat, D., Bennett, B., Noke, K., Donohoe, P., Simmons, D., Kohnen, M., Allan, J., Telnaes, N., Horstad, I., 1996. Molecular indicators of secondary oil migration distances. Nature (Lond.) 383, 593–597.
Le Blanc, L., April 1991. Buried rift methane proposed as tool to locate hydrocarbons. Offshore 23–29.
Levorsen, A.I., 1967. Geology of Petroleum. Freeman, San Francisco.
Lewan, M.D., 1985. Evaluation of petroleum generation by hydrous pyrolysis experimentation. Phil. Trans. Roy. Soc. Lond. 315, 123–134.
Lewan, M.D., 1987. Petrographic study of primary petroleum migration in the Woodford Shale and related rock units. In: Doligez, B. (Ed.), Migration of Hydrocarbons in Sedimentary Basins. Editions Technip, Paris, pp. 113–130.
Link, T.A., 1957. Whence came the hydrocarbons? Am. Assoc. Pet. Geol. Bull. 41, 1387–1402.
Lopatin, N.V., 1971. Temperature and geologic time as factors in coalification. Izv. Akad. Nauk SSSR.Ser. Geol. 3, 95–106.
Li, M., Taisheng, G., Xuepmg, Z., Taijun, Z., Rong, G., Zhenrong, D., 1982. Oil basins and subtle traps in the eastern part of China. In: Halbouty, M.T. (Ed.), The Deliberate Search for the Subtle Trap. Am. Assoc. Pet. Geol., Tulsa, OK, pp. 287–316.
MacDonald, G.J., 1983. The many origins of natural gas. J. Pet. Geol. 5, 341–362.
Magara, K., 1977. Petroleum migration and accumulation. In: Hobson, G.D. (Ed.), Developments in Petroleum Geology. Applied Science Pubs. Banking, pp. 83–126.
Magara, K., 1979. Structured water and its significance in primary oil migration. Bull. Can. Petl. Geol. 27, 87–93.
Magara, K., 1981. Possible primary migration of oil globules. J. Pet. Geol. 3, 325–331.
Magoon, L.B., 1988. The petroleum system–a classification scheme for research, exploration, and resource assessment, Chapter 5–Mapping the Petroleum System–An Investigative Technique 67. In: Magoon, L.B. (Ed.), 1870. Petroleum systems of the United States, USGS Bulletin, pp. 2–15.
Magoon, L.B., Dow, W.G. (Eds.), 1994. The Petroleum System—from Source to Trap. Am. Assoc. Pet Geol., Tulsa, OK. Mem. No. 60.

Magoon, L.B., Dow, W.G., 2000. Mapping the petroleum system—an investigative technique to explore the hydrocarbon fluid system. In: Mello, M.R., Katz, B.J. (Eds.), Petroleum systems of South Atlantic margins. Amer. Assoc. Petrol. Geol. Memoir 73, pp. 53–68.

Mann, U., 1994. An integrated approach to the study of primary petroleum migration. Spec. Publ.—Geol. Soc. Lond. 78, 233–260.

Marchand, A., Libert, P.A., Combaz, Z., 1969. Essai de caractérisation physico-chimiquc de la diagenesis de quelques roches organique, biologiquement homogènes. Rev. Inst. Pet. 24, 3–20.

McAuliffe, C.D., 1979. Oil and gas migration—chemical and physical constraints. AAPG Bull. 63, 761–781.

McTavish, R., 1978. Pressure retardation of vitrinite diagenesis, offshore north west Europe. Nature (Lond.) 271, 648–650.

Meissner, F.F., 1980. Examples of abnormal pressure produced by hydrocarbon generation (abs.). Am. Assoc. Pet. Geol. Bull. 64, 749.

Meissner, F.F., 1991. Origin and migration of oil and gas. In: Gluskoter, H.J., Rice, D.D., Taylor, R.B. (Eds.), Economic Geology, The Geology of North America, vol. P-2. Geological Society of America, U.S., Boulder, CO, pp. 225–240.

Mendele'ev, D., 1877. Entstehung und Vorkommen des Mineralols. Dtsch. Chem. Ges. Ber. 10, 229.

Mendele'ev, D., 1902. The Principles of Chemistry. In: 2nd Engl, vol. 1. Collier, New York (translated from the 6th Russian edition).

Miller, T.W., 1995. New insights on natural hydraulic fractures induced by abnormally high pore pressures. AAPG Bull. 79, 1005–1018.

Momper, J.A., 1978. Oil migration limitations suggested by geological and geochemical considerations. In: Roberts, W.H., Cordell, R.J. (Eds.), Physical and Chemical Constraints on Petroleum Migration. Am. Assoc. Pet. Geol., Tulsa, OK, pp. B1–B60. Course Notes, No. 8.

Momper, J.A., 1981. Oil Expulsion – A Consequence of Oil Generation. Slide-Tape Series in Geology. Am. Assoc. Pet. Geol., Tulsa, OK.

Morakhovsky, V.N., 1995. Bitumen as a result of retrograde metamorphism processes in crystalline complexes of Scandia region. In: PreCambrian of Europe: Stratigraphy, Structure, Evolution and Mineralization. Russian Academy of Sciences, St. Petersburg, pp. 73–74.

Mueller, G., 1963. Properties of extraterrestrial hydrocarbons and theory of their genesis. In: Proc.—World Pet. Congr. 6, Sect. 1, Pap. 29, pp. 1–14.

Nazipov, A.K., Iskanderov, D.B., 1995. Investigation of Precambrian basement in Novoyel-hovskaya superdeep well N 20009. In: PreCambrian of Europe: Stratigraphy, Structure, Evolution and Mineralization. Russian Academy of Sciences, St. Petersburg, p. 80.

Nelson, P., 2009. Pore-throat sizes in sandstones, tight sandstones, and shales. Am. Assoc. Pet. Geol. 93, 329–340.

North, F.K., 1979. Episodes of source-sediment deposition. Part 1. J. Pet. Geol. 2, 199–218.

North, F.K., 1980. Episodes of source-sediment deposition. Part 2. The episodes in individual close-up. J. Pet. Geol. 2, 323–338.

North, F.K., 1982. Review of Thomas Gold's deep-earth-gas hypothesis. Energy Explor. Exploit. 1, 105–110.

North, F.K., 1985. Petroleum Geology. Allen & Urwin Inc. 607.

Palciauskus, V.V., Domenico, P.A., 1980. Microfracture development of compacting sediments: relation to hydrocarbon-maturation kinetics. AAPG Bull. 64, 927–937.

Parrish, J.T., 1982. Upwelling and petroleum source beds, with reference to Paleozoic. AAPG Bull. 66, 750–774.

Parrish, J.T., 1995. Paleogeography of C_{org}-rich rocks and the preservation versus productivity controversy. Stud. Geol. (Tulsa. Okla.) 40, 1–20.

Pederson, T.F., Calvert, S.F., 1990. Anoxia vs. productivity: what controls the formation of organic-carbon rich sediments and sedimentary rocks? AAPG Bull. 74, 454–466.

Pepper, A.S., Corvi, P.J., 1995a. Simple kinetic models of petroleum formation. Part 1: oil and gas generation from Kerogen. Mar. Pet. Geol. 12, 291–319.

Pepper, A.S., Corvi, P.J., 1995b. Simple kinetic models of petroleum formation. Part 2: gas cracking. Mar. Pet. Geol. 12, 321–340.

Pepper, A.S., Corvi, P.J., 1995c. Simple kinetic models of petroleum formation. Part 3: modelling an open system. Mar. Pet. Geol. 12, 417–452.

Peters, K.E., 1986. Guidelines for evaluating petroleum source rocks using programmed pyrolysis. AAPG Bull. 70, 318–329.

Petersil'ye, Ya A., 1962. Origin of hydrocarbon gases and dispersed bitumens of the Khibina alkalic massif (translated from Russian). Geochemistry 1, 14–30.

Peyve, A.V., 1956. General characteristics and spatial characteristics of deep faults. Izv. Akad. Nauk SSSR 1, 90–106.

Philippi, H.W., 1975. On the depth, time and mechanism of petroleum migration. Geochim. Cosmochim. Acta 29, 1021–1049.

Pillinger, C., 1987. Meteorites: nearer yet further away. Nature (Lond.) 326, 445–447.

Porfir'ev, V.B., 1974. Inorganic origin of petroleum. AAPG Bull. 58, 3–33.

Powers, M.C., 1967. Fluid-release mechanisms in compacting marine mudrocks and their importance in oil exploration. Am. Assoc. Pet. Geol. Bull. 51, 1240–1254.

Pratt, W.E., 1942. Oil in the Earth. University of Kansas Press, Lawrence.

Price, L.C., 1976. Aqueous solubility of petroleum as applied to its origin and primary migration. AAPG Bull. 60, 213–244.

Price, L.C., 1978. New evidence for hot, deep origin and migration of petroleum. In: Roberts, W.H., Cordell, R.J. (Eds.), Physical and Chemical Constraints on Petroleum Migration. Am. Assoc. Pet. Geol., Tulsa, OK, pp. F9–F10. Course Notes, No. 9.

Price, L.C., 1980. Shelf and shallow basin oil as related to hot-deep origin of petroleum. J. Pet. Geol. 3, 91–115.

Price, L.C., 1981a. Primary petroleum migration by molecular solution: consideration of new data. J. Pet. Geol. 4, 89–101.

Price, L.C., 1981b. Aqueous solubility of crude oil to 400°C and 2,000 bars pressure in the presence of gas. J. Pet. Geol. 4, 195–223.

Price, L.C., 1983. Geologic time as a parameter in organic metamorphism and vitrinite reflectance as an absolute paleogeothermometer. J. Pet. Geol. 6, 5–38.

Pusey, W.C., 1973a. How to evaluate potential gas and oil source rocks. World Oil 176 (5), 71–75.

Pusey, W.C., 1973b. Paleotemperatures in the Gulf Coast using the ESR-kerogen method. Trans. Gulf Coast Assoc. Geol. Soc. 23, 195–202.

Pushkarev, Y.D., 1995. Gas-petroliferous potential of the Precambrian basement under European platforms according to radiogenic isotope geochemistry of oils and bitumens. In: Pre-Cambrian of Europe: Stratigraphy, Structure, Evolution and Mineralization. Russian Academy of Sciences, St. Petersburg, pp. 90–91.

Rankin, A.R., 1990. Fluid inclusions associated with oil and ore in sediments. In: Ala, M. (Ed.), Seventy-five Years of Progress in Oil Field Technology. Balkema, Rotterdam, pp. 113–124.

Roberts, S.J., Nunn, J.A., 1995. Episodic fluid expulsion from geopressured sediments. Mar. Pet. Geol. 12, 195–204.

Roberts, W.H., Cordell, R.J., 1980. Problems of petroleum migration. Stud. Geol. (Tulsa, Okla.) 10, 1–273.

Robinson, N., Eglinton, G., Brassell, S.C., Gowar, A.P., Parnell, J., 1986. Hydrocarbon compositions of bitumens associated with igneous intrusions and hydrothermal deposits in Britain. In: Advances in Organic Geochemistry. Pergamon, Oxford, pp. 145–152.

Robinson, R., 1963. Duplex origin of petroleum. Nature (Lond.) 199, 113–114.

Robinson, W.E., 1976. Origin and characteristics of Green River oil shale. In: Chilingarian, G.V., Yen, T.F. (Eds.), Oil Shale. Elsevier, Amsterdam, pp. 61–80.

Roedder, E., 1984. Fluid inclusions. Rev. Mineral 12, 1–644.

Royden, L., Sclater, J.G., Von Herzen, R.P., 1980. Continental margin subsidence and heat flow: important parameters in formation of petroleum hydrocarbons. AAPG Bull. 64, 173–187.

Schidlowski, M., Eichmann, R., Junge, C.E., 1974. Evolution des iridischen Sauerstof-Budgets und Entwicklung de Edatmosphare. Umschau 22, 703–707.

Schlanger, S.O., Cita, M.B., 1982. Nature and Origin of Cretaceous Carbon Rich Facies. Academic Press, London.

Schnip, O.A., 1992. Reservoirs in fractured basement on the continental shelf of Vietnam. J. Pet. Geol. 15, 451–464.

Selley, R.C., 1980. Economic basement in the northern North Sea. In: Hardman, R. (Ed.), Symposium on the Reservoir Rocks of the Northern North Sea. Norwegian Petroleum Society, Oslo, pp. 1–22.

Selley, R.C., 1992. Petroleum seepages and impregnations in Great Britain. Mar. Pet. Geol. 9, 226–244.

Showalter, T.T., 1979. Mechanics of secondary hydrocarbon migration and entrapment. AAPG Bull. 63, 723–760.

Snowdon, L.R., 1995. Rock-eval T_{max} suppression: documentation and amelioration. AAPG Bull. 79, 1337–1348.

Staplin, F.L., 1969. Sedimentary organic matter, organic metamorphism, and oil and gas occurrence. Bull. Can. Pet. Geol. 17, 47–66.

Stein, R., 1986. Organic carbon and sedimentation rate—further evidence for anoxic deep-water conditions in the Cenomanian–Turonian Atlantic Ocean. Mar. Pet. Geol. 72, 199–210.

Stoneley, R., 1983. Fibrous calcite veins, overpressures and primary oil migration. AAPG Bull. 67, 1427–1428.

Studier, M.H., Hayatsu, R., Anders, E., 1965. Organic compounds in Carbonaceous chondrites. Science 149, 1455–1459.

Subbotin, S.L., 1966. Upper mantle and inorganic oil. In: Problems in the Origin of Oil. Izd. Nauk Dumka, Kiev, pp. 52–62.

Sugisaki, R., Ido, M., Takeda, H., Isobe, Y., 1983. Origin of hydrogen and carbon dioxide in fault gases and its relation to fault activity. J. Geol. 91, 239–258.

Summerhayes, C., 1987. Organic-rich Cretaceous sediments in the North Atlantic. Spec. Publ. Geol. Soc. Lond. 26, 301–316.

Szatmari, P., 1986. Plate tectonic control of synthetic oil formation. Oil Gas J. September 1, 67–69.

Szatmari, P., 1989. Petroleum formation by Fischer-Tropsch synthesis in plate tectonics. AAPG Bull. 73, 989–998.

Tappan, H., Loeblich, A.R., 1970. Geobiologic implications of fossil phytoplankton evolution and time/space distribution. In: Kosanke, R.M., Cross, S.T. (Eds.), Palynology of Late Cretaceous and Early Tertiary. Michigan State University, Geol. Soc., East Lansing, pp. 247–340.

Tissot, B.P., 1977. The application of the results of organic chemical studies in oil and gas exploration. In: Hobson, G.D. (Ed.), Developments in Petroleum Geology, vol. 1. Applied Science Publishers, London, pp. 53–82.

Tissot, B.P., 1979. Effects on prolific petroleum source rocks and major coal deposits caused by sea level changes. Nature (Lond.) 277, 377–380.

Tissot, B.P., Welte, D.H., 1978. Petroleum Formation and Occurrence: A New Approach to Oil and Gas Exploration. Springer-Verlag, Berlin.

Tissot, P.B., Welte, D.H., 1984. Petroleum Formation and occurrence: A New Approach to Oil and Gas Exploration, 2nd Edn. Springer Verlag, Berlin. 530 pp.

Vail, P.R., Mitchum, R.M., Thompson, S., 1977. Seismic stratigraphy and global changes of sea level. Part 3. Mem.—Am. Assoc. Pet. Geol. 26, 63–82.

Van Hinte, J.E., 1978. Geohistory analysis—application of micropaleontology in exploration geology. AAPG Bull. 62, 201–202.

Verweij, J.M., 1993. Hydrocarbon Migration Systems Analysis. Elsevier, Amsterdam.

Vlierbloom, F.W., Collini, B., Zumberge, J.E., 1986. The occurrence of petroleum in sedimentary rocks of the meteor impact crater at Lake Siljan, Sweden. Org. Geochem. 10, 153–161.

Volkman, J.K., 1988. Biological marker compounds as indicators of the depositional environment of petroleum source rocks. Spec. Publ.—Geol. Soc. Lond. 40, 103–122.

Waples, D.W., 1980. Time and temperature in petroleum exploration. AAPG Bull. 64, 916–926.

Waples, D.W., 1981. Organic Geochemistry for Exploration Geologists. CEPCO, Minneapolis, MN.

Waples, D.W., 1983. Reappraisal of anoxia and organic richness with emphasis on Cretaceous of North Atlantic. AAPG Bull. 67, 963–978.

Waples, D.W., 1994. Modelling of sedimentary basins and systems. AAPG Mem. 60, 307–322.

Waples, D.W., Kamat, H., Suizu, M., 1992a. The art of maturity modelling. Part 1. Finding a satisfactory geologic model. AAPG Bull. 76, 31–46.

Waples, D.W., Kamat, H., Suizu, M., 1992b. The art of maturity modelling. Part 2: alternative models and sensitivity analysis. AAPG Bull. 76, 47–66.

Welte, D.H., 1972. Petroleum exploration and organic geochemistry. J. Geochem. Explor. 1, 117–136.

Welte, D.H., Yukler, M.A., 1984. Petroleum origin and accumulation in basin evolution—a quantitative model. AAPG Mem. 35, 27–78.

Welte, D.H., Hagemann, H.W., Hollerbach, A., Leythaeuser, D., Staht, W., 1975. Correlation between petroleum and source rock. Proc.—World Pet. Congr. 9, 179–191.

White, D.A., Gehman, H.M., 1979. Methods of estimating oil and gas resources. AAPG Bull. 63, 2183–2203.

White, D.E., Waring, G.A., 1963. Volcanic emanations. Geol. Surv. Prof. Pap. (U.S.) 440-K, 1–29.

Zahnle, K., Grinspoon, D., 1990. Comet dust as a source of aminoacids at the Cretaceous/Tertiary boundary. Nature (Lond.) 348, 157–160.

Zielinski, G.W., Bruchhausen, P.M., 1983. Shallow temperature and thermal regime in the hydrocarbon province of Tierra del Fuego. AAPG Bull. 67, 166–177.

Selected Bibliography

For accounts of the origin, composition and diagenesis of petreum source rocks see:
Hue, A.-Y. (Ed.), 1995. Paleogeography, Paleoclimates, and Source Rocks. Am. Assoc. Pet. Geol, Tulsa, OK. Stud. Geol. No. 40.
Katz, B.J., 1995. Petroleum Source Rocks. Springer-Verlag, Berlin.
For detailed accounts of the generation and migration of petroleum, see:
Hunt, J.M., 1996. Petroleum Geochemistry and Geology, second ed. Freeman, New York.
Verweij, J.M., 1993. Hydrocarbon Migration Systems Analysis. Elsevier, Amsterdam.

CHAPTER

6

The Reservoir

Chapter 1 showed that one of the seven essentials prerequisites for a commercial accumulation of hydrocarbons is the existence of a reservoir. Theoretically, any rock may act as a reservoir for oil or gas. In practice, the sandstones and carbonates contain the major known reserves, although fields do occur in shales and diverse igneous and metamorphic rocks. For a rock to act as a reservoir it must possess two essential properties: It must have pores to contain the oil or gas, and the pores must be connected to allow the movement of fluids; in other words, the rock must have permeability.

This chapter is concerned with the nature of hydrocarbon reservoirs, that is, with the internal properties of a trap. It begins by describing porosity and permeability and discussing their relationship with sediment texture. This discussion is followed by an account of how postdepositional (diagenetic) changes may diminish or enhance reservoir performance. Following this discussion is a section on the vertical and lateral continuity of reservoirs and the calculation of oil and gas reserves. The chapter concludes with an account of the various ways of producing hydrocarbons from a reservoir.

6.1 POROSITY

6.1.1 Definition and Classification

Porosity is the first of the two essential attributes of a reservoir. The pore spaces, or voids, within a rock are generally filled with connate water, but contain oil or gas within a field. Porosity is either expressed as the void ratio, which is the ratio of voids to solid rock, or, more frequently, as a percentage:

$$\text{Porosity}(\%) = \frac{\text{volume of voids}}{\text{total volume of rock}} \times 100.$$

Porosity is conventionally symbolized by the Greek lowercase letter phi (ϕ). Pores are of three morphological types: catenary, cul-de-sac, and closed (Fig. 6.1). Catenary pores are those that communicate with others by more than one throat passage. Cul-de-sac, or dead-end, pores have only one throat passage connecting with another pore. Closed pores have no communication with other pores.

Catenary and cul-de-sac pores constitute effective porosity, in that hydrocarbons can emerge from them. In catenary pores hydrocarbons can be flushed out by a natural or

http://dx.doi.org/10.1016/B978-0-12-386031-6.00006-0

FIGURE 6.1 The three basic types of pores.

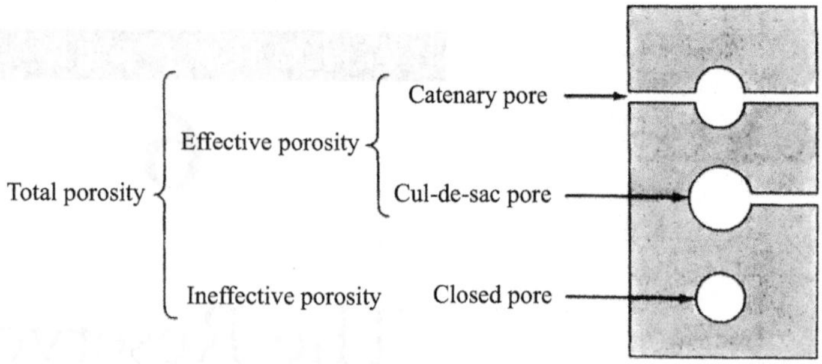

TABLE 6.1 Classification of the Different Types of Porosity Found in Sediments

Time of formation	Type	Origin
Primary or depositional	Intergranular, or interparticle Intragranular, or intraparticle	Sedimentation
	Intercrystalline Fenestral	Cementation
Secondary or postdepositional	Vuggy Moldic	Solution
	Fracture	Tectonics, compaction, dehydration, diagenesis

artificial water drive. Cul-de-sac pores are unaffected by flushing, but may yield some oil or gas by expansion as reservoir pressure drops. Closed pores are unable to yield hydrocarbons (such oil or gas having invaded an open pore subsequently closed by compaction or cementation). The ratio of total to effective porosity is extremely important, being directly related to the permeability of a rock.

The size and geometry of the pores and the diameter and tortuosity of the connecting throat passages all affect the productivity of the reservoir. Pore geometry and continuity are difficult to analyze. In some studies pores are filled by a fluid, which then solidifies, and the rock is then dissolved by acid to reveal casts of the pores. Collins (1961) used molten lead in sandstones, and Wardlaw (1976) used epoxy resin in carbonates.

Several schemes have been drawn up to classify porosity types (e.g., Robinson, 1966; Levorsen, 1967; Choquette and Pray, 1970). Two main types of pore can be defined according to their time of formation (Murray, 1960). Primary pores are those formed when a sediment is deposited. Secondary pores are those developed in a rock some time after deposition (Table 6.1). Primary pores may be divided into two subtypes: interparticle (or intergranular) and intraparticle (or intragranular). Interparticle pores are initially present in all sediments. They are often quickly lost in clays and carbonate sands because of the combined effects of compaction and cementation. Much of the porosity found in sandstone reservoirs is

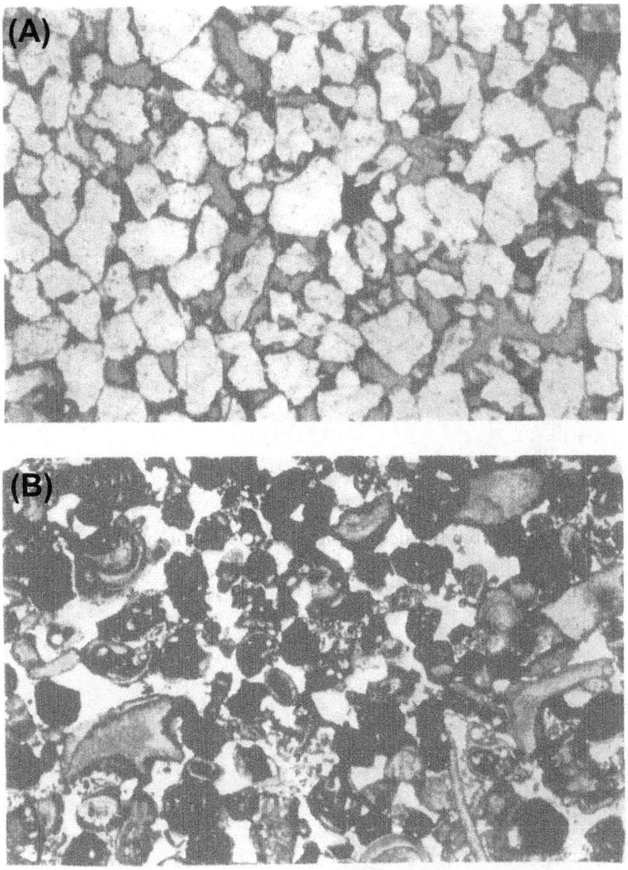

FIGURE 6.2 Thin sections illustrating the different types of primary porosity. (A) Intergranular porosity in Middle Jurassic Brent sandstone (North Sea, United Kingdom). Field of view is approximately 1 cm. (B) Intergranular and intragranular (within fossils) porosity in skeletal pelletal limestone. Field of view is about 1 cm.

preserved primary intergranular porosity (Fig. 6.2(A)). Intraparticle pores are generally found within the skeletal grains of carbonate sands (Fig. 6.2(B)) and are thus often cul-de-sac pores. Because of compaction and cementation they are generally absent in carbonate reservoirs.

Secondary pores are often caused by solution. Many minerals may be leached out of a rock, but, volumetrically, carbonate solution is the most significant. Thus solution-induced porosity is more common in carbonate reservoirs than in sandstone reservoirs. A distinction is generally made between moldic and vuggy porosity. Moldic porosity is fabric selective (Fig. 6.3), that is, only the grains or only the matrix has been leached out. Vugs, by contrast, are pores whose boundaries cross-cut grains, matrices, and/or earlier cement. Vugs thus tend to be larger than moldic pores (Fig. 6.4). With increasing size vuggy porosity changes into cavernous porosity. Cavernous pores are those large enough to cause the drill string to drop by half a meter or to contain one crouched mud-logger. Examples of cavernous porosity are known from the Arab D Jurassic limestone of the Abqaiq field of Saudi Arabia (McConnell, 1951) and from the Fusselman limestone of the Dollarhide field of Texas (Stormont, 1949), both having cavernous pores up to 5 m high.

FIGURE 6.3 Portland limestone (Upper Jurassic) showing biomoldic porosity due to solution of shells. Portland Island, Dorset, United Kingdom.

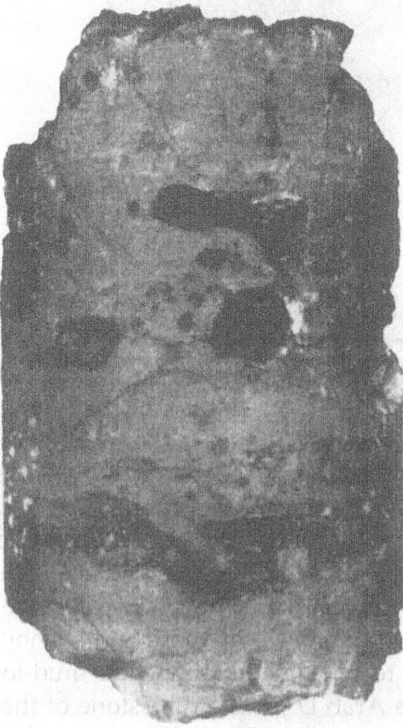

FIGURE 6.4 Core of Zechstein (Upper Permian) dolomite showing vuggy porosity (North Sea, United Kingdom).

Secondary porosity can also be caused not by solution but by the effects of cementation. Fenestral pores occur where there is a "primary or penecontem-poraneous gap in rock framework larger than grain-supported interstices" (Tebbutt et al., 1965). Strictly speaking, therefore, such pores would appear to be primary rather than secondary. In fact, fenestral pores are characteristic of lagoonal pelmicrites in which dehydration has caused shrinkage and buckling of the laminae. This type of fabric has been described from Triassic lagoonal carbonates of the Alps, where it is termed *loferite* (Fischer, 1964), and also from Paleocene dolomite pellet muds from Libya (Conley, 1971).

Intercrystalline porosity, which refers to pores occurring between the crystal faces of crystalline rocks, is a far more important type of secondary porosity than is fenestral porosity. Most recrystallized limestones generally possess negligible porosity. Crystalline dolomites, on the other hand, are often highly porous, with a friable, sugary (saccharoidal) texture. This type of intercrystalline porosity occurs in secondary dolomites that have formed by the replacement of calcite. Dolomitization causes a 13% shrinkage of the original bulk volume, thereby increasing the porosity. Intercrystalline pores tend to be polyhedral, with sheet-like pore throats in contrast to the more common tubular ones (Wardlaw, 1976).

Fracture porosity is the last major type of pore to consider. It is extremely important not so much because it increases the storage capacity of a reservoir but because of the degree to which it may enhance permeability. Fractures are rare in unconsolidated, loosely cemented sediments, which respond to stress by plastic flow. They may occur in any brittle rock, not only sandstones and limestones but also shales, igneous, and metamorphic rocks (Stearns and Friedman, 1972). Because fractures are much larger than are most pores, they are seldom amenable to study from cores alone (Fig. 6.5). Furthermore, the process of coring itself may fracture the rock. Such artifacts must be distinguished from naturally occurring pores. Kulander et al. (1979) have outlined several significant criteria that can be used for this purpose. Fractures may also be recognized from wireline logs, seismic data, and the production history of a well. Cycle skip on sonic logs can be caused by fractures (and other phenomena). A random bag o' nails motif on a dipmeter can be caused by fractures (and other phenomena). Fractures can also be "seen" by the borehole imaging tools (Fig. 6.6). Anomalously low velocities in seismic data may be due to fractures, although again this effect may be due to other types of porosity and/or the presence of gas. High initial production followed by a rapid pressure drop and rapid decline in flow rate in production well tests is often indicative of fracture porosity.

Brittle rocks tend to fracture in three geological settings (Fig. 6.7). Fracture systems may dilate where strata are subjected to tension on the crests of anticlines and the nadirs of synclines. Detailed outcrop studies have demonstrated the correlation between fracture intensity and orientation and the structural style and trend of the anticline (e.g., Harris et al., 1960). Fractures often occur adjacent to faults, and in some cases a field may be directly related to a major fault and its associated fractures (Fig. 6.8). Fracture porosity is also found beneath unconformities, especially in carbonates, where the fractures may be enlarged by karstic solution, as in the Castellan field of offshore Spain. Fracture porosity also occurs in sandstones; for example, in the Buchan field of the North Sea, in which a fractured Devonian sandstone reservoir lies beneath the Cimmerian unconformity (Butler et al., 1976).

Basement rocks sometimes serve as petroleum reservoirs because fracture porosity is sufficiently abundant for them to flow. These reservoirs are commonly dual porosity systems

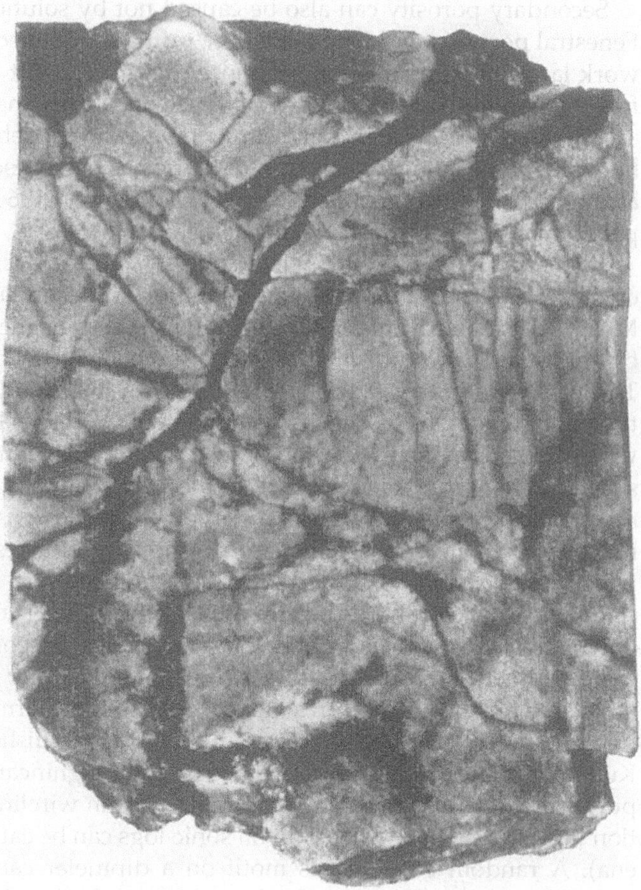

FIGURE 6.5 Fractured core of Gargaf Group sandstone (Cambro-Ordovician). Sirte basin, Libya.

where solution porosity has formed from the leaching of unstable mineral grains. This may be due to hydrothermal alteration where magma intrudes source rocks, as in the Dineh-Bi-Keyah field in Apache County, Arizona (McKenny and Masters, 1968; Biederman, 1986). Alternatively the solution porosity may be due to weathering where basement is unconformably onlapped by source rocks, as in the Augila field of the Sirte basin, Libya (Williams, 1972). In such fields the importance of fracturing is not so much because it enhances porosity, but because it dramatically improves permeability.

Figure 6.9 summarizes the relationship between porosity and permeability for the different types of pore systems.

6.1.2 Porosity Measurement

Porosity may be measured in three ways: directly from cores, indirectly from geophysical well logs, or from seismic data, as discussed earlier in Chapter 3. The last two of these

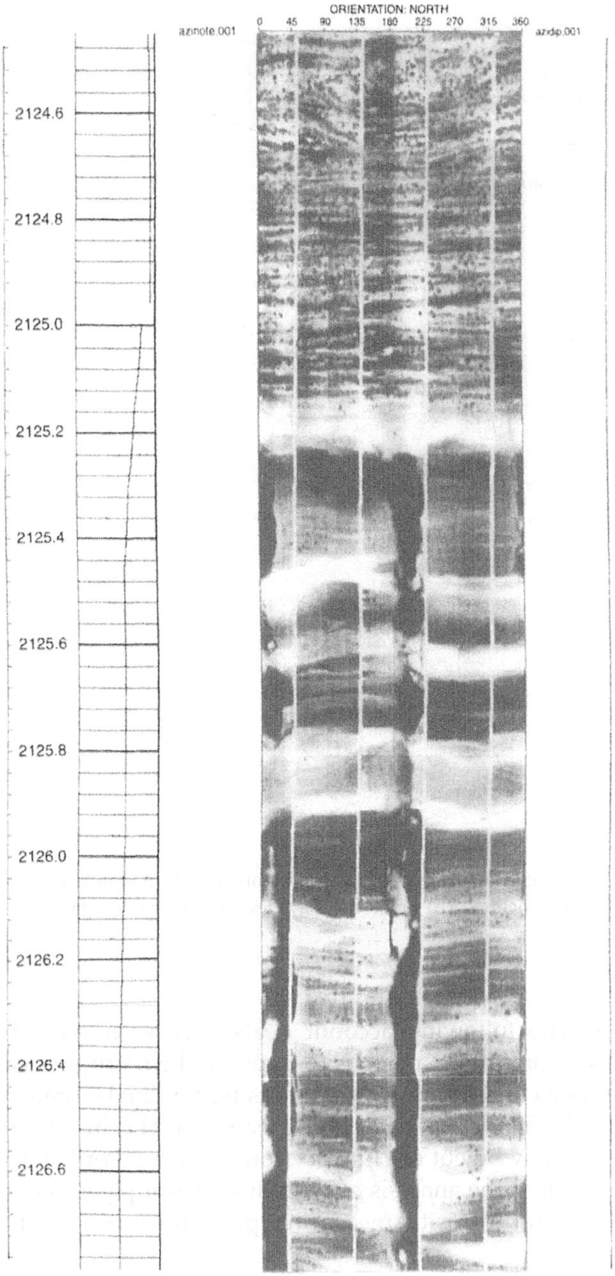

FIGURE 6.6 Borehole Formation Micro-imager (FMI) image showing fracture porosity at sequence boundary between fractured bedded limestone overlain by unfractured laminated limestone. *Courtesy of Schlumberger.*

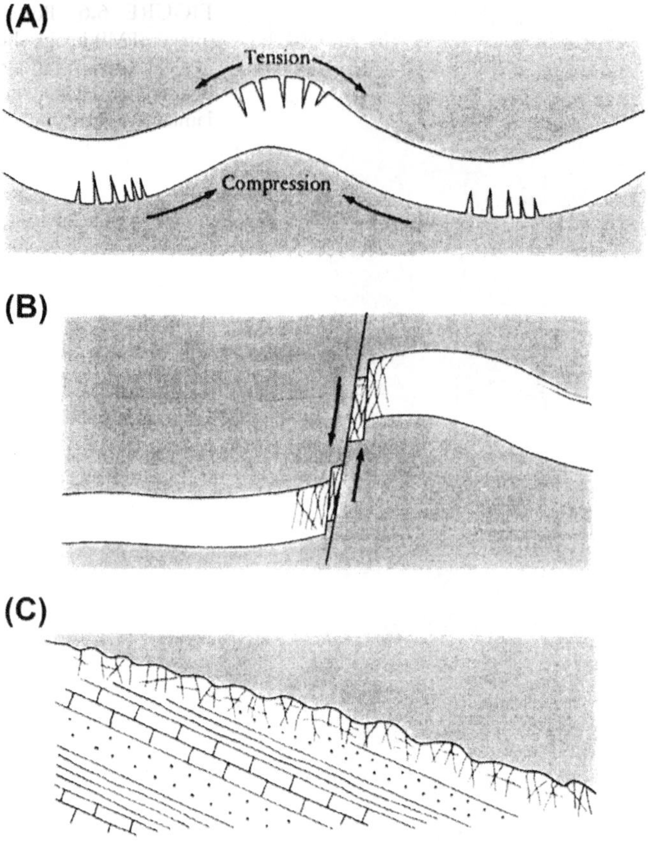

FIGURE 6.7 Illustrations of the various ways in which fracture porosity is commonly found. (A) Fractures may develop on the crests of anticlines and the nadirs of synclines. (B) Fractures may develop adjacent to faults. (C) Fractures may occur beneath unconformities.

methods are discussed in this chapter. The following account deals only with porosity measurement from cores. The main reason for cutting a core is to measure the petrophysical properties of the reservoir. The porosity of a core sample may be measured in the laboratory using several methods (Anderson, 1975; Monicard, 1981). For homogeneous rocks, like many sandstones, samples of only 30 mm^3 or so may be cut or chipped from the core. For heterogeneous reservoirs, including many limestones, the analysis of a whole core sample is generally necessary. Several of the more common porosity measuring procedures are briefly described in the following sections.

6.1.2.1 Washburn–Bunting Method

One of the earliest and simplest methods of porosity measurement is the gas expansion technique described by Washburn and Bunting in 1922. The basic apparatus is shown in Fig. 6.10. Air within the pores of the sample is extracted when a vacuum is created by

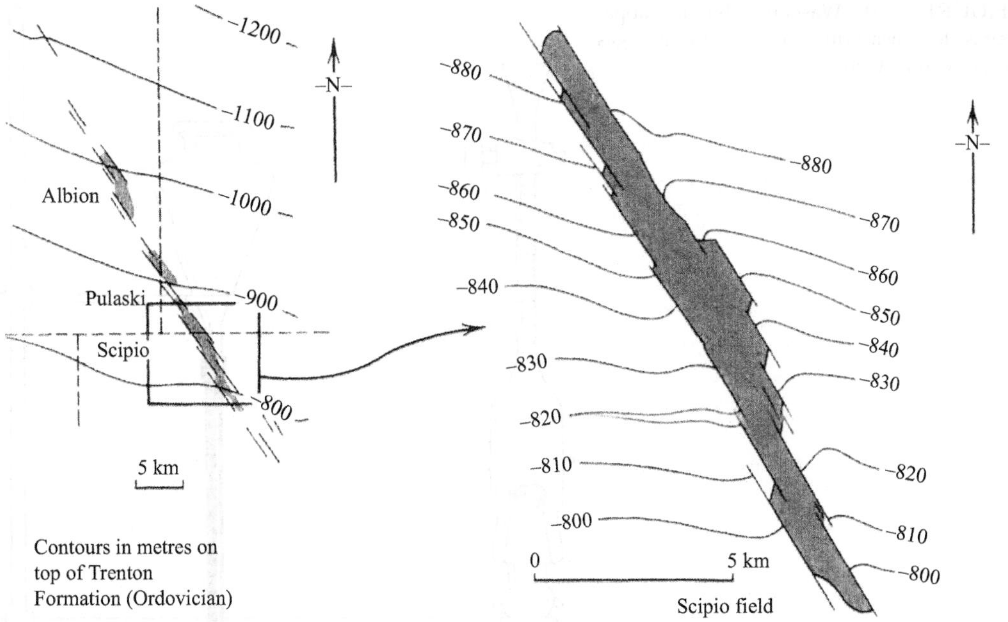

FIGURE 6.8 Map of the Scipio–Pulaski–Albion trend of fields. Production comes from fractured Trenton (Ordovician) dolomite. Right-hand map is a detail of the boxed area in the left-hand map. *From Levorsen (1967). Used with permission.*

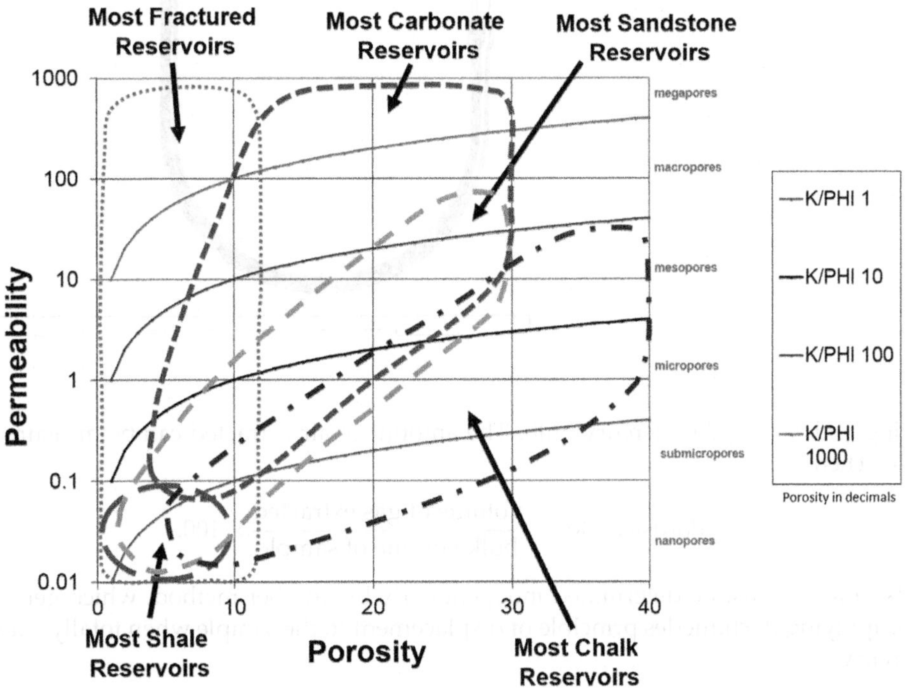

FIGURE 6.9 Graph to illustrate the relationship between porosity and permeability for the different types of reservoirs. Note that fracturing will enhance permeability dramatically for any type of reservoir. The contours represent a constant K/PHI ratio and divide the plot into areas of similar pore types. Data points that plot along a constant ratio have similar flow characteristics. *Modified from Selley (1988).*

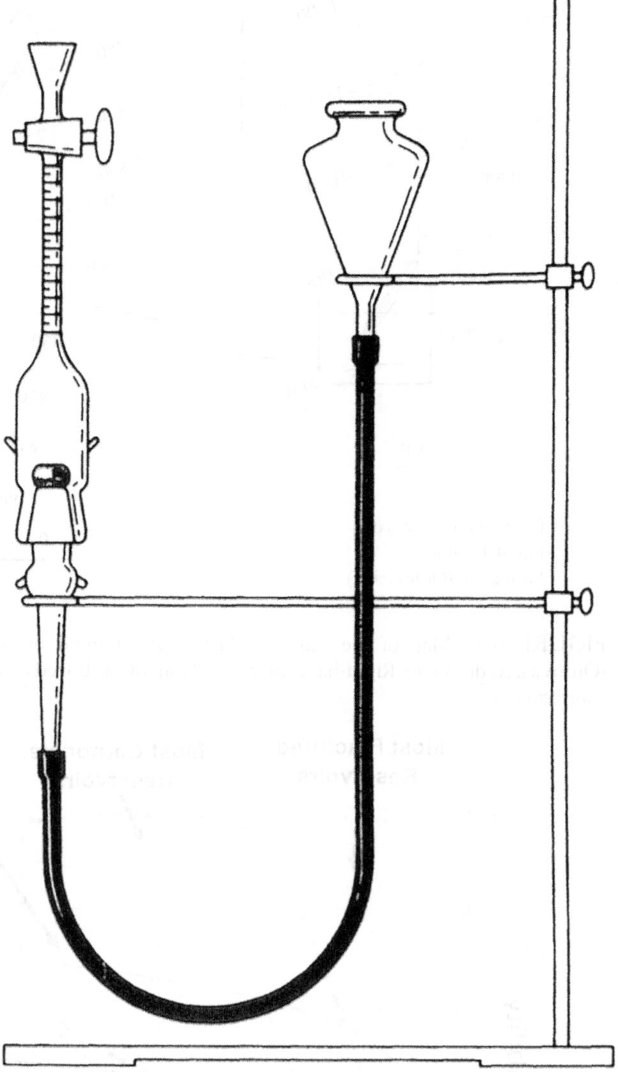

FIGURE 6.10 Washburn–Bunting apparatus for measuring porosity by the gas expansion method.

lowering and raising the mercury bulb. The amount of air extracted can be measured in the burette. Then:

$$\text{Porosity (\%)} = \frac{\text{volume of gas extracted}}{\text{bulk volume of sample}} \times 100.$$

Bulk volume must be determined independently by another method, which generally involves applying Archimedes principle of displacement to the sample when totally submerged in mercury.

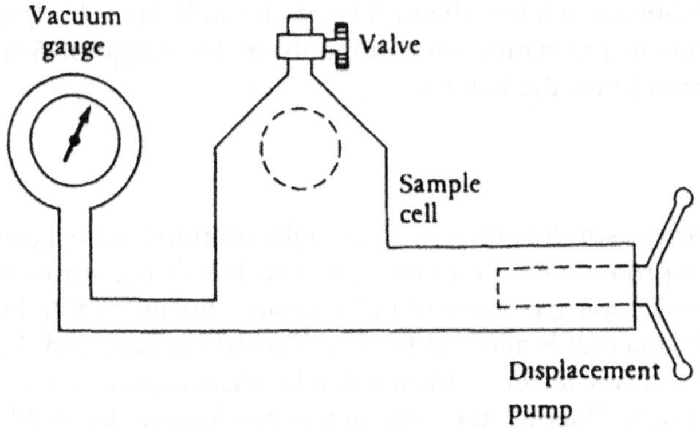

FIGURE 6.11 Boyle's law porosimeter.

6.1.2.2 Boyle's Law Method

Boyle's law (pressure × volume = constant) can be applied to porosity measurement in two ways. One way is to measure the pore volume by sealing the sample in a pressure vessel, decreasing the pressure by a known amount, and measuring the increase in volume of the contained gas. Conversely, the grain volume can be measured and, if the bulk volume is known, porosity can be determined. The Ruska porosimeter is a popular apparatus based on this technique (Fig. 6.11).

6.2 PERMEABILITY

6.2.1 Fundamental Principles

The second essential requirement for a reservoir rock is permeability. Porosity alone is not enough; the pores must be connected. Permeability is the ability of fluids to pass through a porous material. The original work on permeability was carried out by H. Darcy (1856), who studied the flow rates of the springs at Dijon in France. His work was further developed by Muskat and Botset (1931), Botset (1931), and Muskat (1937). They formulated Darcy's law as follows:

$$Q = \frac{K(P_1 - P_2)A}{\mu L},$$

where

Q = rate of flow;
K = permeability;
$(P_1 - P_2)$ = pressure drop across the sample;
A = cross-sectional area of the sample;
L = length of the sample;
μ = viscosity of the fluid.

The unit of permeability is the Darcy. This is defined as the permeability that allows a fluid of 1 centipoise (cP) viscosity to flow at a velocity of 1 cm/s for a pressure drop of 1 atm/cm.

Because most reservoirs have permeabilities much less than a Darcy, the millidarcy (md) is commonly used. Average permeabilities in reservoirs are commonly in the range of 5 to 500 md. Permeability is generally referred to by the letter *K*.

6.2.2 Permeability Measurement

The permeability of a reservoir can be measured in three ways. On the grandest scale it can be measured by means of a drill stem or production test. In this test, a well is drilled through the reservoir. Casing is set and perforated, tubing is run within the casing, the interval to be tested is sealed off with packers, and the interval is allowed to flow. The rate of flow and the drop in pressure at the commencement and conclusion of the test can be measured, and reservoir fluid can be recovered at the surface. Thus all the parameters are known to enable permeability to be calculated from Darcy's law. This type of test is, of course, in many ways the most obvious—and significant. The advantages and limitations of production testing are beyond the scope of this text.

The second way of measuring permeability is from wireline logs. It has long been possible to identify permeable zones in a qualitative way from Spontaneous Potential (SP) and caliper logs, but only recently has it been possible to quantify permeability from logs with any degree of reliability.

The third way of measuring permeability is by means of a permeameter (Fig. 6.12). The version illustrated here is for a fixed laboratory based rig that forces gas through a prepared rock sample. It is conventional to cut plugs from the whole core at intervals of half a meter or so. Whole core analysis is more time-consuming and expensive. It is, however, essential for rocks with heterogeneous pore systems, such as fractured and vuggy reservoirs, and heterogeneous rock fabrics, such as conglomerate and biolithite.

Mobile permeameters are also available. The "probe" permeameter, also called the "mini-" permeameter was first developed nearly half a century ago (Dykstra and Parsons, 1950). It was little used until the 1970s (Weber et al., 1972), since then it has undergone a renaissance. The probe permeameter involves placing a nozzle on a flat rock surface, ensuring that there is an effective seal around the aperture, and then pumping a fluid, generally nitrogen gas, into the rock and measuring the rate of flow. The mini-permeameter has several great advantages over the conventional one. It is sufficiently light and portable that it can be carried around and used on cores in a core store, or on rock outcrops in the field (Hurst and Goggin, 1995). It is fast to operate, and can thus take many readings over a small area of rock, thus testing for small-scale reservoir heterogeneities.

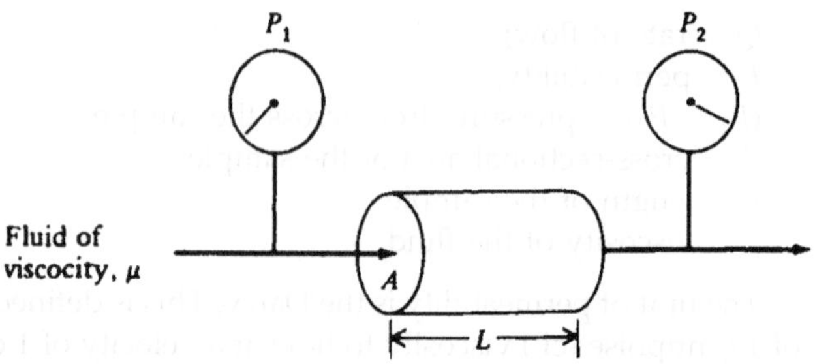

FIGURE 6.12 Basic arrangement for the measurement of permeability. This sketch illustrates a rig suitable for measuring a core plug within a laboratory. "Mini-" or "probe" permeameters are now also widely used in which gas is pumped through a sealed nozzle on a flat rock surface. For further details see text.

Hurst and Rosvoll (1991) report that average permeabilities measured with a minipermeameter were comparable to those obtained by whole core analysis, but both are lower than the permeability measured from plugs cut from the same core.

Shale and mudrock reservoirs are exceeding fine-grained and fracture prone. Data for core porosity, permeability, grain-density, and core saturations can therefore be inaccurate (Luffel and Guidry, 1992). A special sampling and measurement technique referred to as Gas Research Institute (GRI) (Crushed Shale) analysis was developed by Core Lab and the Gas Research Institute for these fine-grained rocks (Core Laboratories, 2014). The method determines the bulk density of the fine-grained rock after which samples are reduced to small particles, 0.5–0.85 mm in diameter. Coring- and sampling-induced fractures are eliminated and pore pathways are shortened. Pore liquid removal and quantification by distillation extraction is then performed. Dry grain volume determination finishes the measurement sequence and provides porosity, grain density, and core saturations. GRI "crushed" core analysis method provides the following: bulk density, grain density, total interconnected porosity, gas filled porosity, core saturations (Sw, So, Sg), and matrix permeability to gas.

6.2.3 Interpretation of Permeability Data

Darcy's law is only valid when there is no chemical reaction between the fluid and the rock and when only one fluid phase completely fills the pores. The situation is far more complex for mixed oil and gas phases, although a Darcy-type equation is assumed to apply. Darcy's law is also only valid for a uniform type of pore system. In dual porosity systems, where, for example fractures and vugs occur, more complex relationships exist than can be accurately expressed by Darcy's law.

A fourth problem is due to the fact that core plugs are commonly contaminated with petroleum and detritus from the drilling mud. Thus they need to be cleaned before measurement can take place. Sadly this cleaning process may also modify the original petrophysical characteristics of the specimen, particularly the clay minerals.

Flow rate depends on the ratio of permeability to viscosity. Thus gas reservoirs may be able to flow at commercial rates with permeabilities of only a few millidarcies, whereas oil reservoirs need minimal permeabilities of the order of tens of millidarcies. For this reason permeability is measured in the laboratory using an inert gas rather than a liquid. Under some conditions—small pore dimensions, very low gas densities, and relatively large mean free paths of gas molecules—the normal condition of zero velocity and no slip at solid surface may not be met. This is the Klinkenberg effect, and a correction is necessary to recalculate the permeability of the rock to air measured in the laboratory to the permeability it would exhibit for liquid or high-density gas (Klinkenberg, 1941). This correction may range from 1% or 2% for high-permeability rocks to as much as 70% for low-permeability rocks.

Permeability is seldom the same in all directions within a rock. Vertical permeability is generally far lower than permeability horizontal to the bedding. Permeability is thus commonly measured from plugs cut in both directions.

When a single fluid phase completely saturates the pore space, permeability is referred to as absolute or specific and is given the dimension L2. The effective permeability refers to saturations of less than 100%. The terms K_w, K_g, and K_o are used to designate the effective permeability with respect to water, gas, and oil, respectively. Effective permeability ranges between 0 and K at 100% saturation, but the sum of permeabilities to two or three phases is always less than one. Relative permeability is the ratio of the effective permeability for a particular fluid at a given saturation to a base permeability. The relative permeability ranges from 0.0 to 1.0. Thus for oil,

$$K_{ro} = \frac{K_o}{K};$$

for gas,

$$K_{rg} = \frac{K_g}{K};$$

and for water,

$$K_{rw} = \frac{K_w}{K}.$$

where

K = absolute permeability (or permeability at irreducible wetting phase saturation);
K_r = relative permeability;
K_g = effective permeability at 100% gas saturation;
K_w = effective permeability at 100% water saturation;
K_o = effective permeability at 100% oil saturation.

Figure 6.13 shows the typical relative permeability relationship for different saturations. It is necessary to consider the relative saturations of how the oil and gas are distributed, and the

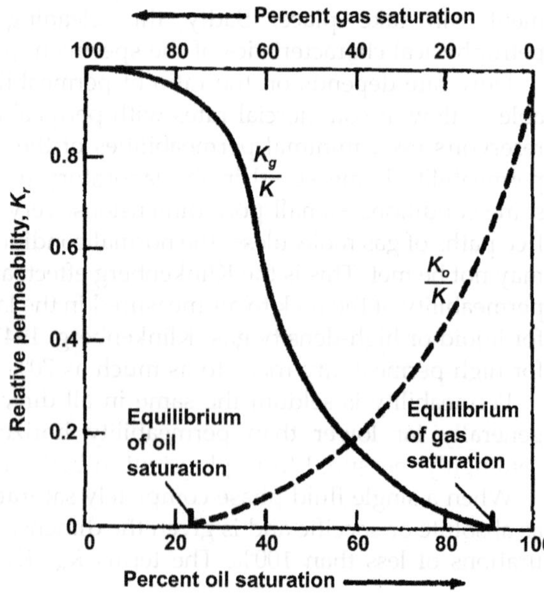

FIGURE 6.13 Typical curves for the relative permeabilities of oil and gas for different saturations.

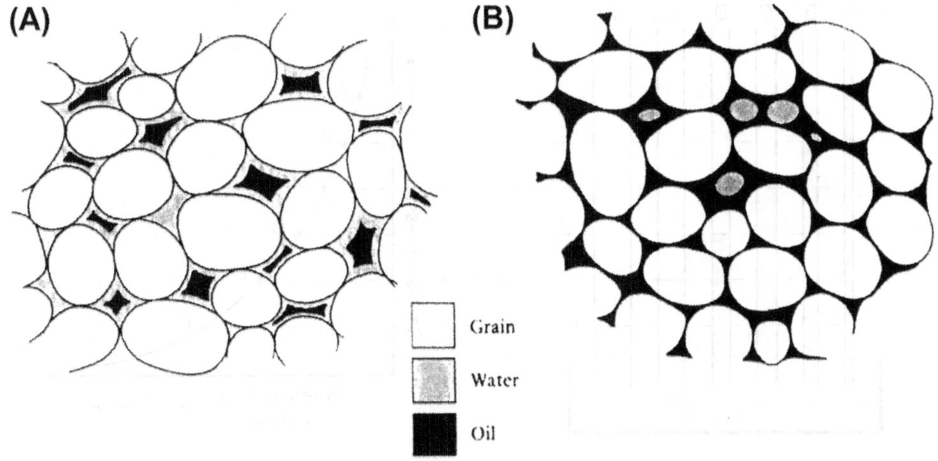

FIGURE 6.14 The concept of wettability in reservoirs. (A) A water-wet reservoir (common). (B) An oil-wet reservoir (rare).

presence or absence of water. Most reservoirs are water-wet, having a film of water separating the pore boundaries from the oil. Few reservoirs are oil-wet, totally lacking a film of connate water at pore boundaries (Fig. 6.14). A mixed, or intermediate wettability condition may also occur.

6.3 CAPILLARY PRESSURE

The consideration of the wettability of pores leads us to the concept of capillarity, the phenomenon whereby liquid is drawn up a capillary tube (Fig. 6.15). The capillary pressure is the difference between the ambient pressure and the pressure exerted by the column of liquid. Capillary pressure increases with decreasing tube diameter (Fig. 6.16). Translated

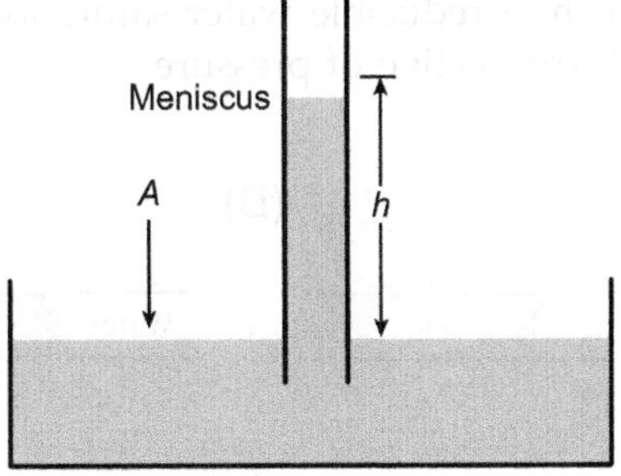

FIGURE 6.15 Capillary tube in a liquid-filled tank. The pressure on the water level (A) equals the pressure due to the hydrostatic head of water (h) minus the capillary pressure across the meniscus.

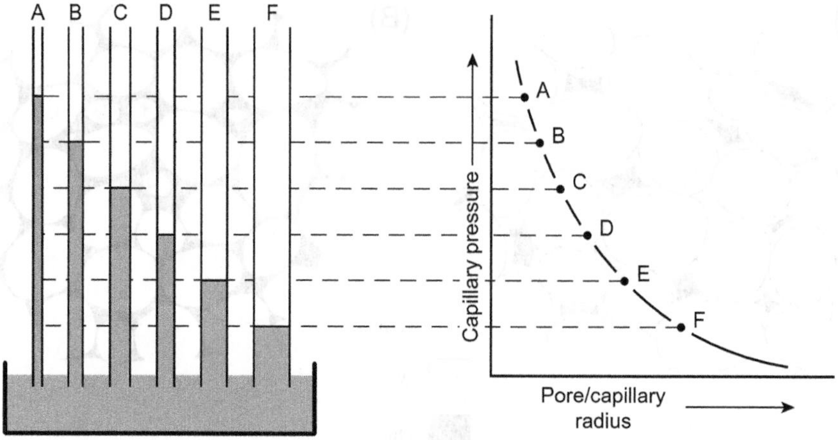

FIGURE 6.16 Capillary tubes of various (exaggerated) diameters showing that the heights of the liquid columns are proportional to the diameters of the tubes.

into geological terms, the capillary pressure of a reservoir increases with decreasing pore size or, more specifically, pore throat diameter. Capillary pressure is also related to the surface tension generated by the two adjacent fluids—it increases with increasing surface tension. In water-wet pore systems the meniscus is convex with respect to water; in oil-wet systems it is concave (Fig. 6.17). For a mathematical analysis of capillary pressure, see Muskat (1949), Dickey (1979), and Chierici (1994). For an account of how capillary pressure analysis may be applied to petroleum reservoirs, see Vavra et al. (1992).

Reservoirs are commonly subjected to capillary pressure tests in which samples with 100% of one fluid are injected with another (gas, oil, water, and mercury may be used). The pressure at which the injected fluid begins to invade the reservoir is the displacement pressure. As pressure increases, the proportions of the two fluids gradually reverse until the irreducible saturation point is reached, at which no further invasion by the second fluid is possible at any pressure. Data from these analyses are plotted as capillary pressure curves (Fig. 6.18). Curve 1 is typical of a good quality reservoir—porous and permeable. Once the entry, or displacement, pressure has been exceeded, fluid invasion increases rapidly for a minor pressure increase until the irreducible water saturation is reached. At this point no further water can be expelled irrespective of pressure.

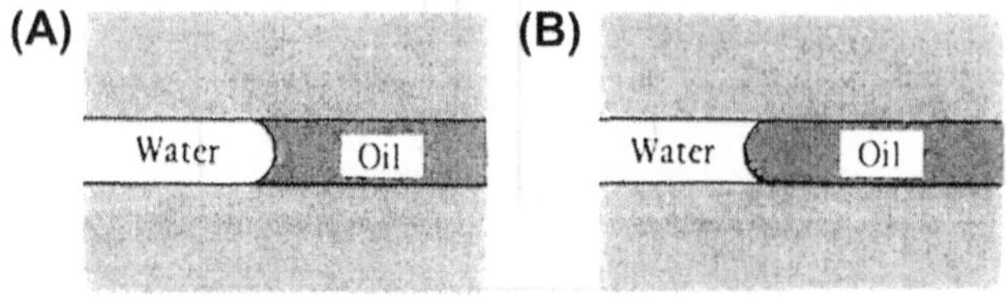

FIGURE 6.17 Cross-sections of capillary tube/pores showing meniscus effect for (A) oil-wet and (B) water-wet reservoirs. Water-wet reservoirs are the general rule.

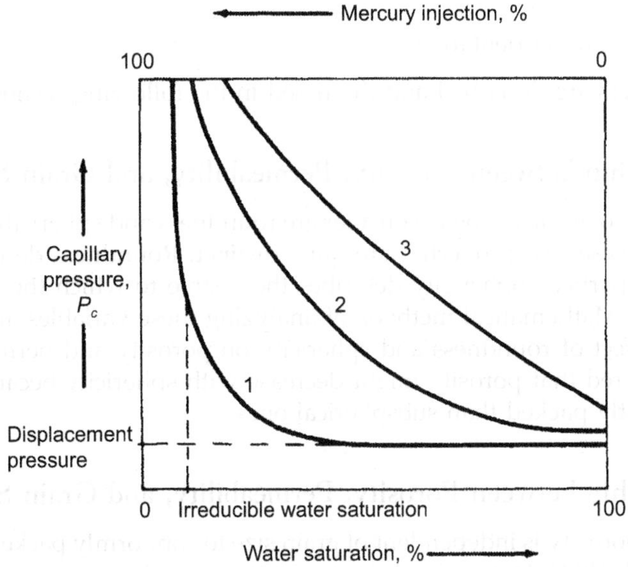

FIGURE 6.18 Capillary pressure curves for various reservoirs: (1) a clean, well-sorted sand with uniform pore diameters, (2) an intermediate quality reservoir, and (3) a poor quality reservoir with a wide range of pore diameters. For full explanation see text.

Curve 2 is for a poorer quality reservoir with a higher displacement pressure and higher irreducible water content. Curve 3 is for a very poor quality reservoir, such as a poorly sorted sand with abundant matrix and hence a wide range of pore sizes. Displacement pressure and irreducible water saturation are therefore both high, and water saturation declines almost uniformly with increasing pressure.

6.4 RELATIONSHIP BETWEEN POROSITY, PERMEABILITY, AND TEXTURE

The texture of a sediment is closely correlated with its porosity and permeability. The texture of a reservoir rock is related to the original depositional fabric of the sediment, which is modified by subsequent diagenesis. This diagenesis may be negligible in many sandstones, but in carbonates it may be sufficient to obliterate all traces of original depositional features. Before considering the effects of diagenesis on porosity and permeability, the effects of the original depositional fabric on these two parameters must be discussed. The following account is based largely on studies by Krumbein and Monk (1942), Gaithor (1953), Rogers and Head (1961), Potter and Mast (1963), Chilingar (1964), Beard and Weyl (1973), Pryor (1973), and Atkins and McBride (1992). The textural parameters of an unconsolidated sediment that may affect porosity and permeability are as follows:

Grain shape (roundness, sphericity)
Grain size

Sorting
Fabric (packing, grain orientation)

These parameters are described and discussed in the following sections.

6.4.1 Relationship between Porosity, Permeability, and Grain Shape

The two aspects of grain shape to consider are roundness and sphericity (Powers, 1953). As Fig. 6.19 shows, these two properties are quite distinct. Roundness describes the degree of angularity of the particle. Sphericity describes the degree to which the particle approaches a spherical shape. Mathematical methods of analyzing these variables are available.

Data on the effect of roundness and sphericity on porosity and permeability are sparse. Fraser (1935) inferred that porosity might decrease with sphericity because spherical grains may be more tightly packed than subspherical ones.

6.4.2 Relationship between Porosity, Permeability, and Grain Size

Theoretically, porosity is independent of grain size for uniformly packed and graded sands (Rogers and Head, 1961). In practice, however, coarser sands sometimes have higher porosities than do finer sands or vice versa (e.g., Lee, 1919; Sneider et al., 1977). This disparity may be due to separate but correlative factors such as sorting and/or cementation.

Permeability declines with decreasing grain size because pore diameter decreases and hence capillary pressure increases (Krumbein and Monk, 1942). Thus a sand and a shale may both have porosities of 10%; whereas the former may be a permeable reservoir, the latter may be an impermeable cap rock.

6.4.3 Relationship between Porosity, Permeability, and Grain Sorting

Porosity increases with improved sorting. As sorting decreases, the pores between the larger, framework-forming grains are infilled by the smaller particles. Permeability decreases with sorting for the same reason (Fraser, 1935; Rogers and Head, 1961; Beard and Weyl, 1973). As mentioned earlier, sorting sometimes varies with the grain size of a particular

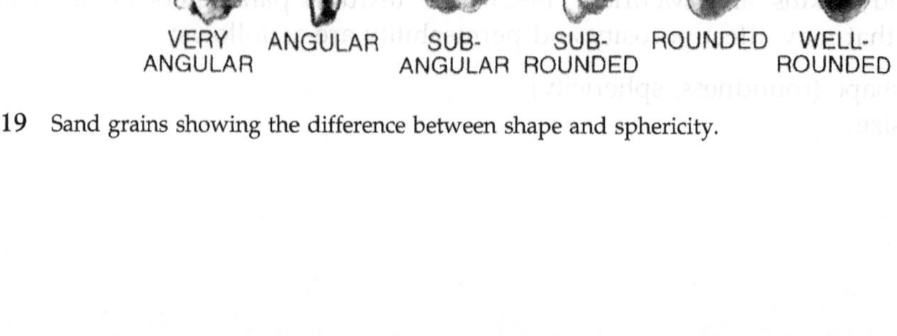

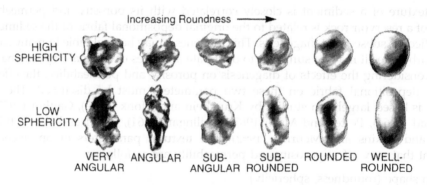

FIGURE 6.19 Sand grains showing the difference between shape and sphericity.

reservoir sand, thus indicating a possible correlation between porosity and grain size. Fig. 6.20 summarizes the effects of sorting and grain size on porosity and permeability in unconsolidated sand.

6.4.4 Relationship between Porosity, Permeability, and Grain Packing

The two important characteristics of the fabric of a sediment are how the grains are packed and how they are oriented. The classic studies of sediment packing were described by Fraser

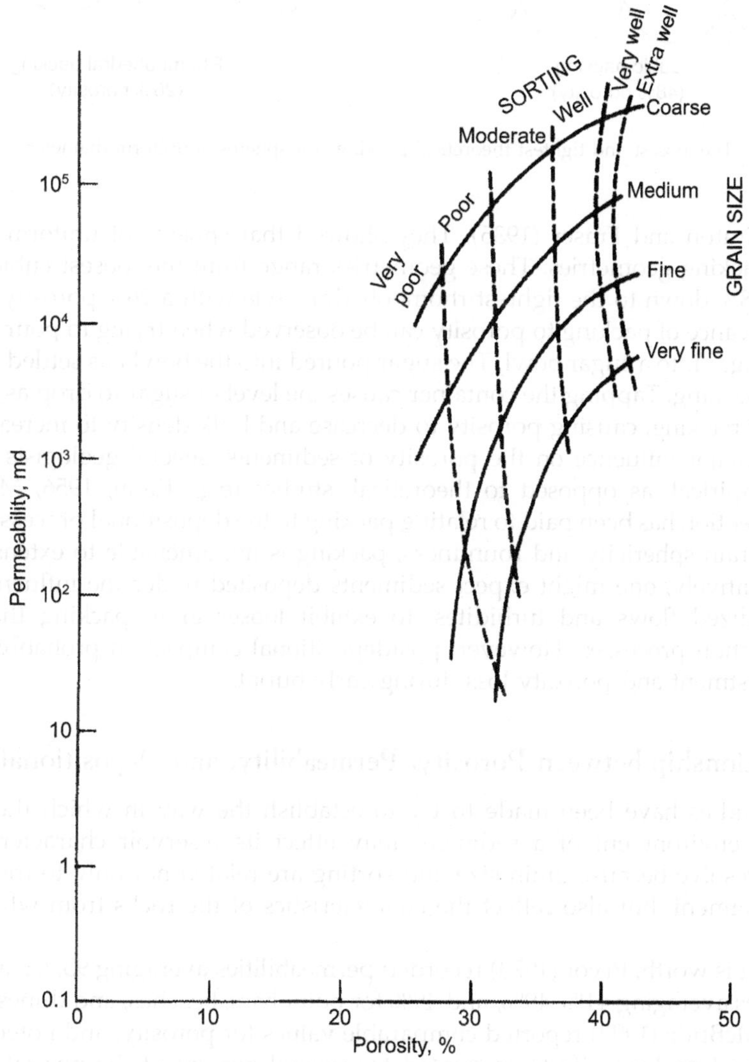

FIGURE 6.20 Graph of porosity against permeability showing their relationship with grain size and sorting for uncemented sands. *After Beard and Weyl (1973), Nagtegaal (1978).*

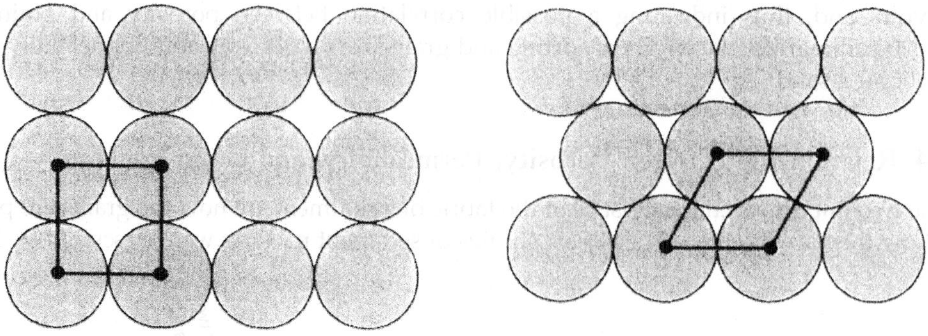

FIGURE 6.21 The loosest and tightest theoretical packings for spheres of uniform diameter.

(1935) and Graton and Fraser (1935). They showed that spheres of uniform size have six theoretical packing geometries. These geometries range from the loosest cubic style with a porosity of 48% down to the tightest rhombohedral style with a 26% porosity (Fig. 6.21).

The significance of packing to porosity can be observed when trying to pour the residue of a packet of sugar into a sugar bowl. The sugar poured into the bowl has settled under gravity into a loose packing. Tapping the container causes the level of sugar to drop as the grains fall into a tighter packing, causing porosity to decrease and bulk density to increase. Packing is obviously a major influence on the porosity of sediments; several geologists have tried to carry out empirical, as opposed to theoretical, studies (e.g., Kahn, 1956; Morrow, 1971). Particular attention has been paid to relating packing to the depositional process (e.g., Martini, 1972). Like grain sphericity and roundness, packing is not amenable to extensive statistical analysis. Intuitively, one might expect sediments deposited under the influence of gravity, such as fluidized flows and turbidites, to exhibit looser grain packing than those laid down by traction processes. However, postdepositional compaction probably causes rapid packing adjustment and porosity loss during early burial.

6.4.5 Relationship between Porosity, Permeability, and Depositional Process

Several studies have been made to try to establish the way in which the depositional process and environment of a sediment may effect its reservoir characteristics. This is not easy to resolve because grain size and sorting are related not only to the final depositional environment, but also reflect the characteristics of the rocks from which they were derived.

For what it is worth, Pryor (1973) recorded permeabilities averaging 93, 68, and 54 darcies, and porosities averaging 41%, 49%, and 49% for point bars, beaches, and dunes, respectively. Atkins and McBride (1992) reported comparable values for porosity, and noted that trapped air bubbles and packing effects accounted for several percent of the porosity. There is no doubt that porosity and permeability are rapidly reduced due to packing adjustments and compaction early on during burial.

6.4.6 Relationship between Porosity, Permeability, and Grain Orientation

The preceding analysis of packing was based on the assumption that grains are spherical, which is generally untrue of all sediments except oolites. Most quartz grains are actually prolate spheroids, slightly elongated with respect to their C crystallographic axis (Allen, 1970). Sands also contain flaky grains of mica, clay, shell fragments, and other constituents. Skeletal carbonates have still more eccentric grain shapes. Thus the second element of fabric, namely, orientation, is perhaps more significant to porosity and permeability than packing is. The orientation of grains may have little effect on porosity, but a major effect on permeability.

Most sediments are stratified, the layering being caused by flaky grains, such as mica, shells, and plant fragments, as well as by clay laminae. Because of this stratification the vertical permeability is generally considerably lower than the horizontal permeability. The ratio of vertical to horizontal permeability in a reservoir is important because of its effect on coning as the oil and gas are produced. Variation in permeability also occurs parallel to bedding. In most sands the grains generally show a preferential alignment within the horizontal plane. Grain orientation can be measured by various methods (Sippel, 1971). Studies of horizontally bedded sands have shown that grains are elongated parallel to current direction (e.g., Shelton and Mack, 1970; von Rad, 1971; Martini, 1971, 1972). For cross-bedded sands the situation is more complex because grains may be aligned parallel to the strike of foresets due to gravitational rolling. Figure 6.22 shows that permeability will be greatest parallel to grain orientation, since this orientation is the fabric alignment with least resistance to fluid movement (Scheiddegger, 1960).

Studies of the relationship between fabric and permeability variation have produced different results on both small and large scales. Potter and Pettijohn (1977) have reviewed the conflicting results of a number of case histories, some of which show a correlation with grain orientation and some of which do not. It is necessary to consider not only small-scale permeability variations caused by grain alignment but also the larger variations caused by sedimentary structures.

Grain-size differences cause permeability variations far greater than those caused by grain orientation. Thus in the cross-bedded eolian sands of the Leman field (North Sea), horizontal permeabilities measured parallel to strike varied from as much as 0.5 to 38.5 md between

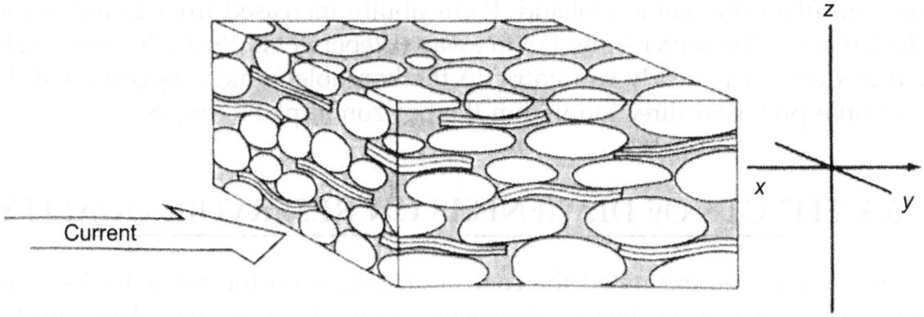

FIGURE 6.22 Block diagram of sand showing layered fabric with grains oriented parallel to current. Generally, $K_x > K_y > K_z$.

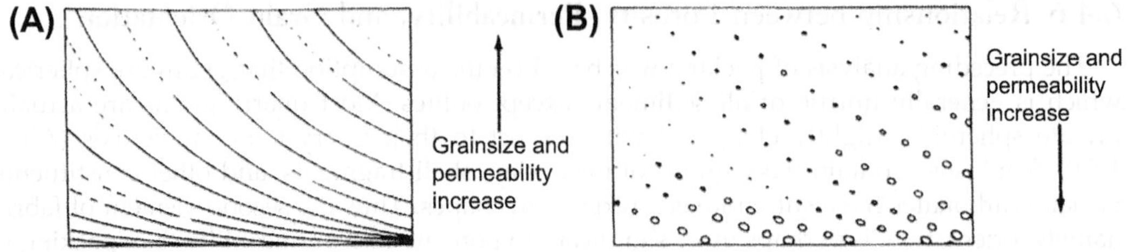

FIGURE 6.23 Permeability variations for (A) downward-fining and (B) downward-coarsening avalanche cross-beds.

adjacent foresets. This range in permeabilities is attributable to variations in grain size and sorting. Similarly, because eolian cross-beds generally show a decreasing grain size from foreset to toeset, permeability diminishes downward through each cross-bedded unit (Van Veen, 1975). Conversely, in many aqueously deposited cross-beds, avalanching causes grain size to increase downward. Thus permeability also increases down each foreset for the reason previously given (Fig. 6.23).

Detailed accounts of permeability variations within sedimentary structures, mainly cross-bedding, have been given by Weber (1982) and Hurst and Rosvoll (1991). The latter study reported 16,000 mini-permeameter readings. These showed that that there was greater variation of permeability found within sedimentary structures than between them.

On the still larger scale of whole sand bodies, grain-size-related permeabilities often have considerable variations (Richardson et al., 1987). When discussing the use of the SP log as a vertical profile of grain size it was pointed out that channels tend to have upward-fining grain-size profiles, and thus upward-decreasing permeability. By contrast barrier bar and delta mouth bar sands have upward-coarsening grain-size profiles, and thus upward-increasing permeability. These vertical changes in permeability are also often accompanied by commensurate variations in porosity. Thus there is a strong sedimentological control over the vertical and lateral variation of porosity and permeability within petroleum reservoirs on a hierarchy of scales, from variations within sedimentary structures, variations within sand bodies, and variations within formations, due to sand body trend (Fig. 6.24).

An example of the scale and significance of these variations is provided by a study of a Holocene sand-filled channel in Holland. Permeability increased from 25 md at the channel margin to 270 md in the center some 175 m away (Weber et al., 1972). Pumping tests showed that the drawdown was nearly concentric to the borehole, which suggests that the permeability had little preferred direction within the horizontal plane (Fig. 6.25).

6.5 EFFECTS OF DIAGENESIS ON RESERVOIR QUALITY

The preceding section examined the control of texture on the petrophysics of unconsolidated sediment. Once burial begins, however, many changes take place, most of which diminish the porosity and permeability of a potential reservoir. These changes, collectively referred to as *diagenesis*, are numerous and complex. The following account specifically

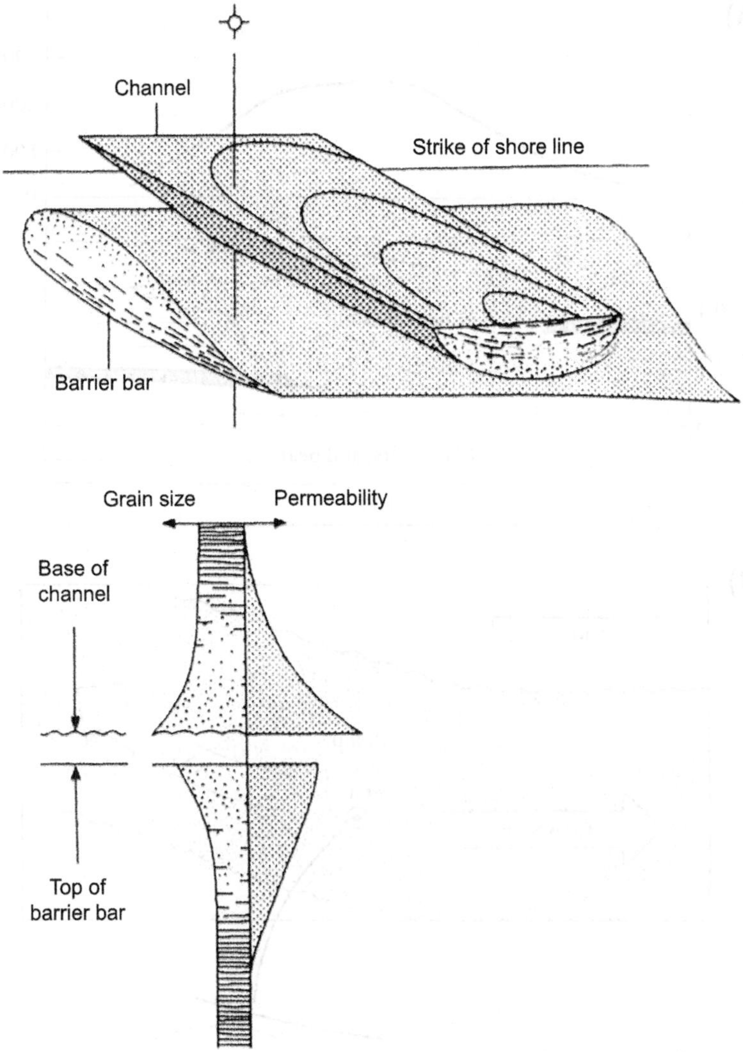

FIGURE 6.24 Diagram to show the permeability variations and trends in sand bodies. Channels often have upward-fining grain-size profiles, so permeability may decline upward within a channel. Barrier sands often have upward-coarsening grain-size profiles, so they often show an upward-increasing permeability. On a regional scale channels tend to trend down the paleoslope, whereas barrier bar sands will parallel it. This hierarchy of permeability variations effects the flow of fluids within petroleum reservoirs. *From Selley (1988).*

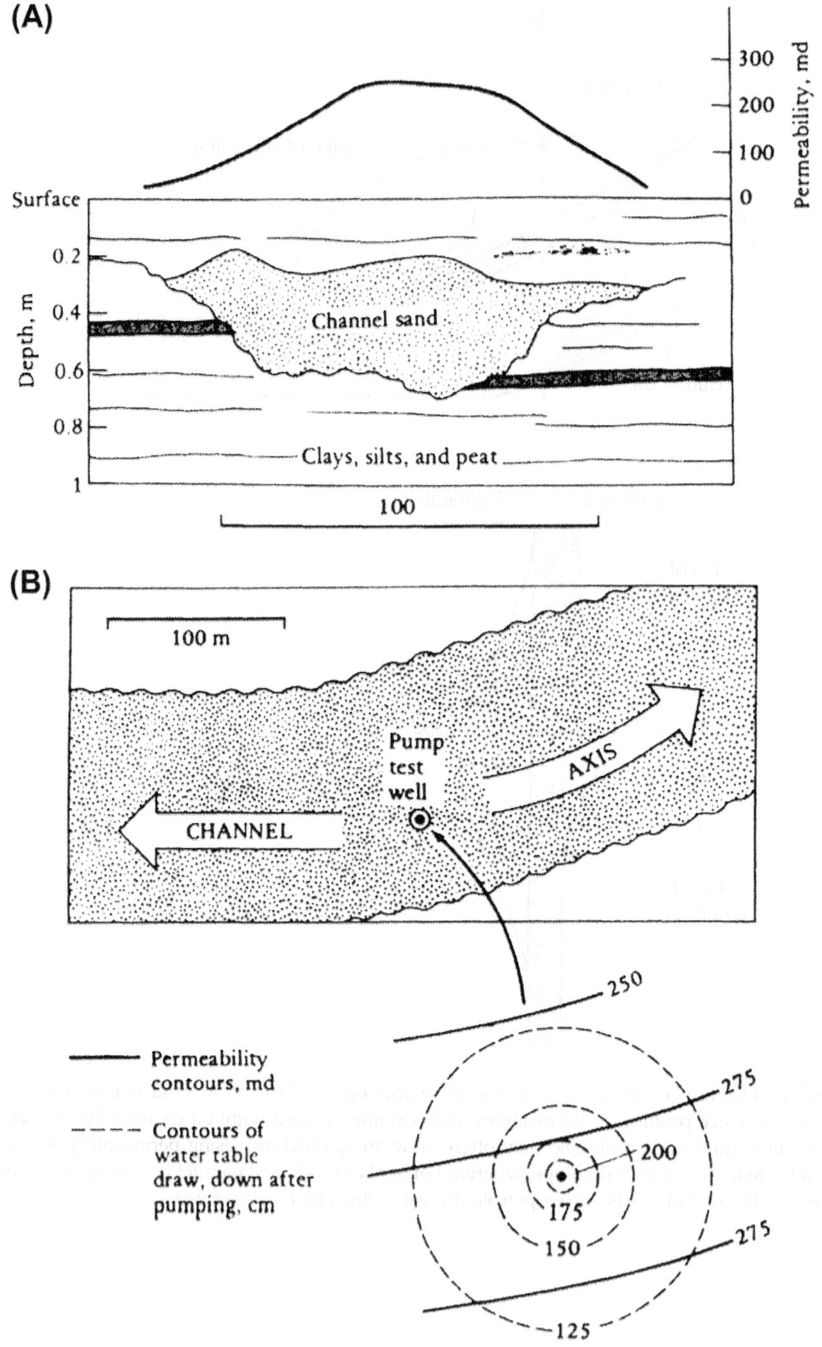

FIGURE 6.25 Permeability variations in a Holocene Dutch channel sand: (A) a cross-section showing how permeability increases toward the channel axis and (B) maps showing how contours of the water level drop are concentric around the pump test well. *After Weber et al. (1972).*

focuses on the effect of diagenesis on reservoir quality. More detailed accounts are found in Blatt et al. (1980), Selley (1988), and Friedman et al. (1992). The diagenesis of sandstone and carbonate reservoirs is considered in turn.

6.5.1 Effects of Diagenesis on Sandstone Reservoirs

The effects of diagenesis on sandstone reservoirs include the destruction of porosity by compaction and cementation, and the enhancement of porosity by solution. In this section the factors that control regional variations of reservoir quality are discussed first. Then the details of cementation and solution are described, and the section concludes with an account of diagenesis associated with petroleum accumulations.

6.5.1.1 Regional Variations on Sandstone Reservoir Quality

Studies of recent sands, such as those of Pryor (1973), show that they have porosities of some 40–50% and permeabilities of tens of darcies. Most sandstone reservoirs, however, have porosities in the range of 10–20% and permeabilities measurable in millidarcies. Although fluctuations do occur, the porosity and permeability of reservoirs decrease with depth. This relationship is of no consequence to the production geologist or reservoir engineer concerned with developing a field, but it is important for the explorationist who has to decide the greatest depth at which commercially viable reservoirs may occur.

The porosity of a sandstone at a given depth can be determined if the porosity gradient and primary porosity are known (Selley, 1978):

$$\phi^{D} = \phi^{P} - GD,$$

where

ϕ^{D} = porosity at a given depth;
ϕ^{P} = primary porosity at the surface;
G = porosity gradient (%ϕ/km);
D = burial depth.

Reviews of porosity gradients and the factors controlling them have been given by Selley (1978) and Magara (1980). The main variables that affect porosity gradients are the mineralogy and texture of the sediments and the geothermal and pressure regimes to which they are subjected. The more mineralogically mature a sand is, the better its ability to retain its porosity (Dodge and Loucks, 1979). Taking the two extreme cases, chemically unstable volcaniclastic sands tend to lose porosity fastest, and the more stable pure quartz sands tend to have the lowest gradients (Fig. 6.26).

Texture also affects the gradient: Poorly sorted sands with abundant clay matrix compact more and lose porosity faster than do clean, well-sorted sands (Rittenhouse, 1971). A similar effect is noted in micaceous sand, as, for example, in the Jurassic of the North Sea (Hay, 1977). Geothermal gradients also affect the porosity gradient of sand, as shown in Fig. 6.27. Because the rate of a chemical reaction increases with temperature, the higher the geothermal gradient, the faster the rate of porosity loss. Pressure gradient also affects porosity. Abnormal pressure preserves porosity, presumably by decreasing the effect of compaction (Atwater and Miller, 1965).

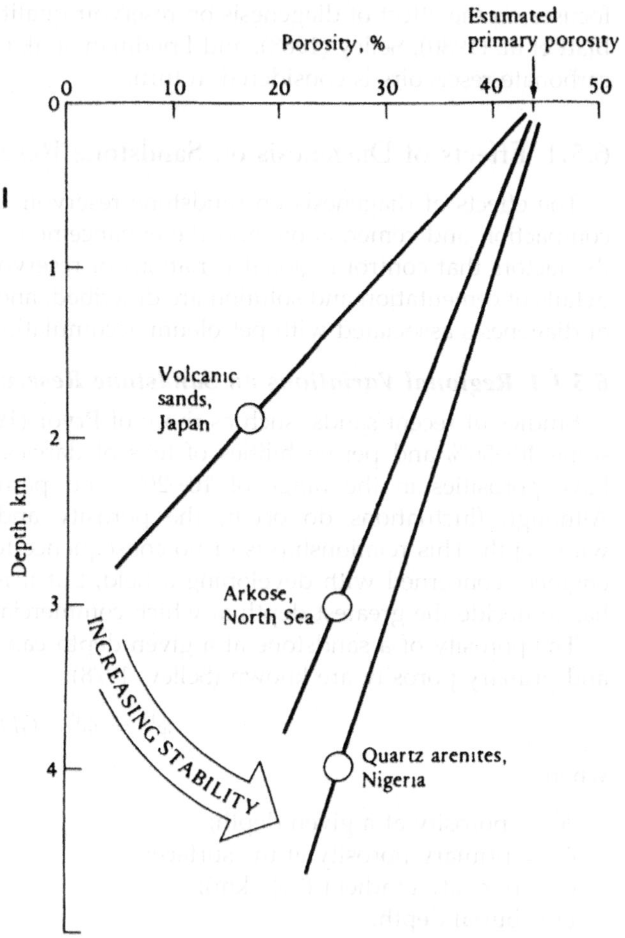

FIGURE 6.26 Graph showing how the porosity gradient decreases with increasing mineralogical maturity. *After Nagtegaal (1978).*

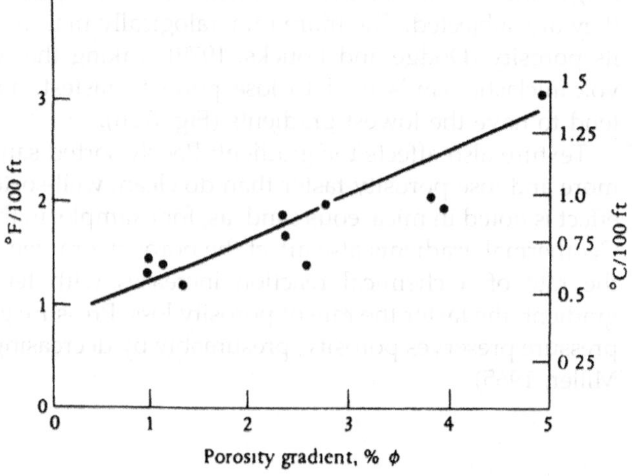

FIGURE 6.27 Graph of geothermal gradient versus porosity for various sandstones showing how they tend to increase together. *From Selley (1978).*

Oil and gas have been commonly observed to preserve porosity in sands (e.g., Fuchtbauer, 1967; O'Brien and Lerche, 1988). Once petroleum enters a trap, the circulation of connate water is diminished, and further cementation inhibited. Some studies show, however, that cementation in a petroleum-saturated reservoir is not completely inhibited, but continues at a reduced rate (Saigal et al., 1993). Gregory (1977) demonstrated the inhibiting effect of petroleum on porosity destruction for the Tertiary sediments of the Gulf Coast, USA. This study shows that water-wet sands have lower porosities than petroleum-saturated sands for the same depth, and that gas preserves porosity more effectively than oil (Fig. 6.28).

The linear rate of porosity loss with depth shown by Fig. 6.28 has been observed in many basins. This has led to the view that porosity is lost at a slow and steady rate, concomitant with burial, and implies that there is little circulation of connate fluid (e.g., Bloch, 1991; Bloch and Helmold, 1995; Cazier et al., 1995; Walderhaug, 1996).

This steady-state model contrasts markedly with the "throbbing basins" model, which is based on evidence that cementation occurs in response to the episodic expulsion of hot water from overpressured shales. The evidence for these "hot Hushes" is based on studies of fluid inclusions and isotope data (e.g., Lynch, 1996). For further accounts on this controversy, see Horbury and Robinson (1993).

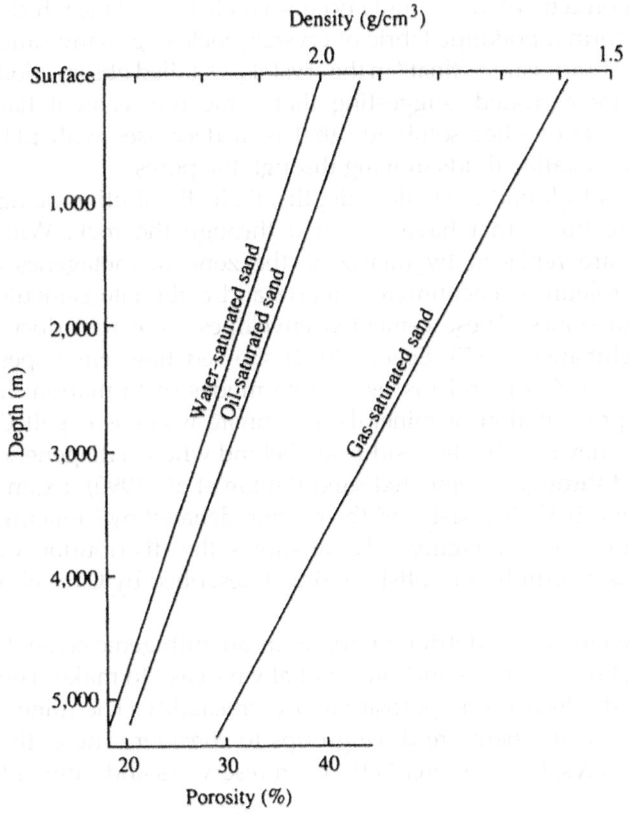

FIGURE 6.28 The graph of porosity and density plotted against depth for Tertiary sandstones in the U.S. Gulf Coast, from data in Gregory (1977). Note how porosity diminishes linearly with increasing depth. These data show how petroleum preserves reservoir porosity, gas being more effective than oil.

6.5.1.2 Porosity Loss by Cementation

On a regional scale the porosity of sandstones is largely controlled by the factors previously discussed (mineralogy, texture, geothermal and pressure gradient). Within a reservoir itself porosity and permeability vary erratically with primary textural changes and secondary diagenetic ones. The diagenesis of sandstones is a major topic beyond the scope of this text. It is treated in depth by McDonald and Surdam (1984), Friedman et al. (1992), and Morse (1994). The following account deals concisely with those aspects particularly relevant to petroleum geology.

Diagenetic changes in a sandstone reservoir include cementation and solution, which are now discussed in turn. A small amount of cementation is beneficial to a sandstone reservoir because it prevents sand from being produced with the oil. The presence of sand in the oil not only damages the reservoir itself but also the production system. Extensive cementation is deleterious, however, because it diminishes porosity and permeability. Many minerals may grow in the pores of a sandstone, but only three are of major significance: quartz, calcite, and the authigenic clays.

Quartz is a common cement. It generally grows as optically continuous overgrowths on detrital quartz grains (Fig. 6.29(A)). The solubility of silica increases with pH, so silica cements occur where acid fluids have moved through the pores.

Calcium carbonate is another common cement. It generally occurs as calcite crystals, which, as they grow from pore to pore, may form a poikilitic fabric of crystals enclosing many sand grains (Fig. 6.29(B)). The grains frequently appear to "float" in the crystals. Detailed observation often shows that grain boundaries are corroded, suggesting that some replacement has occurred. Calcite solubility is the reverse of silica solubility; that is, it decreases with pH. Thus calcite cementation is the result of alkaline fluids moving through the pores.

Quartz and carbonate cements are both found at shallow depths, their distribution being dependent on the history of the pore fluids that have migrated through the rock. With increasing depth carbonate cements are replaced by quartz as the zone of metagenesis approaches. At the depths where petroleum is encountered quartz and carbonate cements commonly occur in sands adjacent to shales. These cemented envelopes were first documented by Fothergill (1955) and Fuchtbauer (1967). Selley (1992) showed how envelopes could be identified on wireline logs (Fig. 6.30), and discussed two modes of formation. It is possible that they result from the precipitation of minerals as connate water is expelled from the adjacent shales. Alternatively they may be the residue left behind where a diagenetic front of acid connate water has moved through a cemented sand (Sultan et al., 1990). Examples have also been described from the U.S. Gulf Coast, and their origin debated by Moncure et al. (1984) and Sullivan and McBride (1991). Figure 6.31 illustrates the distribution of cemented envelopes in the Campos basin turbidites of offshore Brazil described by Carvalho et al. (1995).

Clay may be present in a sandstone either as a detrital matrix or as an authigenic cement. As clays recrystallize and alter during burial, this distinction is not always easy to make. The presence of clay in a reservoir obviously destroys its porosity and permeability. The mineralogy of clays is very complex, but basically there are three groups to consider. These, the kaolinitic, illitic, and montmorillonitic clays, have different effects on reservoirs and different sources of formation.

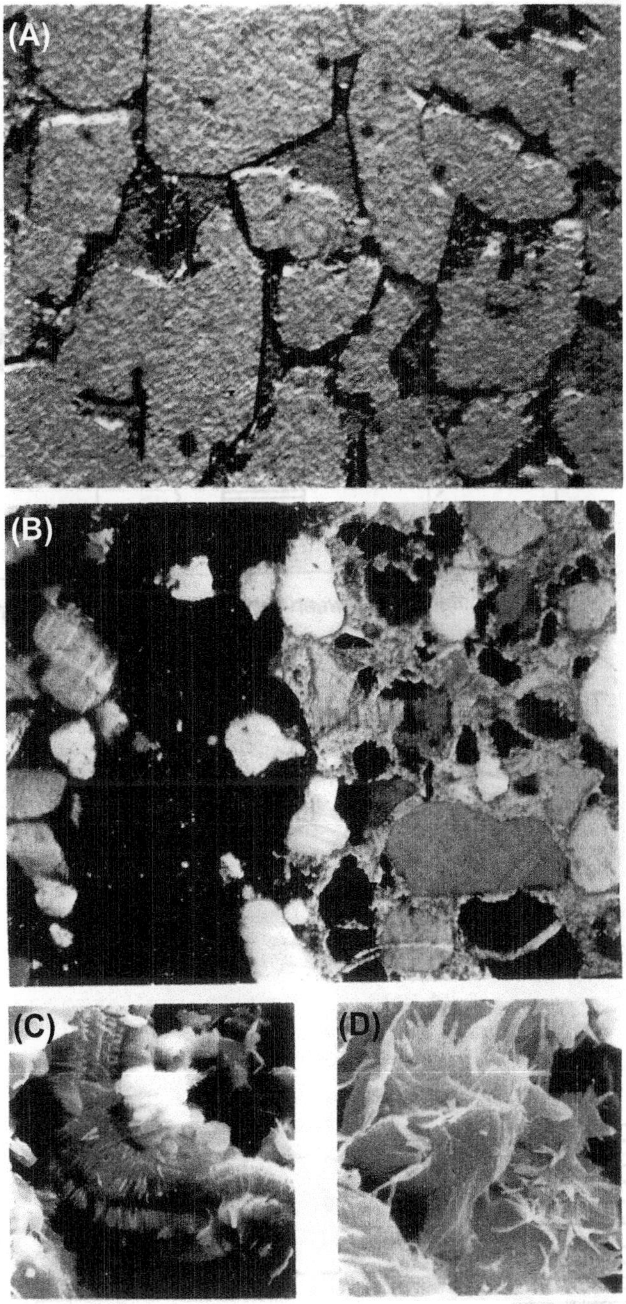

FIGURE 6.29 Photomicrographs of various types of sandstone cement: (A) silica cement in optical continuity on detrital quartz grains, (B) calcite cement with poikilitic fabric of large crystals enclosing corroded quartz grains. (C) authigenic kaolin crystals within pores, and (D) authigenic illite showing fibrous habit. (A) and (B) are thin sections photographed in ordinary and polarizing light, respectively. (C) and (D) are scanning electron micrographs.

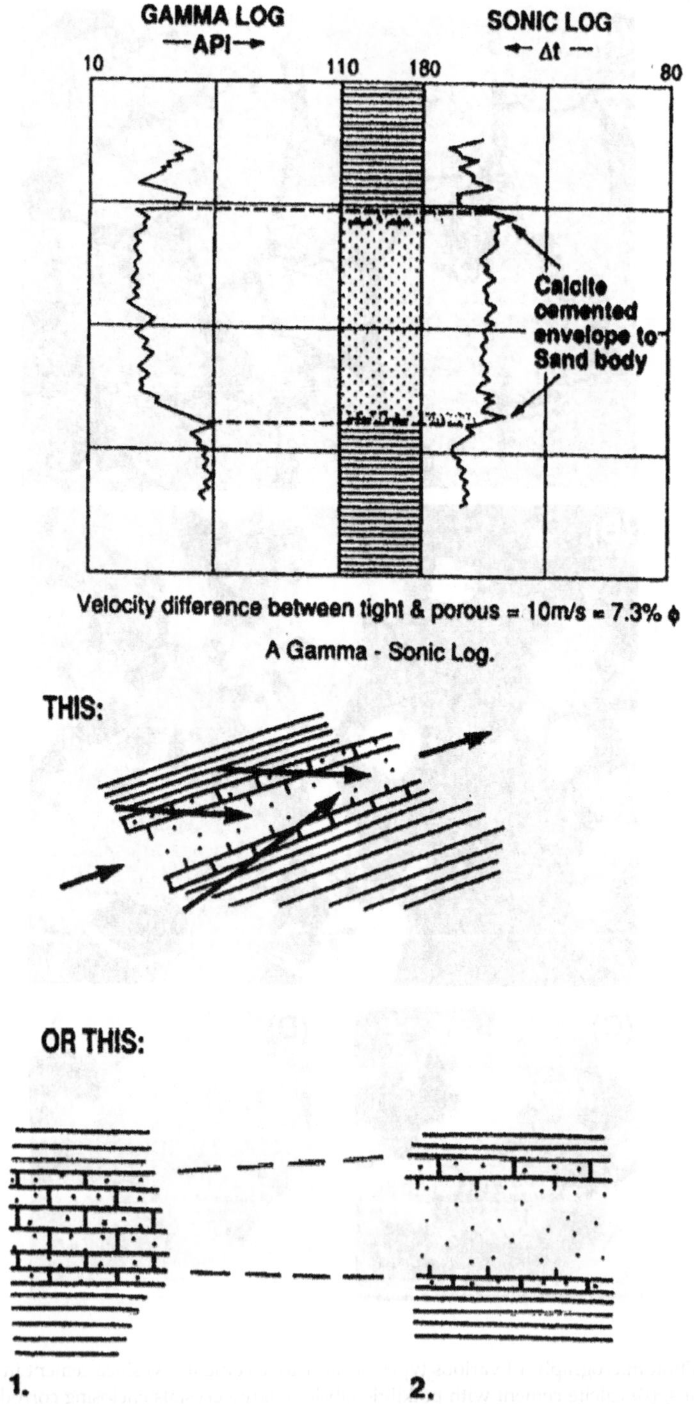

FIGURE 6.30 Gamma sonic log from a well in the UK North Sea to illustrate a Paleocene sand with cemented envelope, and illustrations that show the envelope may result either from the precipitation of minerals as connate waters flow into the sand from compacting shale or from a diagenetic front of acid fluid leaching a passage through an earlier cemented sand. *From Selley (1992).*

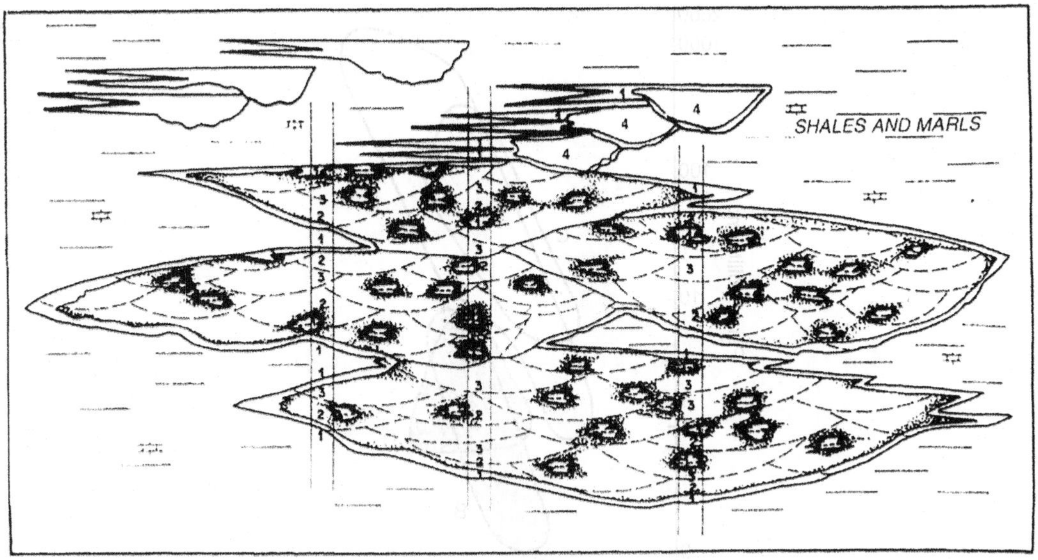

FIGURE 6.31 Diagram to show the distribution of cemented envelopes in Campos basin (Cretaceous) turbidite reservoirs, offshore Brazil. *Reprinted from Carvalho et al. (1995); pp. 226–244, with kind permission from Elsevier Science Ltd.*

Kaolinite generally occurs as well-formed, blocky crystals within pore spaces (Fig. 6.29(C)). This crystal habit diminishes the porosity of the reservoir, but may have only a minor effect on permeability. Kaolinite forms and is stable in the presence of acid solutions. Therefore it occurs as a detrital clay in continental deposits, and as an authigenic cement in sands that have been flushed by acidic waters, such as those of meteoric origin.

Illitic clay is quite different from kaolin. Authigenic illite grows as fibrous crystals, which typically occur as furlike jackets on the detrital grains (Fig. 6.29(D)). These structures often bridge over the throat passages between pores in a tangled mass. Thus illitic cement may have a very harmful effect on the permeability of a reservoir. This effect is demonstrated in Fig. 6.32, which is taken from studies of the Lower Permian Rotliegende sandstone of the southern North Sea by Stalder (1973) and Seemann (1979). Illitic clays typically form in alkaline environments. They are thus the dominant detrital clay of most marine sediments and occur as an authigenic clay in sands through which alkaline connate water has moved.

The montmorillonitic, or smectitic, clays are formed from the alteration of volcanic glass and are found in continental or deep marine deposits. They have the ability to swell in the presence of water. Reservoirs with montmorillonite are thus very susceptible to formation damage if drilled with a conventional water-based mud and must therefore be drilled with an oil-based mud. When production begins, water displaces the oil, causing the montmorillonitic clays to expand and destroy the permeability of the lower part of the reservoir.

Kaolinite, illite, and montmorillonite may all be found in shallow reservoirs, depending on the source material and the diagenetic history. With increasing burial the kaolinites and montmorillonites alter to illite, the collapse of montmorillonites being a possible cause of overpressure, and related to the expulsion of petroleum as discussed earlier. As catagenesis merges into metagenesis, illite crystallinity increases and they evolve into the hydromicas and micas.

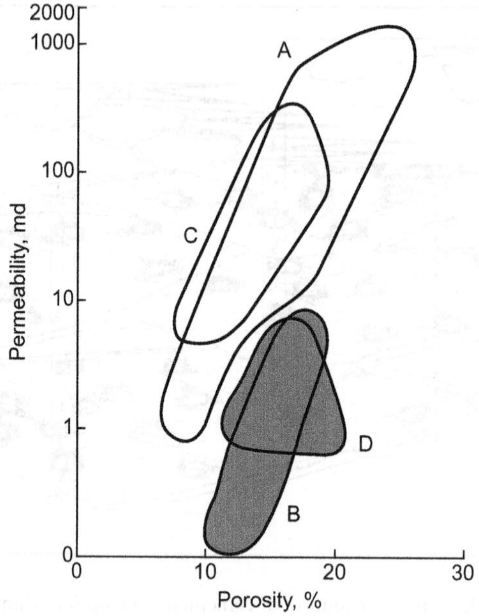

FIGURE 6.32 Porosity–permeability data for illite (B and D) and kaolin (A and C) cemented Rotliegende reservoir sands of the North Sea. *After Stalder (1973), Seemann (1979).*

6.5.1.3 Porosity Enhancement by Solution

Cementation reduces the porosity and permeability of a sand. In some cases, however, solution of cement or grains can reverse this trend. Schmidt et al. (1977) have outlined the petrographic criteria for the recognition of secondary solution porosity in sands. It generally involves the leaching of carbonate cements and grains, including calcite, dolomite, siderite, shell debris, and unstable detrital minerals, especially feldspar. Leached porosity in sands is generally associated with kaolin, which both replaces feldspar and occurs as an authigenic cement.

The fact that carbonate is leached out of the sand and the predominance of kaolin indicate that the leaching was caused by acidic solutions. There are two sources of acidic leaching: epidiagenesis or, more specifically, weathering due to surface waters (Fairbridge, 1967) and decarboxylation of kerogen (Schmidt et al., 1977).

Meteoric water rich in carbonic and humic acids weathers sandstones, and other rocks, at the earth's surface. In many cases kaolinization and leaching generate solution porosity that is enhanced by fracturing. Usually, the weathering-induced porosity is destroyed during reburial by normal diagenesis. The secondary porosity may only be preserved within hydrocarbon reservoirs; the preservative effect of hydrocarbons on porosity has already been noted. Thus many geologists from Hea (1971) to Al-Gailani (1981) and Shanmugam (1990) attributed subunconformity secondary porosity to the effects of epidiagenesis. These views are supported by isotope and fluid inclusion analyses that show that cements formed concomitant with leaching in low temperatures and salinities. These results are consistent with reports of stratified freshwaters preserved beneath unconformities, suggesting that a palaeoaquifer has been preserved (e.g., Macauley et al., 1992).

Examples of subunconformity sand reservoirs with secondary porosity include the Sarir sand of Libya (which formed the basis for Hea's study), the Cambro-Ordovician sand of Hassi Messaoud, Algeria (Balducci and Pommier, 1970), the sub-Cimmerian unconformity Jurassic sands of the North Sea (Bjorlykke, 1984; Selley, 1984, 1990), and Prudhoe Bay, Alaska (Shanmugam and Higgins, 1988).

An alternative source for the acid fluid is the decarboxylation of coal or kerogen, causing the expulsion of solutions of carbonic acid (Schmidt et al., 1977; Loucks et al., 1979). According to this theory, acidic solutions are expelled from a maturing source rock ahead of petroleum migration. These acid fluids generate secondary solution porosity in the reservoir beds. This mechanism explains solution porosity, which has been observed in sandstones that have not undergone uplift and subaerial exposure (Jamison et al., 1980).

6.5.1.4 Diagenesis Associated with Petroleum Accumulations

Sandstone petroleum reservoirs may have undergone a complex diagenetic history, as outlined earlier, before the emplacement of the petroleum. Porosity and permeability are often higher in hydrocarbon reservoirs than in the underlying water zones. As discussed in the preceding section, this discrepancy may be due to epidiagenesis, where the reservoir lies beneath an unconformity. In such instances the changes in porosity and permeability may be above or beneath the oil/water contact. As already noted, however, the presence of hydrocarbons inhibits cementation by preventing connate water from continuing to move through the trap. Flow may continue in the underlying water zone, however, so porosity and permeability are reduced by cementation after oil or gas has migrated into the trap.

Postmigration silica cement in water zones beneath traps has been described from the Silurian Clinton sand of Ohio (Heald, 1940) and the Jurassic reservoirs of the Gifhorn trough, Germany (Philip et al., 1963). Calcite and anhydrite postdate migration in the Permian Lyons sand of the Denver basin (Levandowski et al., 1973), and from the Fulmar sands of the North Sea (Saigal et al., 1993). Many other instances of postmigration cementation are known.

In extreme cases postmigration cementation may be so pervasive as to act as a seat seal, giving rise to "frozen-in" paleotraps, as in the Raudhatain field of Kuwait (Al-Rawi, 1981). Because of postmigration cementation it is important to cut cores below the oil/water contact of a field. If a water-drive production mechanism is anticipated, then the permeability of the aquifer must be known. Data must not be extrapolated from the reservoir downward.

The situation may be still further complicated by partial emigration of petroleum from a trap. This may occur though leakage at the crest, in which case the paleopetroleum/water contact parallels the modern one. In some cases, however, the tectonic tilting of a trap after petroleum invasion generates a tilted paleopetroleum/water contact. This occurs in the Morecombe gas field in the Irish Sea. Here an upper illite-free and a lower platy-illite cemented zone are separated by a sloping surface that cross-cuts the present gas/water contact, and is interpreted as a paleogas/water contact that has since tilted. Quartz and carbonate cement are more prevalent beneath than above the present-day gas/water contact (Ebbern, 1981; Woodward and Curtis, 1987; Stuart, 1993). The Morecambe field thus has four different types of reservoirs in terms of their diagenetic history, porosity, and permeability. Try explaining that to a petroleum reservoir engineer.

The previous pages show that the porosity and permeability of a reservoir is the result of its original texture, on which may have been imprinted diverse diagenetic processes, some of which may enhance, but most of which will diminish, its reservoir quality. Fig. 6.33 is thus a very naive attempt to summarize the potential diagenetic pathways of sandstones and the resultant reservoir quality.

6.5.2 Effects of Diagenesis on Carbonate Reservoirs

The carbonate rocks include the limestones, composed largely of calcite ($CaCO_3$), and the dolomites, composed largely of the mineral of the same name ($CaMg(CO_3)_2$). As reservoirs,

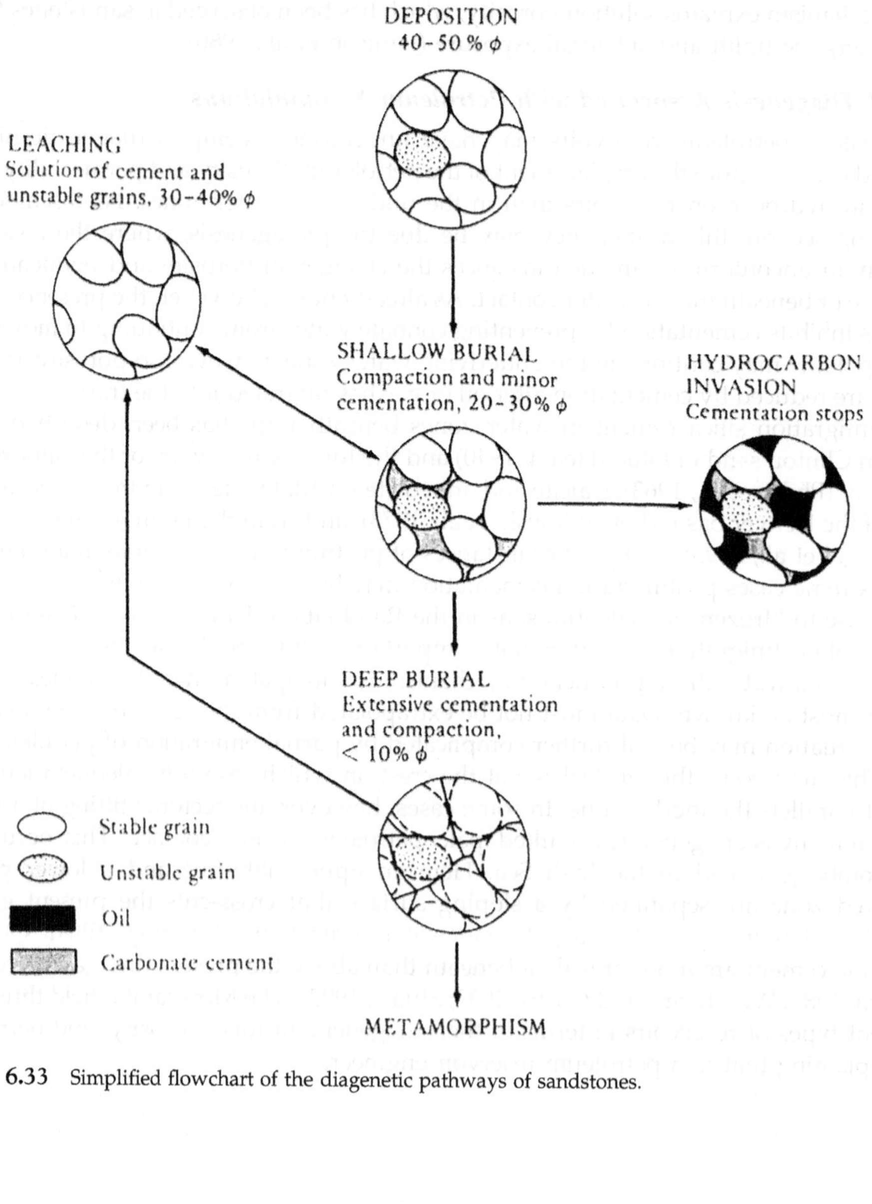

FIGURE 6.33 Simplified flowchart of the diagenetic pathways of sandstones.

carbonates are as important as sandstones, but their development and production present geologists and engineers with a different set of problems. Silica is chemically more stable than calcite. Thus the effects of diagenesis are more marked in limestones than in sandstones. With sandstone reservoirs the main problem is to establish the original sedimentary variations, working out the depositional environment and paleogeography, and using these to predict reservoir variation as the field is drilled. The effects of diagenesis are generally subordinate to primary porosity variations, which is seldom true of carbonate reservoirs.

When first deposited, carbonate sediments are highly porous and permeable and are inherently unstable in the subsurface environment. Carbonate minerals are thus dissolved and reprecipitated to form limestones whose porosity and permeability distribution are largely secondary in origin and often unrelated to the primary porosity. Thus with carbonate reservoirs, facies analysis may only aid development and production in those rare cases where the diagenetic overprint is minimal.

Having discussed the general effects of diagenesis on carbonate reservoirs, it is now appropriate to consider them in more detail. Major texts covering this topic have been published by Reeckmann and Friedman (1982), Roehl and Choquette (1985), Tucker and Wright (1990), and Jordan and Wilson (1994) and Lucia (1999). Shorter accounts that specifically relate carbonate diagenesis to porosity evolution have been given by Purser (1978) and Longman (1980). The following description is based on these references, concentrating on those aspects of diagenesis that significantly affect reservoir quality. Carbonate reefs, sands, and muds are considered separately.

6.5.2.1 Diagenesis and Petrophysics of Reefs

Reefs (used in the broadest sense of the term) are unique among the sediments because they actually form as rock rather than from the lithification of unconsolidated sediment. Modern reefs are reported with porosities of 60–80%. Because reefs are formed already lithified, they do not undergo compaction as do most sediments. Porosity may thus be preserved. However, porosity is only likely to be preserved when the reef is maintained in a static fluid environment, such as when it is rapidly buried by impermeable muds. Where a steady flow of alkaline connate water passes through the reef, porosity may gradually be lost by the formation of a mosaic sparite cement. Where acid meteoric water flows through the reef, the reef may be leached and its porosity enhanced.

Before a reef is finally buried, sea level may have fluctuated several times. Thus several diagenetic phases of both cementation and solution may have been superimposed on the primary fabric and porosity of the reef. It is not surprising, therefore, that there may be no correlation between the reservoir units that an engineer delineates in a reef trap and the facies defined by geologists. This phenomenon is illustrated by the Intisar reefs of Libya, described in Chapter 7 (Brady et al., 1980).

6.5.2.2 Diagenesis and Petrophysics of Lime Sands

Like reefs, lime sands may have high primary porosities (Enos and Sawatsky, 1979). Because they are unconsolidated, however, they begin to lose porosity quickly on burial. Mud compacts and shells fracture as the overburden pressure increases. Like reefs, lime sands may be flushed by various fluids. Acid meteoric waters may leach lime sands and increase porosity, whereas alkaline fluids may cause them to become cemented. Early cementation takes various distinctive forms, which may be recognized microscopically. These forms include fibrous, micritic, and microcrystalline cements formed in the spray zone (beach rock) and on the sea floor (hard ground). These early cements diminish porosity somewhat and may diminish permeability considerably where they bridge pore throats. Early cementation, however, may enhance reservoir potential because it lithifies the sediment, and, as with reefs, the effect of compaction is then diminished. Later diagenesis depends on the fluid environment. Acidic meteoric water may enhance porosity by leaching; alkaline connate water may cause a mosaic of sparite crystals to infill the pores. Where there is no fluid flow, porosity may he preserved and hydrocarbons may invade the reservoir.

Carbonate sands that have not undergone early shallow cementation have less chance of becoming reservoirs. As the sands are buried, they lose porosity rapidly by compaction; although again they are susceptible to secondary leaching when invaded by acid meteoric water during later epidiagenesis. Figure 6.34 summarizes the diagenetic pathways for carbonate sands.

6.5.2.3 Diagenesis and Petrophysics of Lime Muds

Recent carbonates are largely composed of the mineral aragonite, the orthorhombic variety of calcium carbonate. Aragonite is unstable in the subsurface, where it reverts to the hexagonal isomorph calcite. This polymorphic reaction causes an 8% increase in bulk volume, which results in a loss of porosity. Coupled with compaction, this porosity loss means that most ancient lime mudstones are hard, tight, and splintery rocks. They have no reservoir potential unless fractured or dolomitized. A significant exception to this general rule occurs when the lime mud is calcitic rather than aragonitic. The coccolithic oozes, which occur in Cretaceous and younger rocks, are of original calcitic composition. Lacking the aragonite–calcite reaction, these lime muds remain as friable chalks with porosities of 30–40%. Ordinarily, they have very low permeabilities because of their small pore diameters. Chalks are not usually considered to be reservoirs and may even act as cap rocks. In certain circumstances, however, they can act as reservoirs. For example, fractures may enhance their permeability as in the Austin Chalk (Upper Cretaceous) of Texas. In the similar Cretaceous Chalk of the North Sea, reservoirs occur in the Ekofisk and associated fields of offshore Norway. Here overpressure has inhibited porosity loss by compaction, whereas fracturing over salt domes has increased permeability (Scholle, 1977). These chalk reservoirs are thus dual porosity systems in which petroleum occurs both within microporosity between the coccoliths, and within the cross-cutting fractures.

Because of their uniformity, porosity gradients of lime muds can be constructed analogously to those drawn for sandstones (Scholle, 1981). These porosity gradients can seldom be prepared for the more heterogeneous calcarenites, although there are exceptions (see Schmoker and Halley, 1982).

6.5.2.4 Dolomite Reservoirs

The second main group of carbonate reservoirs includes the dolomites. Dolomite can form from calcite (dolomitization) and vice versa (dedolomitization or calcitization) as shown in the following reaction:

$$2\,CaCO_3 + Mg^{2+} = CaCO_3MgCO_3 + Ca^{2+}.$$

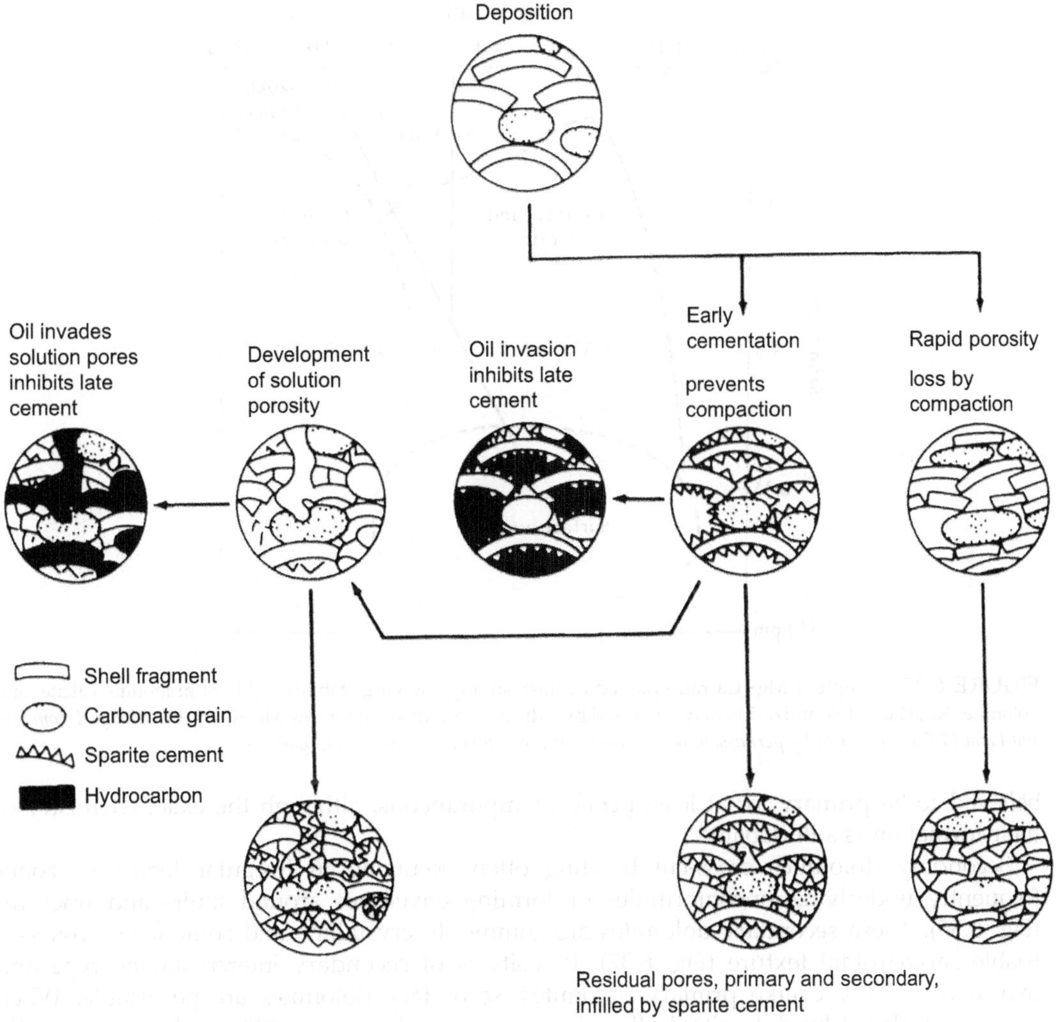

FIGURE 6.34 Simplified flowchart of the diagenetic pathways of lime sand.

The direction in which this reaction moves is a function not only of the Mg–Ca ratio but also of salinity (Fig. 6.35).

A distinction is made between primary and secondary dolomites. Primary dolomites are generally bedded and often form laterally continuous units. Characteristically, they occur in sabkha (salt marsh) carbonate sequences, passing down into fecal pellet lagoonal muds and up into supratidal evaporites. Such dolomites are generally cryptocrystalline, often with a chalky texture. They may thus be porous, but, because of their narrow pore diameter, lack permeability. Examples of primary dolomites in sabkha cycles have been described from the Jurassic Arab-Darb formation of Abu Dhabi (Wood and Wolfe, 1969). Such dolomites are

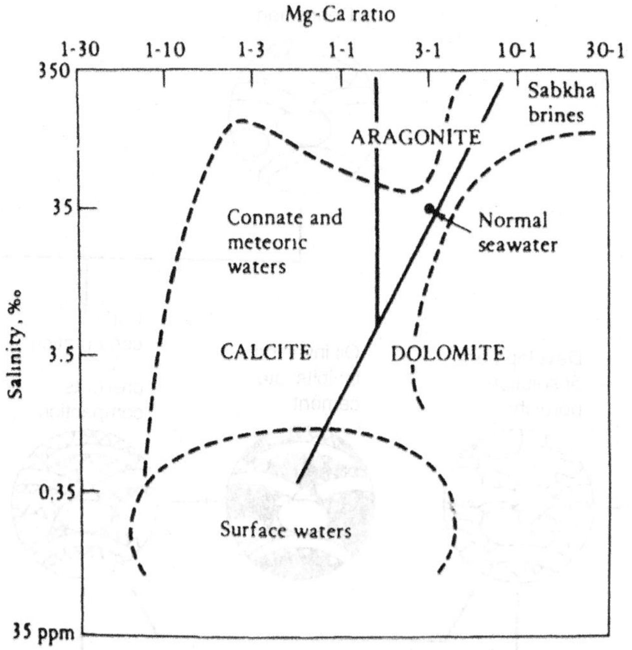

FIGURE 6.35 Graph of Mg–Ca ratio plotted against salinity showing stability fields of aragonite, calcite, and dolomite. Note how dolomitization may be caused by a drop in salinity even for low Mg–Ca ratios. *Modified from Folk and Land (1974); reprinted by permission of the American Association of Petroleum Geologists.*

believed to be primary, or at least penecontemporaneous, although the exact chemistry of their formation is still debated.

Secondary dolomites cross-cut bedding often occurring in irregular lenses or zones frequently underlying unconformities or forming envelopes around faults and fractures (Fig. 6.36). These secondary dolomites are commonly crystalline and sometimes possess a friable saccharoidal texture (Fig. 6.37). Porosity is of secondary intercrystalline type and may exceed 30%. Unlike primary dolomites, secondary dolomites are permeable. When calcite is replaced by dolomite, bulk volume is reduced by some 13%, and hence porosity increases correspondingly. Secondary dolomites beneath unconformities sometimes undergo fracturing. Sometimes this is so extensive that they collapse into breccias (Fig. 6.38).

Thus secondary dolomites are often important reservoirs, as in the Devonian reefs and carbonate platforms of Alberta (Toomey et al., 1970; Luo and Machel, 1995). Regionally extensive subunconformity dolomites occur, for example, on the Wisconsin, San Marcos, and Cincinnati arches of the United States (see Badiozamani, 1973; Rose, 1972; Tedesco, 1994; respectively). These regionally extensive arches host numerous petroleum accumulations where limestones have undergone dolomitization, solution, and fracturing for tens of meters deep beneath truncating unconformities. Petroleum reservoirs in secondary dolomite are very complex, and it may be difficult to assess their reserves and to characterize the reservoir. In the Lisburne field at Prudhoe Bay, Alaska, for example, a limestone bears the dolomitization overprint of two superimposed unconformities (Jameson, 1994).

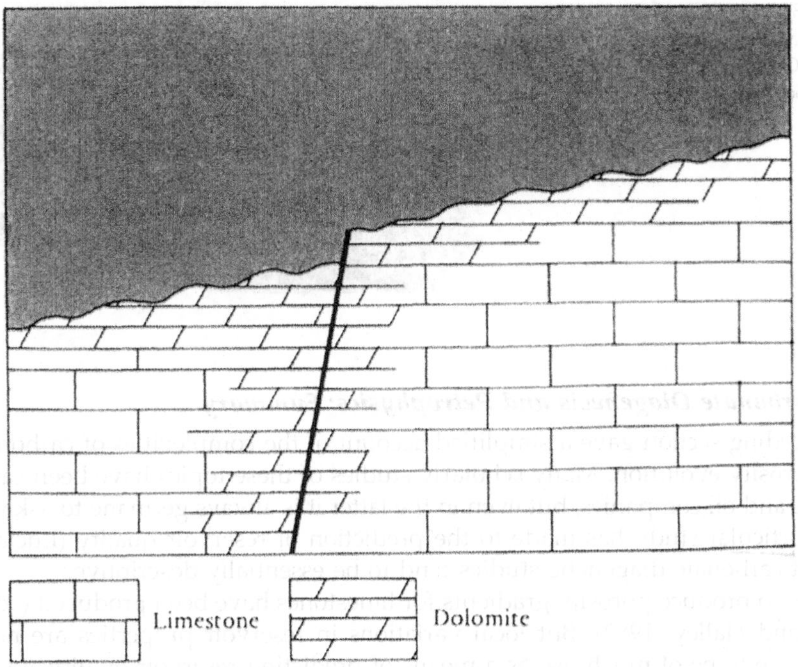

FIGURE 6.36 Cross-section showing how secondary dolomites are often related to faults and unconformities.

FIGURE 6.37 Thin section of a secondary dolomite reservoir, showing coarsely crystalline fabric. Zechstein (Upper Permian). North Sea, United Kingdom.

FIGURE 6.38 Collapsed breccia of Zechstein (Permian) dolomite reservoir from beneath the Cimmerian unconformity. Auk Field, North Sea, United Kingdom. *Courtesy of Shell UK.*

6.5.2.5 Carbonate Diagenesis and Petrophysics: Summary

The preceding section gave a simplified account of the complexities of carbonate diagenesis and porosity evolution. Many scholarly studies of these topics have been carried out in universities and oil companies, but even in the latter it is always germane to ask what contribution a particular study has made to the prediction of reservoir quality (Lucia and Fogg, 1990). Most carbonate diagenetic studies tend to be essentially descriptive.

Attempts to produce porosity gradients for limestones have been produced (Scholle, 1981; Schmoker and Halley, 1982). But local variations in reservoir properties are normally too rapid for these to be of much use as a means of predicting reservoir quality. As with sandstones, secondary solution porosity may be attributed to unconformity-related epidiagenesis, or to the decarboxylation of source rocks (Mazzullo and Harris, 1992). Epidiagenetic processes may enhance porosity and permeability beneath unconformities. Stress-release fractures increase permeability, enabling meteoric water to invade the rock. This results in dolomitization, in the development of solution porosity, and, in extreme cases, in the formation of paleo-karst, which may collapse to form a heterogeneous carbonate rubble. These types of carbonate reservoir are characteristic of truncation stratigraphic traps. Specific field examples are cited in Chapter 7. Sequence stratigraphy is now widely applied to carbonate sequences as a means of identifying unconformities, and thus predicting horizons of solution porosity and dolomitization (Loucks and Sarg, 1993; Saller et al., 1994).

Jardine et al. (1977) and Wilson (1975, 1980) have reviewed the relationships between carbonate reservoir quality, facies, and diagenesis. Wilson (1980) recognizes the following six major carbonate reservoir types, ranging from those whose porosity is largely facies controlled to those with extensive diagenetic overprints.

1. Shoals with primary porosity still preserved; for example, the Arab D of Saudi Arabia and the Smackover of Louisiana.
2. Buildups with primary porosity preserved; for example, some Alberta Devonian reefs and the Golden Lane atoll of Mexico.
3. Forereef talus with primary porosity preserved; for example, the Poza Rica and Reforma fields of Mexico.
4. Pinchout traps where grainstones with preserved porosity or intercrystalline dolomitic porosity are sealed up-dip by sabkha evaporites; for example, the San Andres of West Texas.

5. Subunconformity reservoirs, for which porosity is largely secondary due to solution, fracturing, and dolomitization; for example, the Natih and Fahud fields of Oman.
6. Calcitic chalks with permeability enhanced by fractures; for example, the Ekofisk and associated fields of the North Sea.

Finally, Fig. 6.9 shows the relationship between porosity, permeability, and the various types of pore found in carbonate and other reservoirs.

6.5.3 Atypical and Fractured Reservoirs

Some 90% of the world's oil and gas occur in sandstone or carbonate reservoirs. The remaining 10% occur in what may therefore be termed atypical reservoirs, which range from various types of basement to fractured shale. As discussed earlier, theoretically any rock can be a petroleum reservoir if it is both porous and permeable. Atypical reservoirs may form by two processes: weathering and fracturing.

The role of weathering in the formation of solution porosity in sandstones and carbonate has already been noted. The weathering of certain other rocks has a similar effect. Minerals weather at different rates; so polym締minerallic crystalline rocks can form a porous veneer as the unstable minerals weather out, leaving a granular porous residue of stable mineral grains. This situation usually occurs in granites and gneisses, where the feldspars leach out to leave an unconsolidated quartz sand. This granite wash is a well site geologist's nightmare. The transition from arkose via in situ granite wash to unweathered granite may be long and gentle.

A number of fields produce from weathered basement reservoirs. In the Panhandle–Hugoton field of Texas and Oklahoma the productive granite wash zone is some 70 m thick, although this zone ranges from red shale, via arkose, into fractured granite (Pippin, 1970). The Augila field of the Sirte basin, Libya, also produces from weathered and fractured granite. Some wells in this field flowed at more than 40,000 BOPD (barrels oil per day) (Williams, 1972). The Long Beach and other fields of California produce from fractured Franciscan (Jurassic) schists (Truex, 1972).

Fracturing can turn any brittle rock into a reservoir. Porous but impermeable rocks can be rendered permeable by fracturing as, for example, in chalk and shale. Even totally nonporous rocks may become reservoirs by fracturing. Fractures can be identified from wireline logs, from visual inspection of cores, and from production tests. Parameters of great importance include fracture intensity, which controls porosity and permeability, and fracture orientation, which affects the isotropy of the reservoir (Pirson, 1978; pp. 180–206). Fracture intensity may be expressed by the fracture intensity index (FII):

$$\mathrm{FII} = \frac{(\phi_t - \phi_m)}{(1 - \phi_m)},$$

where ϕ_t = total porosity and ϕ_m = porosity not due to fractures. More recently Narr (1996) has shown how the probability of a borehole intersecting a fracture may be expressed as

$$P_t = \frac{D}{S_{av}},$$

where P_f = probability of a fracture being intercepted, D = diameter of the borehole, and S_{av} = average fracture spacing.

Because fractures are commonly vertical, fractured reservoirs are particularly amenable to production by means of deviated horizontal wells. This approach has been notably successful in the Austin Chalk of Texas where seismic is used to locate faults and adjacent acoustically slow fractured zones, both of which may be penetrated by horizontal wells.

Fractured reservoirs present particular problems of production, the main danger being that oil in the fractures may be rapidly produced and replaced by water, thus preventing recovery from the rest of the pores (Aguilera and Van Poollen, 1978). Examples of reservoirs that produce almost entirely from fractures include the Miocene Monterey cherts of California, which produce oil in the Santa Maria and other fields. Recovery rates are low, of the order of 15,400 barrels per acre (Biederman, 1986).

Shales, which are generally only cap rocks, may themselves act as reservoirs when fractured. Some shale source rocks produce free oil from fractures with no associated water. An example of this phenomenon is provided by the Cretaceous Pierre shale of the Florence field, Colorado (McCoy et al., 1951). Production is erratic, although one well in this field is reported to have produced 1.5 million barrels of oil. Gas production from fractured shale is more common, with many fields producing from Paleozoic shales in Kentucky, Kansas, and elsewhere in the eastern United States. A more detailed account of shale gas is given in Chapter 9. Further details on anomalous and fractured reservoirs can be found in Nelson (1987). Because of advancements in technology (e.g., horizontal drilling and multistage fracture stimulation) and excellent production results, shales are now seen as major reservoirs for oil and gas (a significant change from when the second edition of this book was written in 1998).

6.6 RESERVOIR CONTINUITY

Once an oil or gas field has been discovered, its reserves and the optimum method for recovering them must be established. A detailed knowledge of reservoir continuity is a prerequisite for solving both these problems. Few traps contain reservoirs that are uniform in thickness, porosity, and permeability; most are heterogeneous to varying degrees. Fluid flow in a reservoir is controlled by the amount and type of reservoir heterogeneity (Weber, 1992). Bed continuity, the presence of baffles to flow, and permeability distribution all influence fluid flow. Common reservoir heterogeneity types include faults (sealing and nonsealing), boundaries of genetic units, permeability zonation within genetic units, baffles within genetic units, laminations and cross-bedding, microscopic heterogeneity, and fracturing (open and closed). Thus a reservoir is commonly divided into the gross pay and the net pay intervals. The gross pay is the total vertical interval from the top of the reservoir down to the petroleum:water contact. The net pay is the cumulative vertical thickness from which petroleum may actually be produced. In many fields the net pay may be considerably less than the gross pay. The difference between gross and net pay is due to two factors: the primary porosity with which the reservoir was deposited and the diagenetic processes that have destroyed this original porosity by cementation or enhanced it by solution. There are, however, three main processes that control reservoir continuity: depositional barriers, diagenetic barriers, and tectonic barriers (Haldersen et al., 1987).

6.6.1 Depositional Barriers

Schemes to classify and quantify the lateral continuity of sands have been produced by Potter (1962), Polasek and Hutchinson (1967), Pryor and Fulton (1976), and Harris and Hewitt (1977). Figure 6.39 is based on the scheme proposed by Potter (1962). Two major groups of sand may be recognized according to their lateral continuity: sheets and elongate bodies. Sheets are more or less continuous with length:width ratios of 1:1.

Sheet sands occur in many environments, ranging from turbidite fans and crevasse-splays to coalesced channel sands in braided alluvial plains. Sheets may be discontinuous because of nondeposition or later erosion, with the sands being locally replaced by shales. Discontinuous sheets grade into belt sands, which, although still extensive, contain many local elon gate vacuoles.

Of the elongate sands with length:width ratios greater than 3:1, the best known are ribbons, or shoestrings (Rich, 1923), which are generally deposited in barrier bar environments. Dendroids are bifurcating shoestrings and include both fluvial tributary channel sands and deltaic and submarine fan distributary channels. Pods are isolated sands with length:width ratios of less than 3:1. They include some tidal current sands and some eolian dunes.

Considered vertically, sands may be differentiated into isolated and stacked reservoirs (Harris and Hewitt, 1977). Stacked sands are those that are in continuity with one another either laterally or vertically. The latter are often referred to as multistorey sands (Fig. 6.40). Pryor and Fulton (1976) have produced a more rigorous analysis of sand body continuity

Name		Length-width ratio
Sheet		≏ 1.1
Elongate	Belt	Sheet with holes
	Dendroid	> 3–1 bifurcating
	Ribbon, or shoestring	> 3–1
	Pod	< 3–1

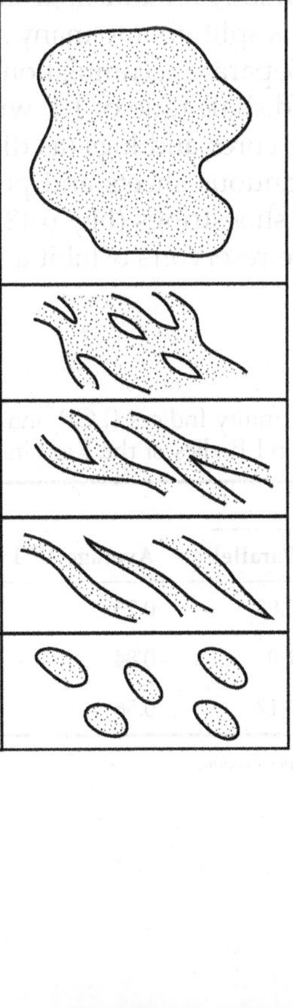

FIGURE 6.39 Nomenclature of sand body geometry. *After Potter (1962).*

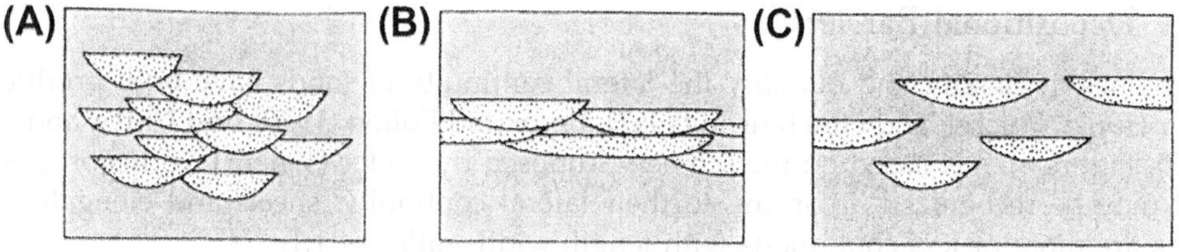

FIGURE 6.40 Descriptive terms for vertical sand body continuity: (A) vertically stacked (multistorey), (B) laterally stacked, and (C) isolated. *After Harris and Hewitt (1977).*

with a lateral continuity index (LCI) and a vertical continuity index (VCI). The LCI is calculated by constructing a series of cross-sections, measuring maximum and minimum sand body continuities, and dividing that length by the length of the section. The VCI is calculated by measuring maximum sand thickness for each well and dividing it into the thickness of the thickest sand. Pryor and Fulton (1976) applied these indices to detailed core data on the Holocene sands of the Rio Grande. Table 6.2 summarizes their data.

Weber (1982), however, approached the problem from the opposite direction, compiling data on the continuity of shale beds for different environments. Many fields have reservoirs with alternating productive sand and nonproductive shale. In these cases detailed correlation is essential both to select the location of wells during development drilling and to predict how fluids will move as the field is produced. These fluids include not only naturally occurring oil and gas but any water or gas that may be injected to maintain pressure and enhance production (Fig. 6.41). The Handil field of Indonesia illustrates just how heterogeneous a reservoir can be (Verdier et al., 1980). It is essentially an anticlinal trap with some faulting. The gross pay is some 1600 m thick; however, it is split up into many separate accumulations with their own gas caps and oil:water contacts. Separate accumulations are partly due to faults, but also occur because the formation has a sand:shale ratio of 1:1, with individual units being 5–20 m thick. Subsurface facies analysis using cores and logs (as discussed in Chapter 3) shows that the productive sands include discontinuous sheets and pods of marine sands interbedded with, and locally cut into by, channel shoestrings (Fig. 6.42). This arrangement is not an isolated case; many other shallow marine reservoirs exhibit a similar complexity, notably those of the Niger delta (Weber, 1971).

TABLE 6.2 Maximum Lateral Continuity Indices (LCI) and Vertical Continuity Indices (VCI) for Holocene Sand Bodies of the Rio Grande Delta

	Average LCI			
Environment	**Perpendicular**	**Parallel**	**Average VCI**	**Average sand thickness (m)**
Fluvial	0.49	0.83	0.75	5–7
Fluviomarine	1.0	1.0	0.84	7–8
Pro-delta	0.3	0.17	0.56	1.5

From Pryor and Fulton (1976). Reprinted with permission.

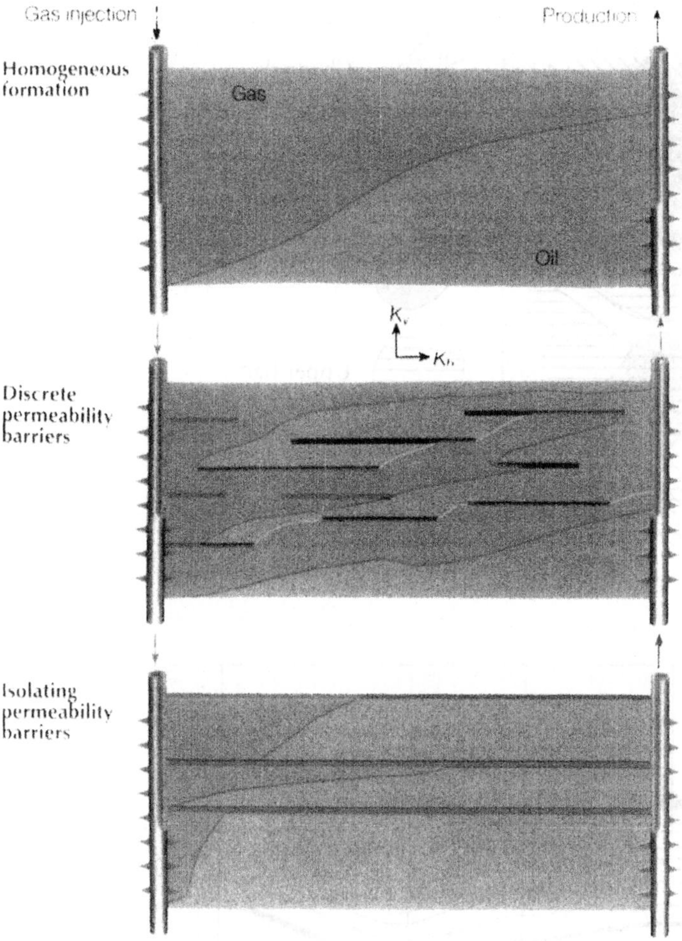

FIGURE 6.41 Illustrations showing the effects of shale continuity on reservoir performance. In homogenous reservoir (upper) the sweep efficiency is low. When the reservoir has discrete zones separated by permeability barriers the efficiency is improved (middle). Isolating permeability barriers leads to the most efficient sweep. *From Ayan et al. (1994); courtesy Schlumberger Oilfield Review.*

Though geologists always like to attempt to use their art to aid reservoir engineers, there comes a time when their services are dispensed with and statistical methods are used. These range from stochastic modeling to fractal analysis (see Martin, 1993; Stolum, 1991; respectively).

6.6.2 Diagenetic Barriers

Diagenetic barriers within reservoirs were extensively discussed earlier in the chapter and need little further elaboration. These are due to the effects of cementation, and may occur in both carbonate and terrigenous reservoirs. They may be related to diagenetic fronts and to petroleum:water contacts.

One particular type of diagenetic barrier that has attracted cause for concern is the occurrence of discontinuous horizons of carbonate cements that sometimes break up into isolated

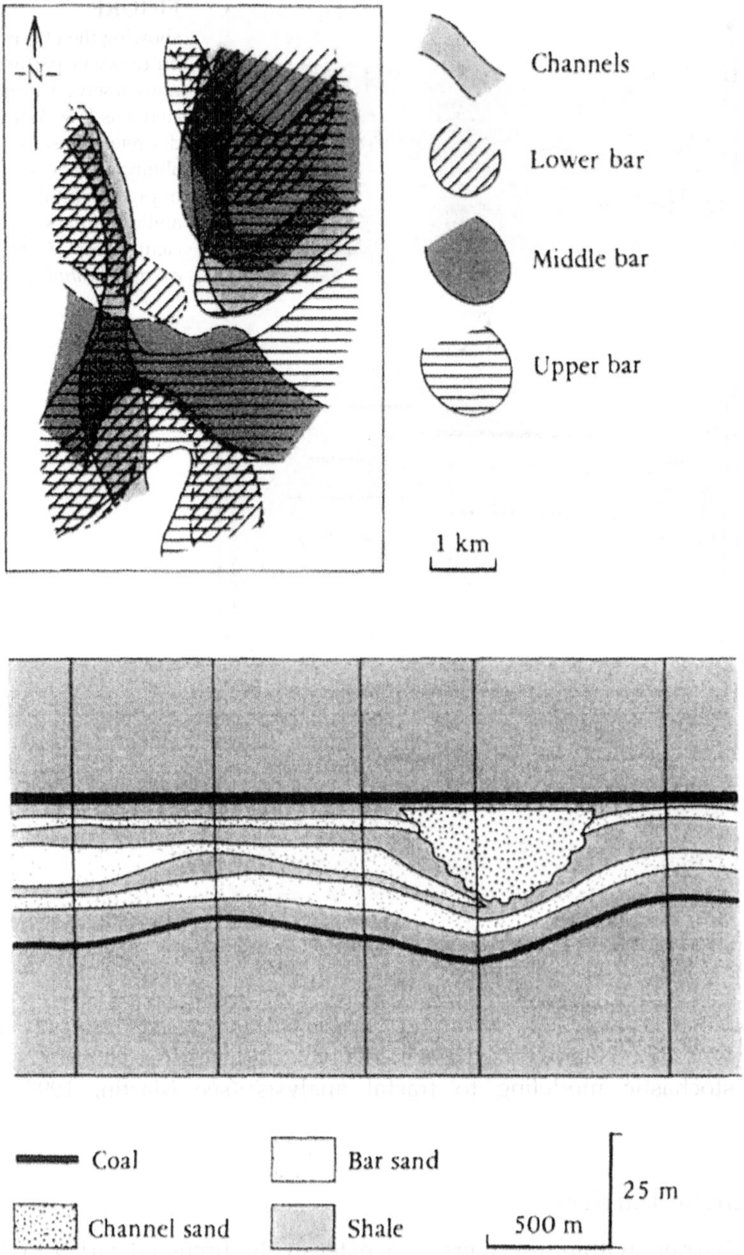

FIGURE 6.42 Map and cross-section of part of the Handil field, Indonesia, indicating the degree of vertical and lateral continuity of reservoirs found in shallow fluviomarine sands. *Modified from Verdier et al. (1980); reprinted by permission of the American Association of Petroleum Geologists.*

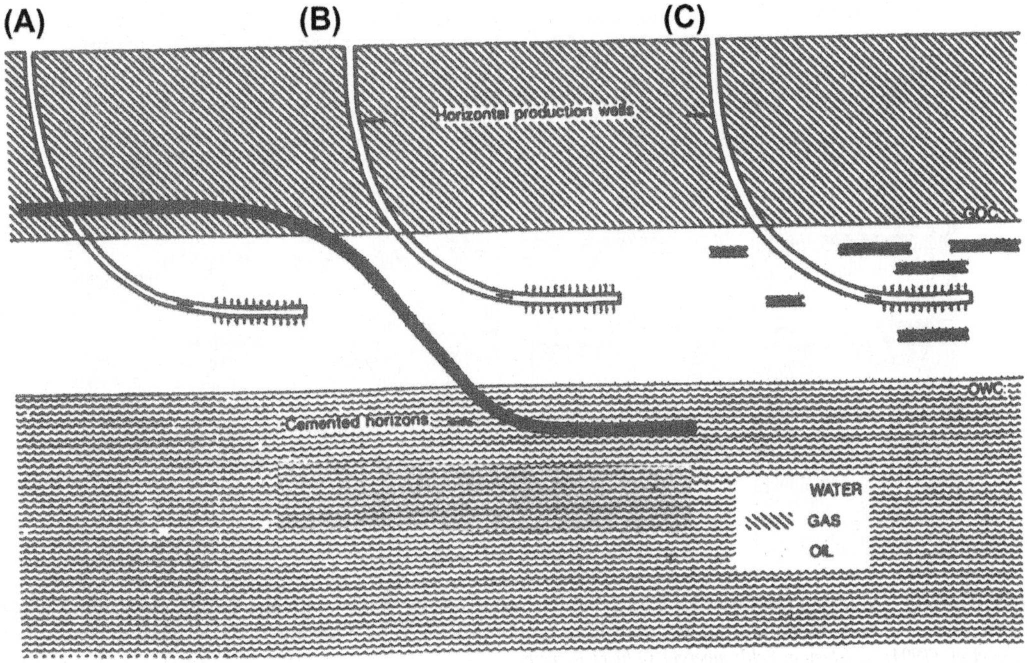

FIGURE 6.43 Illustration showing how horizontal wells may be used to overcome the problems of continuous discordant (left) and discontinuous concordant (right) carbonate cemented horizons within a reservoir. *From Gibbons et al. (1993).*

concretions colloquially referred to as "doggers." These are particularly common in the Jurassic sandstones of northwest Europe, where they are sufficiently abundant to give the eponymous name Dogger to the Middle Jurassic. Within the Jurassic reservoirs of the North Sea great attention has been paid to their significance as permeability barriers (Hurst, 1987; Gibbons et al., 1993). Once they have been identified it may be possible to drill wells to produce petroleum at maximum efficiency (Fig. 6.43).

6.6.3 Structural Barriers

Faults are a common structural barrier within petroleum reservoirs. Faults are sometimes permeable and allow fluid movement; sometimes they are sealed and inhibit it. Open faults are beneficial if they allow petroleum to migrate from a source rock into a reservoir, and are beneficial if they are impermeable and prevent the escape of petroleum from a trap. Within a trap, however impermeable faults inhibit effective petroleum production. Much research has been done on the imaging of faults, and on trying to establish whether they are open or closed. Figure 6.44 shows how faults with a throw of less than 12 m may be below the present limits of seismic resolution. For subseismic scale faults their frequency can be modeled using various statistical methods, including fractal analysis (Childs et al., 1990; Heath et al., 1994; Walsh et al., 1994).

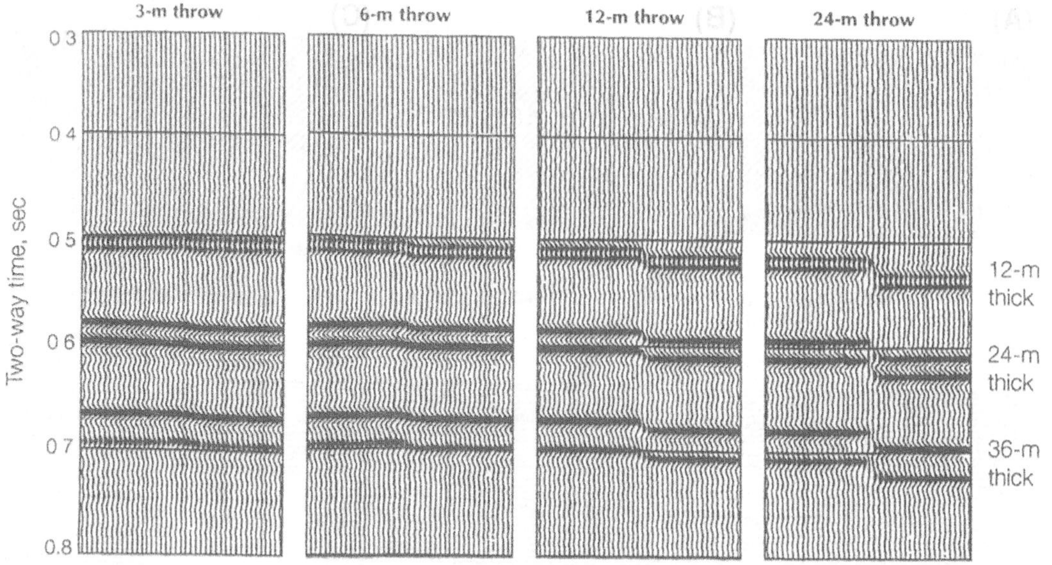

FIGURE 6.44 Modeled responses for a 48-m seismic wave imaging faults of diverse throws crosscutting beds of diverse thickness. Note that a fault of less than 12 m cannot be imaged by a seismic signal $\lambda/4$ bed thickness. *From Aston et al. (1994); courtesy of Schlumberger Oilfield Review.*

The determination of whether a fault is open or closed depends on many variables, including the physical properties of the rocks, the fluid pressures, and the juxtaposition of permeable sands and impermeable shales. Cunning computer programs are available, such as Badley's FAPS (Fault Analysis Software), that attempt to predict the frequency and permeability of small-scale faulting. This is discussed later in Chapter 7, in the context of fault traps.

6.7 RESERVOIR CHARACTERIZATION

Once an accumulation of petroleum has been discovered it is essential to characterize the reservoir as accurately as possible in order to calculate the reserves and to determine the most effective way of recovering as much of the petroleum as economically as possible (Lucia and Fogg, 1990; Lake et al., 1991; Worthington, 1991; Haldersen and Damsleth, 1993). The principal goal of reservoir characterization is to obtain higher recoveries with fewer wells in better positions at minimum cost through optimization (Slatt, 2006). Reservoir characterization first involves the integration of a vast amount of data from seismic surveys, from geophysical well logs, and from geological samples (Fig. 6.45). Note that the data come in a hierarchy of scales, from the megascopic and mesoscopic to the microscopic. It is important to appreciate both the scale and the reliability of the different data sets. For example, the problems of reconciling porosity and permeability data from logs and rock samples has already been mentioned.

The first aim of reservoir characterization is to produce a geological model that honors the available data and can be used to predict the distribution of porosity, permeability, and fluids

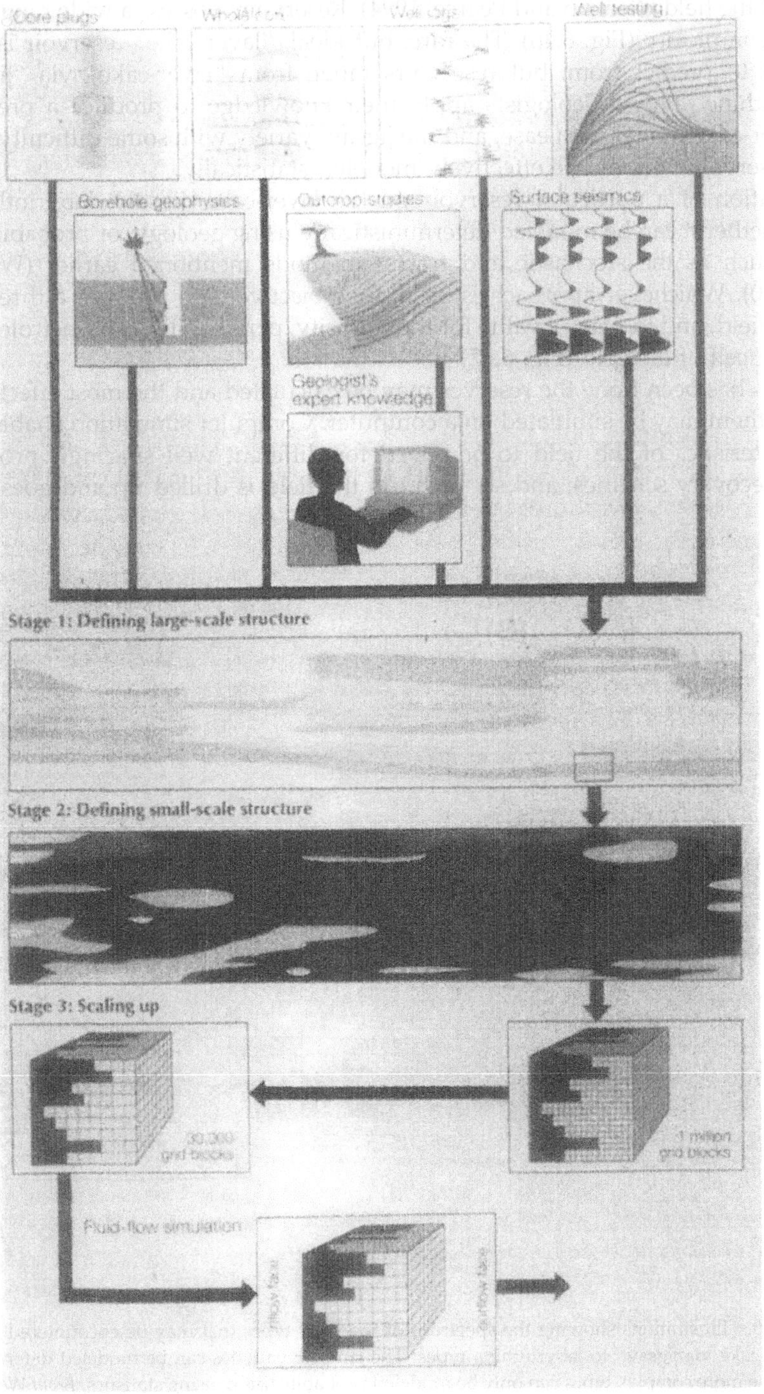

FIGURE 6.45 Diagram showing the stages involved in reservoir characterization. Note how it requires the integration of geophysical, geological, and reservoir engineering data, and involves consideration of data reliability on a range of scales: macroscopic, mesoscopic, and microscopic. *From Corvey and Cobley (1992); courtesy of Schlumberger Oilfield Review.*

throughout the field (Geehan and Pearce, 1994). Reservoirs possess a wide range of degrees of geometric complexity (Fig. 6.46). The rare, but ideal, "layer-cake" reservoir is the easiest to model and to predict from, but reservoirs range from "layer-cake" via "jigsaw puzzle" to "labyrinthine" types. Geologists apply their knowledge to produce a predictive model for the layer-cake model with ease, and the jigsaw variety with some difficulty. But the labyrinthine reservoir can only be effectively modeled statistically.

The location of a particular reservoir on the layer-cake–jigsaw–labyrinthine spectrum decides whether it can be modeled deterministically using geology, or probabilistically using statistics, such as the stochastic and fractal methods mentioned earlier (Weber and van Guens, 1990). Whichever approach is used, the objective is to produce a three-dimensional grid of the field, and to place a value for the porosity, permeability, and petroleum saturation within each cell of the grid (Fig. 6.47).

Once this has been done the reserves may be calculated and the most effective method of producing them may be simulated on a computer. Computer simulation enables the production characteristics of the field to be tested for different well spacings, production rates, enhanced recovery schemes, and so forth. As the field is drilled up and goes on stream an

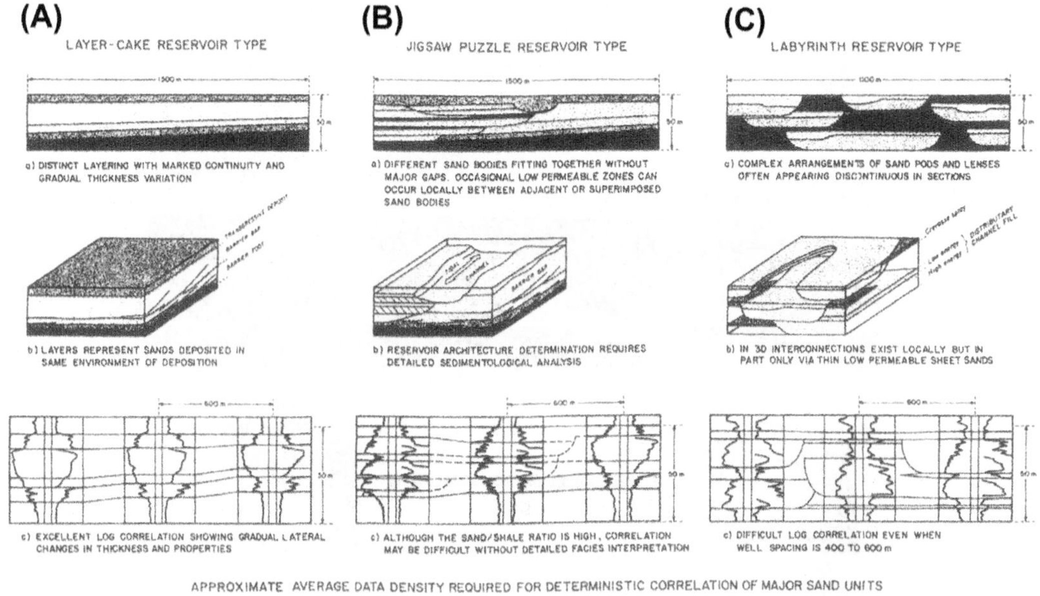

WELL PATTERN	SPACING, m	WELLS/km²
RECTANGULAR	1000	1
TRIANGULAR	1200	0.8
RANDOM		1.3

WELL PATTERN	SPACING, m	WELLS/km²
RECTANGULAR	600	3
TRIANGULAR	800	2
RANDOM		4

WELL PATTERN	SPACING, m	WELLS/km²
RECTANGULAR	200	25
TRIANGULAR	300	13
RANDOM		32

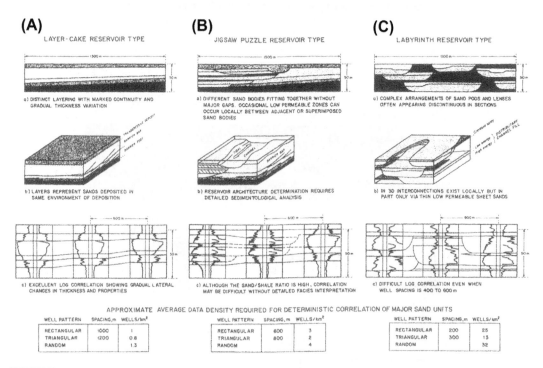

FIGURE 6.46 Illustrations showing the spectrum of reservoir types that may be encountered ranging from the simplest layer-cake via jigsaw, to labyrinthine types. The simpler varieties can be modeled deterministically using geology, but the more complex types can only be modeled probabilistically using statistics. *From Weber and van Guens (1990).*

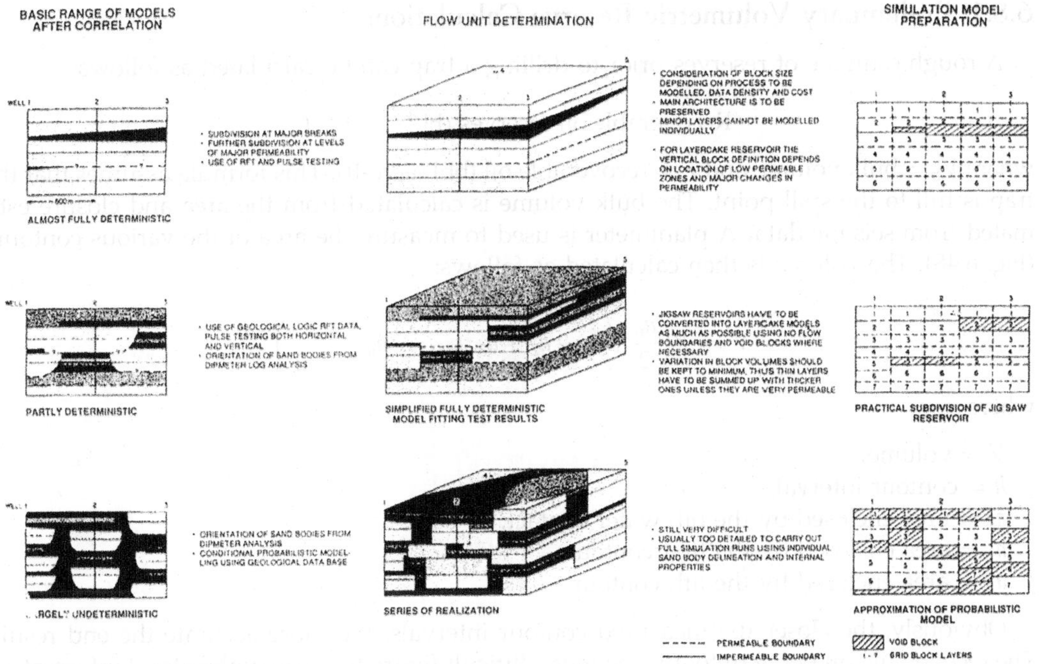

FIGURE 6.47 Illustration showing how reservoirs may be gridded into a series of cells for which a value for the porosity, permeability, and petroleum saturation must be given. Then it is possible to produce an accurate assessment of the reserves and a data for computer simulation to establish the most effective way of producing the petroleum from the reservoir. *From Weber and van Guens (1990).*

iterative process constantly updates and revises the reservoir grid, and enables progressively more accurate production scenarios to be tested.

Reservoir characterization leads to an incremental improvement in production. The improvements come about because of better understanding of the geologic complexities of fields.

6.8 RESERVE CALCULATIONS

Estimates of possible reserves in a new oil or gas held can be made before a trap is even drilled. The figures used are only approximations, but they may give some indication of the economic viability of the prospect. As a proven field is developed and produced, its reserves are known with greater and greater accuracy until they are finally depleted. The calculation of reserves is more properly the task of the petroleum reservoir engineer, but since this task is based on geological data, it deserves consideration here. Several methods are used to estimate reserves, ranging from crude approximations made before a trap is tested to more sophisticated calculations as hard data become available.

6.8.1 Preliminary Volumetric Reserve Calculations

A rough estimate of reserves prior to drilling a trap can be calculated as follows:

$$\text{Recoverable oil reserves (bbl)} = V_b \cdot F,$$

where V_b = bulk volume and F = recoverable oil (bbl/acre-ft). This formula assumes that the trap is full to the spill point. The bulk volume is calculated from the area and closure estimated from seismic data. A planimeter is used to measure the area of the various contours (Fig. 6.48). The volume is then calculated as follows:

$$V = h\left(\frac{a_o}{2} + a_1 + a_2 + a_3 + \ldots + a_{n-1} + \frac{a_n}{2}\right)$$

where

V = volume;
h = contour interval;
a_o = area enclosed by the oil/water contact;
a_1 = area enclosed by the first contour;
a_n = area enclosed by the nth contour.

Obviously, the closer the measured contour intervals, the more accurate the end result. The recoverable oil per acre-foot is the most difficult figure to assess unless local information is available from adjacent fields. It may be calculated according to an estimate of the average porosity of the reservoir (say, 10–30%) and an estimate of the recovery factor (generally 30% or more for sands, and 10–20% for carbonate reservoirs). The recovery factor will vary according to well spacing, reservoir permeability, fluid viscosity, and the effectiveness of the drive mechanism. Generally, several 100 bbl/acre-ft of oil may be recovered.

This method of calculating reserves is very approximate, but may be the only one available before bidding for acreage. It may also be useful for ranking the order in which to drill a number of prospects.

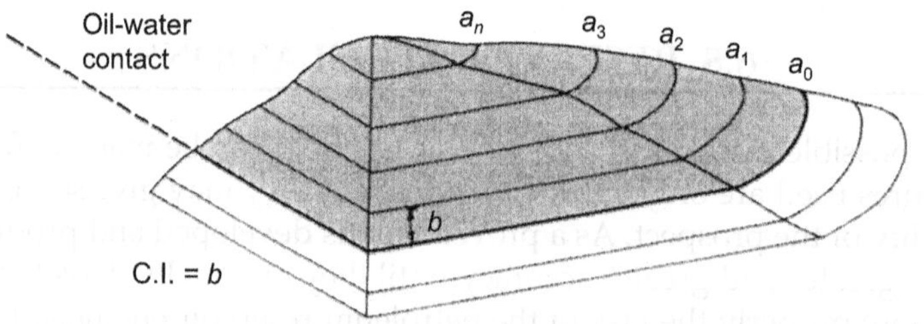

FIGURE 6.48 Isometric sketch of a trap showing how a planimeter survey measures the areas above selected contours. For additional explanation see text.

6.8.2 Postdiscovery Reserve Calculations

Once a field has been discovered, accurate reservoir data become available and a more sophisticated formula may be applied:

$$\text{Recoverable oil (bbl)} = \frac{7758\ V\phi(1 - S_w)R}{\text{FVF}},$$

where

V = the volume (area × thickness);
7758 = conversion factor from acre-feet to barrels;
ϕ = porosity (average);
S_w = water saturation (average);
R = recovery factor (estimated); and
FVF = formation volume factor.

These variables are now discussed in more detail. As wells are drilled on the field, the seismic interpretation becomes refined so that an accurate structure contour map can be drawn. Log and test data establish the oil:water contact and hence the thickness of the hydrocarbon column (Dahlberg, 1979).

The porosity is calculated from wireline logs calibrated from core data, and the water saturation is calculated from the resistivity logs. The recovery factor is hard to estimate even if the performances of similar reservoirs in adjacent fields are available. Approximate values have been given previously.

The FVF converts a stock tank barrel of oil to its volume at reservoir temperatures and pressures. It depends on oil composition, but this dependence can generally be approximated by calculating the FVF's dependence on the solution gas:oil ratio (GOR) and oil density (API gravity). The FVF ranges from 1.08 for low GORs and heavy crudes to values of more than 2.0 for volatile oils and high GORs. The GOR is defined as the volume ratio of gas and liquid phase obtained by taking petroleum from one equilibrium pressure and temperature, in the reservoir, to another, at the surface, via a precisely defined path. For a restrictive set of subsurface conditions it may be calculated from the following equation:

$$\text{Gas:oil ratio (in the reservoir)} = \frac{Q_g}{Q_o} = \frac{\mu_o K_g}{\mu_g K_o}$$

where

Q = flow rate at reservoir temperatures and pressures;
μ = viscosity at reservoir temperatures and pressures;
K = effective permeability;
g = gas; and
o = oil.

Figure 6.49 shows the relationship between FVF and GOR. A more accurate measurement of FVF can be made in the laboratory. Ideally, a sample of the reservoir fluid is collected in a pressure bomb and reheated to the reservoir temperature. Temperature and pressure are

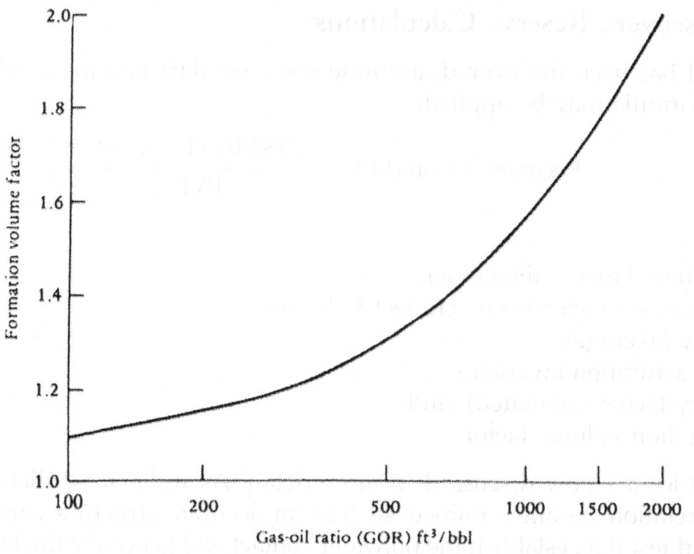

FIGURE 6.49 Graph showing the relationship between FVF and the gas:oil ratio (GOR).

reduced to surface temperature and pressure so that the respective volumes of gas and oil can be measured accurately. This procedure is referred to as pressure–volume–temperature analysis.

As a field is produced, several changes take place in the reservoir. Pressure drops and the flow rate diminishes. The GOR may also vary, although this depends on the type of drive mechanism, as discussed in the next section. These changes are sketched in Fig. 6.50. The changes may be formulated in the material balance equation, which, as its name suggests, equates the volumes of oil and gas in the reservoir at virgin pressure with those produced and remaining in the reservoir at various stages in its productive life. At its simplest level, the material balance equation is a variant of the law of conservation of mass:

$$\text{Weight of hydrocarbons originally in reservoir} =$$
$$\text{Weight of produced hydrocarbons} +$$
$$\text{Weight of hydrocarbons retained in reservoirs.}$$

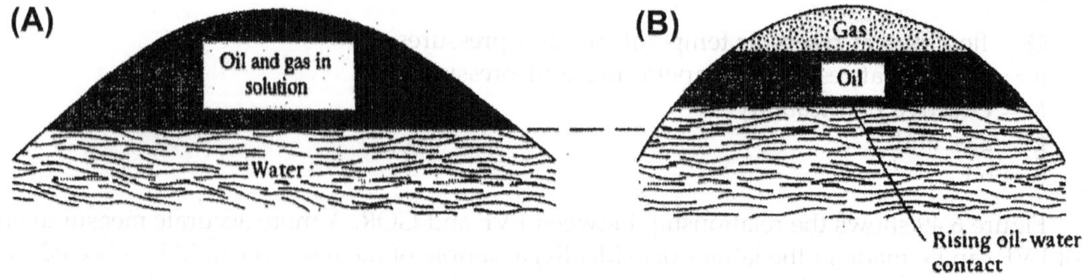

FIGURE 6.50 The changes within a reservoir that may be caused by production: (A) the situation at virgin pressure before production and (B) the situation with depleted pressure after commencement of production.

If all the weights are expressed as stock tank barrels of oil and stock tank cubic feet of gas, then weight can be substituted by volume.

The mass balance equation has several complex variations. Although its solution is a job for the petroleum reservoir engineer, much of the data on which it is based are provided by geologists.

6.9 PRODUCTION METHODS

As mentioned in the review of the mass balance equation in the previous section, hydrocarbons may be produced by several mechanisms. Three natural drive mechanisms can cause oil or gas to flow up the well bore: water drive, gas drive, and gas solution (or dissolved gas) drive.

6.9.1 Water Drive

In water drive production, oil or gas trapped within a reservoir may be viewed as being sealed within a water-filled U-tube (Fig. 6.51). When the tap is opened, oil and gas will flow from the reservoir because of the hydrostatic head of water (Fig. 6.52). Not all aquifers have a continuous recharge of the earth's surface. The degree of water encroachment and pressure maintenance depends on the size and productivity of the aquifer. Note that as the field is produced, water invades the lower part of the trap to displace the oil (Fig. 6.53); only in the most uniform reservoirs does the oil:water contact rise evenly. Because adjacent beds seldom have the same permeability, water encroachment advances at different rates, giving rise to fingering. A further complication may be caused by coning of the water adjacent to boreholes. The extent of coning depends on the rate of production and the ratio between vertical and horizontal permeability.

When a field with a water drive mechanism is produced, the reservoir pressure drops in inverse proportion to the effectiveness of the recharge from the aquifer. Generally, little change occurs in the GOR. With an effective water drive, the flow rate remains constant

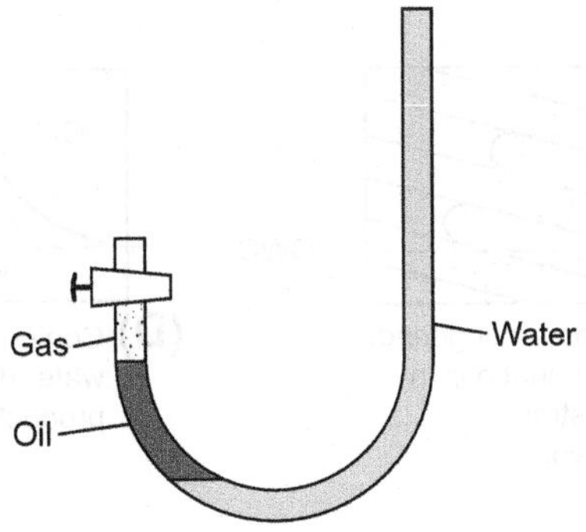

FIGURE 6.51 Sketch showing the principle of water drive petroleum production by analogy with a water-filled U-tube. When the valve is opened (i.e., the trap drilled into), the hydrostatic head of water causes the oil and gas flow.

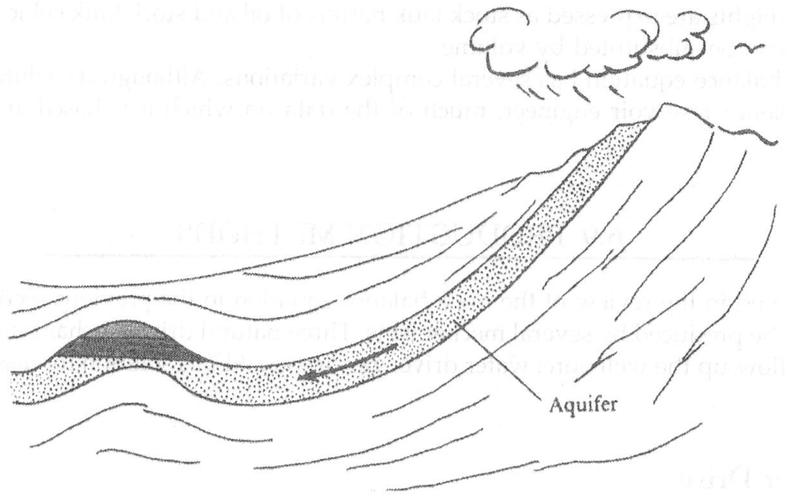
Aquifer

FIGURE 6.52 A typical geological setting for a water drive mechanism. As the field is produced, petroleum is displaced by water from the aquifer.

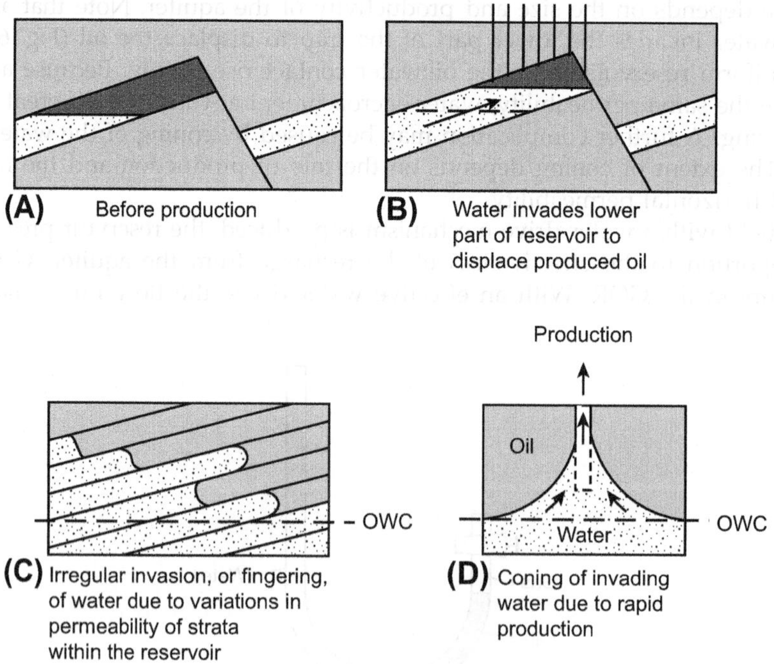
(A) Before production
(B) Water invades lower part of reservoir to displace produced oil
Production
Oil
OWC
Water
OWC
(C) Irregular invasion, or fingering, of water due to variations in permeability of strata within the reservoir
(D) Coning of invading water due to rapid production

FIGURE 6.53 Water drive mechanism (A) before production and (B) during production. (C) and (D) illustrate potential irregularities in the rising oil/water contact.

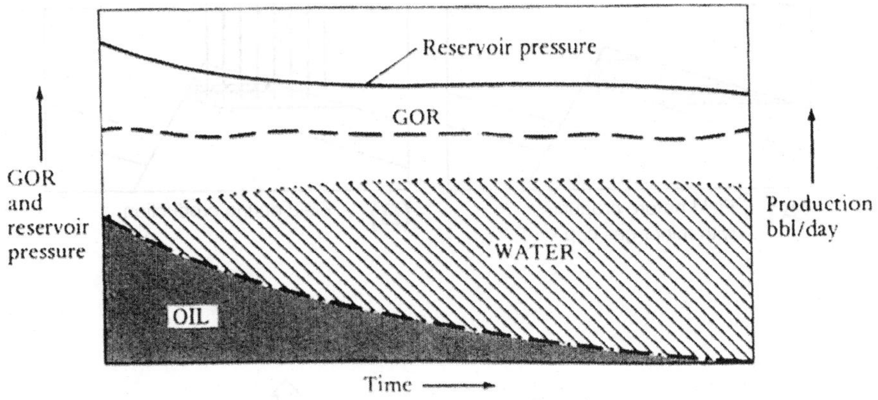

FIGURE 6.54 Typical production history for a water drive field. For explanation see text.

during the life of the fluid, but oil production declines inversely with an increase in water production. These trends are illustrated in Fig. 6.54. The water drive mechanism is generally the most effective, with a recovery factor of up to 60%.

6.9.2 Gas Cap Drive

A second producing mechanism is the gas cap drive, in which the field contains both oil and gas zones. As production begins, the drop in pressure causes gas dissolved in the oil to come out of the solution. This new gas moves up to the gas cap and, in so doing, expands to occupy the pores vacated by the oil. A transitional zone of degassing thus forms at the gas:oil contact. Drawdown zones may develop adjacent to boreholes in a manner analogous to, but the reverse of, coning at the oil/water contact (Fig. 6.55).

The production history of gas cap drive fields is very different from that of water drive fields. Pressure and oil production drop steadily, while the ratio of gas to oil naturally increases (Fig. 6.56). The gas cap drive mechanism is generally less effective than the water drive mechanism, with a recovery factor of 20–50%.

6.9.3 Dissolved Gas Drive

The third type of production mechanism is the dissolved gas drive, sometimes called the solution gas drive. This type of drive occurs in oil fields that initially have no gas cap. Production is analogous to a soda fountain. As production begins, pressure drops and gas bubbles form in the oil and expand, forcing the oil out of the pore system and toward the boreholes. As the gas expands, it helps to maintain reservoir pressure. Initially, the gas bubbles are separated. As time passes, they come together and may form a continuous free-gas phase, which may accumulate as a gas cap in the crest of the reservoir. This point is termed the critical gas saturation, and care should be taken to prevent it from being reached. It may be avoided either by maintaining a slow rate of production or by reinjecting the produced gas to maintain the original reservoir pressure.

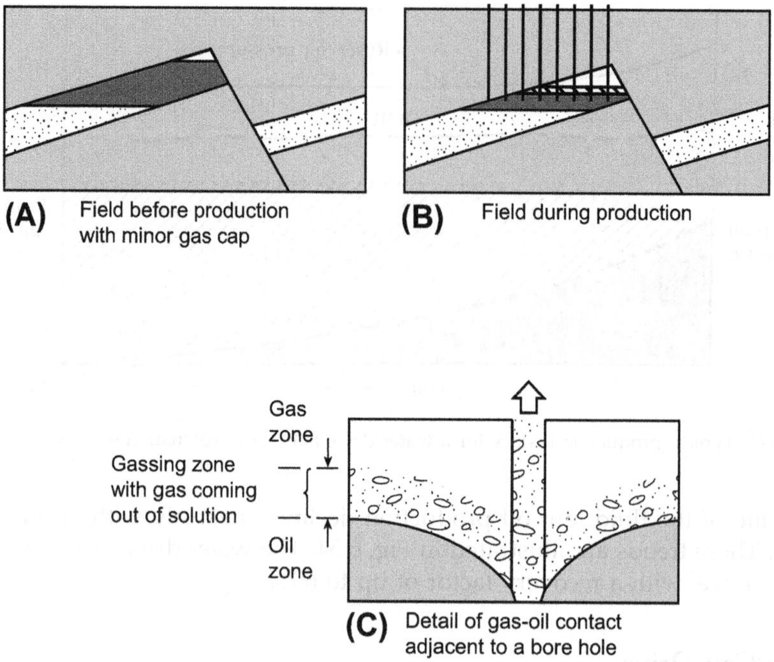

FIGURE 6.55 The gas expansion drive mechanism. A transitional zone develops at the gas/oil contact as pressure drops and the gas separates from the oil. Note the drawdown effect, which may develop adjacent to boreholes (C).

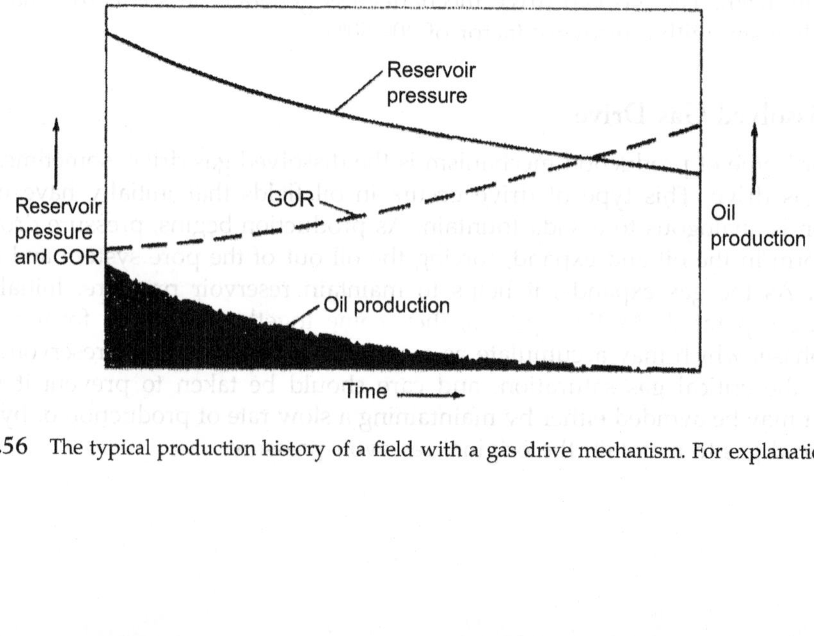

FIGURE 6.56 The typical production history of a field with a gas drive mechanism. For explanation see text.

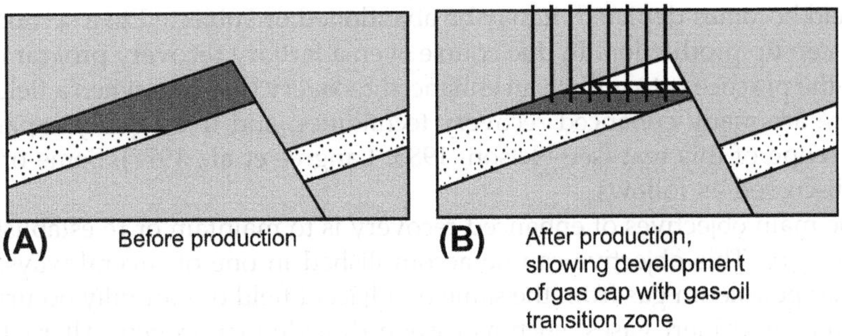

FIGURE 6.57 A field with a gas expansion drive mechanism. If the pressure drops sufficiently to cross the critical gas saturation point, then a free-gas cap may form (B).

Figure 6.57 illustrates what happens in a reservoir with a gas expansion drive, and Fig. 6.58 shows the typical production history. This drive mechanism is generally considered to be the least efficient of the three, with a recovery factor in the range of 7–15%.

6.9.4 Artificial Lift and Enhanced Recovery

Not all fields produce by natural drive mechanisms, and even these natural drive mechanisms do not recover all the oil. Artificial methods are used to produce oil from fields lacking natural drive, and enhanced recovery methods increase the recoverable reserves.

A well will flow oil to the surface it the static pressure at the bottom of the well exceeds the pressure of the column of mud and the frictional effect of the borehole. In many shallow fields with low reservoir pressure and in depleted fields, the wells do not flow spontaneously. In such instances the oil is pumped to the surface using the nodding donkey or bottom hole pumps described in an earlier chapter.

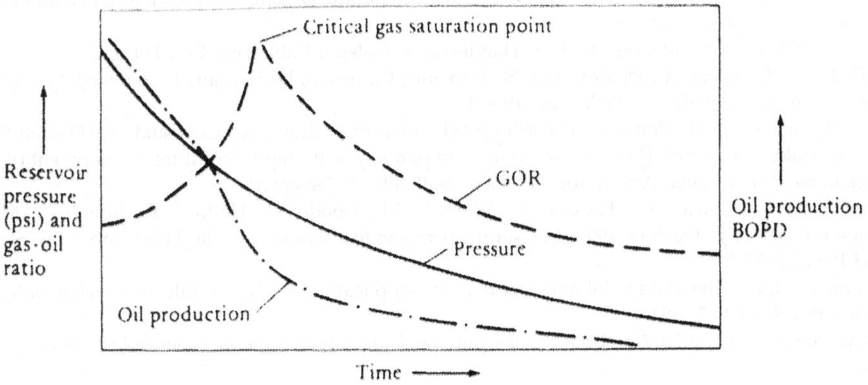

FIGURE 6.58 Graph illustrating the production history of a field with a gas expansion drive. Note how the GOR increases until the critical gas saturation point is reached; then it drops off sharply as the gas separates out.

Once a field becomes depleted, it may be abandoned or subjected to a secondary recovery program to step up production. In due course even a tertiary recovery program may be initiated. Today the practice is to initiate an enhanced recovery program when a field first goes on stream. There are many enhanced recovery techniques, and this rapidly expanding field is beyond the scope of this text (see Nelson, 1985; Langres et al., 1972). Some of the methods are briefly discussed as follows.

One of the main objectives of enhanced recovery is to maintain or re-establish the original reservoir pressure. This objective can be accomplished in one of several ways. Gas may be injected, either petroleum gas from the same or adjacent field or naturally occurring or industrially manufactured inert gases, such as carbon dioxide or nitrogen. Alternatively, liquids may be injected; these liquids may be seawater or connate waters from adjacent strata. Careful chemical analysis and the treatment of injected waters is essential to prevent and monitor unwanted chemical changes taking place within the reservoir. Some waters damage permeability by precipitating salts in pore spaces and causing clays to swell (Giraud and Neu, 1971). Seawater must be treated to destroy bacteria, since they degrade oil and increase its viscosity. More elaborate enhanced recovery methods involve the injection of detergents (micellar floods) to emulsify heavy oil and move it to the surface. Electric radiators have been used to mobilize oil from shales and from conventional petroleum reservoirs (Pizarro and Trevisan, 1990). For these and further details of enhanced recovery, see de Haan (1995).

Nuclear explosions have been used in attempts to enhance petroleum recovery in both the United States and the former USSR (see Howard and King, 1983; Orudjev, 1971; respectively). What better way to end this chapter than with a bang?

References

Aguilera, R., Van Poollen, H.K., 1978. Geologic aspects of naturally fractured reservoirs explained. Oil Gas J. 47–51. December 18.

Al-Gailani, M.B., 1981. Authigenic mineralization at unconformities: implication for reservoir characteristics. Sediment. Geol. 29, 89–115.

Allen, J.R.L., 1970. The systematic packing of prolate spheroids with reference to concentration and dilatency. Geol. Mijnbouw 49, 211–220.

Al-Rawi, M., 1981. Geological Interpretation of Oil Entrapment in the Dubair Formation, Raudhatain Field. Soc. Pet. Eng.. Prepr. No. 9591, pp. 149–158.

Anderson, G., 1975. Coring and Core Analysis Handbook. Petroleum Publishing Co., Tulsa, OK.

Aston, C.P., Bacon, B., Mann, A., Moldoveanu, N., Deplante, C., Ireson, D., Sinclair, T., Redekop, G., April 1994. 3D seismic design. Schlumberger Oilfield Rev. 19–32.

Atkins, J.E., McBride, E.F., 1992. Porosity and packing of Holocene river, dune and beach sands. AAPG Bull. 92, 339–355.

Atwater, G.I., Miller, E.E., 1965. The effect of decrease in porosity with depth on future development of oil and gas reserves in South Louisiana. Am. Assoc. Pet. Geol. Bull. 49, 334 (abstract).

Ayan, C., Colley, N., Cowan, G., Ezekwe, E., Wannell, M., Goode, P., Halford, F., Joseph, J., Mongini, G., Obondoko, G., Pop, J., October 1994. Measuring permeability anisotropy: the latest approach. Schlumberger Oilfield Rev. 24–35.

Badiozamani, K., 1973. The dorag dolomitization model-application to the middle ordovician of wisconsin. J. Sediment. Petrol. 43, 965–984.

Balducci, A., Pommier, G., 1970. Cambrian oil field of Hassi Messaoud, Algeria. Mem.—Am. Assoc. Pet. Geol. 14, 477–488.

Beard, D.C., Weyl, P.K., 1973. The influence of texture on porosity and permeability of unconsolidated sand. Am. Assoc. Pet. Geol. Bull. 57, 349–369.

Biederman, E.W., 1986. Atlas of Selected Oil and Gas Reservoir Rocks from North America. Wiley, New York.

Bjorlykke, K., 1984. Formation of secondary porosity: how important is it? AAPG Mem. 37, 277–286.
Blatt, H., Middleton, G.V., Murray, R., 1980. Origin of Sedimentary Rocks, second ed. Prentice Hall, New Jersey.
Bloch, S., 1991. Empirical prediction of porosity and permeability in sandstones. AAPG Bull. 75, 1145–1160.
Bloch, S., Helmold, K.P., 1995. Approaches to predicting reservoir quality in sandstones. AAPG Bull. 79, 97–115.
Botset, H.G., 1931. The measurement of permeability of porous aluminum discs of water and oils. Rev. Sci. Instrum. 2, 84–95.
Brady, T.J., Campbell, N.D.J., Maher, C.E., 1980. Intisar "D" oil field, Libya. AAPG Mem. 30, 543–564.
Butler, M., Phelan, M.L., Wright, A.W., 1976. The Buchan field, evaluation of a fractured sandstone reservoir. In: Trans. Eur. Symp., Soc. Prof. Well Log Anal., 4th, London, pp. 1–18.
Carvalho, M.V.H., De Ros, L.F., Gomes, N.S., 1995. Carbonate cementation patterns and diagenetic reservoir facies in the campos basin cretaceous turbidites, eastern offshore Brazil. Mar. Pet. Geol. 12, 741–758.
Cazier, E.C., Haywood, A.B., Espinoza, G., Velandia, J., Mugniot, J.-F., Peel, W.G., 1995. Petroleum geology of the Cusiana Field, Llanos Basin, Foothills, Columbia. AAPG Bull. 79, 1444–1463.
Chierici, G.L., 1994. Principles of Petroleum Reservoir Engineering, vols 2. Springer-Verlag, Berlin.
Childs, C., Walsh, J.J., Waterson, J., 1990. A method for estimation of the density of fault displacements below the limits of seismic resolution in reservoir formations. In: Buller, A.T. (Ed.), North Sea Oil and Gas Reservoirs II. Graham & Trotman, London, pp. 309–318.
Chilingar, G.V., 1964. Relationship between porosity, permeability and grainsize distribution of sands and sandstones. In: Van Straaten, L.M.J.U. (Ed.), Deltaic and Shallow Marine Deposits. Elsevier, Amsterdam, pp. 71–75.
Choquette, P.W., Pray, L.C., 1970. Geologic nomenclature and classification of porosity in sedimentary carbonates. Am. Assoc. Pet. Geol. Bull. 54.
Collins, R.E., 1961. Flow of Fluids through Porous Media. Reinhold, New York.
Conley, C.D., 1971. Stratigraphy and lithofacies of lower paleocene rocks, Sirte Basin, Libya. In: Gray, C. (Ed.), Symposium on the Geology of Libya. University of Libya, Tripoli, pp. 127–140.
Core Laboratories, 2014. GRI (Crushed Shale) Analysis. Core Laboratories. http://www.corelab.com/ps/gri-analysis.
Corvey, P., Cobley, T., January 1992. Reservoir characterization using expert knowledge, data and statistics. Schlumberger Oilfield Rev. 25–39.
Dahlberg, E.C., 1979. Hydrocarbon reserve estimation from contour maps. Bull. Can. Pet. Geol. 27, 94–99.
Darcy, H., 1856. Les fontaines publiques de la Ville de Dijon. V. Dalmont, Paris.
de Haan (Ed.), 1995. New Developments in Enhanced Oil Recovery. Spec. Publ.—Geol. Soc. London, vol. 84, pp. 1–286.
Dickey, P.A., 1979. Petroleum Development Geology. Petroleum Publishing Co., Tulsa, OK.
Dodge, M.M., Loucks, R.G., 1979. Mineralogic composition and diagenesis of tertiary sandstones along texas gulf coast. AAPG Bull. 63, 440.
Dykstra, H., Parsons, R.L., 1950. The prediction of oil recovery by waterflood. In: Secondary Recovery, second ed. Am. Pet. Inst., New York, pp. 160–174.
Ebbern, J., 1981. The geology of the Morecombe gas fields. In: Illing, L.V., Hobson, G.D. (Eds.), Petroleum and the Continental Shelf of North-West Europe. Heyden Press, London, pp. 485–493.
Enos, P., Sawatsky, L.H., 1979. Pore space in holocene carbonate sediments. AAPG Bull. 63, 445.
Fairbridge, R., 1967. Phases of diagenesis and authigenesis. In: Larsen, G., Chilingar, G.V. (Eds.), Diagenesis in Sediments. Elsevier, Amsterdam, pp. 19–89.
Fischer, A.G., 1964. The lofer cyclothems of the alpine triassic. Kans. Univ. Geol. Surv. Bull. 169, 107–150.
Folk, R.L., Land, L.S., 1974. Mg/Ca ratio and salinity: two controls over crystallization of dolomite. AAPG Bull. 59, 60–68.
Fothergill, C.A., 1955. The cementation of oil reservoir sands. In: Proc.—World Pet. Congr. 4, Sect, vol. 1, pp. 300–312.
Fraser, H.J., 1935. Experimental study of the porosity and permeability of clastic sediments. J. Geol. 43, 910–1010.
Friedman, G.M., Sanders, J.E., Kopaska-Merkel, D.C., 1992. Principles of Sedimentary Deposits. Macmillan, New York.
Fuchtbauer, H., 1967. Influence of different types of diagenesis on sandstone porosity. In: Proc.—World Pet. Congr., vol. 7, pp. 353–369.
Gaithor, A., 1953. A study of porosity and grain relationships in sand. J. Sediment. Petrol. 23, 186–195.
Geehan, G.W., Pearce, A.J., 1994. Geological reservoir heterogeneity databases and their application to integrated reservoir description. In: Aason, J. (Ed.), North Sea Oil and Gas Reservoirs III. Norwegian Petroleum Institute, Trondheim, pp. 131–140.

Gibbons, K., Hellem, T., Kjemperud, A., Nio, S.D., Vebenstad, K., 1993. Sequence architecture, facies development and carbonate-cemented horizons in the troll field reservoir, offshore Norway. In: Spec. Publ.—Geol. Soc. London, vol. 69, pp. 1–32.

Giraud, A., Neu, J.M., 1971. Injection d'eau: interaction entre l'eau injectée et le reservoir. Rev. Inst. Fr. Pet. 26, 1009–1028.

Graton, L.C., Fraser, H.J., 1935. Systematic packing of spheres, with particular reference to porosity and permeability. J. Geol. 43, 785–909.

Gregory, A.R., 1977. Aspects of rock physics from laboratory and log data that are important to seismic interpretation. Mem.—Am. Assoc. Pet. Geol. 26, 15–46.

Haldersen, H.H., Damsleth, E., 1993. Challenges in reservoir characterization. AAPG Bull. 77, 541–551.

Haldersen, H.H., Chang, D.M., Begg, S.H., 1987. Discontinuous vertical permeability barriers: a challenge to engineers and geologists. In: Kleppe, J. (Ed.), North Sea Oil and Gas Reservoirs. Graham & Trotman, London, pp. 127–151.

Harris, D.G., Hewitt, C.H., 1977. Synergism in reservoir management. The geologic perspective. J. Pet. Technol. 761–770.

Harris, J.F., Taylor, G.L., Walper, J.L., 1960. Relation of deformational fractures in sedimentary rocks to regional and local structure. Am. Assoc. Pet. Geol. Bull. 44, 1853–1873.

Hay, J.T.C., 1977. The thistle oilfield. In: Mesozoic Northern North Sea Symposium. Norwegian Petroleum Society, Oslo, pp. 1–20. Paper II.

Hea, J.P., 1971. Petrography of the Paleozoic—Mesozoic sandstones of the Southern Sirte basin, Libya. In: Gray, C. (Ed.), The Geology of Libya. University of Libya, Tripoli, pp. 107–125.

Heald, K.C., 1940. Essentials for oil pools. In: De Golyer, E.L. (Ed.), Elements of the Petroleum Industry. Am. Inst. Min. Metall. Eng., New York, pp. 26–62.

Heath, A.E., Walsh, J.J., Waterson, J., 1994. Estimation of the effects of sub-seismic faults on effective permeabilities in sandstone reservoirs. In: Aason, J. (Ed.), North Sea Oil and Gas Reservoirs III. Norwegian Petroleum Institute, Trondheim, pp. 173–183.

Horbury, A.D., Robinson, A.G. (Eds.), 1993. Diagenesis and Basin Development. Am. Assoc. Pet. Geol., Tulsa, OK. Stud. Geol. No. 36.

Howard, J.H., King, J.E., 1983. Nuclear frac could be feasible for carbonate reservoirs. Oil Gas J. 84–92. December 12.

Hurst, A., 1987. Problems of reservoir characterization in some North sea sandstone reservoirs solved by the application of microscale geological data. In: Kleppe, J., Berg, E.W., Buller, A.T., Hjelmeland, O., Torsaeter, O. (Eds.), North Sea Oil and Gas Reservoirs. Graham & Trotman, London, pp. 153–167.

Hurst, A., Goggin, D., 1995. Probe permeametry: an overview and bibliography. AAPG Bull. 79, 463–473.

Hurst, A., Rosvoll, K.J., 1991. Permeability in sandstones and their relationship to sedimentary structures. In: Lake, L.W., Carroll, H.B., Wesson, T.C. (Eds.), Reservoir Characterization. Academic Press, San Diego, CA, pp. 166–196.

Jameson, J., 1994. Models of porosity formation and their impact on reservoir description, Lisburne Field, Prudhoe Bay, Alaska. AAPG Bull. 78, 1651–1678.

Jamison, H.C., Brockett, L.D., McIntosh, R.A., 1980. Prudhoe Bay: a 10-year perspective. AAPG Mem. 30, 289–314.

Jardine, D., Andrews, D.P., Wishart, J.W., Young, J.W., July 1977. Distribution and continuity of carbonate reservoirs. J. Pet. Technol. 873–885.

Jordan, C.F., Wilson, J.L., 1994. Carbonate reservoir rocks. AAPG Mem. 60, 141–158.

Kahn, J.S., 1956. The analysis and distribution of the properties of packing in sand size sediments. J. Geol. 64, 385–395.

Klinkenberg, L.J., 1941. The permeability of porous media to liquids and gases. Drill. Prod. Pract. 200–213.

Krumbein, W.C., Monk, G.D., 1942. Permeability as a function of the size parameters of unconsolidated sands. Am. Inst. Min. Metall. Eng. 1–11. Tech. Publ. 1492.

Kulander, B.R., Barton, C.C., Dean, S.L., 1979. Fractographic distinction of coring—induced fractures from natural cored fractures. AAPG Bull. 63, 482.

Lake, L.W., Carroll, H.B., Wesson, T.C., 1991. Reservoir Characterization, II. Academic Press, San Diego, CA.

Langres, C.L., Robertson, J.O., Chilingarian, G.V., 1972. Secondary Recovery and Carbonate Reservoirs. Elsevier, Amsterdam.

Lee, C.H., 1919. Geology and Groundwaters of the Western Part of the San Diego County, California. Water Supply Invig., Washington. Pap. No. 446.

Levandowski, D.W., Kaley, M.E., Silverman, S.R., Smalley, R.G., 1973. Cementation in Lyons sandstone and its role in oil accumulation, Denver Basin, Colorado. Am. Assoc. Pet. Geol. Bull. 57, 2217–2244.

Levorsen, A.I., 1967. The Geology of Petroleum. Freeman, Oxford.

Longman, M.W., 1980. Carbonate diagenetic textures from near surface diagenetic environments. AAPG Bull. 64, 461–487.

Loucks, R.G., Sarg, J.F., 1993. Carbonate sequence stratigraphy—recent developments and applications. AAPG Mem. 57, 1–545.

Loucks, R.G., Dodge, M.M., Galloway, W.W., 1979. Reservoir quality in tertiary sandstones along Texas Gulf Coast. AAPG Bull. 63, 488.

Lucia, F.J., Fogg, G.E., 1990. Geologic/stochastic mapping of heterogeneity in a carbonate reservoir. J. Pet. Technol. 42, 1298–1303.

Lucia, F.J., 1999. Carbonate Reservoir Characterisation. Springer-Verlag, Berlin, 226 pp.

Luffel, D.L., Guidry, F.K., 1992. New core analysis methods for measuring rock properties of Devonian shale. J. Pet. Technol. 4, 1184–1190.

Luo, P., Machel, H.G., 1995. Pore size and pore throat types in a heterogenous dolostonc reservoir, devonian grosmont formation western canada sedimentary basin. AAPG Bull. 79, 1698–1720.

Lynch, F.L., 1996. Mineral/water interaction, fluid flow, and Frio sandstone diagenesis: evidence from the rocks. AAPG Bull. 80, 486–504.

Macaulay, C.Y., Haszeldine, R.S., Fallick, A.E., 1992. Diagenetic pore waters stratified for at least 35 million years: magnus oil field, North sea. AAPG Bull. 76, 1625–1634.

Magara, K., 1980. Comparison of porosity-depth relationships of shale and sandstone. J. Pet. Geol. 3, 175–185.

Martin, J.H., 1993. A review of braided fluvial hydrocarbon reservoirs: the petroleum engineer's perspective. In: Spec. Publ.—Geol. Soc. London, vol. 75, pp. 333–367.

Martini, L.P., 1971. Grain size orientation and paleocurrent systems in the Thorold and Grimsby sandstones (Silurian), Ontario and New York. J. Sediment. Petrol. 41, 425–434.

Martini, L.P., 1972. Studies of microfabrics: an analysis of packing in the Grimsby sandstone (Silurian), Ontario and New York state. In: Int. Geol. Congr., Rep. Sess., 24th, Montreal, 1972, Sect, vol. 6, pp. 415–423.

Mazzullo, S.J., Harris, P.M., 1992. Mesogenctic dissolution: its role in porosity development in carbonate reservoirs. AAPG Bull. 76, 607–620.

McConnell, P.C., December 20, 1951. Drilling and production techniques that yield nearly 850,000 barrels per day in Saudi Arabia's fabulous Abqaiq field. Oil Gas J. 197.

McCoy, A.W., Sielaeff, R.L., Downs, G.R., Bass, N.W., Maxson, J.H., 1951. Types of oil and gas traps in the rocky mountains. Am. Assoc. Pet. Geol. Bull. 35, 1000–1037.

McDonald, D.A., Surdam, R.C. (Eds.), 1984. Clastic Diagenesis. Am. Assoc. Pet. Geol., Tulsa, OK. Mem. No. 37.

McKenny, J.W., Masters, J.W., 1968. Dineh-bi-Keyah field, Apache county, Arizona. Am. Assoc. Pet. Geol. Bull. 52, 2045–2057.

Moncure, G.K., Lahann, R.W., Seibert, R.M., 1984. Origin of secondary porosity and cement distribution in a sandstone/shale sequence from the Frio formation (Oligocene). AAPG Mem. 37, 151–162.

Monicard, R.P., 1981. In: Properties of Reservoir Rocks: Core Analysis. Technip, Paris.

Morrow, N.R., 1971. Small scale packing heterogeneities in porous sedimentary rocks. Am. Assoc. Pet. Geol. Bull. 55, 514–522.

Morse, D.G., 1994. Siliciclastic reservoir rocks. AAPG Mem. 60, 121–139.

Murray, R.C., 1960. Origin of porosity in carbonate rocks. J. Sediment. Petrol. 30, 59–84.

Muskat, M., 1937. Flow of Homogenous Fluids through Porous Media. McGraw-Hill, New York.

Muskat, M., 1949. Physical Principles of Oil Production. McGraw-Hill, New York (reissued by IHRDC, Boston, 1981).

Muskat, M., Botset, H.G., 1931. Flow of gas through porous materials. Physics 1, 27–47.

Nagtegaal, P.J.C., 1978. Sandstone-framework instability as a function of burial diagenesis. J. Geol. Soc. London 135, 101–105.

Narr, W., 1996. Estimating average fracture spacing in subsurface rock. AAPG Bull. 80, 1565–1586.

Nelson, R.A., 1985. Geological Analysis of Naturally Fractured Reservoirs. Gulf Publishing, Houston.

Nelson, R.A., 1987. Fractured reservoirs: turning knowledge into practice. J. Pet. Technol. 39, 407–414.

O'Brien, J.J., Lerche, I., June 1988. The preservation of primary porosity through hydrocarbon entrapment during burial. SPE Form. Eval. 295–299.

Orudjev, S.A., 1971. Underground nuclear explosions to stimulate oil field development. Water Air Conserv. Pet. Ind. Paper SP3b.

Philip, W., Droug, H.J., Haddenbach, H.G., Jankowsky, W., 1963. The history of migration in the Gifhorn trough (N.W. Germany). In: Proc.—World Pet. Congr., vol. 6, pp. 547–742.

Pippin, L., 1970. Panhandle-Hugoton field, Texas–Oklahoma–Kansas—the first fifty years. Mem.—Am. Assoc. Pet. Geol. 14, 204–222.

Pirson, S.J., 1978. Geologic Well Lag Analysis, second ed. Gulf Publishing Co., Houston, TX.

Pizarro, J.O.S., Trevisan, O.V., 1990. Electrical heating of oil reservoirs: numerical simulation and field test results. J. Pet. Technol. 42, 1320–1325.

Polasek, T.L., Hutchinson, C.A., 1967. Characterization of non-uniformities within a sandstone reservoir from a fluid mechanics point of view. In: Proc.—World Pet. congr., vol. 7 (2), pp. 397–407.

Potter, P.E., 1962. Late Mississippian sandstones of Illinois Basin. Circ.—Illt State Geol. Surv. 340.

Potter, P.E., Mast, R.F., 1963. Sedimentary structures, sand shape fabrics and permeability. J. Geol. 71, 441–471.

Potter, P.E., Pettijohn, P.J., 1977. Paleocurrents and Basin Analysis. Springer-Verlag, New York.

Powers, M.C., 1953. A new roundness scale for sedimentary particles. J. Sediment. Petrol. 23, 117–119.

Pryor, W.A., 1973. Permeability—porosity patterns and variations in some holocene sand bodies. Am. Assoc. Pet. Geol. Bull. 57, 162–189.

Pryor, W.A., Fulton, K., 1976. Geometry of Reservoir Type Sand Bodies in the Holocene Rio Grande Delta and Comparison with Ancient Reservoir Analogs. Soc. Pet. Eng.. Prepr. 7045, pp. 81–92.

Purser, B.H., 1978. Early diagenesis and the preservation of porosity in jurassic limestones. J.Pet. Geol. 1, 83–94.

von Rad, U., 1971. Comparison between "magnetic" and sedimentary fabric in graded and crosslaminated sand layers, Southern California. Geol. Rundsch. 60, 331–354.

Reeckmann, A., Friedman, G.M., 1982. Exploration for Carbonate Petroleum Reservoirs. Wiley, New York.

Rich, J.L., 1923. Shoestring sands of eastern Kansas. Am. Assoc. Pet. Geol. 7, 103–113.

Richardson, J.G., Sangree, J.B., Sneider, R.M., 1987. Permeability distribution in reservoirs. J. Pet. Technol. 39, 1197–1199.

Rittenhouse, G., 1971. Mechanical compaction of sands containing different percentages of ductile grains: a theoretical approach. Am. Assoc. Pet. Geol. 52, 92–96.

Robinson, R.B., 1966. Classification of reservoir rocks by surface texture. Am. Assoc. Pet. Geol. 50, 547–559.

Roehl, P.O., Choquette, P.W., 1985. Carbonate Petroleum Reservoirs. Springer-Verlag, New York.

Rogers, J.J., Head, W.B., 1961. Relationship between porosity median size, and sorting coefficients of synthetic sands. J. Sediment. Petrol. 31, 467–470.

Rose, P.R., 1972. Edwards Group, Surface and Subsurface, Central Texas. Bur. Econ. Geol., University of Texas, Austin.

Saigal, G.C., Bjorlykke, K., Larter, S., 1993. The effects of oil emplacement on diagenetic processes—examples from the fulmar reservoir sandstones, central North Sea. AAPG Bull. 76, 1024–1033.

Saller, A.H., Budd, D.A., Harris, P.M., 1994. Unconformities and porosity development in carbonate strata: ideas from the Hedberg Conference. AAPG Bull. 78, 857–872.

Scheiddegger, A.E., 1960. The Physics of Flow through Porous Media. Macmillan, New York.

Schmidt, V., McDonald, D.A., Platt, R.L., 1977. Pore geometry and reservoir aspects of secondary porosity in sandstones. Bull. Can. Pet. Geol. 25, 271–290.

Schmoker, J.W., Halley, R.B., 1982. Carbonate porosity versus depth. A predictable relation for south Florida. AAPG Bull. 66, 2561–2570.

Scholle, P.A., 1977. Chalk diagenesis and its relation to petroleum exploration: oil from chalks, a modern miracle. AAPG Bull. 61, 982–1009.

Scholle, P.A., November 1981. Porosity prediction in shallow vs. deepwater limestones. J. Pet. Technol. 2236–2242. November, 33.

Seemann, U., 1979. Diagenetically formed interstitial clay minerals as a factor in Rothegende sandstone reservoir quality in the North Sea. J. Pet. Geol. 1 (3), 55–62.

Selley, R.C., 1978. Porosity gradients in North Sea oil-bearing sandstones. J. Geol. Soc. London 135, 119–132.

Selley, R.C., 1984. Porosity evolution of truncation traps: diagenetic models and log responses. In: Finstad, K., Selley, R.C. (Eds.), Proceedings of the Norwegian Offshore North Sea Conference. Norwegian Petroleum Society, Oslo. Paper G3.

Selley, R.C., 1988. Applied Sedimentology. Academic Press, London.

Selley, R.C., 1990. Porosity evolution of truncated sandstone reservoirs. In: Ala, M. (Ed.), Seventy-Five Years of Progress in Oil Field Technology. Balkema, Amsterdam, pp. 103–112.

Selley, R.C., 1992. The third age of log analysis: application to reservoir diagenesis. In: Spec. Publ.—Geol. Soc. London, vol. 65, pp. 377–388.

Shanmugam, G., 1990. Porosity prediction in sandstones using erosional unconformities. AAPG Mem. 49, 1–23.

Shanmugam, G., Higgins, J.B., 1988. Porosity enhancement from chert dissolution beneath neocomian unconformity: ivishak formation, North Slope Alaska. AAPG Bull. 72, 523–535.

Shelton, J.W., Mack, D.E., 1970. Grain orientation in determination of paleocurrents and sandstone trends. Am. Assoc. Pet. Geol. 54, 1108–1119.

Sippel, R.F., 1971. Quartz grain orientations. I. The photometric method. J. Sediment. Petrol. 41, 38–59.

Slatt, R.M., 2006. Stratigraphic Reservoir Characterization for Petroleum Geologists, Geophysicists and Engineers. Elsevier, p. 478.

Sneider, R.M., Richardson, F.H., Paynter, D.D., Eddy, R.E., Wyant, I.A., 1977. Predicting reservoir rock geometry and continuity in pennsylvanian reservoirs, Elk city field, Oklahoma. J. Pet. Technol. 29, 851–866.

Stalder, P.J., 1973. Influence of crystallography habit and aggregate. Structure of authigenic clay minerals on sandstone permeability. Geol. Mijnbouw 52, 217–220.

Stearns, D.W., Friedman, M., 1972. Reservoirs in fractured rocks. Mem.—Am. Assoc. Pet. Geol. 16, 82–106.

Stolum, H.H., 1991. Fractal heterogeneity of clastic reservoirs. In: Lake, L.W., Carroll, H.B., Wesson, T.C. (Eds.), Reservoir Characterization 11. Academic Press, San Diego, CA, pp. 579–612.

Stormont, D.H., 1949. Huge caverns encountered in Dollarhide field. April 7 Oil Gas J. 66–68.

Stuart, I.A., 1993. The geology of the North morecombe bay gas field, East Irish Sea Basin. In: Parker, J.R. (Ed.), Petroleum Geology of Northwest Europe. Geol. Soc. London, London, pp. 883–898.

Sullivan, K.B., McBride, E.F., 1991. Diagenesis of sandstones at shale contacts and diagenetic heterogeneity, Frio Formation, Texas. AAPG Bull. 75, 121–138.

Sultan, R., Ortoleva, P., Depasquale, F., Tartaglia, P., 1990. Bifurcation of the Ostwald–Liesegang supersaturation nucleation-depletion cycle. Earth Sci. Rev. 129, 163–173.

Tebbutt, G.E., Conley, C.D., Boyd, D.W., 1965. Lithogenesis of a distinctive carbonate fabric. Contrib. Geol. 4 (1), 1–13.

Tedesco, S.A., 1994. Integrated exploration locates Cincinnati arch dolomite breccias. Oil Gas J. 86–90. November 28.

Toomey, D.F., Mountjoy, E.W., Mackenzie, W.S., 1970. Upper Devonian (Frasnian) algae and foraminifera from the ancient wall carbonate complex, Jasper National Park, Alberta, Canada. Can. J. Earth Sci. 7, 946–981.

Truex, J.N., 1972. Fractured shale and basement reservoir long beach unit, California. Am. Assoc. Pet. Geol. 50, 1931–1938.

Tucker, M.E., Wright, V.P., 1990. Carbonate Sedimentology. Blackwell, Oxford.

Van Veen, F.R., 1975. Geology of the leman gas-field. In: Woodland, A.E. (Ed.), Petroleum and the Continental Shelf of North-West Europe, vol. 1. Applied Science Publishers, Barking, UK, pp. 223–232.

Vavra, C.L., Kaldi, J.G., Sneider, R.M., 1992. Geological applications of capillary pressure: a review. AAPG Bull. 76, 840–850.

Verdier, A.C., Oki, T., Atik, S., 1980. Geology of the Handil field (East Kalimantan, Indonesia). AAPG Mem. 30, 399–422.

Walderhaug, O., 1996. Kinetic modelling of quartz cementation and porosity loss in deeply buried sandstone reservoirs. AAPG Bull. 80, 731–745.

Walsh, J.J., Watterson, J., Yielding, G., 1994. Determination and interpretation of fault size populations: procedures and problems. In: Aason, J. (Ed.), North Sea Oil and Gas Reservoirs III. Norwegian Petroleum Institute, Trondheim, pp. 141–155.

Wardlaw, N.C., 1976. Pore geometry of carbonate rocks as revealed by pore casts and capillary pressure. AAPG Bull. 60, 245–257.

Washburn, E.W., Bunting, E.N., 1922. Determination of porosity by the method of gas expansion. J. Am. Ceram. Soc. 5, 48–112.

Weber, K.J., 1971. Sedimentological aspects of oil fields in the Niger delta. Geol. Mijnbouw 50, 559–576.

Weber, K.J., 1982. Influence of common sedimentary structures on fluid flow in reservoir models. J. Pet. Technol. 34, 665–672.

Weber, K.J., van Guens, L.C., 1990. Framework for constructing clastic reservoir simulation models. J. Pet. Technol. 42, 1248–1297.

Weber, K.J., Eijpe, R., Leijnse, D., Moens, C., 1972. Permeability distribution in a Holocene distributary channel-fill near Leerdam (the Netherlands). Geol. Mijnbouw 51, 53–62.

Weber, K., 1992. Reservoir modeling for simulation purposes: part 10. Reservoir engineering methods. AAPG Special Volumes Dev. Geol. Reference Manual 531–535.

Williams, J.J., 1972. Augila field, Libya: depositional environment and diagenesis of sedimentary reservoir and description of igneous reservoir. Mem.—Am. Assoc. Pet. Geol. 16, 623–632.

Wilson, J.L., 1975. Carbonate Facies in Geologic History. Springer-Verlag, Berlin.

Wilson, J.L., 1980. A review of carbonate reservoirs. Mem.—Can. Soc. Pet. Geol. 6, 95–117.

Wood, G.V., Wolfe, M.J., 1969. Sabkha cycles in the Arab/Darb Formation off the Trucial Coast of Arabia. Sedimentology 12, 165–191.

Woodward, K., Curtis, C.J., 1987. Predictive modelling for the distribution of production-constraining illites—Morecombe Gas Field, Irish Sea, offshore UK. In: Brooks, J., Glennie, K. (Eds.), Petroleum Geology of North West Europe. Graham & Trotman, London, pp. 205–216.

Worthington, P.F., 1991. Reservoir characterization at the mesoscopic scale. In: Lake, L.W., Carroll, H.B., Wesson, T.C. (Eds.), Reservoir Characterization. Academic Press, San Diego, CA, pp. 123–165.

Selected Bibliography

For details of the deposition and diagenesis of reservoir rocks, see:

Friedman, G.M., Sanders, J.E., Kopaska-Merkel, D.C., 1992. Principles of Sedimentary Deposits. Macmillan, New York (This major text deals not only with petrography but also with wider aspects of reservoir rocks).

Lucia, F.J., 1990. Carbonate Reservoir Characterization. Springer, Berlin.

Selley, R.C., 2000. Applied Sedimentology, second ed. Academic Press, London.

For further details on the petroleum engineering aspects of reservoirs, see:

Archer, J.S., Wall, C.G., 1986. Petroleum Engineering Principles and Practice. Graham & Trotman, London.

Ashton, M. (Ed.), 1992. Advances in Reservoir Geology. Geol. Soc. London, London. Spec. Publ. No. 69.

Chierici, G.L., 1994. Principles of Petroleum Reservoir Engineering, vols 2. Springer-Verlag, Berlin (This shows how to do petroleum reservoir engineering by math alone, without complicating matters with geology).

Cubitt, J.M., England, W.A. (Eds.), 1995. The Geochemistry of Reservoirs. Geol. Soc. London, London. Spec. Publ. No. 86.

Timmerman, E.H., 1982. Practical Reservoir Engineering, vols 1 and 2. Penn Well Publishing Company, Tulsa, OK.

CHAPTER

7

Traps and Seals

7.1 INTRODUCTION

In the early days of oil exploration in the United States, no specific legislation governed the exploration and exploitation of petroleum. Initially, the courts applied the game laws, which stated that oil and gas were fugacious (likely to flee away), moving from property to property, and ultimately owned by the man on whose land they were trapped (Dott and Reynolds, 1969).

The term trap was first applied to a hydrocarbon accumulation by Orton (1889): "...stocks of oil and gas might be trapped in the summits of folds or arches found along their way to higher ground." A detailed historical account of the subsequent evolution of the concept and etymology of the term trap is found in Dott and Reynolds (1969).

As discussed in Chapter 1, a trap is one of the seven essentials requisites for a commercial accumulation of oil or gas. Levorsen (1967) gave a concise definition of a trap as "the place where oil and gas are barred from further movement." This definition needs some qualification. Explorationists in general and geophysicists in particular search for hydrocarbon traps. Perhaps it would be more accurate to say that they search for potential traps. Only after drilling and testing is it known whether the trap contains oil or gas. In other words, a trap is still a trap whether it is barren or productive.

7.2 NOMENCLATURE OF A TRAP

Many terms are used to describe the various parameters of a trap. These terms are defined as follows and illustrated with reference to an anticlinal trap, the simplest type (Fig. 7.1). The highest point of the trap is the crest, or culmination. The lowest point at which hydrocarbons may be contained in the trap is the spill point; this lies on a horizontal contour, the spill plane. The vertical distance from crest to spill plane is the closure of the trap. A trap may or may not be full to the spill plane, a point of both local and regional significance. Note that in areas of monoclinal dip the closure of a trap may not be the same as its structural relief (Fig. 7.2). This situation is particularly significant in hydrodynamic traps. The zone immediately beneath the petroleum is referred to as the bottom water, and the zone of the reservoir laterally adjacent to the trap as the edge zone (Fig. 7.1).

Elements of Petroleum Geology
http://dx.doi.org/10.1016/B978-0-12-386031-6.00007-2

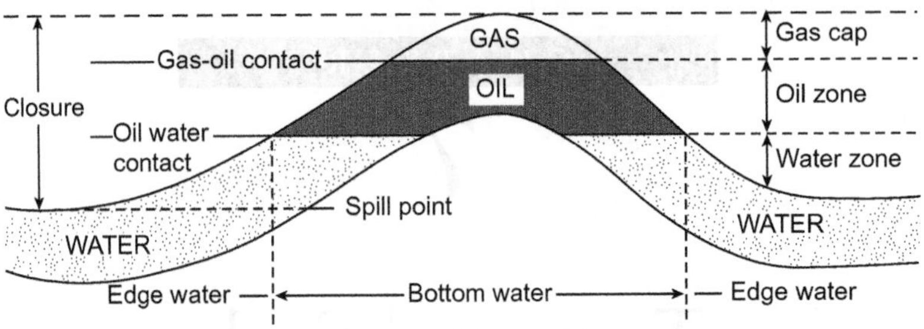

FIGURE 7.1 Cross section through a simple anticlinal trap.

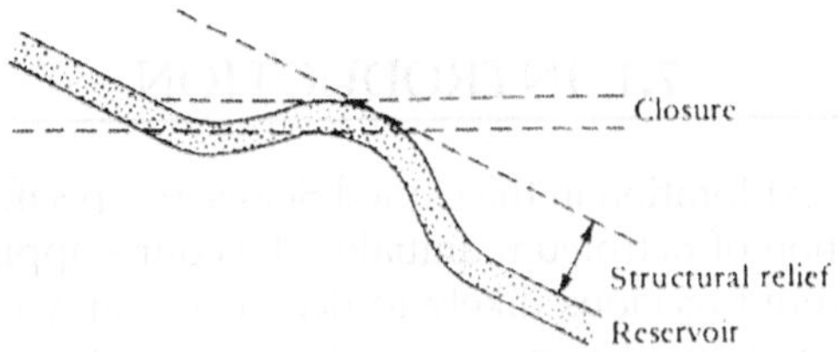

FIGURE 7.2 Cross section through a trap illustrating the difference between closure and structural relief.

Within the trap the productive reservoir is termed the pay. The vertical distance from the top of the reservoir to the petroleum/water contact is termed gross pay. This thickness may vary from only 1 or 2 m in Texas to several hundred meters in the North Sea and Middle East. All of the gross pay does not necessarily consist of productive reservoir, however, so gross pay is usually differentiated from net pay. The net pay is the cumulative vertical thickness of a reservoir from which petroleum may be produced. Development of a reservoir necessitates mapping the gross:net pay ratio across the field.

Within the geographic limits of an oil or gas field there may be one or more pools, each with its own fluid contact. Pool is an inaccurate term, dating back to journalistic fantasies of vast underground lakes of oil; nonetheless, it is widely used. Each individual pool may contain one or more pay zones (Fig. 7.3).

7.3 DISTRIBUTION OF PETROLEUM WITHIN A TRAP

A trap may contain oil, gas, or both. The oil:water contact (commonly referred to as OWC) is the deepest level of producible oil. Similarly, the gas:oil contact (GOC) or gas:water contact, as the case may be, is the lower limit of producible gas. The accurate evaluation of these surfaces is essential before the reserves of a field can be calculated, and their establishment is one of the main objectives of well logging and testing.

Where oil and gas occur together in the same trap, the gas overlies the oil because the gas has a lower density. Whether a trap contains oil and/or gas depends both on the chemistry and level of maturation of the source rock (see Chapter 5) and on the pressure and temperature of the reservoir itself. Fields with thick oil columns may show a more subtle

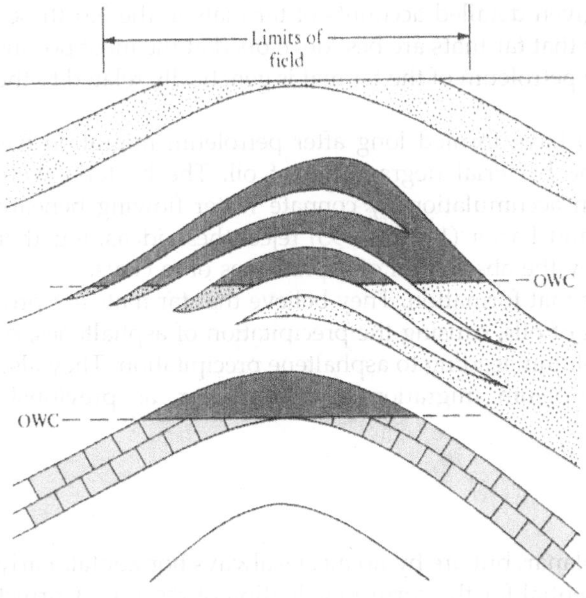

FIGURE 7.3 Cross section through a field illustrating various geological terms. This field contains two pools, that is, two separate accumulations with different oil:water contacts (OWC). In the upper pool the net pay is much less than the gross pay because of nonproductive shale layers. In the lower pool the net pay is equal to the gross pay.

gravity variation through the pay zone. Boundaries between oil, gas, and water may be sharp or transitional. Abrupt fluid contacts indicate a permeable reservoir; gradational ones indicate a low permeability with a high capillary pressure. Not only does a gross gravity separation of gas and oil occur within a reservoir, but more subtle chemical variations may also exist. The petroleum geochemistry of a trap is of concern to reservoir engineers trying to discover the most effective way of producing its petroleum. The petroleum geochemistry of a trap is of interest to petroleum explorationists, since it reveals its source and migration history, parameters that may point to the discovery of new adjacent fields (Cubitt and England, 1995).

Beyond the gross stratification of gas and oil, it is now known that petroleum layering may occur on a scale down to an order of some 10 m (Larter and Aplin, 1995). Particular interest focuses on vertical variations of the water saturation, since this obviously affects the producibility of the petroleum, and on establishing whether the water is free or bound, using tools such as NMR log discussed in Chapter 6.

7.3.1 Tar Mats

Some oil fields have a layer of heavy oil, termed a *tar mat*, immediately above the bottom water. Notable examples include fields such as Prudhoe Bay, Alaska, and Sarir, Libya (see Jones and Speers (1976) and Lewis (1990), respectively): Tar mats are also sometimes associated with late pyrite cementation. Wireline log interpretation of tar mats provides petrophysicists with some of their biggest challenges. Tar mats are very important to identify and understand because they impede the flow of water into a reservoir when the petroleum is produced.

Wilhelms and Larter (1994a,b) have given detailed accounts of tar mats in the North Sea fields and elsewhere. These studies show that tar mats are best developed at the most porous and permeable parts of the reservoir. The petroleum of the tar mat is genetically related to the supradjacent accumulation.

Traditionally tar mats are believed to have formed long after petroleum migration has ceased, and have been attributed to the bacterial degradation of oil. The bacteria were brought into contact with the petroleum accumulation by connate water flowing beneath the petroleum:water contact. Wilhelms and Larter (1994b, 1995) reject these ideas, together with the suggestion that tar mats form by the absorption of asphaltenes onto clays.

They propose two mechanisms for tar mat formation. They believe that tar mats are produced either by the thermal degradation of oils, causing the precipitation of asphaltenes, or by the increased gas solution in the oil column, leading to asphaltene precipitation. They also believe that tar mats form during petroleum migration, not long after, as previously supposed.

7.3.2 Tilted Fluid Contacts

Fluid contacts in a trap are generally planar, but are by no means always horizontal. Early recognition of a tilted fluid contact is essential for the correct evaluation of reserves. Correct identification of the cause of the tilt is necessary for the efficient production of the field. There are several causes of tilted fluid contacts. They may occur where a hydrodynamic flow of the bottom waters leads to a displacement of the hydrocarbons from a crestal to a flank position. This displacement can happen with varying degrees of severity (Fig. 7.4). The presence of this type of tilted OWC can be established from pressure data, which will show a slope in the potentiometric surface. Many, but by no means all, hydrodynamically tilted fields occur above sea level. Where these fields are shallow, the occurrence of tar mats is not unusual because of the degradation of oil due to the movement of water beneath the oil zone.

In some fields the OWC has tilted as a result of production, presumably because of fluid movement initiated by the production of oil from an adjacent field. This phenomenon has been recorded, for example, from the Cairo Field of Arkansas (Goebel, 1950). An alternative explanation for a sloping fluid contact is that a trap has been tilted after hydrocarbon invasion, and the contact has not moved. Considered on its own, this theory is unlikely because, within the geological timescale, the oil and/or gas have ample time to adjust to a new horizontal level.

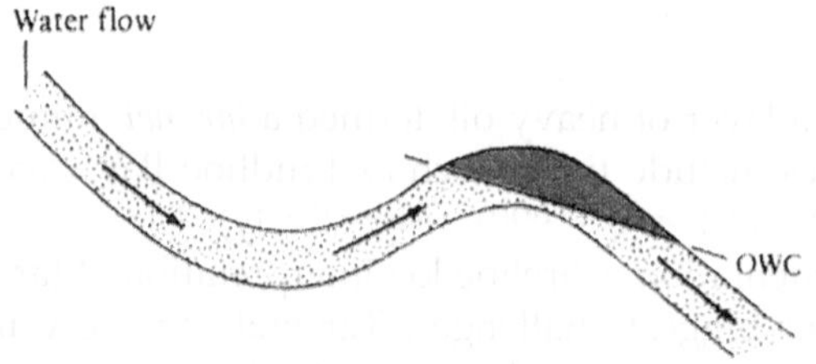

FIGURE 7.4 Cross section through a trap showing tilted oil:water contact (OWC) due to hydrodynamic flow.

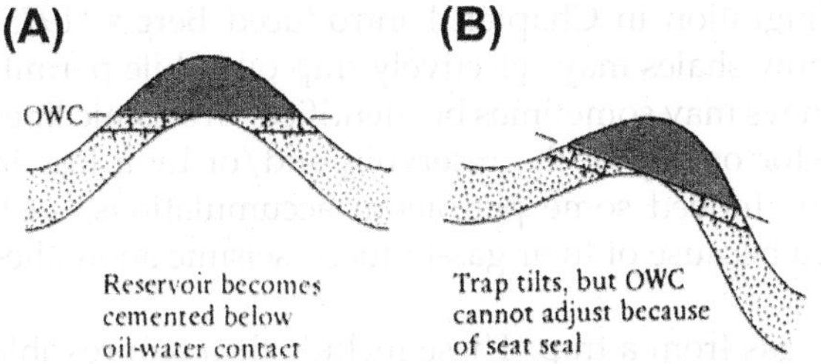

FIGURE 7.5 Cross section through a trap showing how a tilted oil:water contact (OWC) may be caused by (A) cementation of the water zone followed by (B) tilting.

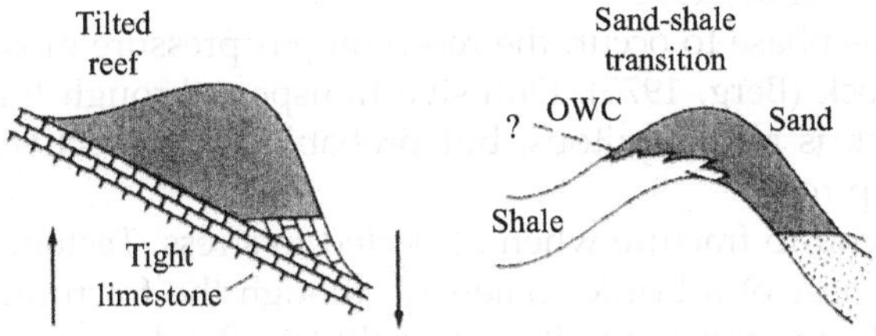

FIGURE 7.6 Cross sections through traps showing how apparently tilted oil:water contacts (OWC) may be caused by facies changes.

A tar mat may decrease permeability to such an extent that if a trap is tilted, the OWC may be unable to adjust to the new horizontal datum. Alternatively, cementation can continue in the reservoir beneath an oil zone while it is halted in the trap itself. This cementation seldom provides a seat seal sufficiently tight to restrain the hydrocarbons from later movement. It may, however, diminish permeability sufficiently to provide some interesting production problems: the updip edge of the field having anomalously low permeability, and the water zone on the downdip side having anomalously high permeability (Fig. 7.5).

A third possible cause of a tilted OWC may be a change in facies. Theoretically, a change in grain size across the reservoir causes a tilted contact. With declining grain size, capillary pressure increases, allowing a rise in the OWC. In practice, the effect of this rise is likely to be negligible, at the most only a few meters (Yuster, 1953). On the other hand, where the change is actually lithological, the lower contact of the field may be tilted. In this situation the underlying lithology, either a shale or basement, is generally impermeable. In such cases, the tilted lower surface of the reservoir is not truly an OWC, but a seat seal (Fig. 7.6).

7.4 SEALS AND CAP ROCKS

For a trap to have integrity it must be overlain by an effective seal. Any rock may act as a seal as long as it is impermeable (Downey, 1994). Seals will commonly be porous, and may in fact be petroleum saturated, but they must not permit the vertical migration of petroleum from the trap. Shales are the commonest seals, but evaporites are the most effective. Shales are commonly porous, but because of their fine grain size have very high capillary forces that prevent fluid flow.

The discussion on primary petroleum migration in Chapter 4 introduced Berg's (1975) concept of the capillary seal. Berg showed how shales may selectively trap oil, while permitting the upward migration of gas. Gas chimneys may sometimes be identified on seismic lines either by a velocity pull-down of the reflector on top of the reservoir, and/or by a loss in seismic character in the overlying reflectors. Indeed some petroleum accumulations, such as the Ekofisk Field, are sometimes identified because of their gas-induced seismic anomalies (Van den Berg and Thomas, 1980).

Several mechanisms allow the leakage of gas from a trap. These include the compressible Darcy flow of a free gas phase, the transport of dissolved gas in aqueous solution along adjacent aquifers under hydrodynamic conditions and the diffusive transport through the water-saturated pore space of the cap rock (Krooss et al., 1992).

For compressible Darcy flow of a free gas phase to occur, the reservoir gas pressure must exceed the capillary pressure in the cap rock (Berg, 1975). Diffusive transport through the water-saturated pore space of the cap rock is an ubiquitous, but probably least effective, method of gas transportation through a cap rock.

With increasing induration shales will tend to fracture when subjected to stress. Tectonic movements may thus destroy the effectiveness of a brittle shale seal, though the fractured shale may then, of course, serve as a petroleum reservoir in its own right (see the discussion of shale gas in Chapter 9).

7.5 CLASSIFICATION OF TRAPS

Hydrocarbons may be trapped in many different ways. Several schemes have been drawn up to attempt to classify traps (e.g., Clapp, 1910, 1929; Lovely, 1943; Hobson and Tiratsoo, 1975; Biddle and Wielchowsky, 1994). Most trap classificatory schemes are based on the geometry of the trap, but Milton and Bertram (1992) use the seal as the classificatory parameter. Two major genetic groups of trap are generally agreed on: structural and stratigraphic. A third group, combination traps, is caused by a combination of processes. Agreement breaks down, however, when attempts are made to subdivide these groups.

Table 7.1 presents a classification of hydrocarbon traps. The table is based on information previously cited in this chapter, and can only be regarded as a crude attempt to pigeonhole such truly fugacious entities as traps. The table has no intrinsic merit other than to provide a framework for the following descriptions of the various types of hydrocarbon traps.

Structural traps are those traps whose geometry was formed by tectonic processes after the deposition of the beds involved. According to Levorsen (1967), a structural trap is "one whose upper boundary has been made concave, as viewed from below by some local deformation, such as folding, or faulting, or both, of the reservoir rock. The edges of a pool occurring in a structural trap are determined wholly, or in part, by the intersection of the underlying water table with the roof rock overlying the deformed reservoir rock." Basically, therefore, structural traps are caused by folding and faulting.

A second group of traps is caused by diapirs, where salt or mud have moved upward and domed the overlying strata, causing many individual types of trap. Arguably, diapiric traps are a variety of structural traps; but since they are caused by local lithostatic movement, not regional tectonic forces, they should perhaps be differentiated.

TABLE 7.1 Crude Classification of Hydrocarbon Traps Based on Previous Schemes Cited in the Text

I	Structural traps—caused by tectonic processes
	Fold traps: Compressional anticlines; Compactional anticlines
	Fault traps
II	Diapiric traps—caused by flow due to density contrasts between strata
	Salt diapirs
	Mud diapirs
III	Stratigraphic traps—caused by depositional morphology or diagenesis (for detailed classification see Table 7.2.)
IV	Hydrodynamic traps—caused by water flow
V	Combination traps—caused by a combination of two or more of the above processes

Stratigraphic traps are those traps whose geometry is formed by changes in lithology. The lithological variations may be depositional (e.g., channels, reefs, and bars) or postdepositional (e.g., truncations and diagenetic changes). Hydrodynamic traps occur where the downward movement of water prevents the upward movement of oil, thus trapping the oil without normal structural or stratigraphic closure. Such traps are rare. The final group, combination traps, is formed by a combination of two or more of the previously defined genetic processes.

The various types of traps—structural, diapiric, stratigraphic, hydrodynamic, and combination—are described at greater length and illustrated with examples in the following sections.

7.6 STRUCTURAL TRAPS

As previously stated, the geometry of structural traps is formed by postdepositional tectonic modification of the reservoir. Table 7.1 divides structural traps into those caused by folding and those caused by faulting. These two classifications are now considered in turn.

7.6.1 Anticlinal Traps

Anticlinal, or fold, traps may be subdivided into two classes: compressional anticlines (caused by crustal shortening) and compactional anticlines (developed in response to crustal tension).

7.6.1.1 Compressional Anticlines

Anticlinal traps caused by compression are most likely to be found in, or adjacent to, subductive troughs, where there is a net shortening of the earth's crust. Thus, fields in such traps are found within, and adjacent to, mountain chains in many parts of the world.

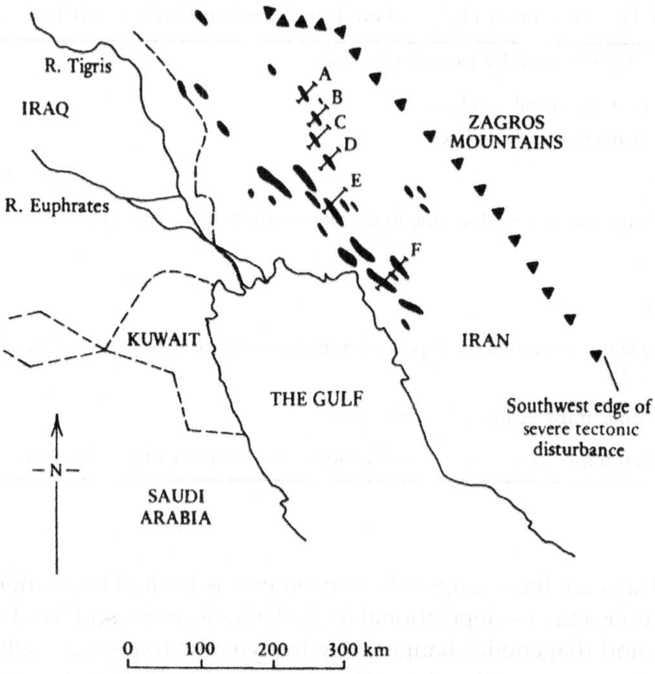

FIGURE 7.7 Map showing the location of the folded anticlinal traps of Iran. For cross section see Fig. 7.8.

One of the best known oil provinces with production from compressional anticlines occurs in Iran (Fig. 7.7). Many such fields are found in the foothills of the Zagros Mountains. Sixteen of these fields are in the "giant" category, with reserves of more than 500 million barrels of recoverable oil or 3.5 trillion cubic feet of recoverable gas (Halbouty et al., 1970). These fields have been described in considerable detail over the years (Lees, 1952; Falcon, 1958, 1969; Slinger and Crichton, 1959; Hull and Warman, 1970; Colmann-Sadd, 1978). The main producing horizon is the Asmari limestone (Lower Miocene), a reservoir with extensive fracture porosity. Flow rates and productivity are immense, with some individual wells having flowed 50 million barrels (No. 7-7, Masjid-i-Suleiman Field). The cap rock is provided by evaporites of the lower Fars Group (Miocene), whose disharmonic folding makes it difficult to extrapolate from the surface to the reservoirs. The traps themselves lie to the southwest of the main Zagros Mountain thrust belt. Individual anticlines are up to 60 km in length and some 10–15 km wide (Fig. 7.8). The axial planes of the folds pass downward into thrust faults, which die out in a zone of decollement within the underlying Hormuz salt (Precambrian?).

A second major hydrocarbon province that contains compressional anticlinal traps occurs in the Tertiary basins of California. Here, a number of fault-bounded troughs are infilled by thick regressive sequences in which organic-rich basinal muds are overlain by turbidites capped by younger continental beds. These sediments have locally undergone tight compressive folding associated with the transcurrent movement of the San Andreas Fault system (Barbat, 1958; Schwade et al., 1958; Simonson, 1958). Many of the fields are associated with faulting:

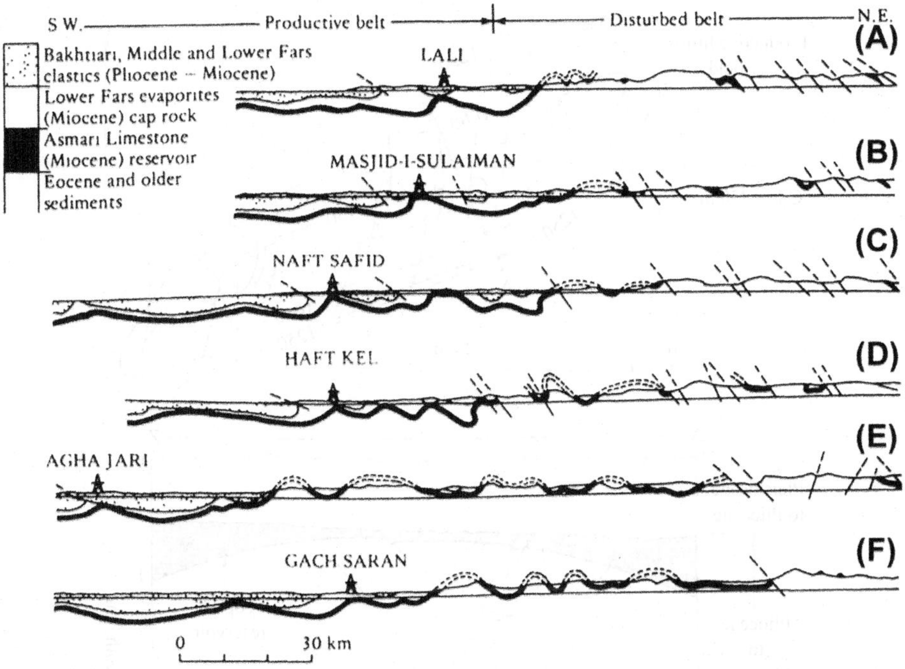

FIGURE 7.8 Southwest to northeast true-scale cross sections through some of the folded structures of Iran. Locations shown in Fig. 7.7. *After Falcon (1958), reprinted by permission of the American Association of Petroleum Geologists.*

normal, reversed, and strike slip (Fig. 7.9). The Long Beach-Wilmington Field of the Los Angeles Basin is a giant field in a compressional anticline, cross-cut by normal faults perpendicular to the fold axis (Mayuga, 1970).

Folds are often involved in thrusting within the mountain chains themselves. Hydrocarbons may be trapped in anticlines above thrust planes and in reservoirs sealed beneath the thrust. A major play of this type occurs in the eastern Rocky Mountains, including the Turner Valley field of Alberta and the Painter Valley Reservoir field of Wyoming (Fig. 7.10). Such fields are extremely difficult to find and develop because of the problems of seismic interpretation due to complex faulting and steeply dipping beds. With recent improvements in seismic technology, however, this task is becoming easier, opening up previously neglected areas to exploration.

7.6.1.2 Compactional Anticlines

A second major group of anticlinal traps is formed not by compression but by crustal tension. Where crustal tension causes a sedimentary basin to form, the floor is commonly split into a mosaic of basement horsts and grabens. The initial phase of deposition infills this irregular topography. Throughout the history of the basin, the initial structural architecture usually persists, controlling subsequent sedimentation. Thus anticlines may occur in the sediment cover above deep-seated horsts (Fig. 7.11). Closure may be enhanced both by compaction and sedimentation. Differential sedimentation of clays increases

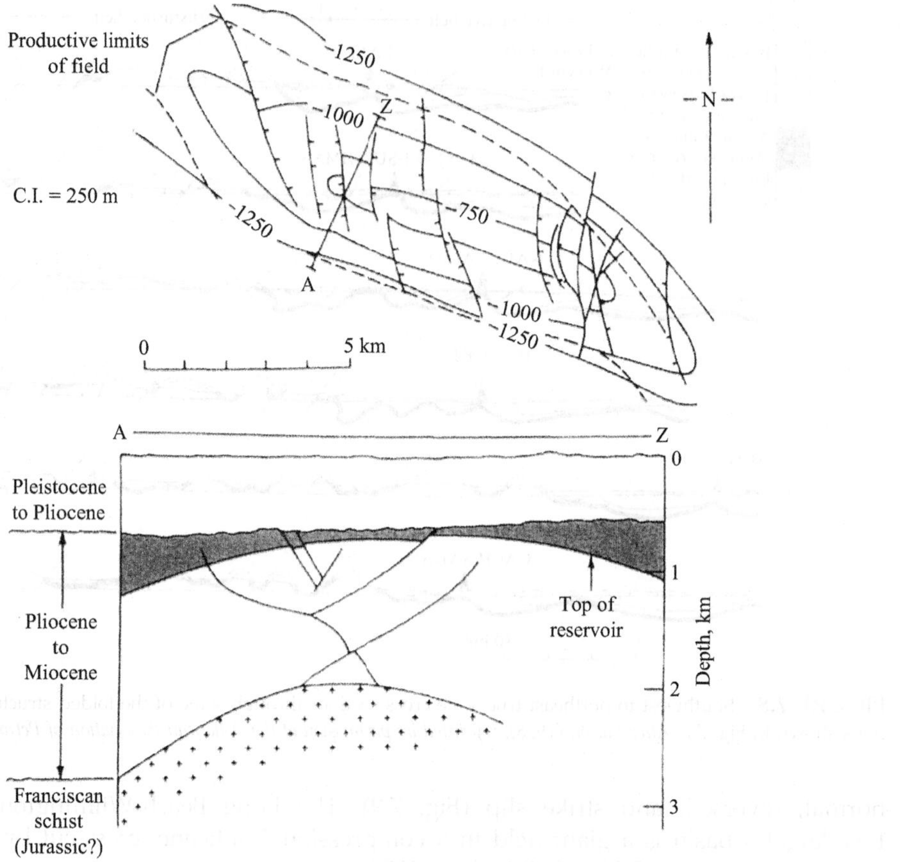

FIGURE 7.9 The Wilmington oil field, Long Beach, California. (Upper) Structure contour map on top of the Ranger Zone, just below the crest of the reservoir. (Lower) Southwest–northeast cross section along the line A–Z. *Modified from Mayuga (1970), reprinted by permission of the American Association of Petroleum Geologists.*

the amplitude of the fold because, although the percentage of compaction is constant for crest and trough, the actual amount of compaction is greater for the thicker flank sediment (Fig. 7.12).

Differential depositional rates also enhance structural closure. Carbonate sedimentation tends to be higher in shallow, rather than deep, water; so shoal and reefal facies may pass off-structure into thinner increments of basinal lime mud. Similarly, terrigenous shoal sands may develop on the crests of structures and pass down flank into deeper water muds. Thus, reservoir quality often diminishes down the flank of such structures.

Good examples of oil fields trapped in compactional anticlines occur in the North Sea. Here, Paleocene deep-sea sands are draped over Mesozoic horsts (Blair, 1975). These fields include the Forties, Montrose, Maureen, and East Frigg fields (Fig. 7.13).

The traps of compactional and compressional anticlines are very different. As just discussed, compactional folds may have considerable variations in reservoir facies across

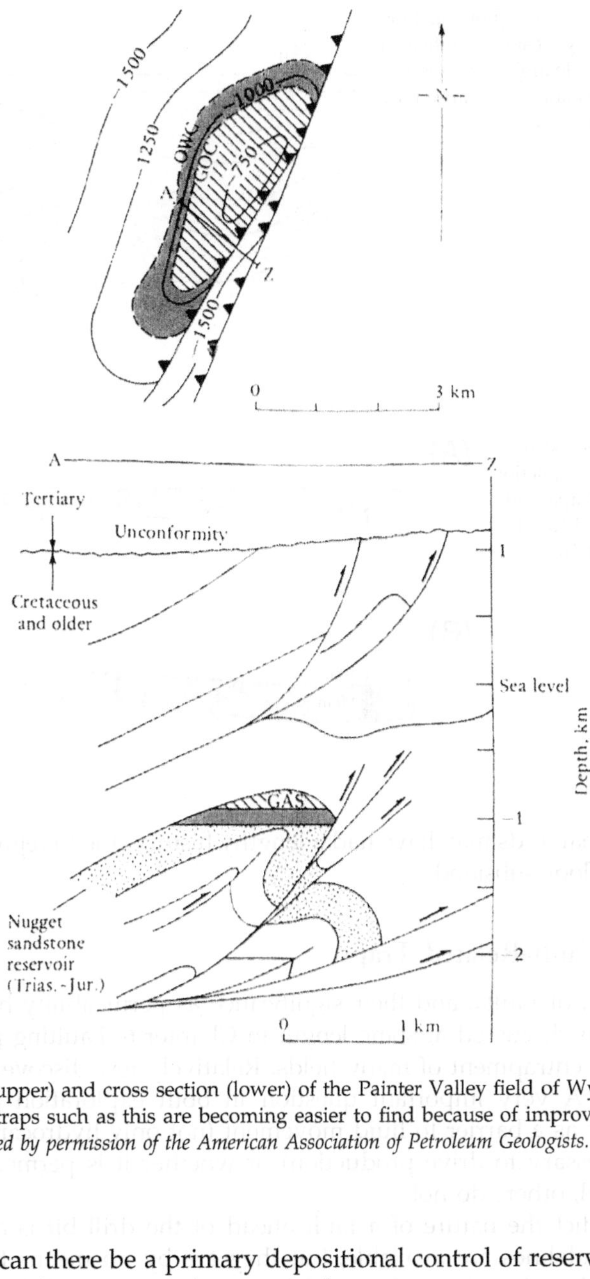

FIGURE 7.10 Map (upper) and cross section (lower) of the Painter Valley field of Wyoming. Thrust-associated compressional anticline traps such as this are becoming easier to find because of improving seismic data. *Modified from Lamb (1980), reprinted by permission of the American Association of Petroleum Geologists.*

structure. Not only can there be a primary depositional control of reservoir quality, but later diagenetic changes can be extensive, because such structures are prone to subaerial exposure, leaching, and, in extreme cases, truncation.

Whereas compressional folds are generally elongated perpendicular to the axis of crustal shortening, compactional folds are irregularly shaped, reflecting the intersection of fault trends in the basement. Compressional folds generally form in one major tectonic event,

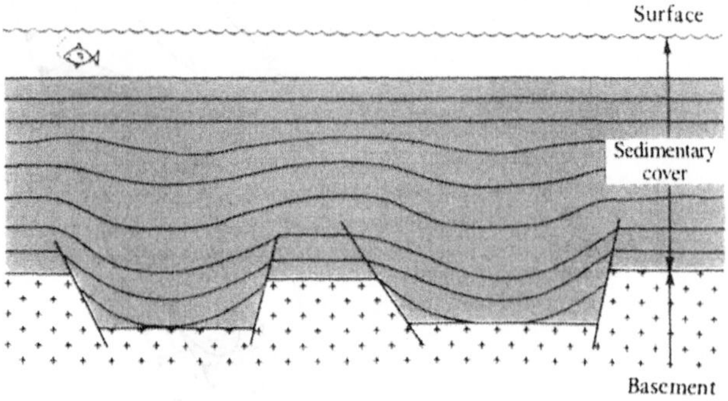

FIGURE 7.11 Cross section showing how basement block faulting causes anticlinal structures in sediments; closure decreases upward. These drape anticlines are caused by tension rather than compression.

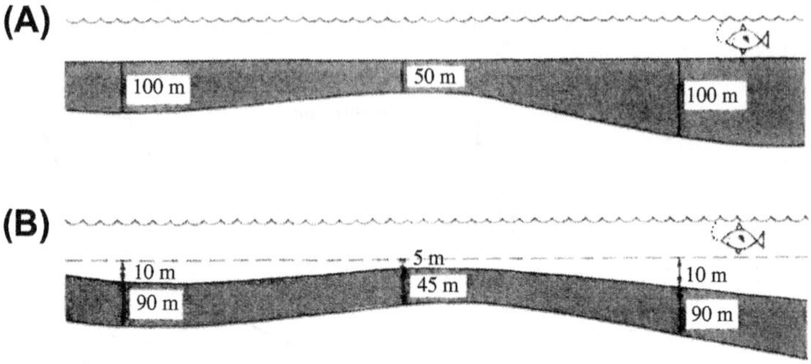

FIGURE 7.12 Cross sections showing how burial compaction enhances closure on drape anticlines. (A) Deposition and (B) after a uniform compaction of 10%.

whereas compactional folds may have had a lengthy history due to rejuvenation of basement faults as the basin floor subsided.

7.6.2 Fault and Fault-Related Traps

The identification of faults, and their significance as permeability barriers within petroleum reservoirs, was discussed at some length in Chapter 6. Faulting plays an indirect but essential role in the entrapment of many fields. Relatively few discovered fields are caused solely by faulting. A very important question in both exploration and development is whether a fault acts as a barrier to fluid movement (not only hydrocarbons but also water, which may be necessary to drive production) or whether it is permeable. The problem is that some faults seal, others do not.

Attempts to predict the nature of a fault ahead of the drill bit is a very active area of research. A few guidelines are available, but they are by no means foolproof. Where the throw of the fault is less than the thickness of the reservoir, it is unlikely to seal. Faults in brittle rocks are less likely to seal than those in plastic rocks. In lithified rocks faults may be accompanied by extensive fracturing, which may be permeable; indeed, some fields are caused solely by fracture porosity adjacent to a fault (refer back to Fig. 6.8). Sometimes, however, fractures may have undergone later cementation.

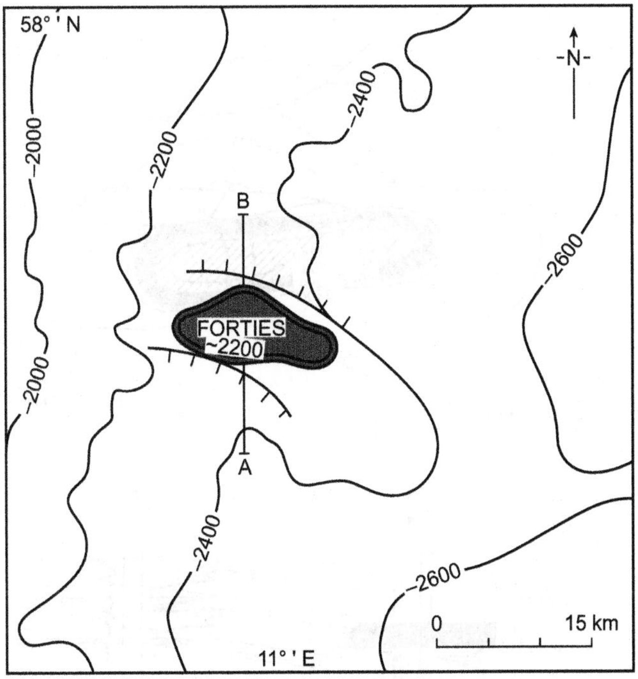

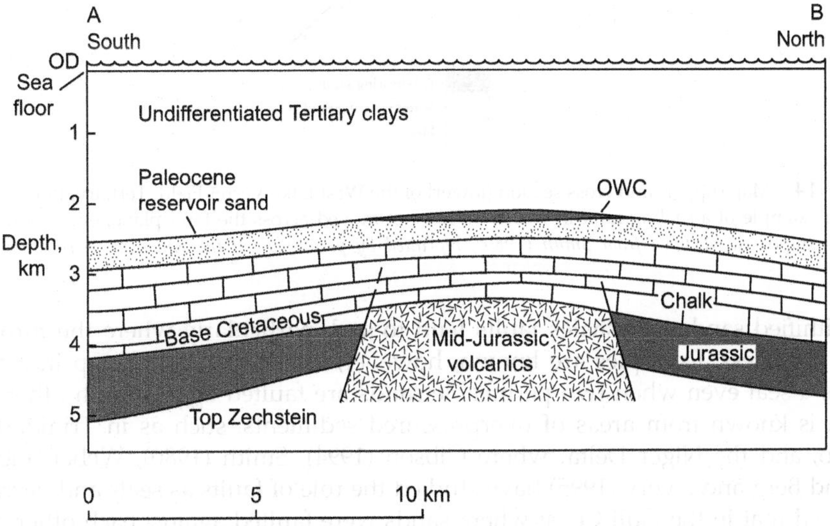

FIGURE 7.13 Map (upper) and cross section (lower) of the Forties field of the North Sea. This field is essentially a compactional anticline draped over an old basement high.

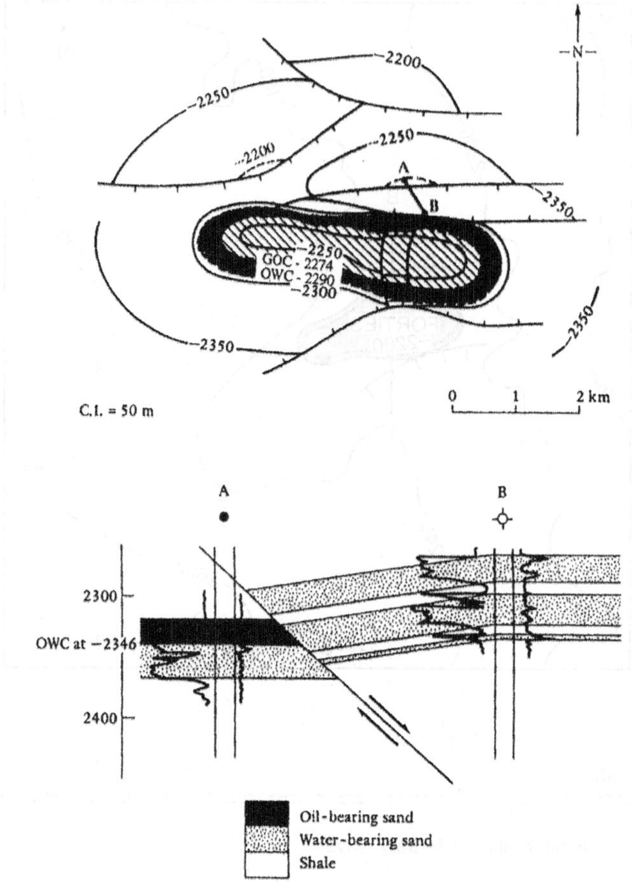

FIGURE 7.14 Map (upper) and cross section (lower) of the West Lake Verret Field, Tertiary, Louisiana. This field provides an example of a sealing fault in which oil has not moved across the fault plane, even though permeable sands are juxtaposed. *Modified from Smith (1980), reprinted by permission of the American Association of Petroleum Geologists.*

In unlithified sands and shales faults tend to seal, particularly where the throw exceeds reservoir thickness. Examples are known, however, where clay caught up in a fault plane can act as a seal even when two permeable sands are faulted against each other. This phenomenon is known from areas of overpressured sediments, such as in Trinidad, the Gulf of Mexico, and the Niger Delta, where Gibson (1994), Smith (1980), Weber and Daukoru (1975), and Berg and Avery (1995) have studied the role of faults as seals and barriers. Smith (1980) noted that in the Gulf Coast where sands were faulted against each other, the probability of the fault sealing increased with the age difference of the two sands. Figure 7.14 shows cross sections of the West Lake Verret Field in which hydrocarbon and water-bearing sands are faulted against one another. The faults do not separate sands of different facies or capillary displacement pressure, so Smith (1980) concluded that the faults sealed by virtue of impermeable material smeared along the fault planes. Particular attention is paid to the

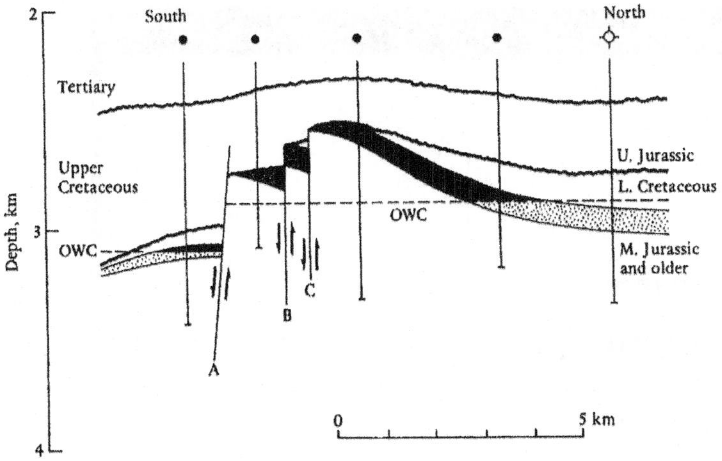

FIGURE 7.15 Cross section through the Piper Field of the North Sea showing how within the same field (A) some faults seal and (B and C) others do not. *After Williams et al. (1975).*

sand:shale ratio of the section that is faulted. This may be used in probabilistic estimations of whether faults are open or closed (Gibson, 1994). Figure 7.15 shows a complexly faulted field in which the OWCs indicate that some faults are conduits, whereas others are seals.

Because of the importance of predicting the extent to which permeable reservoir beds are juxtaposed or offset by faults, sophisticated computer packages are now available to image such scenarios (Fig. 7.16).

Bailey and Stoneley (1981) have shown that there are eight theoretical geometries for fault traps, assuming that faults do not separate juxtaposed permeable beds (Fig. 7.17). Six of these geometries may be valid traps provided that there is also closure in both directions parallel to the fault plane. Although many fields are trapped by a combination of faulting and other features, pure fault traps are rare. Figure 7.18 illustrates one example of a simple faulted trap. Fault and fault-related traps may be conveniently categorized according to whether transverse or tensional forces operate.

7.6.2.1 Traps Related to Transverse Faults

Transverse faults give rise to several distinctive types of petroleum traps (Wilcox et al., 1973). Transverse movement of basement blocks takes place along wrench faults. These movements are expressed in the overlying sediment cover in a trend of en echelon folds. Fold axes are oblique to the wrench fault and indicate its direction of movement. In some instances, fold axes may be offset by faults (Fig. 7.19). In cross section, wrench faults split upward into low-angle faults. This phenomenon is sometimes referred to *as flower structure* (Gregory, in Harding and Lowell (1979)). The Newport-Inglewood fault in the Ventura basin of California provides a classic example of petroleum entrapment in en echelon folds developed along a wrench fault (Harding, 1973).

Few wrench faults have no vertical movement. Indeed, sedimentary basins often form by rapid subsidence of one side of a wrench fault. In such settings, the folds may be present only on the basinward side of the fault and may form structural noses where they are truncated (Fig. 7.19). Examples of this type of trap occur where the southwestern flank of the San Joaquin Basin is truncated by the San Andreas Fault, California (Harding, 1974).

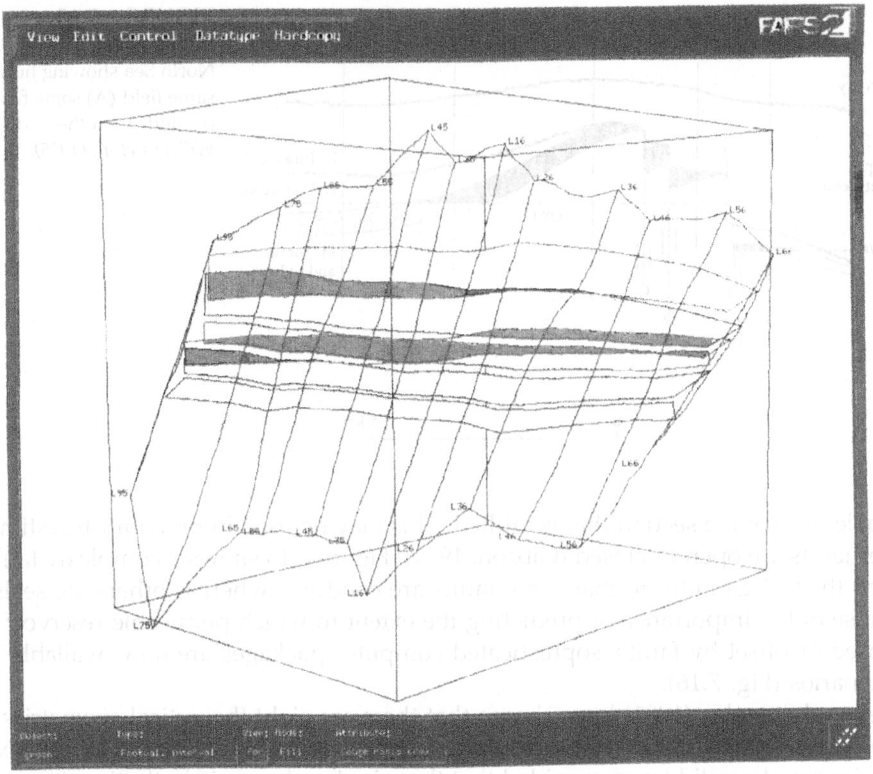

FIGURE 7.16 Computer-generated illustration to show where permeable beds are in communication across a fault plane. This is an example of the FAPS software used for the 3D display and analysis of faults developed by Badley Earth Sciences Ltd. in collaboration with the Fault Analysis Group, University of Liverpool, United Kingdom.

7.6.2.2 Traps Related to Tensional Faults

A particularly important group of traps is associated with tensional faults. In some faults the throw increases incrementally downward. This demonstrates that fault movement was synchronous with deposition. Such faults are thus termed growth faults. In plan view the fault trace is frequently curved and when viewed from the downthrown side, it is concave. Similarly, in profile the fault angle diminishes downward.

There are two types of growth fault. One type is basement-related and tends to be long in plan view and to have had a lengthy history controlling facies and sedimentation rate (Shelton, 1968). A good example of this type occurs on the flank of the Sarir Field in Libya (Fig. 7.20). One of the best known large-scale growth faults is the Vicksburg flexure, which extends for some 500 km around the Gulf Coast of Texas. The maximum increase in sediment across the fault is of the order of 1500 m near the Mexican border. Most of this thickening occurs in the Vicksburg Group (Oligocene); considerably less thickening occurs in the overlying Frio Group (Miocene).

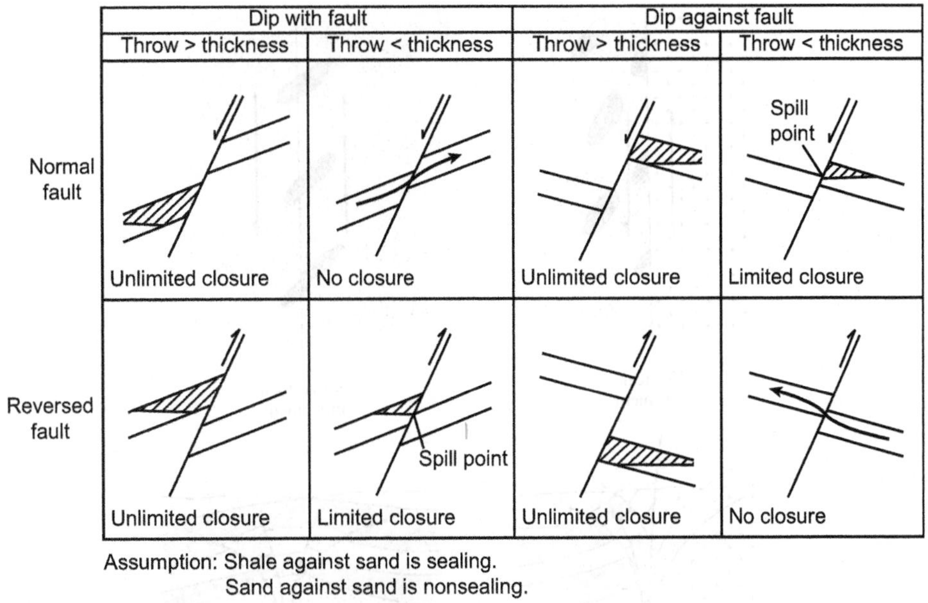

FIGURE 7.17 The eight theoretical configurations of petroleum traps associated with faulting. These configurations are drawn on the assumption that oil can move across, but not up, the fault plane when permeable sands are juxtaposed. *After Bailey and Stoneley (1981), reprinted with permission from Black-well Science Ltd.*

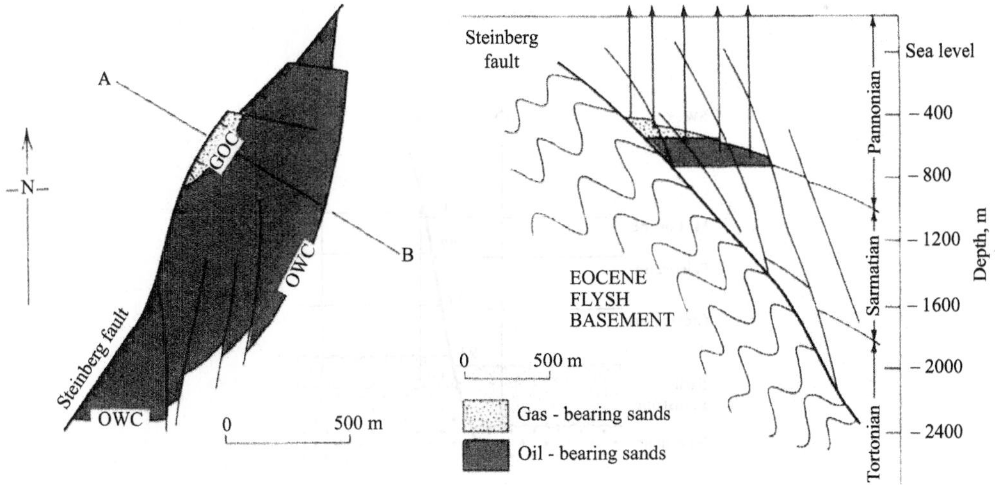

FIGURE 7.18 Map (left) and cross section (right) through a fault trap in the Gaiselberg Field, Austria. *Modified from Janoschek (1958), reprinted by permission of the American Association of Petroleum Geologists.*

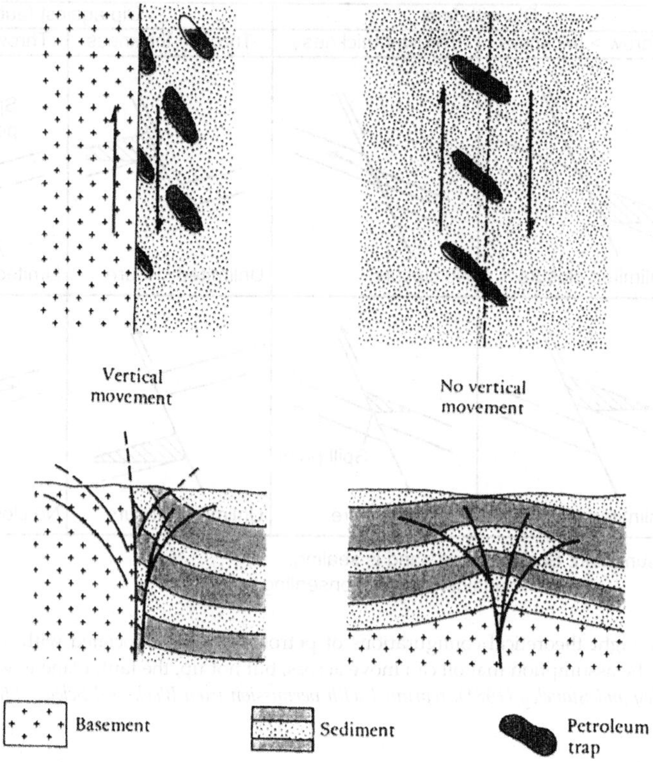

FIGURE 7.19 Maps (upper) and cross sections (lower) showing the types of petroleum traps associated with wrench faults. For examples and explanation see text.

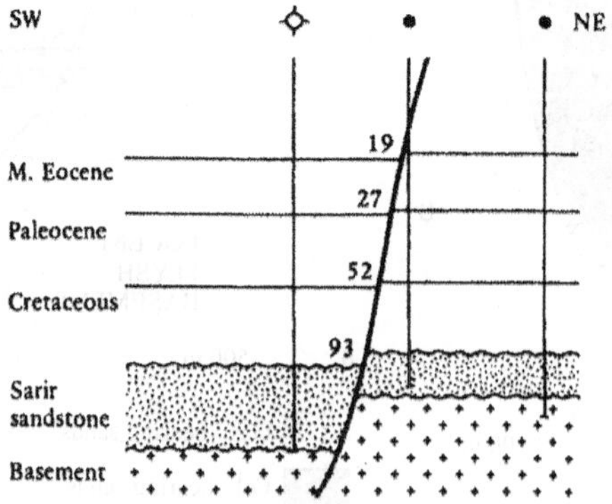

FIGURE 7.20 Cross section through a basement-related growth fault in the Sarir Field, Libya. The numbers show the throw (in meters) of the fault for various markers and demonstrate an incremental increase downward. *Modified from Sanford (1970), reprinted by permission of the American Association of Petroleum Geologists.*

Not only do formations thicken across the fault toward the Gulf but the percentage of net sand increases too. This suggests that subsidence on the downthrown side of the fault formed a natural sediment trap. Characteristically, there is a local reversal of the easterly regional dip of strata adjacent to the fault plane, with rollover anticlines developed on its downthrown side (Fig. 7.21). The strata dip toward the fault to fill the space caused by separation along the plane of the fault. The dip reversal of the anticlines is often enhanced by antithetic faults downthrown and dipping in toward the major fault.

Oil and gas are trapped both in the rollover anticlines and in sand pinchout stratigraphic traps on both upthrown and downthrown sides of the Vicksburg Fault (Halbouty, 1972). It has been estimated that some 3 billion barrels of oil and 20 trillion cubic feet of gas are trapped adjacent to the Vicksburg flexure, partly in pure fault traps, but more commonly in rollover anticlines and pinchouts.

Traced eastward into Louisiana, the Vicksburg Fault disappears beneath the clastic wedge of the Mississippi Delta. Growth-fault-related traps dominate this hydrocarbon province too, but they are different in scale and genesis. Individual faults are seldom more than a few kilometers in length; although with their curved traces, scalloped fault patterns can also occur. These faults are not basement-related, but pass downward to die out as horizontal shear planes either in overpressured shales or in the deeper Louann Salt (Triassic–Jurassic).

Similarly, in the Tertiary sediments of the Niger Delta, growth faults play a major role in the migration and entrapment of oil and gas (Evamy et al., 1978). A detailed analysis by Weber et al. (1980) was based on observation of the fault-associated fields and laboratory experiments. They concluded that in most instances growth faults were not sealing because they were associated not only with clay gangue but also with slivers of permeable sand. Migration of hydrocarbons appears to occur both up the fault plane and across it, from overpressured mature source shales into normally pressured (or at least lower-pressured) sands. In the upper part of a rollover anticline, traps may be filled to the spill point; but where sands are faulted against overpressured shale, traps may be filled to below the spill point (Fig. 7.22).

7.6.3 Relationship between Structural Traps and Tectonic Setting

The classification and account of structural traps just given are essentially descriptive and static; that is, the traps were not examined in their tectonic context. Structural traps do not occur at random. The types and distribution of structural traps are closely related to the regional tectonic setting and history of the region in which they are found.

A genetic classification of structural traps has been developed by Harding and Lowell (1979). They note that structural traps can be grouped according to the tectonic forces operating, according to whether the basement or only the detached cover is involved, and according to their habitat with respect to tectonic plates. This last classification is covered in Chapter 8. Table 7.3 presents a genetic grouping of structural traps. It shows that a major distinction is made between regions where basement and cover are attached and regions where basement and cover are detached (generally because of the presence of intervening evaporites or overpressured clays). Compressive and tensional forces can operate in both situations; the former occur at convergent plate boundaries, the latter at divergent ones. This type of genetic grouping of structural traps is more useful than a purely descriptive one. It is an exploration tool that predicts the type of structural trap to be anticipated in any given tectonic setting.

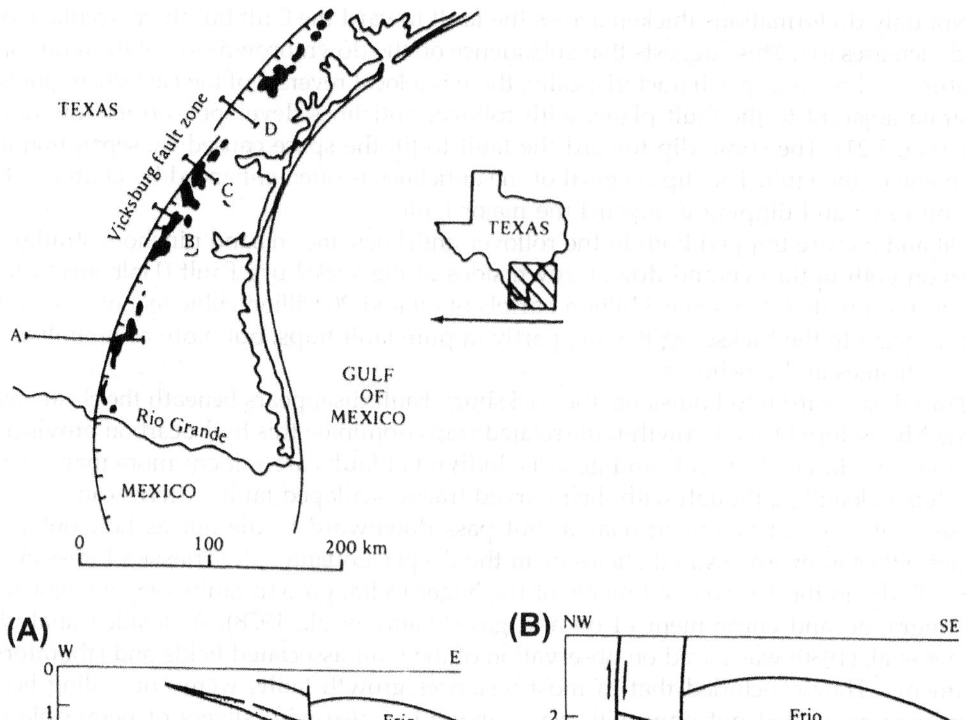

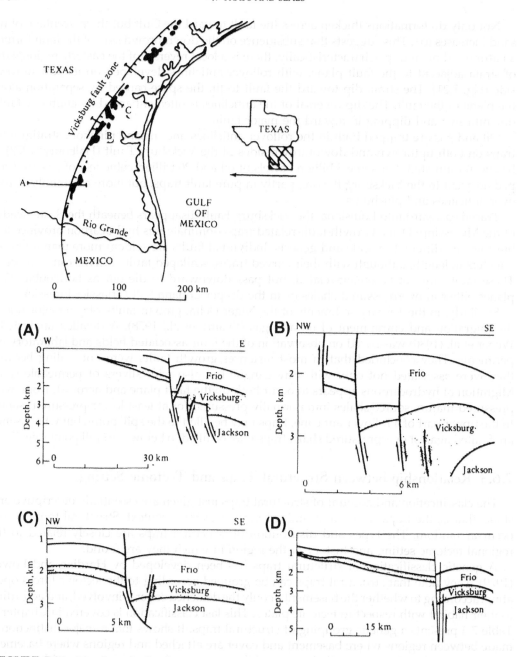

FIGURE 7.21 Map (upper) of and cross sections (lower) through the Vicksburg growth fault of South Texas. Traps associated with this structure contain estimated reserves of 3 billion barrels of oil and 20 trillion cubic feet of gas. Note how formations thicken basinward across the fault, the rollover anticlines and antithetic minor faults dipping toward the major fault. Production comes from sands in both the Vicksburg (Oligocene) and Frio (Eocene) Groups. The locations of cross-sections A, B, C & D are shown on the map. *Modified from Stanley (1970), reprinted by permission of the American Association of Petroleum Geologists.*

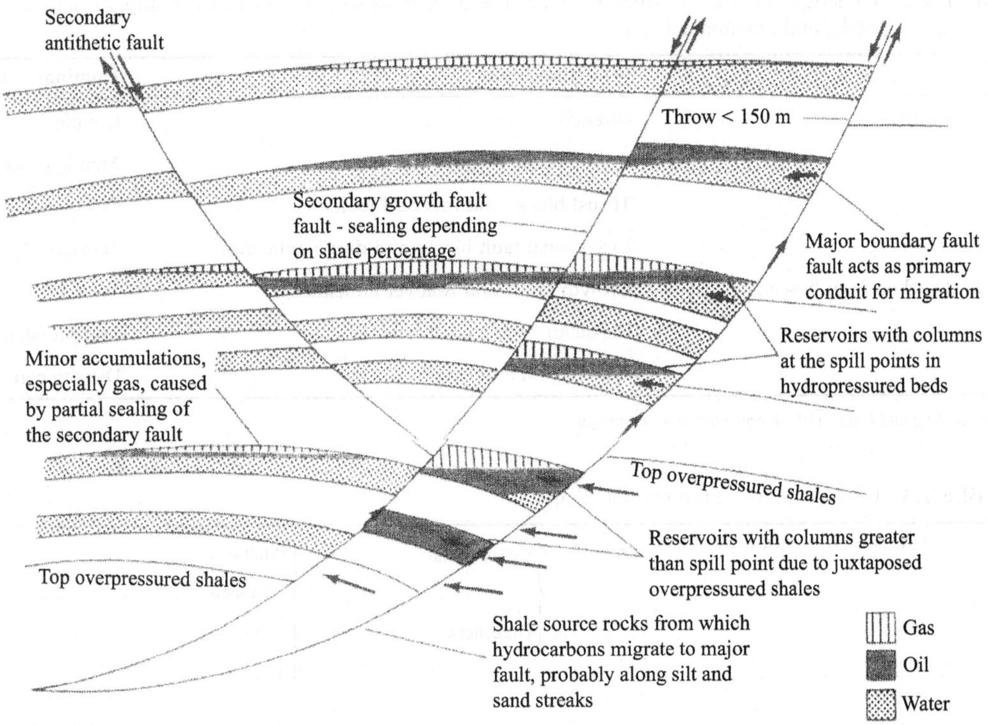

FIGURE 7.22 The mode of accumulation of oil and gas in Niger Delta growth fault traps. Unlike basement-related growth faults, this type shears out horizontally into overpressured shales. *After Weber et al. (1980).*

7.7 DIAPIRIC TRAPS

Diapiric traps are produced by the upward movement of sediments that are less dense than those overlying them. In this situation the sediments tend to move upward diapirically and, in so doing, may form diverse hydrocarbon traps. Such traps cannot be regarded as true structural traps, since tectonic forces are not required to initiate them (although in some cases they may do so). Similarly, diapirically related traps are not initiated by stratigraphic processes, although in some cases they may be caused by depositional changes across the structure. Diapiric traps are generally caused by the upward movement of salt or, less frequently, overpressured clay.

7.7.1 Salt Domes

Salt has a density of about 2.03 g/cm^3. Recently deposited clay and sand have densities less than that of salt. As the clay and sand are buried, however, they compact, losing porosity and gaining density. Ultimately, a burial depth is reached when sediments are denser than salt. Depending on a number of variables, this point may occur between about 800 and

TABLE 7.2 Genetic Grouping of Structural Traps Based on Basement–Cover Relationship, Structural Style, and Dominant Force

	Structural Style	Dominant Force
Basement involved	Wrench Fault	Couple
	Regional paleohigh	Mantle processes
	Thrust blocks and reversed faults	Compression
	Extensional fault blocks and drape anticlines	Tension
Cover detached from basement	Growth faults and rollover anticlines	
	Decollement-related structures	Compression
	Diapirs (salt and clay)	Density contrast

From Harding and Lowell (1979), reprinted with permission.

TABLE 7.3 Classification of Stratigraphic Traps

I.	Unassociated with unconformities	Depositional	Pinchouts
			Channels
		Diagenetic	Bars
			Reefs
			Porosity and/or permeability transition
II.	Associated with unconformities	Supraunconformity	On lap
			Strike valley
		Subunconformity	Channel
			Truncation

From Rittenhouse (1972), reprinted with permission.

1200 m (Fig. 7.23). When this point is reached, the salt will tend to flow up through the denser overburden. This movement may be triggered tectonically, and the resultant structures may show some structural alignment. In other instances, however, the salt movement is apparently random. The exact mechanics of diapiric movement has been studied by observation, experiment, and mathematical calculation over many years (e.g., Halbouty, 1967; Berner et al., 1972; Bishop, 1978; Alsop et al., 1995; Jackson et al., 1996).

In some salt structures the overlying strata are only updomed, whereas in others the salt actually intrudes its way upward; the latter are referred to as piercement structures. In some instances the salt may actually reach the surface, forming solution sinks in humid climates and salt glaciers in arid climates (like Iran) (Kent, 1979). Salt movement, or *halokinesis*, plays an important role in the entrapment of oil and gas in the US Gulf Coast, Iran, and the Arabian Gulf and the North Sea. Oil and gas may be trapped by salt movement in many ways (Fig. 7.24). In the simplest cases subcircular anticlines may trap hydrocarbon over the crest of a salt dome. Notable examples of this type include Ekofisk and associated fields of offshore

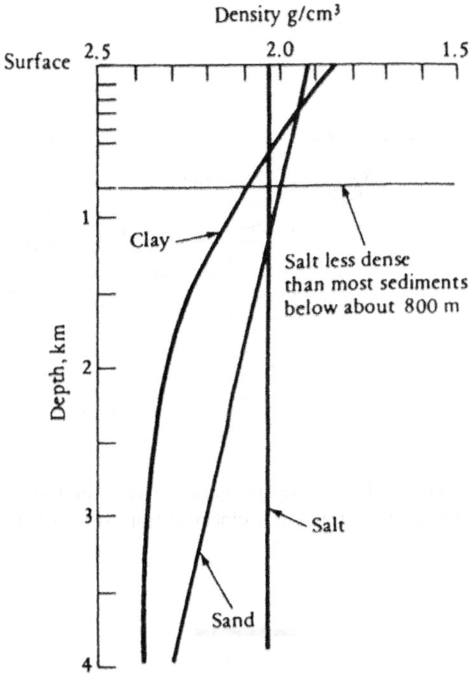

FIGURE 7.23 Density–depth curves for sand, clay, and salt. The graph shows that salt is less dense than other sediments below about 800 m, and salt movement may therefore be anticipated once this burial depth has been reached.

Norway and Denmark (Fig. 7.25). The crestal dome may be complicated by radial faults or a central graben. Around the flank of the dome, oil or gas can be trapped by faults, both sediment against sediment and sediment against salt, and by stratigraphic truncation, pinchout, and onlap. Some salt domes are pear or mushroom shaped in cross section, and petroleum is trapped beneath the peripheral overhang zone.

As the salt moves upward, a cap of diagenetically produced limestone, dolomite, and anhydrite often develops (Kyle and Posey, 1991). This cap may contain oil and gas in fractures. When a salt dome moves up close to the earth's surface, its crest may be dissolved by groundwater. Overlying sediments may collapse into the space thus formed, so giving rise to solution-collapse breccias. Ultimately, these too may act as petroleum reservoirs.

In the Arabian Gulf today coral atolls grow above salt domes (e.g., Das Island), and analogues containing oil and gas are present in the subsurface. When a salt dome moves upward, a concentric rim-syncline may form where the adjacent salt once was. In some instances the rim-syncline may be infilled by sand. Subsequent salt movement and shale compaction may result in structural closure of the sand. This phenomenon is sometimes referred to as a sedimentary anticline (i.e., of atectonic origin) or turtle-back (Fig. 7.26). These structures are essentially residual highs caused by adjacent salt moving into domes. The Bryan Field of Mississippi is one example of a turtle-back trap (Oxley and Herliny, 1972).

From the preceding account it is apparent that traps associated with salt domes can be very complex indeed. Not only are there many different trap situations but any one salt

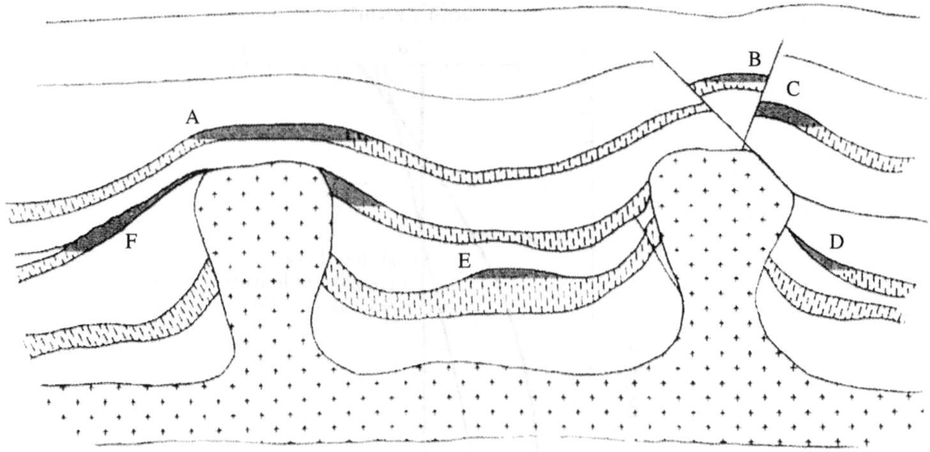

FIGURE 7.24 Crustal cross section illustrating the various types of trap that may be associated with salt movement: (A) domal trap; (B and C) fault traps; (D) pinchout trap; (E) turtle-back or sedimentary anticline; and (F) truncation trap.

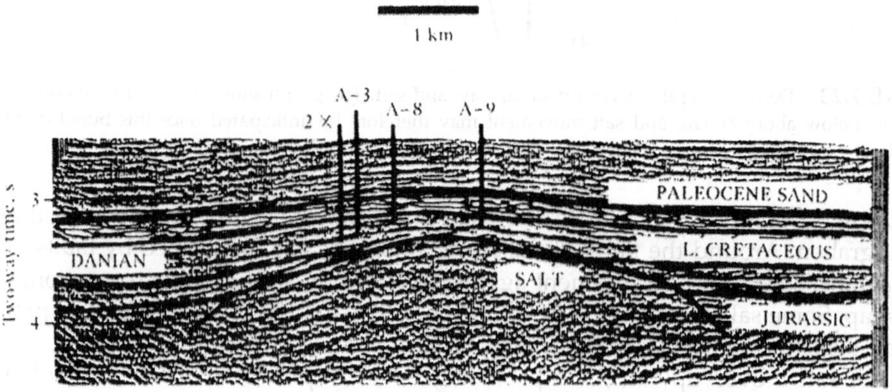

FIGURE 7.25 Seismic cross section through the Cod Field in the Norwegian sector of the North Sea. This structure is an example of a salt dome trap. Production comes from Paleocene deep-sea sands. *From Kessler et al. (1980).*

dome may host many separate traps of different types. Although the total reserves pertaining to one diapir can be vast, they may be contained in many separate accumulations. Each accumulation may have its own fluid and pressure characteristics. Individual oil columns may be high because of steeply dipping beds; individual pressures may be high because of the forces set up by the salt movement itself. A salt dome can be easily located by gravity or seismic methods. Once found, however, the associated traps may be too small to be delineated by present-day seismic resolution.

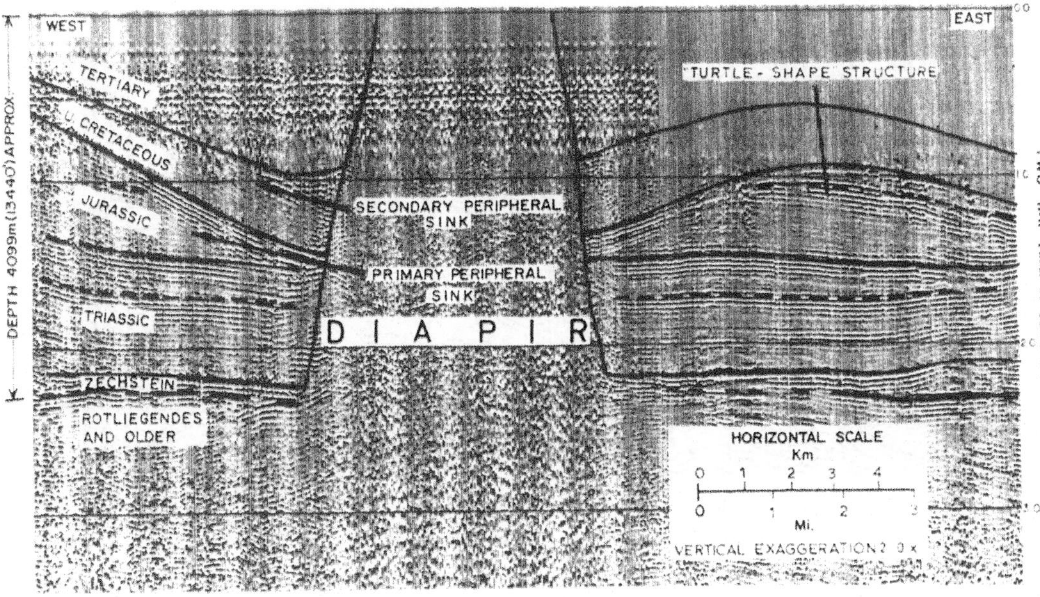

FIGURE 7.26 Seismic line through a Zechstein (Upper Permian) salt dome in the southern North Sea, showing rim-syncline due to salt withdrawal and associated turtle-back structure. *From Christian (1969).*

7.7.2 Mud Diapirs

The foregoing account deals largely with diapirs formed by salt. Diapiric mud structures also exist, and they too may generate hydrocarbon traps. Overpressure and overpressured shales were discussed earlier. By its very nature, an overpressured shale has a higher porosity and therefore a lower density than does normally compacted clay.

As already discussed, prodelta clays may be overpressured because of rapid burial beneath an advancing prism of deltaic sediment. This sediment thus becomes unstable, tending to slump seaward over the clays beneath it. This slumping is associated with growth faulting.

Sometimes diapirs of overpressured clay intrude the younger, denser cover, and, like salt domes, these mud lumps may even reach the surface. Mud diapirs are known from the Mississippi, Niger, Mackenzie, and other recent deltas and are less common, but not absent, from pre-Tertiary deltas (Fig. 7.27). Many of these mud diapirs have hydrocarbons trapped in ways analogous to those previously described for salt domes (e.g., the Beaufort Sea of Arctic Canada).

7.8 STRATIGRAPHIC TRAPS

Another major group of traps to be considered are the stratigraphic traps, whose geometry is due to changes in lithology. Such changes may be caused by the original deposition of the rock, as with a reef or channel. Alternatively, the change in lithology may be postdepositional, as with a truncation or diagenetic trap.

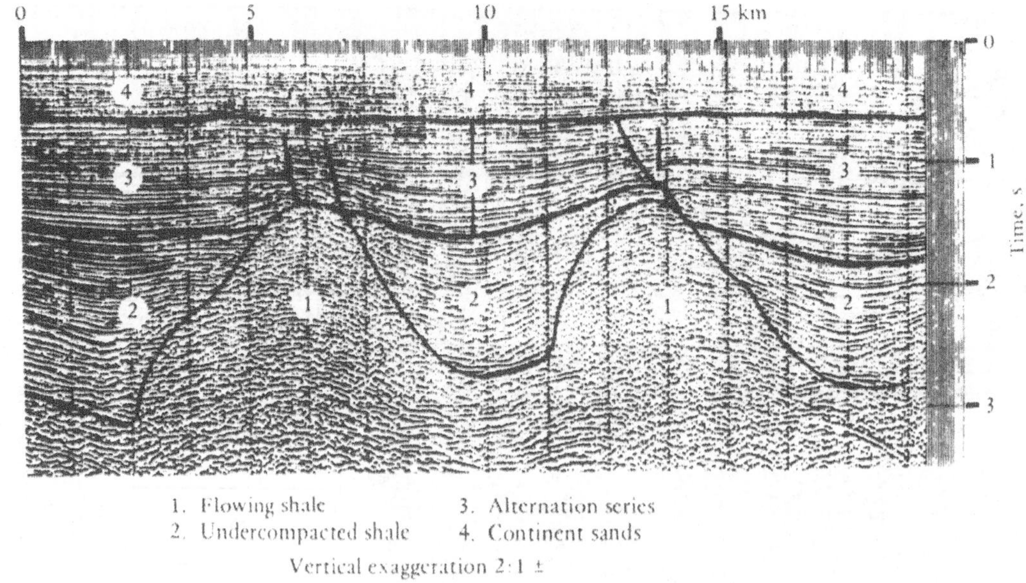

FIGURE 7.27 Cross section illustrating overpressured clay diapirs in a regressive deltaic sequence. *From Dailly (1976), with permission from the Canadian Society of Petroleum Geologists.*

The concept of the stratigraphic trap was first enunciated by Carll (1880) when he realized that the entrapment of oil in the Venango sands of Pennsylvania could not be explained by the then-prevailing anticlinal theory. Clapp (1917), in his second classification of oil fields, included lenticular sand and pinchout traps. The term *stratigraphic trap* was first coined by Levorsen in his presidential address to the American Association of Petroleum Geologists at Tulsa in 1936. He defined a stratigraphic trap as "one in which the chief trap-making element is some variation in the stratigraphy, or lithology, or both, of the reservoir rock, such as a facies change, variable local porosity and permeability, or an up-structure termination of the reservoir rock, irrespective of the cause" (Levorsen, 1967). For reviews of the evolution of the term and concept of the stratigraphic trap, see Dott and Reynolds (1969) and Rittenhouse (1972).

Stratigraphic traps are less well known and harder to locate than structural traps; their formation processes are even more complex. Nonetheless, as with structural traps a broad classification of different types can be made. Table 7.2 shows a scheme based on that of Rittenhouse (1972). Like most classifications it has its limitations, since many fields represent transitional steps between clearly defined types. It does, however, provide a convenient framework for the following account of stratigraphic traps. Major sources of data on stratigraphic traps are found in King (1972), Busch (1974), and Conybeare (1976).

7.8.1 Stratigraphic Traps Unrelated to Unconformities

Table 7.2 shows that a major distinction can be made between stratigraphic traps associated with unconformities and those occurring within normal conformable sequences. This distinction itself is somewhat arbitrary, since some types of trap, such as channels and reefs, occur both at unconformities and within conformable sequences. Those traps that do not necessarily require an unconformity are considered in this section. In this group the major distinction is between traps due to deposition and traps due to diagenesis. The depositional traps (the facies change traps of Rittenhouse) include channels, bars, and reefs. Diagenetic traps are due to porosity and permeability changes caused by solution and cementation. These various types of traps are discussed and illustrated as follows.

7.8.1.1 Channel Traps

Many oil and gas fields are trapped within different types of channels. Before examining examples of these fields, some background information is necessary. A channel is an environment for the transportation of sand, which may or may not include sand deposition. Thus, whereas a barrier island will always be made of sand, channels are frequently clay plugged. This situation is not necessarily bad, because the channel fill may act as a permeability barrier and thus trap hydrocarbons in adjacent porous beds. Therefore, finding a channel is not a guarantee of finding a reservoir. Also, channels occur both cut into unconformities and within conformable sequences (although it could be argued that, by definition, a channel is a *prima facie* case for the existence of some kind of depositional break).

Many good examples of channel stratigraphic traps occur in the Cretaceous basins along the eastern flanks of the Rocky Mountains, from Alberta down through Montana, Wyoming, Colorado, and New Mexico (Harms, 1966). These traps occur both cut into a major sub-Cretaceous unconformity and within the Cretaceous beds. The South Glenrock oil field of Wyoming serves as a typical example illustrating many of the characteristics of these fields (Curry and Curry, 1972). The South Glenrock field occurs in a syncline, which partly underlies a thrust block of Precambrian basement. The field contains oil trapped in both barrier bar and fluvial channel reservoirs. Only the latter is considered here. The channel has a width of some 1500 m and a maximum depth of some 15 m. It has been mapped for a distance of more than 15 km and shows a meandering shape (Fig. 7.28). The channel is partly infilled by sand and partly clay plugged. The SP curve on some wells suggests upward-fining point bar sequences. The channel is cut into the marine Skull Creek shale and is overlain by the marine Mowry shale. Immediately above the channel sand are thin, marine bar sand reservoirs. There can be no doubt about the channel origin of the lower reservoir, and, with its meandering geometry, there can be little argument as to its fluvial environment. This example demonstrates two main points. First note the scale: With a thickness of 15 m and a width of 1500 m, such reservoirs are not the hosts to giant oil accumulations. Second, because of the partial clay plug, only part of the channel is actually reservoir.

Oil is also trapped in channels other than fluvial ones. Perhaps, one of the best examples is Busch's (1961, 1971) illustration of oil fields in the deltaic distributary channels of Oklahoma (Fig. 7.29).

In the old days, channel traps could only be found by serendipity because it was so hard to predict the geometry and trend of sands in the subsurface. With modern seismic surveys, however, channels can be mapped on 2D and 3D surveys. Figure 7.30 illustrates a channel subtly imaged on a 2D seismic line. 3D examples are much more dramatic (see, for example, Brown, 1985; Rijks and Jauffred, 1991).

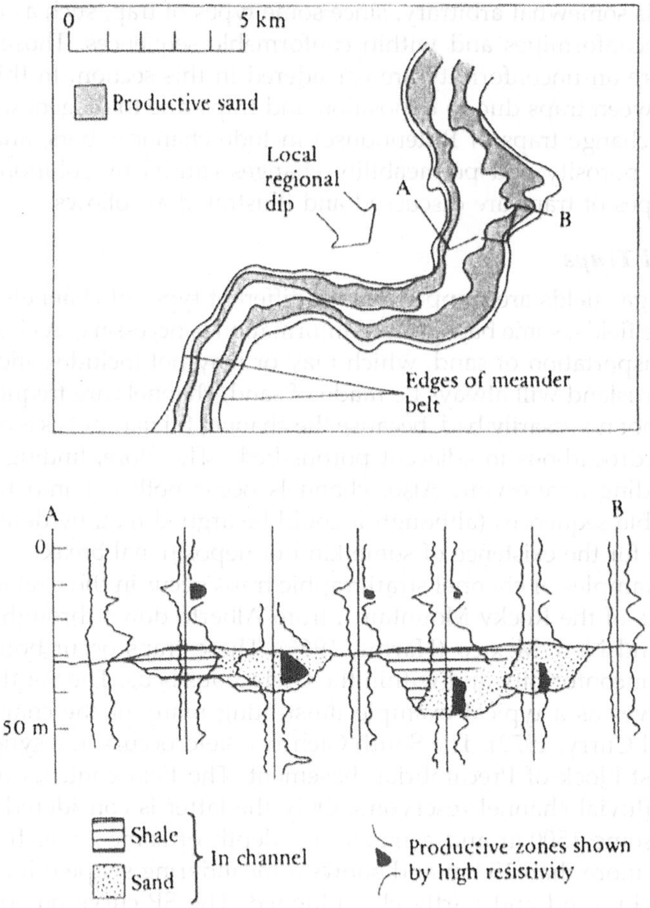

FIGURE 7.28 Map and cross section of the Cretacous South Glenrock oil field, Powder River Basin, Wyoming. Note the small dimensions of the reservoir and that not all of the channel contains sand. *Modified from Curry and Curry (1972), reprinted by permission of the American Association of Petroleum Geologists.*

7.8.1.2 Barrier Bar Traps

Marine barrier bar sands often make excellent reservoirs because of their clean, well-sorted texture. Coalesced barrier sands may form blanket sands within which oil may be structurally trapped. Sometimes, however, isolated barrier bars may be totally enclosed in marine and lagoonal shales. These barrier bars may then form shoestring stratigraphic traps parallel to the paleoshoreline. Many examples of this type of trap are known, but one of the classics is the Bisti field described by Sabins (1963, 1972). This field occurs in Cretaceous rocks of the San Juan Basin, New Mexico. Three stacked sand bars, with an aggregate thickness of only about 15 m, occur totally enclosed in the marine Mancos shale. The field is about 65 km long and 7 km wide. In some wells the three sands merge totally and cannot be

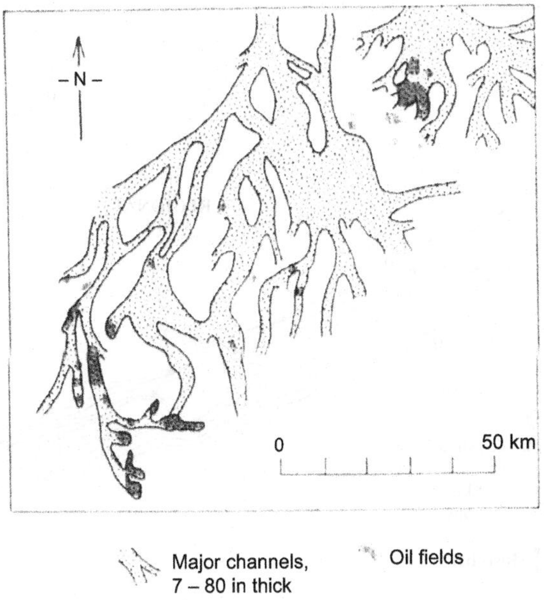

FIGURE 7.29 Map of the Booch delta, Seminole Country, Oklahoma. This celebrated study shows how oil is stratigraphically trapped in the axes of a major Pennsylvanian (Upper Carboniferous) delta distributary channel system. This type of map can only be constructed with imagination and ample well control. *Modified from Busch (1961), reprinted by permission of the American Association of Petroleum Geologists.*

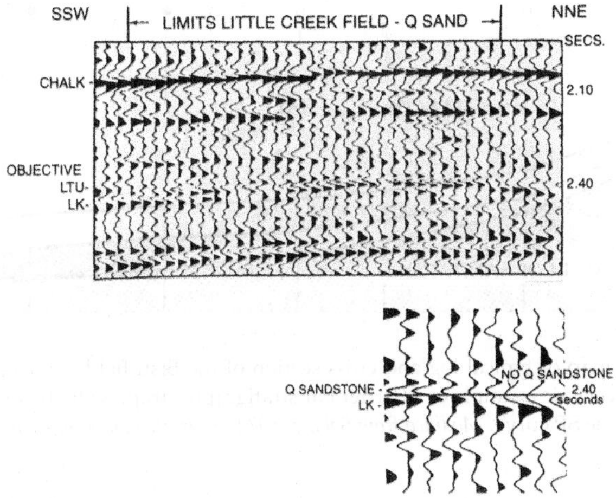

FIGURE 7.30 Seismic lines through the Little Creek Field, Mississippi. The discovery well was located to test a closure at the stippled Lower Tuscaloosa horizon (LTU) on the upper line. It located the Little Creek Field serendipitously. The extent of the field was subsequently determined by carefully mapping one negative amplitude event (seismic line lower right). *(From Werren et al. (1990), with permission from Springer-Verlag.)* This is a subtle trap. 3D seismic surveys show channels more dramatically (e.g., Brown, 1985; Rijks and Jauffred, 1991).

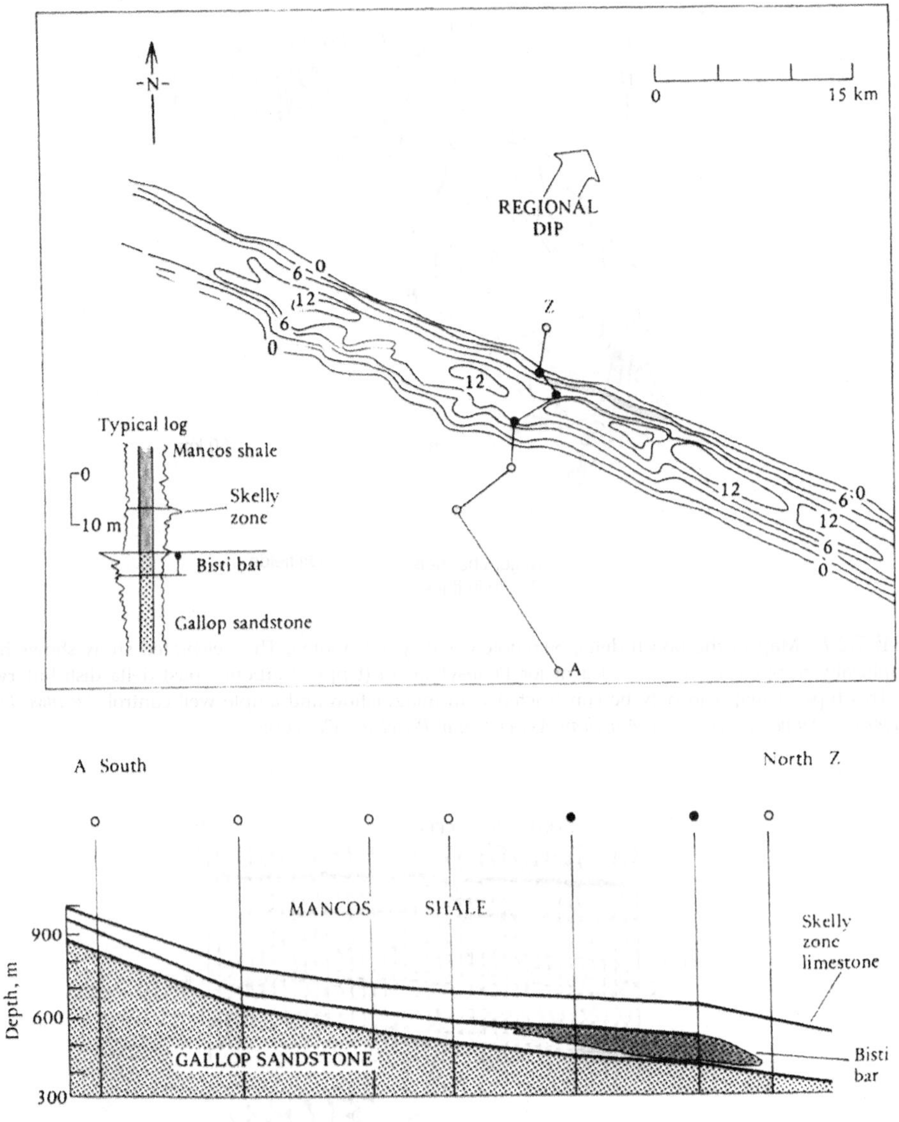

FIGURE 7.31 Isopach map of typical log and cross section of the Bisti field (Cretaceous) of the San Juan Basin, New Mexico. This field is a classic example of a barrier bar stratigraphic trap. Note the regressive upward-coarsening grain-size motif shown on the SP curve. *Modified from Sabins (1972), reprinted by permission of the American Association of Petroleum Geologists.*

separated. SP logs show typical upward-coarsening bar sand motifs in some of the wells (Fig. 7.31).

Barrier bar sand traps occur in many of the Rocky Mountain Cretaceous basins, ranging from the Viking sands of Alberta and Saskatchewan in the north (Evans, 1970; Cant, 1992, p. 13), via the Powder River Basin of Montana and Wyoming (Asquith, 1970; Woncik, 1972),

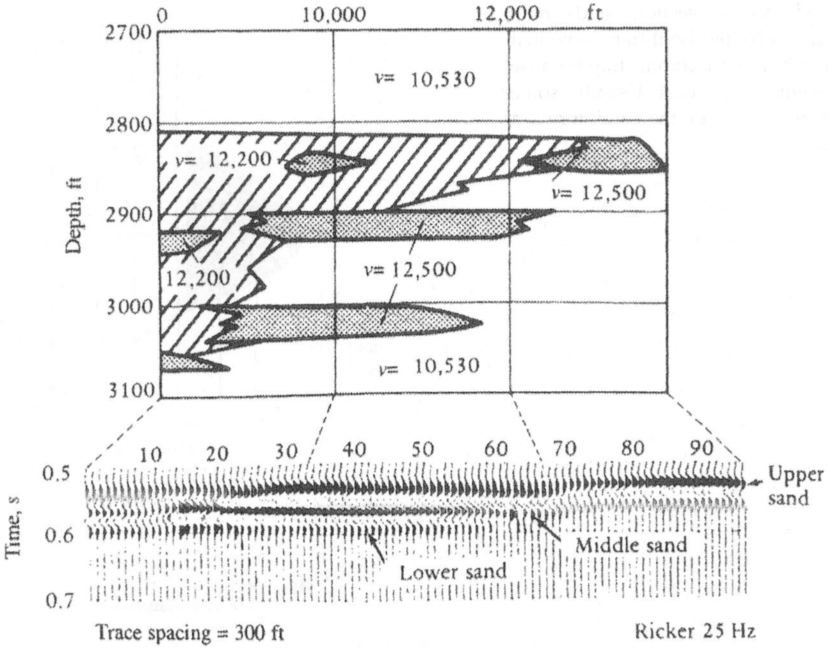

FIGURE 7.32 Geological model (upper) and computer-simulated seismic lines (lower) for regressing coastal barrier island sands. *From Meckel and Nath (1977), namely Fig. 27, p. 434.*

to the Denver Basin, Colorado (Tobison, 1972; Weimer and Davis, 1977) and the San Juan Basin of New Mexico (Hollenshead and Pritchard, 1961; Sabins, 1963, 1972).

As with channel traps, in the old days barrier bar fields were normally only found by serendipity. Today, however, cunning seismic techniques can be used to locate such subtle features (Fig. 7.32).

7.8.1.3 Pinchout Traps

Isolated barrier bar shoestring stratigraphic traps like those of the Bisti field are rare. Generally, a regressing barrier island deposits a sheet of sand. This sand may form a continuous reservoir, although in some instances shale permeability barriers may separate successive progradational events. Where these sheet sands pass updip into lagoonal or intertidal shales, they may give rise to pinchout, or feather edge, traps. Note that for these traps to be valid, they also need some closure in both directions along the paleostrike. This closure may be stratigraphic (where, as shown in the example in Fig. 7.33, the shoreline has an embayment) or structural, in which case the field should more properly be classified as a combination trap, rather than a stratigraphic trap.

Barrier bar sands, sheets, and shoestrings often occur as an integral part of major regressive–transgressive cycles. Sands are thickest and best developed in the regressive phase, and tend to occur only as discrete shoestrings during the transgressions. Examination of the Rocky Mountain Cretaceous basins shows that the transgressive sands tend to make the best traps.

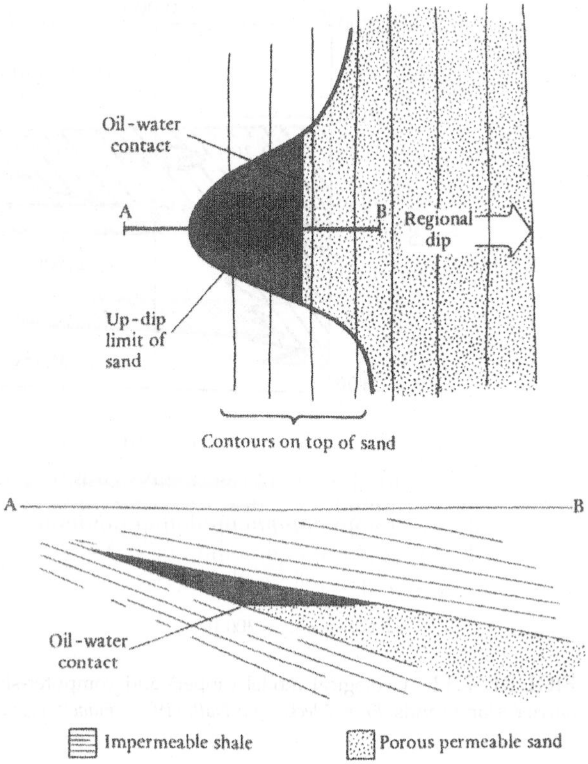

FIGURE 7.33 Cross section and map showing a stratigraphic pinchout trap. Note that this example is a pure stratigraphic trap because of the embayment of the coast. Usually, some structural closure on top of the sand forms a combination trap.

The regressive sands tend to lack updip seals as they pass shoreward into channel sands. The transgressive sands, by contrast, pass updip into sealing lagoonal and tidal flat shales (Mackenzie, 1972) (Fig. 7.34).

More complex pinchout traps can occur where both barrier bar and channel sands are in fluid communication with one another. The Bell Creek field (Cretaceous) of the Powder River Basin is an example of this type of trap (Berg and Davies, 1968; McGregor and Biggs, 1972). It lies on a basin margin where stratigraphic traps can be mapped in fairways of fluvial channels, and deltaic and bar sands (Fig. 7.35). The Bell Creek field is one of the few stratigraphic traps of "giant" status, with reserves of over 200 million barrels. It occurs at the mouth of the major river estuary in a complex of productive channel and barrier bar sands that interfinger with nonproductive marsh and lagoon muds and sands (Fig. 7.36). This complexity of reservoir geometry has lead to the occurrence of several GOC and OWC.

7.8.1.4 Reefs

Reefs, or carbonate buildups (to use a nongenetic term), have long been recognized as one of the most important types of stratigraphic traps. Modern reefs have been intensely studied by biologists and geologists (e.g., Maxwell, 1968; Jones and Endean, 1973), and accounts of ancient reefs are found in most sedimentology textbooks (see also Laporte (1974)).

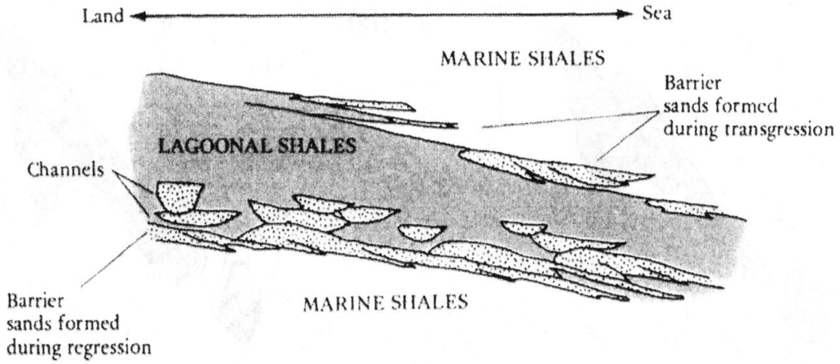

FIGURE 7.34 Cross section of regressive–transgressive shoreline deposits showing how barrier sands deposited during transgression can form stratigraphic traps due to updip lagoonal shale seal. Barrier sands formed during regressions lack updip seals, being in communication with channel sands. *Modified from Mackenzie (1972), reprinted by permission of the American Association of Petroleum Geologists.*

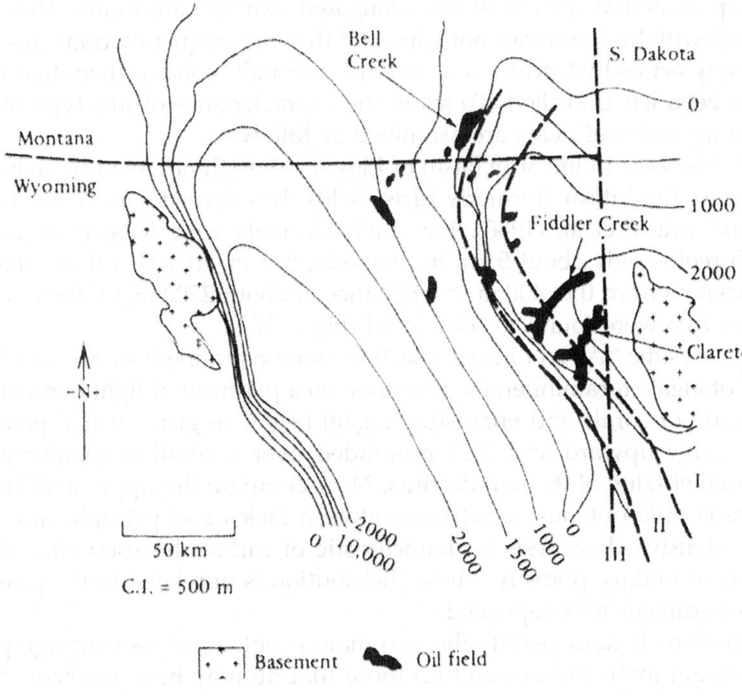

FIGURE 7.35 Map of the Powder River Basin of Wyoming and Montana, USA, contoured on top of the Fall River Sandstone (Cretaceous) showing productive fairways of stratigraphic traps: (I) fluvial channels, (II) deltaic distributary channel systems, and (III) marine bar sands. *Modified from Woncik (1972) and McGregor and Biggs (1970), reprinted by permission of the American Association of Petroleum Geologists.*

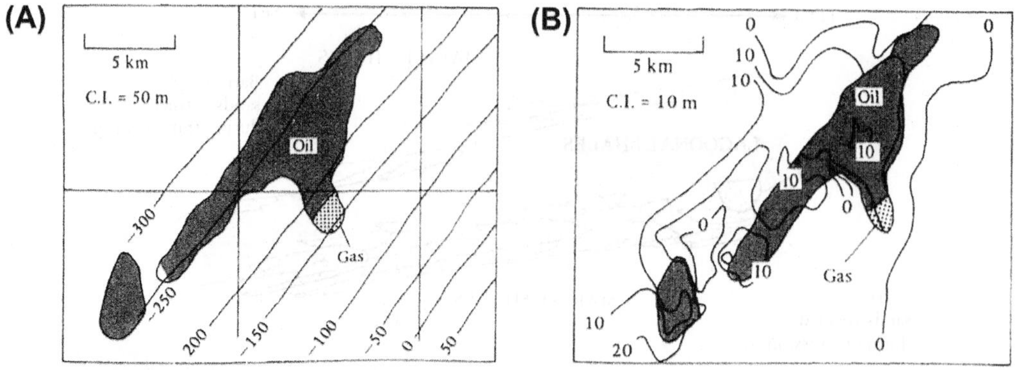

FIGURE 7.36 Maps of the Bell Creek giant stratigraphic trap (for location, see Fig. 7.30). (A) Structure contour map on top of the Muddy sand (the main reservoir) showing lack of structural control on the field boundaries. *(Modified from McGregor and Biggs (1970), reprinted by permission of the American Association of Petroleum Geologists.)* (B) Gross sand isopach. Basically an up-dip pinchout trap, the Bell Creek reservoir consists of a complex of marine shoreface and fluviodeltaic channel sands with multiple fluid contacts.

Reefs develop as domal (pinnacle) and elongated (barrier) antiforms. They grow a rigid stony framework with high primary porosity, and they are frequently transgressed by marine shales, which may act as hydrocarbon source rocks. Small wonder, then, that few identified reef traps have been left undrilled. To show the characteristics of this type of stratigraphic trap, some examples of reef fields are examined as follows.

Many reef fields occur in the Sirte Basin of Libya, although not all of these have been published. One group, the Intisar (formerly Idris) fields, has been described by Terry and Williams (1969) and Brady et al. (1980). Five pinnacle reefs were located in a concession of 1880 km^2. Each reef is only about 5 km in diameter, but up to 400 m thick; that is, the reefs build up in thickness from 0 to 400 m in a distance of about 2500 m. Of the five reefs located and tested, only two were found to contain oil (Fig. 7.37).

Figure 7.38 shows the "A" reef in cross section, demonstrating how the reef began to form as a biostrome of algal–foraminiferal wackestone on a platform of tight nonreef limestone. A reef (largely made of corals and encrusting algae) began to grow at one point on the biostrome. It first grew upward and then prograded over a coralline biomicrite, which had formed as a forereef talus of its own detritus. Now compare the upper and lower halves of Fig. 7.38. There is little obvious correlation between facies and petrophysics. As discussed in Chapter 6, extensive diagenesis is characteristic of carbonate reservoirs. Therefore, it is common to find secondary porosity whose distribution is unrelated to the primary porosity with which the sediment was deposited.

Thus the Intisar reefs demonstrate the two main problems of reef stratigraphic traps: (1) Not all reefs contain hydrocarbons and (2) those that do may have reservoir characteristics unrelated to depositional facies. Nonetheless, many reef hydrocarbon provinces exist around the world, notably in the Arabian Gulf, Western Canada, and Mexico.

In common with sandstone stratigraphic traps, reefs were very hard to locate in the subsurface until the advent of geophysics. Gravity surveys were an early effective exploration

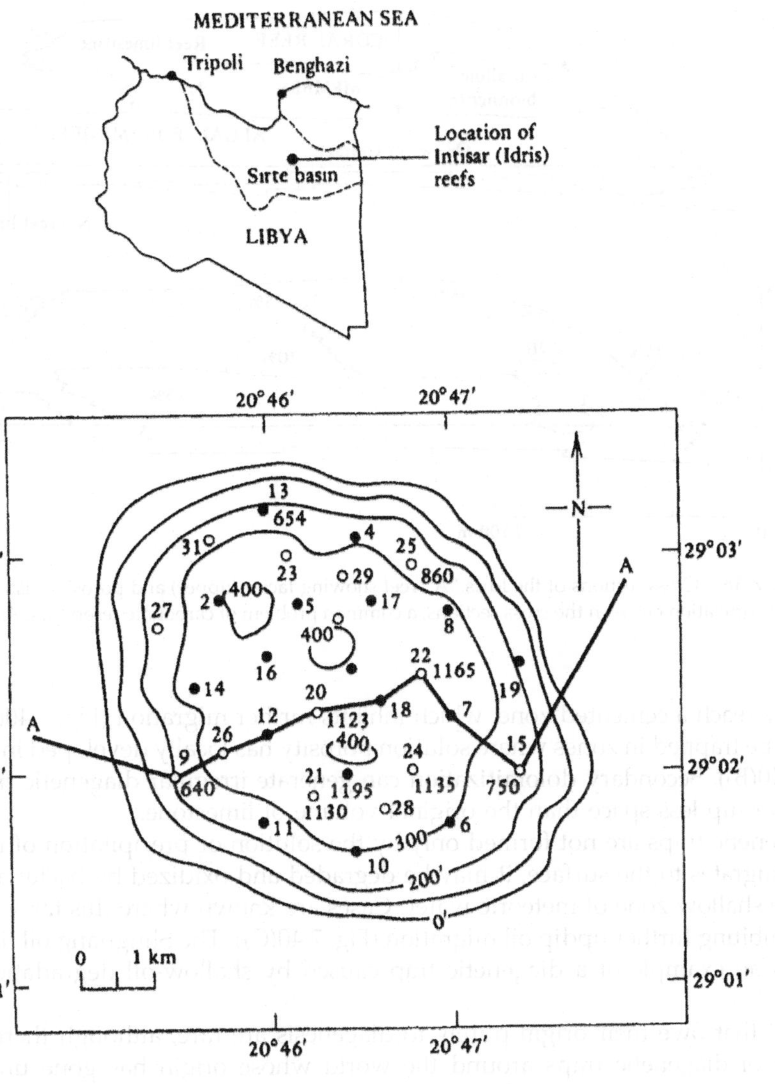

FIGURE 7.37 Isopach map of the Idris "A" reef, Sirte Basin, Libya. *After Terry and Williams (1969).*

tool (e.g., Ferris, 1972), but are now surplanted by 2D and 3D seismic surveys (see Fig. 7.39 and Plate 3).

7.8.1.5 Diagenetic Traps

Diagenesis plays a considerable role in controlling the quality of a reservoir within a trap. As discussed in Chapter 6, solution can enhance reservoir quality by generating secondary porosity, whereas cementation can destroy it. In some situations diagenesis can actually generate a hydrocarbon trap (Rittenhouse, 1972). Oil or gas moving up a permeable carrier

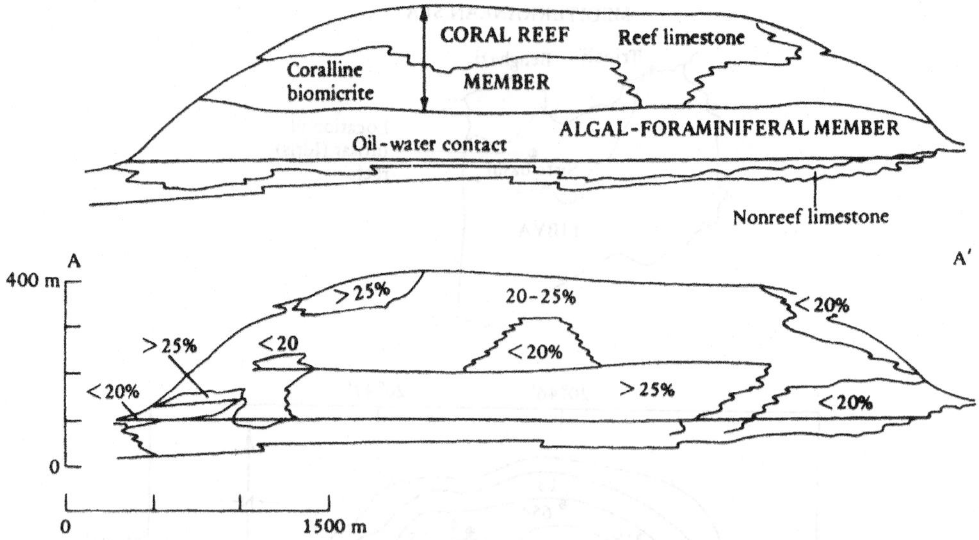

FIGURE 7.38 Cross sections of the Idris "A" reef showing facies (upper) and porosity distribution (lower). Note the lack of correlation between the cross sections, a common problem of carbonate reservoirs. *After Terry and Williams (1969).*

bed may reach a cemented zone, which inhibits further migration (Fig. 7.40(A)). Conversely, oil may be trapped in zones where solution porosity has locally developed in a cemented rock (Fig. 7.40(B)). Secondary dolomitization can generate irregular diagenetic traps as the dolomite takes up less space than the original volume of limestone.

Diagenetic traps are not formed only by the solution or precipitation of mineral cements. As oil migrates to the surface, it may be degraded and oxidized by bacterial action if it reaches the shallow zone of meteoric water. Cases are known where this tarry residue acts as a seal, inhibiting further updip oil migration (Fig. 7.40(C)). The Shuguang oil field in the Liaohe Basin is an example of a diagenetic trap caused by shallow-oil degradation (Ma Li et al., 1982).

Traps that owe their origin purely to diagenesis are rare, although there are probably a number of diagenetic traps around the world whose origin has gone unrecognized, and many are yet to be found (Wilson, 1977). Many traps, however, are due to a combination of diagenesis and one or more other causes. This type of origin is particularly true of the sub-unconformity traps discussed in the next section.

7.8.2 Stratigraphic Traps Related to Unconformities

The channel, bar, reef, and diagenetic traps just described can occur both in conformable sequences and adjacent to unconformities. The first three types often overlie unconformities, whereas diagenetically assisted traps underlie them.

The role of unconformities in the entrapment of hydrocarbons has been remarked by geologists from Levorsen (1934, 1964) to Chenoweth (1972) and Bushnell (1981). Unconformities

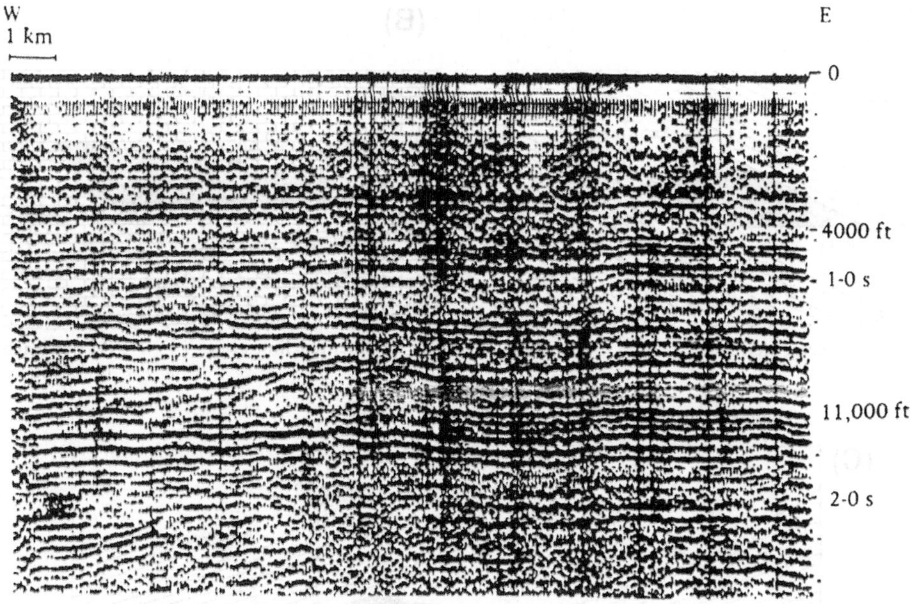

FIGURE 7.39 Seismic line through a limestone reef or buildup, clearly visible as a characterless lense at 11,000 feet. Note regular reflectors of back reef lagoonal sediments to the east, and basinal reflectors onlapping the reef from the west. *From Fitch (1976).*

facilitate the juxtaposition of porous reservoirs and impermeable shales that may act as source and seal. A large percentage of the known global petroleum reserves are trapped adjacent to the worldwide unconformities and source rocks of Late Jurassic to mid-Cretaceous age. Many of these reserves are held in structural and combination traps, as well as the pure stratigraphic traps, which are described as follows. As already shown (Table 7.2), unconformity-related traps can be divided into those that occur above the unconformity and those that occur below it. These two types of unconformity traps are now described in turn.

7.8.2.1 Supraunconformity Traps

Stratigraphic traps that overlie unconformities include reefs and various types of terrigenous sands. These traps may be divided into three classes according to their geometry: sheet, channel, and strike valley. Shallow marine or fluvial sands may onlap a planar unconformity. A stratigraphic trap may occur where these sands are overlain by shale and where the subunconformity rocks are also impermeable. In many ways, therefore, these onlap traps are similar to the pinchout traps described previously. Note, in particular, that both traps require stratigraphic permeability changes or structural closure in both directions along the paleostrike for the trap to be valid. The Cut Bank field of Montana is an example of an onlap stratigraphic trap (Fig. 7.41). Here the Lower Cretaceous Cut Bank sand unconformably onlaps Jurassic shales and is itself onlapped by the Kootenai shale (Shelton, 1967). It contains recoverable reserves of more than 200 million barrels of oil and 300 billion cubic feet of gas. In this case the reservoir is a fluvial sand, but in other onlap traps the reservoir is of marine origin.

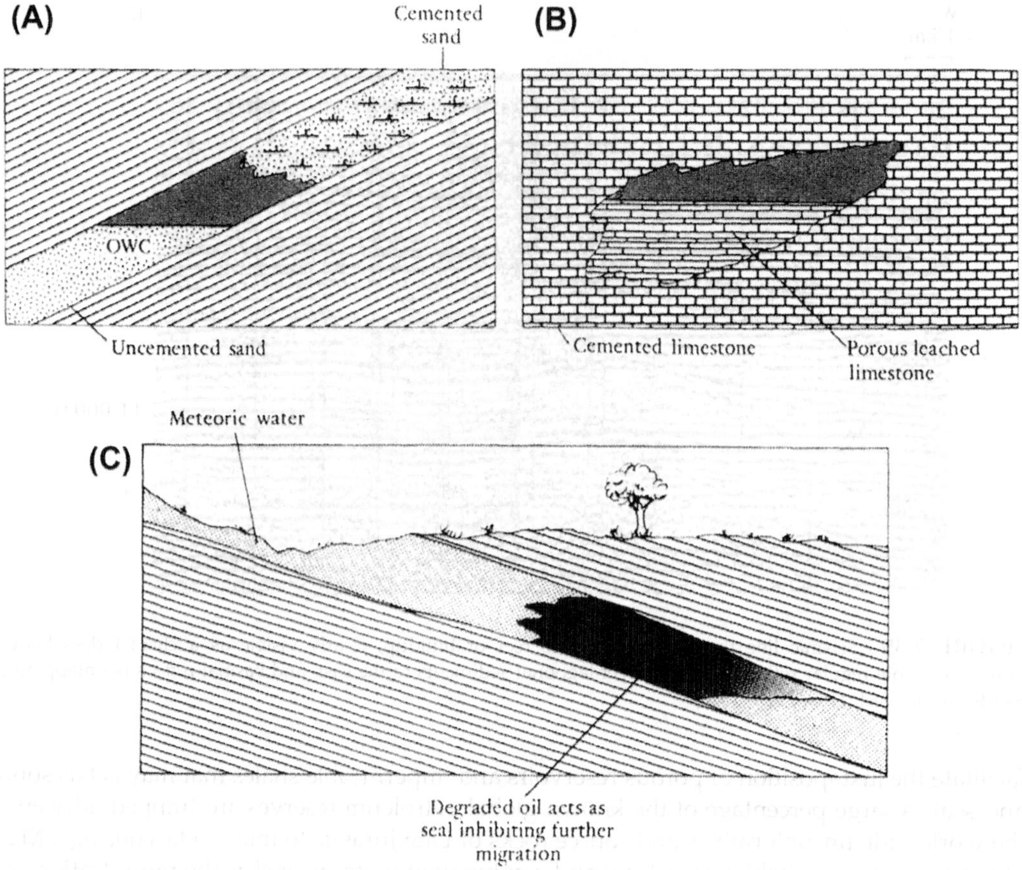

FIGURE 7.40 Configurations for diagenetic traps caused by (A) cementation, (B) solution, and (C) shallow-oil degradation.

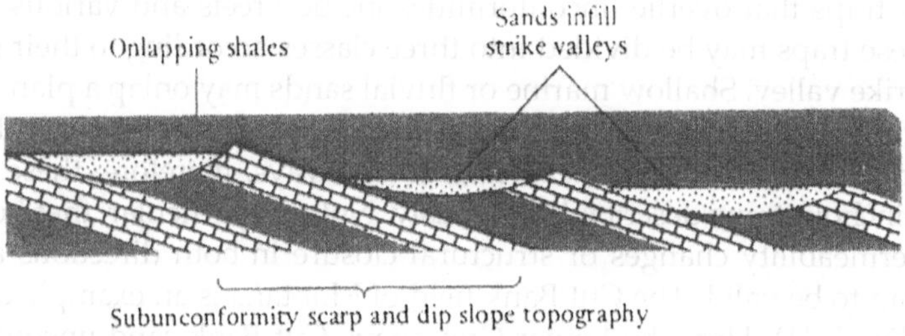

FIGURE 7.41 Cross section showing the occurrence of strike valley sands. Examples of oil trapped in these sands are known from the Pennsylvanian of Oklahoma and the Cretaceous of New Mexico.

Where an unconformity is irregular, sand often infills valleys cut into the old land surface (valley fill traps). The location and trend of these valleys may be related to the resistance to weathering of the various strata that once cropped out at the old land surface, basement tectonics, or other causes. This process gives rise to paleogeomorphic traps (Martin, 1966; Wood and Hopkins, 1992). The two main groups of paleogeomorphic traps are valley fill (channel) and strike valley. Rivers draining a land surface may incise valleys into the bedrock, and these valleys may then be infilled with alluvium (both porous sand and impermeable shale). Pinchouts and facies changes within the valley fill can make predicting of reservoir facies difficult (Biddle and Wielchowsky, 1994). The Glen Rock field, Wyoming, has already been described as an example of a fluvial channel stratigraphic trap. There is little difference between valleys cut in unconformities and those within conformable sequences. The Fiddler Creek and Clareton fields (Fig. 7.35) are examples of paleogeomorphic channel traps. Many of the Cretaceous Muddy Formation fields in the Powder River Basin and Pennsylvanian Morrow Formation of the Mid-Continent produce from incised valleys. There are many other such traps, ranging from fluvial channels to deep submarine ones like those of Miocene age in California (Martin, 1963).

Where alternating beds of hard and soft rocks are weathered and eroded, the soft strata form strike valleys between the resistant ridges of harder rock. Fluvial, and occasionally marine, sands within the strike valleys may be blanketed by a transgressive marine shale. Oil and gas may be stratigraphically trapped in the strike valley sands (Fig. 7.41). This type of trap was first described from the sub-Pennsylvanian unconformity of Oklahoma. Some of these sands are 60–70 km in length, yet only 1 or 2 km wide. Local thickening of the reservoir occurs where primary consequent valleys intersect strike valleys (Busch, 1959, 1961, 1974; Andresen, 1962). Other examples have been identified in the basal Niobrara (Upper Cretaceous) sands of the San Juan Basin, New Mexico (McCubbin, 1969). These sands are shorter and wider than the Oklahoma sands. In both cases, however, the sand bodies are generally less than 30 m thick.

7.8.2.2 Subunconformity Traps

Stratigraphic traps also occur beneath unconformities where porous permeable beds have been truncated and overlain by impermeable clay. In many instances a seat seal is also provided by impermeable strata beneath the reservoir. As with pinchouts and onlaps, some closure is needed in both directions along the paleostrike. This closure may be structural or stratigraphic, but for many truncation traps it will be provided by the irregular topography of the unconformity (Fig. 7.42).

Most, if not all, truncation traps have had their reservoir quality enhanced by epidiagenesis. This secondary solution, porosity induced by weathering, is particularly well known in limestones, but also occurs in sands and basement. Weathering of limestones ranges from minor moldic and vuggy porosity formation to the generation of karstic and collapse breccia zones of great reservoir potential. One noted example is the Casablanca field of offshore Spain (Watson, 1982). Figure 6.38 shows a Zechstein dolomite collapse breccia due to pre-Cretaceous weathering of the Auk field in the North Sea (Brennand and Van Veen, 1975). Subunconformity solution porosity is important in many sandstone reservoirs, such as the Sarir Group of Libya and the Brent Sand of the North Sea.

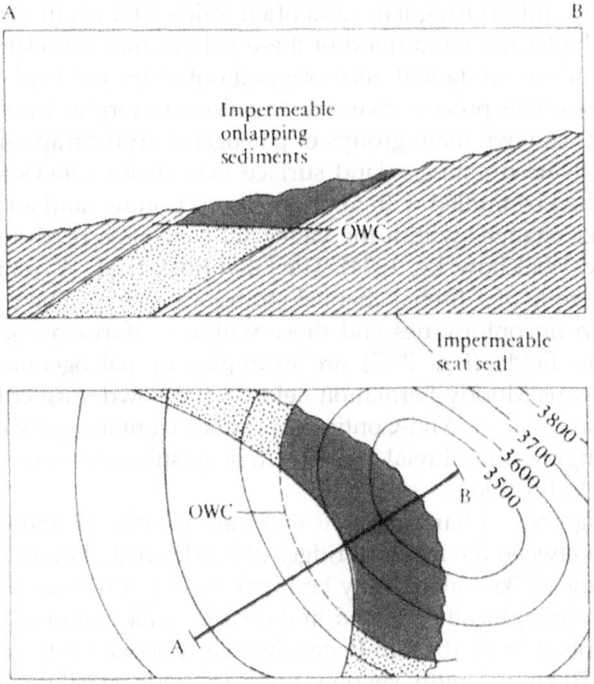

FIGURE 7.42 Cross section (upper) and map (lower) illustrating the geometry of a truncation trap. The contours drawn on the unconformity surface of the map define a buried hill.

As discussed earlier, a number of fields produce from basement rocks, notably in Vietnam and China. In almost every case the reservoir is unconformably overlain by shales that have acted as source and seal. Production comes from fractures and solution pores, where unstable minerals (generally mafics and occasionally feldspar) have weathered out. A notable example of this type of trap is the Augila field of Libya, where one well produced 40,000 bopd from granite (Williams, 1968, 1972). This remarkable stratigraphic trap also produces from sands and carbonates that onlap a buried granite hill (Fig. 7.43).

7.8.2.3 Unconformity-Related Traps: Conclusion

The preceding sections describe the various ways in which unconformities may trap hydrocarbons. It must be stressed that few traps are simple and monocausal. The number of combination traps in which an unconformity is combined with folding or faulting far outweighs simple unconformity traps. Many unconformity traps produce from both onlapping and subcropping reservoirs, as in the Augila field just noted. Similarly, as unconformities converge and merge upstructure, reservoirs that are both onlap and subcrop traps form. The East Texas field is one such example (Fig. 7.44). This giant field, with estimated recoverable reserves of 5600 million barrels, has a length of some 75 km and a maximum width of 25 km. It produces from the Cretaceous Woodbine sand, which unconformably overlies the Washita Group and is itself truncated by the Austin Chalk (Halbouty, 1994).

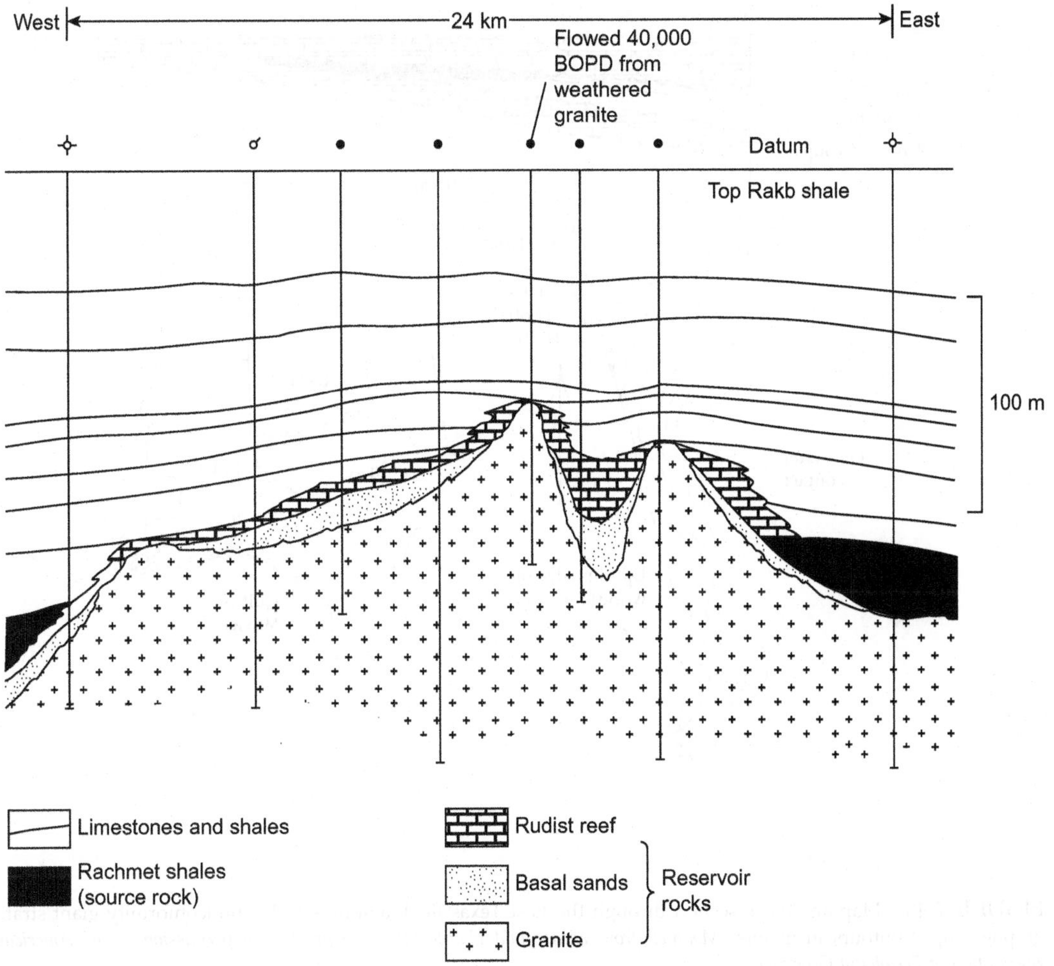

FIGURE 7.43 Cross section through the Augila field of Libya. This field is a complex trap that produces partly from sands and reefal carbonates and partly from fractured and weathered granite. *Modified from Williams (1968, 1972).*

There has been considerable debate as to whether shales above or below an unconformity provide the oil for adjacent reservoirs. With modern geochemical techniques of matching crudes and their parent kerogen, this problem can generally be solved. In some cases the source rock is clearly beneath the unconformity. A notable example of this condition is the Hassi Massaoud field, in which Cambro-Ordovician sands contain oil derived from the Tannezuft Shale (Silurian) on a large paleohigh sealed by Triassic evaporites (Balducci and Pommier, 1970). Another example is the Brent field and its associates in the North Sea. Here the early Cretaceous Cimmerian unconformity seals many traps ranging in age from Jurassic to Devonian. All are sourced by the Upper Jurassic Kimmeridge clay.

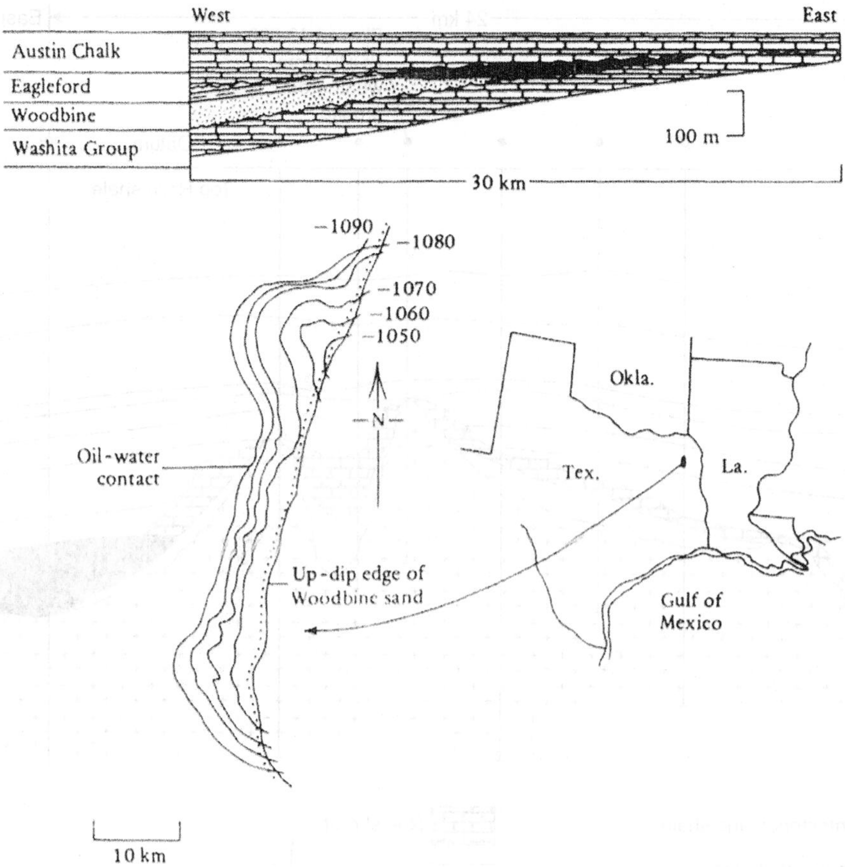

FIGURE 7.44 Map and cross section through the East Texas field, a supra- and subunconformity giant stratigraphic trap. Contours in meters. *Modified from Minor and Hanna (1941), reprinted by permission of the American Association of Petroleum Geologists.*

Conversely, in many fields shales above an unconformity act as source and seal. This situation occurs, for example, in Prudhoe Bay, Alaska (Morgridge and Smith, 1972; Jones and Speers, 1976) and Sarir field, Libya (Sanford, 1970). That an impermeable shale should act as both source and seal may seem incomprehensible at first. Actually, two explanations can account for this phenomenon. The oil trapped in a structural culmination need not necessarily have moved downward in a physical sense, but may have moved up the flank of the structure from adjacent lows, and in so doing moved downward only in a stratigraphic sense. Also, any good engineer can prove mathematically that fluid can move downward if there is a sufficient pressure differential.

7.8.3 Relationship between Stratigraphic Traps and Sedimentary Facies

In the same way that structural traps are related to tectonic style, stratigraphic traps are often related to sedimentary environments. A genetic grouping of stratigraphic traps and facies may be more useful as an aid to prospect prediction than the formal classification used for the preceding descriptive section. This approach needs to be qualified. As discussed in Chapter 6, the distribution of porosity is often unrelated to facies in carbonate reservoirs. Thus, apart from the obvious environmental control of reefs, carbonate stratigraphic traps are seldom facies related. Diagenetic processes and the position of unconformities are generally more significant than earlier environmental controls.

Turning to sandstones, Chapter 6 noted that reservoir quality is generally facies related and the effects of diagenesis are usually less significant. Pinchout traps, whether due to onlap or truncation, can occur in a blanket sand deposited in any environment. All they require is the necessary structural tilt and an adequate seal. Many varieties of sandstone stratigraphic trap have distributions that are largely facies related, as shown in Table 7.4. Channel traps may, of course, be found in almost any environment from fluvial to deep marine, as already noted. Strike valley sands are generally of continental origin. Growth-fault-related traps, such as rollover anticlines, and clay diapir traps are not stratigraphic in origin, but their occurrence is closely related to environment. They are restricted to rapidly deposited regressive wedges of deltaic sands and overpressured clay. Diapiric traps are commonly present to the basinward side of growth faults. In deep basinal settings stratigraphic traps may occur in submarine channels and in submarine fans. The latter include updip pinchouts as well as closed structures where fan paleotopography is preserved.

This brief review shows that many stratigraphic traps are related to depositional environment. This correlation obviously aids the prediction of the type of trap to be anticipated in a particular sedimentary facies.

TABLE 7.4 Relationship between Stratigraphic Traps and Sedimentary Environments of Sandstone Reservoirs

Environment		Trap Type
Continental	Eolian	Pinchout
	Fluvial	Channel
		Strike valley
Coastal	Barrier bar	Shoestring
	Delta	Channel
		Crevasse·splay and mouth bar
		Growth-fault-related
		Diapir-related
Deep marine		Submarine fan pinchout or paleotopographic closure
		Submarine channel

7.9 HYDRODYNAMIC TRAPS

The third group of traps to consider, in addition to structural and stratigraphic ones, is the hydrodynamic traps. In these traps hydrodynamic movement of water is essential to prevent the upward movement of oil or gas. The concept was first formulated by Hubbert (1953) and embellished by Levorsen (1966). The basic argument is that oil or gas will generally move upward along permeable carrier beds to the earth's surface except where they encounter a permeability barrier, structural or stratigraphic, beneath which they may be trapped.

When water is moving hydrodynamically down the permeable beds, it may encounter upward-moving oil. When the hydrodynamic force of the water is greater than the force due to the buoyancy of the oil droplets, the oil will be restrained from upward movement and will be trapped within the bed without any permeability barrier. Levorsen illustrated this concept by using the analogy of corks in a glass tube below which water was flowing. The corks would tend to accumulate where the tube was restricted, causing a local increase in the fluid potential gradient.

In the real world this increase might occur where there was a local reversal of dip or facies change, or even a local fluctuation in the potentiometric gradient (Fig. 7.45). Hubbert (1953) expounded the theory behind the concept of the hydrodynamic trap. The role of the hydrodynamic flow in causing tilted OWCs was discussed earlier in this chapter. Although this phenomenon occurs in many fields, pure hydrodynamic traps are very rare. One such example, however, is the Wheat field of the Delaware Basin, West Texas, described by Adams (1936). This field occurs in a gentle flexure in monoclinally dipping beds, which have closure but not vertical relief (Fig. 7.46). Other examples have been described from the Paris Basin, France, from Poland (see Heuillon (1972); Zawiska (1986); respectively), and from Canada (Eremako and Michailov, 1974).

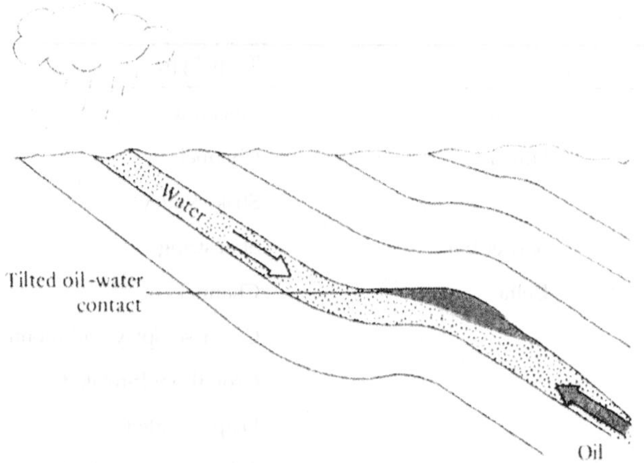

FIGURE 7.45 Crustal cross section showing a pure hydrodynamic trap. There is no vertical structural relief. Oil migrating upward is trapped in a monocline by the downward flow of water.

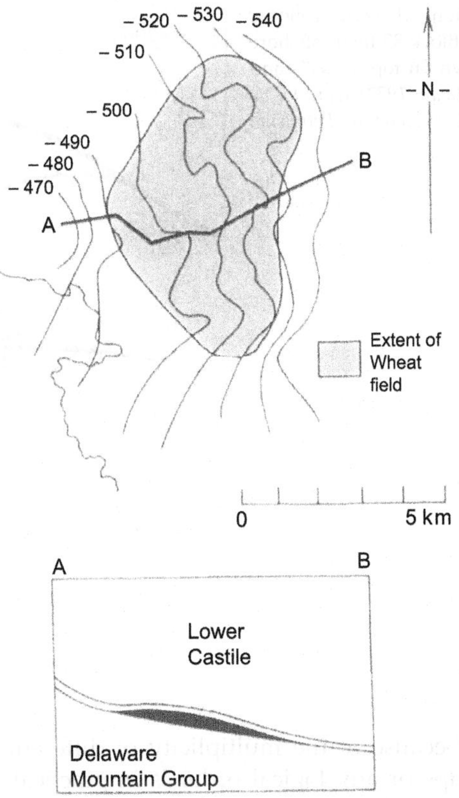

FIGURE 7.46 Map (upper) and cross section (lower) through the Wheat field, Delaware Basin, Texas. This field is an example of a hydrodynamic trap. The structure contours on the map are drawn on top of the Wheat field reservoir. *Modified from Adams (1936), reprinted by permission of the American Association of Petroleum Geologists.*

An obvious tilted oil/water contact points to the role of hydrodynamic flow. If there was no flow and pressures were hydrostatic, oil could not be trapped because the trap lacks four-way structural closure. Traps like the Wheat field are very rare and are not regarded as a prime target for exploration, although the significance of hydrodynamic flow in shifting fields down the flank of structures must always be remembered.

7.10 COMBINATION TRAPS

Many oil and gas fields around the world are not due solely to structure or stratigraphy or hydrodynamic flow, but to a combination of two or more of these forces. Such fields may properly be termed *combination traps*. Most of these traps are caused by a combination of structural and stratigraphical processes. Structural–hydrodynamic and stratigraphic–hydrodynamic traps are rare. Examples are known from the Rocky Mountain Cretaceous basins where Hubbert developed his theories of hydrodynamic flow, and from Indonesia

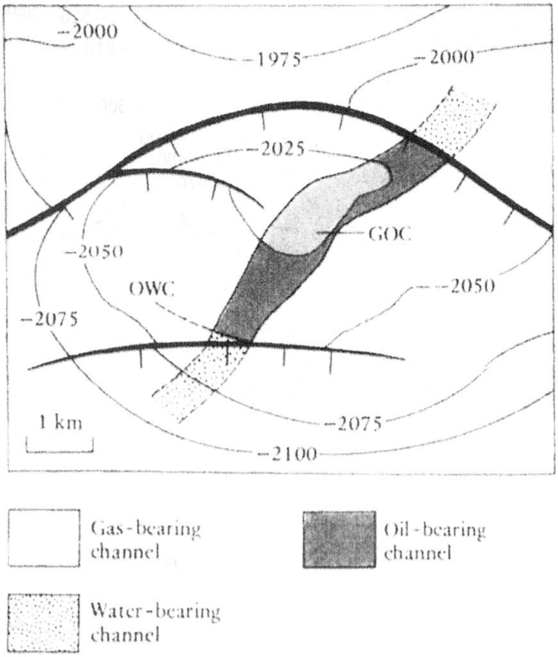

FIGURE 7.47 Combination channel rollover anticline trap, the Main Pass Block 35 field, offshore Louisiana. Contours are drawn on top of "G" sand (CI = 25 m). *Modified from Hartman (1972), reprinted by permission of the American Association of Petroleum Geologists.*

(Cockroft et al., 1987). Because of the multiplicity of different types of combination traps, discussing them in groups or any logical order is impractical. One or two examples must suffice.

On a small scale, oil may be trapped in shoestring sands (channels or bars) that cross-cut anticlines (Fig. 7.47). As pointed out earlier, pinchout, onlap, and truncation traps all require closure, which is very often structural, along the strike. Likewise, folded and/or faulted beds may be sealed by unconformities to form another group of traps. Examples of folded and faulted unconformity traps are now discussed in turn.

The Prudhoe Bay field on the North Slope of Alaska is a fine example of a combination trap (Morgridge and Smith, 1972; Jones and Speers, 1976; Jamison et al., 1980; Bushnell, 1981). Here a series of Carboniferous, Permian, Triassic, Jurassic, and basal Cretaceous sediments were folded into a westerly plunging anticlinal nose. This structure was truncated progressively to the northeast and overlain by Cretaceous shales, which act as source and seal to the trap. Oil and gas are trapped in the older beds, which subcrop the unconformity. The main reservoir is the fluvial Sadlerochit Sandstone (Triassic). Additional closure is provided by major faults on the northern and southwestern side of the structure (Fig. 7.48).

Fault unconformity traps characterize the Jurassic Brent province of the northern North Sea. The reservoir is the Middle Jurassic deltaic Brent Sandstone Group, which is overlain by Upper Jurassic shales. These include the Kimmeridge clay formation, which is the source rock. Late Jurassic–Early Cretaceous Cimmerian movement created numerous fault blocks, which were tilted, truncated, and unconformably overlain by Cretaceous shales. The resultant traps, which include fields such as Brent, Statford, Murchison, Hutton, Thistle, and Piper

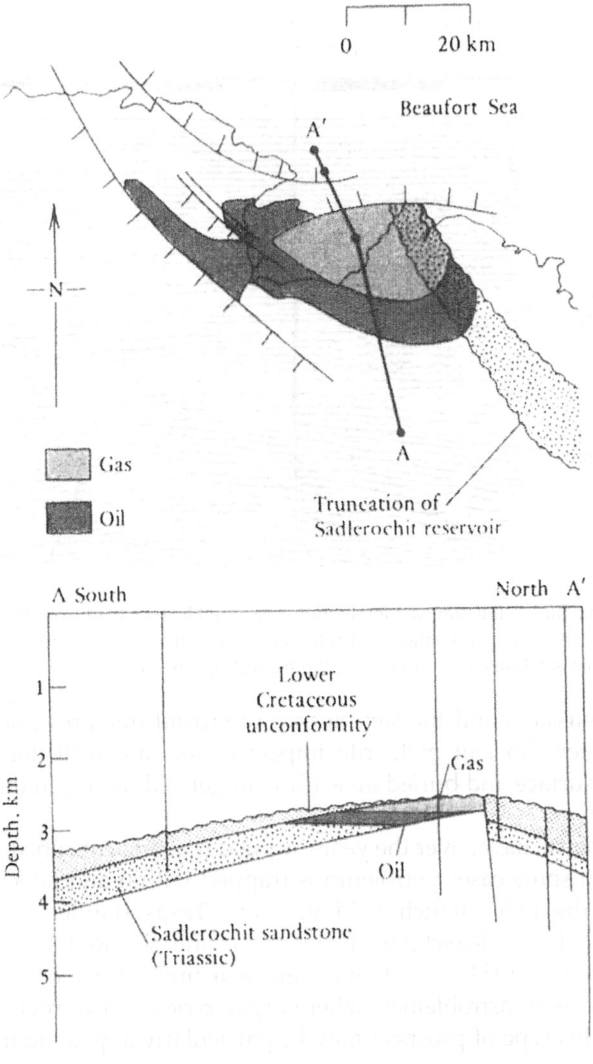

FIGURE 7.48 Map (upper) and cross section (lower) of the Prudhoe Bay field of Alaska. *Modified from Jamison et al. (1980), reprinted by permission of the American Association of Petroleum Geologists.*

(Fig. 7.15), are thus all combination fault unconformity traps with varying degrees of truncation of the reservoir. Such combination traps can normally be found by seismic surveys since both the fault and the truncating unconformity are usually detectable (Fig. 7.49).

7.10.1 Astrobleme Traps

Finally, mention must be made of one very rare, but spectacular, type of combination trap, namely, the astrobleme, or meteorite impact crater. Gold's theory for the origin of petroleum

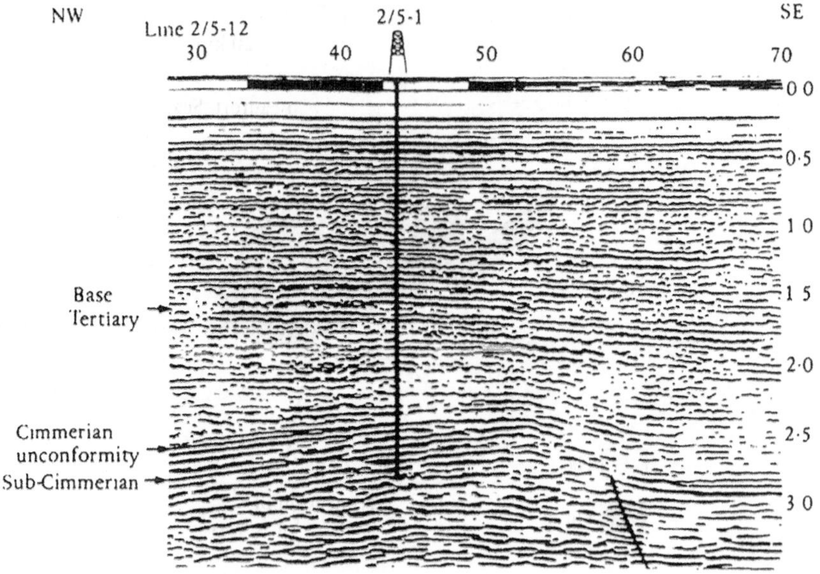

FIGURE 7.49 Seismic line to illustrate a combination trap, the Heather field from the Northern North Sea *(from Gray and Barnes (1981))*. This is a combination of a tilted fault block that has been truncated by the Cimmerian unconformity. The reservoir is a Jurassic sand beneath the truncating unconformity.

by earthquake outgassing, and the significance of crustal fractures due to meteorite impact was discussed earlier. Ancient meteorite impact craters are well documented around the globe, both on the surface and buried beneath younger sediments, in the subsurface (Jenyon, 1990, pp. 363–368).

Numerous geologists have, over the years, argued that ancient subcircular structures are of meteoritic origin. In some cases petroleum is trapped within and adjacent to them. Notable examples include the Lyles Ranch field in south Texas (LeVie, 1986), the petroliferous Avak structure in Alaska (Kirschner et al., 1992), and the Red Wing Creek Field of the Williston Basin (Parson, 1974). No doubt there is a multiplicity of combination traps that might be associated with astroblemes, where organic-rich source rocks pre- or postdate the impact structure. This type of prospect may be particularly appealing to adherents of Gold's theory of the origin of petroleum.

Thus the list of combination traps is endless. The preceding examples illustrate some of their variety and complexity.

7.11 TRAPS: CONCLUSION

7.11.1 Timing of Trap Development Relative to Petroleum Migration and Reservoir Deposition

The time of trap formation relative to petroleum migration is obviously extremely important. If traps predate migration, they will be productive. If they postdate migration, they will be barren. Questions concerning a prospect should include these: Which horizons are known or presumed to be source rocks? Can time–burial depth curves be used to determine the time of petroleum generation? If the prospect is structural, did the fold or fault form before or after migration? In the case of truncation traps the source rock may underlie or overlie the unconformity (as in Hassi Messaoud and Prudhoe Bay, respectively). It is important, though, to establish that a truncation trap was thoroughly sealed before petroleum generation began.

Postmigration structural movement may also be relevant. Faults can open to allow petroleum to undergo further migration. Structural closure may tighten. This tightening is not harmful by itself, unless it is accompanied by crestal fractures, which increase the permeability of the seal. Uplift and erosion may breach the crests of traps. Regional tilting may trigger extensive secondary migration of petroleum because the spill points of traps may be altered.

The time factor is also important, although less so, when considering the relationship between the deposition of the reservoir and the formation of structural traps. At its simplest level syndepositional structural growth will affect sedimentation and hence reservoir characteristics; postdepositional structures, on the other hand, may not correlate with facies variations within the reservoir. In a terrigenous basin with synchronous structural growth, channels and fans of fluvial, deltaic, or submarine origin will develop in the lows. Simultaneously, winnowed marine shoal sands may be present on the structural highs, although, where too high, truncation may be present. In carbonate basins oolite shoals and reefs may be anticipated on structural highs, but, for reasons already discussed, porosity is often unrelated to facies in carbonate sediments.

The situation may be rather different in areas where structural traps postdate sedimentation. Once sediments are lithified, they respond to stress by fracturing. Thus in both sandstone and carbonate reservoirs, porosity and permeability due to fractures may be closely related to structure. Fracture intensity, and thus reservoir performance, may be enhanced over the crests of folds and the apices of diapirs, as well as adjacent to faults.

This review of the timing of trap formation relative to reservoir deposition and petroleum generation emphasizes its importance. Generally, the relationship between structural movement and petroleum migration is more important than whether structures are syn- or postdepositional.

7.11.2 Relative Frequency of the Different Types of Traps

This chapter has covered a great deal of ground. Therefore, it is appropriate to close with an attempt to arrange the various types of traps in some sort of order of importance. A great aid to this classification has been a global analysis of the trapping mechanisms of known giant oil fields by Moody (1975). His study showed that by far the majority of giant oil fields are anticlines, followed, a long way behind and in order of decreasing importance, by combination traps, reefs, pinchouts, truncations, salt domes, and faults (Fig. 7.50).

These interesting data deserve careful analysis. They only concern giant fields, which are defined as those with more than 500 million barrels of recoverable reserves. This chapter has shown that stratigraphically trapped oil tends to occur in small fields because of the limited extent of channel, bar, and reef reservoirs. Also, these data pertain to oil fields, not gas fields, although, intuitively, they would probably be similar.

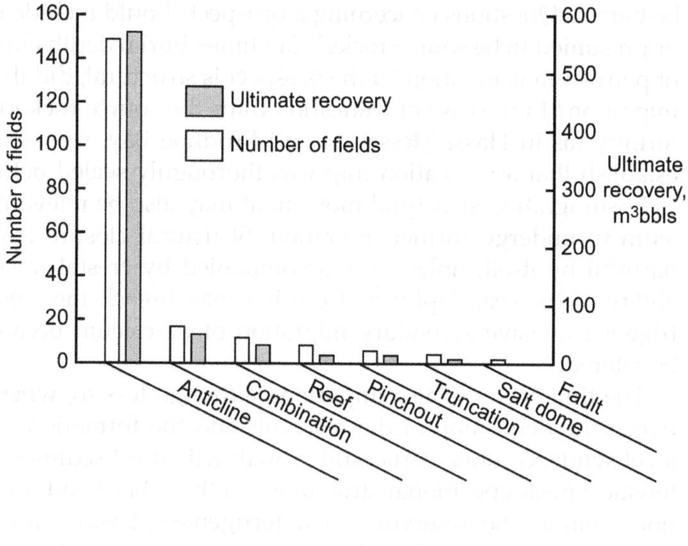

FIGURE 7.50 Histograms showing the mode of entrapment of giant oil fields around the world. *After Moody (1975).*

Note also that these figures were compiled some time ago. With the steady improvement in seismic geophysics, the percentage of known subtle stratigraphic traps may now have increased in proportion to anticlines. Most importantly, these figures record the entrapment of known giant oil fields. The various percentages of the different types of giant trap that actually exist under the earth might be completely different.

These figures, therefore, reflect mankind's ability to find oil, not the total number of fields in the world. Finding oil in anticlines is easy. They may be mapped at the surface or detected seismically in the subsurface. The concept of the anticlinal trap is a simple one to grasp for the managers, accountants, engineers, and farmers who may actually make the decision to drill, or not, based on a geologist's recommendation.

Stratigraphic traps, on the other hand, are harder to locate. Few can simply be picked off a brightly colored seismic section; most require an integration of seismic, log, and real rock data with sophisticated geological concepts. An explorationist would therefore find it harder to develop a stratigraphic trap prospect, and harder still to explain it to the lay audience, which may hold the purse strings. Pratt (1942) spoke very truly, though now politically incorrect, when he said that "oil is found in the minds of men."

References

Adams, J.E., 1936. Oil pool of open reservoir type. Am. Assoc. Pet. Geol. Bull. 20, 780–796.

Alsop, G.I., Blundell, D.J., Davison, I., 1995. Salt tectonics. Spec. Publ. Geol. Soc. London 100, 1–320.

Andresen, M.J., 1962. Paleodrainage patterns: their mapping from subsurface data, and their paleogeographic value. Am. Assoc. Pet. Geol. Bull. 46, 398–405.

Asquith, D.O., 1970. Depositional topography and major marine environments, late Cretaceous, Wyoming. Am. Assoc. Pet. Geol. Bull. 54, 1184–1224.

Bailey, R.J., Stoneley, R., 1981. Petroleum: entrapment and conclusion. In: Tarling, D.H. (Ed.), Economic Geology and Geotectonics. Blackwell, Oxford, pp. 73–97.

Balducchi, A., Pommier, G., 1970. Cambrian oil field of hassi massaoud, Algeria. Mem. Am. Assoc. Pet. Geol. 14, 477–488.

Barbat, F.W., 1958. The Los Angeles basin area, California. In: Weeks, L.G. (Ed.), The Habitat of Oil. Am. Assoc. Pet. Geol., Tulsa, OK, pp. 62–77.

Berg, R.R., 1975. Capillary pressures in stratigraphic traps. AAPG Bull. 59, 939–956.

Berg, R.R., Avery, A.H., 1995. Sealing properties of tertiary growth faults, Texas Gulf Coast. AAPG Bull. 79, 375–393.

Berg, R.R., Davies, D.K., 1968. Origin of lower cretaceous muddy sandstone at Bell creek field, Montana. Am. Assoc. Pet. Geol. Bull. 52, 1888–1898.

Berner, H., Ramsberg, H., Stephansson, O., 1972. Diapirism in theory and experiment. Tectonophysics 15, 197–218.

Biddle, K.T., Wielchowsky, C.C., 1994. Hydrocarbon traps. AAPG Mem. 60, 219–223.

Bishop, R.S., 1978. Mechanism for emplacement of piercement diapirs. AAPG Bull. 62, 1561–1584.

Blair, D.G., 1975. Structural styles in North Sea oil and gas fields. In: Woodland, A.W. (Ed.), Petroleum and the Continental Shelf of Northwest Europe, vol. I. Applied Science Publishers, London, pp. 327–338.

Brady, T.J., Campbell, N.D.H., Maher, C.E., 1980. Intisar 'D' oil field, Libya. AAPG Mem. 30, 543–564.

Brennand, T.P., Van Veen, F.R., 1975. The Auk oil field. In: Woodland, A.W. (Ed.), Petroleum and the Continental Shelf of Northwest Europe. Applied Science Publishers, London, pp. 275–284.

Brown, A.R., 1985. 3D interpretation of seismic data. AAPG Mem. 42, 1–341.

Busch, D.A., 1959. Prospecting for stratigraphic traps. Am. Assoc. Pet. Geol. Bull. 43, 2829–2843.

Busch, D.A., 1961. Prospecting for stratigraphic traps. In: Peterson, J.A., Osmond, J.C. (Eds.), Geometry of Sandstone Bodies. Am. Assoc. Pet. Geol., Tulsa, OK, pp. 220–232.

Busch, D.A., 1971. Genetic units in delta prospecting. Am. Assoc. Pet. Geol. Bull. 55, 1137–1154.

Busch, D.A., 1974. Stratigraphic Traps in Sandstones—Exploration Techniques. Mem. No. 21. Am. Assoc. Pet. Geol., Tulsa, OK.

Bushnell, H., January 12, 1981. Unconformities—key to N. slope oil. Oil Gas J. 112–118.

Cant, D.J., 1992. The stratigraphic and palaeogeographic context of shoreline-shelf reservoirs. In: Rhodes, E.G., Moslow, T.F. (Eds.), Marine Clastic Reservoirs. Springer-Verlag, Berlin, pp. 3–38.

Carll, J.F., 1880. The geology of the oil regions of Warren, Venango, Clarion and Butler Counties. Pa. Geol. Surv. 3, 1–482.

Chenoweth, P.A., 1972. Unconformity traps. Mem. Am. Assoc. Pet. Geol. 15, 42–46.

Christian, H.E., 1969. Some observations on the initiation of salt structures of the Southern British North Sea. In: Ion, D. (Ed.), The Exploration for Petroleum in Europe and North Africa. Inst. Pet., London, pp. 231–250.

Clapp, F.G., 1910. A proposed classification of petroleum and natural gas fields based on structure. Econ. Geol. 5, 503–521.

Clapp, F.G., 1917. Revision of the structural classification of the petroleum and natural gas fields. Geol. Soc. Am. Bull. 28, 553–602.

Clapp, F.G., 1929. Role of geologic structure in the accumulation of petroleum. In: Structure of Typical American Oil Fields, vol. 2. Am. Assoc. Pet. Geol., Tulsa, OK, pp. 667–716.

Cockroft, P.J., Edwards, G.A., Phoa, R.S.K., Reid, H.W., 1987. Applications of pressure analysis and hydrodynamics to petroleum exploration in Indonesia. Proc. Annu. Conv. Indones. Pet. Assoc. 16, 87–107.

Colmann-Sadd, S.P., 1978. Fold development in Zagros simply-folded belt, southwest Iran. AAPG Bull. 62, 984–1003.

Conybeare, C.E.B., 1976. Geomorphology of Oil and Gas Fields in Sandstone Bodies. Elsevier, Amsterdam.

Spec. Publ. No. 86. In: Cubitt, J.M., England, W.A. (Eds.), 1995. The Geochemistry of Reservoirs. Geol. Soc. London, London.

Curry, W.H., Curry III, W.H., 1972. South Glenrock oil field, Wyoming: pre discovery thinking and postdiscovery description. Mem. Am. Assoc. Pet. Geol. 16, 415–427.

Dailly, G.C., 1976. A possible mechanism relating progradation, growth faulting, clay diapiristn and overthrusting in a regressive sequence of sediments. Bull. Can. Pet. Geol. 24, 92–116.

Dott, R.H., Reynolds, M.J., 1969. Sourcebook for Petroleum Geology. Mem. No. 5. Am. Assoc. Pet. Geol., Tulsa, OK.

Downey, M.A., 1994. Hydrocarbon seal rocks. AAPG Mem. 60, 159–164.

Eremako, N.A., Michailov, I., 1974. Hydrodynamic pools at fault. Bull. Can. Pet. Geol. 22, 106–118.

Evamy, D.D., Haremboure, J., Kamerling, P., Knapp, W.A., Molloy, F.A., Rowlands, P.H., 1978. Hydrocarbon habitat of Tertiary Niger delta. AAPG Bull. 62, 1–39.

Evans, W.E., 1970. Imbricate linear sand bodies of Viking Formation in Dodsland-Hoosier area of southwestern Sakatchewan, Canada. AAPG Bull. 54, 469–486.

Falcon, N.L., 1958. Position of oil fields of southwest Iran with respect to relevant sedimentary basins. In: Weeks, L.G. (Ed.), The Habitat of Oil. Am. Assoc. Pet. Geol., Tulsa, OK, pp. 1279–1293.

Falcon, N.L., 1969. Problems of the relationship between surface structure and deep displacements illustrated by the Zagros Range. Spec. Publ. Geol. Soc. London 3, 9–22.

Ferris, C., 1972. Use of gravity meters in search for traps. Mem. Am. Assoc. Pet. Geol. 16, 252–270.

Fitch, A.A., 1976. Seismic Reflection Interpretation. Borntraeger, Berlin.

Gibson, R.G., 1994. Fault-zone seals in siliciclastic strata of the Columbus Basin, Offshore Trinidad. AAPG Bull. 78, 1372–1385.

Goebel, L.A., 1950. Cairo field, Union County, Arkansas. Am. Assoc. Pet. Geol. Bull. 34, 1954–1980.

Gray, W.D.T., Barnes, G., 1981. The heather oil field. In: Illing, L.V., Hobson, G.D. (Eds.), Petroleum Geology of the Continental Shelf of Northwest Europe. Heyden, London, pp. 335–341.

Halbouty, M.T., 1967. Salt Domes, Gulf Region, United States and Mexico. Gulf Publishing Co., Houston, TX.

Halbouty, M.T., 1972. Rationale for deliberate pursuit of stratigraphic, unconformity and paleogeomorphic traps. Am. Assoc. Pet. Geol. Bull. 56, 537–541.

Halbouty, M.T., 1994. East Texas field. In: Foster, N.H., Beaumont, E.A. (Eds.), Stratigraphic Traps II. Am. Assoc. Pet. Geol., Tulsa, OK, pp. 189–206.

Halbouty, M.T., Meyerhoff, A.A., King, R.E., Dott, R.H., Klemme, H.D., Shabad, T., 1970. World's giant oil and gas field, geologic factors affecting their formation and basin classification. Part I. Giant oil and gas fields. Mem. Am. Assoc. Pet. Geol. 14, 502–556.

Harding, T.P., 1973. Newport-Inglewood fault zone, Los Angeles basin, California. Am. Assoc. Pet. Geol. Bull. 57, 97–116.

Harding, T.P., 1974. Petroleum traps associated with wrench faults. AAPG Bull. 58, 1290–1304.

Harding, T.P., Lowell, J.D., 1979. Structural styles, their plate-tectonic habitats, and hydrocarbon traps in petroleum provinces. AAPG Bull. 63, 1016–1058.

Harms, J.C., 1966. Valley fill, western Nebraska. Am. Assoc. Pet. Geol. Bull. 50, 2119–2149.

Hartman, J.A., 1972. "G2", channel sandstone, Mainpass Block 35 field, offshore Louisiana. Am. Assoc. Pet. Geol. Bull. 56, 554–558.

Heuillon, B., 1972. Oil fields of Neocomian of Paris Basin, France. Mem. Am. Assoc. Pet. Geol. 10, 599–609.

Hobson, G.D., Tiratsoo, E.N., 1975. Introduction to Petroleum Geology. Scientific Press, Beaconsfield, IA.

Hollenshead, C.T., Pritchard, R.L., 1961. Geometry of producing Mesaverde sandstones, San Juan basin. In: Peterson, J.A., Osmond, J.C. (Eds.), Geometry of Sandstone Bodies. Am. Assoc. Pet. Geol., Tulsa, OK, pp. 98–118.

Hubbert, M.K., 1953. Entrapment of petroleum under hydrodynamic conditions. Am. Assoc. Pet. Geol. Bull. 37, 1454–2026.

Hull, C.E., Warman, H.R., 1970. Asmari oil fields of Iran. Mem. Am. Assoc. Pet. Geol. 14, 428–437.

Jackson, M.P.A., Roberts, D.G., Snelson, S., 1996. Salt tectonics: a global perspective. AAPG Mem. 65, 1–454.

Jamison, H.C., Brockett, L.D., McIntosh, R.A., 1980. Prudhoe Bay: a 10-year perspective. AAPG Mem. 30, 289–314.

Janoschek, R., 1958. The Inner Alpine Vienna Basin, an example of a small sedimentary area with rich oil accumulation. In: Weeks, L.G. (Ed.), The Habitat of Oil. Am. Assoc. Pet. Geol., Tulsa, OK, pp. 1134–1152.

Jones, H.P., Speers, R.G., 1976. Permo-triassic reservoirs on Prudhoe Bay field, North slope, Alaska. Mem. Am. Assoc. Pet. Geol. 24, 23–50.

Jenyon, M.K., 1990. Oil and Gas Traps. Wiley, Chichester, UK.

Jones, O.A., Endean, R., 1973. Biology and Geology of Coral Reefs, 2 vols. Academic Press, London.

Kent, P., 1979. The emergent Hormuz salt plugs of southern Iran. J. Pet. Geol. 2, 117–144.

Kessler, L.G., Zang, R.D., Englehorn, J.A., Eger, J.D., 1980. Stratigraphy and sedimentology of a Palaeocene submarine fan complex, Cod Field, Norwegian North Sea. In: The Sedimentation of North Sea Reservoir Rocks, vol. 7. Norwegian Petroleum Society, Oslo, pp. 1–19.

Mem. No. 16. In: King, R.E. (Ed.), 1972. Stratigraphic Oil and Gas Fields—Classification, Exploration Methods, and Case Histories. Am. Assoc. Pet. Geol., Tulsa, OK.

Kirschner, C.E., Grantz, A., Mullen, M.W., 1992. Impact origin of the Avak Structure, Arctic Alaska and genesis of the Barrow gas fields. AAPG Bull. 76, 651–679.

Krooss, B.M., Leythauser, D., Schaefer, R.G., 1992. The quantification of diffusive hydrocarbon losses through cap rocks of natural gas reservoirs—a reevaluation. AAPG Bull. 76, 403–406.

Kyle, J.R., Posey, H.H., 1991. Halokinesis, cap rock development and salt dome mineral resources. In: Melvin, J.L. (Ed.), Evaporites, Petroleum and Mineral Resources. Elsevier, Amsterdam, pp. 413–476.
Lamb, C.F., 1980. Painter reservoir field—giant in Wyoming thrust belt. AAPG Bull. 64, 638–644.
Spec. Publ. No. 18. In: Laporte, L.F. (Ed.), 1974. Reefs in Time and Space. Soc. Econ. Paleontol Mineral., Tulsa, OK.
Larter, S.R., Aplin, A.C., 1995. Reservoir geochemistry: methods, applications and opportunities. Spec. Publ. Geol. Soc. London 86, 5–32.
Lees, G.M., 1952. Foreland folding. Q. J. Geol. Soc. London 108, 1–34.
LeVie, D.S., 1986. South Texas' Lyles ranch field: production from an astrobleme? Oil Gas J. 135–138.
Levorsen, A.I., 1934. Relation of oil and gas pools to unconformities in the mid Continent region. In: Wrather, R.E., Lahee, F.H. (Eds.), Problems of Petroleum Geology. Am. Assoc. Pet. Geol., Tulsa, OK, pp. 761–784.
Levorsen, A.I., 1964. Big geology for big needs. Am. Assoc. Pet. Geol. Bull. 48, 141–156.
Levorsen, A.I., 1966. The obscure and subtle trap. Am. Assoc. Pet. Geol. Bull. 50, 2058–2067.
Levorsen, A.I., 1967. Geology of Petroleum, second ed. Freeman, San Francisco.
Lewis, C.J., 1990. Sarir field. In: Beaumont, E.A., Foster, N.H. (Eds.), Treatise of Petroleum Geology. Atlas of Oil and Gas Fields. Structural Traps II. Am. Assoc. Pet. Geol., Tulsa, OK, pp. 253–267.
Lovely, H.R., 1943. Classification of oil reservoirs. Am. Assoc. Pet. Geol. Bull. 27, 224–237.
Li, M., Taisheng, G., Xueping, Z., Taiju, Z., Rong, G., Zhenrong, D., 1982. Oil basins and subtle traps in the eastern part of China. In: Halbouty, M.T. (Ed.), The Deliberate Search for the Subtle Trap. Am. Assoc. Pet. Geol., Tulsa, OK, pp. 287–316.
Mackenzie, D.B., 1972. Primary stratigraphic traps in sandstone. Mem. Am. Assoc. Pet. Geol. 16, 47–63.
Martin, D.B., 1963. Rosedale channel: evidence for late Miocene submarine erosion in Great Valley of California. Am. Assoc. Pet. Geol. Bull. 47, 441–456.
Martin, R., 1966. Paleogeomorphology and its application to exploration for oil and gas (with examples from Western Canada). Am. Assoc. Pet. Geol. Bull. 50, 2277–2311.
Maxwell, W.G.H., 1968. Atlas of the Great Barrier Reef. Elsevier, Amsterdam.
Mayuga, M.N., 1970. California's giant—Wilmington oil field. Mem. Am. Assoc. Pet. Geol. 14, 158–184.
McCubbin, D.G., 1969. Cretaceous strike-valley sandstone reservoirs, northwestern New Mexico. Am. Assoc. Pet. Geol. Bull. 53, 2114–2140.
McGregor, A.A., Biggs, C.A., 1970. Bell Creek field, Montana: a rich stratigraphic trap. Mem. Am. Assoc. Pet. Geol. 14, 128–146.
McGregor, A.A., Biggs, C.A., 1972. Bell Creek field, Montana: a rich stratigraphic trap. Mem. Am. Assoc. Pet. Geol. 16, 367–375.
Meckel, L.D., Nath, A.K., 1977. Geological consideration of stratigraphic trap modelling. Mem. Am. Assoc. Pet. Geol. 26, 417–438.
Milton, N.J., Bertram, G.T., 1992. Trap-styles—a new classification based on sealing surfaces. AAPG Bull. 76, 983–999.
Minor, H.E., Hanna, M.A., 1941. East Texas oil field. In: Stratigraphic Type Oil Fields. Am. Assoc. Pet. Geol., Tulsa, OK, pp. 600–640.
Moody, J.D., 1975. Distribution and geological characteristics of giant oil fields. In: Fischer, A.G., Judson, S. (Eds.), Petroleum and Global Tectonics. Princeton University Press, Princeton, NJ, pp. 307–320.
Morgridge, D.L., Smith, W.B., 1972. Geology and discovery of prudhoe Bay field, eastern arctic slope, Alaska. Mem. Am. Assoc. Pet. Geol. 16, 489–501.
Orton, E., 1889. The Trenton limestone as a source for petroleum and inflammable gas in Ohio and Indiana. U.S. Geol. Surv. Annu. Rep. 8, 475–662.
Oxley, M.L., Herliny, D.E., 1972. The Bryan field—a sedimentary anticline. Geophysics 37, 59–67.
Parson, E.S., 1974. The Red Wing creek field, North Dakota—an extraterrestrial hydrocarbon trap. Am. Assoc. Pet. Geol. 58, 910.
Pratt, W.E., 1942. Oil in the Earth. University of Kansas Press, Lawrence.
Rijks, E.J.H., Jauffred, J.C.E.M., September 1991. Attribute extraction: an important application in any detailed 3D interpretation study. Geophys. Leading Edge Explor. 11–19.
Rittenhouse, G., 1972. Stratigraphic trap classification. Mem. Am. Assoc. Pet. Geol. 16, 14–28.
Sabins, F.F., 1963. Anatomy of a stratigraphic trap. Am. Assoc. Pet. Geol. Bull. 47, 193–228.

Sabins, F.F., 1972. Comparison of bisti and Horseshoe Canyon stratigraphic traps, San Juan basin, New Mexico. Mem. Am. Assoc. Pet. Geol. 10, 610–622.

Sanford, R.M., 1970. Sarir oil field, Libya—desert surprise. Mem. Am. Assoc. Pet. Geol. 14, 449–476.

Schwade, I.T., Carlson, S.A., O'Flynn, J.B., 1958. Geologic environment of Cuyuma Valley oil fields, California. In: Weeks, L.G. (Ed.), The Habitat of Oil. Am. Assoc. Pet. Geol., Tulsa, OK, pp. 78–98.

Shelton, J.W., 1967. Stratigraphic models and general criteria for recognition of alluvial, barrier bar and turbidity current sand deposits. Am. Assoc. Pet. Geol. Bull. 51, 2441–2460.

Shelton, J.W., 1968. Role of contemporaneous faulting during basinal subsidence. Am. Assoc. Pet. Geol. Bull. 52, 399–413.

Simonson, R.R., 1958. Oil in the San Joachim Valley. In: Weeks, L.G. (Ed.), The Habitat of Oil. Am. Assoc. Pet. Geol., Tulsa, OK, pp. 99–112.

Slinger, F.C.P., Crichton, 1959. The geology and development of the Gachsaran field, southwest Iran. Proc. World Pet. Congr. 5, 349–375. Sect. 1.

Smith, D.A., 1980. Sealing and non-sealing faults in Louisiana Gulf Coast salt basin. AAPG Bull. 64, 145–172.

Stanley, T.B., 1970. Vicksburg fault zone, Texas. Mem. Am. Assoc. Pet. Geol. 14, 301–308.

Terry, C.E., Williams, J.J., 1969. The Idris "A" bioherm and oilfield, Sirte basin, Libya. In: Hepple, P. (Ed.), The Exploration for Petroleum in Europe and North Africa. Inst. Pet., London, pp. 31–48.

Tobison, N.H., 1972. Boxer field, Morgan County, Colorado, 1972. Mem. Am. Assoc. Pet. Geol. 10, 383–388.

Van den Berg, E., Thomas, O.T., 1980. Ekofisk: first of the giant oil fields in Western Europe. AAPG Mem. 30, 195–224.

Watson, H.J., 1982. Casablanca field, Offshore Spain, a palcogeomorphic trap. In: Halbouty, M.T. (Ed.), The Deliberate Search for the Subtle Trap, Am. Assoc. Pet. Geol. Mem., vol. 32, pp. 237–250.

Weber, K.J., Daukoru, E., 1975. Petroleum geology of the Niger delta. Trans. World Pet. Congr. 9 (2), 209–221.

Weber, K.J., Mandl, G., Pilaar, W.F., Lehner, F., Precious, R.G., 1980. The role of faults in hydrocarbon migration and trapping in Nigerian growth fault structures. In: Offshore Technol. Conf. Pap., vol. 3356, pp. 2643–2651.

Weimer, R.J., Davis, T.L., 1977. Stratigraphic and seismic evidence for late Cretaceous faulting, Denver basin, Colorado. Mem. Am. Assoc. Pet. Geol. 26, 277–300.

Werren, E.G., Shew, R.D., Adams, E.R., Stancliffe, R.J., 1990. Meander-belt reservoir geology, mid-dip Tuscaloosa, Little Creek field, Mississippi. In: Barwis, J.H. (Ed.), Sandstone Petroleum Reservoirs. Springer-Verlag, New York, pp. 85–108.

Wilcox, R.E., Harding, T.P., Seely, D.R., 1973. Basic wrench tectonics. Am. Assoc. Pet. Geol. Bull. 57, 74–96.

Wilhelms, A., Larter, S.R., 1994a. Origin of tar mats in petroleum reservoirs. Part 1. Introduction and case studies. Mar. Pet. Geol. 11, 418–441.

Wilhelms, A., Larter, S.R., 1994b. Origin of tar mats in petroleum reservoirs. Part 2. Formation mechanisms for tar mats. Mar. Pet. Geol. 11, 442–456.

Wilhelms, A., Larter, S.R., 1995. Overview of the geochemistry of some tar mats from the North Sea and USA: implication for tar mat origin. Spec. Publ. Geol. Soc. London 86, 87–102.

Williams, J.J., 1968. The stratigraphy and igneous reservoirs of the Augila field, Libya. In: Barr, T.F. (Ed.), Geology and Archeology of Northern Cyrenaica, Libya. Pet. Explor. Soc., Libya, pp. 197–206.

Williams, J.J., 1972. Augila field, Libya: depositional environment and diagenesis of sedimentary reservoir and description of igneous reservoir. Mem. Am. Assoc. Pet. Geol. 16, 623–632.

Williams, J.J., Connor, D.C., Peterson, K.E., 1975. The Piper oil-field, U. K. North Sea: a fault-block structure with Upper Jurassic beach-bar reservoir sands. In: Woodland, A.W. (Ed.), Petroleum and the Continental Shelf of North West Europe, vol. 1. Applied Science Publishers, London, pp. 363–378.

Wilson, H.H., 1977. 'Frozen-in' hydrocarbon accumulations in diagenetic traps—exploration targets. AAPG Bull. 61, 483–491.

Woncik, J., 1972. Recluse field, Campbell County, Wyoming. Mem. Am. Assoc. Pet. Geol. 16, 376–382.

Wood, J.M., Hopkins, J.C., 1992. Traps associated with paleovalleys and interfluves in an unconformity bounded sequence: lower Cretaceous glauconitic member, Southern Alberta, Canada. AAPG Bull. 76, 904–926.

Yuster, S.T., 1953. Some theoretical considerations of tilted water tables. Trans. Am. Inst. Min. Metall. Eng. 198, 149–153. Tech. Pap. 3564.

Zawiska, L., June 1986. Hydrodynamic conditions of hydrocarbon accumulation exemplified by the Carboniferous formation in the Lublin Synclinorium, Poland. Form. Eval. Soc. Pet. Eng. 286–294.

Selected Bibliography

For major compilations of oil field case histories, see:

Mem. No. 14. In: Halbouty, M.T. (Ed.), 1970. Geology of Giant Petroleum Fields. Am. Assoc. Pet. Geol., Tulsa, OK.

Mem. No. 30. In: Halbouty, M.T. (Ed.), 1980. Giant Oil and Gas Fields in the Decade: 1968–1978. Am. Assoc. Pet. Geol., Tulsa, OK.

Mem. No. 16. In: King, R.E. (Ed.), 1972. Stratigraphic Oil and Gas Fields. Am. Assoc. Pet. Geol., Tulsa, OK.

For an elegant review of the relationship between trap types, sedimentary basins, and plate tectonics, see:

Harding, T.P., Lowell, J.D., 1979. Structural styles, their plate tectonic habitats, and hydrocarbon traps in petroleum provinces. AAPG Bull. 63, 1016–1058.

CHAPTER

8

Sedimentary Basins and Petroleum Systems

8.1 BASIC CONCEPTS AND TERMS

A sedimentary basin is an area of the earth's crust that is underlain by a thick sequence of sedimentary rocks. Hydrocarbons commonly occur in sedimentary basins and are absent from intervening areas of igneous and metamorphic rocks (North, 1971). This fundamental truth is one of the cornerstones of the sedimentary–organic theory for the origin of hydrocarbons. (This theory is in opposition to the cosmic–igneous theory discussed in Chapter 5.) Therefore it is important to direct our attention not only to the details of traps and reservoir rocks but also to the broader aspects of sedimentary basin analysis. Before acquiring acreage in a new area, and long before attempting to locate drillable prospects, it is necessary to establish the type of basin to be evaluated and to consider what productive fairways it may contain and where they may be extensively located. This chapter describes the various types of basin with reference to examples from around the world and discusses the relationship between the genesis and evolution of a basin and its hydrocarbon potential.

First, however, some of the basic terms and concepts must be defined. A sedimentary basin is an area on the earth's surface where sediments have accumulated to a greater thickness than they have in adjacent areas. No clear boundary exists between the lower size limit of a basin and the upper limit of a syncline. Most geologists would probably take the view that a length of more than 100 km and a width of more than 10 km would be a useful dividing line. Most sedimentary basins cover tens of thousands of square kilometers and may contain more than 5 km of sedimentary fill. Note that a sedimentary basin is defined as an area of thick sediment, with no reference to its topography. A sedimentary basin may occur as part of a mountain chain, beneath a continental peneplain, or in an ocean. Conversely, a present-day ocean basin need not necessarily qualify as a sedimentary basin; indeed, many are floored by igneous rocks with only a veneer of sediment.

This distinction between topographic and sedimentary basins needs further elaboration. Both types of basin have a depressed basement. Sedimentary basins may or may not have

Elements of Petroleum Geology
http://dx.doi.org/10.1016/B978-0-12-386031-6.00008-4

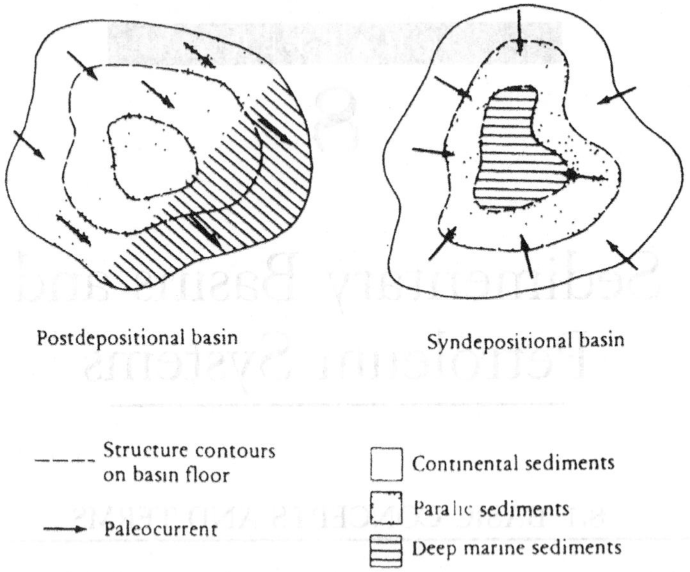

FIGURE 8.1 The differences between syndepositional and postdepositional sedimentary basins.

been marked topographic basins during their history. Many basins are infilled with continental and shallow marine sediments, and totally lack deep-sea deposits.

Similarly, a distinction needs to be made between syndepositional and postdepositional basins. Most sedimentary basins indicate that subsidence and deposition took place simultaneously. This simultaneous occurrence is shown by facies changes and paleocurrents that are concordant with structure. On the other hand, in some basins paleocurrent directions and facies are discordant with and clearly predate the present structure (Fig. 8.1). This is particularly characteristic of intracratonic basins, as is shown later. The distinction between these two types of basins is critically important in petroleum exploration because of the need for traps to have formed before hydrocarbon generation and migration. Stratigraphic traps are generally formed before migration, except for rare diagenetic traps. Structural traps may predate or postdate migration, and establishing the chronology correctly is essential.

A further important distinction must be made between topography and sediment thickness. When examining regional isopach or isochron maps, it is tempting to assume that they are an indication of the paleotopography of the basin. This is by no means always true. The depocenter (area of greatest sediment thickness) is not always found in the topographic nadir of the basin, but may frequently be a linear zone along the basin margin. This is true of terrigenous sediments, where maximum deposition may take place along the edge of a delta. Sediments thin out from the delta front both up the basin margin and also seaward. Similarly, in carbonate basins most deposition takes place along shelf margins, where organisms thrive in shallow, well-oxygenated conditions with abundant nutrients. Thus reefs and skeletal and oolite sands thin out toward basin margin sabkhas and basinward into condensed sequences of lime mud.

Many studies have shown that a depocenter may migrate across a basin. The topographic center of the basin need not necessarily move with it. Examples of this phenomenon have been documented from Gabon, the Maranhao Basin of Brazil, and Iraq (Belmonte et al., 1965; Mesner and Woodridge, 1964; Ibrahim, 1979; respectively). Note that the thickness of each of the formations measured at outcrop should not be added to determine the overall thickness of sediment within a basin. This measurement can only be made from drilling or geophysical data (Fig. 8.2).

Now that basins have been considered in time and profile, they may be viewed in plan. The term *basin* has two interpretations. In the broadest sense, as already defined, a sedimentary basin is an area of the earth's surface underlain by sediments. In a narrower sense basins may be subdivided into true basins; those that are subcircular in plan and those that are elongate (troughs). Embayments, lacking centripetal closure, are basins that open out into larger basins (Fig. 8.3).

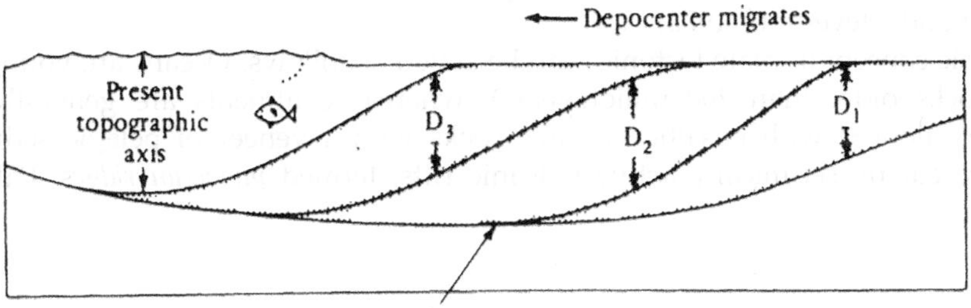

FIGURE 8.2 Cross-section illustrating migrating basin depocenters. Note how measuring the apparent thickness of each unit at the surface, and summing them, will give an erroneous overall thickness of the basin fill.

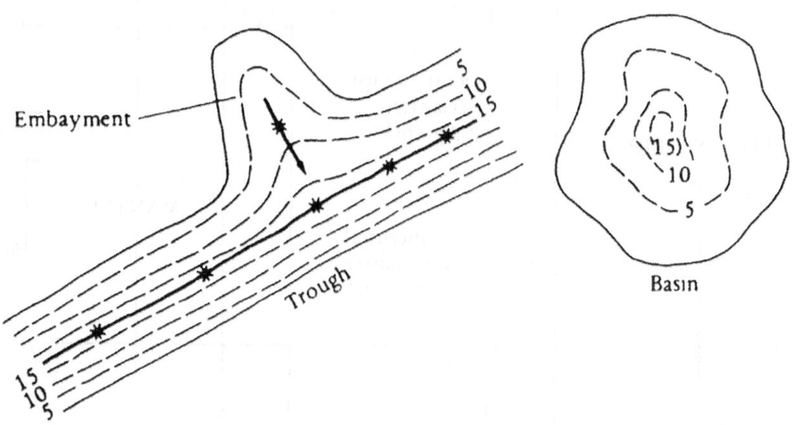

FIGURE 8.3 Basins, defined as areas of the earth's surface underlain by sediments, may be subdivided into true basins, embayments, and troughs. Contours are in kilometers.

8.2 MECHANISMS OF BASIN FORMATION

Sedimentary basins form part of the earth's crust, or lithosphere; they are generally distinguishable from granitic continental and basaltic oceanic crust by their lower densities and slower seismic velocities. Beneath these crustal elements is the more continuous subcrustal lithosphere. The crust is thin, dense, and topographically low across the ocean basins, but thick, of lower density, and, consequently, of higher elevation over the continents (Fig. 8.4). The lithosphere is made up of a series of rigid plates, which overlie the denser, yet viscous, asthenosphere.

The lithospheric plates drift slowly across the asthenosphere. Knowledge of plate tectonics is of fundamental importance in understanding sedimentary basins. Detailed exposition of this topic is beyond the scope of this text, but a brief summary is necessary before considering the mechanics of basin formation. Further details are found in Seyfert and Sirkin (1973), Fischer and Judson (1975), Tarling and Runcorn (1973), Davies and Runcorn (1980), Tarling (1981), and Allen and Allen (1990). A skeptical review of these ideas can be gained from Meyerhoff and Meyerhoff (1972).

The basic concept of plate tectonics can be stated as follows. Oceans are young (generally lacking rocks older than 200 million years), whereas continents are generally far older. Oceans are floored with basaltic volcanic rocks with a veneer of pelagic sediments. The oceans are cut by seismically active volcanic rifts, termed *midocean ridges*. Paleomagnetic

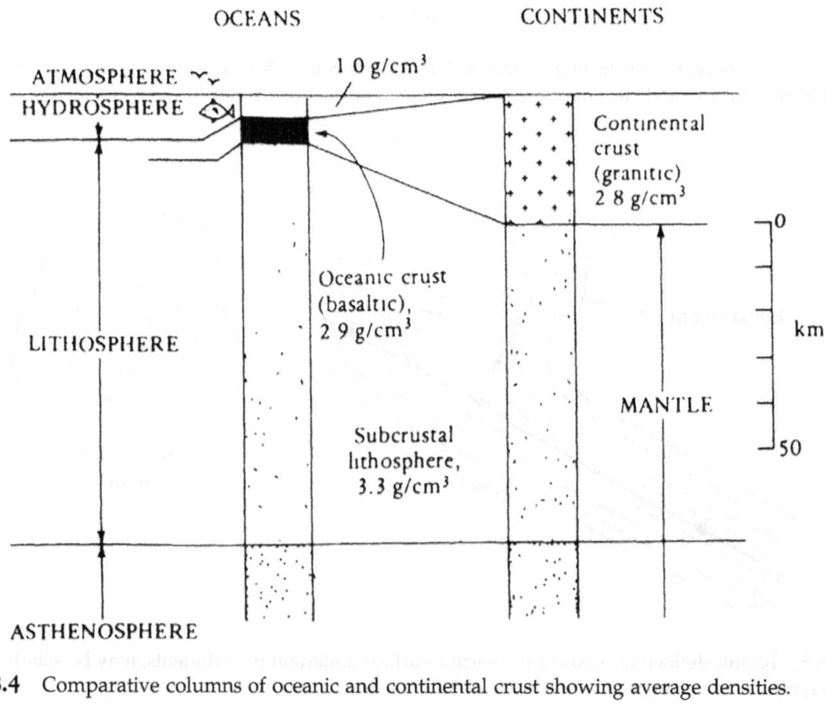

FIGURE 8.4 Comparative columns of oceanic and continental crust showing average densities.

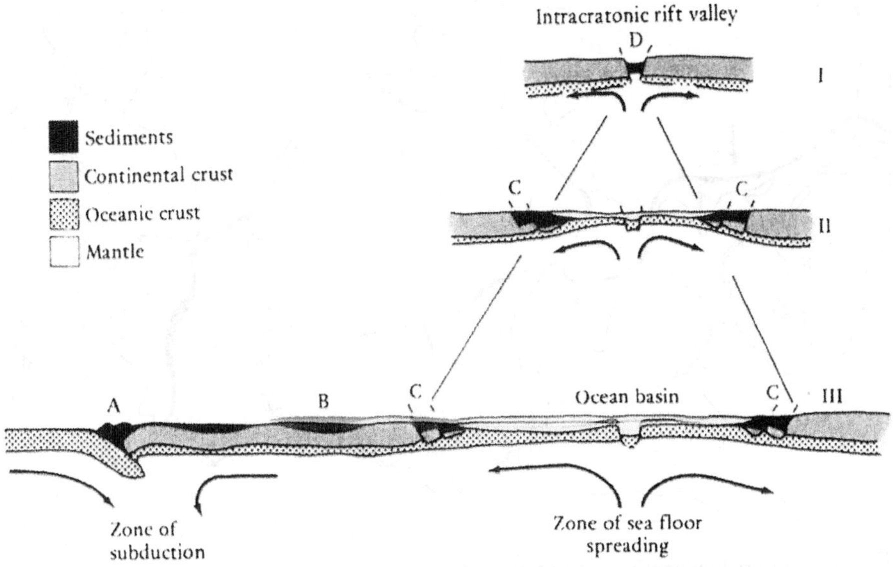

FIGURE 8.5 Cross-sections illustrating the basic concepts of plate tectonics, showing how basins form in response to crustal movement driven by convection cells in the mantle. (I) An axis of sea floor spreading develops, in this instance, beneath continental crust. Updoming occurs and a rift valley is formed (D). The East African rifts are a modern example. (II) Crustal separation causes the rift to split into two continental margin basins (C) separated by a widening ocean. (III) Concomitant with the formation of new oceanic crust at the spreading ridge, crust is drawn down into the mantle at zones of subduction (A). In these areas deep basins undergo extensive tectonism as their sediments are compressed by the converging plates. Intracratonic sag basins develop intermittently on areas of continental crust (B).

reversals and age dating show that rocks become progressively older away from the ridges and toward the continental margins. The midocean ridges can be traced landward into sediment-infilled rifts within continental granitic crust. The evidence suggests, therefore, that new crust, largely of basaltic composition, forms where tension and upwelling occur along these zones of sea floor spreading. Simultaneously, crust is drawn down into the asthenosphere at complementary linear features known as *zones of subduction*. These zones appear as folded troughs of sediment within or adjacent to continental masses and as volcanic island arcs within the oceans. Figure 8.5 shows the process of crustal gestation and digestion, and Fig. 8.6 shows the recognized plate boundaries of the earth.

Although there is general unanimity on the identification of the major plate boundaries, details of some of the smaller ones (microplates) are still somewhat unclear. Three types of plate boundaries are recognized: trailing, subductive, and transcurrent. Trailing, or rift, boundaries occur where new crust forms and plates diverge. Subductive boundaries occur where plates converge. Some plate boundaries are transcurrent where two plates move past each other. Transcurrent plate boundaries are marked by extensive transform faulting accompanied by deep basins and thrust belts of local extent but great complexity (Crowell, 1974; Dickinson and Seely, 1979).

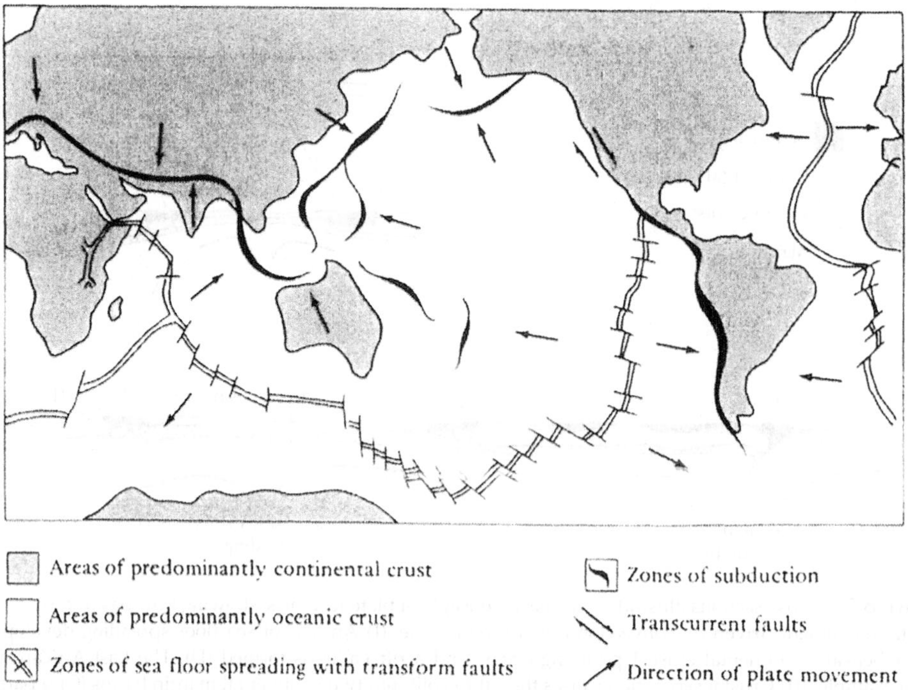

FIGURE 8.6 Map of the Earth showing approximate distribution of oceanic and continental crust and plate boundaries. *After Heirtzler (1968), Vine (1970), others.*

Basins can form in four main ways (Fischer, 1975). Three of these processes are summarized in Fig. 8.7. One major group of basins, the rift basins, forms as a direct result of crustal tension at the zones of sea floor spreading. A second major group of basins occurs as a result of crustal compression at convergent plate boundaries. A third type of basin can form in response not to lateral forces but to vertical crustal movements. For reasons not fully understood, phase changes can take place beneath the lithosphere. These changes may take the form of localized cooling, and therefore contraction, resulting in a superficial hollow, which becomes infilled by sediment. Conversely, the lithosphere may locally heat up and expand, causing an arching of the crust. Erosion of this zone will then occur. Sometimes this crustal doming is a precursor to rifting and drifting. Alternatively, subsequent cooling and subsidence result in the formation of an intracratonic hollow, which may be infilled with sediment.

A fourth mechanism of basin formation is simple crustal loading due to sedimentation. This process poses a "chicken-and-egg" problem, however. Basins of this type require an initial depression in the crust before deposition may begin. Thus, loaded basins characterize continental margins where a prograding delta can initiate and maintain the depression of adjacent oceanic crust.

Basins formed as a result of crustal thinning and rifting are of particular interest to the petroleum industry because they are an important habitat for petroleum. Many theories

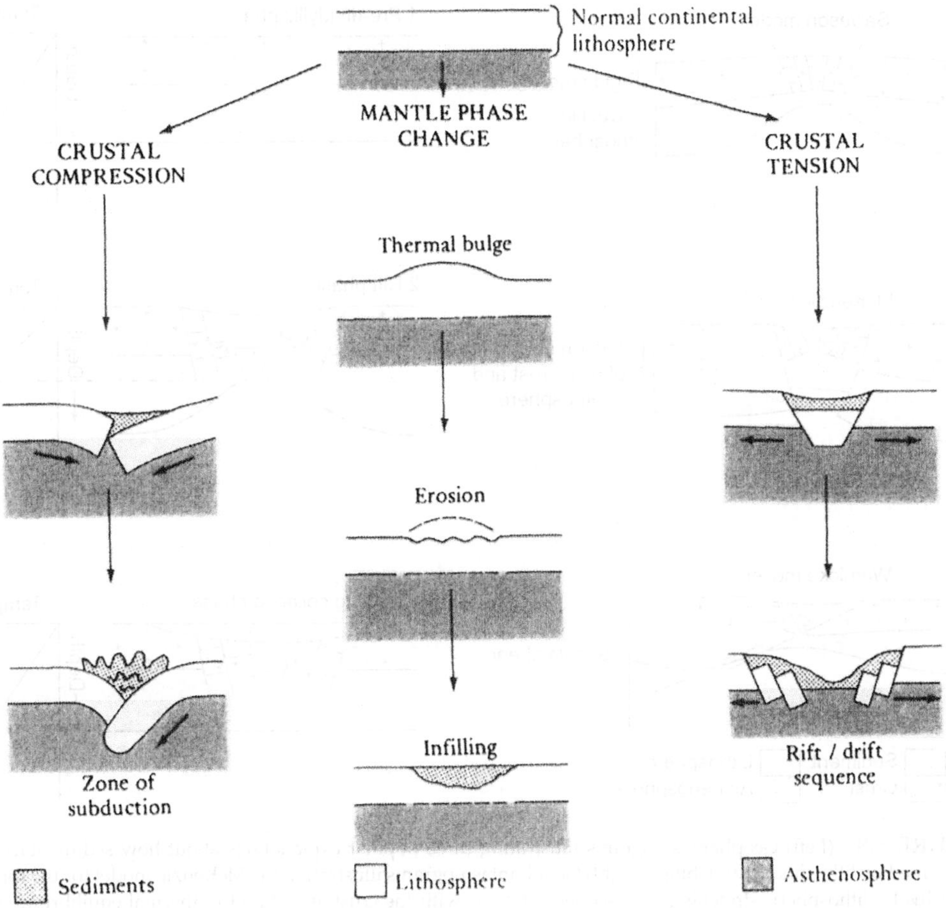

FIGURE 8.7 Cross-sections showing the various types of basin formation discussed in the text. *After Fischer (1975).*

have been advanced to explain their formation. A useful review of these can be found in Allen and Allen (1990). Of the many models proposed, three are particularly significant:

1. Salveson (1976, 1979) proposed a model of passive crustal separation, in which the continental crust was deemed to deform by brittle failure, while the subcrustal lithosphere is thinned by ductile necking. This model was largely based on studies of the Red Sea and Gulf of Suez rift system.
2. McKenzie (1978) proposed a model that assumed that both the crust and subcrustal lithosphere deformed by brittle failure. This model was largely based on studies of the North Sea Basin.
3. Wernicke (1981, 1985) proposed a model for crustal thinning by means of simple shear, in which a low-angle fault extends from the surface right through the lithosphere. This model was largely based on studies of the basin and range tectonic province of North America.

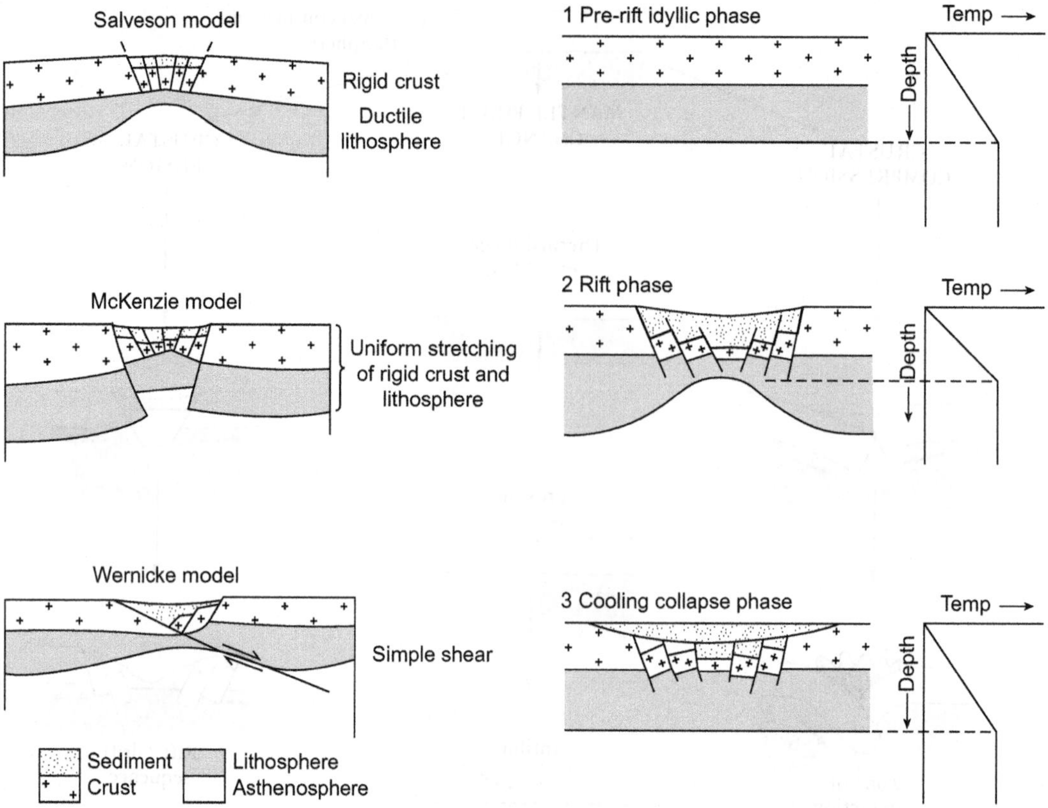

FIGURE 8.8 (Left) Geophantasmograms illustrating three popular explanations about how sedimentary basins are formed by lithospheric stretching. (Right) Geophantasmograms illustrating the McKenzie model for the formation of basins by lithospheric stretching. The sequence begins with the crust at rest and in thermal equilibrium. Crustal thinning and uplift of the asthenosphere are associated with high heat flow and the formation of a rift basin. Subsequent cooling and shrinkage causes the crust to collapse gently resulting in the "steer's head" basin form. Thermal equilibrium is finally reestablished.

These three models are illustrated in Fig. 8.8 on the left. The McKenzie model has received particular interest in the oil industry because it offers a means of predicting the history of heat flow in a sedimentary basin. This is a prerequisite to the accurate modeling of petroleum generation. In the McKenzie model rifting commences on a level surface that is in thermal equilibrium. As the crust thins and rifting develops, the heat flux increases and the temperature of the shallow rocks rises. After rifting has ceased, the crust cools, shrinks, and collapses. Sedimentation continues, but now infills a gently subsiding basin. Faults die out at the top of the syn-rift sediments. The crust returns to thermal equilibrium. The resultant basin is colloquially referred to as a "steer's head" basin, because it is reminiscent of a Texas Longhorn, or Highland Cattle (Fig. 8.8, right).

McKenzie demonstrated mathematically that the heat flow within a basin was related to the amount of crustal stretching, termed the β value. The higher the rate of stretching, the higher the heat flux during the initial phase of rifting (Fig. 8.9, upper). The β value for a basin

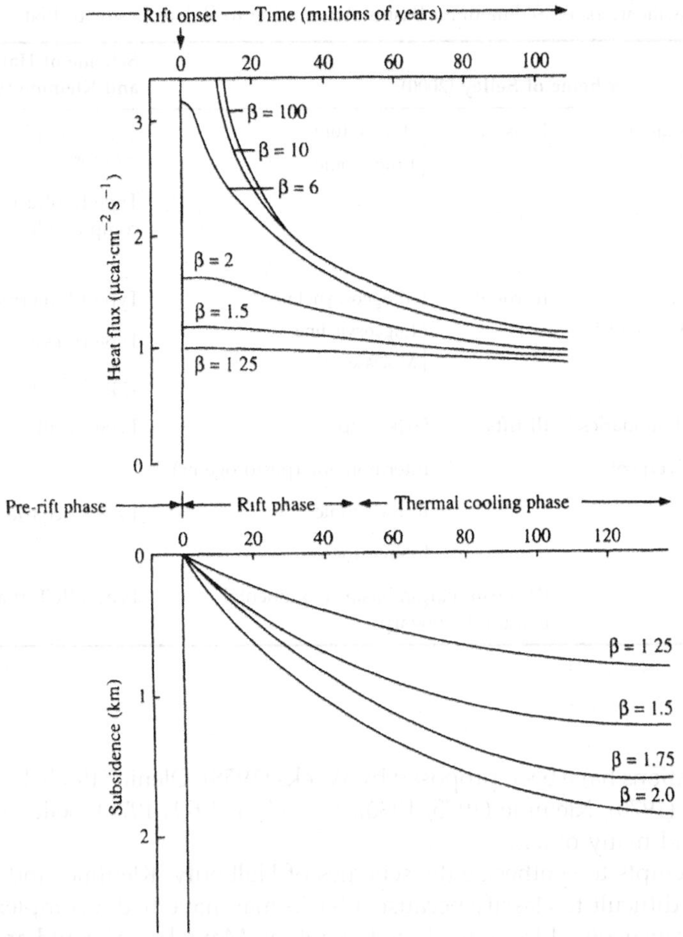

FIGURE 8.9 Illustration of the McKenzie model for basin formation by lithospheric stretching. (Upper) Graph of heat flux against time for various β factors. (Lower) Graph of subsidence against time for various β factors. *Developed from McKenzie (1978).*

may be discovered by constructing a burial history curve for the basin and comparing it with known curves calculated for given β factors (Fig. 8.9, lower). The accurate prediction of the history of heat flow in a sedimentary basin is a prerequisite to the accurate modeling of petroleum generation (Dore et al., 1991; Helbig, 1994).

8.3 CLASSIFICATION OF SEDIMENTARY BASINS

Many schemes have been proposed to classify sedimentary basins. The early schemes were largely descriptive. Today, with the current understanding of plate tectonics, it is now possible to devise schemes that are not only descriptive but also genetic (Busby and Ingersoll,

TABLE 8.1 Classifications of Sedimentary Basins Attempting to Relate Basins to Plate Tectonics

Scheme of Selley (2000)			Scheme of Halbouty et al. (1970) and Klemme (1975, 1980)
Cratonic suite associated with crustal stability	I basins	Intracratonic Epicratonic	Type I simple, saucer-shaped interior Type II intracontinental composite foreland shelf
Geosynclinal suite at convergent plate boundaries	II troughs	Miogeosuyncline Eugcosyncline Molasse	Type VI intermontane Type IV extracontinental downwarp Type VII intermontane
Transcurrent plate boundaries	III rifts	Strike-slip	Type III rifts
Rift-drift suite at divergent plate boundaries		Intermontane (postorogenic)	
		Intracratonic	Type IV coastal graben pull-apart
		Intercratonic	
	IV ocean margin basins (continental margin downwarp)		Type VIII Tertiary deltas

1995). Classifications have been proposed by Weeks (1958), Olenin (1967), Uspenskaya (1967), Halbouty et al. (1970), Klemme (1975, 1980), Perrodon (1971, 1978), Selley (2000), Allen and Allen (1990), and many others.

Table 8.1 attempts to synthesize the schemes of Halbouty, Klemme, and Selley. Sedimentary basins are difficult to classify because a basin may have had a complex history, during which it may have evolved from one type to another. Many basins could arguably be placed in more than one class. No great weight should be attached to Table 8.1. Its main merit is that it shows how basins can be linked to their genesis, providing a logical framework for the ensuing description of the various types of basins (Fig. 8.10).

8.4 CRATONIC BASINS

Cratonic basins are essentially subcircular basins that lie wholly or dominantly on granitic continental crust. The floors of such basins may be broken into a mosaic of horsts and grabens, but major rifting is absent. The genesis of circular sag basins, such as those of Africa, has long attracted attention. Several modes of origin have been postulated. One of the most popular mode proposes that thermal doming over a mantle "hot spot" generates doming of the crust, leading to the erosion of uplifted crustal rocks, followed by cooling and crustal collapse, initially into a rift, followed by gentle sag subsidence (Allen and Allen, 1990).

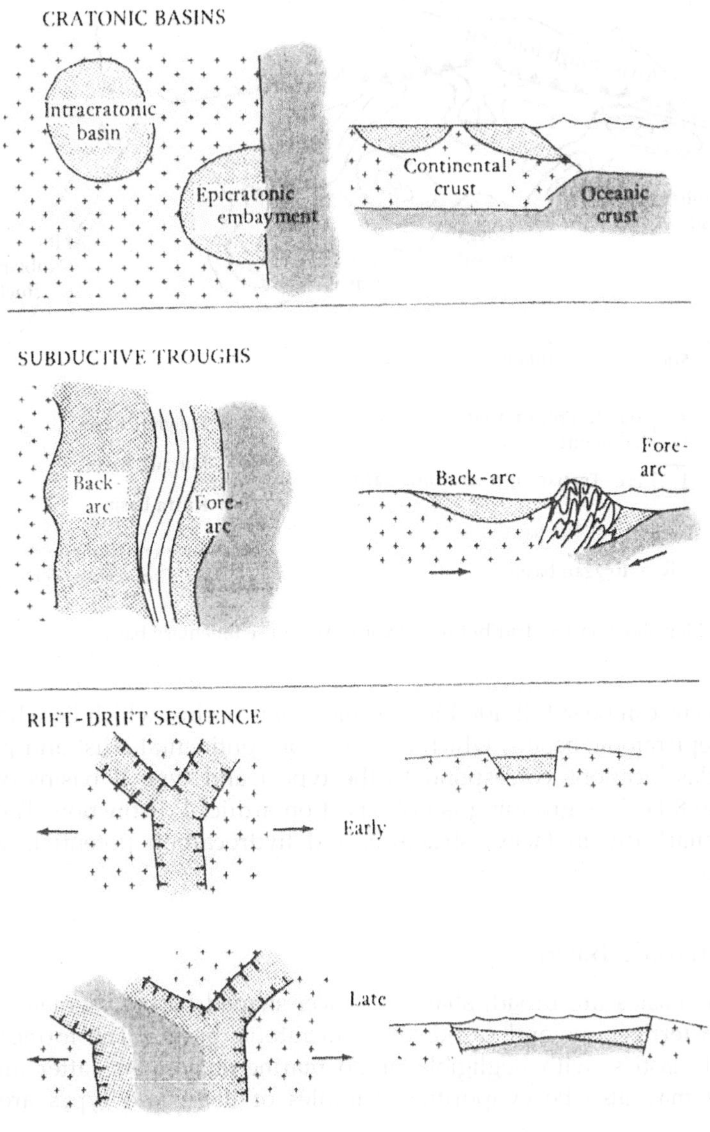

FIGURE 8.10 Geophantasmograms showing the geometry of the various types of basins. Beware of this figure. Note that transitions occur between the different categories and that a basin may evolve from one type to another.

More recently it has been suggested that crustal sags may result from "cold spots" due to mantle cooling, resulting in downwelling and a dignified sagging of the crust (Hartley and Allen, 1994). This model implies that sag basins may lack a precursor rift basin, and an early high heat flux, important considerations when modeling basins for petroleum generation studies.

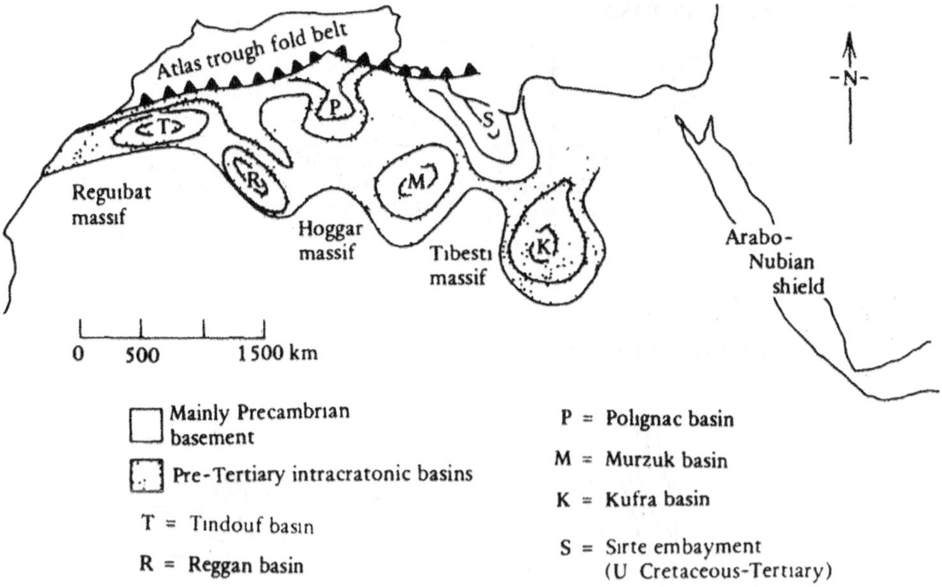

FIGURE 8.11 Map showing the distribution of North African sedimentary basins.

Cratonic basins can be subdivided into intracratonic basins, which lie wholly on continental crust, and epicratonic basins, which lie partly on continental crust and partly on oceanic crust. These classifications correspond to the type I and type II basins of Halbouty and Klemme (Table 8.1). This grouping is not based on artificial distinction. These two types of basins differ markedly in facies, structure, and hydrocarbon potential, as the following account shows.

8.4.1 Intracratonic Basins

Intracratonic basins are broad, shallow, saucer-shaped basins. A major division can be made between terrigenous and carbonate intracratonic basins. The former are dominated by continental clastics, with negligible or no marine shales; the latter are more marine, although they may also be evaporitic. Examples of these two types are described and discussed.

A series of intracratonic basins occurs in North Africa between the Atlantic Ocean and the Red Sea. These basins are separated from one another by ridges of Precambrian igneous and metamorphic basement and tend to plunge northward toward the Mediterranean (Fig. 8.11). These basins show a remarkably uniform Paleozoic stratigraphy, but their characters become distinctly different in the Mesozoic. The Murzuk and Kufra basins of southern Libya are examples of intracratonic basins (Fig. 8.12). They are both floored with the widespread Pan-Saharan Paleozoic sequence. Names vary from basin to basin, and facies' boundaries are diachronic, but the stratigraphy is remarkably uniform from Arabia to Africa (see Selley (1996, 1997a), respectively). A blanket of braided alluvial sands, several hundred meters

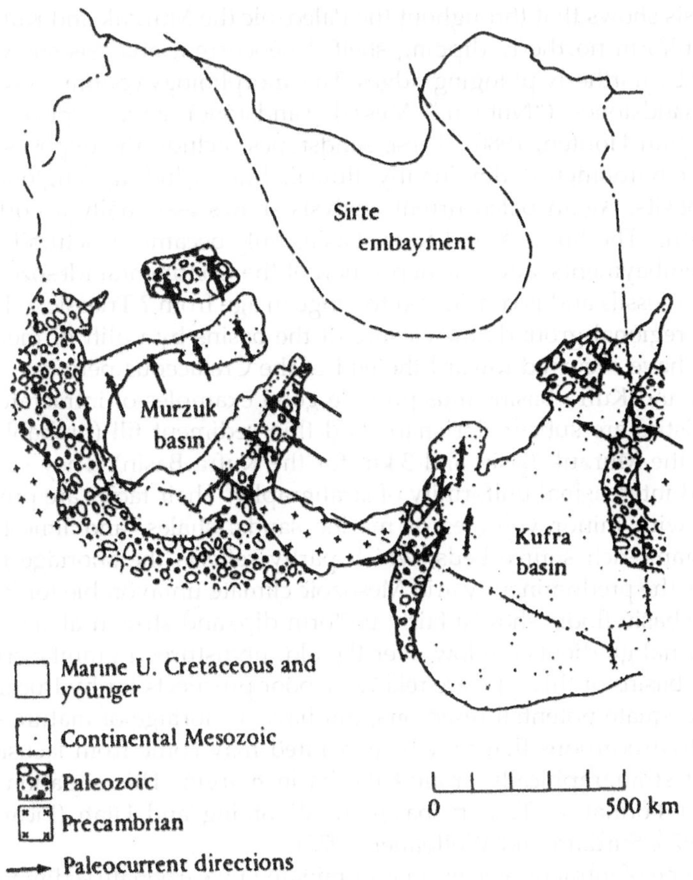

FIGURE 8.12 Map of Libya showing the Murzuk and Kufra intracratonic basins. Note how paleocurrent data show that basin subsidence postdated sedimentation.

thick, is overlain by a thinner, but still uniform, blanket of marine shoal sands. The precise age of these sands is uncertain because of a lack of fossils, but they are generally referred to the Cambro-Ordovician. These beds are succeeded by a marine graptolitic shale: the Tannezuft shale (Silurian) of Algeria and Libya and the Arenig shale of the Khreim Group in Jordan. This shale is an organic-rich oil source rock in northwest Algeria, but becomes thinner, siltier, and less organic when traced south and east toward the craton. A major regression then deposited the predominantly deltaic Acacus sandstone (Silurian) and the predominantly fluvial Tadrart sandstone (Devonian) in North Africa, and the equivalent Khreim Group and Al Jouf sandstone of Arabia. Overlying Carboniferous marine limestones and shales in Algeria and northwestern Libya pass southeastward into paralic sands and shales and finally to fluvial sands in the southeastern Kufra basin. After a major regression at the end of the Carboniferous, the sea has never again returned to the Murzuk and Kufra basins (Lestang, 1965; Klitzsch, 1970).

Facies analysis shows that throughout the Paleozoic the Murzuk and Kufra basins lay on a more or less uniform northerly dipping shelf. Paleocurrent analysis shows that the basins were separated by northerly plunging ridges. This morphology continued while the continental Mesozoic sandstones ("Nubian," Messak sandstone) were deposited (McKee, 1965; Klitzsch, 1972; Van Houten, 1980). These sandstones include the deposits of a wide range of continental environments: dominantly fluvial, but including fanglomerate, lacustrine, and eolian deposits. Again paleocurrent analysis shows essentially a northerly paleoslope across both basins. The Murzuk and Kufra basins only became structurally enclosed basins, as opposed to embayments, after the deposition of the Continental Mesozoic. This deposit is largely barren of fossils and is considered to range in age from ? Triassic to Lower Cretaceous (Wealden). On regional grounds the closure of the basins by uplift of their northern edges would seem to have occurred toward the end of the Cretaceous period.

The Murzuk and Kufra basins thus provide good examples of intracratonic basins. Their main characteristics are subcircular shape and thin sediment fill (probably of the order of some 2 km for the Murzuk Basin and 3 km for the Kufra Basin). They show a remarkable intrabasinal and interbasinal uniformity of stratigraphy. Their facies are predominantly nonmarine sands, with minor volumes of marine sands, shales, and limestones. There is a shortage of organic-rich source beds, which partly reflects the shortage of marine shales, and also reflects the predominantly arid Mesozoic climate unfavorable for lacustrine oil shale deposition. The basin floors show a fairly uniform dip, and structural anomalies are largely absent. Geothermal gradients are low over the old, undisturbed granitic crust.

Intracratonic basins of this type are relatively poor prospects for hydrocarbon exploration. They contain adequate potential reservoirs, but have a shortage of mature source rocks and structure. The hydrocarbons that may be generated may come from lacustrine source beds and be trapped stratigraphically around the basin margin. This situation is illustrated by the Green River Formation Tertiary basins of Wyoming and Utah (Picard, 1967; Eugster and Surdam, 1973; Surdam and Wolfbauer, 1975).

The second type of intracratonic basin is dominated by carbonate sedimentation. This variation is not due to an underlying structural difference, but rather due to climatic and other factors, and transitions between the two types are present. The Williston and Michigan basins of North America are good examples of carbonate intracratonic basins. The Williston Basin occupies parts of Saskatchewan, Montana, and North Dakota. It has a diameter of some 400 km and contains some 3 km of sediment, ranging in age from Cambrian to Tertiary (Darling and Wood, 1958; Smith et al., 1958; Dallmus, 1958; Harding and Lowell, 1979).

The Paleozoic sequence begins with a basal Cambrian marine shoal sand, followed by largely marine shales and shallow water limestones, with sabkha evaporites and red beds in the Devonian and Lower Carboniferous. A more or less uniform Paleozoic stratigraphy becomes regionally varied in the Mesozoic. Triassic and Jurassic sediments pinch out toward the basin margin, and a major sub-Cretaceous unconformity cuts across earlier rocks down to and including basement. This process plays a major part in the sealing of subcrop truncation traps. More than a kilometer of Cretaceous and Tertiary shales and clastics was then deposited in shallow marine and continental environments (Fig. 8.13). These beds, although not themselves significantly petroliferous, played an important part not only as a seal but also as a blanket cover, which enabled the Paleozoic source shales to mature and generate oil and gas.

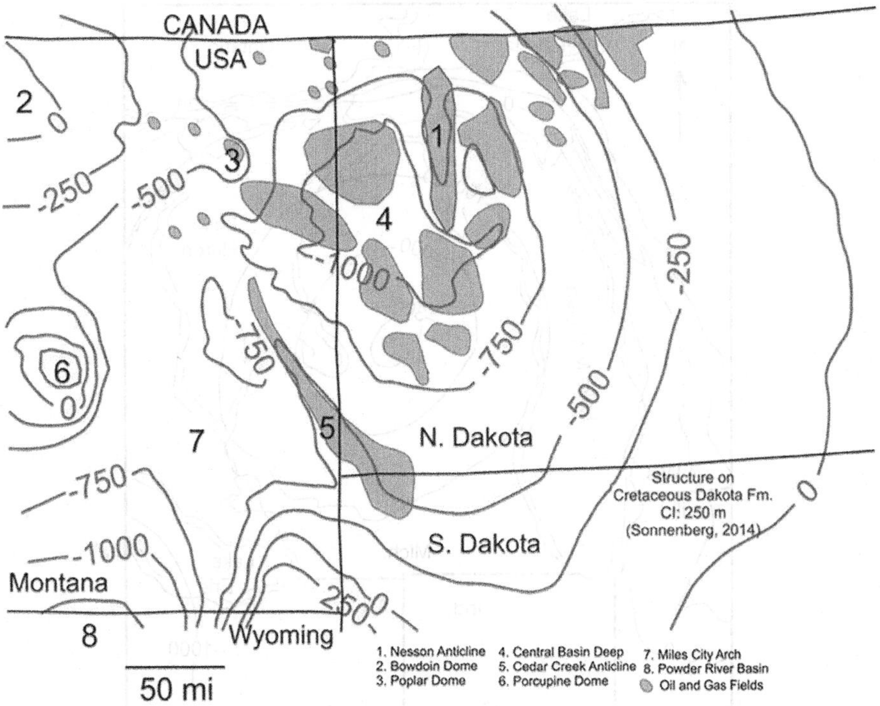

FIGURE 8.13 Structure contour map on the top of the Cretaceous Dakota Formation of the Williston Basin (contours in 250-m intervals). This basin is a good example of a closed intracratonic basin, with a fill of more than 3 km of shallow marine and continental sediments.

The Williston Basin is a major oil province and contains giant oil or gas fields in the Late Devonian–Early Mississippian Bakken Formation. Prior to the Bakken play which started in 2000, it produced oil and gas from numerous relatively small accumulations. The Bakken play production occurs mainly in the basin center. Traps are of three main types: (1) a number of broad regional arches, such as the Miles City arch and Porcupine dome; (2) several positive trends, which are related to basement faults, such as the Cedar Creek and Nesson anticlines; and (3) basin center continuous (discussed in Chapter 9). Warps and faults trend north–south or northwest–southeast. Oil emigrating from the source beds in the deeper part of the basin has been trapped in pre-Cretaceous reservoirs in anticlinal, truncation, and combination traps (structural closure plus unconformity) beneath both the Cretaceous and Jurassic erosion surfaces. The basin center area is a continuous unconventional accumulation in the Bakken and Three Forks formations.

Another good example of an intracratonic carbonate basin is the Michigan Basin to the southeast of the Williston Basin (Cohee and Landes, 1958; Delwig and Evans, 1969; Mesolella et al., 1974). This basin is also subcircular in plan, with a sediment thickness of some 5 km. Unlike the Williston Basin, the section consists largely of Lower Paleozoic rocks, with a veneer of Devonian to Jurassic strata. Facies are largely shallow marine with a basal Cambrian shoal sand overlain by shales, carbonates, and evaporites (Fig. 8.14). Particular

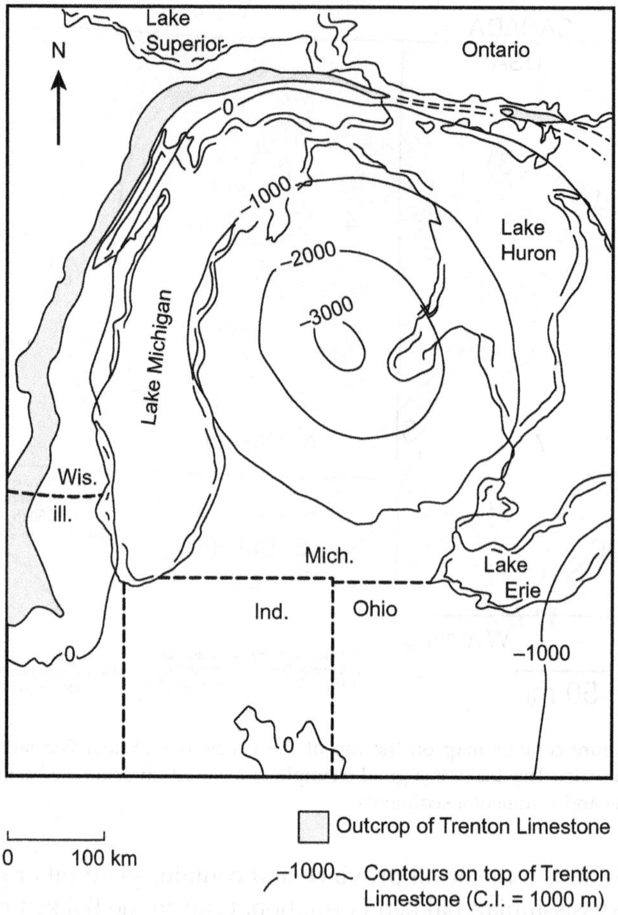

FIGURE 8.14 Structure contour map on the top of the Trenton Limestone (Ordovician) showing the shape of the intracratonic Michigan Basin of the Great Lakes region of North America. *Modified from Cohee and Landes (1958); reprinted by permission of the American Association of Petroleum Geologists.*

interest has centered on the Silurian rocks. These consist of basinal carbonates rimmed by, in turn, pinnacle reefs on the platform slope, a barrier reef, and platform backreef carbonates. This depositional topography was infilled with Upper Silurian evaporites and minor carbonates.

Like the Williston Basin, the Michigan Basin is not a major hydrocarbon province, and does not contain a single known giant oil or gas field. Similarly, however, it contains many hundreds of small oil and gas fields. These are mainly trapped on the myriad pinnacle reefs and on culminations on the concentric barrier reef. Smaller reserves have also been found, ranging from the Ordovician Trenton limestone to the basal Upper Carboniferous (Pennsylvanian) sands.

8.4.2 Epicratonic Embayments

Epicratonic embayments are basins that lie on the edge of continental crust. They are not true closed basins, but plunge toward major oceanic areas floored with basaltic crust. Epicratonic embayments correspond broadly to type II intracontinental composite foreland shelf basins of Halbouty et al. (1970). As with intracratonic basins, a major distinction can be made between dominantly terrigenous and dominantly carbonate-filled embayments. The Tertiary Gulf Coast of the United States and the Niger Delta embayment illustrate a terrigenous basin, and the Sirte embayment of Libya illustrates a carbonate-filled basin.

The Gulf Coast embayment of the southern United States is a major embayment containing some 15 km of sediment (Fig. 8.15). Basement is overlain by the Louann salt of Jurassic age (?) (Murray, 1960; Wilhelm and Ewing, 1972; Antoine, 1974; Dow, 1978). This salt is succeeded by a series of prograding wedges, which range in age from Cretaceous to Recent. Each wedge is composed of a thin up-dip section of fluvial sands, which thickens seaward into deltaic sands and muds. These deltaic sands and muds thin seaward, in turn, into deep marine clays of the Gulf of Mexico. These sediments contain major reserves of oil and gas (21.5 billion barrels of recoverable oil and 17,000 ft^3 of recoverable gas, according to Ivanhoe (1980)). These reserves occur in a series of fairways that become young toward the Gulf (Fig. 8.16). Within each fairway, production occurs where the oil window intersects the breakup zone of interfingering slope mud source beds and deltaic sands. Hydrocarbon generation has been aided by rapid sedimentation and hence overpressuring. This process has resulted in abnormally low geothermal gradients over the gulf depocenter as heat builds up in the "devil's kitchen" far below (Jones, 1969). Within the productive fairways hydrocarbons occur in a variety of traps. These traps include the rollover anticlines associated with the Vicksburg flexure (Section 7.5.1.2), rollover anticlines associated with local growth faults originating in the overpressured clays, and diapiric traps due to both Louann salt domes and younger mud diapirs.

The Niger embayment of West Africa is in many ways analogous to the Mississippi embayment. It, too, plunges from continental to oceanic crust, but the subsidence that initiated sedimentation is clearly related to rifting of the African craton as the Atlantic ocean developed. The embayment merges up-dip into the Benue and Chari rifts, which extend into Niger and Chad (Fig. 8.17). Like the Mississippi, the Niger embayment contains a series of prograding clastic wedges, which range in age from Upper Cretaceous to Recent. Three diachronous formations are recognized: the predominantly fluvial Benin Formation, the deltaic sands and shales of the Agbada Formation, and the slope muds of the Akata Formation (Short and Stauble, 1967; Weber and Daukoru, 1975; Evamy et al., 1978; Avbobo, 1978; Reijers, 1997). Like the Mississippi delta the clays are overpressured, giving rise to growth faults and rollover anticlines, which form the major traps (Section 7.5.1.2). The Niger delta does not show a series of seaward younging productive fairways, although oil and gas are regularly distributed within each major growth fault structural unit. Abnormally low geothermal gradients are again encountered over the depocenter, the top of the oil window rising from some 5 km in the middle to 3 km around the edge. The Mississippi and Niger embayments are both characterized by high gas:oil ratios and waxy low-sulfur crudes. These characteristics probably reflect the high humic content of their kerogen.

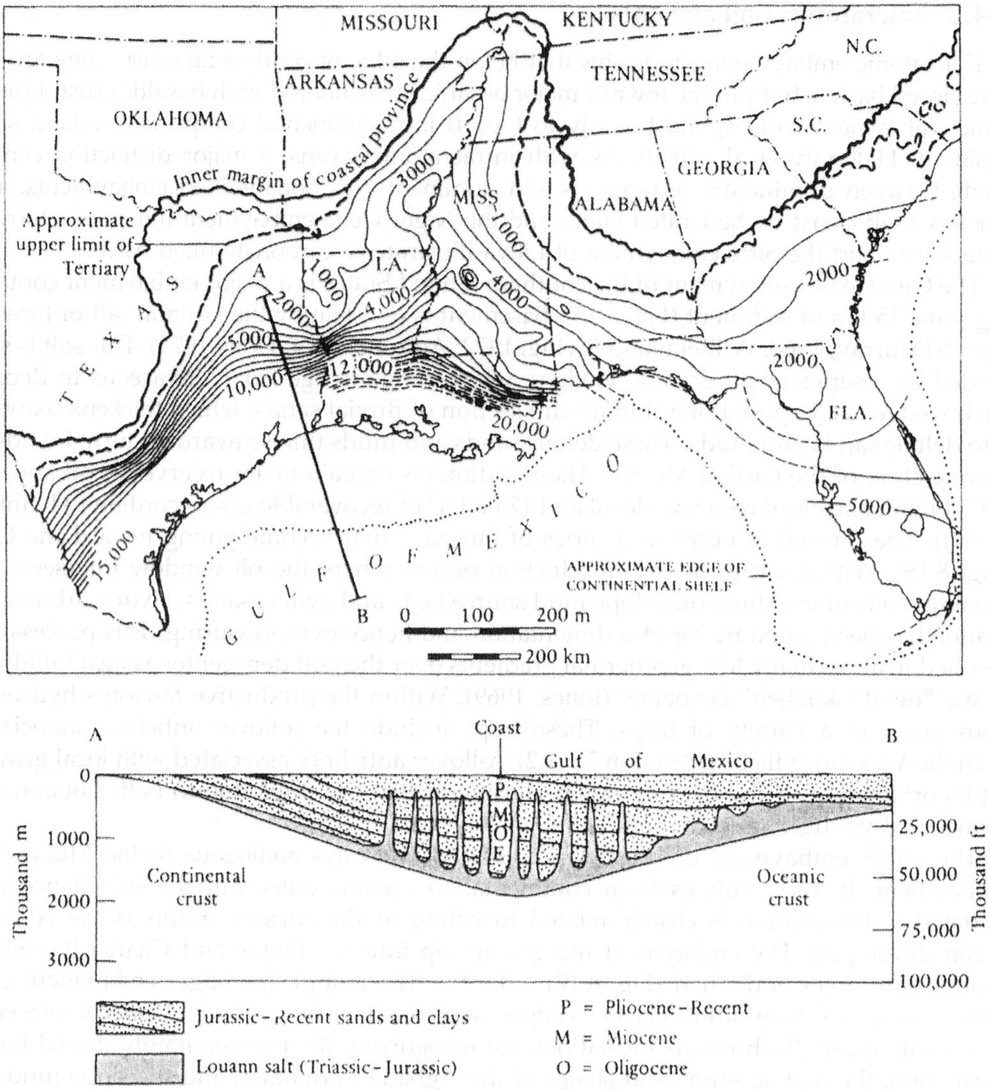

FIGURE 8.15 Map and cross-section of the northern Gulf of Mexico coastal basin. *Modified from Murray (1960); reprinted by permission of the American Association of Petroleum Geologists.*

Many other clastic epicratonic embayments occur around the world, especially in southeast Asia and the Canadian Arctic (Bruce and Parker, 1975). However, few are as well documented and apparently prolific as those just described.

Not all epicratonic embayments are terrigenous; some have a predominantly carbonate fill. The Sirte embayment is an example of this type (Conant and Goudarzi, 1967; Grey, 1971; Salem and Busrewil, 1981; Selley, 1997b). Paleocurrent analysis shows that from the Cambrian

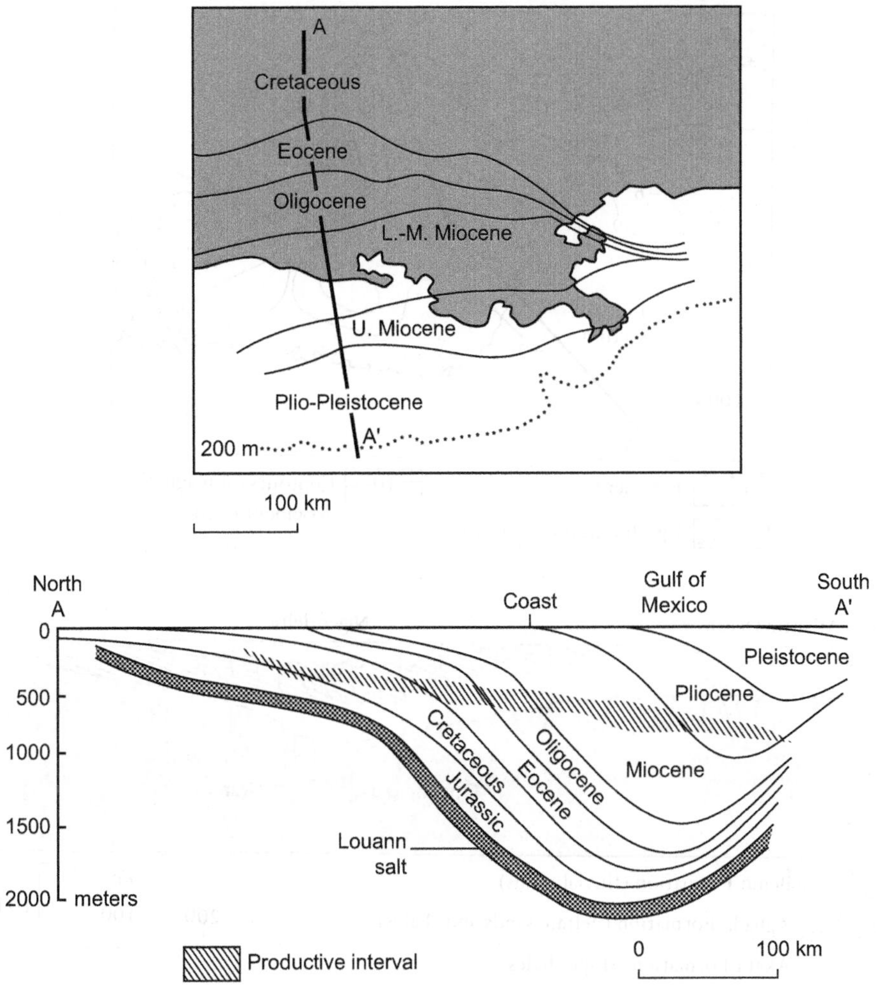

FIGURE 8.16 (Upper) Map showing the productive fairways of the Gulf Coast embayment. (Lower) Cross-section showing the progradational nature of the basin fill and its relation to hydrocarbon production. *Modified from Dow (1978); reprinted by permission of the American Association of Petroleum Geologists.*

to the early Cretaceous the area now occupied by the Sirte embayment was a northerly plunging ridge that separated the Murzuk and Kufra embayments, as they then were. This arch collapsed in the mid-Cretaceous. The Sirte unconformity directly overlies Precambrian basement in the center of the embayment and progressively younger rocks away from the basin axis. A locally developed basal sand is overlain on tilted fault blocks by Upper Cretaceous to Paleocene reefs and, in the troughs, by organic-rich shales that locally onlap and overlie the highs. The Lower Eocene consists of up to a kilometer of interbedded evaporites and carbonates. Carbonate sedimentation continued in the Middle Eocene, to be succeeded by shallow marine and fluvial sands and shales from the Oligocene to Recent (Fig. 8.18).

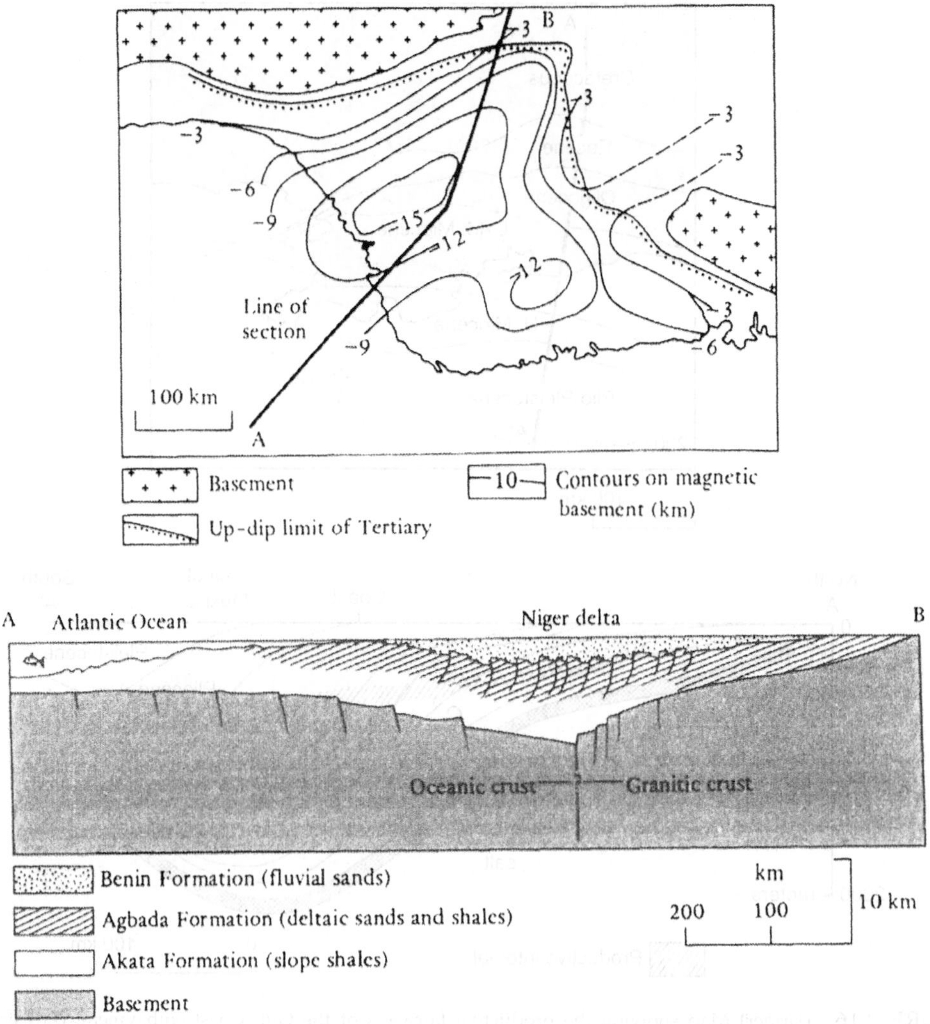

FIGURE 8.17 (Upper) Map of the Niger Delta showing depth to magnetic basement. (Lower) Cross-section along the line A–B. *Modified from Evamy et al. (1978); reprinted by permission of the American Association of Petroleum Geologists.*

The Sirte embayment contains three main productive horizons. Many fields are combination traps on structural highs sealed by the Sirte unconformity. Reservoirs range in age, from Precambrian granite (Aguila) through various sandstones that range in age from Cambrian to Lower Cretaceous (Messla and Sarir). The second and most important play involves production from Upper Cretaceous and Paleocene reefal carbonates on the crests of fault blocks (Zelten, Waha, Dahra, etc.). A third, minor play occurs in the Oligocene sands, where the Gialo field is in the giant category (Fig. 8.19). Overall reserves of the Sirte embayment have been estimated at 30 billion barrels of recoverable oil and 32 trillion cubic feet of gas (Ivanhoe, 1980). It contains more than 13 fields in the giant category (Halbouty et al., 1970).

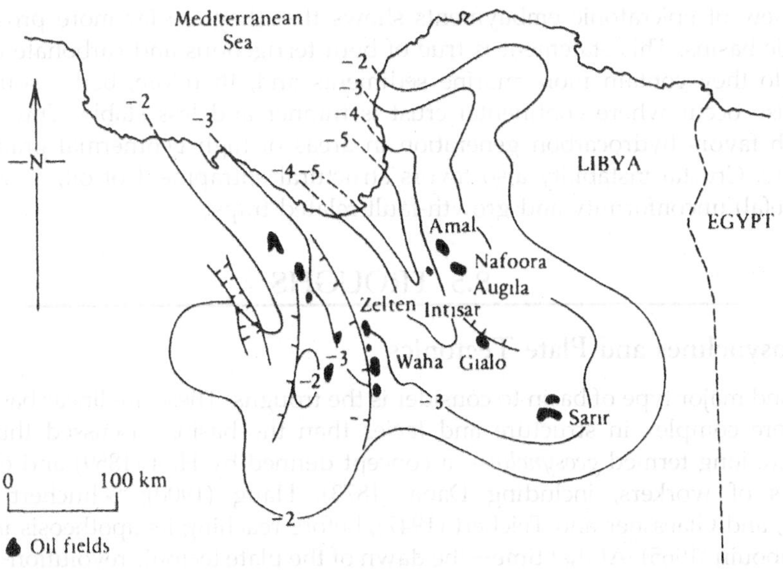

FIGURE 8.18 Map of the Sirte embayment showing distribution of oil fields (black) and structure contours (kilometers) on the Sirte unconformity (pre-Upper Cretaceous). *Modified from Sanford (1970); reprinted by permission of the American Association of Petroleum Geologists.*

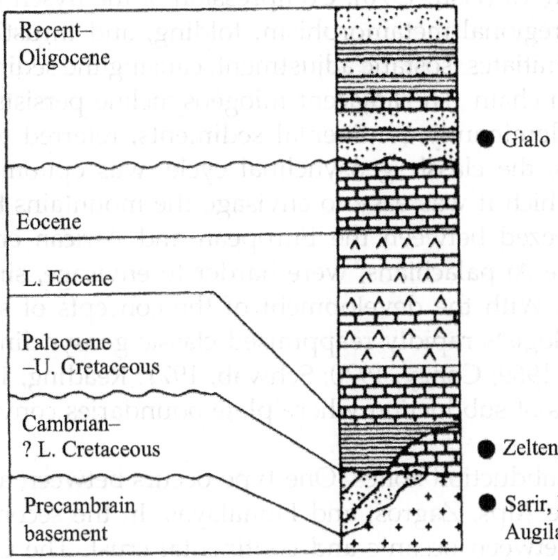

FIGURE 8.19 Summary stratigraphy of the Sirte embayment showing the main reservoir formations. The Upper Cretaceous–Paleocene shales provide the main source rock.

This review of epicratonic embayments shows that they are far more prospective than intracratonic basins. This statement is true of both terrigenous and carbonate embayments. Not only do they contain more marine sediments and, therefore, better source potential but they also occur where continental crust is thinner and less stable. Thus heat flow is high, which favors hydrocarbon generation in areas of high geothermal gradients due to overpressure. Crustal instability also favors structural entrapment of oil, as well as stratigraphic (reefal) unconformity and growth-fault-related traps.

8.5 TROUGHS

8.5.1 Geosynclines and Plate Tectonics

The second major type of basin to consider is the troughs. These are linear basins far larger and far more complex in structure and facies than the basins discussed thus far. These troughs were long termed *geosynclines,* a concept defined by Hall (1859) and elaborated on by a series of workers, including Dana (1873), Haug (1900); Schuchert (1923), Kay (1944,1947), and Glaessner and Teichert (1947), before reaching its apotheosis in the mighty work of Aubouin (1965). At that time—the dawn of the plate tectonic revolution—the concept of the geosyncline could be summarized as follows. Geosynclines consist of two parallel troughs. One, the miogeosyncline, which lies on continental crust and consists of an oceanward-thickening wedge of shallow marine limestones, sandstones, and shale. This trough is separated from the second trough, the eugeosyncline, by the miogeanticlinal ridge. The eugeosyncline is deeper than the miogeosyncline and is infilled largely by deep-water sediments. Initially, these sediments may be bathyal muds, but as the geosyncline evolves, turbidite sands infill it from a rising arc of islands on its oceanward side. This wedge of clastics has been referred to as *flysch* (Hsu, 1970). Compression of the flysch trough is accompanied by igneous intrusion, regional metamorphism, folding, and thrusting. Each phase of compression, or orogenesis, initiates isostatic adjustment, causing the sediments of the trough to rise and form a mountain chain. The adjacent miogeosyncline persists and is filled by a postorogenic wedge of predominantly continental sediments, referred to as molasse (Van Houten, 1973). This process, the classic geosynclinal cycle, was epitomized for European geologists by the Alps, in which it was easy to envisage the mountains being formed from a trough of sediments squeezed between the European and African cratons. Continental margin geosynclines, like the Appalachians, were harder to envisage, since one side of the vice was apparently absent. With the development of the concepts of sea floor spreading and crustal subduction, geologists rapidly reappraised classic geosynclinal theory (Ahmad, 1968; Mitchell and Reading, 1969; Coney, 1970; Schwab, 1971; Reading, 1972). Geosynclines are now interpreted as zones of subduction where plate boundaries converge, and the term has fallen into disuse.

There are three types of subduction zones. One type occurs between areas of continental crust, as, for example, in the Alps, Zagros, and Himalayas. In the second type the trough develops at the boundary between oceanic and continental crust. The Cordillera of North America, the Andes, and the Banda Arc of Southeast Asia illustrate this type. The third type is where subduction occurs between two plates of essentially oceanic crust, as, for example, in the Japanese and New Zealand arcs.

These three types of subduction zone may form part of an evolutionary sequence in which an ocean closes as plates of oceanic crust converge, finally resulting in the juxtaposition of continental crustal blocks. Figure 8.20 illustrates these different types of subduction zones. Note the change in terminology: back-arc, arc, and fore-arc broadly correspond to the old miogeosyncline, geanticline, and eugeosyncline, respectively.

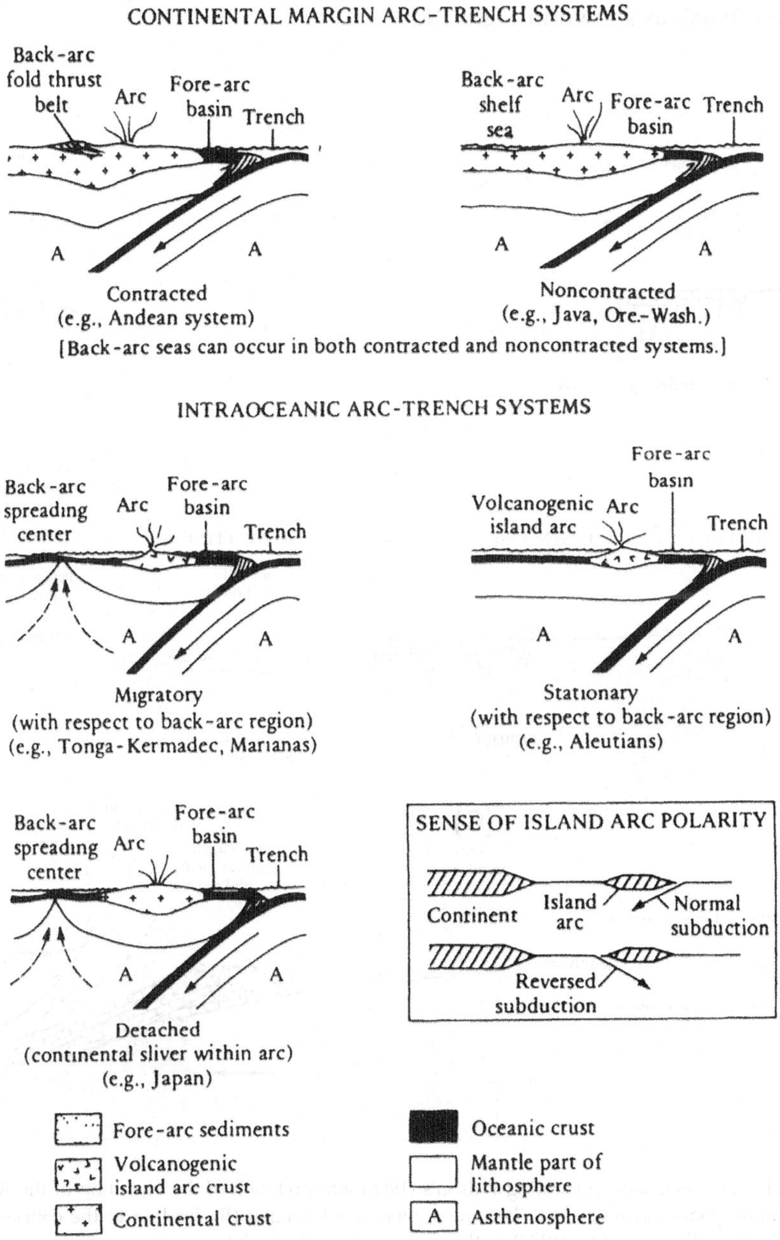

FIGURE 8.20 Cross-sections illustrating the various types of zones of subduction and their associated sedimentary troughs. *From Dickinson and Seely (1979), namely Fig. 1, p. 4.*

This later reinterpretation of sedimentary troughs has been complicated by the fact that some troughs that should theoretically be zones of subduction actually contain rifts with flat-lying sediments, which suggests a lack of compression, if not tension. The Peruvian trench is one such embarrassing example (Scholl et al., 1968). This phenomenon has led to the idea that plate movement may occasionally reverse, so that subductive zones become divergent, albeit for intermittent periods. The Alps have been interpreted in such a way by Wilson (1966, 1968), as shown in Fig. 8.21.

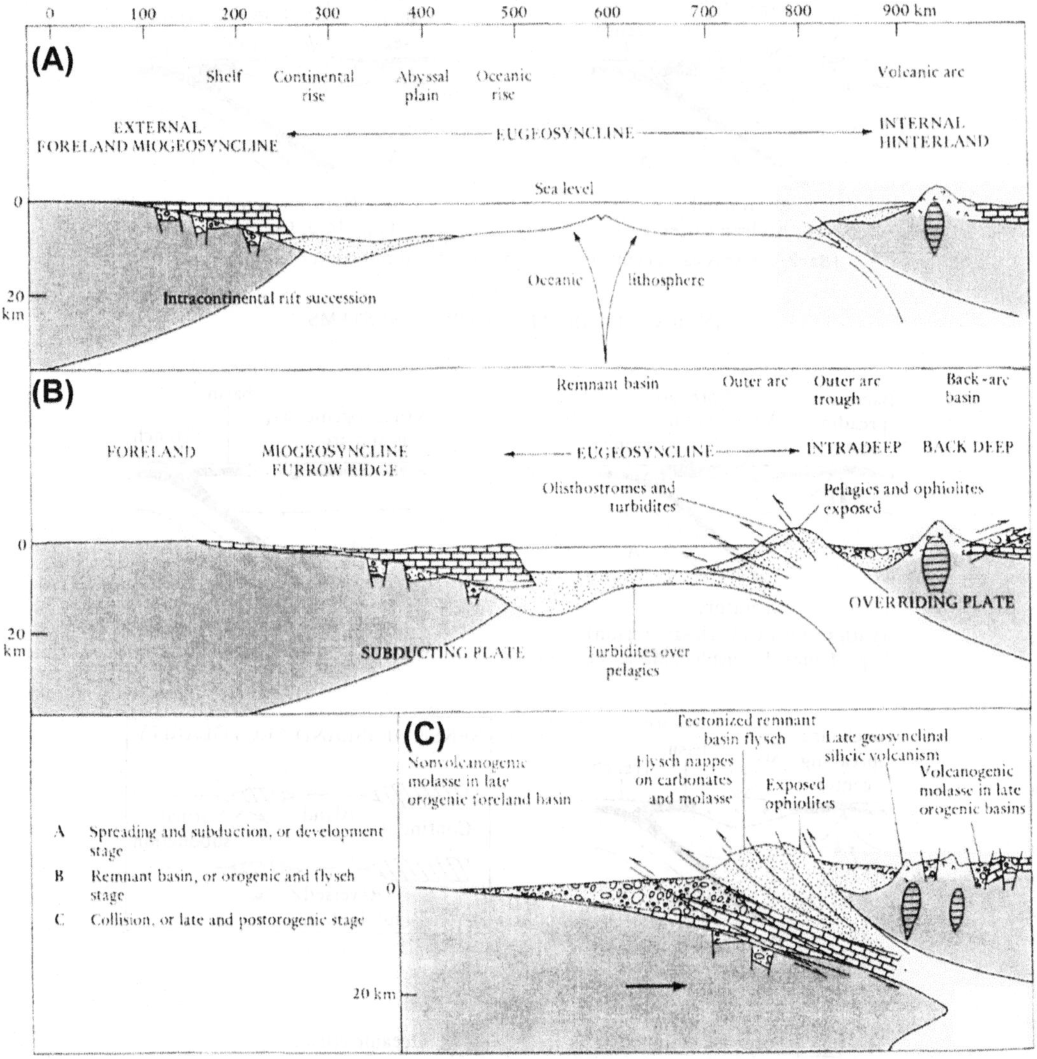

FIGURE 8.21 Cross-sections illustrating Wilson's (1968) interpretation of the evolution of the Alps. An original sea-floor-spreading phase (A) reverses to become a zone of subduction (B), leading to the collision of continental plates (C). *After Mitchell and Reading (1978), with permission from Blackwell Science Ltd.*

8.5.2 Back-Arc Troughs

The complexity of facies, structure, and history makes it difficult to generalize about the petroleum potential of subductive troughs. Consider first the back-arc, or miogeosynclinal, troughs. These troughs are asymmetric shelves whose sediments thicken toward the arc. This type of basin can also be regarded as an elongated epicratonic embayment and is more or less synonymous with Klemme's extracontinental downwarp. The deposits of these basins are largely shallow marine shales, carbonates (often reefal), and mature tidal shelf sands, with perhaps a feather edge of nonmarine sediments. Between major subductive phases, extensive source rocks may be deposited along the basin's seaward margin. Orogenic movements leave these back-arc basins terminated on their outer side by a thrust belt.

The Arabian Gulf, Sumatra, and Western Canada illustrate this type of basin. A map and cross-section of the Arabian Gulf is shown in Fig. 7.8. For further details see Kamen-Kaye (1970) and Murris (1980,1981).

Figure 8.22 illustrates the salient features of the Western Canada Basin. This basin contains a wedge of sediments that thickens westward to some 5 km until it is abruptly truncated by the Rocky Mountain thrust belt. The earlier Paleozoic sediments include sands, carbonates (with spectacular Devonian reefs), organic-rich shales, and evaporites. Major oil and gas production occurs in the Devonian reefs (Illing, 1959; Barss et al., 1970; Evans, 1972; Klovan, 1974) and in the Viking and Cardium (Cretaceous) shallow marine sands. Recoverable reserves are estimated at 16.3 billion barrels of oil and 94 trillion cubic feet of gas (Ivanhoe, 1980).

The source of terrigenous detritus is important in basins such as the Western Canada Basin. For most of the time the sands are produced by the slow weathering of the craton, followed by extensive reworking and deposition mainly in high-energy marine environments. These sands are thus mineralogically and texturally mature and have good porosity and permeability. As the arc rises, however, it begins to shed detritus into the back-arc basin. These later sands are often mineralogically immature, especially if derived from volcanic rocks. Rapidly deposited in fluviodeltaic environments, their mineralogical and textural immaturity may render them poorer reservoirs than the earlier shelf-derived sands.

Southeast Asia provides another example of back-arc basins and their relationship to fore-arc basins and plate boundaries (Schuppli, 1946; Haile, 1968; Crostella, 1977; Ranneft, 1979; Bowen et al., 1980; Wood, 1980). Here the continental crust of the Sunda Shelf moves westward toward the Indian Ocean plate, forming a convergent subductive plate boundary. Simultaneously, northward movement of the Australian plate is causing subduction on the southern margin of the Sunda plate. A number of back-arc basins separate the shelf from a volcanic arc (Fig. 8.23). These basins are infilled with a series of Eocene–Recent prograding clastic wedges in the Sumatra and Borneo basins and with carbonates over the more stable southeastern part of the shelf. These basins contain more than 10 fields with recoverable reserves in excess of 8 billion barrels of oil and 24 trillion cubic feet of gas (St. John, 1980). The back-arc basins are separated from the fore-arc basins by a volcanic arc. The latter are infilled with marine sediment, derived largely from the volcanic arc. The fore-arc basins are gradually being subducted between the Indian and Sunda plates. Thrust slices thus appear locally as an outer nonvolcanic island arc (e.g., the Andaman and Nicobar Island chains). Hydrocarbon production from the fore-arc basins is negligible for several reasons,

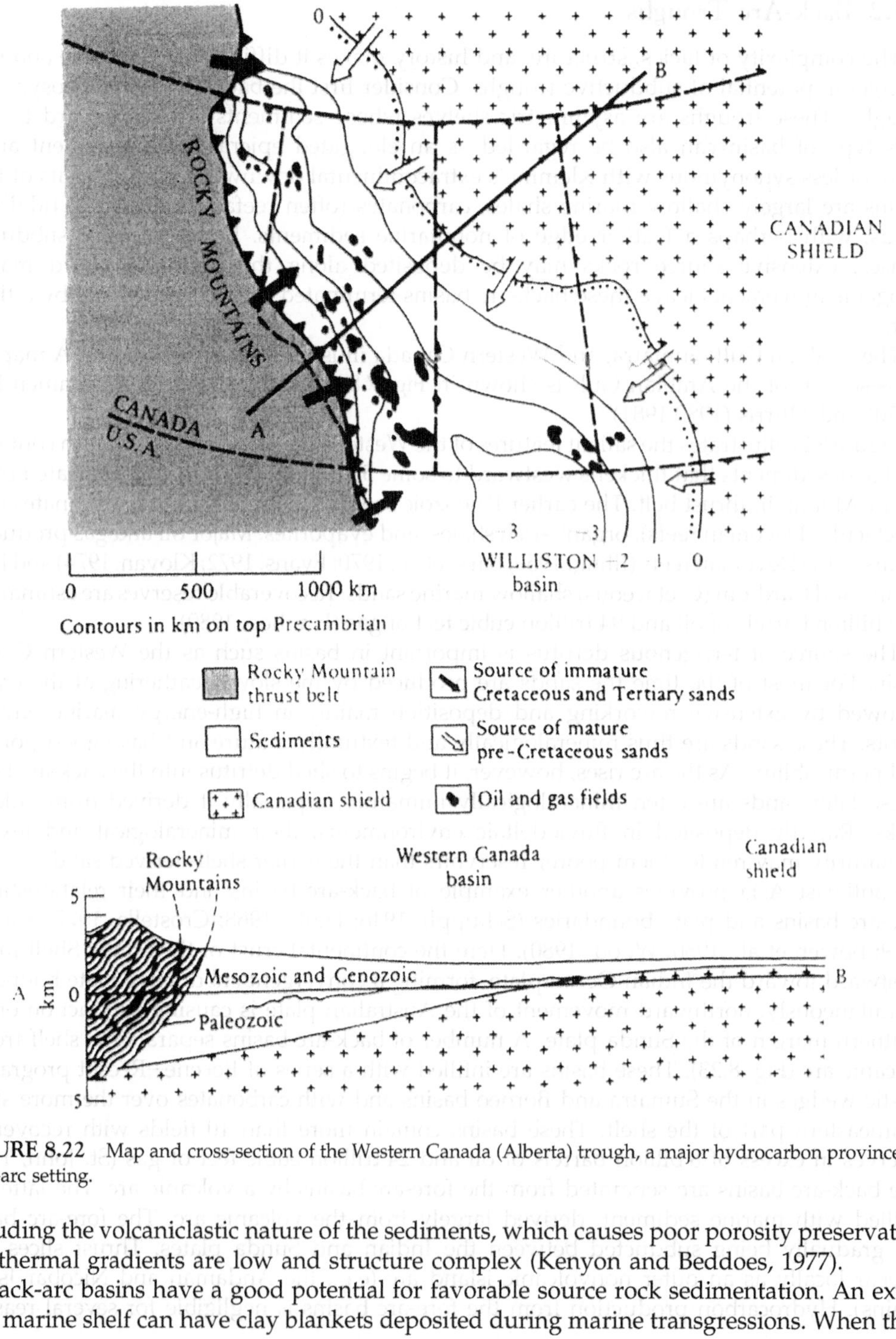

FIGURE 8.22 Map and cross-section of the Western Canada (Alberta) trough, a major hydrocarbon province in a back-arc setting.

including the volcaniclastic nature of the sediments, which causes poor porosity preservation. Geothermal gradients are low and structure complex (Kenyon and Beddoes, 1977).

Back-arc basins have a good potential for favorable source rock sedimentation. An extensive marine shelf can have clay blankets deposited during marine transgressions. When these

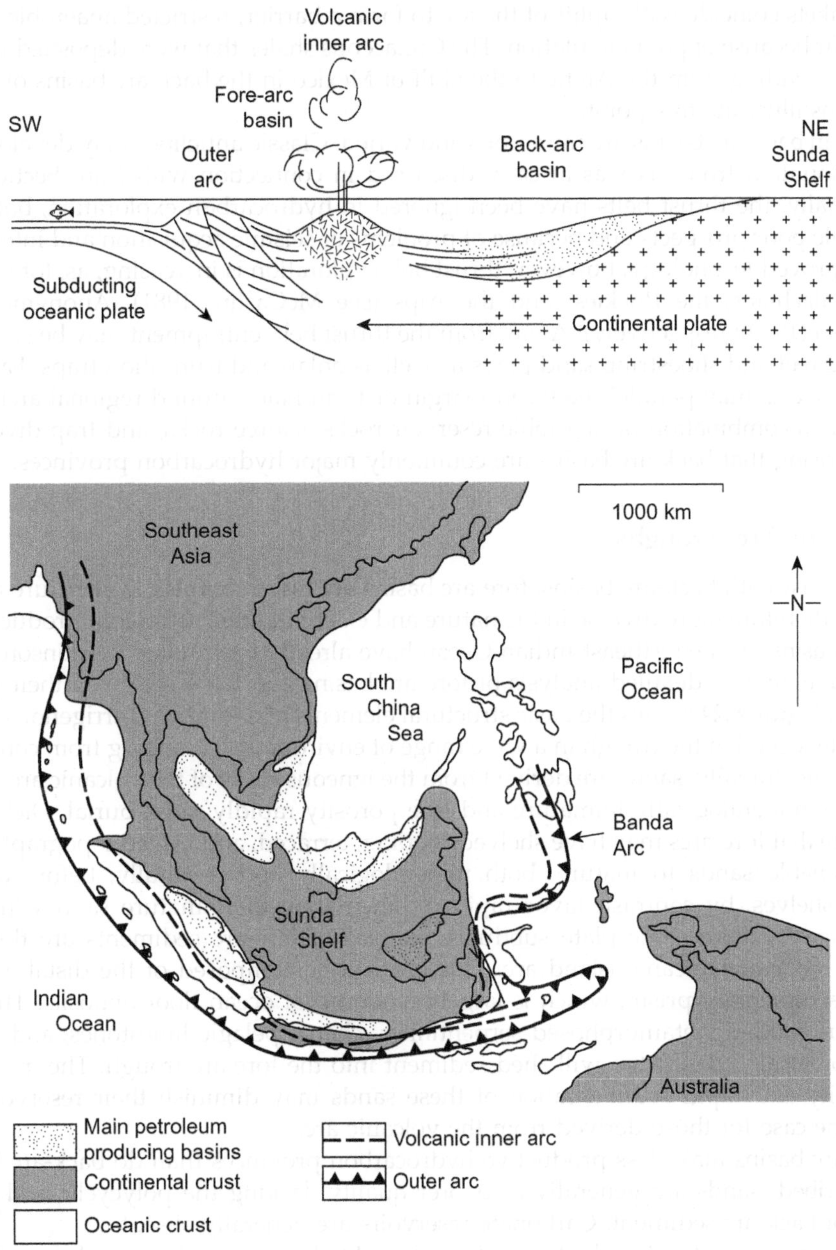

FIGURE 8.23 Cross-section and map of Southeast Asia showing the relationship between fore-arc and back-arc troughs. The back-arc troughs are the main oil-producing basins. *After Haile (1968), Crostella (1977), Ranneft (1979), Bowen et al. (1980), Wood (1980).*

clay blankets coincide with uplift of the arc to form a barrier, restricted anaerobic conditions may occur because of poor circulation. The Cretaceous shales that were deposited in the great seaway extending from the Arctic to the Gulf of Mexico in the back-arc basins of the Rocky Mountains illustrate this point.

Traps in back-arc basins are numerous and varied. Classic anticlines may develop adjacent to the mountain front, as was already discussed in connection with Iran (Section 7.6.1.1). Traditionally, the thrust belts have been ignored in hydrocarbon exploration, both because of possible poor prospects and because of problems of seismic acquisition and interpretation. With improved technology, however, thrust belt exploration is increasing, as, for example, in the Appalachians, the Rockies, and the Alps (see McCaslin (1981), Anonymous (1980), Bachmann (1979); respectively). Away from the thrust belt, entrapment may be stratigraphic, including reef and shoestring sand plays as well as onlap and truncation traps. Fairways for these prospects may parallel the basin margin or form halos around regional arches.

With this combination of favorable reservoir rocks, source rocks, and trap diversity, it is not surprising that back-arc basins are commonly major hydrocarbon provinces.

8.5.3 Fore-Arc Troughs

In contrast with back-arc basins, fore-arc basins are more complex in structure and facies. They are therefore more diverse in the nature and extent of their petroleum productivity. The fore-arc basins of the northeast Indian Ocean have already been noted. Dickinson and Seely (1979) have given a detailed analysis of fore-arc basins and have reviewed their petroleum potential. Figure 8.24 shows the basic structural elements of a fore-arc. Terrigenous sediments may be deposited in the trough in a wide range of environments, ranging from continental to deep marine. Initially, sands are derived from the igneous rocks of the volcanic arc. They thus tend to be mineralogically immature and lose porosity rapidly upon burial. Dickinson and Seely note that fore-arcs may have shelved, sloping, terraced, and ridged topographies. Broad shelves enable sands to mature, both mineralogically and texturally, before deposition. Narrow shelves, by contrast, favor rapid sedimentation of immature sand with minimal reworking. As the oceanic plate subducts, wedges of fore-arc sediments are thrust up to form an outer nonvolcanic island arc. This arc will be composed of the distal edge of the fore-arc sedimentary prism, which is largely made up of ocean floor deposits. These thrust slices thus contain metamorphosed serpentinites, cherts, pelagic limestones, and turbidites. As this outer arc rises, it too will shed sediment into the fore-arc trough. The mineralogical immaturity and rapid sedimentation of these sands may diminish their reservoir quality, as was the case for those derived from the volcanic arc.

Fore-arc basins make less productive hydrocarbon provinces than do back-arc basins. As just described, sands are generally of poorer quality, lacking the polycyclic and reworked history of back-arc sediment. Carbonate reservoirs are generally absent.

Source rocks may be abundant and of good quality because of the prevalence of deep and often restricted sea floor conditions. However, locating the main area of oil generation may be difficult. Geothermal gradients are often low in fore-arcs because the cool sediment is subducted, which depresses the isotherms. Those hydrocarbons that are generated are more commonly trapped in structural fold and fault traps than in stratigraphic ones. Extensive structural deformation may cause traps to be small and hard to develop.

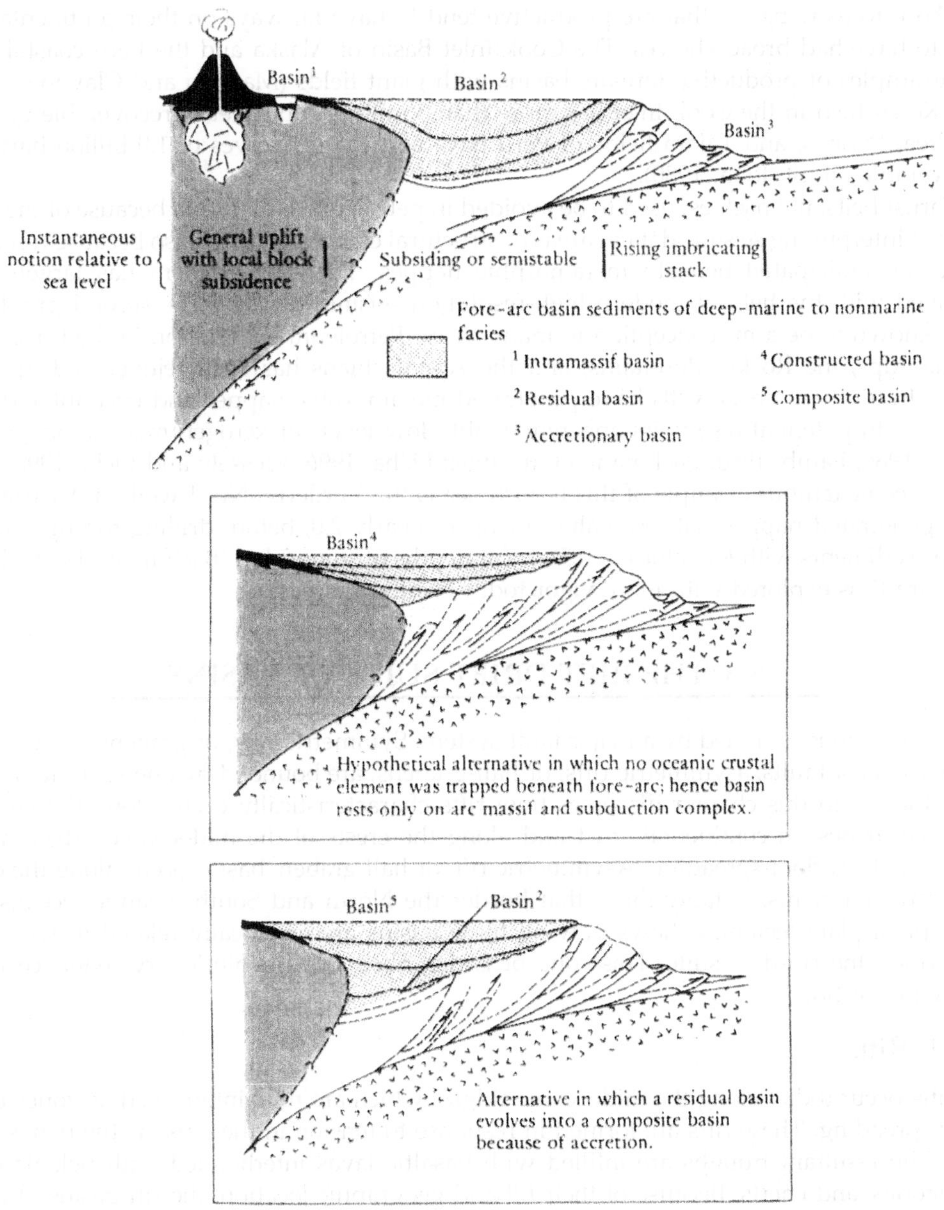

FIGURE 8.24 Cross-section through a fore-arc trough that developed where a continental plate has overriden a subducting oceanic plate. *From Dickinson and Seely (1979), namely Fig. 3. p. 7.*

Those fore-arc basins that are productive tend to have fairways on their continental side and to have had broad shelves. The Cook Inlet Basin of Alaska and the Peru coastal basin are examples of productive fore-arc basins with giant fields (Magoon and Claypool, 1981). The Kenai field in the Cook Inlet has more than 5 million cubic feet of recoverable gas. The La Brea, Parinas, and Talara fields of Peru have aggregate reserves of 1.0 billion barrels of recoverable oil (St. John, 1980).

Thrust belts, themselves, used to be avoided in petroleum exploration because of the problems of interpreting seismic data in areas of structural complexity, and also because reservoirs were not anticipated beneath metamorphic nappes. The first problem has largely been resolved with the help of modern high-resolution seismic surveys. The second problem is now known to be a misconception in many cases. Petroleum exploration in the thrust belts of the Alps, the Rocky Mountains, and the Appalachians has been rejuvenated in recent years. In all these areas wells have penetrated metamorphic nappes and encountered sediments with potential reservoirs and remarkably low levels of kerogen maturation (Anonymous, 1980; Lamb, 1980; Bachmann et al., 1982; Picha, 1996; Wessely and Liebl, 1996).

A specific famous example of this occurrence is the Vorderiss No. 1 well, in Austria. This well penetrated nappes with R_o values of up to nearly 2.0, before drilling through to sub-nappe sediments with R_o values of 0.5–0.6 at depths of some 5 km (Bachmann, 1979). Thrust belts are thus explored with more vigor today than in the past.

8.6 THE RIFT-DRIFT SUITE OF BASINS

A rift basin is bounded by a major fault system. Symmetric rifts, or grabens, are bounded by two sets of faults; asymmetric rifts, or half-grabens, are bounded by one set of faults. The introduction to this chapter discusses how rifts characteristically occur along the crests of regional arches on continental crust and along the crests of the midoceanic ridges, which are axes of sea floor spreading. Asymmetric rift, or half-graben, basins occur along the edges of many continents, notably those that border the North and South Atlantic Oceans. The concept of plate tectonics shows how all these basins are genetically related in what may be termed the rift-drift suite. This concept is described in its evolutionary sequence in the following section.

8.6.1 Rifts

Rifts occur today along the midoceanic ridges, which are now interpreted as zones of sea floor spreading. These rifts are formed in response to tension in the crust as the plates separate. The resultant troughs are infilled with basaltic lavas interbedded with pelagic clays, limestones, and cherts. Because of their fill and geographic location, the rift basins of midocean ridges are not attractive areas for hydrocarbon exploration.

A number of rift basins occur on continental crust, including the Rhine Valley of Germany and the Baikal rift of Russia (Illies and Mueller, 1970; Salop, 1967). Both of these basins cross-cut areas of arched crust and show a tendency to radiate into minor rifts at both ends of the main rift. They are infilled with up to 5 km of sediment and have igneous extrusives associated with them (Fig. 8.25). Equally well known, and in many ways similar, are the great rift valleys of Africa. Central Africa is now known to be crossed by rifts that, extending inland

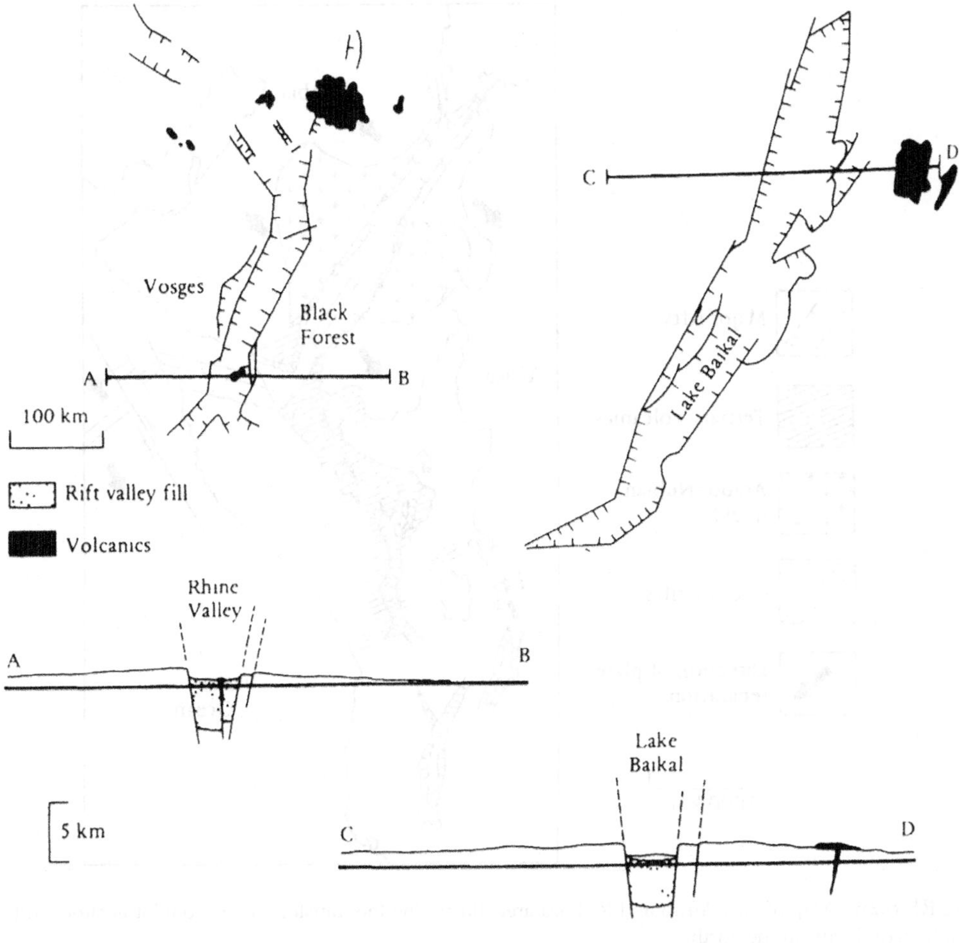

FIGURE 8.25 Maps and cross-sections of the Rhine rift of Germany (left) and the Lake Baikal rift of the Soviet Union (right). *After Selley (2000).*

from Nigeria, Mozambique, and Somalia, intersect in Sudan, Chad, and Niger. These rifts are of Cretaceous and early Tertiary age. Largely blanketed by younger sediments, they have only recently become known as a result of petroleum exploration in these countries. More conspicuous is a series of younger rifts that occur in eastern Africa (Unesco, 1965; Baker et al., 1972; Darrcott et al., 1973; Veevers, 1981). These rifts are very similar to the Rhine and Baikal rifts. They show crustal doming with local reversal of drainage. Major rifts bifurcate at their terminations, are volcanically and seismically active, and are infilled with volcanic, fluvial, and lacustrine sediments of Miocene to Recent age. As rifts are traced northeastward toward the Omo Depression on the Ethiopian coast of the Red Sea, rift initiation began earlier (Eocene), basaltic lava outpourings are more extensive in time and space, and the sediments include extensive evaporite deposits (Fig. 8.26).

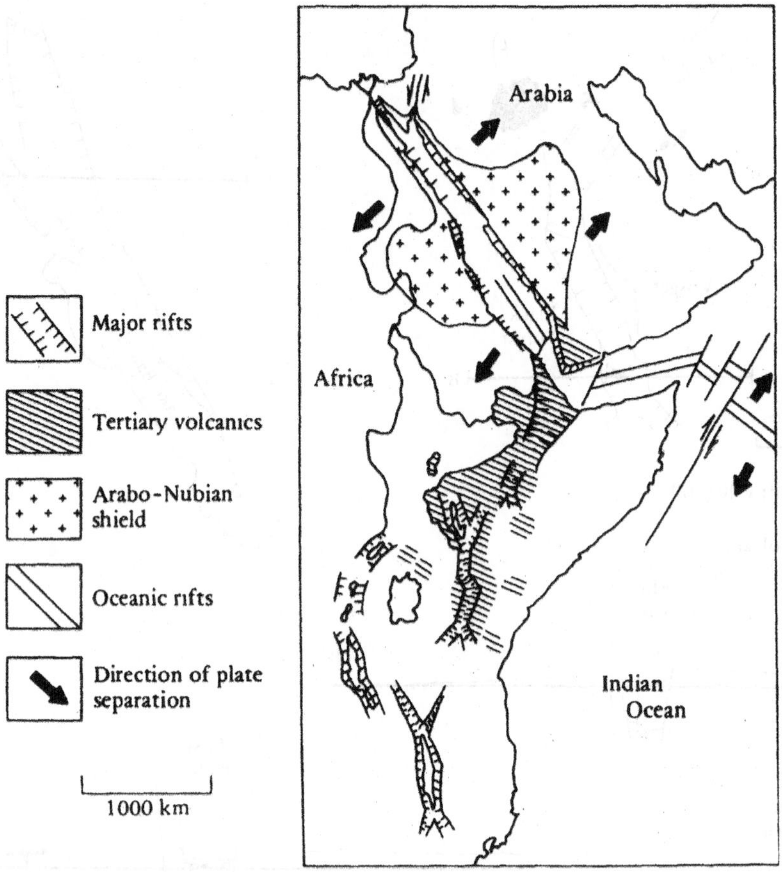

FIGURE 8.26 Map of East Africa and Red Sea area illustrating the transformation from intracratonic rifts in the south to oceanic rifts in the north.

The Red Sea is itself a complex rift that provides a genetic link between the intracratonic rifts just discussed and the ocean margin rifts (Heybroek, 1965; Lowell and Genik, 1972; Lowell et al., 1975; Thiebaud and Robson, 1979). The margins of the Red Sea are two parallel-sided half-grabens, whose major faults downthrow seaward. On the upthrown sides of the faults the granites of the Arabo-Nubian shield crop out with occasional veneers of basaltic lavas (Fig. 8.27). Within the coastal basins Paleozoic Nubian sandstones and Cretaceous and Eocene limestones are locally truncated on numerous fault blocks. This horst and graben floor to the coastal basins is onlapped by Miocene evaporites, reefal limestones, and younger sands. Many oil fields have been discovered in Tertiary carbonates and sandstones and in Nubian sandstone reservoirs in the fault blocks of the Gulf of Suez (Fig. 8.28).

The deeper central part of the Red Sea consists of basaltic oceanic crust with a veneer of young deep marine deposits. It contains a longitudinal axial rift, which can be traced, offset by many transform faults, into a midoceanic rift in the Indian Ocean. The Red Sea thus

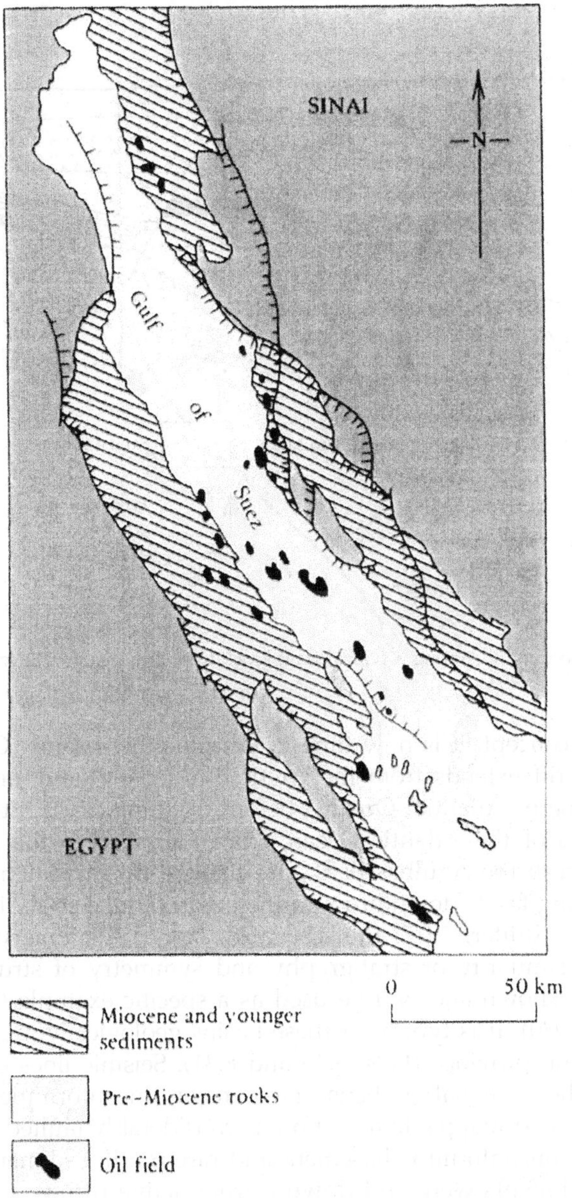

FIGURE 8.27 Map of the Gulf of Suez rift basin showing location of faults and oil fields. *After Thiebaud and Robson (1979).*

provides the link between the intracratonic rifts of East Africa and the oceanic rifts of the mid-ocean ridges. The Red Sea may be regarded as an incipient ocean basin in which a rift in the Arabo-Nubian shield separated into two half-graben basins, between which new oceanic basaltic crust is now forming.

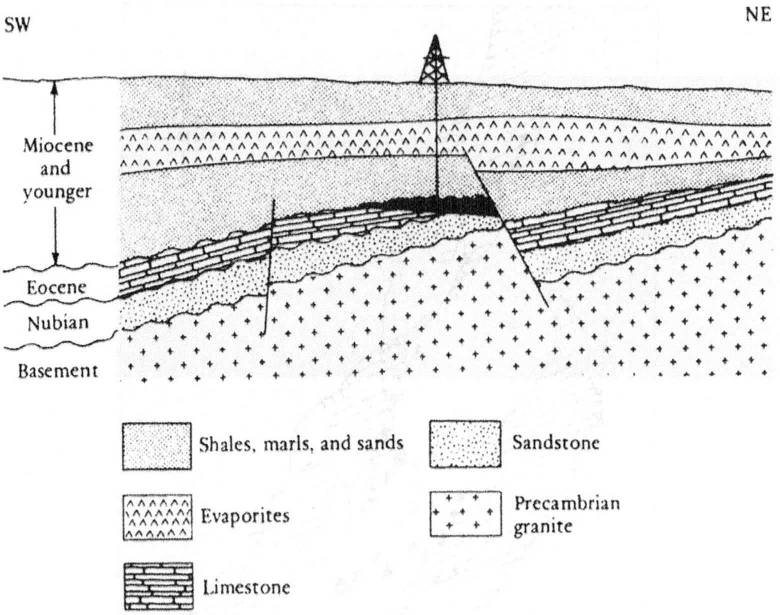

FIGURE 8.28 Cross-section illustrating a typical fault block trap in the Gulf of Suez.

Armed with this concept, it is now time to examine the Atlantic Ocean. A well-defined ridge with an axial rift extends from the Arctic, through Iceland, and down through the northern and southern Atlantic Oceans before running east into the Indian Ocean (Fig. 8.6). Both sides of the Atlantic Ocean, from Labrador to the Falkland Islands and from the Barents Sea to the Agulhas Bank, are flanked by asymmetric half-graben basins, whose major bounding faults downthrow to the ocean (Drake et al., 1968; Burk and Drake, 1974; Lehner and Dc Ruiter, 1977; Pratsch, 1978; Hoc, 1979; Emery, 1980). These basins show a remarkable similarity of stratigraphy and symmetry of structure (Fig. 8.29). The Gabon Basin is well known and will be used as a specific example (Belmonte et al., 1965; Brink, 1974; Vidal, 1980). It is typical of these basins geologically, but unusual in that it is a modest hydrocarbon province (Figs. 8.30 and 8.31). Seismic lines of the Atlantic coastal basins show that they are half-grabens. Two major unconformities may be discerned. One, generally the lowest mappable reflection, is considerably faulted. This marks the onset of rifting. Below this unconformity, basement and rare prerift sediments occur.

The overlying sediments were laid down during active rifting. They consist largely of continental clastics, but, as the fault blocks subside progressively toward the continent, a barrier develops between the rift and new oceanic crust. Behind this barrier sapropelic lacustrine shales may form in humid climates. Within the Cocobeach Group of the Gabon Basin these shales have generated oil. In arid climates, however, evaporites may form in the back basin; these evaporites are common in the Atlantic coastal basins. As the ocean opened from north to south, the onset of evaporite development becomes young southward: Permian in northern Europe, Triassic in the Georges Bank and Senegal basins, Jurassic in the

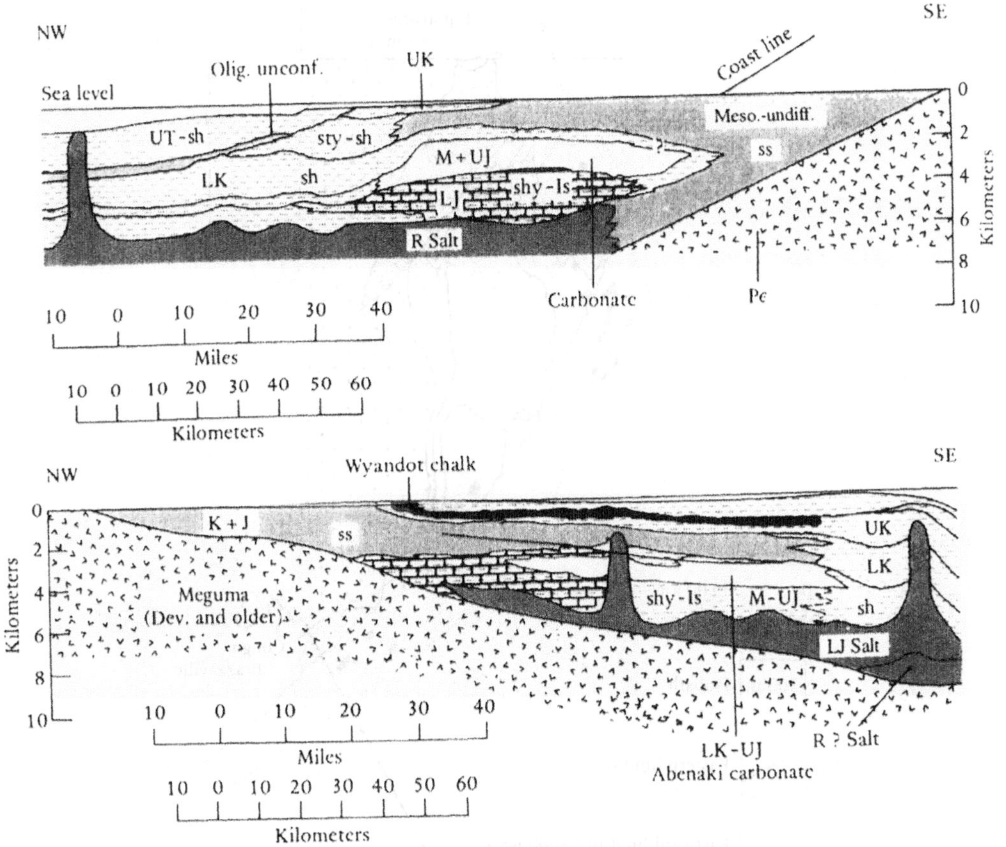

FIGURE 8.29 Cross-sections through the Moroccan (upper) and Scotian (lower) continental shelves showing the symmetry of structure and similarity of stratigraphy of these two Atlantic margin half-graben basins. *After Bhat et al. (1975) with permission from the Canadian Society of Petroleum Geologists.*

Gulf of Mexico, and Aptian in the Brazilian, Gabon, Cuanza, and Congo basins (Evans, 1978).

The second major unconformity overlies the evaporites. It marks the end of the rifting phase and the onset of drifting as the Atlantic Ocean widened. Rift faults generally die out at this surface, except at the basin margin. The overlying sediments start with an organic-rich shale, deposited as the sea invaded the incipient Atlantic Ocean. This shale is then overlain by a major regressive wedge. Sometimes this wedge is composed of carbonates, with marked thinning from a reefal shelf edge into basinal marls. Alternatively, there are terrigenous progrades in which basinal shales pass shoreward and upward into turbidites and paralic and continental deposits.

Basins of this half-graben type, with rift-drift sequences similar to those just described, are not unique to the Atlantic Ocean, but also occur around other opening oceans. The Exmouth Plateau of northwest Australia is such an example (Exxon and Wilcox, 1978). Collectively,

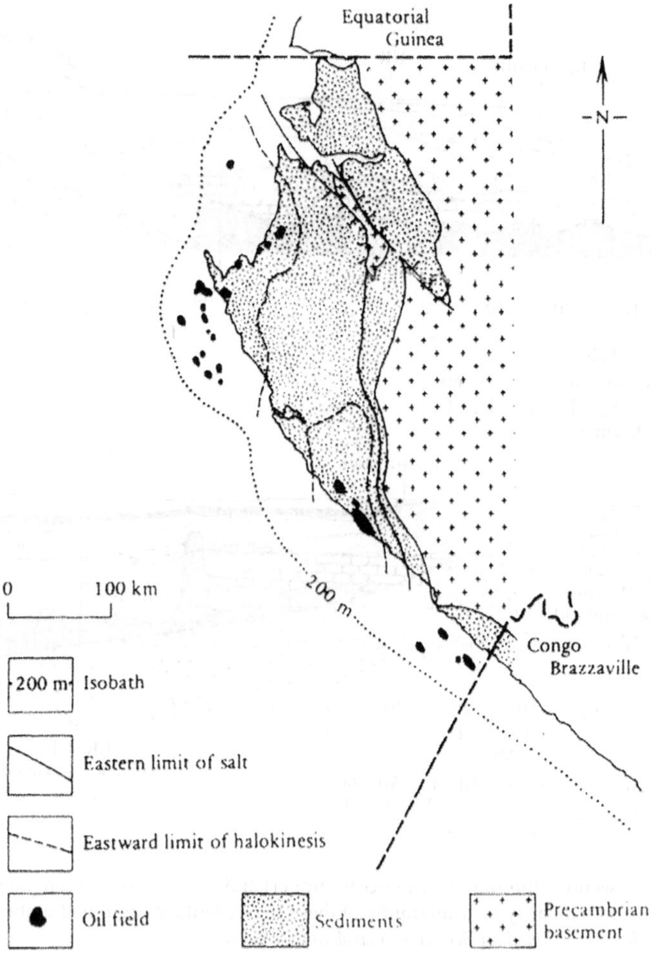

FIGURE 8.30 Geological map of the Gabon basin, compiled from various sources.

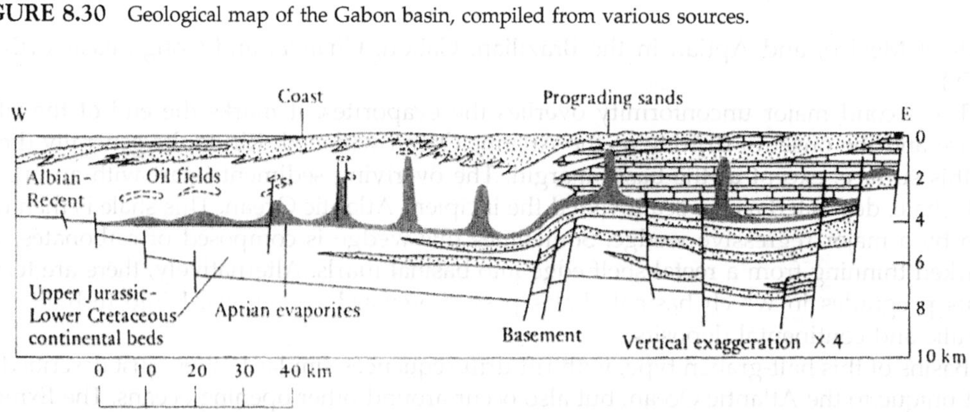

FIGURE 8.31 Cross-section of the Gabon basin. *Modified from Brink (1974) reprinted by permission of the American Association of Petroleum Geologists.*

such coasts are referred to as passive, or trailing, continental margins, in contrast to the active, or Pacific, continental margins where subductive arcs occur.

The preceding review shows that the Atlantic coastal basins evolved from intracratonic rifts, which were initiated on axes of crustal divergence and incipient sea floor spreading. For many years Atlantic-type coastal basins could only be explored along their less prospective up-dip edges. Within recent years offshore exploration has shown that many of these edges contain considerable reserves of oil and gas. Traps include tilted fault blocks (sealed by evaporites or the drift-onset unconformity), salt domes, and anticlines draped over basement horsts. For reasons still not clear, most of the productive basins appear to be in the Southern Hemisphere.

8.6.2 Failed Rift Basins: Aulacogens

There is one last type of rift basin to consider. Earlier, intracratonic rifts were noted to develop a triradiate pattern, often with triple-rift, or triple-R, junctions at the end of each major rift. As the crust draws apart, only two rifts out of each triple-R junction actually separate and become ocean margin half-grabens. One rift has failed to open. These aborted rifts, also called *failed arms* or *aulacogens*, are prime targets for petroleum exploration and deserve to be examined closely. The Benue Trough of Nigeria, which has already been mentioned, is the failed arm of a triple-rift junction that developed on the site of the present Niger Delta. When the old Pangean southern continent began to break up, rifts developed from the Benue trough north into Niger and Chad. Simultaneously, rifts developed along the Tibesti-Sirte Arch, which separated the Murzuk and Kufra basins (Section 8.4.1). The Sirte embayment, which was introduced earlier as an epicratonic embayment, developed along this axis and could therefore be regarded as a complex failed rift.

One of the best known failed rifts occurs in the North Sea of Europe. When the European, Greenland, and North American plates began to separate, a triple-R junction developed somewhere to the northeast of Scotland. Two of these arms opened to form the Norwegian Sea and the Atlantic Ocean, which are both flanked by half-graben basins. The southeastern branch of the triple-R junction subsided but failed to open, providing the North Sea oil province. This province is described in many papers in volumes edited by Woodland (1975), Finstad and Selley (2000,1977), Illing and Hobson (1981), Brooks and Glennie (1987), Glennie (1990), and Parker (1993). Short reviews are given by Ziegler (1975) and Selley (1976).

Figures 8.32 and 8.33 illustrate the main structural features of the North Sea as seen in plan and sections. Rifting began in the Permian and continued throughout the Triassic, with the deposition of fluvial and eolian sands and evaporites. A major transgression in the Jurassic resulted in the deposition of organic-rich shales and paralic sands. As rifting continued, submarine fault scarps along the rift margins poured conglomerates and turbidites onto the basin floor. At the end of the Jurassic a major unconformity, the Cimmerian event, marked the end of rifting and the onset of drifting as the Atlantic Ocean opened to the west. Lower Cretaceous sands and shales onlap pre-Cimmerian fault blocks. These blocks have considerable structural relief and locally underwent crestal erosion. Quiescence in the Late Cretaceous allowed the widespread deposition of coccolithic chalk. Renewed rifting in the Paleocene caused chalk turbidites and melanges to be shed into the Central Graben, following which a major delta complex prograded east and southeastward from Scotland. Submarine channel

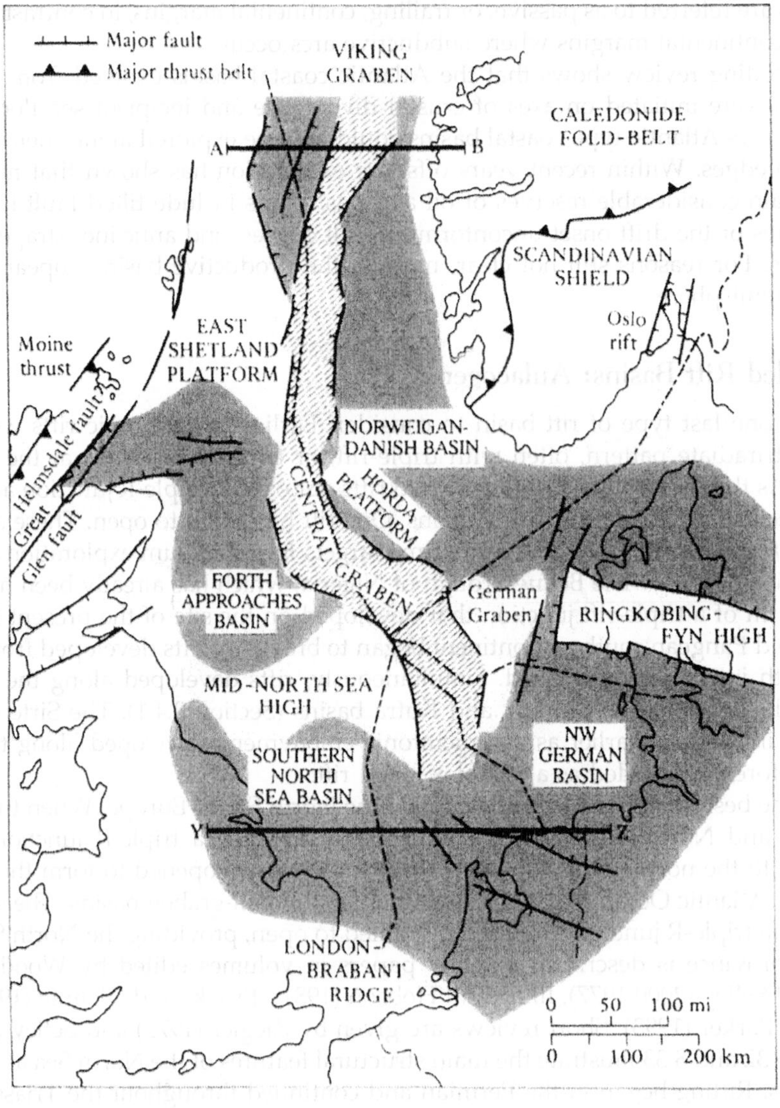

FIGURE 8.32 Map of the North Sea showing the main structural elements. This major hydrocarbon province is a classic example of a failed rift system.

sands and turbidite fans were deposited at the foot of the delta complex. The basin continued to be infilled by marine clays and occasional shallow marine sands up to the present day.

The North Sea contains four major hydrocarbon plays. In the southern North Sea Basin gas occurs in block-faulted anticlines beneath the Upper Permian Zechstein salt. Lower Permian eolian dune sands are the reservoir, and underlying Carboniferous coal beds provide the source for the gas. This play contains several gas fields in the giant category, including

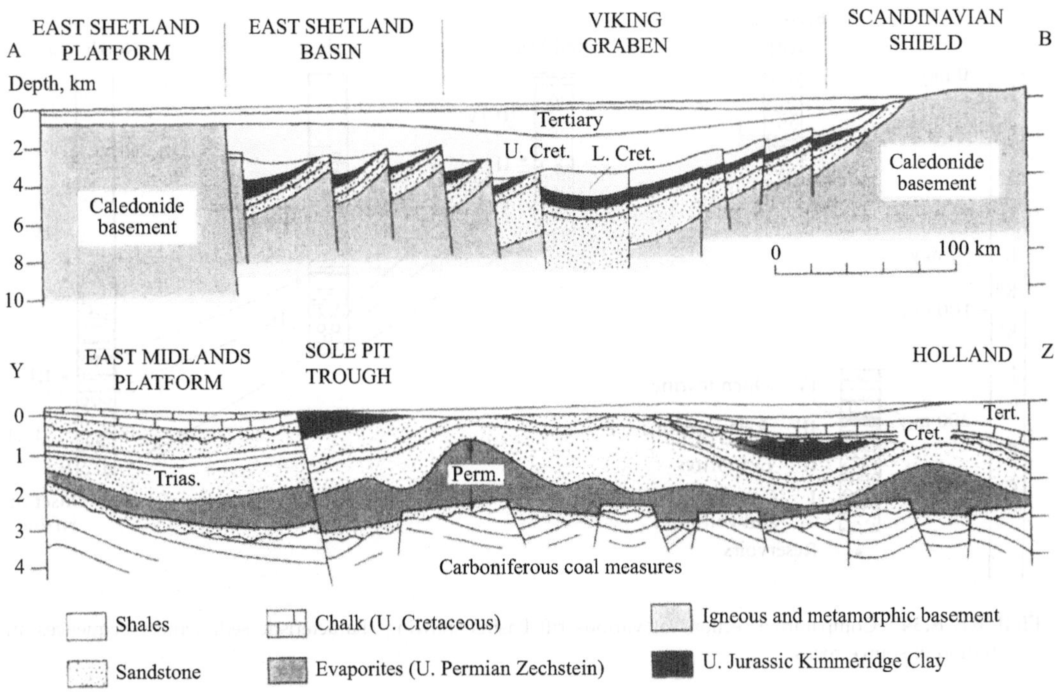

FIGURE 8.33 Cross-sections of the North Sea. Locations shown in Fig. 8.32.

Groningen, Hewett, Leman, and Indefatigable. The second major play occurs in Jurassic sands in tilted fault blocks sealed by Cretaceous shales and limestones that onlap the Cimmerian unconformity. The Brent, Statfjord, Piper, and Heather fields are of this type (refer back to Figs. 7.15, 7.48, and 7.49). The third major play occurs in southwestern offshore Norway. Here Ekofisk and associated fields produce from fractured, overpressured Cretaceous chalk reservoirs domed over Permian salt structures (refer back to Fig. 7.25). Finally, production comes from Paleocene deep-sea sands draped over pre-Cimmerian horsts. The Montrose, Frigg, and Forties fields are of this type (refer back to Fig. 7.13).

The Jurassic, Cretaceous, and Paleocene oil and gas fields are all believed to have been largely sourced from Jurassic shales. Estimates of North Sea reserves vary widely, but Ivanhoe (1980) cites recoverable reserves of 20 billion barrels of oil and 40 trillion cubic feet of gas.

Several conditions have made the North Sea rift basin a major hydrocarbon province. Excellent reservoirs are provided by polycyclic sands, often with subunconformity-leached porosity. Thick, rich source beds within the rift axis interfinger with and underlie the reservoirs. Traps are many and varied, including horsts, combination fault block truncations, salt domes, and compactional anticlines. Geothermal gradients, although now near average, were once abnormally high, enhancing hydrocarbon generation and migration.

This shows why rift basins in general, and failed rifts in particular, may be major petroleum provinces (Lambiase, 1994). Schneider (1972) has shown that a regular sequence of

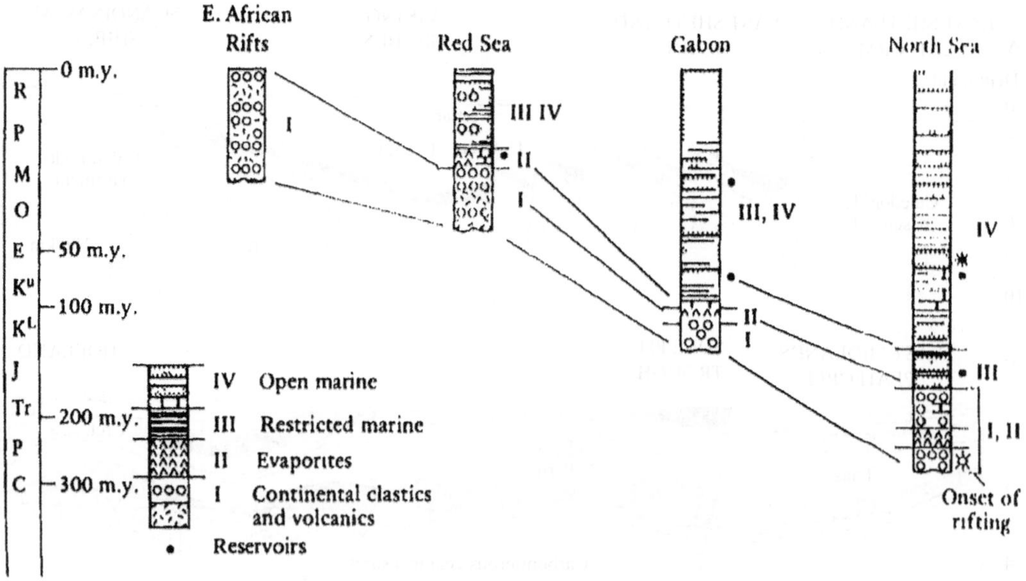

FIGURE 8.34 Comparative sections of various rift basins showing characteristic sedimentary sequences and distribution of source rocks.

facies tends to occur within the rift-drift suite of basins. Stage one, when the rift was still above sea level, consists of continental clastics, which are often associated with volcanics. The subsiding rift floor ultimately reaches sea level. This condition favors evaporite formation as the trough surface oscillates above and below the sea. As the rift floor is finally submerged, evaporites are overlain by organic-rich marine muds deposited in the restricted trough. Finally, as the rift dilates into an open sea, carbonate shelves and prograding clastic wedges build out over the old rift floor onto oceanic crust. Figure 8.34 shows how this ideal sequence applies to the examples of the rift-drift sequence of basins just described.

8.7 STRIKE-SLIP BASINS

We have now reviewed the different types of basins associated with convergent and divergent plate boundaries. Some plate margins are transcurrent and develop a particular type of rift basin. Transcurrent plate margins are defined by major wrench faults. Where plates move past one another, however, the movement is seldom wholly parallel; an oblique component commonly causes crustal compression. Similarly, compressive phases may alternate with phases of separation. These changes occur both in time and space. Thus, for example, the divergent plate boundary of the Red Sea can be traced northeastward to the Dead Sea rift. The transcurrent nature of the Dead Sea faults has been understood from biblical to recent times (see Zechariah 14.4 and Quennell (1958); respectively). By contrast the San Andreas transverse fault system of California passes into compressive plate boundaries at both ends (Fig. 8.6).

Transform fault systems give rise to very distinctive types of rift basin. These rift basins are generally very deep, subside rapidly, and have rates of high heat flow. The Dead Sea rift is such an example. It is some 15 km wide and 150 km long. Left lateral displacement has been estimated in the order of 70–100 km (Wilson et al., 1983). More than 6 km of subsidence has occurred since the Miocene period. The Dead Sea valley has long been noted for its petroleum seeps, occasionally giving rise to floating blocks of asphalt (Nissenbaum, 1978). Commercial quantities of oil have yet to be found within the margins of the rift.

By contrast the strike-slip basins associated with the transcurrent fault systems of California are very petroliferous (Fig. 8.35). These rifts are mainly of late Tertiary age. They are characterized by thick sequences (more than 10 km) of rapidly deposited clastics in which abyssal shales (source rocks) pass up through thick submarine fans (reservoirs) into paralic and continental deposits. The basins are often asymmetric, with alluvial and submarine conglomerates developed adjacent to the active boundary faults (Crowell, 1974. Petroleum is trapped in

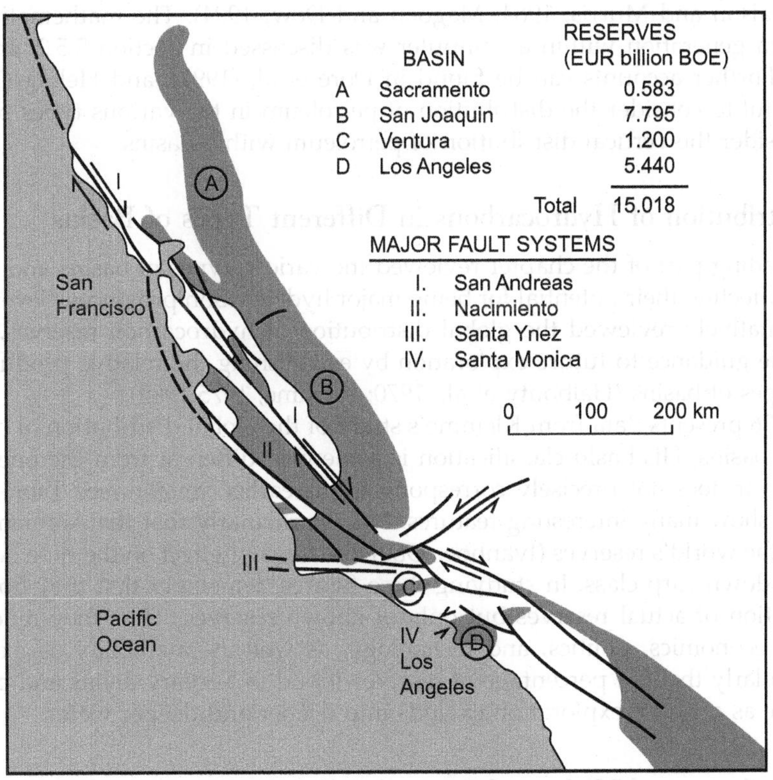

FIGURE 8.35 Map of part of California showing major strike-slip basins and their associated faults.

en echelon and flower-structure faulted anticlines (see Section 7.6.2); examples are given in Weeks (1958). Stratigraphic entrapment in submarine channels and fans is less common (Section 7.6.2). Estimated ultimate recoverable reserves for the Californian basins is in excess of 15 billion barrels of oil equivalent (St. John, 1980). Additional information on strike-slip basins can be found in Allen and Allen (1990), pp. 115–138).

8.8 SEDIMENTARY BASINS AND PETROLEUM SYSTEMS

The concept of the petroleum system was introduced in Chapter 5, where it was discussed in terms of how petroleum systems were related to the volumes of petroleum generated from source rocks. Recall that a petroleum system was defined as "a dynamic petroleum generating and concentrating physico-chemical system functioning in geologic space and time" (Demaison and Huizinga, 1994). It is now appropriate to integrate the concept of the petroleum system with sedimentary basins, a topic that has been intensively studied in recent years (Demaison and Murris, 1984; Magoon and Dow, 1994). The mathematical modeling of petroleum generation within a computer was discussed in Section 5.5.2, and illustrated in Plate 5. Further accounts can be found in Dore et al. (1991), and Helbig (1994). First it may be useful to consider the distribution of petroleum in the various types of basins, and then to consider the vertical distribution of petroleum within basins.

8.8.1 Distribution of Hydrocarbons in Different Types of Basins

The preceding part of the chapter reviewed the various types of basins and analyzed the conditions affecting their potential for being major hydrocarbon provinces. Several geologists have quantitatively reviewed the global distribution of hydrocarbon reserves. This review may provide guidance to future exploration by establishing the relative productivity of the different types of basins (Halbouty et al., 1970; Klemme, 1975,1980).

Figure 8.36 presents data from Klemme's study of the global distribution of hydrocarbons in different basins. His basin classification is somewhat different from the one used in this chapter, and it does not precisely correspond to his earlier version (see Table 8.1), but the figure does show many interesting features. Note particularly that the Arabian Gulf, which has 38% of the world's reserves (Ivanhoe, 1980), has a major effect on the type IV Continental borderland downwarp class. In studying these figures, remember that they do not indicate the distribution of actual reserves but only of known reserves. Thus they reflect factors of geography, economics, politics, and technology, as well as our ability as explorationists. Note particularly that the percentage of reserves found in Tertiary deltas and coastal basins will increase as offshore exploration extends into deeper and deeper water.

8.8.2 Distribution of Hydrocarbons within Basins

Oil and gas tend to occur in sedimentary basins in a regular pattern. Considered vertically, oil gravity decreases with depth. Heavy oils tend to be shallow and, with increasing depth, pass down into light oils, condensate, and finally gas, until the point at which hydrocarbons and porosity are absent. Hunt (1979) took all the API data from the 1975 International

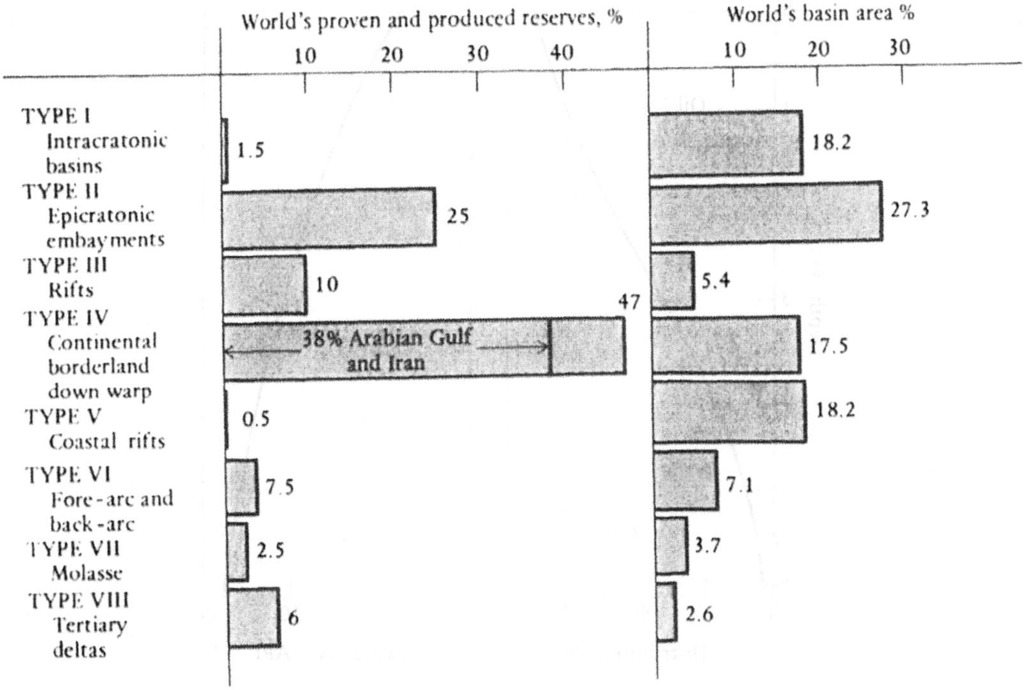

FIGURE 8.36 Histograms comparing the surface area of different types of sedimentary basins and their currently known petroleum reserves. *Based on data from Klemme (1980).*

Petroleum Encyclopedia to produce the graphs shown in Fig. 8.37. These graphs bear out the statement just made. Although oil API generally increases with depth (density decreases), many local conditions may disrupt this pattern, for example, the existence of several source beds in a basin, hydrocarbon flow along faults, flushing and degradation, and uplift and erosion. A noted anomaly is provided by the Niger Delta, where gas commonly occurs at shallower depths than oil (Unomah, 1993).

Oils tend to become lighter not only downward but also laterally toward a basin center. Typically, heavy oils occur around basin margins, and condensate and gas in the center. The cause, or causes, of these vertical and lateral variations of oil and gas are of considerable importance.

An ingenious explanation was advanced by Gussow (1954) and has since gained wide acceptance as Gussow's principle. This theory may be stated briefly as follows. Consider a sedimentary basin, up whose flanks extend continuous, permeable reservoir beds that contain many traps. Consider also that the hydrocarbons, oil and gas, emigrate from the "devil's kitchen" in the deep center of the basin up the flanks. On reaching the first trap, gas may displace all the oil, which will be forced below the spill point and up to the next trap, where the process will be repeated. Finally, all the gas will be retained, so there will be a trap with a gas cap and an oil column. The trap will be full to spill point, and oil forced below the spill point will emigrate up, filling trap after trap until all the oil is contained. The

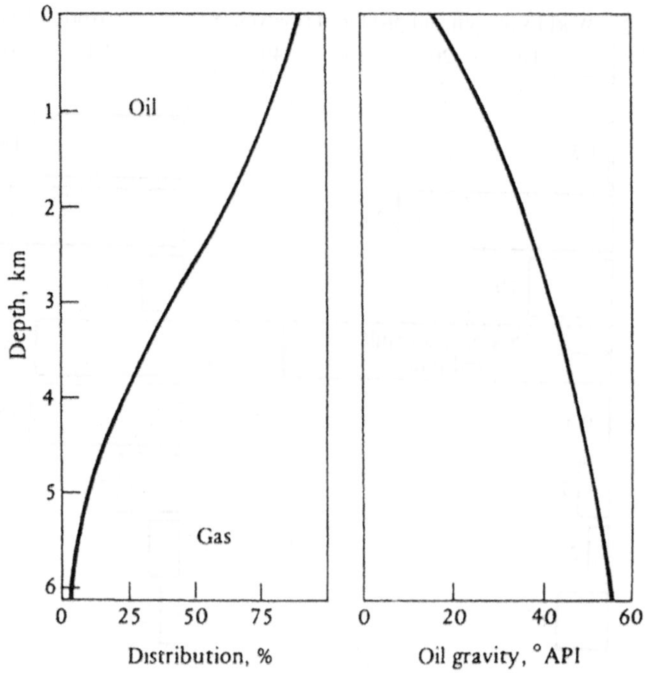

FIGURE 8.37 Vertical variation in the distribution of oil and gas and in oil gravity. These curves were calculated from data in the 1975 International Petroleum Encyclopedia. These global trends have many local exceptions. *From Hunt (1979), used with permission.*

final trap will not be full to spill point, and any additional traps that may be structurally higher will be barren (Fig. 8.38).

It has already been pointed out that some fields show a gravity segregation of oil (Section 7.2.1). Therefore it is not hard to envisage Gussow's principle explaining not only the gross distribution of oil and gas in a basin but also the less obvious variations in oil gravity. Gussow established his principle from work on the Bonnie-Glenn-Wizard Lake Devonian reef complex of Alberta, across a distance of about 160 km. Other excellent examples of differential entrapment have been noted from the Niagaran reefs of Michigan (Gill, 1979) and from the Mardin Group of Anatolia (Erdogan and Akgul, 1981).

There are problems, however, in applying Gussow's principle over large distances and in basins with discontinuous reservoirs. Thus Bailey et al. (1974) showed how oil and gas distribution can be mapped with great continuity right across the Alberta Basin, where many reservoirs are isolated reefs and where unconformities provide regionally extensive permeability barriers. The zonation of gas, light oil, and heavy oil from basin center to margin may be due to a combination of thermal maturation and degradation by meteoric water.

Many other basins around the world exhibit a regular gravity zonation of hydrocarbons, yet lack continuity of reservoir from trap to trap. The chalk fields of the Ekofisk area of the North Sea, for example, show an increase in API gravity toward the Central Graben, yet lack any regional permeability.

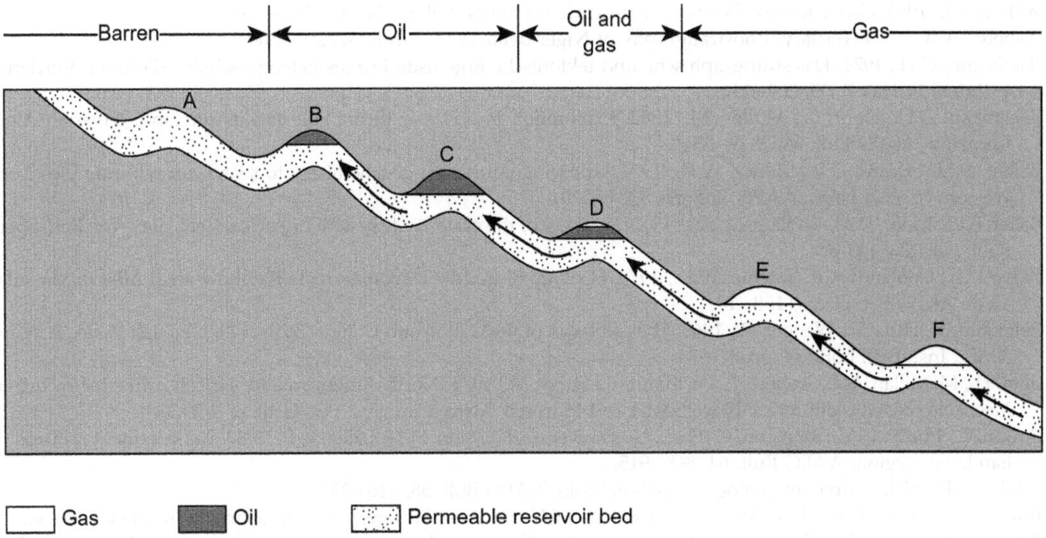

FIGURE 8.38 Cross-section through a hypothetical sedimentary basin with a laterally continuous permeable stratum folded into a multitude of traps. This figure illustrates Gussow's principle.

Thus the differentiation of oil and gas found in a basin is related to two factors. One factor is the level of thermal maturation of the source beds: gas being generated at higher temperatures and therefore at greater depths than oil (see Section 7.3.3). The shallowest traps around a basin rim are likely to have been flushed by meteoric water, as can be checked from their salinity. The shallowest oil is likely to be heavy where oil moving up from the basin center has been degraded by contact with meteoric water. The second factor is the regional distribution of permeable carrier formations, a factor that is termed the "impedance" of the petroleum system as discussed earlier.

Thus Gussow's principle must be applied with care when considering the basinwide distribution of oil, condensate, and gas. It is particularly useful, however, when trying to understand local variations in the distribution of different gravities of petroleum. This is illustrated by Fig. 8.38. If trap C was the first to be drilled and it was full to spill point, then both traps B and D will be prospective. It may be anticipated that trap D will be full to spill point, but one can only speculate about the thickness of the oil column in trap B. Consider, on the other hand, that trap B had been drilled first and had been found not to be full to spill point. It would be sound policy to drill trap C, but trap A should be farmed out at the earliest opportunity, since it is unlikely to have received any oil.

References

Ahmad, F., 1968. Orogeny, geosynclines and continental drift. Tectonophysics 5, 177–189.

Allen, P.A., Allen, J.R., 1990. Basin Analysis Principles and Applications. Blackwell, Oxford.

Anonymous, July 1980. Overthrust belt action hot in and out of fairway. Oil Gas J. 123–125.

Antoine, J.W., 1974. Continental margins of the Gulf of Mexico. In: Burk, C.A., Drake, C.I. (Eds.), The Geology of Continental Margins. Springer-Verlag, New York, pp. 683–693.

Aubouin, J., 1965. Geosynclines. Developments in Geotectonics, vol. I. Elsevier, New York.
Avbobo, A.A., 1978. Tertiary lithostratigraphy of Niger delta. AAPG Bull. 62, 295–306.
Bachmann, G.H., 1979. Das stratigraphische und tektonische Ergenisde Erdges tiefengufschulb—Bohring. Vorderiss 1. Erdoel-Erdgas-Z. 95, 209–214.
Bachmann, G.H., Dohr, G., Muller, M., 1982. Exploration in a classic thrust belt and its foreland: Bavarian Alps, Germany. AAPG Bull. 66, 2529–2542.
Bailey, N.J.L., Evans, C.R., Milner, W.D., 1974. Applying petroleum geochemistry to search for oil: examples from Western Canada basin. AAPG Bull. 58, 2284–2294.
Baker, B.H., Mohr, P.A., Williams, L.A.J., 1972. Geology of the eastern rift system of Africa. Geol. Soc. Am. Bull. Spec. Publ. No. 136. 145 pp.
Barss, D.L., Copland, A.B., Ritchie, W.D., 1970. Geology of middle Devonian reefs Rainbow area, Alberta, Canada. Mem. Assoc. Pet. Geol. 14, 8–19.
Belmonte, Y., Hirtz, P., Wenger, R., 1965. The salt basin of the Gabon and Congo (Brassaville). In: Salt Basins Around Africa. Inst. Petl, London, pp. 55–78.
Bhat, H., McMillan, N.J., Aubert, J., Porthault, B., Surin, M., 1975. North American and African drift—the record in Mesozoic coastal plain rocks. Nova Scotia and Morocco. Mem. Can. Soc. Pet. Geol. 4, 375–390.
Bowen, C., Purdy, G.M., Johnston, C., Shor, G., Lawver, L., Hartano, H.M.S., Jezek, P., 1980. Arc-continent collision in Banda Sea region. AAPG Bull. 64, 868–915.
Brink, A.H., 1974. Petroleum geology of Gabon basin. AAPG Bull. 58, 216–235.
Brooks, J., Glennie, K.W., 1987. Petroleum Geology of North West Europe, vol. 2. Graham & Trotman, London.
Bruce, C., Parker, E.H., 1975. Structural features and hydrocarbon deposits in the Mackenzie delta. Proc. World Pet. Congr. 9 (2), 251–261.
Burk, C.A., Drake, C.L. (Eds.), 1974. The Geology of Continental Margins. Springer-Verlag, New York.
Busby, C., Ingersoll, R., 1995. Tectonics of Sedimentary Basins. Blackwell, Oxford.
Cohee, G.V., Landes, K.K., 1958. Oil in the Michigan basin. In: Weeks, L.G. (Ed.), The Habitat of Oil. Am. Assoc. Pet. Geol, Tulsa, OK, pp. 473–493.
Conant, L.C., Goudarzi, G.H., 1967. Stratigraphic and tectonic framework of Libya. Am. Assoc. Pet. Geol. Bull. 51, 719–730.
Coney, P.J., 1970. The geotectonic cycle and the new global tectonics. Geol. Soc. Am. Bull. 81, 739–747.
Crostella, A., 1977. Geosynclines and plate tectonics in Banda arcs, eastern Indonesia. AAPG Bull. 61, 2063–2081.
Crowell, J.C., 1974. Origin of late Cenozoic basin in southern California. Spec. Publ. Soc. Econ. Paleontol. Mineral. 22, 190–204.
Dallmus, K.F., 1958. Mechanics of basin evolution and its relation to the habitat of oil. In: Weeks, L.G. (Ed.), The Habitat of Oil. Am. Assoc. Pet. Geol., Tulsa, OK, pp. 883–931.
Dana, J.D., 1873. On some results of the earth's contraction from cooling, including a discussion of the origin of mountains and the nature of the earth's interior. Am. J. Sci. 5, 423–443, 6, 6–14, 104–115, 161–172.
Darling, G.B., Wood, P.W.J., 1958. Habitat of oil in the Canadian portion of the Williston basin. In: Weeks, L.G. (Ed.), Habitat of Oil. Am. Assoc. Pet. Geol., Tulsa, OK, pp. 129–148.
Darrcott, B.W., Girdler, R.W., Fairhead, J.D., Hall, S.A., 1973. The East African rift system. In: Tarling, D.H., Runcorn, S.K. (Eds.), Implications of Continental Drift to the Earth Sciences. Academic Press, London, pp. 757–766.
Davies, P.A., Runcorn, S.A., 1980. Mechanics of Continental Drift and Plate Tectonics. Academic Press, London.
Delwig, L.F., Evans, R., 1969. Depositional processes in Salina salt of Michigan, Ohio, and New York. Am. Assoc. Pet. Geol. Bull. 53, 949–956.
Demaison, G., Huizinga, H.J., 1994. Genetic classification of petroleum systems using three factors: charge, migration and entrapment. AAPG Mem. 60, 73–92.
Demaison, G., Murris, R.J., 1984. Petroleum geochemistry and basin evaluation. Mem. Am. Assoc. Pet. Geol. 35.
Dickinson, W.R., Seely, D.R., 1979. Structure and stratigraphy of forearc regions. AAPG Bull. 63, 2–31.
Dore, A.G., Augustson, J.H., Hermanrud, C., Stewart, D.J., Sylta, O. (Eds.), 1991. Basin Modelling: Advances and Applications. Elsevier, Amsterdam.
Dow, W.C., 1978. Petroleum source beds on continental slopes and rises. AAPG Bull. 62, 1584–1606.
Drake, C.L., Ewing, J.T., Stokand, H., 1968. The continental margin of the United States. Can. J. Earth Sci. 5, 99–110.
Emery, K.O., 1980. Continental margins—classification and petroleum prospects. AAGP Bull. 64, 297–315.

Erdogan, L.T., Akgul, A., 1981. An oil migration and re-entrapment model for the Mardin group reservoirs of Southeast Anatolia. J. Pet. Geol. 4, 57–75.
Eugster, H.P., Surdam, R.C., 1973. Depositional environment of the Green River formation of Wyoming. A preliminary report. Geol. Soc. Am. Bull. 84, 1115–1120.
Evamy, B.D., Harembourne, J., Kamerling, P., Knapp, W.A., Molloy, F.A., Rowlands, P.H., 1978. Hydrocarbon habitat of tertiary Niger delta. AAPG Bull. 62, 1–39.
Evans, H., 1972. Zama—a geophysical case history. Spec. Publ. Am. Assoc. Pet. Geol. 10, 440–452.
Evans, R., 1978. Origin and significance of evaporites in basins around Atlantic margin. AAPG Bull. 62, 234.
Exxon, N.F., Wilcox, J.B., 1978. Geology and petroleum potential of Exmouth plateau area off western Australia. AAPG Bull. 62, 40–72.
Finstad, K.G., Selley, R.C. (Eds.), 1975. Proceedings of the Jurassic Northern North Sea Symposium. Norsk Petroleumsforening, Stavanger.
Finstad, K.G., Selley, R.C. (Eds.), 1977. Proceedings of the Second Jurassic Northern North Sea Symposium. Norwegian Petroleum Society, Oslo.
Fischer, A.G., 1975. Origin and growth of basins. In: Fischcr, A.G., Judson, S. (Eds.), Petroleum and Global Tectonics. Princeton University Press, Princeton, NJ, pp. 47–82.
Fischer, A.G., Judson, S. (Eds.), 1975. Petroleum and Global Tectonics. Princeton University Press, Princeton, NJ.
Gill, D., 1979. Differential entrapment of oil and gas in Niagaran pinnacle—reef belt of Northern Michigan. AAPG Bull. 63, 608–620.
Glaessner, M.F., Teichert, C., 1947. Geosynclines: a fundamental concept in geology. Am. J. Sci. 245, 465–482, 571–591.
Glennie, K.W., 1990. Introduction to the Petroleum Geology of the North Sea, third ed. Black-well, Oxford.
Grey, C., 1971. The Geology of Libya. University of Libya, Tripoli.
Gussow, W.C., 1954. Differential entrapment of oil and gas: a fundamental principle. Am. Assoc. Pet. Geol. Bull. 38, 816–853.
Haile, N.S., 1968. Geosynclinal theory and the organizational pattern of the north west Borneo geosyncline. Q. J. Geol. Soc. London 124, 171–194.
Halbouty, M.T., Meyerhoff, A.A., King, R.E., Dott, R.H., Klemme, H.D., Shabad, T., 1970. World's giant oil and gas fields, geologic factors affecting their formation and basin classification. Mem. Am. Assoc. Pet. Geol. 14, 502–555.
Hall, A.J., 1859. Natural History of New York, vol. 3. Appleton & Coy, New York.
Harding, T.P., Lowell, J.D., 1979. Structural styles, their plate-tectonic habitats, and hydrocarbon traps in petroleum provinces. AAPG Bull. 63, 1016–1058.
Hartley, R.W., Allen, P.A., 1994. Interior cratonic basins of Africa: relation to continental break-up and role of mantle convection. Basin Res. 6, 95–113.
Haug, E., 1900. Les geosynchnaux et les aires continentales. Geol. Soc. Fr. Bull. 28, 617–711.
Heirtzler, J.R., 1968. Sea floor spreading. Sci. Am. 219, 60–70.
Helbig, K. (Ed.), 1994. Modelling the Earth for Oil Exploration. Elsevier, Amsterdam.
Heybroek, F., 1965. The red sea Miocene evaporite basin. In: The Salt Basins Around Africa. Inst. Pet., London, pp. 17–40.
Hoc, A., 1979. Continental Margins Geological and Geophysical Research Needs and Problems. Natl. Acad. Sci., Washington, DC.
Hsu, K.J., 1970. The meaning of the world flysch, a short historical search. Spec. Pap. Geol. Assoc. Can. 7, 1–11.
Hunt, J.M., 1979. Petroleum Geochemistry and Geology. Freeman, San Francisco.
Ibrahim, M.W., 1979. Shifting depositional axes of Iraq: an outline of geosyncline history. J. Pet. Geol. 2, 181–197.
Illies, J.H., Mueller, S.T., 1970. Graben Problems. Slucheizerbart'sche Verlagsbuchhandlung, Stuttgart.
Illing, L.V., 1959. Deposition and diagenesis of some upper Paleozoic carbonate sediments in Western Canada. Proc. World Pet. Congr. 5, 23–52.
Illing, L.V., Hobson, G.D., 1981. Petroleum Geology of the Continental Shelf of Northwest Europe. Heydcn, London.
Ivanhoe, L.F., June 30, 1980. World's giant petroleum provinces. Oil Gas J. 146–147.
Jones, P.H., 1969. Hydrodynamics of geopressure in the northern Gulf of Mexico. J. Pet. Technol. 17, 803–810.
Kamen-Kaye, M., 1970. Geology and productivity of Persian Gulf synclinorium. Am. Assoc. Pet. Geol. Bull. 54, 2371–2394.
Kay, M., 1944. Geosynclines in continental development. Science 99, 461–462.

Kay, M., 1947. Geosynclinal nomenclature and the craton. Am. Assoc. Pet. Geol. Bull. 31, 1289–1293.

Kenyon, C.S., Beddoes, K., 1977. Geothermal gradient map of southeast Asia. South East Asia Pet. Explor. Soc. 1–50.

Klemme, H.D., 1975. Geothermal gradients, heat flow, and hydrocarbon recovery. In: Fischer, A.F., Judson, S. (Eds.), Petroleum and Global Tectonics. Princeton University Press, Princeton, NJ, pp. 251–306.

Klemme, H.D., 1980. Petroleum basins—classification and characteristics. J. Pet. Geol. 3, 187–207.

Klitzsch, E., 1970. Die Strukturgeschichte der Zentral Sahara. Geol. Rundsch. 59, 459–527.

Klitzsch, E.H., 1972. Problems of continental Mesozoic strata of southwestern Libya. In: Whiteman, A.J., Dessauvagie, T.F.J. (Eds.), African Geology. Ibadan University, Nigeria, pp. 483–493.

Klovan, J.E., 1974. Development of Western Canadian Devonian reefs and comparison with Holocene analogues. AAPG Bull. 58, 787–799.

Lamb, C.F., 1980. Painter reservoir field—giant in Wyoming thrust belt. AAPG Bull. 64, 638–644.

Spec. Publ. No. 80. In: Lambiase, J.J. (Ed.), 1994. Hydrocarbon Habitat in Rift Basins. Geol. Soc. London, London.

Lehner, P., Dc Ruiter, P.A.C., 1977. Structural history of Atlantic margin of Africa. AAPG Bull. 61, 961–981.

Lestang, J. De, 1965. Das Palaozoikum am Rande des Afro-Arabischen Gondwanakontinentes. Z. Dtsch. Geol. Ges. 117, 479–488.

Lowell, J.D., Genik, G.J., 1972. Sea floor spreading and structural evolution of the southern Red Sea. Am. Assoc. Pet. Geol. Bull. 56, 247–259.

Lowell, J.D., Genik, G.J., Nelson, T.H., Tucker, P.M., 1975. Petroleum and plate tectonics of the southern Red Sea. In: Fischer, A.G., Judson, S. (Eds.), Petroleum and Global Tectonics. Princeton University Press, Princeton, NJ, pp. 129–156.

Magoon, L.B., Claypool, G.E., 1981. Petroleum geology of Cook Inlet basin – an exploration model. AAPG Bull. 65, 1043–1062.

Magoon, L.B., Dow, W.G. (Eds.), 1994. The Petroleum System—from Source to Trap. Mem. Am. Assoc. Pet. Geol., vol. 60.

McCaslin, E., January 12, 1981. Hot overthrust belt exploration continues. Oil Gas J. 113–114.

McKee, E.D., 1965. Origin of Nubian and similar sandstones. Geol. Rundsch. 52, 551–587.

McKenzie, D.P., 1978. Some remarks on the development of sedimentary basins. Earth Planet. Sci. Lett. 40, 25–32.

Mesner, J.C., Woodridge, L.C.P., 1964. Maranhao Paleozoic basin and Cretaceous coastal basins. Am. Assoc. Pet. Geol. Bull. 48, 1475–1512.

Mesolella, K.J., Robinson, J.D., McCormick, L.M., Ormiston, A.R., 1974. Cyclic deposition of Silurian carbonates and evaporites in Michigan basin. AAPG Bull. 58, 34–62.

Meyerhoff, A.A., Meyerhoff, H.A., 1972. The new global tectonics: major inconsistencies. Am. Assoc. Pet. Geol. Bull. 56, 269–336.

Mitchell, A.H., Reading, H.G., 1969. Continental margins, geosynclines, and ocean floor spreading. J. Geol. 77, 629–646.

Mitchell, A.H.G., Reading, H.G., 1978. Sedimentation and tectonics. In: Reading, H.G. (Ed.), Sedimentary Environments and Facies. Blackwell, Oxford, pp. 439–476.

Murray, G.E., 1960. Geologic framework of gulf coastal province of United States. In: Shepard, F.P. (Ed.), Recent Sediments, Northwestern Gulf of Mexico. Am Assoc. Pet. Geol, Tulsa, OK, pp. 5–33.

Murris, R.J., 1980. Middle East: stratigraphic evolution and oil habitat. AAPG Bull. 64, 597–618.

Murris, R.J., 1981. Middle East: stratigraphic evolution and oil habitat. Geol. Mijnbouw 60, 467–480.

Nissenbaum, A., 1978. Dead Sea asphalts—historical aspects. AAPG Bull. 62, 837–844.

North, F.K., 1971. Characteristics of oil provinces: a study for students. Bull. Can. Pet. Geol. 19, 601–658.

Olenin, V.B., 1967. The principles of classification of oil and gas basins. Aust. Oil Gas J. 13, 40–46.

Parker, J.R. (Ed.), 1993. Petroleum Geology of Northwest Europe, vol. 2. Geol. Soc. London, London.

Perrodon, A., 1971. Classification of sedimentary basins: an essay. Sci. Terre. 16, 193–227.

Perrodon, A., 1978. Coup d'oil surles provinces géantes d'hydrocarbures. Rev. Inst. Fr. Pet. 33, 493–513.

Picard, M.D., 1967. Paleocurrents and shoreline orientations in Green river formation (Eocene). Raven Ridge and red wash areas, Northeastern Uinta Basin, Utah. Am. Assoc. Pet. Geol. Bull. 5, 383–392.

Picha, F.J., 1996. Exploring for hydrocarbons under thrust belts—a challenging new frontier in the Carpathians. AAPG Bull. 80, 1547–1564.

Pratsch, J.C., 1978. Future hydrocarbon exploration on continental margins and plate tectonics. J. Pet. Geol. 1, 95–105.

Quennell, A.M., 1958. The structural and geomorphic evolution of the Dead Sea rift. Q. J. Geol. Soc. London 114, 1–24.

Ranneft, T.S.M., 1979. Segmentation of island arcs and application to petroleum geology. J. Pet. Geol. 1, 35–53.

Reading, H.G., 1972. Global tectonics and the genesis of flysch successions. In: Int. Geol. Congr., Rep. Sess., 24th, Montreal, 1972, Sect. 6, pp. 59–66.

Reijers, T.J.A., 1997. Niger delta basin. In: Selley, R.C. (Ed.), The Sedimentary Basins of Africa. Elsevier, Amsterdam.

Salem, M., Busrewil, M.T. (Eds.), 1981. The Geology of Libya, vol. 3. Academic Press, London.

Salop, L.I., 1967. Geology of the Baikal Region, vol. 2. Izd Nedva, Moscow.

Salveson, J.O., 1976. Variations in the oil and gas geology of rift basins. In: 5th Pet. Explor. Semin., Egyptian General Petroleum Corporation, Cairo, pp. 15–17.

Salveson, J.O., 1979. Variations in the geology of rift basins—a tectonic model. In: Rio Grande Rift: Tecton. Magmat. [Sel. Pap. Int. Symp.], Santa Fe, 1978, pp. 11–28.

Sanford, R.M., 1970. Sarir oil field, Libya—desert surprise. Mem. Am. Assoc. Pet. Geol. 14, 449–476.

Schneider, E.D., 1972. Sedimentary evolution of rifted continental margins. Mem. Geol. Soc. Am. 132, 109–118.

Scholl, D.W., Von Huenc, R., Ridlon, J.B., 1968. Spreading of the ocean floor: undeformed sediments in the peru-chile trench. Science 159, 869–871.

Schuchert, C., 1923. Sites and nature of the North American geosynclines. Geol. Soc. Am. Bull. 34, 151–230.

Schuppli, H.M., 1946. Geology of the oil basins of East Indian archipelago. Am. Assoc. Pet. Geol. Bull. 30, 1–22.

Schwab, F.L., 1971. Geosynclinal compositions and the new global tectonics. J. Sediment. Pet. 41, 928–938.

Selley, R.C., 1976. The habitat of North sea oil. Proc. Geol. Assoc. London 87, 359–388.

Selley, R.C., 1996. Ancient Sedimentary Environments, fourth ed. Chapman & Hall, London.

Selley, R.C., 1997a. The sedimentary basins of North Africa: stratigraphy. In: Selley, R.C. (Ed.), The Sedimentary Basins of Africa. Elsevier, Amsterdam.

Selley, R.C., 1997b. The Sirte basin of Libya. In: Selley, R.C. (Ed.), The Sedimentary Basins of Africa. Elsevier, Amsterdam.

Selley, R.C., 2000. Applied Sedimentology, 2nd Edition. Academic Press, London.

Seyfert, C.K., Sirkin, L.A., 1973. Earth History and Plate Tectonics. Harper & Row, New York.

Short, K.C., Stauble, A.J., 1967. Outline of geology of Niger delta. Am. Assoc. Pet. Geol. Bull. 51, 761–779.

Smith, G.W., Summers, G.F., Wellington, D., Lee, J.L., 1958. Mississippian oil reservoirs in Williston basin. In: Weeks, L.G. (Ed.), Habitat of Oil. Am. Assoc. Pet. Geol, Tulsa, OK, pp. 149–177.

St John, B., 1980. Sedimentary Basins of the World and Giant Hydrocarbon Accumulations. Am. Assoc. Pet. Geol, Tulsa, OK.

Surdam, R.C., Wolfbauer, C.A., 1975. Green river formation, Wyoming: a playa-lake complex. Geol. Soc. Am. Bull. 86, 335–345.

Tarling, D.H. (Ed.), 1981. Economic Geology and Geotectonics. Blackwell, London.

Tarling, D.H., Runcorn, S.K., 1973. Implications of Continental Drift to the Earth Sciences, vols. 1 and 2. Academic Press, London.

Thiebaud, C.E., Robson, D.A., 1979. The geology of the area between Wadi Warden and Wadi Gharandal, East Clysmic Rift, Sinai, Egypt. J. Pet. Geol. 1, 63–75.

Unesco, S., 1965. East African Rift System, Upper Mantle. Committee, University College, Nairobi.

Unomah, G.I., June 14, 1993. Why gas is trapped stratigraphically shallower than oil in Niger Delta. Oil Gas J. 53–55.

Uspenskaya, N. Yu, 1967. Principles of oil and gas territories subdivisions and the classification of oil and gas accumulations. Proc. World Pet. Congr. 7 (2), 961–969.

Van Houten, F.B., 1973. Meaning of molasse. Geol. Soc. Am. Bull. 84, 1973–1976.

Van Houten, F.B., 1980. Latest Jurassic—Cretaceous regressive facies, northeast African craton. AAPG Bull. 64, 857–867.

Veevers, J.J., 1981. Morphotectonics of rifted continental margins in embryo (East Africa), youth (Africa-Arabia) and maturity (Australia). J. Geol. 89, 57–82.

Vidal, J., 1980. Geology of the Grondin field. AAPG Mem. 30, 577–590.

Vine, F.J., 1970. The geophysical year. Nature (London) 227, 1013–1017.

Weber, K.J., Daukoru, E., 1975. Petroleum geology of the Niger delta. Proc. World Pet. Congr. 9 (2), 209–221.

Weeks, L.G., 1958. Factors of sedimentary basin development that control oil occurrence. Am. Assoc. Pet. Geol. Bull. 32, 1093–1160.

Wernicke, B., 1981. Low-angle normal faults in the Basin and range Province; nappe tectonics in an extending orogen. Nature (London) 291, 645–648.
Wernicke, B., 1985. Uniform-sense normal shear of the continental lithosphere. Can. J. Earth Sci. 22, 108–125.
European Association of Geophysical Explorationists Spec. Publ. No. 5. In: Wessely, G., Liebl, W. (Eds.), 1996. Oil and Gas in Alpidic Thrustbelts and Basins of Central and Eastern Europe. Geol. Soc. London, London.
Wilhelm, O., Ewing, M., 1972. Geology and history of the Gulf of Mexico. Geol. Soc. Am. Bull. 83, 575–600.
Wilson, J.E., Kashai, E.L., Croker, P., June 20, 1983. Hydrocarbon potential of Dead Sea rift valley. Oil Gas J. 147–154.
Wilson, J.T., 1966. Did the Atlantic close and then re-open? Nature (London) 211, 676–681.
Wilson, J.T., 1968. Static or mobile earth: the current scientific revolution. Proc. Am. Philos. Soc. 112, 309–320.
Wood, P.W.J., July 21, 1980. Hydrocarbon plays in tertiary S. E. Asia basins. Oil Gas J. 90–96.
Woodland, A.W. (Ed.), 1975. Petroleum and the Continental Shelf of North West Europe, vol. 1. Applied Science Publishers, London.
Ziegler, P.A., 1975. Geologic evolution of North Sea and its tectonic framework. AAPG Bull. 59, 1073–1097.

Selected Bibliography

Allen, P.A., Allen, J.R., 1990. Basin Analysis Principles and Applications. Blackwell, Oxford.
Busby, C., Ingersoll, R., 1995. Tectonics of Sedimentary Basins. Blackwell, Oxford.
Demaison, G., Murris, R.J., 1984. Petroleum geochemistry and basin evaluation. Mem. Am. Assoc. Pet. Geol. 35.
Dore, A.G., Augustson, J.H., Hermanrud, C., Stewart, D.J., Sylta, O. (Eds.), 1991. Basin Modelling: Advances and Applications. Elsevier, Amsterdam.
Helbig, K. (Ed.), 1994. Modelling the Earth for Oil Exploration. Elsevier, Amsterdam.
Magoon, L.B., Dow, W.G. (Eds.), 1994. The Petroleum System—from Source to Trap. Mem. Am. Assoc. Pet. Geol, vol. 60.

CHAPTER

9

Nonconventional Petroleum Resources

9.1 INTRODUCTION

The petroleum geologist is largely concerned with exploring for crude oil and natural gas, and the major part of this book is devoted to this theme. Vast amounts of energy, however, are also locked up in what may loosely be described as nonconventional petroleum resources. These include gas hydrates, tar sands, oil shales, tight oil reservoirs, shale gas, and coal bed methane. These reserves have been hard to unlock because the relatively low cost of producing petroleum from conventional sources has inhibited technological research into their extraction. An increase in the cost of producing conventional petroleum provides an incentive to develop methods of producing these nonconventional petroleum resources.

This chapter deals with the plastic and solid hydrocarbons (excluding coal), describing their composition, origin, and distribution, and briefly reviewing production processes. Gas hydrates are not discussed here because they were dealt with at some length in Chapter 2. Shale gas and coal bed methane are also described in this chapter.

9.2 PLASTIC AND SOLID HYDROCARBONS

Plastic and solid hydrocarbons are common in sedimentary rocks of diverse ages in many parts of the world. They are distinct from crude oils, which are liquid at normal temperatures and pressures, although many of these hydrocarbons are viscous, and their viscosity decreases with increasing temperature. Fig. 9.1 classifies the solid hydrocarbons.

Mankind's first exposure to hydrocarbons was from surface occurrences of solid and viscous tars rather than from liquid crude oil, which was not encountered in large quantities until boreholes were drilled. The heavy hydrocarbons are colloquially termed *pitch, asphalt, bitumen,* or *tar,* and the words are often used interchangeably. Although the first three terms are synonymous, tar is, strictly speaking, a dark brown, viscous liquid produced by the destructive distillation of organic matter—wood, coal, or shale. This distillation may occur naturally in the subsurface or artificially by man. Stockholm, or Archangel, tar was collected from the cavities below the "meilers" in which charcoal was burnt. This tar was used extensively for the wooden ships of the maritime nations of northern Europe.

Elements of Petroleum Geology
http://dx.doi.org/10.1016/B978-0-12-386031-6.00009-6

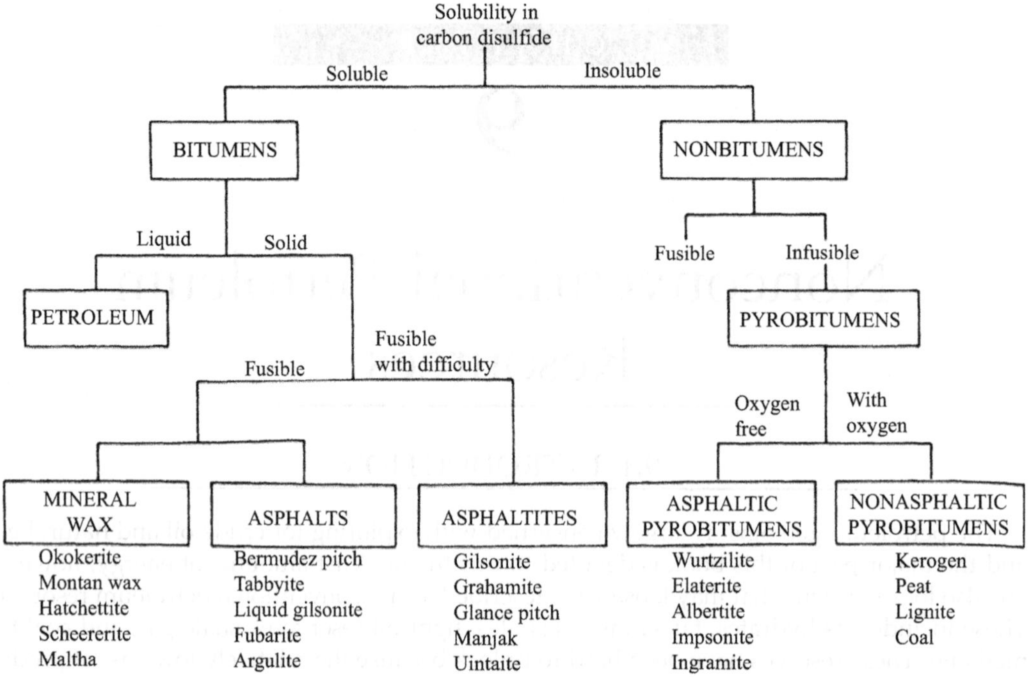

FIGURE 9.1 Terminology and classification of plastic and solid hydrocarbons. *After Abraham (1945).*

The following sections describe the mode of occurrence, the composition, and the economic significance of solid hydrocarbons.

9.2.1 Occurrence of Plastic and Solid Hydrocarbons

The solid and heavy, viscous hydrocarbons occur as lakes or pools on the earth's surface and are disseminated in veins and pores in the subsurface. Two genetically distinct modes of occurrence are recognized: (1) inspissated deposits and (2) secondary deposits.

9.2.1.1 Inspissated Deposits

Inspissated deposits are heavy hydrocarbons from which the light fraction has been removed. This situation can occur where an accumulation of liquid oil has been brought to the surface of the earth by a combination of migration, coupled with uplift and erosion. As this process takes place, the oil will be subjected to flushing by meteoric water, leading to oxidation and bacterial degradation. The lighter hydrocarbons may come out of solution as the pressure drops and may be lost to the surface through fractures as the cap rock is breached. Thus the oil accumulation is proportionally enriched in the viscous, heavy hydrocarbons (Connan and Van der Weide, 1974).

Surface occurrences of asphalt are not only fossilized unroofed oil fields but may also be produced from seepages escaping from oil fields in the subsurface. These hydrocarbon seeps can occur in many ways, but are principally found where faults and fractures allow hydrocarbons to emigrate up through the cap rocks of reservoirs (Fig. 9.2). The seeps may be of mud, brine, gas, or oil, although oil has a tendency to become tarry for the reasons already outlined.

Care must be taken to distinguish petroleum seeps from the iridescent films of ferric hydroxide that are sometimes found floating on puddles, to the confusion of novice petroleum explorationists. The criteria to differentiate genuine petroleum from ferric hydroxide scum were laid down by Craig, in Wade (1915) as follows:

Films of ferric hydroxide	Films of petroleum
1. Iridescent reds, red-browns, and blues	Iridescent greens, pinks, and purples
2. Dries with a semimetallic luster	Never does
3. Breaks into angular fragments	Deforms into plastic convolutions
4. Crystalline flakes may sink	Film always floats

Hydrocarbon seeps are known from all over the world and are characteristic of many, but by no means all, major hydrocarbon provinces. Notable examples occur in Oklahoma, Venezuela, Trinidad, Burma, Iran, Iraq, and elsewhere in the Middle East. The numerous references to pitch and bitumen in the Bible have already been noted. In the Dead Sea, blocks of pitch that are sufficiently large for men to stand on them are occasionally exuded from fault planes (Nissenbaum, 1978; Spiro et al., 1983). The seepages on the shores of the Caspian Sea

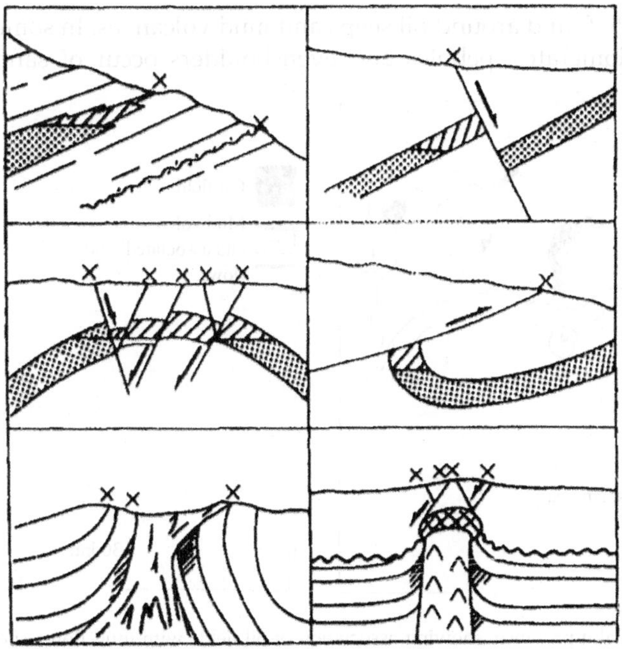

FIGURE 9.2 Sketch cross-sections illustrating the various ways in which oil seeps may occur. (Top left) Outcropping carrier beds and unconformities. (Middle left) Ineffectively sealed anticlinal traps. (Bottom left) Associated with mud volcanoes. (Top right) Via open normal faults. (Middle right) Via open thrust faults. (Bottom right) Over salt diapirs. *From Levorsen, 1967. Used with permission.*

have long attracted attention, especially in the region of Baku (Fig. 9.3). Here, not only oil but also mud and gas pour out on the earth's surface. Spontaneous ignition frequently occurs, so that this region has been called the Land of Eternal Fires. It has been suggested that Baku, and other burning seeps in this area, formed the basis for the Zoroastrian religion, which flourished in Persia for many hundreds of years. In the thirteenth century Marco Polo wrote of one seep in the Baku area: "On the confines toward Geirgene there is a fountain from which oil springs in great abundance, in as much as a hundred shiploads might be taken from it at one time" (translated by Yule, 1871).

Petroleum seeps also occur offshore, notably in the Santa Barbara Channel of California (where the oil companies have been blamed for causing pollution) and off the coasts of Trinidad, Yucatan, and Ecuador. Gas seeps, possibly of both shallow biogenic and thermal origin, are commonly recorded on sparker surveys around the world (refer back to Fig. 3.57), and are a major hazard to drilling and pipeline-laying operations.

Hydrocarbon seeps are, of course, not just a modern phenomenon but are a form of nature's own pollution, which has gone on throughout Phanerozoic time. The Pleistocene seeps of California are well known for their contained vertebrate fauna. Since hydrocarbons are biodegradable, ancient inspissated oil occurs along faults and unconformities. Tarry seeps ranging from Miocene to Pliocene age occur in Iran (Elmore, 1979).

9.2.1.2 *Secondary Deposits*

The extensive flow of oil from an oil seep may give rise to secondary deposits. These deposits are admixtures of sediment and heavy degraded oil. Extensive deposits of secondary petroleum are extremely rare, because of the tendency for oil to be degraded and oxidized in the atmosphere and under water.

Secondary deposits are occasionally found around oil seeps and mud volcanoes. In some rare cases of intraformational conglomerates, pebbles and even boulders occur of sand

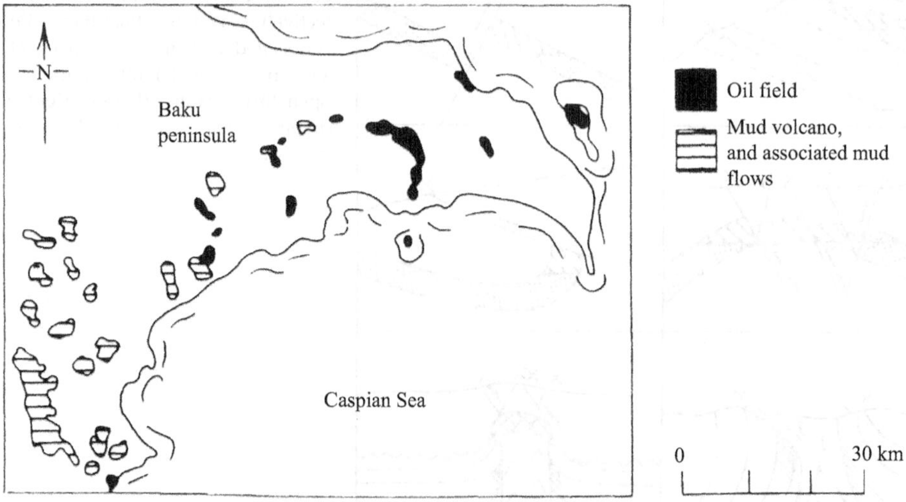

FIGURE 9.3 Map of Aspheron Peninsula, Baku province, showing association of oil fields with mud flows and mud volcanoes. *From Levorsen (1967). Used with permission.*

cemented by dead oil. One such example is known in Lower Cretaceous fluvial channels at Mupe Bay on the southern coast of England (Lees and Cox, 1937; Selley, 1992) (Fig. 9.4). Such secondary deposits are of limited economic significance. They may, however, be extremely important because they help to date the migration of petroleum in an area. The relationship between petroleum migration and trap formation is obviously extremely important.

Seeps have attracted attention as an indicator of subsurface accumulations of petroleum since the earliest days of exploration, and many studies have been carried out to assess their significance (e.g., MacGregor, 1993). It has commonly been remarked that all major petroleum provinces are characterized by surface indications, whereas the converse is not true. For example, more than 173 surface indications of petroleum have been found in the United Kingdom, a country not noted as a major petroleum province (Selley, 1992).

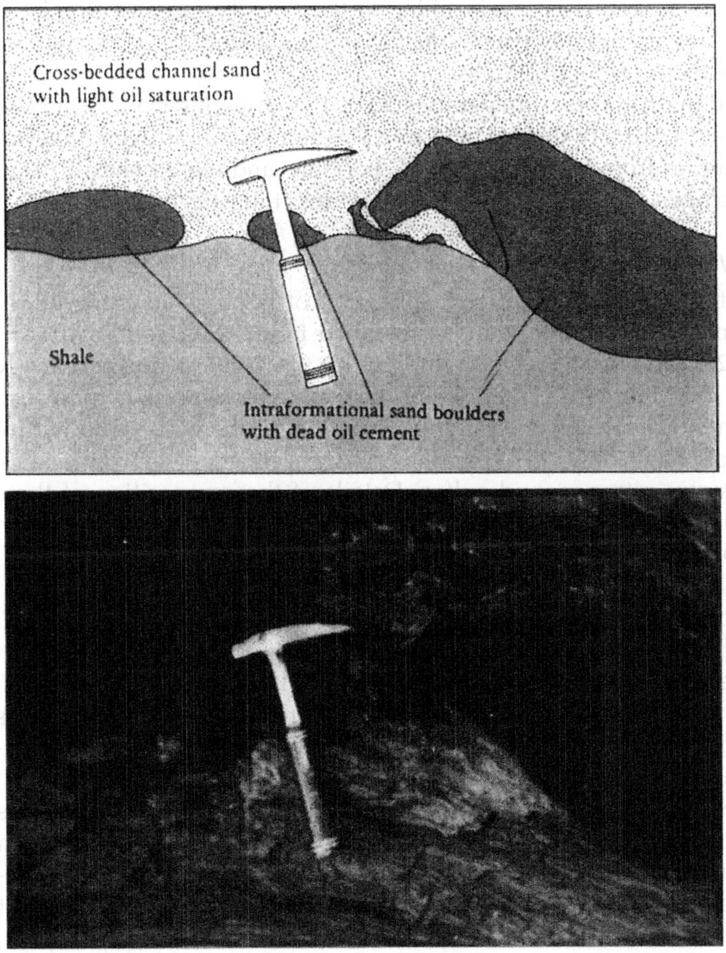

FIGURE 9.4 Sketch and photograph of Wealden (Lower Cretaceous) oil sand Mupe Bay. Dorset. A cross-bedded fluvial channel sand is saturated with sweet light oil. The floor of the channel is lined by boulders of sand cemented with dead oil. The channel was eroding a penecontemporaneous oil seep. This outcrop demonstrates that oil generation and migration had begun by the Early Cretaceous. The light oil saturation testifies to present-day migration to the surface. *See Selley and Stoneley (1987) for further details.*

9.2.2 Composition

When crude oil is dissipated at the surface of the earth, its composition gradually changes as it loses its lighter components and becomes a solid hydrocarbon. Although, as already shown, the terms *pitch, tar,* and *asphalt* have been used synonymously for such material, their composition can be described more specifically. This classification is related to the type of the original crude oil and its degree of inspissation.

Paraffinic oils degenerate to waxy residues, whereas the naphthenic oils move along a more complex path. These two major types of solid hydrocarbon are discussed in the following sections.

9.2.2.1 *Waxy Solid Hydrocarbons*

Waxy solid hydrocarbons are produced by the inspissation of paraffinic crude oils. They are generally plastic, waxy yellow to dark brown substances called *ozokerite* (from the Greek oderiferous wax). This substance occurs as a vein filling in the Boryslav area of Russia, in Utah, and in India. A particular variant known as *hatchettite,* or mineral tallow, occurs in ironstone nodules in the coal measures of South Wales.

Chemically, the waxy solid hydrocarbons are paraffin derivatives, ranging from C_{22} to C_{29}. They contain 84–86% carbon, 14–16% hydrogen, and traces of sulfur and nitrogen. In terms of molecular structure, ozokerites from Utah contain 81% paraffins and naphthenes, 10% aromatics, and 9% oxygen, nitrogen, and sulfur compounds (Hunt et al., 1954).

9.2.2.2 *Asphaltic Solid Hydrocarbons*

Asphaltic solid hydrocarbons are produced by inspissation of naphthenic crude oil. In terms of their elemental chemistry, these compounds are similar to the waxy solid hydrocarbons. They contain 79.5–87.2% carbon, 8.9–13.2% hydrogen, up to 8% sulfur, and commonly traces of nitrogen, oxygen, and inorganic compounds (generally rock-forming minerals). Their molecular composition, however, is quite different from that of the waxy solid hydrocarbons. Asphalts from Utah have only 18% paraffins and naphthenes, 48% aromatics, and 34% compounds of oxygen, nitrogen, and sulfur. Details of the composition of these substances are given by Hobson and Tiratsoo (1981).

As the asphalts are subjected to inspissation, they grade through the asphaltites to asphaltic pyrobitumen. Asphalt (Greek for bitumen) is a solid to semisolid, dark brown to black substance with a specific gravity of less than 1.15 and a melting point of 65–95 °C. It is 10–70% soluble in naphtha and totally soluble in carbon disulfide.

The asphaltites have a specific gravity of less than 1.20 and a melting point of 120–320 °C. They are up to 60% soluble in naphtha and 60–90% soluble in carbon disulfide. Gilsonite is an asphaltite that occurs in fissures in the Tertiary shales of the Uinta Basin of Utah and Colorado. Glance pitch, manjak, and grahamite are other local names for asphaltites.

The asphaltic pyrobitumens are rubbery, brown substances with a specific gravity of less than 1.25. They do not melt, but swell and decompose on heating. They are insoluble in naphtha and only slightly soluble in carbon disulfide. The various names for asphaltic pyrobitumens include albertite, elaterite, impsonite, and wurtzilite (Table 9.1). Some of the best studied deposits of these minerals are associated with the Green River shales (Hunt et al., 1954). Figure 9.5 summarizes the composition and evolution of the solid hydrocarbons.

TABLE 9.1 Characteristics of Some Bitumens and Waxes

		Percent by weight		
Name	**H–C ratio**	**S**	**N**	**Origin**
Ozokerite	1.96–1.89	0.1	0.1	Paraffinic oil
Tabbyite	1.62			↑
Ingranite	1.62			
Wurtzlite	1.6–1.59	4.0	2.2	Increasing degradation of naphthenic oil
Gilsonite	1.47–1.42	0.4	2.8	
Albertite	1.32–1.24	0.7	1.0	
Elaterite				

Chemical data from Hunt et al. (1954). Reprinted with permission.

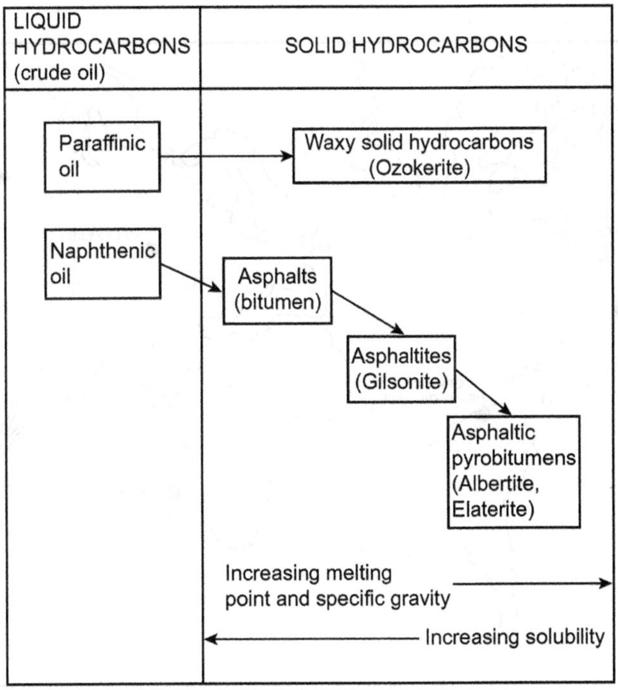

FIGURE 9.5 Diagram illustrating the evolution of the waxy and asphaltic hydrocarbons.

9.3 TAR SANDS

9.3.1 Composition of Tar Sands

Heavy, viscous oil deposits occur at or near the earth's surface in many parts of the world (Fig. 9.6). These deposits have American Petroleum Institute (API) gravities in the range of 5°–15° and typically occur within highly porous sands, generally referred to as tar sands, or bitumen oil sands and heavy oil deposits. As Table 9.2 shows, vast reserves of hydrocarbons are contained in these beds. For many years tar sands have largely been a curiosity. The oil industry lacked the incentive and hence the technology to extract the oil. Global price rises in the 1970s changed this picture. Much research has now been carried out on methods of extraction.

This section examines the composition, occurrence, origin, and extraction of tar sands and associated heavy oils. Further details are given by Phizackerly and Scott (1967) and Hills (1974). The change from normal crude oil, via heavy oil, into tar is a gradational one, both in terms of physical and chemical properties. Physically, the oils become heavier and more viscous. Chemically, the oils tend to contain more inorganic impurities and to be more sulfurous and aromatic. The sulfur content of oil sand varies from about 0.5–5.0%, but occasionally exceeds this range, attaining 6–8% in the La Brea deposits of Trinidad. This value compares with an average of about 1% for normal crudes. Figure 9.7 shows the composition of a number of tar sands compared with light crudes. As shown in the

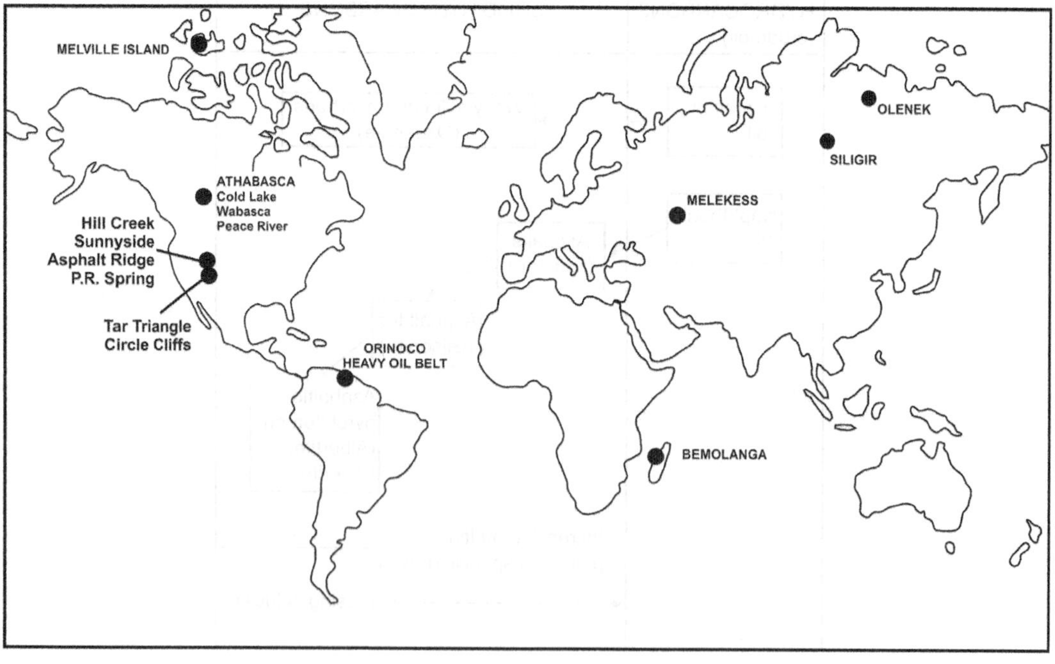

FIGURE 9.6 Location of heavy oil/tar deposits of the world. *Modified from Demaison (1977).*

TABLE 9.2 Data on Some Major Heavy Oil and Natural Bitumen Basins (Billions of Barrels, BBO)

Geologic province	Country	Total original heavy oil in place	Total original natural bitumen in place
Arabian	Bahrain, Iran, Iraq, Jordan, Kuwait, Neutral zone, Oman, Qatar, Saudi Arabia, Syria	842	
Eastern Venezuela	Venezuela, Trinidad, Tobago	593	2090
Maracaibo	Venezuela, Colombia	322	169
Campeche	Mexico	293	0.06
Bohai Gulf	China	141	7.63
Zagros	Iran, Iraq	115	
Campos	Brazil	105	
West Siberia	Russia	88.4	
Tampico	Mexico	65.3	
Western Canada sedimentary	Canada	54.9	2330
Timan-Pechora	Russia	54.9	22
San Joaquin	United States	53.9	<0.01
Putumayo	Colombia, Ecuador	42.4	0.919
Central Sumatra	Indonesia	40.6	
North Slope	United States	37	19
Niger Delta	Cameroon, Equatorial Guinea, Nigeria	36.1	
Los Angeles	United States	33.4	<0.01
North Caspian	Kazakhstan, Russia	31.9	421
Volga–Ural	Russia	26.1	263
Ventura	United States	25.2	0.505
Gulf of Suez	Egypt	24.7	0.5
Northern North Sea	Norway, United Kingdom	22.8	10.9
Gulf Coast	United States	19.7	
Salinas	Mexico	16.6	
Middle Magdalena	Colombia	16.4	
Pearl River	China	15.7	
North Ustyurt	Kazakhstan	15	
Brunei–Sabah	Brunei, Malaysia	14.7	

(Continued)

TABLE 9.2 Data on Some Major Heavy Oil and Natural Bitumen Basins (Billions of Barrels, BBO)—Cont'd

Geologic province	Country	Total original heavy oil in place	Total original natural bitumen in place
Diyarbakir	Syria, Turkey	13.5	
Northwest German	Germany	9.48	
Barinas-Apure	Venezuela, Colombia	9.19	0.38
North Caucasus–Mangyshlak	Russia	8.6	0.06
Cambay	India	8.28	
Santa Maria	United States	8.06	2.03
Central Coastal	United States	8.01	0.095
Big Horn	United States	7.78	
Arkla	United States	7.67	
Moesian	Bulgaria, Moldova, Romania	7.39	
Assam	India	6.16	
Oriente	Peru	5.92	0.25
Molasse	Austria, Germany, Italy, Switzerland	5.79	0.01
Doba	Chad	5.35	
Morondava	Madagascar	4.75	2.21
Florida–Bahama	Cuba, United States	4.75	0.48
Southern North Sea	United Kingdom	4.71	
Durres	Albania	4.7	0.37
Caltanissetta	Italy, Malta	4.65	4.03
Neuquen	Argentina	4.56	
North Sakhalin	Russia	4.46	<0.01
Cabinda	Angola, Congo (Brazzaville), Democratic Republic of Congo (Kinshasa)	4.43	0.363

From Meyer et al. (2007).

figure, compared with normal oils, tar sands tend to be enriched in resins and asphaltenes and impoverished in saturated hydrocarbons. There is little significant difference in their aromatic content.

Two possible explanations have been proposed to account for the differences between normal light crudes and tar sands. The heavy oils may be young and have yet to mature

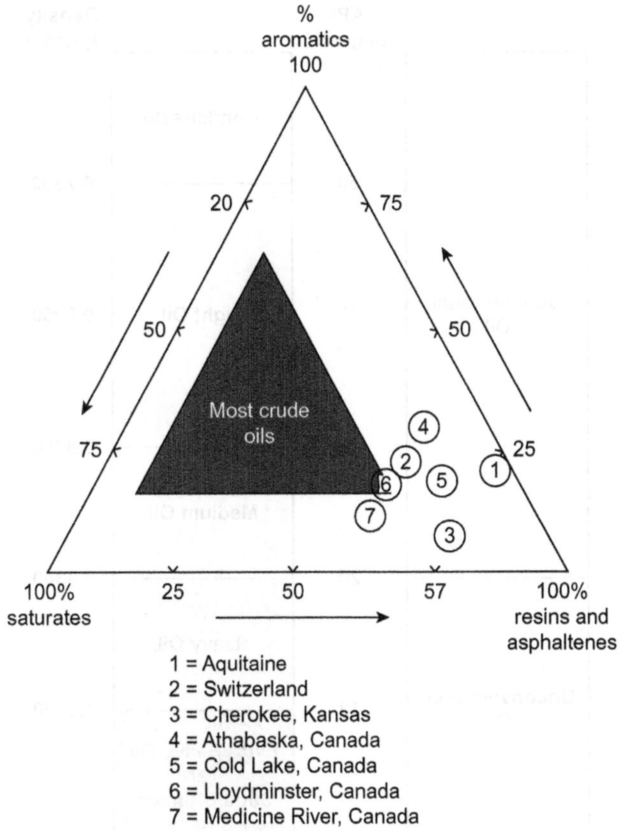

FIGURE 9.7 Triangular diagram showing the composition of various tar sands. *Data from Hills (1974).*

to light oil. Alternatively, the heavy oils may have been produced by the degradation of mature light oil. Many geologists have discussed the chemical evolution of mature oils as the oils emigrate into seeps on shallow traps and are subjected to flushing by meteoric water (e.g., Evans et al., 1971; Byramjee, 1983). Degradation occurs by both organic and inorganic processes. Many of the saturated hydrocarbons are soluble in water, especially the light-end members of the paraffin series. Organic degradation also occurs through bacterial action. Sulfate-reducing bacteria may also be responsible for taking sulfate ions from the meteoric water and combining them with hydrocarbons to form benzothiophenes and dibenzothiophenes. Nonetheless, the origin of heavy oils is controversial.

A classification scheme for conventional oil and unconventional oil based on API gravity is shown in Fig. 9.8. Oils less than 20° gravity in the reservoir are unconventional; oils above 20° gravity but below 50° gravity are conventional. Condensate values range from 50° to 120°.

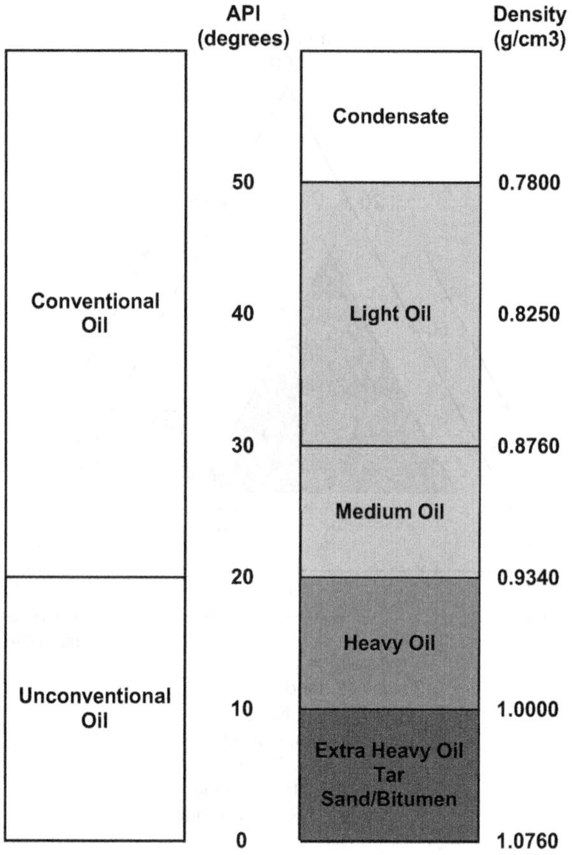

FIGURE 9.8 Classification scheme for oils based on API gravity.

9.3.2 Geological Distribution of Tar Sands

It is perhaps appropriate at this point to consider what light the geological setting of tar sands may throw on their origin. Kappeler (1967) and Demaison (1977) reviewed the mode of occurrence of tar sands around the world. The majority occur where major basins transgress over Precambrian shields. They occur in fluvial or deltaic sands; they rarely occur in marine sands or carbonates. Considered on a smaller scale, tar sands occur in two ways: either in situ in erosionally breached traps or migrated from deep traps into surface seepages. The Bemolanga oil sands of Malagasy are good examples of the former; the La Brea deposits of Trinidad, the Athabasca tar sands of Canada, and the Orinoco Heavy Oil Belt of Venezuela are examples of the latter. These four major tar sand deposits are now described and their possible genesis discussed.

9.3.2.1 The Bemolanga Tar Sands of Malagasy

The Bemolanga oil sands of Malagasy are the third largest such deposits in the world (Fig. 9.9). They occur in the Isalo Group (Triassic) of the Karroo system, covering an area

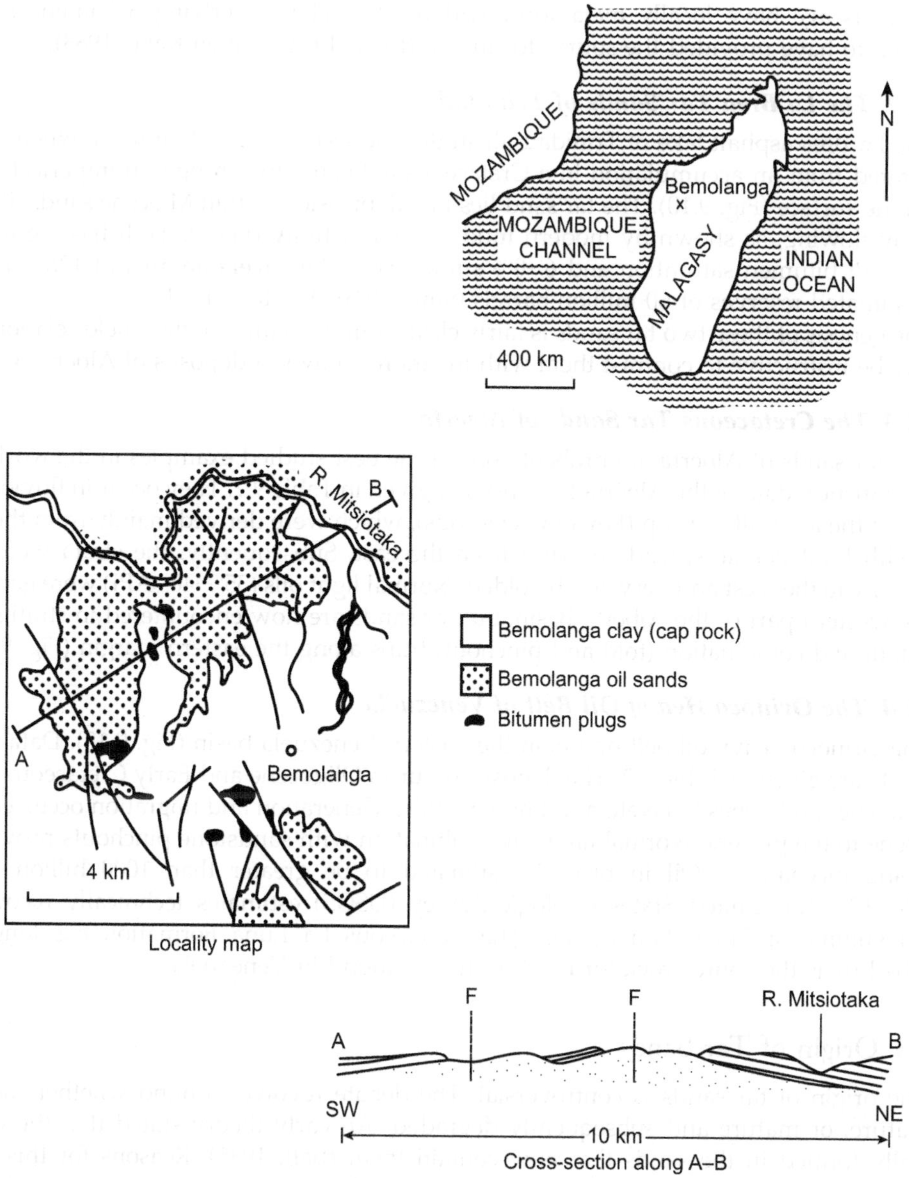

FIGURE 9.9 Maps and sections illustrating the occurrence of the Bemolanga tar sands of Malagasy. *After Kent (1954).*

of some 388 km^2 with an overburden of up to 100 m. The tar sands lie within a complex faulted anticline, where the cap rock of the Bemolanga clay has been removed by recent erosion. The average pay is some 30 m thick, and the in-place reserves are estimated at 1.75 million barrels. The sands are cross-bedded, coarse, and fluvial. Local dykes of

Cretaceous age have locally metamorphosed the tar. The underlying Sakamena shale is believed to have provided the source for the oil (Kent, 1954; Kamen-Kaye, 1983).

9.3.2.2 *The La Brea Tar Sands of Trinidad*

The La Brea asphalt lake of Trinidad illustrates the second type of tar sand, where oil has emigrated from an accumulation in a Cretaceous anticline up through an imperfect seal to reach the surface (Fig. 9.10). The tar sand lies in a depression within Miocene sands; it is still actively flowing, as shown by modern mud volcano activity (Fig. 9.11). It has a gravity of 1°–2° API bitumen saturation and 6–8% sulfur. The lake covers an area of 126 acres and has estimated reserves of 60 million barrels (Suter, 1951; Walters, 1974).

The genesis of these two tar sands is fairly clear from the nature of their field relationships. It may be instructive to contrast them with the more equivocal deposits of Alberta, Canada.

9.3.2.3 *The Cretaceous Tar Sands of Alberta*

The tar sands of Alberta are probably one of the best studied examples in the world. The basic statistical data of the Alberta tar sands are given in Table 9.2. They occur in fluviodeltaic sands of the Manville Group (Lower Cretaceous), which were derived mainly from the west, but with local deltaic spreads of sand from the east. Subsequently, the strata were tilted regionally to the west and very gently folded. Normal light oils (35°–40° API) were generated in the western part of the Alberta Basin, but tar sands are now encountered in stratigraphic pinchout and combination (fold and pinchout) traps along the eastern margin (Fig. 9.12).

9.3.2.4 *The Orinoco Heavy Oil Belt of Venezuela*

The Orinoco heavy oil belt occurs in the Eastern Venezuela basin (Fig. 9.13). Data for the deposits are given in Table 9.2. The deposits occur in Oligocene and Early Cretaceous fluvial and deltaic sandstones (Alayeto and Louder, 1974). Generation and migration occurred from Miocene to the present. Normal faults in combination with sandstone pinchouts provide the trapping mechanism. Oil in place is estimated to be greater than 1000 billion barrels (Table 9.2). The United States Geologic Survey (USGS) estimates technically recoverable mean volume of 513 billion barrels. The Cretaceous La Luna Formation has long been ascribed to as the source rock for much of the oil found in Venezuela.

9.3.3 Origin of Tar Sands

The origin of tar sands is controversial. The debate revolves around whether the oil is immature, or mature and subsequently degraded. An early theory stated that the oil was actually formed in the sands that now contain them (Ball, 1935). Reasons for this theory included the flat-lying nature of the sediments and the apparent absence of normal crudes in the area. This idea has since been shown to be untrue. The Athabaska hydrocarbons are now generally believed to have migrated into their present reservoir in a conventional manner. Many geologists and chemists have suggested that the heavy oils are mature oils that have undergone degradation by meteoric flushing. Deroo et al. (1974) show how these oils can be traced gradationally up the basin margin from deep light oil pools. This change is accompanied by a gradual diminution in water salinity from more than 80 g/l to less than 20 g/l over a distance of some 300 km. Nevertheless, this theory has been disputed

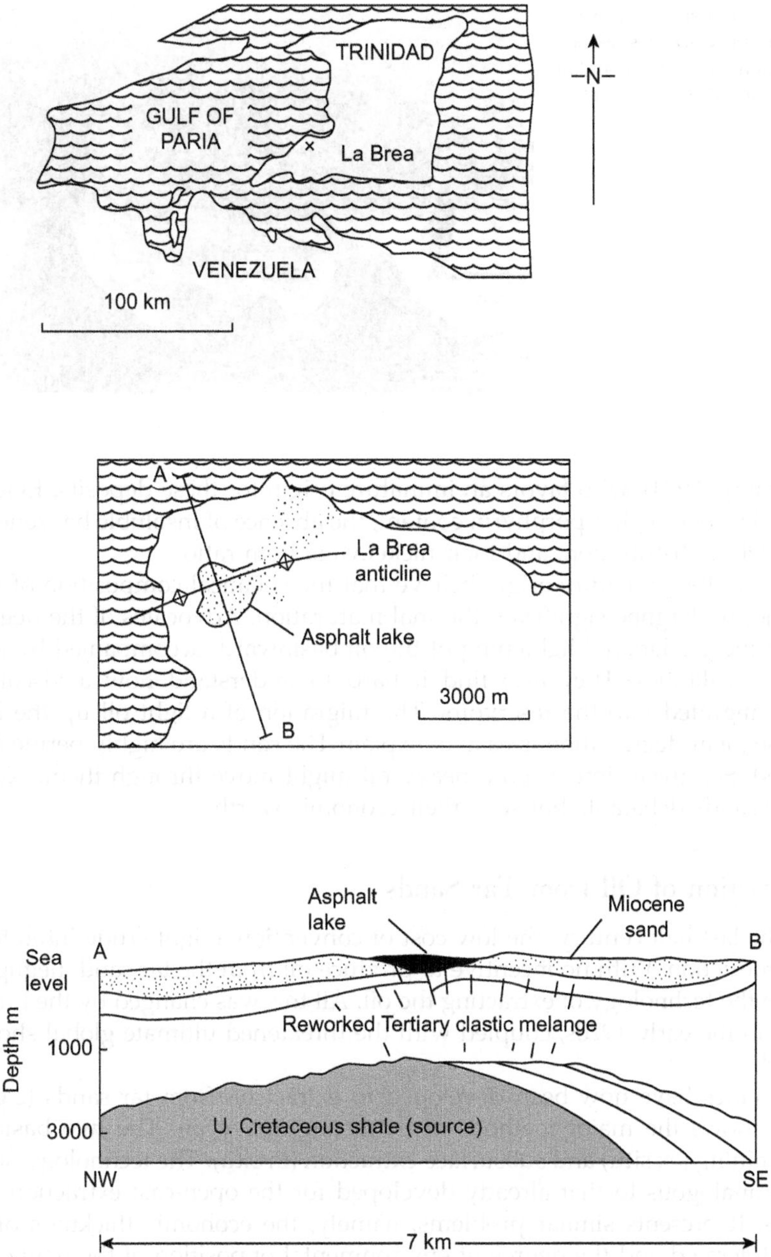

FIGURE 9.10 Maps and sections illustrating the La Brea tar sands of Trinidad. *After Kugler (1961).*

FIGURE 9.11 General view of the petroliferous mud volcanoes associated with Monkeytown. Trinidad Pitch Lake. *From Wall and Sawkins (1860).*

by Montgomery (1974), who favors an immature origin for these deposits. Evidence for this theory includes their higher porphyrin content, the absence of insoluble benzene matter, their molecular weight distribution, and their nickel:vanadium ratio.

Advocates of the immature origin believe that the chemical composition of the oil shows that it has not undergone significant thermal maturation. Advocates of the degradation theory point to the gradational lightening of the oil basinward, accompanied by increased formation water salinities. They also find it hard to understand how a viscous, heavy oil could have migrated into the reservoirs. The migration of a light oil up the basin margin and its subsequent degradation is easy to explain The sands are highly permeable, however (1–5 darcies). So, given time, even a heavy oil might move through them. Not only is the origin of tar sands debatable but so is their economic worth.

9.3.4 Extraction of Oil from Tar Sands

During the last half century, the low cost of conventional light crude inhibited interest in heavy oil production. With no incentive to produce oil from the tar sand, negligible research was put into the technology of extracting the oil. All this was changed by the huge price rises of crude oil in the early 1970s, coupled with the threatened ultimate global shortage of conventional oil.

Many schemes have now been developed to extract oil from tar sands (e.g., Ali, 1974). Figure 9.14 shows the major methods of extracting heavy oil. The two basic approaches are surface mining (ex situ) and subsurface extraction (in situ). The technology of strip mining tar sands is analogous to that already developed for the open-cast extraction of other economic rocks. It presents similar problems, namely, the economic thickness of overburden that can be tolerated and the degree of environmental opposition of the native population.

Once the tar sands have been quarried, the oil contained in them must be extracted. At present the conventional method consists of disaggregation of the sand and separation of the oil by hot water and/or steam. Variations on this technique have been employed on the Alberta tar sands. About 20% of the oil sands reserves in Alberta are recovered by surface mining where the overburden rock is less than 75 m. For the remaining 80% of the reserves buried

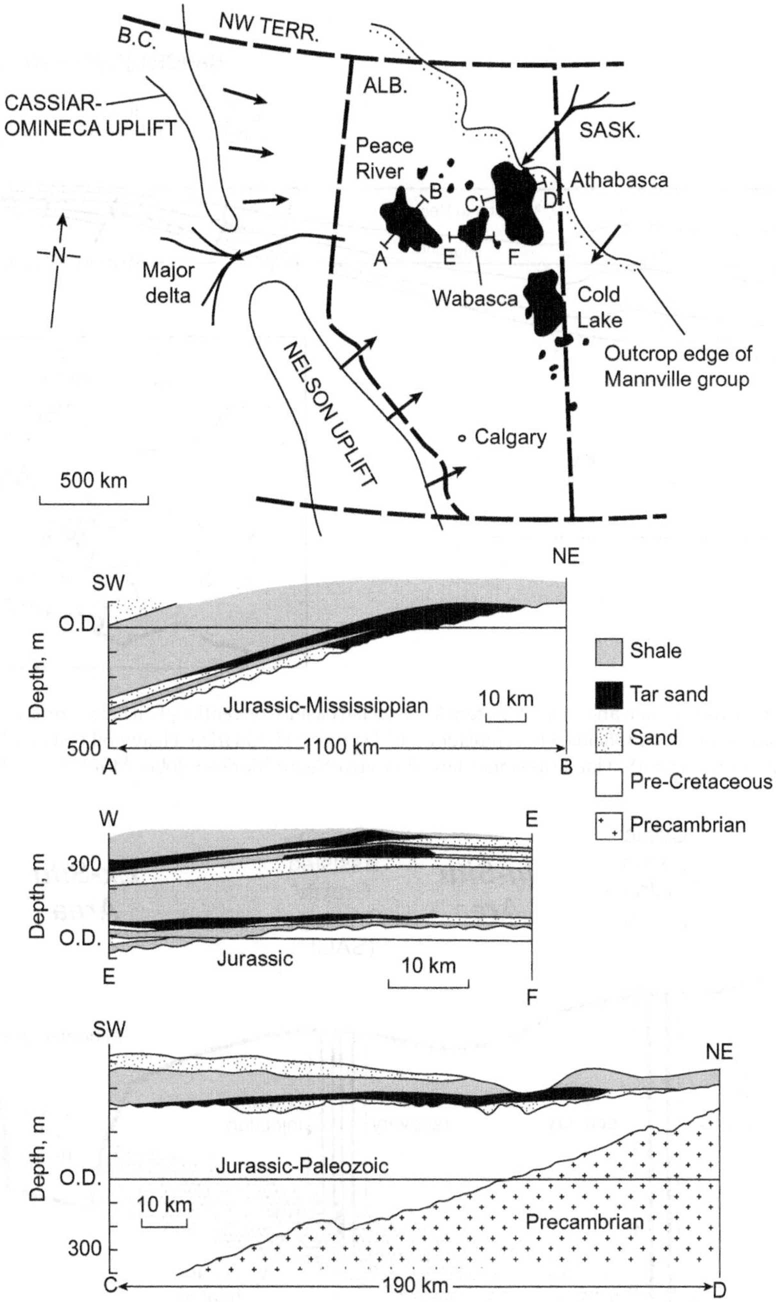

FIGURE 9.12 Map and cross-sections of the Lower Cretaceous tar sands of Alberta. *After Jardine (1974), with permission from the Canadian Society of Petroleum Geologists.*

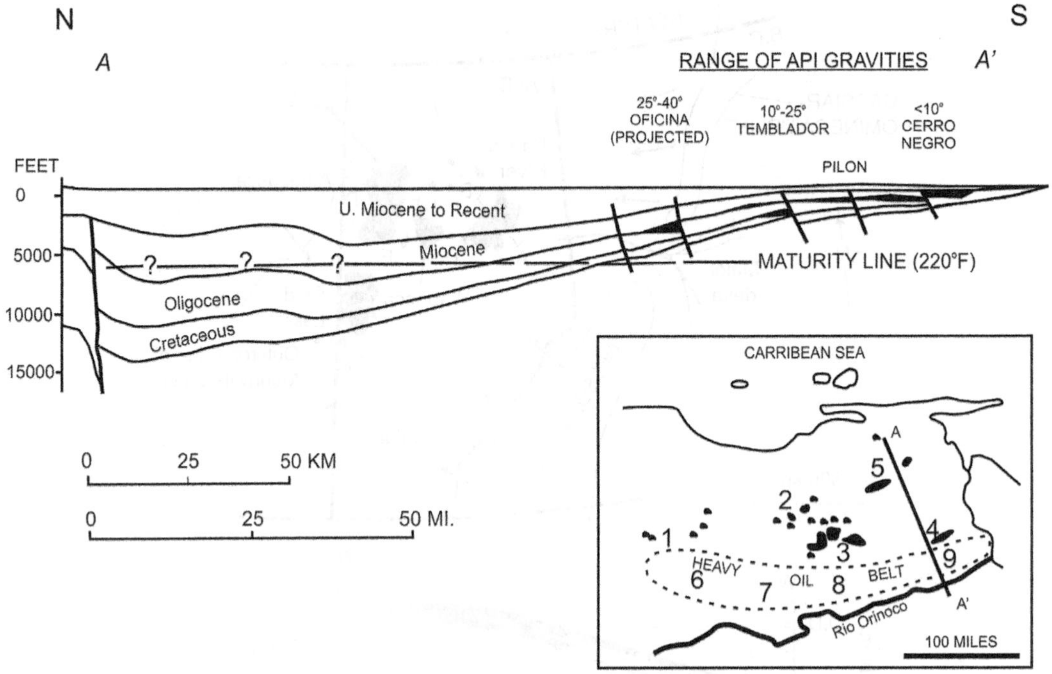

FIGURE 9.13 Cross-section and map of Eastern Venezuelan basin. Conventional oil fields are: (1) Greater Mercedes; (2) Greater Anaco; (3) Greater Oficina; (4) Temblador and Tucupita; (5) Guarico. Heavy oil belt consists of (6) Gorrin-Machete; (7) Altamira-Suata; (8) Hamaca-Santa Clara; (9) Cerro Negro-Morichal-Jobo. *Modified from Demaison (1977).*

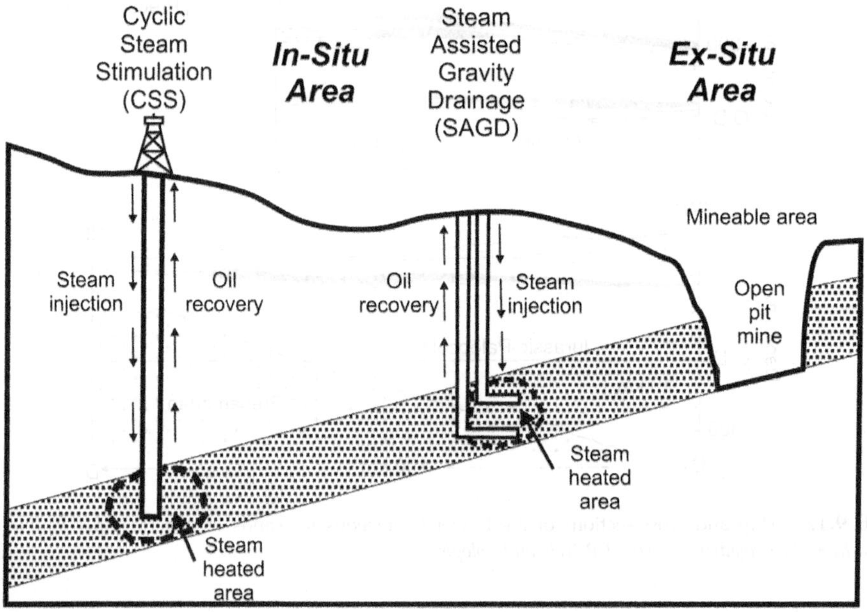

FIGURE 9.14 Methods of extracting oil from Athabasca tar sands. Methods include open pit mining, steam assisted gravity drainage, and cyclic steam stimulation.

at depths greater than 75 m, in situ technologies, largely thermal techniques, are used to extract the bitumen.

Where the overburden is too thick for conventional open-cast mining, in situ extraction methods must be developed. Basically, two types of extraction methods are used: One involves the injection of a solvent to dissolve the oil, whereas the other seeks to reduce the oil's viscosity by heating. Both these methods are inhibited by the tendency for the high permeability of clean sands to decrease in proportion to the level of oil saturation. Great care is thus needed to assess correctly both the petrophysical variations of the sands and their oil saturations (Towson, 1977; Kendall, 1977).

For thermal extraction, heat may either be taken down into the reservoir in hot water or steam or, alternatively, be generated in situ by combustion of the tar. In situ combustion necessitates the introduction of oxygen. All these methods can be accomplished either by the "huff-and-puff" procedure or by monodirectional flooding. In the first method wells are alternately used for injection and production. The second method is akin to conventional oil or gas recovery techniques in that a continuous flood is established from peripheral injection wells to centrally located producers.

Some 1.25 million barrels of oil per day was produced from the Athabasca tar sands in 2006. Almost 44% of Canadian oil production was from oil sands in 2007. It has been estimated that more than 50% of Canada's total crude production could come from these tar sands by 2010 (Newell, 1993).

In situ thermal techniques currently being used include steam-assisted gravity drainage (SAGD), and cyclic steam stimulation (CSS). The quality of the resource and depth determines the recovery method used. It is important to match the recovery technique to the reservoir characteristics.

The most common in situ process is SAGD. In this technique, a pair of horizontal wells is drilled into the bitumen deposit. Steam is injected into the upper well and causes the bitumen to lose its viscosity causing it to flow under gravity to the lower well where it is then pumped to the surface. SAGD was developed in the 1980s by the Alberta Oil Sands Technology and Research Authority (an Alberta crown corporation). SAGD combined with directional drilling technology made this operation economic by the mid-1990s. The operation recovers about 60% of the oil in place. SAGD requires a "cleaner" reservoir with high bitumen saturation.

CSS has been used in the Canadian tar sands since the mid-1980s. In this technique an individual well is cycled through injection, soak, and oil production. High-pressure steam (steam includes both vapor and hot water) is first injected into a well at temperatures of 300–340 °C for a period of weeks to several months. The well is then allowed to soak for days to weeks to allow the heat to soak into the formation and reduce the viscosity of the bitumen. The hot oil is then pumped out of the formation for periods of weeks to months. When production falls off the cycle is repeated. This cycle is repeated many times (upward of 15 cycles). Recovery factors are 20–25%. CSS is not viable for shallow reservoirs that do not have a thick capping shale seal to manage the high steam pressures. CSS works effectively to a broader range of reservoir quality and depths than SAGD.

In situ cold heavy oil production with sand is used in areas where the oil is fluid enough to flow without heating. The oil is pumped out of sands using progressive cavity pumps. It is commonly used in Venezuela and Wabasca, Alberta Oil Sands, southern part of Cold Lake

Oil Sands, and in the Peace River Oil Sands. The technique recovers only about 5–10% of the oil in place. The advantage of this method is that it is cheap compared to other methods. The Disadvantage is disposing of produced sands.

In situ solvent extraction is a newer technique which is showing promise. This technique uses a pair of wells like the SAGD method. Instead of steam, vaporized hydrocarbon solvents are injected into the upper well to dilute bitumen causing the diluted bitumen to flow to the lower well where it is produced. It is more energy efficient than steam injection and partially upgrades the bitumen to oil in the formation. This process is referred to as vapor extraction process (VAPEX). The hydrocarbon solvents can be reused.

CSS, SAGD, and VAPEX are not mutually exclusive. Areas can first be exploited by CSS before going to SAGD production. SAGD can be combined with VAPEX to improve recoveries and improve economics.

At the present time the extraction of oil from tar sands is an economic proposition. Improved technology and global energy costs have made this once uneconomic resource very important in the global energy mix. Other experimental technologies are also being applied to the tar sands. The size of this resource worldwide is enormous and thus much research is being applied to improve recoveries.

9.4 OIL SHALES

Oil shale, also known as kerogen shale, is a fine-grained sedimentary rock that yields oil on heating. It differs from tar sand in more ways than grain size. In tar sands the oil is free and occurs within the pores. In oil shales, however, oil seldom occurs free, but is contained within the complex structure of kerogen, from which it may be distilled. In a whimsical memoir Craig (1915) wrote "To make the matter clear, the relation of malt and hops to beer is somewhat similar to that of oil shales, coal and lignite to petroleum. The malt and hops do not contain beer, but it can be made from them by causing certain chemical changes. Oil can be made from oil shales, coal and lignite, but only by destroying them."

The term "shale oil" is often used for crude oil produced from shales without artificial heating. The term "tight oil" is recommended by the International Energy Agency and World Energy Council for usage over "shale oil" because of the confusion of the terms oil shale and shale oil (Wikipedia, 2014).

Oil shales are widely distributed around the globe and may contain locked within them more energy than in all the presently discovered conventional oil reserves. The world's oil shales may contain some 30 trillion barrels of oil, only about 2% of which is accessible using present-day technology (Yen and Chilingarian, 1976).

Oil shales have been known to man for many centuries. The Mongol hordes used burning oil shale arrowheads to terrify mandarins on the Great Wall of China as long ago as the thirteenth century (Nasiam, 1937). In 1694 the British Crown issued a patent to Martin Eele for the extraction of oil from shale. Subsequently, a considerable extraction industry has developed, although it is now temporarily quiescent.

The following sections review the composition, distribution, origin, and extraction of oil shales. More detailed accounts are found in Dyni (2005), Yen and Chilingarian (1976), Burger (1973), and Russell (1980).

9.4.1 Chemical Composition of Oil Shales

Most oil shales are essentially claystones and siltstones, but bituminous marls and lime mudstones are also known. The basic constituents of oil shales may be grouped as follows:

1. *Inorganic components* (<90%)

Detrital	Quartz	
Detrital	Feldspar	
Detrital	Mica	
Detrital	Accessoryminerals	Authigenic
Detrital	Carbonates	Authigenic
Detrital	Clays	Authigenic
	Pyrite	

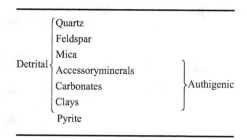

2. *Organic components* (10–27%)

Bitumens (hydrocarbons soluble in CS_2)	≃2%
Kerogens (insoluble in CS_2)	≃8%

The exact amount of organic matter needed before a shale can be classified as an oil shale is arbitrary. A cutoff value of 10% has been cited. With increasing organic content, oil shales grade into the cannel coals. A figure of 27% has been taken as the critical value between oil shales and cannel coals (Shanks et al., 1976).

The inorganic component of oil shales differs little from that of conventional shales. There is a framework of detrital grains of quartz, feldspar, mica, and other accessories. Large amounts of clay are present, both detrital floccules and authigenic crystals. Various carbonate minerals are present, both calcitic skeletal fragments and authigenic calcite, dolomite, ankerite, and siderite. Authigenic pyrite is especially characteristic of oil shales because of its anaerobic origin. The presence of this sulfur causes one of the major problems of oil shale refining.

The important constituent of oil shale is kerogen. In a study of the Scottish Carboniferous oil shales, Stewart (1912) defined kerogen as "the name given to carbonaceous matter in shale which gives rise to crude oil on distillation." The term *kerogen* was suggested by Professor Crum Brown. Subsequently, more rigid definitions have been proposed (see, for example, Dott and Reynolds, 1969). Stewart's original definition of kerogen is now too broad for most authorities, since it would include the bitumen fractions. Currently, kerogen is defined and distinguished from bitumen by its insolubility in carbon disulfide. Further discussion of the chemical origin and significance of kerogen is given in Chapter 5.

Oil shales can be classified as terrestrial, lacustrine, and marine in origin (Fig. 9.15). The terrestrial type consists of lipid-rich organic matter such as resins, spores, waxy cuticles, and corky tissues of roots and stems of terrestrial vascular plants found in coal-forming swamps and bogs. The terrestrial type is also called cannel coal. Lacustrine oil shale is lipid-rich organic matter derived from algae that lived in freshwater or brackish or saline

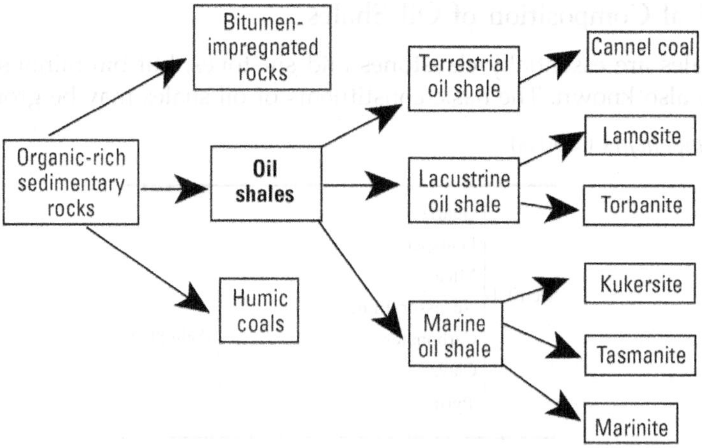

FIGURE 9.15 Classification of oil shales. *From Dyni (2005).*

lakes. Lacustrine oil shales are subdivided into lamonsite and torbanite. Lamonsite is a black oil shale in which the chief component is lamalginite derived from planktonic algae. Torbanite is also known as boghead coal and the organic matter is derived from lipid-rich microscopic plant remains. Marine oil shales are lipid-rich organic matter derived from marine algae, and marine dinoflagellates. Marine oil shales are subdivided into kukersite, tasmanite, and marinite. Marinite is a dark gray to black oil shale consisting of lamalginite and bituminite derived mainly from phytoplankton. They are the most abundant type of oil shale deposit. Kukersite and tasminite are related to specific types of algae from which the organic matter was derived.

The organic precursor of the kerogens in oil shales are hard to determine. However, two particular types of oil shale whose biological origin is well known are torbanite and tasmanite. Torbanite comes from the village of Torban in the Midland Valley of Scotland, where this particular variety of oil shale occurs in Carboniferous coal measures. Torbanites, also termed *bog head* or *cannel coal,* are made of algal matter. They contain the remains of the freshwater blue-green algae *Botryococcus*. Torbanite appears to be synonymous with Kukuserite, an algal oil shale of Ordovician age that occurs in Estonia.

Modern algal deposits analogous to Torbanite have been described from Lake Balkhash, Siberia, and from the Coorong lagoon of South Australia. These occurrences are known as balkashite and coorongite, respectively, naturally. Coorongite is a gelatinous, yellow-green mess, which dries to a dark, rubbery material (Broughton, 1920). Its algal composition can be seen under the microscope.

Tasmanite is a rare variety of oil shale that occurs principally in Tasmania, Australia (the type locality), and in the Brooks Range of North Alaska. It is composed of leiosphores of the algae *Tasmanites.* Although analogous in composition to torbanite, it occurs in marine deposits. Torbanites and tasmanites are relatively rare varieties of oil shale. The source of the organic matter of most oil shales is equivocal, but generally believed to be originally admixtures of algal and terrestrial humic materials.

9.4.2 Distribution of Oil Shales

The following account of the environment, global distribution, and temporal distribution of oil shales is based on surveys by Schlatter (1969), Duncan and Swanson (1965), Cane (1976), and Duncan (1976).

Although carbonaceous shales and hydrocarbon source rocks are known (Section 5.2.2), there appear to be few genuine Precambrian oil shales. In the early Paleozoic, however, siliceous oil shales were deposited on marine shelves throughout the Northern Hemisphere. The Alum shales and Kukersite deposits of the Baltic Shield are examples of these. In the late Paleozoic, oil shales were also deposited on marine shelves in central European Russia and in the central and eastern United States and Canada. The Chattanooga shale is an example of the latter type, and, although it covers some 500,000 km^2, it has a remarkably uniform stratigraphy.

Oil-shale-forming environments appear to have changed significantly at the Devonian–Carboniferous boundary. Paralic deltaic lacustrine oil shales are characteristic of Carboniferous deposits around the world. The Midland Valley of Scotland oil shales, which include torbanite, are of this age and type, as are other examples in France, Spain, South Africa, Australia, and Russia. The Carboniferous Albert oil shale of Nova Scotia, however, is totally lacustrine.

Both paralic and marine shelf oil shales formed during the Permian period. The first type includes the coal measures of the Sidney basin in New South Wales, Australia. Here an area of some 3000 square miles contains rich torbanite deposits. The Permian coal measures of this basin are some 200 m thick. The oil shales occur interbedded with the coal-bearing strata, but are best developed in the west toward the margin of the basin (Cane, 1976).

The Irati oil shale is of similar age. This oil shale covers several hundred thousand square kilometers in Brazil, Argentina, and Uruguay, and the equivalent White Band of the Karroo Series in South Africa (Anderson and McLachlan, 1979) (Fig. 9.16).

In late Mesozoic time many oil shales formed in marine, deltaic, and lacustrine environments. In northwest Europe, in particular, a major phase of oil shale and organic-rich shale deposition began in northern Germany in the Liassic and moved diachronically northward up the North Sea into the East Greenland Mesozoic basin, where organic-rich mud deposition continued into the Cenomanian stage. These deposits include the Kimmeridge oil shale of the Dorset Coast. Here the 2-m-thick "Kimmeridge Coal" was mined as an oil shale for many years. Although it was used locally as a domestic fuel, its high sulfur content was incompatible with the tourist trade. The organic matter of the Kimmeridge clay is a type II mixed kerogen, although the oil shale itself is a type I algal kerogen. Kimmeridgian organic shales are the major source rock of North Sea oil, their richness increasing as they are traced from shelf areas into the grabens of the axial rift system (Barnard and Cooper, 1981; Goff, 1983). Analogous marine platform oil shales occur in Alaska; in Saskatchewan and Manitoba, Canada; and in Nigeria.

The Cenozoic era saw another return to lacustrine oil shale deposition, although marine oil shales do occur. The best known example of lacustrine oil shale deposition is the Green River Formation (Eocene) of Colorado, Utah, and Wyoming (Figs 9.17 and 9.18). In this area oil-shale-bearing rocks cover some 16,500 square miles in a formation up to 700 m thick (Duncan and Swanson, 1965; Knutson, 1981). Total reserves, if they can be recovered, have been estimated at 1.8 trillion barrels (Yen and Chilingarian, 1976).

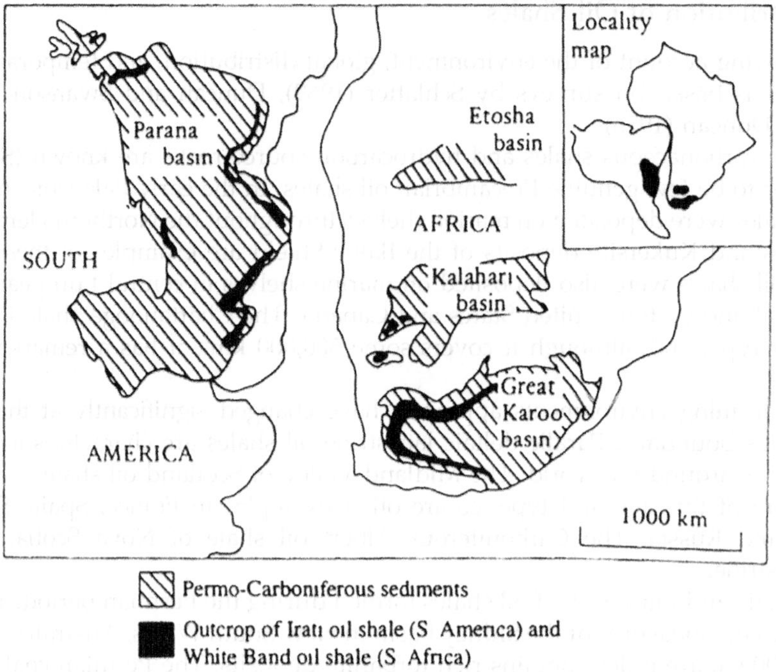

FIGURE 9.16 Outcrop of the Permian oil shales of South America and South Africa. These oil shales were deposited in several vast lakes. *After Anderson and McLachlan (1979).*

Analyses of Green River oil shales yield the following composition (Yen and Chilingarian, 1976).

Kerogen	11.04 %	
Bitumen	2.76 %	
Total organic matter:		14.16%
Dolomite and calcite	43.1 %	
Feldspar	20.7 %	
Micas and clays	12.9 %	
Quartz	8.6 %	
Pyrites	0.86 %	
Total inorganic matter:		85.8 %
TOTAL:		99.96 %

This review shows that the occurrence and distribution of oil shales are widespread in place and time. Three major depositional environments are favorable for oil shale generation: lakes, fluvially dominated deltas, and certain marine shelves.

Torbanites and oil shales of mixed humic and nonmarine algal origin are characteristic of lacustrine and deltaic shales. Shallow marine oil shales, however, are commonly of mixed humic and marine algal composition, including the rarer algally dominated tasmanite variety. Oil shales of both lacustrine and marine types are occasionally associated with autochthonous chemical deposits. The carbonates and evaporites of the Green River

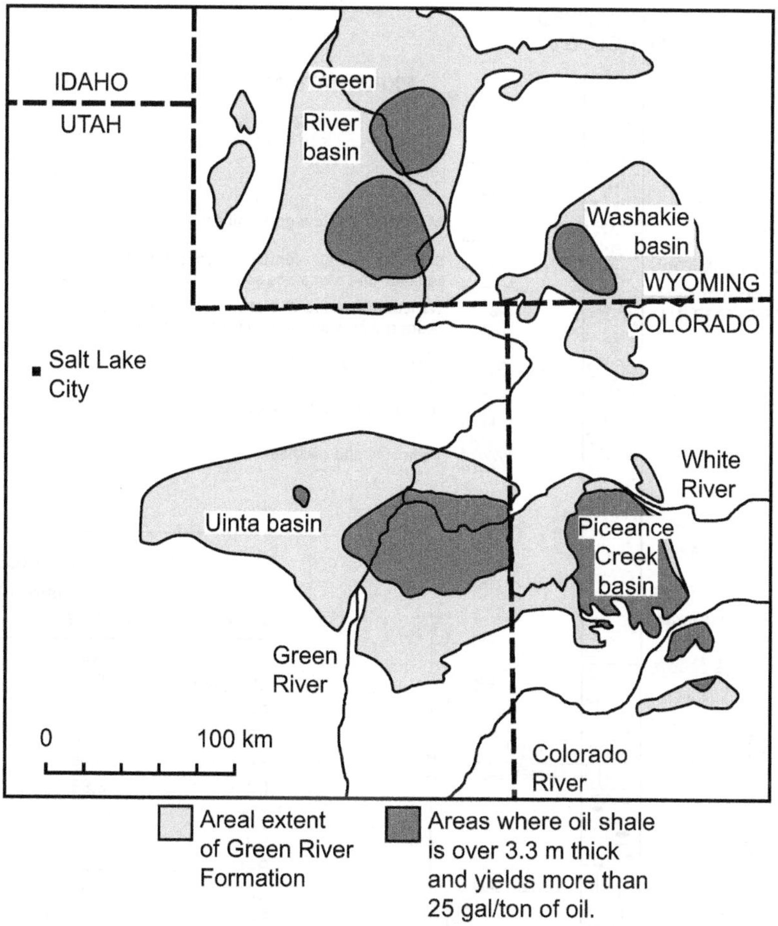

FIGURE 9.17 Outcrops of the Eocene oil-shale-bearing Green River Formation of the central United States. This formation was deposited in a series of lakes occasionally interbedded with evaporites. *Modified from Knutson (1981). Reprinted by permission of the American Association of Petroleum Geologists.*

Formation illustrate the lacustrine types, whereas the phosphates and cherts of the Phosphoria Formation illustrate the marine types.

9.4.3 Extraction of Oil from Oil Shale

The preceding section described the widespread occurrence of oil shales around the globe. The world's reserves of shale oil are estimated to be on the order of 30 trillion barrels, and the United States alone contains more than 700 billion barrels (Yen and Chilingarian, 1976).

The commercial extraction of oil from shale was begun by James Young in Scotland in 1862, 3 years after the drilling of Drakes' well in Pennsylvania. During the next century, oil shale extraction was carried out in England, France, Spain, Sweden, Australia, and South Africa. In all these countries the industry has since largely collapsed.

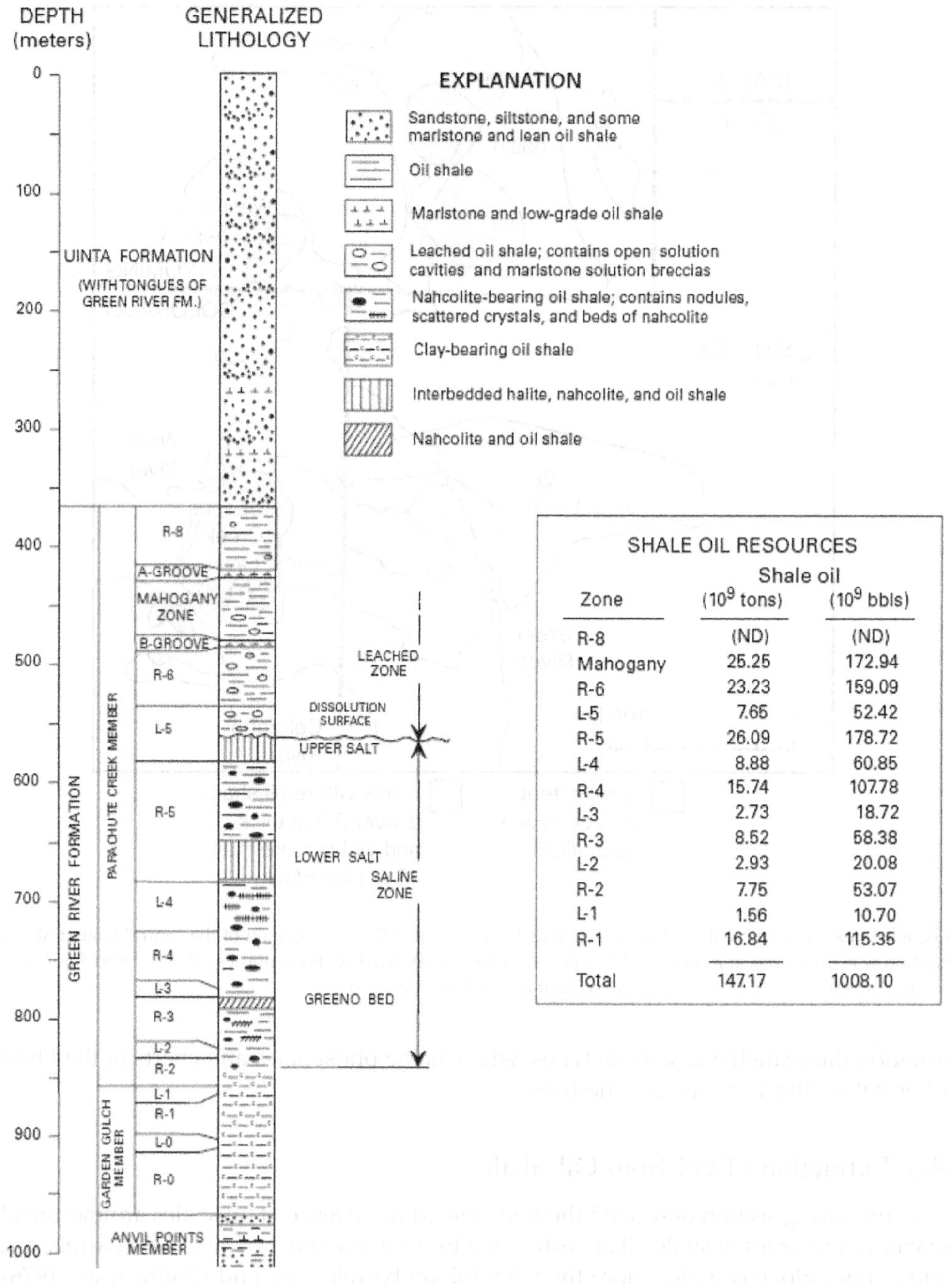

Zone	Shale oil (10^9 tons)	Shale oil (10^9 bbls)
R-8	(ND)	(ND)
Mahogany	25.25	172.94
R-6	23.23	159.09
L-5	7.65	52.42
R-5	26.09	178.72
L-4	8.88	60.85
R-4	15.74	107.78
L-3	2.73	18.72
R-3	8.52	58.38
L-2	2.93	20.08
R-2	7.75	53.07
L-1	1.56	10.70
R-1	16.84	115.35
Total	147.17	1008.10

FIGURE 9.18 Generalized stratigraphic column of the Green River Formation, Piceance Basin, northwestern Colorado. *From Dyni (2005).*

In the state of Parana, Brazil, a 2500 ton/day plant based on the Irati oil shale began production in the mid-1970s. Petrobras has produced 2200 tons/day from a retort at Sao Mateus. A 4400 bbl/day plant has been built, and plans have been laid for a 50,000 bbl/day plant. In the Green River shale areas, about 10 oil extraction plants were once in various stages of planning and development. These plants included both strip mining, room and pillar mining, and in situ retorting. For many years, two plants have operated in the Grand Valley of Colorado, USA.

Russia and China have both had large-scale oil shale extraction industries. In Russia extraction plants operate in the Volga Basin. In China there are plants at Fushun in Manchuria and at Mowming in Kwantung Province (see Prien, 1976; for an historical review).

The reasons for the rise and fall of the oil shale extraction industry are twofold: economics and technology. As technology for extracting oil from boreholes commenced, the rapid expansion of the conventional oil industry largely put oil shale extraction out of business. Crude oil could be found, produced, transported, refined, and marketed at a lower cost than shale oil. Today, however, the oil shale industry is in a similar situation as the tar sand industry. The collapse of oil prices in the early 1980s stopped many tar sand and oil shale extraction projects. But when the cost of conventional oil production increases, the time will return when the extraction cost of shale oil will become comparable to that of conventional crude oil production.

Again, like the tar sands, there are two basic methods of winning oil from shale: (1) by retorting shale quarried at the surface or (2) by underground in situ extraction (Dinneen, 1976). Most of the surface extraction methods are based on the open-cast quarrying of shale, although the "Kimmeridge coal" was mined along conventional adits. The crushed shale is then placed in a retort and heated. Several sources of heat can be used, but the most efficient is by the combustion of gas previously generated from the shale. As the shale is heated (generally in the range of 425–475 °C), oil and gas are driven off by the pyrolysis of the kerogen. Spent shale collects at the bottom of the retort. The oil is refined, the gas reinjected, and the ash cooled and disposed of, it is hoped at least partly, back in the hole whence it came. This method has considerable environmental objections. The mountains of spent oil shale are not very pretty. Furthermore, spent oil shale is rich in radioactive and heavy metals that are liable to leach out and contaminate groundwater supplies, giving rise to monster problems.

In situ methods of retorting are rather different. The two basic problems are similar to those encountered in tar sand extraction. First, an effective method of heating the shale is necessary; second, the rock must be rendered permeable. In Sweden, during the 1940s, the Ljungstrom process was used (Salomonsson, 1951). Closely spaced shallow boreholes were drilled into the oil shale. Heating elements were inserted, and the rock temperature was raised to 400 °C. The oil thus generated was pumped from the wells. The amount of energy used to extract the oil was probably greater than that contained within the oil itself, but because it was derived from renewable hydroelectric power, that did not matter too much.

Sinclair Oil and Gas tried another in situ method in 1953 on the Green River shale of the Piceance Basin (Grant, 1964). A series of wells was drilled into the oil shale and artificially fractured to generate permeability. Compressed air was pumped into certain wells, which were then ignited. Combustion gas from these wells moved through the shale and retorted the oil. This oil was recovered from producing wells.

A slightly different technique was developed by Equity Oil Co. in the same area (Dougan et al., 1970). In the site selected the shales were already permeable due to the leaching of salts. Natural gas was injected and ignited in the wells. Shell Oil Co. has experimented with the

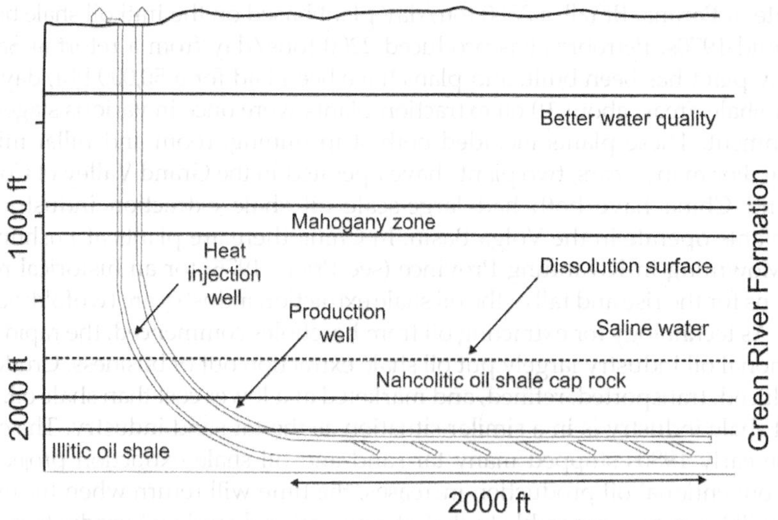

FIGURE 9.19 The AMSO experimental process for extracting oil from oil shale, Piceance Basin, northwest Colorado. *Modified from Burnham et al. (2012).*

injection of hot hydrocarbon solvents, such as carbon disulfide. Further details of shale oil extraction technology are found in Dinneen (1976).

The Piceance Basin of northwest Colorado has enormous potential for economic oil shale development. Several ongoing experimental techniques are being tested for in situ extraction. In situ heating techniques include electric subsurface heaters and freeze walls (Shell Oil Co.), an electrofrac process using electrically conductive fractures (ExxonMobil), radiofrequency waves transmitted to efficiently heat oil shale, in situ heating with fuel cells, and conduction, convection, reflux (CCR) method to heat oil shales (AMSO). The CCR technique combines horizontal wells which are heated by downhole heater or burners aided by refluxing of generated oil and other horizontal or vertical wells for production of the oil. Fig. 9.19 illustrates the CCR (AMSO) method.

Whether the energy budget is in credit in all these methods is debatable. In other words, the amount of energy derived from the shale oil is not always greater than that expended in its extraction. Even if the derived energy is less than the expended energy, these processes may one day become financially viable. This event will happen when the rising cost of conventional crude oil intersects the cost of shale oil. The latter may be expected to decline, in real terms, with improved technology.

9.5 TIGHT OIL RESERVOIRS

Tight oil reservoirs consist of sapropelic organic-rich source beds and adjacent tight reservoirs (sandstones, carbonates, siltstones, etc.) and are located in the oil generation window in a sedimentary basin (Fig. 9.20). This play type can occur updip from basin-center gas plays and downdip from regional water-saturated intervals (Fig. 9.21). The reservoirs can be just

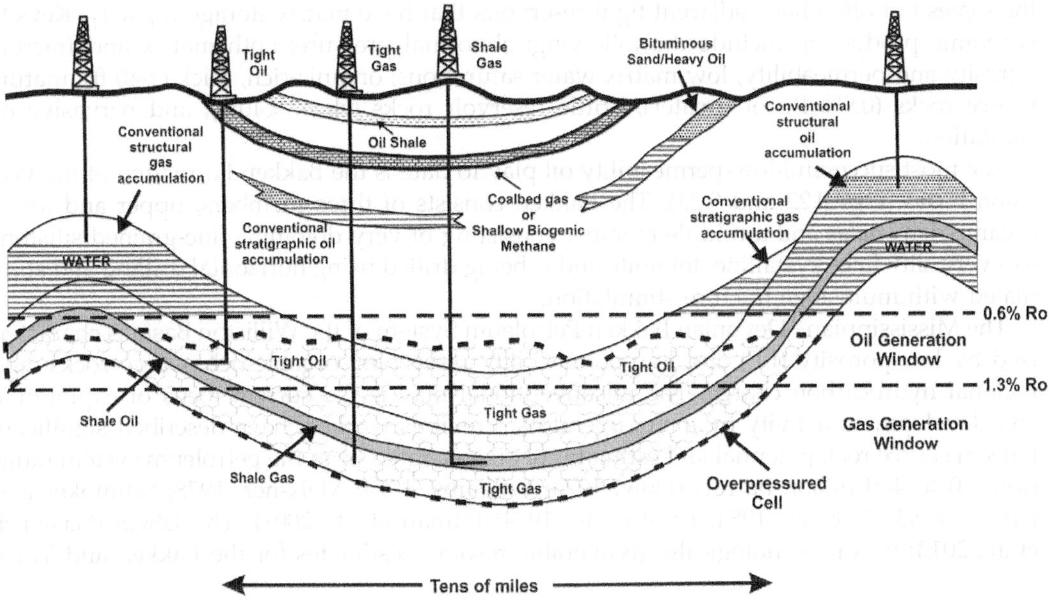

FIGURE 9.20 Diagram illustrating unconventional and conventional reservoirs. Unconventional reservoirs include oil shale, bituminous and/or heavy oil, coalbed methane, shallow biogenic methane reservoirs, tight oil, tight gas, shale gas, and shale oil. Conventional reservoirs have downdip water contacts and are structural, stratigraphic, or combination traps. *Modified from USGS.*

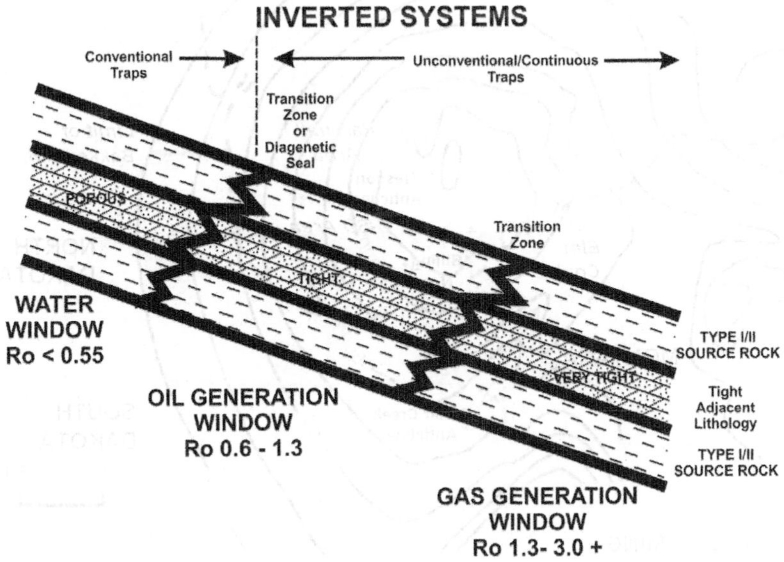

FIGURE 9.21 Inverted hydrocarbon system resulting from sapropelic source rocks being in both the oil- and gas-generating windows. Tight adjacent lithology can be tight sandstone, dolostone, chalk, etc. *Modified from Meckel and Smith (1993).*

the shales but often have adjacent tight reservoirs that have matrix storage capacity. Keys to economic production include the following: abnormal pressure; both matrix and fracture porosity and permeability; low matrix water saturations; organic-rich, thick (>40 ft), mature source rocks (0.9% Ro or greater); brittle reservoir rocks (clays <40%); and pervasive oil saturation.

The most successful low-permeability oil play to date is the Bakken Formation of the Williston Basin (Figs 9.22 and 9.23). The Bakken consists of three members, upper and lower organic-rich shales and a middle member consisting of very dolomitic fine-grained siltstone to a very silty fine-crystalline dolomite and is being drilled using horizontal drilling and stimulated with multistage fracture stimulation.

The Mississippian–Devonian Bakken Petroleum System of the Williston Basin is characterized by low-porosity (LP) and low-permeability reservoirs, organic-rich source rocks, and regional hydrocarbon charge. The unconventional play is the current focus of exploration and development activity by many operators. Previous workers have described significant Bakken source rock potential and estimates of oil generated from the petroleum system range from 10 to 400 billion barrels (Dow, 1974; Williams, 1974; Meissner, 1978; Schmoker and Hester, 1983; Webster, 1984; Price et al., 1984; Pitman et al., 2001). The USGS (Gaswirth et al., 2013) mean technologically recoverable resource estimates for the Bakken and Three

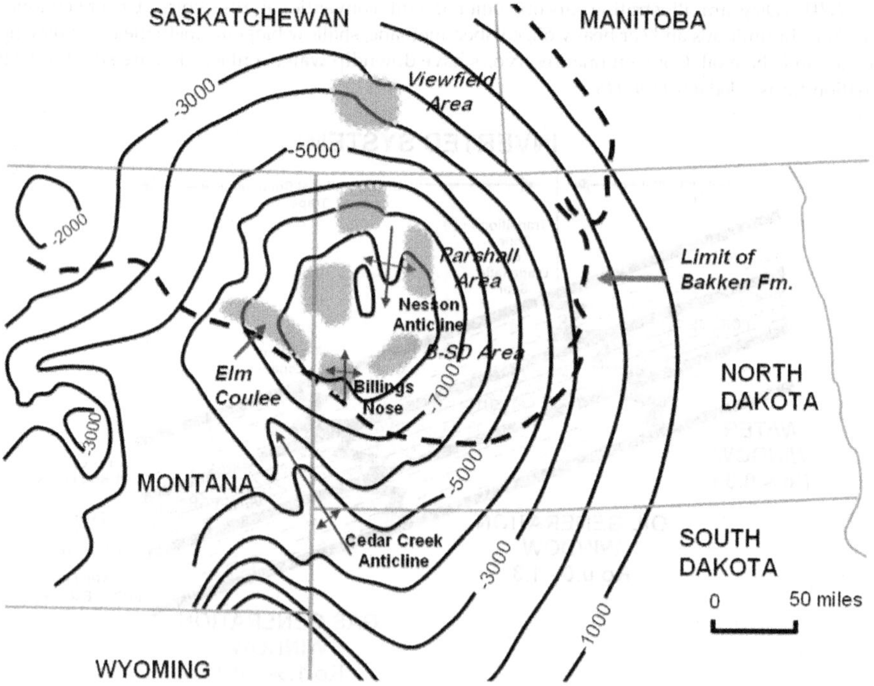

FIGURE 9.22 Index map of Williston Basin showing recent field discoveries (2000 to present) in the Bakken Formation. New Williston Basin fields classified as giants (>100 million barrels ultimate recovery) include Elm Coulee, Parshall, and Viewfield.

Forks formations are 7.37 billion barrels of oil, 6.7 trillion cubic ft of associated/dissolved natural gas, and 527 million barrels of natural gas liquids.

The structure of the Williston Basin at the base of the Mississippian is illustrated by Fig. 9.22. The basin is semicircular and prominent structural features are the Nesson, Billings, Little Knife, and Cedar Creek anticlines. Many of the structural features have a documented ancestral origin and influenced Paleozoic sedimentary patterns (Gerhard et al., 1990). Recurrent movement on Precambrian faults or shear zones is seen elsewhere in the Rock Mountain region. The Nesson anticline is the location of the first oil discoveries in the 1950s. The first oil production on the Nesson anticline was from the Silurian Interlake Formation in 1951 and subsequent oil production was established from the Mississippian Madison Group (the main producer in the basin). The Williston Basin produces mainly oil from several Paleozoic reservoirs (Fig. 9.23). The probable source rock to reservoir rock petroleum systems is illustrated by Fig. 9.23.

The Bakken petroleum system consists of the Bakken Formation, lower Lodgepole, and upper Three Forks (Fig. 26). A petroleum system consists of source beds and all the genetically related hydrocarbon accumulations. The Bakken Formation over most of the Williston Basin consists of three members: (1) upper shale, (2) middle silty dolostone or dolomitic siltstone and sandstone, (3) lower shale. The source beds for the petroleum system are the upper and lower organic-rich Bakken shales. The reservoir rocks for the petroleum system are all the members of the Bakken, the lower Lodgepole, and upper Three Forks.

The Bakken petroleum system is thought to have created a continuous type of accumulation in the deeper parts of the Williston Basin (Nordeng, 2009). A continuous accumulation is

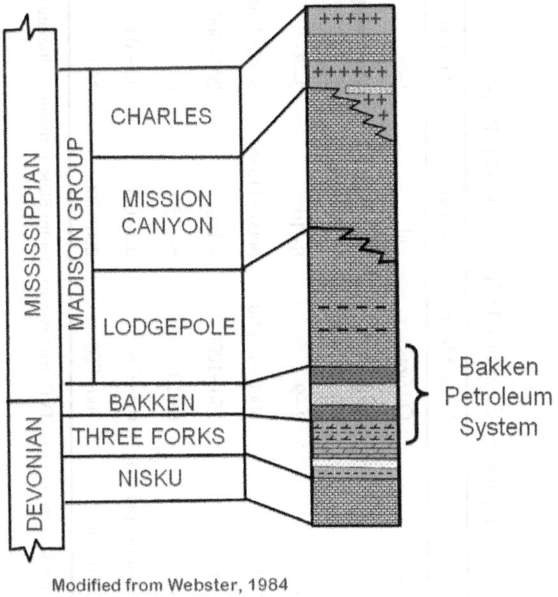

FIGURE 9.23 Stratigraphic column for upper Devonian and lower Mississippian rocks of the Williston Basin. Bakken petroleum system consists of upper and lower organic-rich shales in the Bakken and reservoirs in the Three Forks, Bakken, and lower Lodgepole.

TABLE 9.3 Comparison of Various Tight Oil and Gas Plays

	Vaca Muerta	Bakken	Barnett	Eagle Ford Gas	Eagle Ford Oil	Haynesville	Marcellus
Hydrocarbon	Oil/gas	Oil	Gas	Gas	Oil	Gas	Gas
Age	Jurassic/ Cretaceous	Devonian–Mississippian	Mississippian	Cretaceous	Cretaceous	Jurassic	Devonian
TOC (wt%)	2–9	8–10	3–8	3–6	3–6	2–3	4–6
Depth (TVD ft)	8000–11,000	8000–11,000	6000–9000	11,000–12,000	5000–11,000	10,000–14,000	5000–8500
Thickness (ft)	150–1000+	<140	200–500	200–300	80–175	150–350	50–300
Porosity (%)	6–10	5–7.5	6	9–11	9–11	9–12	6
Pressure gradient (psi/ft)	0.5–0.8	0.5–0.8	0.5–0.6	0.6–0.8	0.5–0.6	0.7–0.89	0.55–0.7
EUR/Well (MBOE)	300–750	200–700	350	1000	200+	1000	700

TOC: total organic carbon, wt.%; TVD: true vertical depth; EUR: estimated ultimate recovery; MBOE: thousand barrels of oil equivalent.

a hydrocarbon accumulation that has some or all of the following characteristics: pervasive hydrocarbon charge throughout a large area; no well-defined oil– or gas–water contact; diffuse boundaries; commonly is abnormally pressured; large in-place resource volume, but low recovery factor; little water production; geologically controlled "sweet spots"; reservoirs commonly in close proximity to mature source rocks; reservoirs with very low matrix permeabilities; and occurrence of water updip from hydrocarbons. The Bakken petroleum system meets all these characteristics.

Many of the reservoirs in the Bakken petroleum system have low permeability. Productive areas or "sweet spots" are localized areas of improved reservoir permeability through natural fracturing or development of matrix permeability, or combination of both.

The Elm Coulee Field was discovered in 2000 with horizontal completions in the middle Bakken. The field is located in the western part of the Williston Basin in northeast Montana (Fig. 9.22). Prior to the horizontal drilling in 2000, the area had scattered vertical well production (marginal to uneconomic) from the Bakken (the Bakken was a secondary objective for wells targeting deeper horizons). Horizontal drilling began in the field in 2000 and to date over 600 wells have been drilled. The estimated ultimate recovery for the field is over 200 million barrels of oil. Cumulative production from the Elm Coulee area from the Bakken up to 2013 is 101 MMBO and 92 BCFG. Horizontal drilling and multistage hydraulic fracture stimulation of the horizontal leg are key technologies that enable a low-permeability reservoir to produce. Stratigraphic trapping plays a key role at Elm Coulee (Sonnenberg and Pramudito, 2009).

The Parshall area on the east flank of the Nesson anticline consists of the Parshall and Sanish fields. The Parshall and Sanish fields were discovered in 2006 with horizontal completion in the middle Bakken (Fig. 9.23). EOG drilled and completed the 1-36 Parshall (sec. 36, T 150N, R90W) for 463 barrels oil per day (BOPD) and 128 thousand cubic feet of gas per day (MCFGPD). Whiting Oil & Gas drilled the Bartleson 44-1H in early 2006 and completed the well for 193 BOPD and 172 MCFD. Through 2013, the contiguous fields have produced approximately 146 MMBO and 84 BCFG from 852 wells completed in the Bakken and Three Forks. These fields illustrated that significant production from the middle Bakken and Three Forks exists in North Dakota. The fields also connect to the Ross Field to the north. Stratigraphic trapping and source bed maturity play key roles at Parshall area.

A comparison of active tight oil and gas plays is shown in Table 9.3. The key ingredient for all the successful plays is starting with a very rich source rock.

9.6 COALBED METHANE

The occurrence of methane gas in coal measures is only too well known because as "fire damp" it is a widespread safety hazard to coal miners. Extensive ventilation systems are required to extract methane from working coal mines. The US Department of Energy has estimated that some 250 MMcf (million cubic feet) of methane is vented daily from US coal mines as a routine safety measure. Thus coalbed methane has now joined the list of nonconventional petroleum resources (Rightmire, 1984; Gayer and Harris, 1996).

Coalbed gas is now being commercially produced from Pennsylvanian (Upper Carboniferous) coal measures in Alabama, and from the Cretaceous coals of the Western interior of the

United States. In the early 1990s some 500 wells were producing more than 28 million cubic meters of gas (Kuuskraa et al., 1992). Coalbed methane is being sought in the United Kingdom and elsewhere around the world.

Unlike shale gas, most coals only produce dry gas, methane. The coalbed gases thus have a lower calorific value than shale gas. Coalbeds normally have higher permeabilities than shales, because they more easily fracture, either naturally or artificially. Thus coalbed methane wells normally have higher flow rates and may therefore justify surface pressuring and pipeline transportation. The parameters that control the methane-generating potential of a coal include its rank, ash content, maceral type, matrix porosity, fracture porosity, pressure, and water content (Kim, 1977; Ayers and Kelso, 1989).

Coalbed methane occurs in water-saturated underground coal seams. Coal acts as both a source rock and reservoir rock for the gas. Gas is retained in coalbeds as sorbed gas and as free gas in fractures (cleats). Water must be drawn off to lower the pressure so methane will desorb from the coal and flow to the wellbore. Reservoirs can be normally or abnormally pressured and the gas can be of both thermal and biogenic origin. Coalbed methane resources are abundant across the United States. The coal rank for the majority of the productive areas is high-volatile bituminous (high capacity to adsorb gas). The best coalbed production comes from the San Juan Basin of Colorado and New Mexico (Fig. 9.24). The lowest coal rank

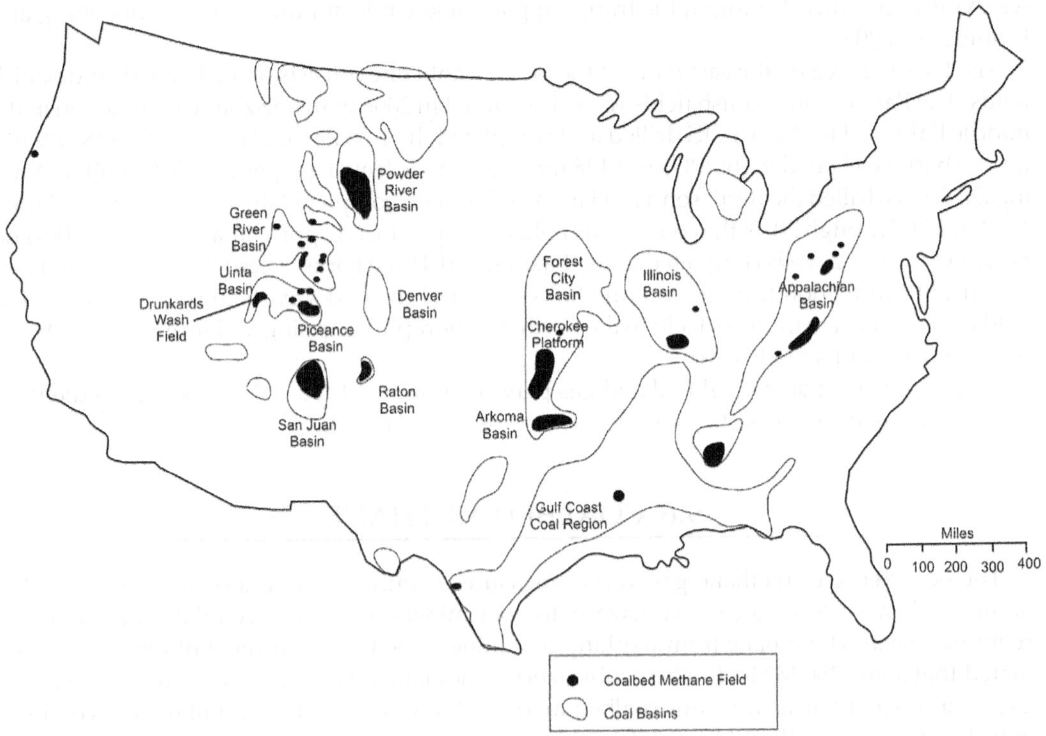

FIGURE 9.24 Map showing US coalbed methane basins and producing fields. *Modified from Eia (2009a)*; http://www.eia.gov/oil_gas/rpd/coalbed_gas.pdf.

TABLE 9.4 Key Properties for 13 US Productive Coalbed Methane Plays in Historic Areas

Coal basin or deposit	State(s)	Coal rank	Major producing coal-bearing formation or group	Typical net coal (ft)	Gas content scf/ton	Well spacing	Recovery factor	Avg. well production (Mcf/day)	Estimated undiscovered resource (Tcf)
San Juan	CO, NM	Hvb	Fruitland Formation	70	430	320	0.80	2000	7.69
Black Warrior	AL, MS	Mvb	Pottsville Formation	25	350	80	0.65	100	5.18
Central Appalachian	WV, VA, KY, TN	Mvb	Norton Formation, New River Formation, Lee Formation, Pocahontas Formation	10	300	80	0.50	80	1.06
Piceance	CO	Hvb	Williams Fork Formation	60	750	40	0.15	140	11.55
Powder River	WY, MT	Subbit.	Fort Union Formation	75	30	80	0.70	200	10.04
Uinta	UT	Hvb	Black Hawk Formation, Ferron SS.	24	400	160	0.50	625	3.81
Arkoma	OK, AR	Hvb	McAlester Formation, Hartshorne Formation	6	550	60–80	0.60	80	3.81
Raton	CO, NM	Hvb	Raton Formation Vermejo Formation	30	350	160	0.55	250	1.88
Northern Appalachian	PA, WV, OH, KY, MD	Hvb	Allegheny group	20	150	60–80	0.50	50	18.37
Cahaba	AL	Hvb	Pottsville Formation	50	260	80			
Cherokee	KS, OK, MO	Hvb	Cabaniss Formation, Krebs Formation	4	200	80	0.60	100	2.93
Greater Green River	WY	Hvb	Mesaverde group	30	350	160	0.60	240	4.66
Illinois	IL, IN, KY	Hvb	Carbondale group	5	150	80		30	2.52
Forest City	KS, IA, MO, NE	Hvb		27	100	40			

Hvb: high volatile bituminous; Mvb: medium volatile bituminous; Subbit.: subbituminous; SS: sandstone.
Modified from Limerick (2004).

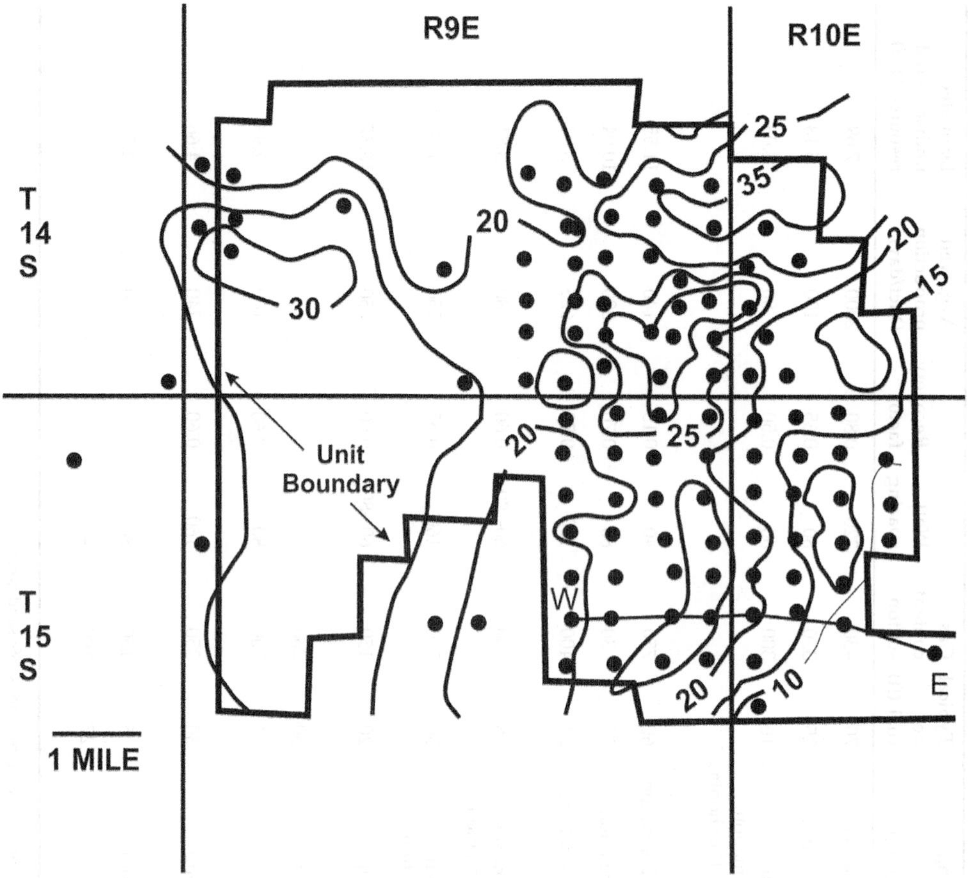

FIGURE 9.25 Map showing isopach of total coal, Drunkard's Wash Field, Utah. *Modified from Lamarre and Burns (1997).*

production comes from the Powder River Basin where the coal rank is subbituminous. The largest coalbed methane resources in the United States are thought to reside in the Rocky Mountain region. The Potential Gas Agency estimates the potential resource of natural gas from coals in the United States to be 166 trillion cubic feet (Tcf). Coalbed methane (CBM) accounts for about 21% of the US unconventional gas production. Key properties for US coalbed methane plays are listed in Table 9.4.

Hydraulic pressure is the major trapping mechanism for CBM (Dallegge and Barker, 2000). Once the pressure is lowered through water production, gas is able to desorb from the coals. Methane and carbon dioxide are the major components of CBM gas.

An example of a coalbed methane field is Drunkard's Wash Field (Utah) (Figs 9.25 and 9.26). Gas in place for the field is estimated to range from 230 to 630 Bcf (billion cubic feet) (Dallegge and Barker, 2000). The field produces from the Ferron member of the Mancos Shale (Lamarre and Burns, 1997). Most wells show the classic coalbed methane production curve with first increasing gas rates as reservoir pressure declines due to production of water

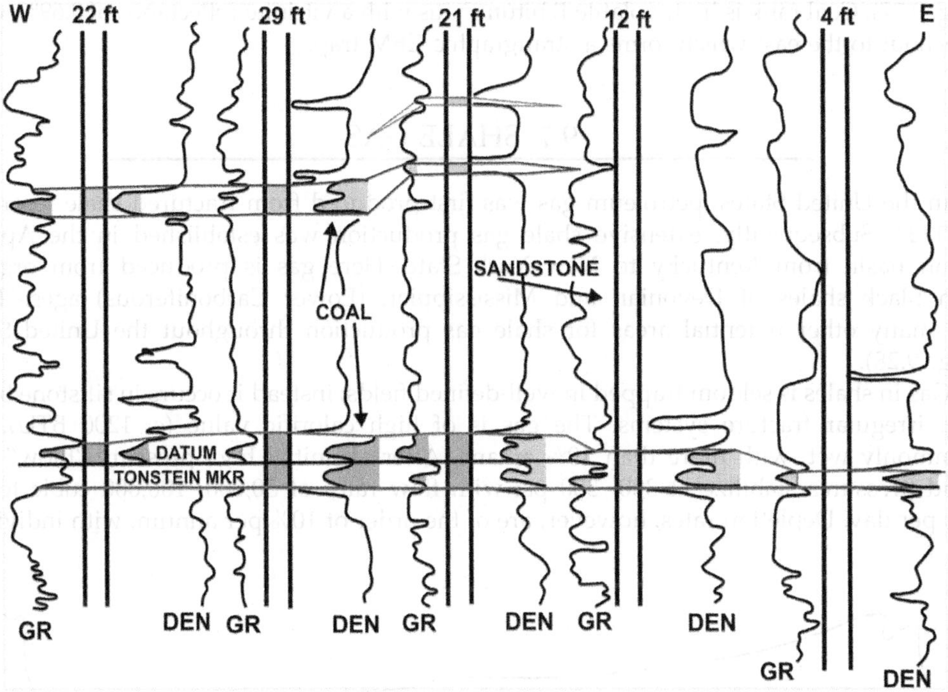

FIGURE 9.26 Stratigraphic cross-section showing Ferron coals and thinning to the east. *Modified from Lamarre and Burns (1997).*

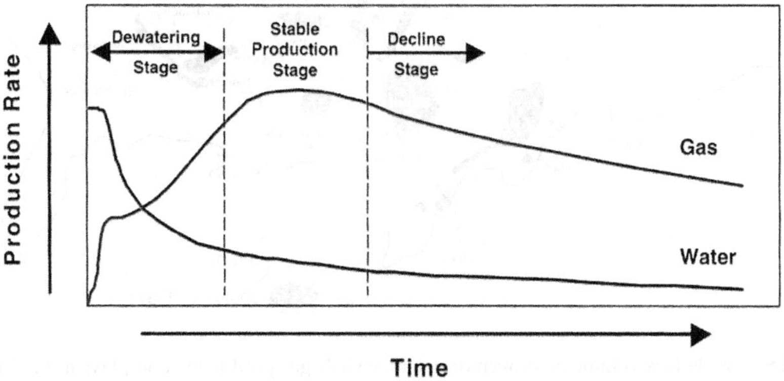

FIGURE 9.27 Typical fluid production profile for a coalbed methane well.

(Fig. 9.27). Coal rank is high-volatile B bituminous with a vitrinite reflectance of 0.69%. Coals pinchout to the east which forms a stratigraphic CBM trap.

9.7 SHALE GAS

In the United States, petroleum gas was first produced from fractured shale reservoirs in 1821. Subsequently, extensive shale gas production was established in the Appalachian Basin from Kentucky to New York State. Here gas is produced from organic-rich black shales of Devonian and Mississippian (Lower Carboniferous) ages. There are many other potential areas for shale gas production throughout the United States (Fig. 9.28).

Gas in shales is seldom trapped in well-defined fields; instead it occurs in siltstone bands and irregular fracture systems. The gas is of high calorific value (c. 1200 BTU), and commonly wet, with more than 10% ethane. After an initial high-pressure "blow" well head pressures stabilize at 300–500 psi with flow rates of 50,000–100,000 cubic feet of gas per day. Depletion rates, however, are of the order of 10% per annum, with individual

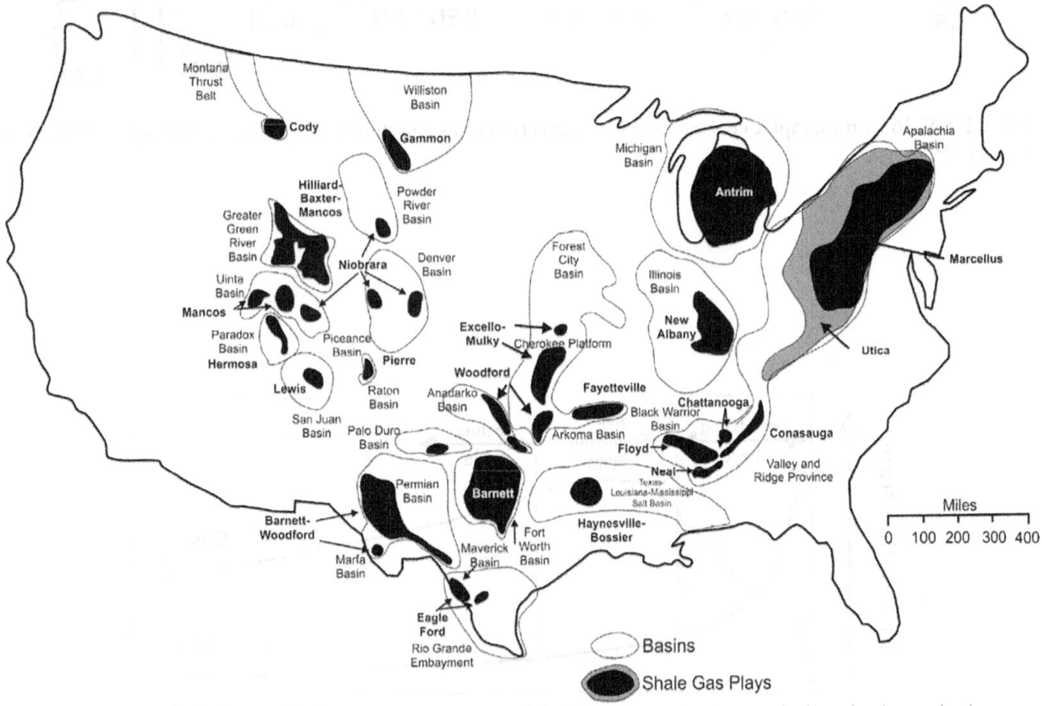

FIGURE 9.28 Shaded areas indicate proven or potential shale gas production and plays in the United States. The Devonian shales of the Appalachians in the eastern United States have been producing gas since 1821. *Modified from Eia (2009).*

wells producing for 40–50 years. Wells are seldom more than 700 m deep, and thus cheap to drill.

Shale gas exploitation is not profitable for major international oil companies. The reserves are too small and the payout time too long. Shale gas is, however, economically viable in the United States for two reasons. The "cottage industry" scale of many small oil companies permits operations with small profit margins. Unlike in the United Kingdom, land ownership includes the mineral rights of a property. Thus ranchers are generally only too pleased to see their natural resources developed.

In the late nineteenth century shale gas was found by random drilling. Shale gas production became widespread across the Appalachian Basin, but became of secondary importance with the establishment of the conventional dogma of drilling for petroleum in sandstone and carbonate reservoirs on structural highs.

Shale gas production, which had continued steadily since 1821, underwent a renaissance during the energy crisis of the late 1970s and early 1980s and continues to increase. US shale gas production has accelerated from 195 Bcf in 1992 to 300 Bcf in 1995 (Reeves et al., 1996).

Extensive research has been undertaken in the exploration for and production of shale gas. It is now known that the regional distribution of shale gas is controlled by the quantity, quality, and level of the maturation of organic matter in the shale formations. Both biogenic and thermogenic shale gas are known to occur (Fig. 9.29). Local concentrations of shale gas occur either in siltstone strata or in natural fracture systems.

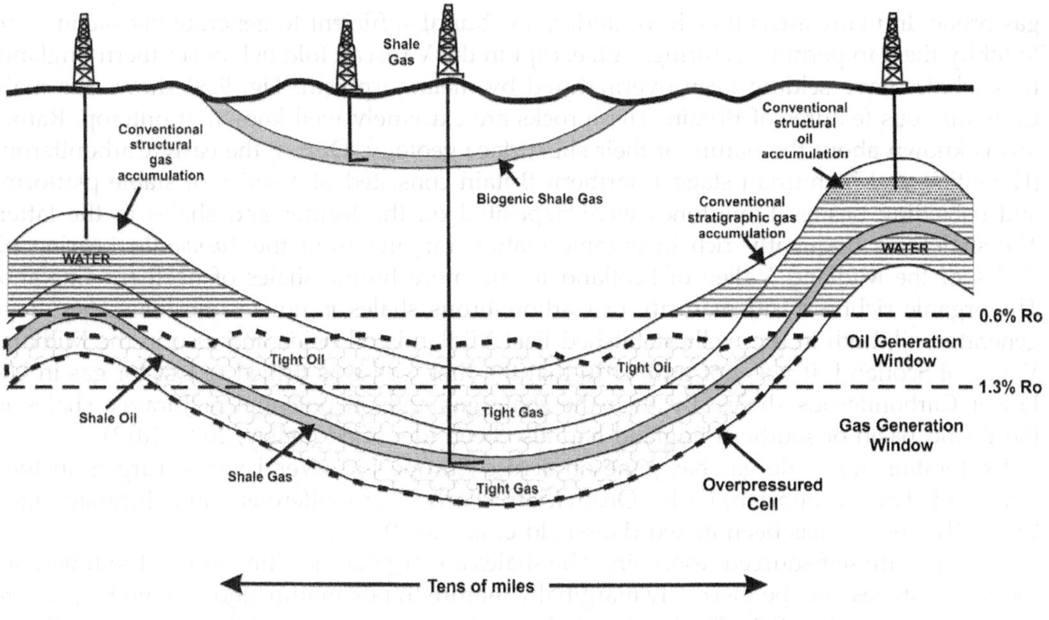

FIGURE 9.29 Scenario for shale gas production. Both biogenic and thermogenic accumulations are known to occur (thermogenic in the deep basin center and biogenic in shallow basin areas). Optimum permeability is found in basin-center shales that are in the gas generation window. Siltstones intermixed with shales can enhance permeabilities. Natural fractures also enhance permeabilities. *Modified from USGS.*

Fracture systems may be best developed where strata are stretched over the crests of anticlines, but they also occur along regional fault and basin hingeline trends. These can be located by remote sensing and seismic surveys. Conventional seismic surveys are inappropriate to locate low-velocity gas-charged shale zones. Specific methods of shooting and processing seismic data have now been developed.

Conventional drilling with a mud-filled hole will seldom locate shale gas. The weight of the mud forces the gas away from the well bore. Such gas as may escape into the drilling mud is recorded as "background gas," with little thought that it may be commercial.

Shale gas exploration and production has undergone a renaissance in the United States in the past 20 years (Zuckerman, 2013). The revolution has been brought about by a combination of horizontal drilling, multiphase hydraulic fracturing, and three-dimensional seismic surveys. Shale gas and shale oil production have had a major impact on the US economy and have led to a major reduction in carbon dioxide emissions as coal-fired power stations are closed down.

The US paradigm for shale gas and oil production is now being replicated across the world, with estimates of country-by-country resources evaluated by the US Energy Information Administration (EIA, 2013).

Over 25 years ago research suggested that an analog of the Appalachian shale gas province might exist in the United Kingdom (Selley, 1987). Fine-grained carbonaceous sediments occur extensively throughout the country. Lower Paleozoic black shales are largely overmature, and Jurassic shales are largely immature, but between these there occur the black shales of the Carboniferous system. These contain considerable organic matter, both oil prone, and gas prone. In many areas they have undergone burial sufficient to generate petroleum, and to lithify them to permit fracturing. But, except in the Variscan fold belt of southern England, these shales have seldom been overmatured by metamorphism. Fig. 9.30 shows the main Carboniferous features of Britain. These rocks are extremely well known at outcrop. Rather less is known about the nature of their subsurface geology. During the early Carboniferous (Dinantian and Namurian stages) northern Britain consisted of a series of stable platforms and subsiding basins. Limestones were deposited on the former and shales in the latter. The shales are frequently rich in organic matter, ranging from the *Tasmanites*-bearing oil shales of the Midland Valley of Scotland to the more humic shales of northern England. The organic richness and maturity of Carboniferous shales is not in question. They have generated oil, both in the well-established East Midlands oil fields and also in the Midland Valley of Scotland. It was not until 2010 that the first well was drilled to test for gas in UK Lower Carboniferous shales. By then the shale gas and oil potential of Jurassic shales in the Weald Basin of southern England had also been identified (Selley, 2005, 2012).

Exploration for shale gas has gone on across Europe for several years. Targets include shales of Late Cambrian/Early Ordovician, Early Carboniferous, and Jurassic ages. Currently, success has been mixed (Horsfield et al., 2012).

Shale gas are self-sourced reservoirs. The shales are organic-rich, fine-grained sedimentary rocks. The shales may be thermally marginally mature to postmature and contain biogenic to thermogenic gas (Fig. 9.29, Table 9.5). Shale has low matrix permeability and generally requires fracturing to provide permeability. Gas is stored in shale in three different manners: adsorbed gas, free gas in fractures and intergranular porosity, and as gas dissolved in kerogen and bitumen. Gas adsorbed on organic material is released as the formation pressure

TABLE 9.5 Key Properties for Six US Productive Shale Gas Plays in Historic Areas

Property	Fayetteville shale	Barnett shale	Ohio shale	Lewis shale	Antrim shale	New Albany shale
Basin	Arkoma	Fort Worth	Appalachian	San Juan	Michigan	Illinois
Age	Mississippian	Mississippian	Devonian	Cretaceous	Devonian	Devonian
Discovery year	2003	1981	1821	1950s	1940	1863
Depth (ft)	3000–5000	7000–8500	2500–6000	4500–6000	500–2500	1000–2500
Gross thickness (ft)	200–300	200–400	300–1000	500–1900	160	180
Bottomhole temp. (°F)	100–150	200	100	130–170	75	80–105
Maturation (Vr %)	1.2–3.0	1.1–1.4	0.6–1.9	1.6–1.9	0.6–0.7	0.6–1.2
Richness (wt% TOC)	2–5	2–5	2–6	0.5–2.5	5–15	5–20
Mineralogy (% nonclay)	30–70	45–70	45–60	50–75	55–70	50–70
Porosity (%)	4–8	3–7	2–6	2–5	5–12	5–12
Gas content (scf/ton)	40–120	30–80	20–100	15–45	40–100	40–80
Pressure gradient (psi/ft)	0.40–0.44	0.5	~0.3 psi/ft	0.22	0.43	~0.43
Gas in place, Bcf/Section	25–65	140–160	5–10	40	6–15	7–10
Avg. vert well EUR (Bcfe)	0.2–0.6	1.4	0.1–0.5	0.1–0.5	0.75	0.25–0.75
Avg. hz well EUR (Bcfe)	0.6–2.0	2.5	NA	NA	NA	NA
Cumulative gas production 1978–2006,[a] (Bcfe)	17	2560	3276	269	2481	50
Historic producing area	Van Buren & Conway Co., AR	Wise Co., TX	Pike Co., KY	San Juan & Rio Arriba Co., NM	Otsego Co., MI	Hamson Co., IN

[a]IHS, State agencies and Independent Estimates.
Vr should be VR: vitrinite reflectance; Vert: vertical Hz: horizontal; NA: not available; Bcfe: billion cubic feet of gas equivalent.
From Hill et al. (2008).

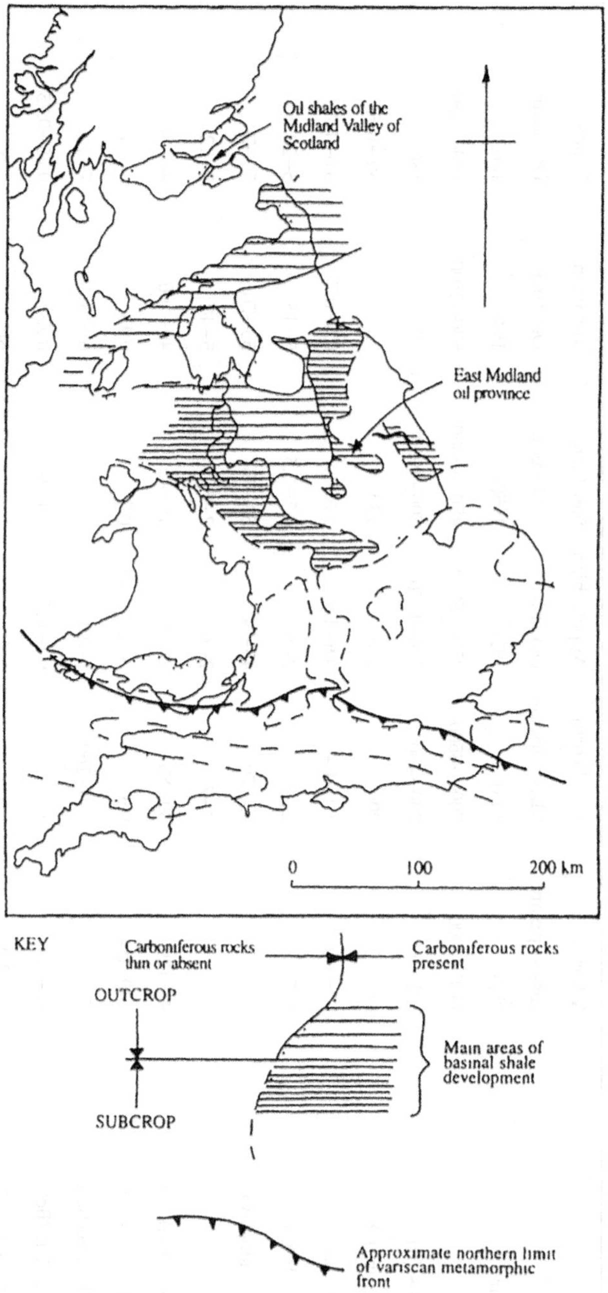

FIGURE 9.30 Areas of potential shale gas production in the United Kingdom. It was not until 2010 that this play was tested. *After Selley (1987).*

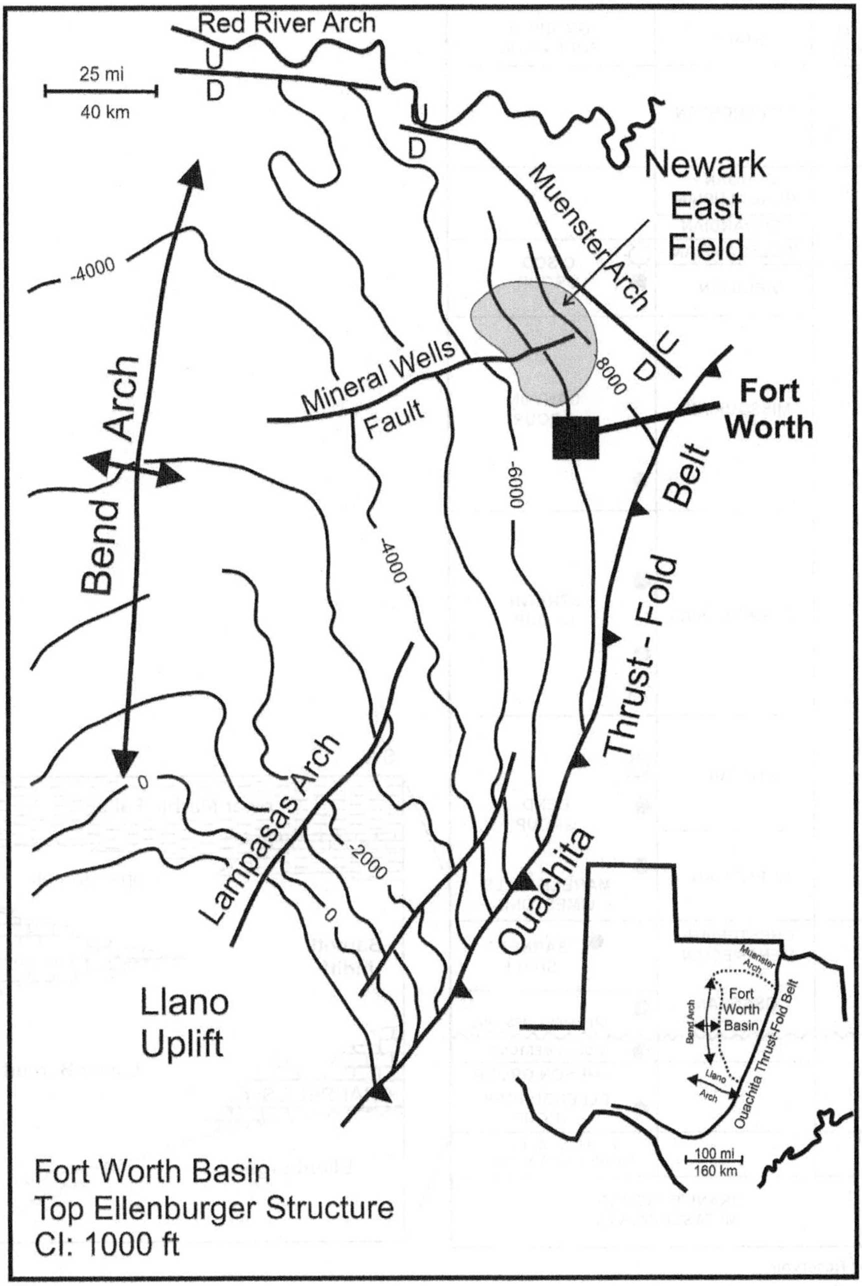

FIGURE 9.31 Structure contour map of Fort Worth Basin showing Newark East Field. *Modified from Montgomery et al. (2005).*

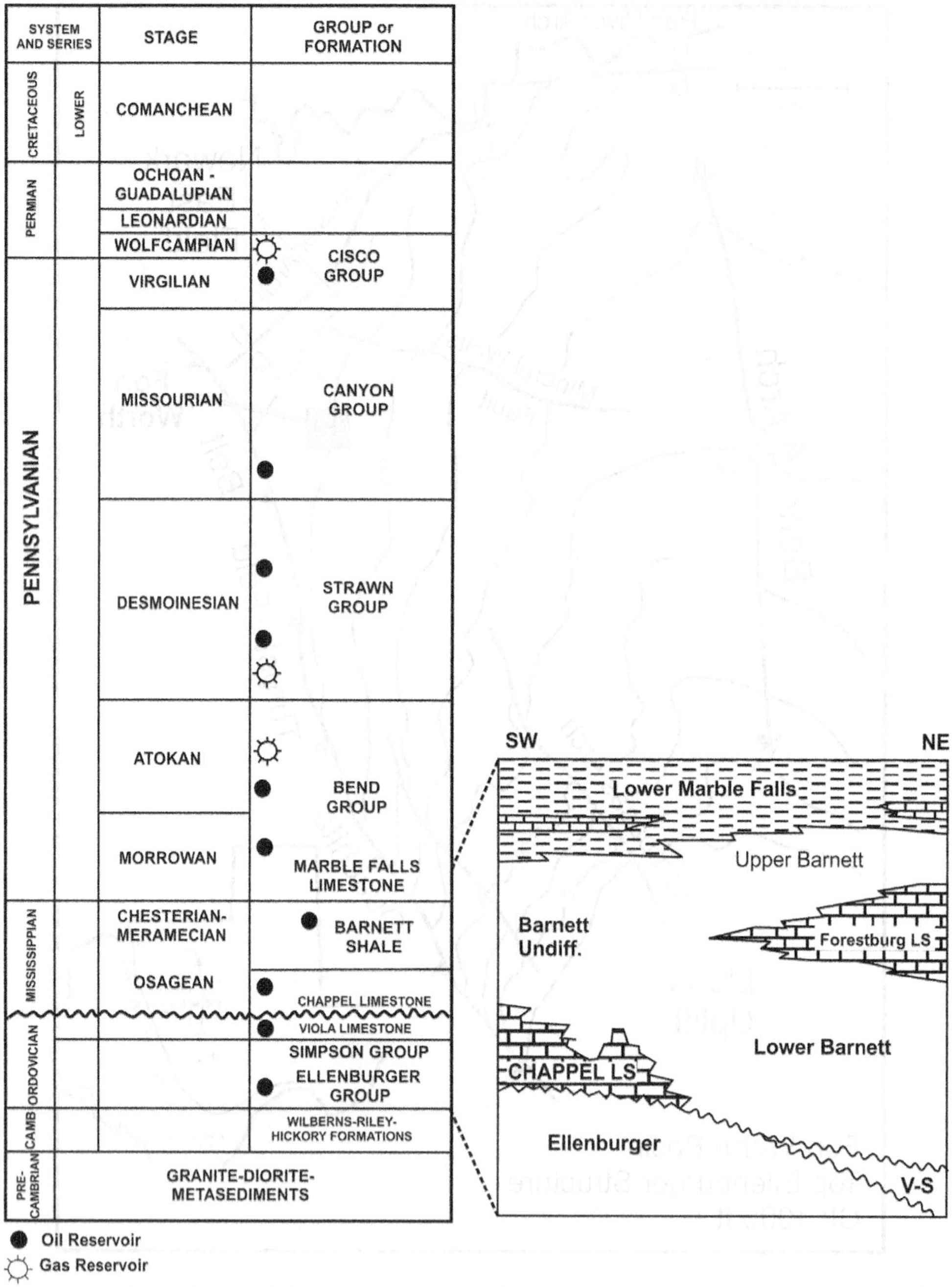

FIGURE 9.32 Stratigraphic column, Fort Worth Basin. Expanded section shows detailed Mississippian stratigraphy. *Modified from Montgomery et al. (2005).*

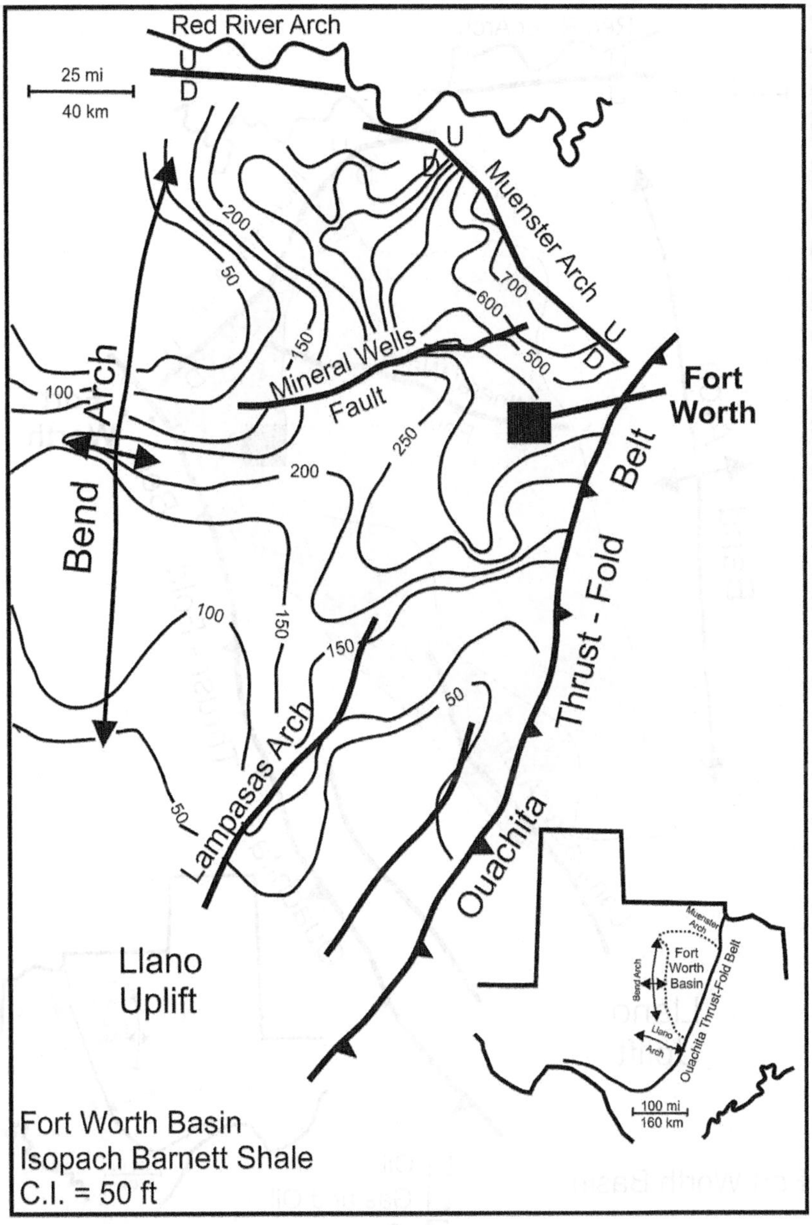

FIGURE 9.33 Isopach of Mississippian Barnett shale. *Modified from Montgomery et al. (2005).*

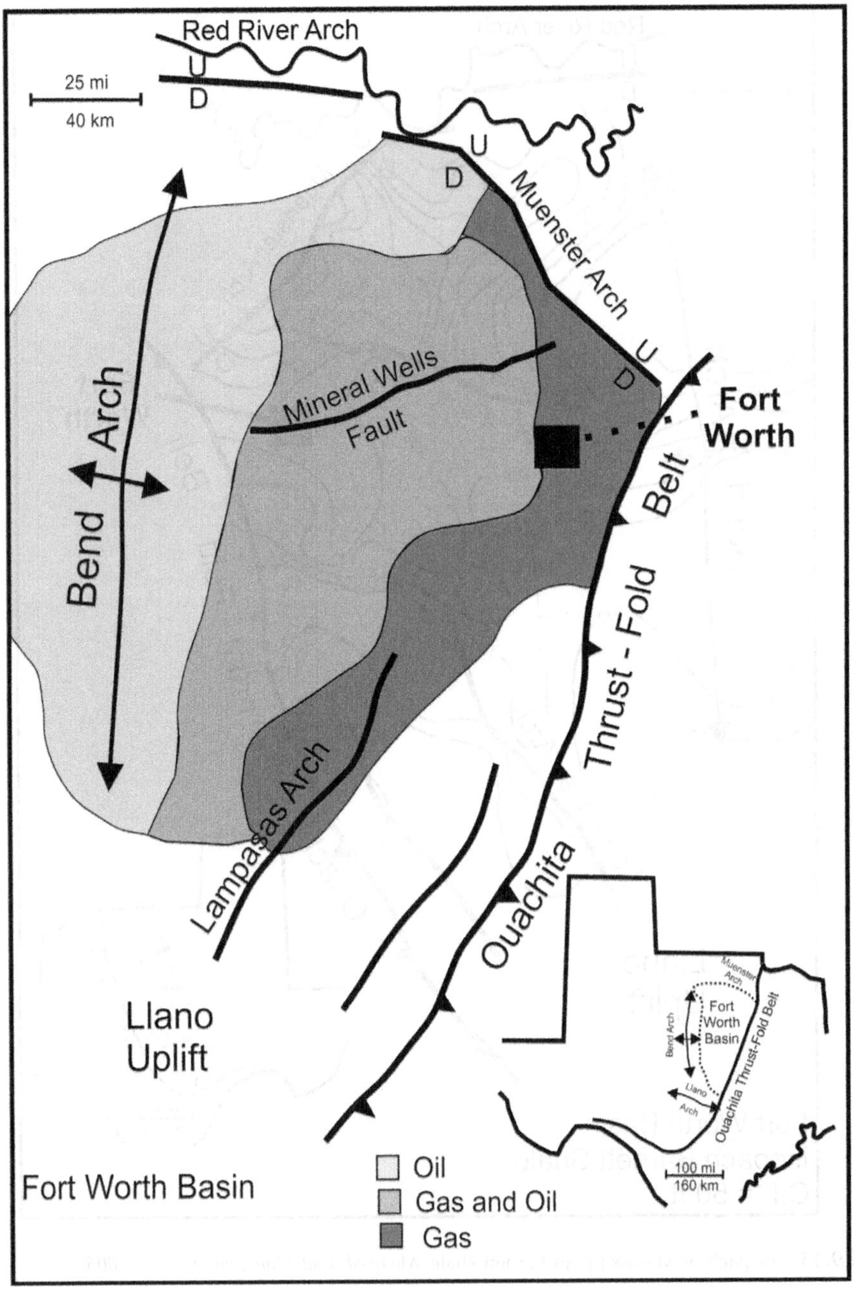

FIGURE 9.34 Generalized maturity map, Barnett Shale, based on hydrocarbon production. *Modified from Montgomery (2005).*

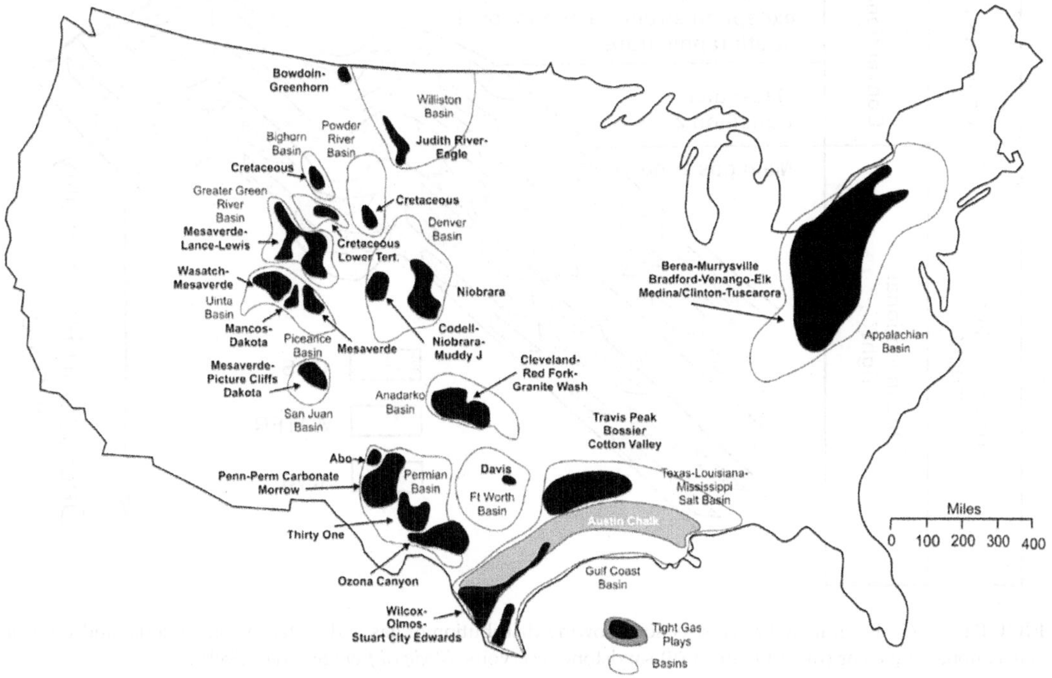

FIGURE 9.35 Major tight gas plays, lower 48 States, United States. *Modified from EIA (2010).*

declines with production. Gas in fractures is produced immediately. Shale plays are being pursued in many basins across the United States. The historic gas shale plays to date are the New Albany (Illinois Basin), Lewis (San Juan Basin), Barnett (Fort Worth Basin), Antrium (Michigan Basin), and Ohio (Appalachian Basin). Gas shales are currently 12% of the US unconventional gas production.

Table 9.5 summarizes important attributes of gas shale plays.

An example of shale gas plays is the Barnett Shale of the Fort Worth Basin (Figs 9.31, 9.32, 9.33). The Barnett is Mississippian in age and the shale serves as the source, seal, and reservoir. The first well completed in the Barnett was in 1981 but extensive development occurred from the mid-1990s to the present.

Most of the production is associated with the Newark East Field located in the bottom of the Fort Worth Basin (Fig. 9.31).

The gross thickness of the Barnett ranges from 50 to 700 ft (Fig. 9.33). The shales are thermally mature with vitrinite reflectance values that range from 1.1% to 1.4%. The total organic carbon content of the shales range from 2–5 wt%. Reservoir pressure gradient is 0.52 psi/ft. The Barnett depth ranges from 7000 to 8500 ft. The shale mineralogy of the nonclay portion ranges from 45% to 70%.

The maturity zones for the Barnett shale are shown in Fig. 9.34. The Barnett illustrates the classic inverted petroleum system with the oil zone occurring structure updip from basin-center gas accumulations.

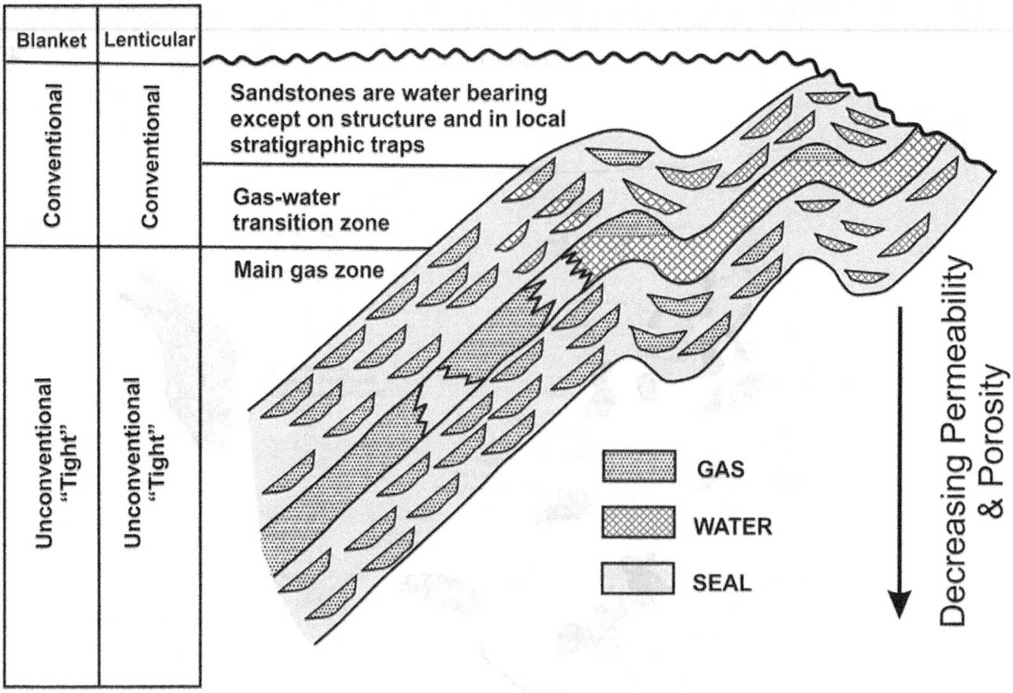

FIGURE 9.36 Diagrammatic cross-section showing distribution of gas and water in conventional and tight unconventional lenticular (L) and blanket (B) sandstone reservoirs. *Modified from Spencer (1989).*

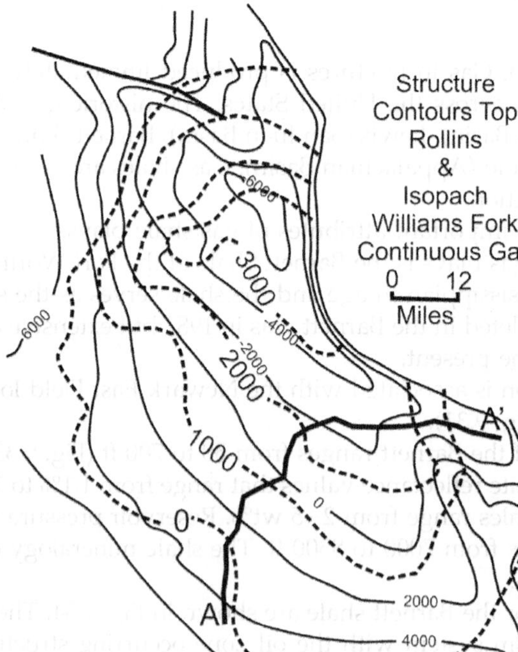

FIGURE 9.37 Structure contour map on top Rollins sandstone and isopach Williams Fork continuous gas shows for the Piceance Basin of Colorado. *Modified from Cumella and Scheevel (2012).*

TABLE 9.6 Comparison of Conventional and Tight Gas Reservoirs

	Conventional gas sandstone	Tight gas blanket and lenticular sandstone	Tight gas blanket siltstone, silty shale (HP reservoir)	Tight gas blanket chalk (HP reservoir)
Porosity (%)	14–25+	3–12+	10–30+ in individual siltstone laminations	<25–45
Porosity type	Primary (intergranular), some secondary	Common secondary (microvug), some intergranular	Dominantly primary, some secondary	Primary
Porosity communication	Good to excellent short pore throats	Poor, relatively long, sheet or ribbonlike capillary system	Good, short pore throats, but gas flow impeded by clays, small size of pores, and high S_w	Excellent, but gas flow impeded by size of pores and high S_w
Relative clay content in pores	Low	High to moderate	Low to high	Low
Geophysical well log interpretation	Generally reliable in low-clay-content reservoirs	Inaccurate; true porosity difficult to determine	Generally unreliable owing to very thin porous laminations and high water saturation	Fair, some problems with deep mud filtrate invasion
Water saturation (%)	25–50	45–70	40–90	30–70
In situ permeability to gas (md)	1.0–500	0.1–0.0005	<0.1	Mostly < 0.1
Capillary pressure	Low	Relatively high	Moderate	Moderate to high
Reservoir rock composition	Abundant quartz, minor feldspar, and rock fragments	Quartz (60–90%), common rock fragments, and some detrital feldspar and mica; may have carbonate cement	Quartz, feldspar, rock fragments, and clay; may have carbonate cement	Silt-size calcium carbonate microtossils, minor clay, and quartz
Grain density (gm/cm^3)	2.65	2.65–2.74; average 2.68–2.71 in siltstone	Probably 2.65–2.701	2.71
Reservoir pressure	Usually normal to underpressure	May be underpressured or overpressured	Underpressured	Underpressured
Recovery of gas in place (%)	75–90	<15–50 estimated for individual reservoirs	Probably low	30–50

Modified from Spencer (1989).

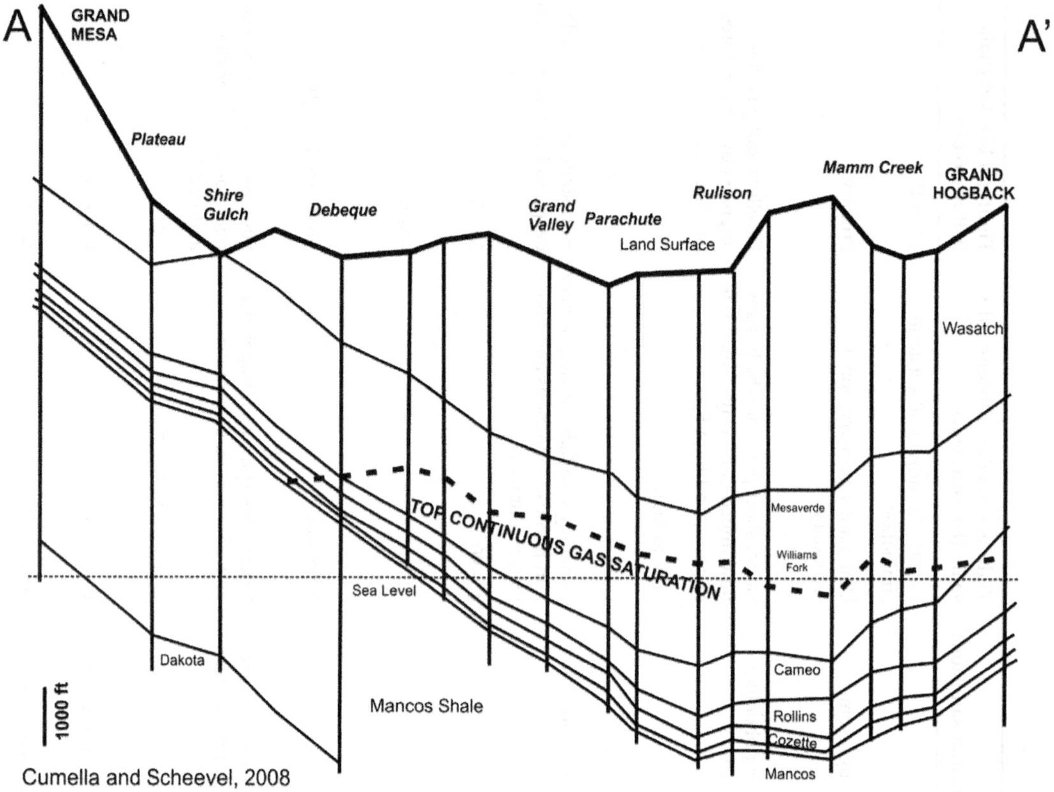

FIGURE 9.38 Cross-section A–A' through Wasatch, Mesaverde, and Mancos intervals, Piceance Basin. Top of gas saturated Williams Fork shown by black dashed line. *Modified from Cumella and Scheevel (2012).* (For interpretation of the references to color in this figure legend, the reader is referred to the online version of this book.)

9.8 TIGHT GAS RESERVOIRS

Tight unconventional gas reservoirs can be subdivided into two types based on porosity low porosity (LP). and low permeability (Spencer, 1989). These types are referred to as high porosity (HP) and major tight gas plays for the United States are illustrated by Fig. 9.35.

The high porosity (HP) type is present in the northern Great Plains and eastern Plains region of the United States. The reservoirs are at shallow burial depths and consist of chalks, siltstones, and very fine-grained sandstones. Although these reservoirs have HP (10–40%), they have low in situ permeability (<0.1 md) because of small pores and pore throats and small grain size. The gas accumulations are structural and the gas produced is thought to be biogenic.

The second type of tight gas reservoirs are referred to as low porosity (LP) type. Such reservoirs have LP (3–12%) and low in situ permeability. Diagenetic alteration has resulted in the reservoirs having small microvug porosity. Pores are interconnected by sheet or cracklike

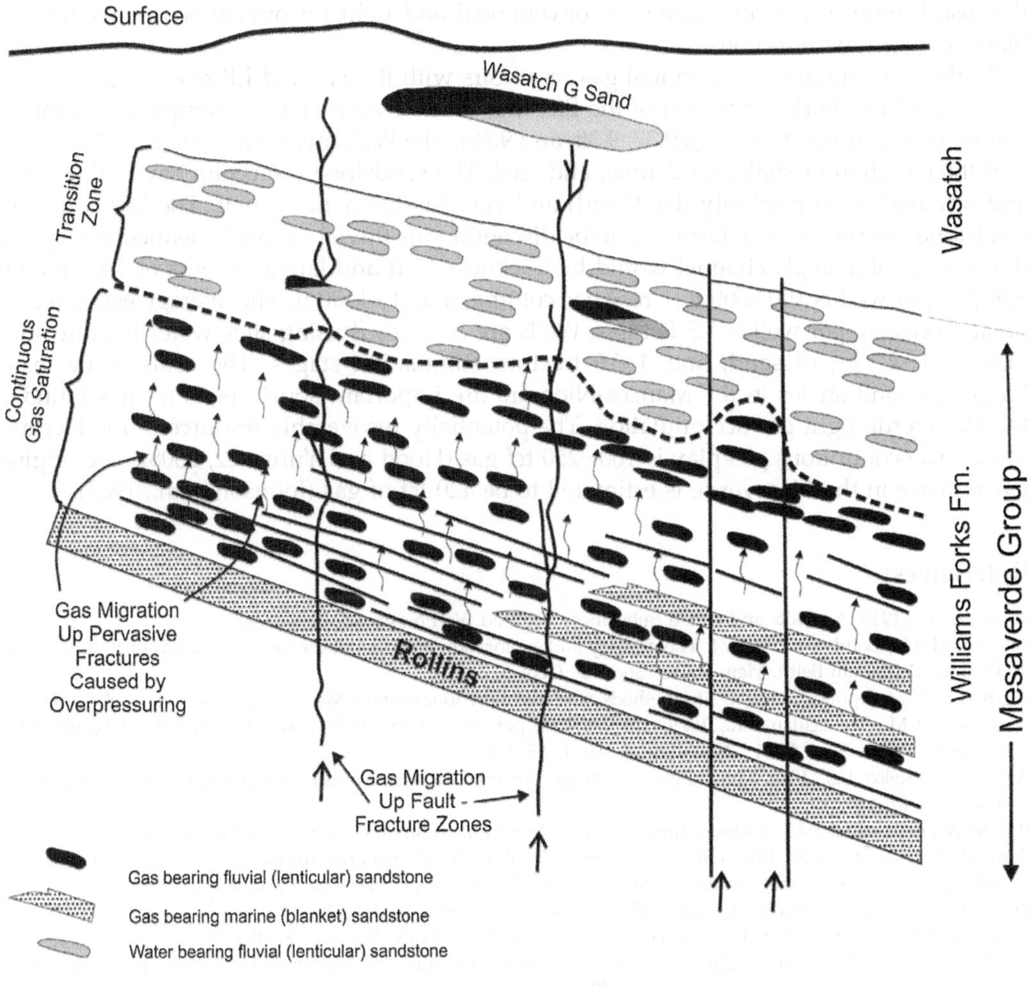

FIGURE 9.39 Schematic illustrating the Piceance Basin basin-center gas accumulation. Source beds are coals in the Mesaverde and Mancos organic-rich shales. *Modified from Cumella and Scheevel (2012).*

capillaries (sometimes referred to as slot pores). The gas accumulations are often referred to as continuous, basin-centered, or deep basin. The gas is of thermogenic origin.

Unconventional LP reservoirs are generally located in the deep central part of basins and have the following characteristics: pervasive gas saturation; commonly lack downdip water; have an updip contact with regional water saturation; have low in situ permeability (<0.1 md) and low matrix porosity (3–12%); are abnormally pressured (either over- or under-pressured with respect to hydrostatic pressure); reservoirs may be single or vertically stacked; natural fracturing enhances production in local areas; gas is of thermal origin; and thus the accumulations are associated with mature source rocks that are either actively generating or have recently ceased generation. Fig. 9.36 shows a diagrammatic cross-section showing the distribution of gas and water in conventional and tight unconventional lenticular and blanket sandstone reservoirs.

Table 9.6 compares conventional gas reservoirs with the HP and LP reservoirs.

The Williams Fork Formation of the Piceance Basin is an excellent example of a tight gas continuous accumulation (Figs 9.37, 9.38, and 9.39). The Williams Fork is a several-thousand-foot thick section of shale, sandstone, and coal. The sandstone reservoirs have LP (8–10%) and low matrix permeability (0.001 md) and very limited aerial extent. The Williams Fork sandstone reservoirs are interpreted to be discontinuous fluvial channels (lenticular). Typical dimensions of a single channel would be thickness 15 ft and lateral extents of 500–1000 ft. Net pay per well is 100–500 ft. Gross gas column is up to 1500 ft. The average estimated ultimate recovery per well is 1.3 Bcf gas. Wells are completed using slick water fracture stimulation, 1–2 ppg of sand, and 4–10 fracture stimulation stages. The coals in the basal Mesaverde and shales in the Mancos-Niobrara are important source beds for gas found in the Mesaverde tight gas accumulation. The potentially recoverable resource of the Piceance Mesaverde continuous gas play is 100–250 tcf gas (Hood and Yurewicz, 2008). The original gas in place in the Mesaverde is estimated to be 420 tcf of gas (Johnson et al., 1987).

References

Abraham, H., 1945. Asphalts and Allied Substances, sixth ed. Van Nostrand, New York.

Alayeto, M.B.E., Louder, L.W., 1974. The geology and exploration potential of the heavy oil sands of Venezuela (the Orinoco Petroleum Belt.). Mem.—Can. Soc. Pet. Geol. 3, 1–18.

Ali, S.M.F., 1974. Application of in situ methods of oil recovery to tar sands. Mem.—Can. Soc. Pet. Geol. 3, 199–211.

Anderson, A.M., McLachlan, I.R., 1979. The oil-shale potential of the early Permian White Band Formation in southern Africa. Spec. Publ.—Geol. Soc. S. Afr. 6, 83–90.

Ayers, W.B., Kelso, B.S., 1989. Knowledge of methane potential for coalbed resources grows, but needs more study. Oil Gas J. 87, 64–67.

Ball, M.W., 1935. Athabaska oil sands: apparent example of local origin of oil. Am. Assoc. Pet. Geol. Bull. 19, 153–171.

Barnard, P.C., Cooper, B.S., 1981. Oils and source rocks of the North Sea area. In: Illing, L.V., Hobson, G.D. (Eds.), Petroleum Geology of the Continental Shelf of North-West Europe. Inst. Pet., London, pp. 169–175.

Broughton, A.C., 1920. Coorongite. Trans. Proc. R. Soc. South Aust. 44, 386–395.

Burger, J., 1973. L'exploitation des pyroschistes on schistes bitumineux. Rev. Ins. Fr. Pet. 28, 315–372.

Burnham, A., Day, R.L., McConaghy, J., Hradisky, M., Coats, D., Smith, P., Wallman, H., LaBrecque, D., Foulkes, J., Allix, Pl, 2012. Initial results from the AMSO RD&D pilot test program. In: 32nd Oil Shale Symposium, Golden, CO.

Byramjee, R.J., July 4, 1983. Heavy crudes and bitumen categorized to help assess resources, techniques. Oil Gas J. 78–82.

Cane, R.F., 1976. The origin and formation of oil shale. In: Yen, T.F., Chilingarian, G.V. (Eds.), Oil Shales. Elsevier, Amsterdam, pp. 27–60.

Connan, J., Van der Weide, B.M., 1974. Diagenetic alteration of natural asphalts. Mem.—Can. Soc. Pet. Geol. 3, 134–147.

Craig, A., 1915. The supposed oil-bearing areas of South Australia. South Aust. Geol. Surv. Bull. 4, 1–54.

Cumella, S., Scheevel, J., 2012. Mesaverde tight gas sandstone sourcing from underlying Mancos-Niobrara shales. Search Discov. article # 10450.

Dallegge, T.A., Barker, C.E., 2000. Coal-Bed Methane Gas-in-Place Resource Estimates Using Sorption Isotherms and Burial History Reconstruction: An Example from the Ferron Sandstone Member of the Mancos Shale. U.S. Geological Survey Professional Paper 1625-B, Utah. Chapter L. 26 p.

Demaison, G.J., 1977. Tar sands and supergiant oil fields. AAPG Bull. V. 63, 1950–1961.

Deroo, G., Tissot, B., McCrossan, R.G., Der, F., 1974. Geochemistry of the heavy oils of Alberta. Mem.—Can. Soc. Pet. Geol. 3, 148–167.

Dinneen, G.V., 1976. Retorting technology of oil shale. In: Yen, T.F., Chilingarian, G.V. (Eds.), Oil Shale. Elsevier, Amsterdam, pp. 181–198.
Dome, K., May 8, 1979. Cold comfort for Alberta's oil industry. Financ. Times, London 32.
Doscher, T.M., 1967. Technical problems in in-situ methods for recovery of bitumen from tar sands. Proc.—World Pet. Congr. 7. Sess. 13, Pap. 6.
Dott, R.H., Reynolds, M.J., 1969. Sourcebook for Petroleum Geology. Mem. No. 5. Am. Assoc. Pet. Geol., Tulsa, OK.
Dougan, P.M., Reynolds, F.S., Root, P.J., 1970. The potential for in situ retorting of oil shale in the Piceance Creek basin of Northwestern Colorado. Q. Colo. Sch. Mines C5 (4), 57–62.
Dow, W.G., 1974. Application of oil-correlation and source-rock data to exploration in Williston Basin. Am. Assoc. Pet. Geol. Bull. 58, 1253–1262.
Duncan, D.C., 1976. Geological setting of oil-shale deposits and world prospects. In: Yen, T.F., Chilingarian, G.V. (Eds.), Oil Shale. Elsevier, Amsterdam, pp. 13–26.
Duncan, D.C., Swanson, V.E., 1965. Organic-rich shale of the United States and world land areas. Geol. Sum Circ. (U.S.) 523, 1–30.
Dyni, J.R., 2005. Geology and Resources of Some World Oil-Shale Deposits. USGS Scientific Investigations report 2005-5294, 42 p.
EIA, 2009a. Coalbed Methane Basin Map accessed 2013. http://www.eia.gov/oil_gas/rpd/coalbed_gas.pdf.
EIA, 2009b. Shale Gas Plays accessed 2013. http://www.eia.gov/oil_gas/rpd/shaleusa2.pdf.
EIA, 2010. Major Tight Gas Plays, Lower 48 States accessed 2013. http://www.eia.gov/oil_gas/rpd/tight_gas.pdf.
EIA, 2013. World Shale Gas Resources. http://www.eia.gov/analysis/studies/worldshalegas/pdf/overview/pdf.
Elmore, R.D., 1979. Miocene-Pliocene syndepositional tar deposit in Iran. AAPG Bull. 63, 444.
Evans, C.R., Rogers, M.A., Bailey, N.J.L., 1971. Evolution and alteration of petroleum in western Canada. Chem. Geol. 8, 147–170.
Finken, R.E., Heldau, R.F., July 17, 1972. Phillips solves Venezuelan tar-belt producing problems. Oil Gas J. 108–114.
Gaswirth, S.B., Marra, K.R., Cook, T.A., Charpentier, R.R., Gautier, D.L., Higley, D.K., Klett, T.R., Lewan, M.D., Lillis, P.G., Schenk, C.J., Tennyson, M.E., Whidden, K.J., 2013. Assessment of undiscovered oil resources in the Bakken and Three Forks Formations, Williston Basin Province, Montana, North Dakota, and South Dakota, 2013: U.S. Geological Survey Fact Sheet 2013–3013, 4 p., http://pubs.usgs.gov/fs/2013/3013/.
Gayer, R., Harris, I., 1996. Coalbed methane and coal geology. Spec. Publ. Geol. Soc. London 109.
Gerhard, L.C., Anderson, S.B., Fischer, D.W., 1990. Petroleum geology of the Williston Basin. In: Leighton, M.W., Kolata, D.R., Oltz, D.T., Eidel, J.J. (Eds.), Interior Cratonic Basins, 51. Am. Assoc. Pet. Geol. Bull, pp. 507–559.
Goff, J.C., 1983. Hydrocarbon generation and migration from Jurassic source rocks in the E. Shetland Basin and Viking Graben of the northern North Sea. J. Geol. Soc. London 140, 445–474.
Grant, B.F., 1964. Retorting oil shale underground—problems and possibilities. Q. Colo. Sch. Mines 59 (3), 39–46.
Hill, D.G., Curtis, J.B., Lillis, P.G., 2008. Update of North America shale-gas exploration and development. In: Hill, D., Lillis, Curtis (Eds.), Rocky Mountain Association of Geologists Guidebook CD, pp. 11–42.
Hills, L.V. (Ed.), 1974. Oil Sands: Fuel of the Future. Mem No. 3. Can. Soc. Pet. Geol., Calgary, Alberta, Canada.
Hobson, G.D., Tiratsoo, N., 1981. Introduction to Petroleum Geology, second ed. Scientific Press, London.
Hood, K.C., Yurewicz, D.A., 2008. Assessing the Mesaverde basin-centered gas play, Piceance Basin, Colorado. In: Cumella, S.P., Shanley, K.W., Camp, W.K. (Eds.), Understanding, exploring, and developing tightgas sandsd 2005 Vail Hedberg Conference: AAPG Hedberg Series, no. 3, p. 87-104.
Horsfield, B., Shulz, H.-M., Kapp, I., 2012. Shale gas in Europe. Search & Discovery Article No. 10380 Am. Assoc. Pet. Geol. 27.
Hunt, J.M., Stewart, F., Dickey, P.A., 1954. Origins of hydrocarbons of Uinta basin, Utah. Am. Assoc. Pet. Geol. Bull. 38, 1671–1698.
Jardine, D., 1974. Cretaceous oil sands of Western Canada. Mem.—Can. Soc. Pet. Geol. 3, 50–67.
Johnson, R.C., Crovelli, R.A., Spenser, C.W., Mast, R.F., 1987. An assessment of gas resources in low permeability sandstones of the Upper Cretaceous Mesaverde Group, Piceance Basin, Colorado: U. S. Geological Survey Open-File Report 87–357, 165 p.
Kamen-Kaye, M., 1983. Mozambique-Madagascar geosyncline. Vol. II. Petroleum geology. J. Pet. Geol. 5, 287–308.
Kappeler, J.H., 1967. Discussion of paper by Phizackerly and Scott. Proc.—World Pet. Congr. 3, 551–571.
Kendall, G.H., 1977. Importance of reservoir description in evaluating in situ recovery methods for Cold Lake heavy oil. Part I. Reservoir description. Bull. Can. Pet. Geol. 25, 314–317.

Kent, P.E., 1954. A Visit to Western Madagascar. B. P. (unpublished report).
Kim, A.G., 1977. Estimating methane content of bituminous coalbeds from absorption data. Rep. Invest.—U.S. Bur. Mines 1–22. RI-8245.
Knutson, C.F., 1981. Oil shale. AAPG Bull. 65, 2283–2289.
Kugler, H.G., 1961. Geological Map and Section of Pitch Lake Morne I'Enfer Area. Pet. Assoc, Trinidad.
Kuuskraa, V.A., Boyer, C.M., Kelafant, J.A., 1992. Hunt for quality basin goes abroad: coalbed gas. Oil Gas J. 90, 49–54.
Lamarre, R., Burns, T.D., 1997. Drunkard's ash unit: coalbed methane production from Ferron coals in east-central Utah. In: Innovative Applications of Petroleum Technology Guidebook. Rocky Mountain Association of Geologists, pp. 47–59.
Lees, G.M., Cox, P.T., 1937. The geological basis of the present search for oil in Great Britain by the D'Arcy Exploration Co., Ltd. Q. J. Geol. Soc. London 93, 156–194.
Limerick, S.H., 2004. Coalbed methane in the United States: a GIS study. Search Discov. Artile # 10066 http://www.searchanddiscovery.com/documents/2004/limerick/index.htm.
MacGregor, D.S., 1993. Relationships between seepage, tectonics and subsurface petroleum reserves. Mar. Pet. Geol. 10, 606–619.
Meckel, L., Smith, 1993. Unpublished Diagram of Inverted Hydrocarbon System.
Meyer, R.F., Attansi, E.D., Freeman, P.A., 2007. Heavy oil and natural bitumen resources in geologic basins of the world. USGS open file report 2007 e1084, 36.
Meissner, F.F., 1978. Petroleum geology of the Bakken formation, Williston Basin, North Dakota and Montana. In: Estelle, D., Miller, R. (Eds.), The Economic Geology of the Williston Basin, 1978 Williston Basin Symposium. Montana Geological Society, pp. 207–230.
Montgomery, D.S., 1974. Geochemistry of the heavy oils of Alberta: discussion. Mem.—Can. Soc. Pet. Geol. 3, 184–185.
Montgomery, S.L., Jarvie, D.M., Bowker, K.A., Pollastro, R.M., 2005. Mississippian Barnett shale, Fort Worth basin, north-central Texas: gas-shale play with multi-trillion cubic foot potential. Am. Assoc. Petrol. Geol. 89, 155–175.
Nasiam, O.W.T., 1937. Early use of hydrocarbons in the Imperial Chinese Empire. In: Professor Li Lo Memorial Volume. University of Beijing, pp. 301–331.
Newell, E., August 2, 1993. Oilsands play larger role in Canadian oil production. Oil Gas J. 25–30.
Nordeng, S.H., 2009. The Bakken petroleum system: an example of a continuous petroleum accumulation. N. D. Dep. Miner. Resour. Newsl. 36 (1), 19–22.
Nissenbaum, A., 1978. Dead Sea asphalts—historical aspects. AAPG Bull. 62, 837–844.
Phizackerly, P.H., Scott, L.O., 1967. Major tar sand deposits of the world. Proc.—World Pet. Congr. 7 (3), 551–571.
Pitman, J.K., Price, L.C., LeFever, J.A., 2001. Diagenesis and Fracture Development in the Bakken Formation, Williston Basin: Implications for Reservoir Quality in the Middle Member. US Geological Survey Professional Paper 1653, 19 p.
Prien, C.H., 1976. Survey of oil shale research in the last three decades. In: Yen, T.F., Chilingarian, G.V. (Eds.), Oil Shale. Elsevier, Amsterdam, pp. 235–267.
Price, L.C., Ging, T., Daws, T., Love, A., Pawlewicz, M., Anders, D., 1984. Organic metamorphism in the Mississippian-Devonian bakken shale North Dakota portion of the Williston Basin. In: Woodward, J., Meissner, F.F., Clayton, J.C. (Eds.), Hydrocarbon Source Rocks of the Greater Rocky Mountain Region. Rocky Mountain Association of Geologists, Denver, pp. 83–134.
Reeves, S.R., Kuustraa, V.A., Hill, D.G., January 22, 1996. New basins invigorate U.S. gas shales play. Oil Gas J. 53–58.
Rightmire, C.T., 1984. Coalbed methane resource. Stud. Geol. (Tulsa, Okla) 17, 1–13.
Russell, P.L., 1980. History of Western Oil Shale. Applied Science Publishers, Barking, UK.
Salomonsson, G., 1951. The Ljungstrom In situ Method for Oil-Shale Recovery. In: Oil Shale and Cannel Coal, vol. 2. Inst. Pet., London, 260–280.
Schlatter, L.E., 1969. Oil shale occurrence in Western Europe. In: Hepple, P. (Ed.), The Exploration for Petroleum in Europe and North Africa. Inst. Pet., London, pp. 251–258.
Schmoker, J.W., Hester, T.C., 1983. Organic carbon in bakken formation, United States portion of Williston Basin. Am. Assoc. Pet. Geol. Bull. 67, 2165–2174.
Selley, R.C., June 15, 1987. British shale gas potential scrutinized. Oil Gas J. 62–64.

Selley, R.C., 1992. Petroleum seepages and impregnations in Great Britain. Mar. Pet. Geol. 9, 226–244.
Selley, R.C., Stoneley, R., 1987. Petroleum habitat in South Dorset. In: Brooks, J., Glennie, K. (Eds.), Petroleum Geology of North West Europe. Graham & Trotman, London, pp. 139–148.
Selley, R.C., 2005. UK shale-gas resources. In: Doré, A.G., Vining, B.A. (Eds.), Petroleum Geology: North-West Europe & Global Perspectives – Proceedings of the 6th Petroleum Geology Conference. Geological Society, London, pp. 707–714.
Selley, R.C., 2012. UK shale gas: the story so far. Mar. Pet. Geol. 31 (1), 100–109. http://dx.doi.org/10.1016/j.marpetgeo.2012.08.017.
Shanks, W.C., Seyfried, W.E., Meyer, W.C., O'Neil, T.J., 1976. Mineralogy of oil shale. In: Yen, T.F., Chilingarian, G.V. (Eds.), Oil Shale. Elsevier, Amsterdam, pp. 81–102.
Sonnenberg, S.A., Pramudito, A., 2009. Petroleum geology of the giant Elm Coulee field, Williston Basin. Am. Assoc. Pet. Geol. Bull. 93, 1127–1153.
Spencer, C.W., 1989. Review of characteristics of low-permeability gas reservoirs in western United States. Am. Assoc. Petrol. Geol. 73, 613–629.
Spiro, B., Welte, D.H., Rullkotter, J., Schaefer, R.G., 1983. Asphalts, oils and bituminous rocks from the Dead Sea area—a geochemical correlation study. AAPG Bull. 67, 1163–1175.
Stewart, D.R., 1912. The Oil Shales of the Lothians. Part III. Mem. Geol. Surv., Scotland.
Suter, H.H., 1951. The general and economic geology of Trinidad. Colon. Geol. Miner. Resour. 2, 271–307.
Towson, D.E., 1977. Importance of reservoir description in evaluating in situ recovery methods for Cold Lake heavy oil. Part II. In situ application. Bull. Can. Pet. Geol. 25, 328–340.
Wade, A., 1915. The supposed oil-bearing areas of South Australia. Geol. Surv. South Aust. 4.
Wall, G.P., Sawkins, J.G., 1860. Report on the Geology of Trinidad. HMSO, London.
Walters, E.J., 1974. Review of the world's major oil sands deposits. Mem.—Can. Soc. Pet. Geol. 3, 240–263.
Webster, R.L., 1984. Petroleum source rocks and stratigraphy of the Bakken formation in North Dakota. In: Woodward, J., Meissner, F.F., Clayton, J.C. (Eds.), Hydrocarbon Source Rocks of the Greater Rocky Mountain Region. Rocky Mountain Association of Geologists, Denver, pp. 57–81.
Wikipedia, 2014. Shale Oil. http://en.wikipedia.org/wiki/Shale_oil. accessed 2014.
Williams, J.A., 1974. Characterization of oil types in Williston basin. Am. Assoc. Pet. Geol. Bull. 58, 1243–1252.
Yen, T.F., Chilingarian, G.V. (Eds.), 1976. Oil Shale. Elsevier, Amsterdam.
Yule, C., 1871. The Book of Ser Marco Polo the Venetian. London.
Zuckerman, G., 2013. The Frackers. Penguin, New York, 404 pp.

Selected Bibliography

Solid hydrocarbons and oil seeps

Hunt, J.M., 1996. Petroleum Geochemistry and Geology, second ed. Freeman, San Francisco.
Tissot, B.P., Welte, D.H., 1978. Petroleum Formation and Occurrence. Springer-Verlag, Berlin (Chapter 5, Petroleum Alteration, covers the degradation of oils to form asphalts).

Tar sands

Chilingarian, G.V., Yen, T.F. (Eds.), 1978. Bitumen Asphalts and Tar Sands. Elsevier, Amsterdam.
Hills, L.V. (Ed.), 1974. Oil Sands: Fuel of the Future. Mem. No. 3. Can. Soc. Pet. Geol, Calgary, Alberta, Canada. This volume contains many papers on the occurrence of tar sands and the technology of extracting oil from them. Most of the papers concentrate on the Athabaska deposits.

Oil shales

Tissot, B.P., Welte, D.H., 1978. Petroleum Formation and Occurrence. Springer-Verlag, Berlin (Chapter 8), pp. 225–236, deals with oil shales.
Yen, T.F., Chilingarian, G.V. (Eds.), 1976. Oil Shale. Elsevier, Amsterdam. This volume contains many papers on the occurrence of oil shales and the technology of extracting oil from them. A large part of the text is concerned with the oil shales of the Green River Formation.

Coal-bed methane:

Gayer, R., Harris, I., 1996. Coalbed Methane and Coal Geology. Spec. Publ. Geol. Soc. London 109.

Rightmire, C.T., Eddy, G.E., Kirr, J.N. (Eds.), 1984. Coalbed Methane Resources of the United States. Stud. Geol., No. 17. Amer. Assoc. Pet. Geol, Tulsa OK.

CHAPTER

10

Conclusions

10.1 PROSPECTS AND PROBABILITIES

The preceding chapters described and discussed the elements of petroleum geology: from its generation and migration, to its entrapment in reservoirs, distribution in basins, and dissipation and degradation at the surface. This chapter considers how these concepts can actually be applied to petroleum exploration and production and reviews the global distribution of reserves and their future supply and demand.

As noted in Chapter 1, many years passed before geology was applied to petroleum exploration, and even now its relevance is occasionally questioned. In particular, several studies purport to show that oil could be found in a new area by drilling wells at random rather than by the application of science. Harbaugh et al. (1977) have reviewed many such studies.

Regarding the random approach to petroleum exploration, three comments are appropriate. First, one should recall the statistician who was drowned trying to cross a river with an average depth of 1 m. Second, one should note that the success of random drilling is likely to increase in proportion to the percentage of productive acreage in a basin (Fig. 10.1). Thus almost anyone (even in a hurry) could find oil in parts of Texas and Oklahoma, but might be less fortunate in offshore Labrador. Petroleum exploration has been likened to betting on horses. One can either pick a name at random or study the racing form. Similarly, drilling locations may be picked at random or by the application of geology in an attempt to reduce the odds. The third point is that no commercial enterprise could embark on a program of random drilling if geological methods might shorten the number of wells drilled before oil is found and produced and a cash flow established. Good geology is essential for a small company, for which a string of dry holes spells catastrophe. Only a state company, with an unlimited supply of taxpayers' money, could adopt a philosophy of random drilling.

10.1.1 Prospect Appraisal

Having established that geology is more appropriate than random drilling, we now turn to prospect evaluation. The two aspects of prospect evaluation are geology and economics. The questions that need to be answered are these: Is the well likely to find oil and gas? If so, are the economics such that it would be commercially viable?

Elements of Petroleum Geology
http://dx.doi.org/10.1016/B978-0-12-386031-6.00010-2

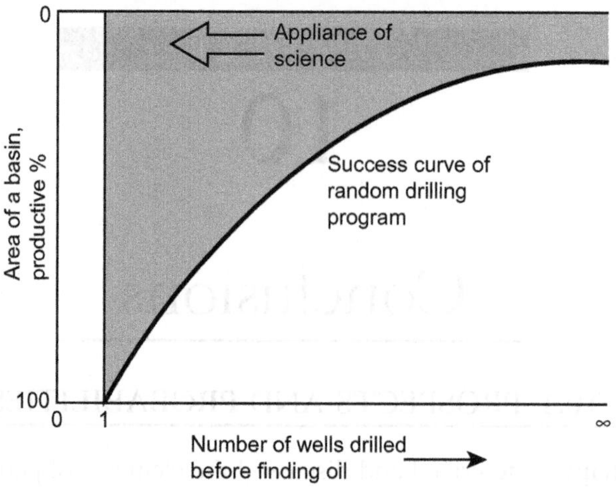

FIGURE 10.1 Graph showing that the smaller the productive area of a basin, the greater the need for the application of science.

10.1.2 Geological Aspects

The probability of a well finding petroleum ranges from 1.0 (petroleum certain to be present) to 0.0 (petroleum certain to be absent). Chapter 1 introduced the five parameters that must be present for a commercial oil or gas field to occur:

1. A source rock
2. A reservoir
3. A trap
4. A cap rock
5. Sufficient heat to generate oil or gas, but not destroy it

Thus instead of subjectively rating the probability of a prospect finding oil or gas on a scale of 1.0 to 0.0, the probability of each condition being fulfilled may be assessed, and the overall probability found from their product:

Probability of a prospect containing oil or gas $= (p_1 \times p_2 \times p_3 \times p_4 \times p_5)$, where

p_1 = probability of a source rock being present
p_2 = probability of a reservoir being present
p_3 = probability of a trap being present
p_4 = probability of a cap rock being present
p_5 = probability of correct thermal maturation occurring

Where several prospects have been identified, they may be analyzed individually, placed in a matrix, their probability of success calculated, and hence their likelihood of finding oil ranked, as shown in Table 10.1. Variations on the preceding technique are infinite, depending on which criteria are considered to be essential for a commercial accumulation to occur (White, 1993).

TABLE 10.1 Matrix for Calculating the Probability of Success for Four Prospects

	Prospects			
Probabilities	**A**	**B**	**C**	**D**
Probability of a source rock	0.9	0.6	0.7	0.3
Probability of a reservoir	1.0	0.7	0.5	0.4
Probability of a trap	0.7	0.6	0.5	0.6
Probability of a seal	0.8	0.5	0.6	0.0
Probability of correct maturation level	0.6	0.4	0.4	0.1
PROBABILITY OF SUCCESS	0.302	0.050	0.042	0.000
	Decreasing probability of success →			

Earlier Chapters (5 and 8) introduced the concept of the petroleum system in terms of petroleum generation and basin type. The petroleum system approach is essential to both an objective appraisal of a prospect and an assessment of the ultimate recoverable reserves of a basin. Basin modeling allows the thermal and subsidence history of a basin to be computed as a means of predicting the volumes and distribution of petroleum in a given area (Lerche, 1990a,b). One such appraisal system developed by Shell (Sluijk and Nederlof, 1984) modeled two geological processes and one technical process, namely:

1. The volume of the "petroleum charge," calculated from the volume, richness, and expulsion factor of source rocks.
2. The trapping and preservation of the petroleum.
3. The efficiency of the recovery system.

Basin modeling packages based on this and similar parameters are now commonplace, and are widely used throughout the industry (Plate 5). But it is important to be able to metamorphose technical success into commercial profitability.

10.1.3 Economic Aspects

The probability of a well-finding petroleum is only one aspect of successful exploration. Since the object of an oil company is to make money, the financial risk involved and the potential profitability must be established, both for individual prospects and for a string of wells. According to the "gamblers' ruin" law, there is a chance of going broke because of a run of bad luck, irrespective of the long-term probability of success. Whether a gambler (or company) is ruined before being commercially successful depends both on the probability of geological success (previously discussed) and four commercial parameters:

1. Potential profitability of venture
2. Available risk investment funds
3. Total risk investment
4. Aversion to risk

Greenwalt (1982) has shown how some of these parameters may be quantified:

$$1 - S = (1 - P_S)N,$$

where S = aversion to risk, P_s = probability of geological success, and N = number of ventures necessary to avoid gamblers' ruin.

Further, and much more elaborate, aids to exploration decision making involve more sophisticated quantification of these commercial parameters. Computer simulation techniques may then be used to aid the decision of whether or not to embark on an exploration venture, and, if so, to determine the amount of risk the investors' finances can tolerate. Such economic considerations lie beyond the field of geology, although they are an extremely important aspect of petroleum exploration. For further details see Newendorp (1976), Megill (1984), and Lerche (1990c).

10.2 RESERVES AND RESOURCES

Now that the probability of finding petroleum in individual prospects has been considered, it is relevant to discuss the probable reserves to be found in sedimentary basins and ultimately the whole world. Chapter 6 briefly reviewed how reserves in a field could be measured, and Chapter 8 showed how the petroleum potential of different types of basins could vary. It is useful to be able to calculate the reserves that a sedimentary basin may contain. First, however, some terms must be defined.

Resources are industrially useful natural materials, not only petroleum but water, minerals, and aggregate. Resources may be defined by two criteria: the economic feasibility of extraction and geological knowledge (Fig. 10.2). Reserves are resources that can be economically extracted from the resource base (McKelvey, 1975; Arnott, 2005). Figure 10.3

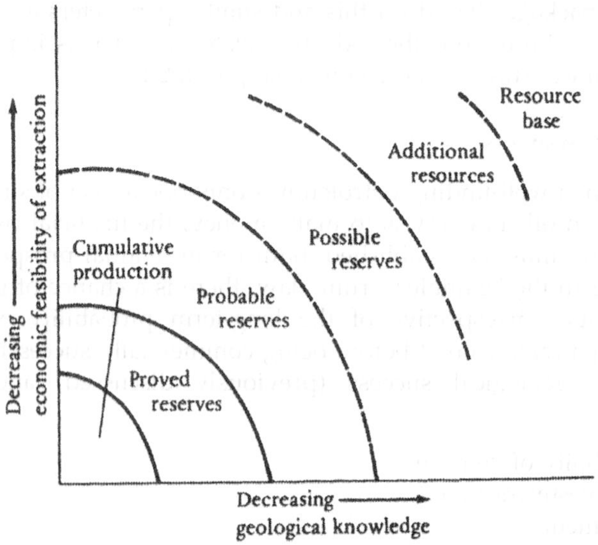

FIGURE 10.2 The concepts of reserves and resources. *After Ion (1979, 1981).*

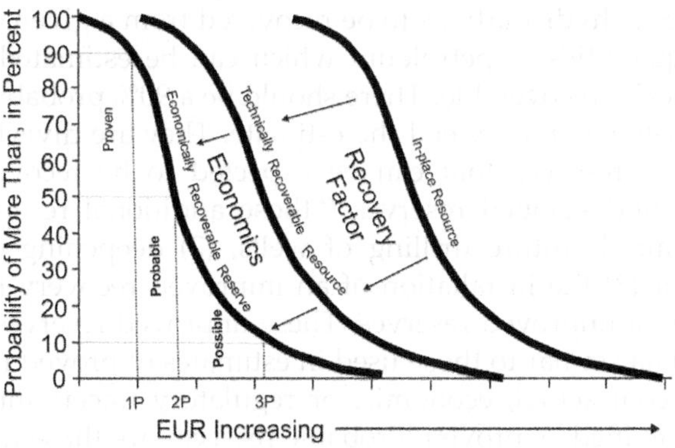

FIGURE 10.3 In-place resource with a recovery factor leads to technically recoverable resource; good economics turn technically recoverable resource into economically recoverable reserves. See text for a discussion on proven, probable, and possible reserves.

illustrates the concept of a recovery factor applied to in-place resources leads to technically recoverable resources. Economics dictate when the technically recoverable resource becomes a reserve. Thus, as the price of oil rises and falls, the oil in a marginally commercial field may be a reserve one day, but only a resource the next. Similarly, as extraction technology improves, oil shale and tar sand resources become reserves.

Reserves are subdivisible into several categories, as shown in Fig. 10.4 (SPE, 2014). Definitions have been proposed jointly by the Society of Petroleum Engineers, the American Petroleum Institute, and the American Association of Petroleum Geologists: "proved

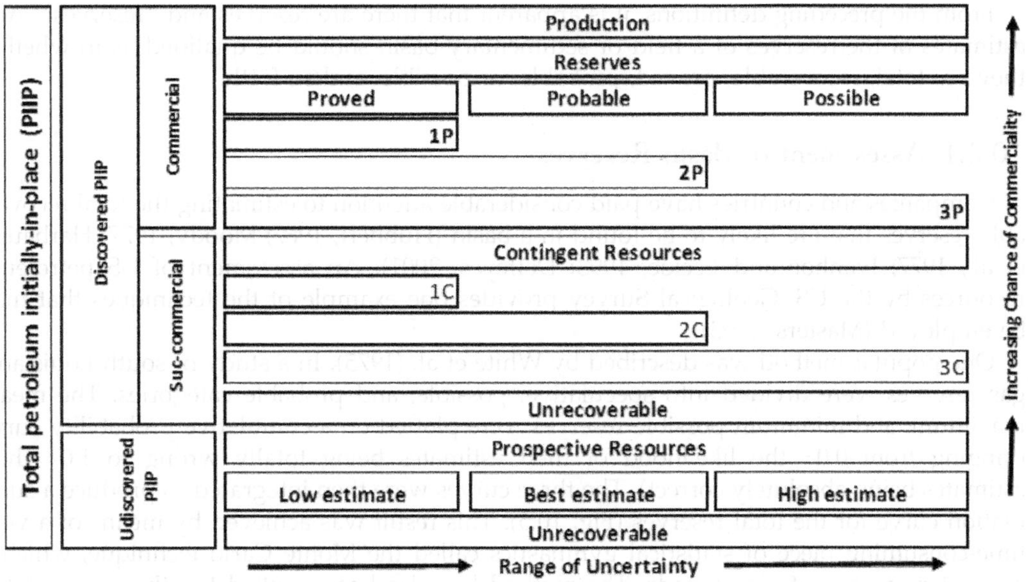

FIGURE 10.4 The different types of reserves and resources (based on Society of Petroleum Engineers (SPE)/American Association of Petroleum Geologists (AAPG) definitions). For definitions see text.

reserves are estimates of hydrocarbons to be recovered from a given date forward." Proven reserves are those quantities of petroleum which can be estimated with reasonable certainty to be economically recoverable. There should be a 90% probability that the quantities actually recovered will equal or exceed the estimate. They are divisible into proved developed reserves ("those reserves that can be expected to be recovered through existing wells") and proved undeveloped reserves ("those additional reserves that are expected to be recovered from: (1) future drilling of wells, (2) deepening of existing wells to a different reservoir, or (3) the installation of an improved recovery project"). Probable reserves are a category of unproved reserves. These unproved reserves are based on geological or engineering data similar to those used in estimates of proved reserves but in relation to which technical, contractual, economic, or regulatory uncertainties preclude such reserves from being classified as proven. Probable reserves are those quantities of petroleum which have lower probability of being recovered than the proved reserves, but higher than the possible reserves. There should be at least a 50% probability that the quantities actually recovered will equal or exceed the probable estimate. Possible reserves are a category of unproved reserves. Possible reserves have a lower probability of being recovered than probable or proved reserves. Technical, contractual, economic, or regulatory uncertainties preclude possible reserves from being classified as proven. Proved reserves are referred to as 1P reserves; proved and probable reserves as 2P reserves; and proved, probable, and possible categories as 3P reserves.

Contingent resources refer to quantities of petroleum that have been estimated as potentially recoverable from known reservoirs but are not commercially recoverable yet. Prospective resources refer to quantities of petroleum that have been estimated as potentially recoverable from undiscovered accumulations by application of future development projects.

From the preceding definitions, it is apparent that there are reserves and "reserves." Any estimates of the reserves of a field or sedimentary basin should be qualified as to whether they are total, recoverable, proved, probable, or possible, and so forth.

10.2.1 Assessment of Basin Reserves

Companies and countries have paid considerable attention to estimating the total recoverable reserves that are likely to be found in a basin (Hubbert, 1967; Moody, 1977; Harbaugh et al., 1977; Ivanhoe and Leckie, 1993; Deffeyes, 2001). An assessment of US petroleum resources by the US. Geological Survey provides one example of the techniques that may be employed (Masters, 1993).

One popular method was described by White et al. (1975). In a study of south Louisiana, gas reserves were divided into speculative, possible, and probable categories. The mean, maximum, and minimum possible reserves were plotted on a cumulative probability curve (ranging from 0.0 = the likelihood of their estimates being totally wrong to 1.0 = their estimates being absolutely correct). The three curves were then integrated to produce a summation curve for the total reserves (Fig. 10.5). This result was achieved by means of a very time-consuming piece of statistical gymnastics called the Monte Carlo technique, which a computer can execute in seconds. The method has also been applied to other areas of the United States (Miller et al., 1975).

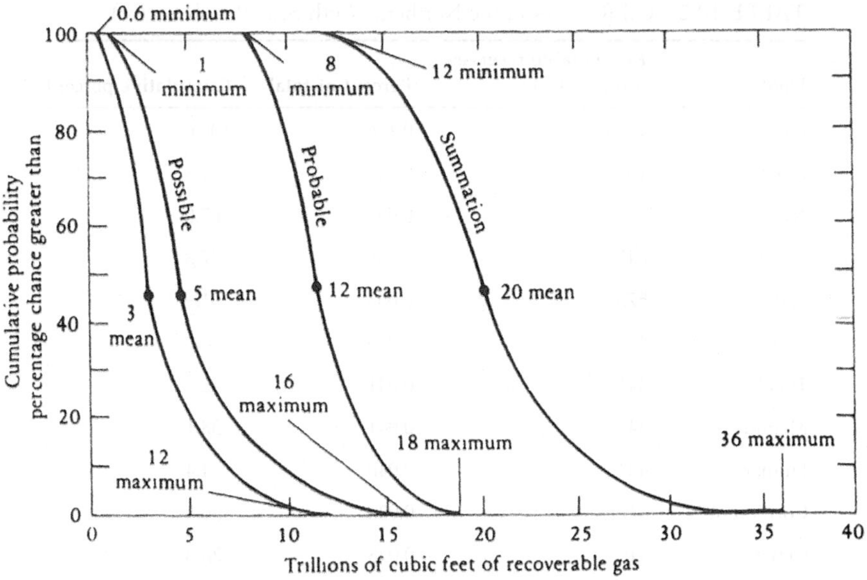

FIGURE 10.5 Estimates of producible gas reserves in south Louisiana. The three left–hand curves show the probable, possible, and speculative gas reserves estimated by Exxon. The right-hand curve is a summation of the three curves produced by a Monte Carlo procedure. *After White et al. (1975).*

Many studies show that the size–frequency distribution of fields in a sedimentary basin approaches a log-normal distribution. That is to say, the number of fields increases with decreasing field size.

When the sizes of fields are plotted on a logarithmic scale against cumulative percent, they approximate to a straight line. Log-normal distributions have been noted for field sizes in northern Louisiana (Kaufman, 1963), Kansas (Griffiths, 1966), the Denver basin (Haun, 1971), and western Canada (Kaufman et al., 1975). Table 10.2 documents reserve estimates for fields discovered in the North Sea up to 1981. Figure 10.5 plots these data as cumulative percentages against log probability. This graph shows that the North Sea broadly conforms to the log-normal distribution of field size reported form other petroliferous basins. This type of information is useful because it gives the probability of discovering a field of a particular size. When the minimum size of an economically viable field is known, so is the probability of finding one. For different basins the slope of the line in Fig. 10.6 will vary, but once the slope has been established from a few fields, the ultimate recoverable reserves of the basin can be estimated.

It is commonly remarked that the reserves of a new field, based on the results of the discovery well, diminish during subsequent development drilling when the complexities of the field become apparent. On a more positive note studies show that fields consistently produce more petroleum than early estimates of recoverable reserves predicted. Indeed Attanasi and Root (1994) point out that important annual additions to US reserves are made, not by drilling wells, but by re-estimating the recoverable reserves of old fields.

TABLE 10.2 Oil Reserves in the Northern North Sea (British Sector)

Field	Recoverable reserves (million bbl)	Percent of total	Cumulative percent
Forties	1918	0.176	100.0
Brent	1610	0.148	81.8
Ninian	1007	0.092	67.0
Piper	647	0.059	57.8
Statfjord	573	0.053	51.9
Beryl	485	0.045	46.6
Thistle	441	0.041	42.1
Magnus	441	0.041	38.1
Fulmar	434	0.040	34.4
Cormorant N	412	0.038	30.4
Claymore	419	0.038	26.6
Murchison	318	0.029	22.8
Dunline	294	0.027	19.9
Brae	294	0.027	17.2
Beryl B	293	0.027	14.5
Hutton NW	276	0.025	11.8
Hutton	257	0.023	9.3
Cormorant S	191	0.018	7.0
Beatrice	154	0.014	5.2
Maureen	154	0.014	3.8
Montrose	89	0.008	2.4
Heather	73	0.007	1.6
Auk	59	0.005	0.9
Argyll	48	0.004	0.4
Total	10,887		

From 1981 data published by the Department of Energy.

10.2.2 Assessment of Global Reserves and Resources

Many estimates of global petroleum reserves and remaining resources have been attempted. This is sometimes done by studying the size and frequency distribution of giant fields (e.g., Halbouty, 1986), and sometimes basin by basin (e.g., Ivanhoe, 1985; Ivanhoe and Leckie, 1993), or by country (EIA, 2013). Most oil companies are continually compiling estimates,

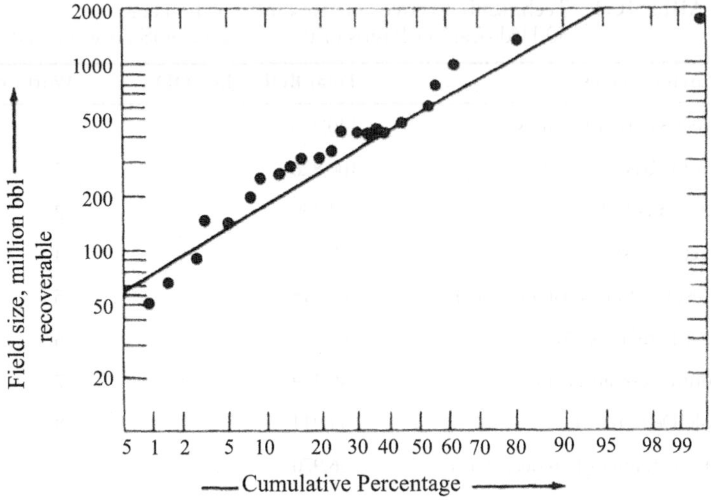

FIGURE 10.6 North Sea, United Kingdom, reserves shown from a plot of field size against cumulative probability based on the data in Table 10.2.

both of the ultimate recoverable reserves of the world, and future demand for oil, gas, and other sources of energy. Needless to say, such estimates differ widely. Table 10.3 shows the estimate of remaining resources by sedimentary basin or province area compiled by the USGS (MMBOE (million barrels oil equivalent), 6000 cubic feet of gas are taken as equal to 1 barrel of oil). These resource estimates do not include unconventional resources.

World conventional reserve estimates are a moving target but are updated on a yearly basis by the EIA and several major oil and gas companies (e.g., BP and Exxon). Figures 10.7–10.9 show EIA oil and gas reserves by country. Figures 10.10 and 10.11 show EIA oil and gas reserves by region. The oil reserves include heavy oil/bitumen deposits from Venezuela and Canada. The largest oil reserves are located in the Middle East with the exception of the heavy oil/bitumen deposits of Venezuela and Canada. Large gas reserves are located in Russia, the Middle East, and the United States.

The world consumes about 90 million barrels of oil a day. World oil reserves indicate that this is about a 50-year supply of conventional oil.

Giant oil and gas fields are defined by those having 500 million barrels or greater of ultimately oil or gas equivalent production. Giants are thought to represent about 40–60% of the world's petroleum reserves and approximately 60% of global production (Höök et al., 2009). There are approximately 1000 giant fields in the world. They are located throughout the world with clusters in the Western Siberian Basin and Persian Gulf. The discovery of giants peaked in the 1960s and 1970s (Fig. 10.12). Giants are located in several basin settings: passive margin (~36%), rift basins (~30%), and collision zones (~20%). Of concern with giant fields is the lower numbers of them being discovered in the past decade. Unconventional resources will play an important role in replacing the declining production from the giant fields.

TABLE 10.3 Technically Recoverable Resource in the Largest 30 Hydrocarbon Basins of the World (USGS Assessments)

Province name	Total BOE (MMBOE)	World rank
West Siberian Basin (Russia)	140,526	1
Santos Basin (Brazil)	100,826	2
Zagros Fold Belt	74,459	3
Arctic Alaska	72,766	4
West-Central Coastal (Africa)	65,245	5
East Barents Basins	61,755	6
South Caspian Basin	49,479	7
Nile Delta Basin	44,944	8
Mesopotamian Foredeep Basin	36,970	9
Morondava	35,037	10
USA Gulf Coast	32,197	11
Ocean	31,903	12
Niger Delta	31,562	13
East Greenland Rift Basins	31,387	14
Red Sea Basin	26,844	15
Mozambique Coastal	26,723	16
Northwest Shelf	26,199	17
Levant Basin	25,160	18
Yenisey-Khatanga Basin	24,920	19
West Greenland-East Canada	24,827	20
Tanzania Coastal	21,950	21
Guyana–Suriname Basin	21,373	22
Campos Basin	21,209	23
Essaouira Basin	20,550	24
Amerasia Basin	19,747	25
North Caspian Basin	15,050	26
Falklands Plateau	14,980	27
Bonaparte Basin	13,316	28
Tarim Basin	12,256	29
North Slope, ANWR	12,113	30

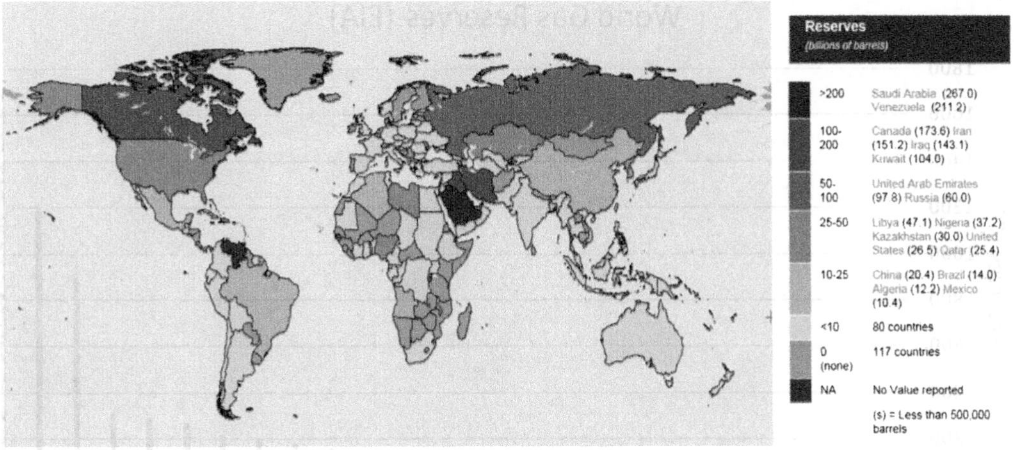

FIGURE 10.7 World oil reserves (billions of barrels). *From EIA (http://www.eia.gov/countries/index.cfm?view=reserves).*

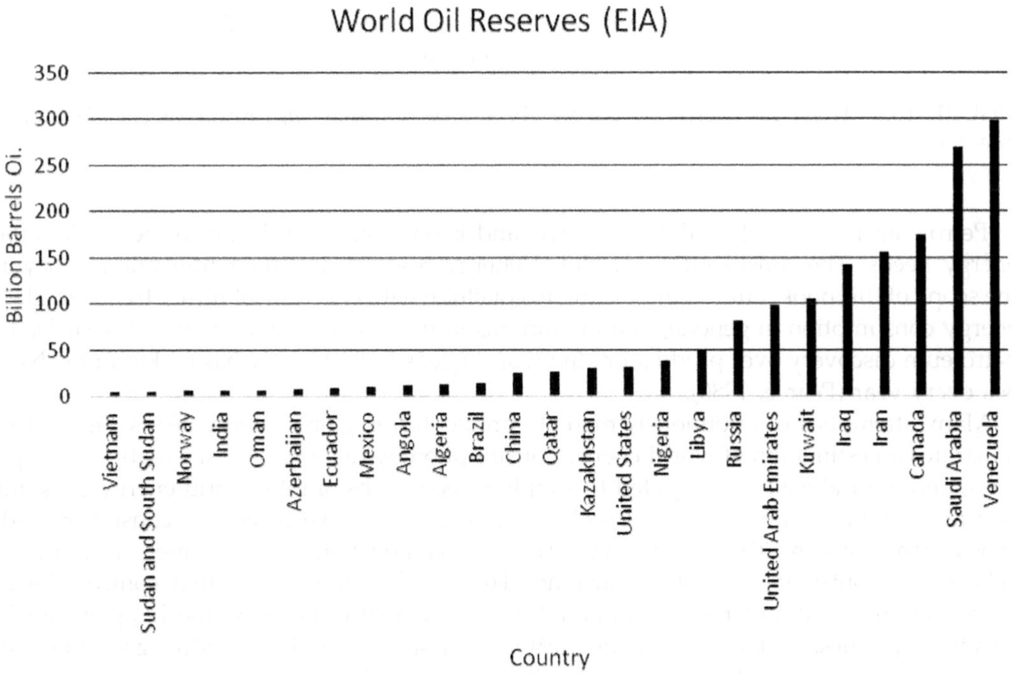

FIGURE 10.8 World oil reserves by country (billions of barrels). *From EIA data. (http://www.eia.gov/countries/index.cfm?view=reserves).*

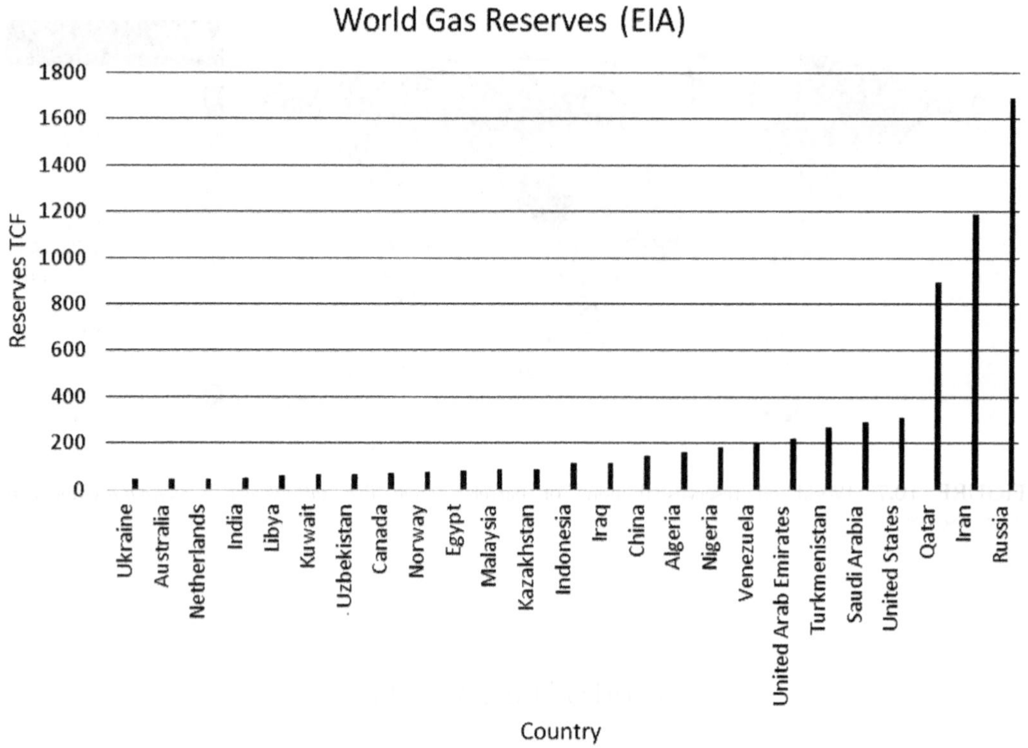

FIGURE 10.9 World gas reserves by country (TCF). *From EIA data. (http://www.eia.gov/countries/index.cfm?view=reserves).*

Petroleum is obviously a finite resource, and is only one contributor to the world's total energy needs. The consideration of global energy reserves and their utilization is beyond the scope of this book, but it is interesting to conclude with one view of future trends in global energy consumption in general, and oil and gas in particular. To maintain a 30-year lead of petroleum discovery over production, the world needs to find a new basin akin to the North Sea every year (Pearce, 1980).

Many attempts are continuously made to predict future global energy needs. Figure 10.13 illustrates an estimation of world energy consumption by fuel type. Fossils fuels (coal, liquid fuels, and natural gas) are projected to supply three-fourths of total world energy consumption in 2040. Petroleum and liquid fuels share of world marketed energy consumption declines from 34% in 2010 to 28% in 2040. The combustion of fossil fuels is popularly believed to contribute to global warming. Though the mechanisms that control climatic change are not well understood, abundant data show that it happens, and happens quickly on a human timescale. Climate change, carbon sequestration, and the gradual greening of the petroleum industry are discussed by Lovell (2009). Global warming, however, may be preferable to another ice age. It could be argued that we should keep the home fires burning, and the internal combustion engine throbbing, lest our grandchildren freeze to death in the dark.

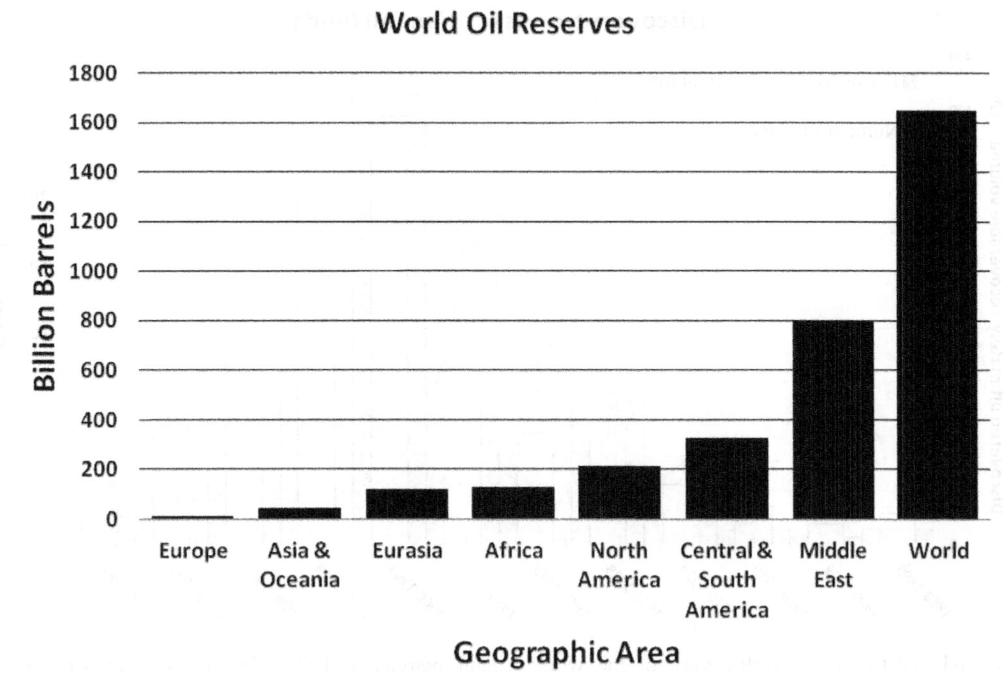

FIGURE 10.10 World oil reserves by geographic area (billion barrels). *Data from EIA.*

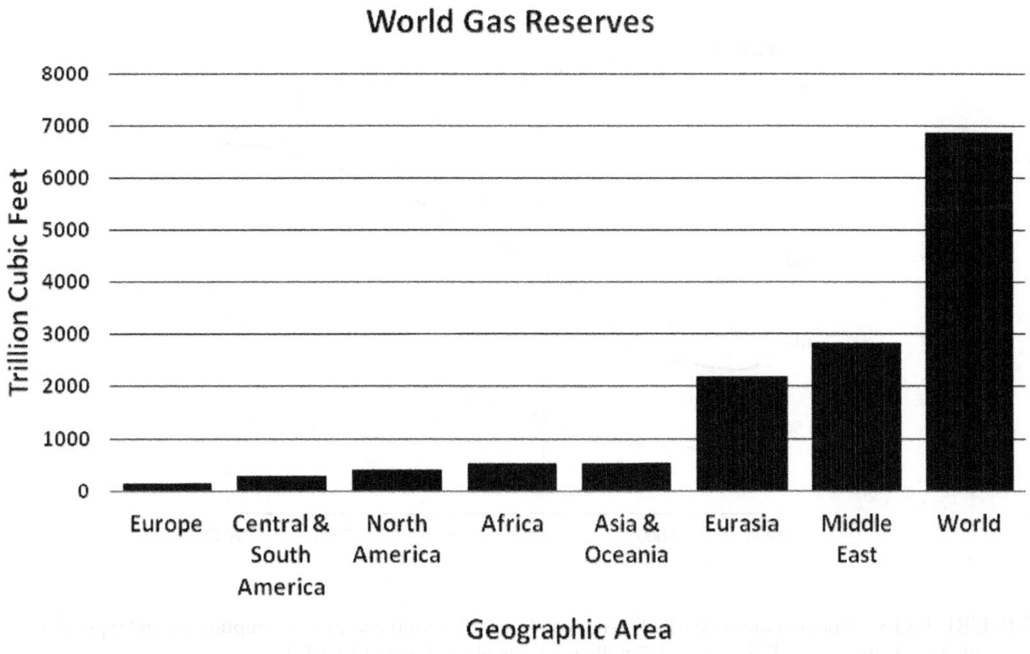

FIGURE 10.11 World gas reserves by geographic area (TCF). *Data from EIA.*

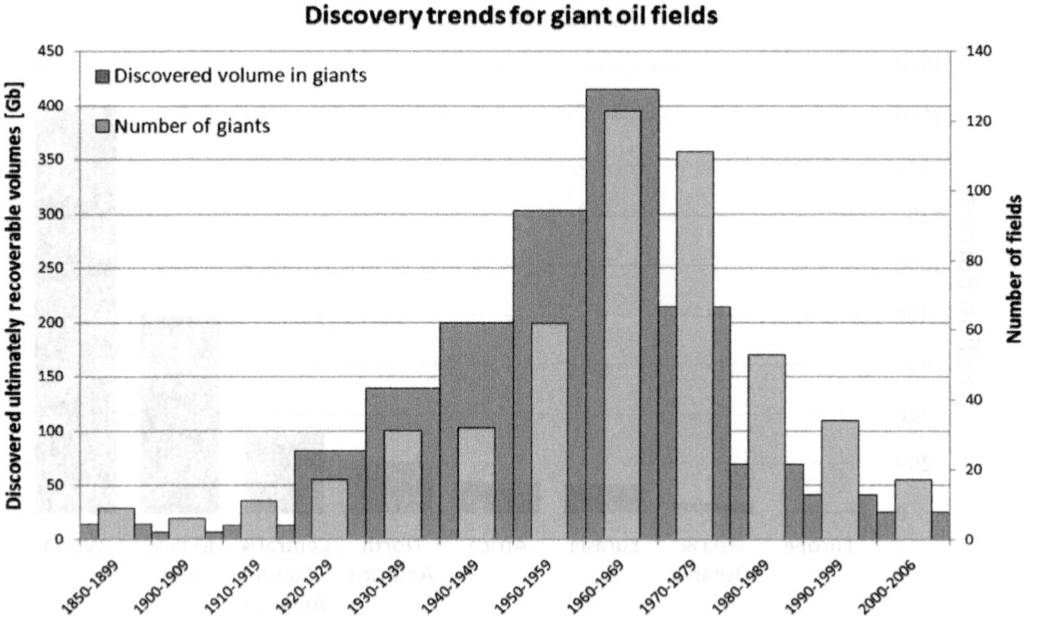

FIGURE 10.12 Rate of discovery of the world's giant petroleum fields. *Note*: GBO = gigabarrels of oil (giga = 10^9). *From Höök et al. (2009).*

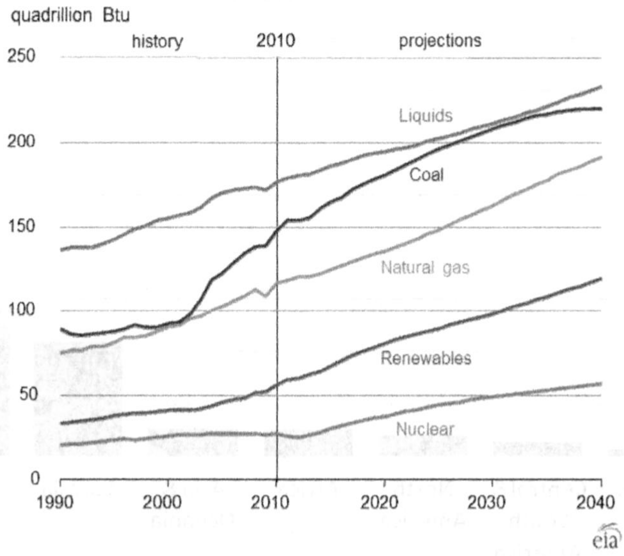

FIGURE 10.13 One of many published scenarios of possible world energy consumption by fuel type, from 1990 to 2040. *Note*: 1 quadrillion BTU equals 172 million barrels of oil. *From EIA (2013).*

References

Arnott, R., 2005. Reserves. In: Selley, R.C., Cocks, L.R.M., Plimer, I. (Eds.), Encyclopedia of Geology, vol. 4, pp. 331–339.

Attanasi, E.D., Root, D.H., 1994. The enigma of oil and gas field growth. AAPG Bull. 78, 321–332.

EIA, 2013. Int. Energy Outlook 2013, 300 p. http://www.eia.gov/forecasts/ieo/pdf/0484 (2013).pdf (accessed 2014).

Fuller, J.G.C.M., 1993. The oil industry today. In: The British Association Lectures. Geological Society of London, London, pp. 9–19.

Greenwalt, W.A., 1982. Determining venture participation. J. Pet. Technol. 33, 2189–2195.

Griffiths, J.C., 1966. Exploration for natural resources. Oper. Res. 14 (2), 189–209.

Halbouty, M.T., 1986. Future petroleum provinces of the world. Mem.—Am. Assoc. Pet. Geol. 40.

Harbaugh, J.W., Doveton, J.H., Davis, J.C., 1977. Probability Methods in Oil Exploration. Wiley, London.

Haun, J.D., 1971. Potential Oil and Gas Resources. Unpublished progress report. Colorado School of Mines, Golden.

Höök, M., Hirsch, R., Aleklett, K., June 2009. Giant oil field decline rates and their influence on world oil production. Energy Policy 37 (6), 2262–2272.

Hubbert, M.K., 1967. Degree of advancement of petroleum exploration in the United States. Am. Assoc. Pet. Geol. Bull. 51, 2207–2227.

Ion, D.C., 1979. World energy supplies. Proc. Geol. Assoc. 90, 193–203.

Ion, D.C., 1981. Availability of World Energy Resources (and Supplements). Graham & Trotman, London.

Ivanhoe, L.F., November 18, 1985. Potential of world's significant oil provinces. Oil Gas J. 164–168.

Ivanhoe, L.F., Leckie, G.G., February 15, 1993. Global oil, gas fields, sizes tallied, analyzed. Oil Gas J. 87–90.

Kaufman, G.M., 1963. Statistical Decision and Related Techniques in Oil and Gas Exploration. Prentice-Hall, Englewood Cliffs, NJ.

Kaufman, G.M., Baker, Y., Kruyt, D., 1975. A probabilistic model of oil and gas discovery. Stud. Geol. (Tulsa, Okla.) 1, 113–142.

Lerche, I., 1990a. Basin Analysis: Quantitative Methods, vol. 1. Academic Press, San Diego.

Lerche, I., 1990b. Basin Analysis: Quantitative Methods, vol. 2. Academic Press, San Diego.

Lerche, I., 1990c. Oil Exploration: Basin Analysis and Economics. Academic Press, San Diego.

Masters, C.D., 1993. U.S. Geological survey petroleum resource assessment procedures. AAPG Bull. 77, 452–454.

McKelvey, V.E., 1975. Concepts of reserves and resources. Stud. Geol. (Tulsa, Okla.) 1, 11–14.

Megill, R.E., 1984. An Introduction to Risk Analysis, second ed. Pennwell, Tulsa, OK.

Miller, B.M., et al., 1975. Geological estimation of undiscovered recoverable oil and gas resources in the United States. Geol. Surv. Circ. (U.S.) 725, 1–78.

Moody, J.D., 1977. Perspectives on energy problems. In: Cameron, V.S. (Ed.), Exploration and Economics of the Petroleum Industry. Bender, New York, pp. 1–16.

Newendorp, P.D., 1976. Decision Analysis for Petroleum Exploration. Petroleum Publishing Co, Tulsa, OK.

Pearce, A.E., 1980. Oil: hydrocarbons or B.T.U.'s. Esso Mag. 113, 3–8.

SPE, 2014. Reserve definitions. http://www.spe.org/industry/reserves.php.

Sluijk, D., Nederlof, M.H., 1984. Worldwide geological experience as a systematic basis for prospect appraisal. AAPG Mem. 35, 15–26.

White, D.A., 1993. Geologic risking guide for prospects and plays. AAPG Bull. 77, 2048–2061.

White, D.A., Garrett, R.W., Marsh, G.R., Baker, R.A., Gehman, H.M., 1975. Assessing regional oil and gas potential. Stud. Geol. (Tulsa, Okla.) 1, 143–159.

Selected Bibliography

For an exposé of statistical techniques applied to prospect evaluation, see:

Harbaugh, J.W., Doveton, J.H., Davis, J.C., 1977. Probability Methods in Oil Exploration. Wiley, London.

For methods of assessing the economics of oil prospects, see:

Megill, R.E., 1984. An Introduction to Risk Analysis, second ed. Pennwell, Tulsa, OK.

Newendorp, P.D., 1976. Decision Analysis for Petroleum Exploration. Petroleum Publishing Co, Tulsa, OK.

For discussions and predictions of future demand/production of oil, gas, and other energy sources, see:

Campbell, C.J., 1991. The Golden Century of Oil 1950–2050, the Depletion of a Resource. Kluwer, Dordrecht, The Netherlands.

Deffeyes, K.S., 2001. Hubbert's Peak. Princeton Univ. Press, 208 pp.

Fuller, J.G.C.M., 1993. The oil industry today. In: The British Association Lectures. Geological Society of London, London, pp. 9–19.

For discussions and predictions about the oil industry and climate change, see:

Lovell, B., 2009. Challenged by Carbon: The Oil Industry & Climate Change. Cambridge University Press, Cambridge, 214 pp.

Index

Note: Page numbers followed by "f" and "t" indicate figures and tables respectively

S

Color Plates

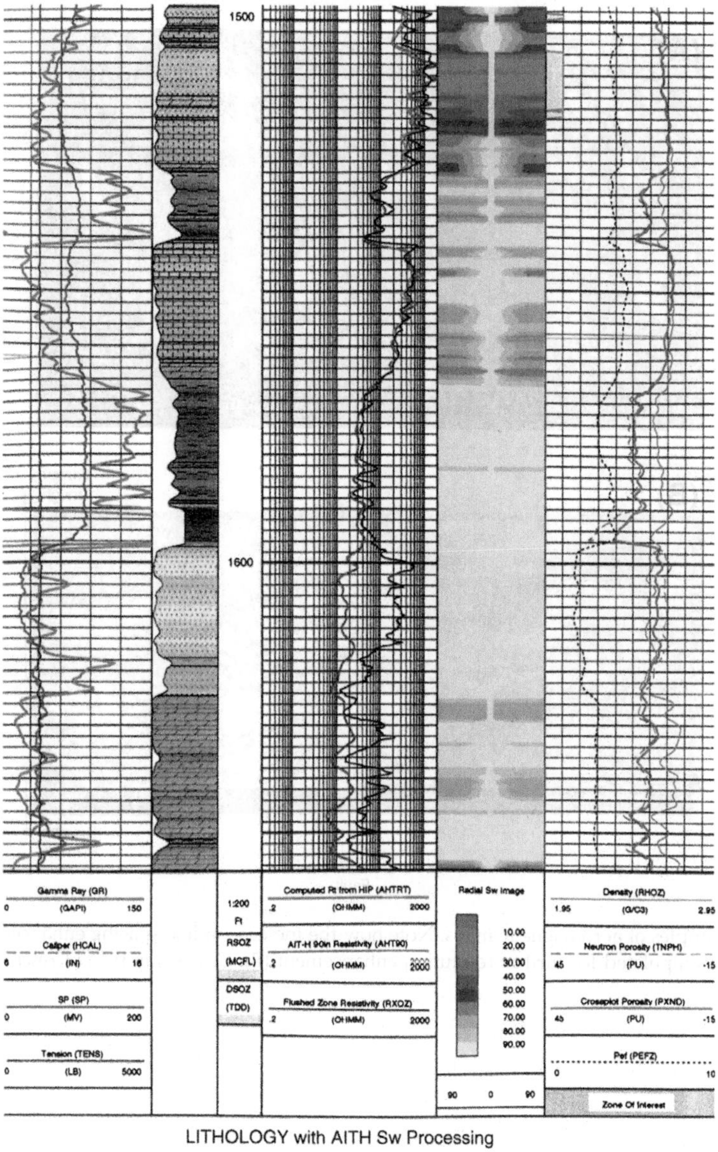

PLATE 3.1 An example of a suite of well logs. Gamma, self-potential, and caliper logs in left-hand column help to indicate lithology. Resistivity logs in track to immediate right of depth column calculate hydrocarbon saturation in right-hand colored column. Porosity logs in the far right column. Reservoir zone at the top of the section is highlighted by pink bar at right of radial S_w log. *Courtesy of Schlumberger Wireline and Testing.*

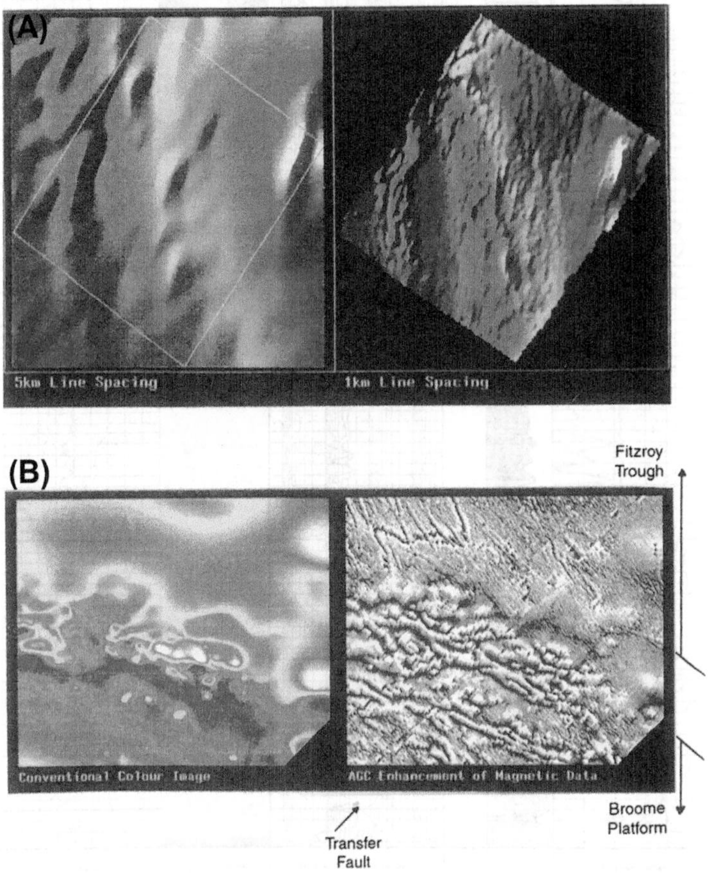

PLATE 3.2 Examples of aeromagnetic maps. Note how the increase in line spacing enhances resolution (A), and how data can be manipulated for further resolution enhancement (B). *Courtesy of World Geoscience.*

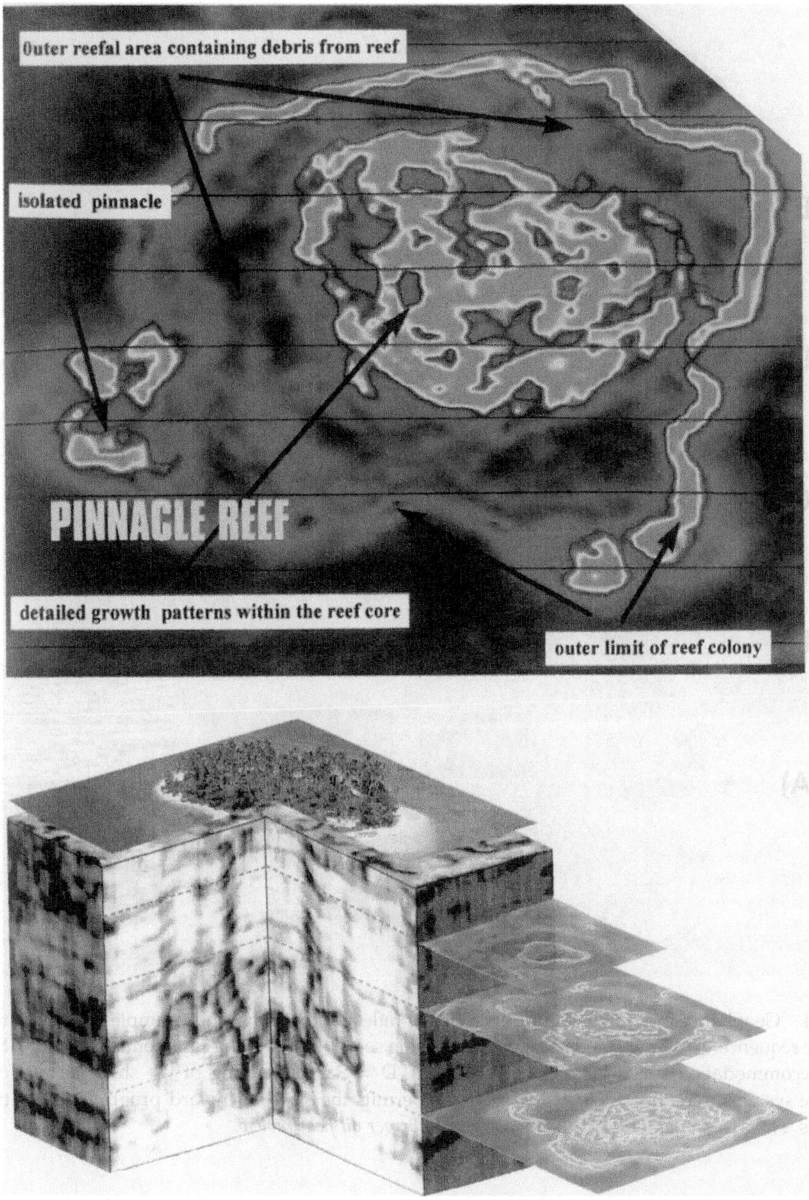

PLATE 3.3 A limestone reef imaged by 3D seismic, with accompanying geophantasmogram. Note how not only horizontal time slices, but any orientation may be selected for display from 3D seismic data. These images were produced by Coherence Cube technology. *Courtesy of GMG Europe Ltd.*

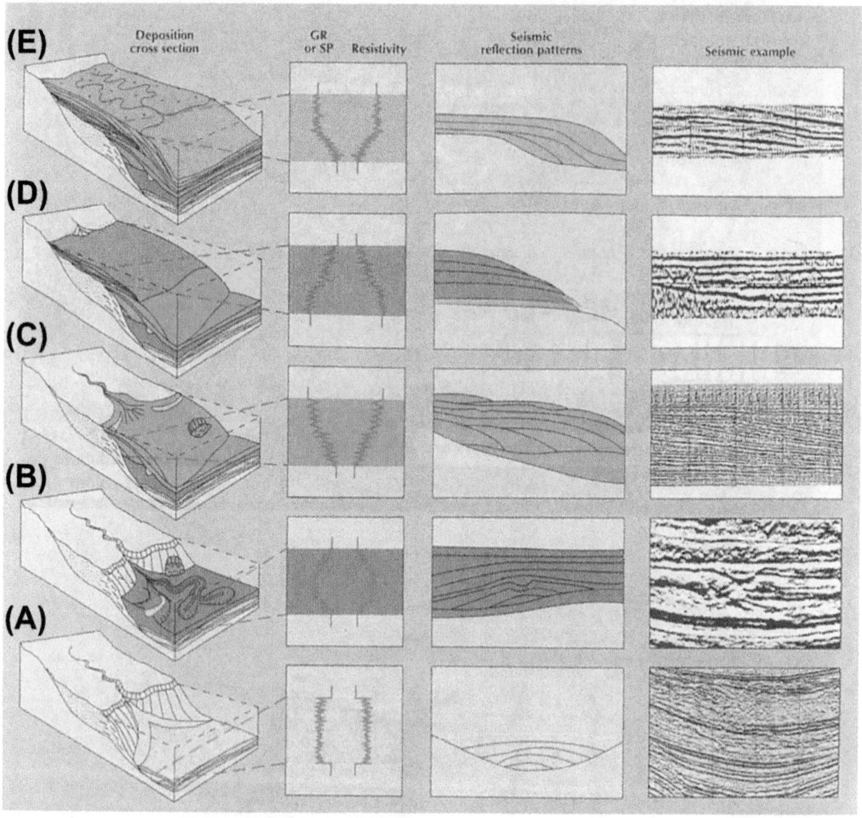

PLATE 3.4 Geophantasmograms, log motifs, seismic reflection patterns, and examples to illustrate a third-order stratigraphic sequence. A = Low stand sea level. B = Rising sea level floods the continental slope. C = Sea floods shelf providing accommodation space for prograding deltas. D = Rapid flooding of the shelf permits deposition of a transgressive systems tract. E = Slow rising sea level permits the slow basinward progradation of thin high stand system tracts. *From Neal et al. (1993); courtesy of Schlumberger oil field review.*

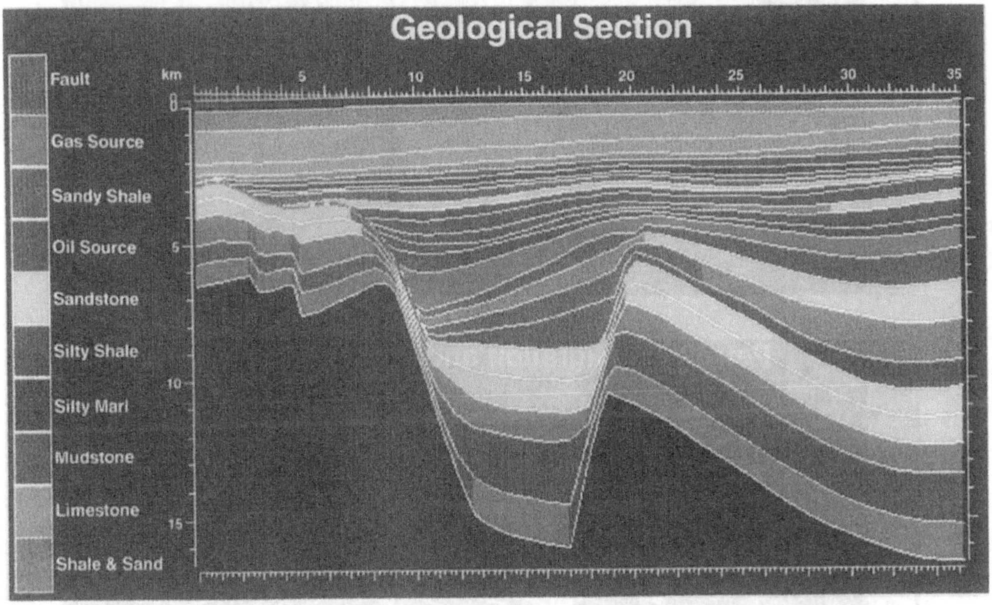

PLATE 5.5 Three computer-generated cross-sections to illustrate how a sedimentary basin may be modeled to predict the distribution of pressure and petroleum. *Courtesy of A. Vear and British Petroleum.*

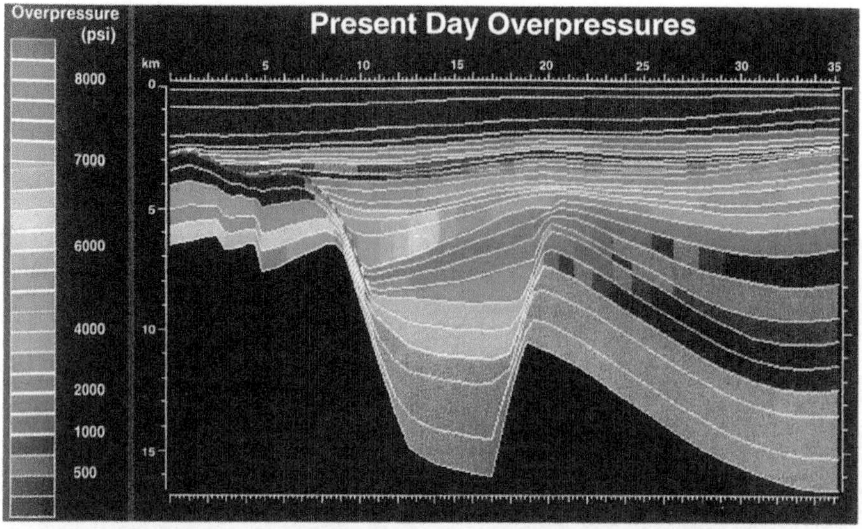

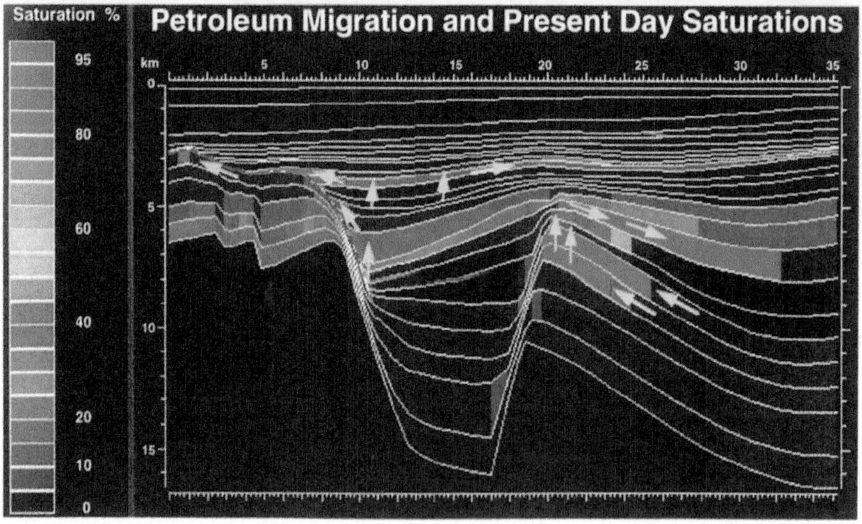

PLATE 5.5 cont'd

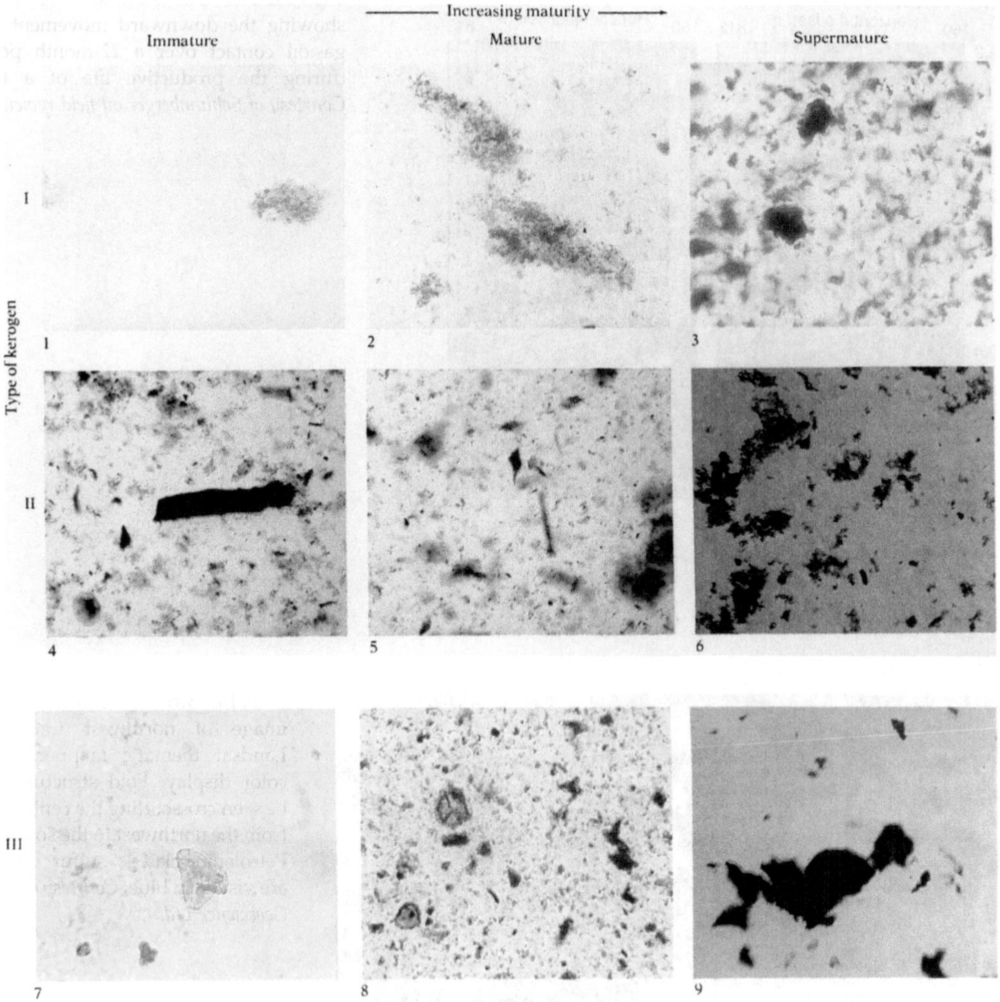

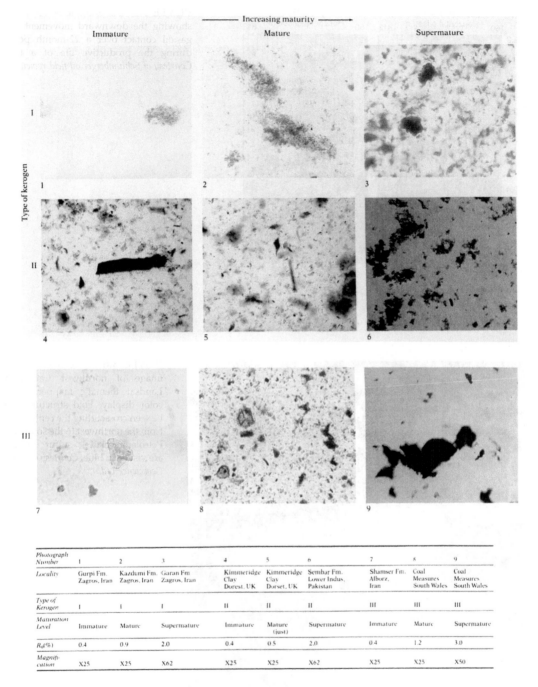

Photograph Number	1	2	3	4	5	6	7	8	9
Locality	Gurpi Fm. Zagros, Iran	Kazdumi Fm. Zagros, Iran	Garan Fm Zagros, Iran	Kimmeridge Clay Dorest, UK	Kimmeridge Clay Dorset, UK	Sembar Fm. Lower Indus, Pakistan	Shamser Fm. Alborz, Iran	Coal Measures South Wales	Coal Measures South Wales
Type of Kerogen	I	I	I	II	II	II	III	III	III
Maturation Level	Immature	Mature	Supermature	Immature	Mature (just)	Supermature	Immature	Mature	Supermature
R_0(%)	0.4	0.9	2.0	0.4	0.5	2.0	0.4	1.2	3.0
Magnification	X25	X25	X62	X25	X25	X62	X25	X25	X50

PLATE 5.6 Photomicrographs of different types of kerogen at various maturation stages. *Courtesy of Kinghorn and Rahman.*

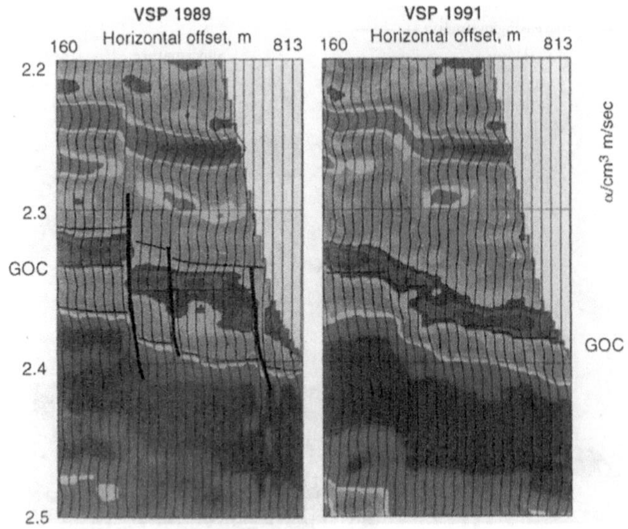

PLATE 3.5 An example of 4D seismic showing the downward movement of a gas:oil contact over a 22-month period during the productive life of a field. *Courtesy of Schlumberger oil field review.*

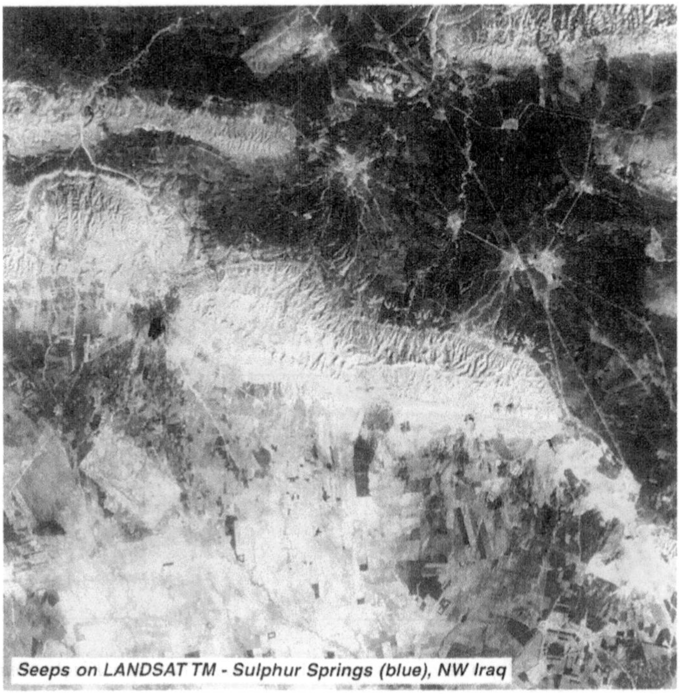

Seeps on LANDSAT TM - Sulphur Springs (blue), NW Iraq

PLATE 3.6 A remotely sensed image of northwest Iraq from Landsat thematic mapper. False-color display. Fold structures can be seen crosscutting the central area from the northwest to the southeast. Petroleum-related sulfur springs are visible in blue. *Courtesy of World Geoscience Ltd.*

Printed by Printforce, the Netherlands